AF805811

Optimization

Optimization is a foundational topic in mathematics, underpinning nearly all of our modern industrial and technological world. Assuming only basic knowledge of linear algebra and calculus, this book provides a rapid, yet thorough, overview of applied mathematical optimization for advanced undergraduates, beginning graduate students, or practitioners in science and engineering. The text opens with an "Optimization Bootcamp", introducing methods at a beginning level, before progressing to deep-dives into advanced topics and research-ready methods. The focus throughout is on modern applications of machine learning, inverse problems, and control. Rich pedagogy includes Python code with simple working examples and advanced case studies. Every section is accompanied by YouTube lectures to encourage interaction with the material. Using intuitive explanations, this book makes the material as simple and interesting as possible, while still having the depth, breadth and precision required to empower use in research and real-world applications.

Steven L. Brunton is the Boeing AI and Data-Driven Engineering Professor of Mechanical Engineering at the University of Washington, where he is the Director of the NSF AI Institute in Dynamic Systems and the Director of the AI Center for Dynamics and Control. His research has been recognized with awards including the Presidential Early Career Award for Scientists and Engineers. Steve is also passionate about teaching math to engineers as an author of five textbooks and through his popular YouTube channel, "eigensteve".

"I would strongly recommend Steve Brunton's Optimization Bootcamp to any beginning student of Applied Math, Engineering, or Machine Learning. The book covers many of the most commonly used optimization methods, with practical examples and problems in different fields, from fitting models to data, to designing mechanical structures. Its integrated Python examples send a clear message to the student: This material is meant to be used."

Stephen Boyd, Stanford University

"Steve Brunton explores optimization with clarity and ambition. Throughout, the book maintains an excellent balance between mathematical insight and practical implementation, with well-chosen examples and Python code that illuminate what is happening beneath the algorithmic surface. This is an accessible text for readers encountering the material for the first time and a valuable reference for researchers wanting to study one of the topics presented in greater depth."

Richard Murray, Caltech

Optimization

A Bootcamp for Machine Learning, Inverse Problems, and Control

STEVEN L. BRUNTON
University of Washington

CAMBRIDGE
UNIVERSITY PRESS

CAMBRIDGE
UNIVERSITY PRESS

Shaftesbury Road, Cambridge CB2 8EA, United Kingdom

One Liberty Plaza, 20th Floor, New York, NY 10006, USA

477 Williamstown Road, Port Melbourne, VIC 3207, Australia

314–321, 3rd Floor, Plot 3, Splendor Forum, Jasola District Centre,
New Delhi – 110025, India

Cambridge University Press is part of Cambridge University Press & Assessment,
a department of the University of Cambridge.

We share the University's mission to contribute to society through the pursuit of
education, learning and research at the highest international levels of excellence.

www.cambridge.org
Information on this title: www.cambridge.org/9781009755863
DOI: 10.1017/9781009755856

© Steven L. Brunton 2026

This publication is in copyright. Subject to statutory exception and to the provisions
of relevant collective licensing agreements, no reproduction of any part may take
place without the written permission of Cambridge University Press & Assessment.

When citing this work, please include a reference to the DOI 10.1017/9781009755856

First published 2026

A catalogue record for this publication is available from the British Library

A Cataloging-in-Publication data record for this book is available from the Library of Congress

ISBN 978-1-009-75586-3 Hardback

Cambridge University Press & Assessment has no responsibility for the persistence
or accuracy of URLs for external or third-party internet websites referred to in this
publication and does not guarantee that any content on such websites is, or will
remain, accurate or appropriate.

For EU product safety concerns, contact us at Calle de José Abascal, 56, 1°, 28003 Madrid,
Spain, or email eugpsr@cambridge.org.

Contents

Preface

I have been scheming to write this book for years. Optimization is such a fundamental topic in mathematics, quietly underpinning nearly all of our modern industrial and technological world. Over time, I've come to realize that optimization is a persistent theme underlying much my work, including machine learning, control theory, and model discovery. However, optimization is often overlooked in engineering applied mathematics curricula, or at least it was in mine. Perhaps this is *because* it is so foundational, much as we often take for granted the concrete and steel all around us, or the production and transportation networks that ensure food and goods arrive. It seems we rarely notice them until they *don't* work. Optimization is one of the best industrialized fields of mathematics, which means that it is constantly working behind the scenes; this also contributes to it being somewhat hidden. For example, fast, robust, and easy-to-use optimization algorithms are the unsung heroes of the machine learning era. However, they are often forgotten in favor of other enabling factors of modern neural networks, advanced computing chips, and abundant data. How, we should ask, are these neural network parameters tuned to fit the data?

This book is designed to provide a rapid, yet reasonably thorough, overview of the field of applied mathematical optimization. This could easily fill several books, but I have opted to streamline and cover about 80% of the key ideas in about 20% of the space. In other words, my goal was to find a Pareto *optimal* balance of topics and depth. In my experience, mathematical optimization is often presented in a form that is either extremely mathematical and abstract, too shallow to be useful, or so narrow that it covers only one topic very well. In this text, I seek a middle ground. As such, this book is not meant to be the definitive compendium on all things optimization, but rather a guide to rapidly become familiar with the main issues and types of optimization, the leading optimization algorithms, and how to use them. The methods discussed in this book have been chosen primarily for their relevance and general applicability. Further, I attempted to present a progression of topics, from basic introductory material that students can apply immediately, to research-level techniques.

An earlier version of this book was called *Optimization Bootcamp*, to reflect the quick and introductory nature of the treatment. However, it has grown into something a bit larger and more serious, so now Chapter 1 is the Optimization Bootcamp, and then we take deep-dive excursions into various topics in the subsequent chapters. This idea of an *Optimization Bootcamp* was inspired by my *Control Bootcamp*, which is a series of videos I created almost ten years ago to help my students quickly get up to speed in control theory without taking several courses, many of which were too mathematically abstract to be immediately useful or intuitive for the practicing engineer. Because this format proved so effective,

I began planning this current book and accompanying video lectures, as I believe that optimization is one of the most useful fields in applied mathematics, up there with linear algebra and calculus. Who knows, maybe there will be a *Linear Algebra Bootcamp* next.

Every section of this book is accompanied by YouTube lectures to make it as easy as possible to interact with the material. Likewise, the book includes extensive Python code with working examples, ranging from very simple to rather advanced case studies. This book only assumes knowledge of basic linear algebra and calculus, so it should be accessible to advanced undergraduates and beginning graduate students. When possible, I will also try to connect to history, as this has always made the material come to life for me.

Topics in this book are motivated by three main application domains: machine learning, control theory, and inverse problems. Although there are many other compelling uses for mathematical optimization, I feature these three topics because they are particularly interesting and relevant in the modern era. They each demonstrate some of the most impressive and relatable benefits of optimization, and they are each areas of intense active research, meaning that they are *living* fields of applied optimization.

I thoroughly enjoyed writing this book, and I tried my best to make this subject come to life for my readers, which are two ways of saying the same thing. This was also a structured excuse to indulge in learning a topic I've wanted to know more about for years. I still have a lot to learn, so if you catch anything off, or have suggestions, please let me know. I wrote this book with the second edition in mind, and I hope you, the reader, will help improve this material through use. The next generation of readers, and I, thank you.

As a parting thought: Optimization might make you rich and famous, but there is no guarantee that it will make you happy. Sometimes optimizing is not optimal. There are times you need to wander and explore and enjoy and fritter, to have no objective, to be unconstrained. With that said, it is your turn to begin optimizing.

Themes and organization of this book. A number of key themes have emerged throughout this book. First, in optimization there is the high-level concept and problem statement, and then there is the algorithmic implementation; the two are often disconnected. Understanding how to formulate and then solve an optimization problem is my primary goal for you when reading this book. In my experience, this skill takes repetition and practice. Second, in the modern era, we are often faced with intractable, nonconvex, nonsmooth, or ill-posed optimization problems. For example, training a neural network or learning a nonlinear control policy is highly nonconvex. One of the most exciting trends in modern optimization is the increasing availability to either solve such nonconvex problems directly, or solve nearby problems, using relaxation and regularization, with more tractable forms. A third and perhaps the most prevalent theme is that of *data-driven applied optimization*. The most exciting emerging techniques involve optimization based on data: finding *optimal* low-dimensional patterns from high-dimensional data, *optimal* sensor placement, machine learning *optimization*, *optimal* control, etc.

Much of the book is organized around the major dichotomies of convex vs nonconvex optimization problems, and constrained vs unconstrained problems. Chapter 1 begins with a substantial introductory overview, the *optimization bootcamp*, introducing key concepts of the gradient, convexity, and constraints. Gradient-based methods are widely used in modern optimization, and they apply to convex and nonconvex problems alike, so we begin Chapter 2 here. Next, we explore two of the largest classes of convex problems: linear programming (Chapter 3) and linear least-squares regression (Chapter 4). This is followed by Chapter 5 where we explore nonsmooth and gradient-free methods, along with global approaches for nonconvex problems. Here, we begin to solve increasingly challenging (and interesting!) problems, as it is increasingly possible to directly tackle nonconvex and nonsmooth problems using proximal methods. Although these methods have less global guarantees, they are quite useful in practice. Next, we cover constraints and duality in Chapter 6, which are topics I've wanted to understand more deeply for about twenty years. I struggled to learn Lagrange multipliers and variational methods to my satisfaction, and so this chapter ended up being a passion project of mine. That said, it is still a work in progress. I then introduce Bayesian optimization and statistical optimization problems in Chapter 7, which will be used in the last three chapters. The last three chapters may be thought of as introductions to entire large application domains of optimization: inverse problems (Chapter 8), control theory (Chapter 9), and machine learning (Chapter 10).

Online material. This book was designed to make extensive use of online material, including videos, codes, data, and homework. All material can be found at:

`eigensteve.com/optimization`

I have recorded and posted video lectures on YouTube for most topics in this book. Supplementary videos are also included for students to fill in gaps in their knowledge. All videos may be found in the various playlists on my YouTube channel @eigensteve :

`https://www.youtube.com/@eigensteve`

In addition, all of the Python code and data used in the book are available for download. The codes posted online are more extensive than those presented in the book, and I included code used to generate publication quality figures. These are available at:

`github.com/dynamicslab/optimizationbook`

How to use this book. The intended audience for this book is advanced undergraduates, beginning graduate students, or practitioners in engineering and science. As such, methods are introduced at a beginning level, but we quickly get to advanced topics and research-ready methods. The perspective taken in this book is that applied optimization is a mathematical toolkit and a common language for solving modern problems in science and engineering.

Most important, I want this book to be fun, interesting, and empowering for researchers new to the field. As always, my goal is to make the material as simple as possible, while still having the depth and breadth required to be useful in research and real-world applications. I hope you enjoy learning about applied optimization as much as I have, and that you seek opportunities to use these tools to make positive change in the world.

Acknowledgments

Writing a book is a lonely, daunting, sometimes seemingly impossible endeavor. But it becomes not only possible, but a joy and a privilege with the support of so many wonderful friends and colleagues.

I am extremely grateful to Prof. Bing Brunton, who painstakingly read the entire book at several stages, providing valuable and detailed feedback throughout. Importantly, she helped me remember my own voice at key points, and encouraged me to channel my passion for teaching. She even lugged around paper copies to mark up on planes and vacations. Bing believed in me, supported this effort wholeheartedly, and there is no way imaginable that this book would have been possible without her. Throughout most of my adult life, she has also helped me remember when *not* to optimize. I am incredibly grateful.

When revising and polishing this book, I had an ace in the hole, Prof. Sasha Aravkin at the University of Washington. Sasha is a long-time friend and colleague. My students and I have always turned to him when we have a tricky optimization problem. I asked Sasha if he was willing to proofread the book, and he not only proofread, but gave me a series of extremely valuable impromptu lectures and working sessions to help shore up gaps in my knowledge. I learned a ton during this process, and I hope I was able to pass along some fraction in the current version. Especially in the chapters on duality, and the sections on proximal algorithms and line search, Sasha's expertise was invaluable. If you find anything particularly interesting or insightful in these sections, you can thank Sasha; any remaining mistakes are entirely my own.

I'd also like to thank Prof. Michelle Hickner, who co-taught a first course on the topic of *Applied Optimization* with me, and who has been my partner in developing an entire new Masters Degree in AI for Engineering. I have learned a lot by working with Michelle, and I don't think I would have had the energy or time to do this if it wasn't for her excellent work.

My amazing lab at University of Washington and my many fantastic collaborators deserve my heartfelt thanks. Getting to work on interesting problems with people you like and respect makes it hardly work at all. I'd especially like to thank my fantastic students and postdocs working in my lab during the writing of this book, not only for teaching me loads of optimization, but also for being patient with my long absences and my longer manic optimization explanations: Sam Ahnert, Hangjun Cho, Sara Ichinaga, Ana Larrañaga Janeiro, Kartik Krishna, Christian Lagemann, Esther Lagemann, Keillan Laurenceau, Brittany Lydon, Sajeda Mokbel, Elania Opusunju, Sam Schneider, Dima

Tretiak, Chace Wilcoxson, Joey Williams, Filippo Zacchei, and Nick Zolman. I'd also like to express my gratitude for extensive interactions and conversations with my incredible collaborators Jean-Christophe Loiseau and Jared Callaham. There are far too many others to name, and so I will suffice to say that I thank you all.

I am fortunate to have an incredible following of brilliant scientists and engineers on YouTube and X/Twitter. Towards the end of my first draft, I solicited feedback and was overwhelmed by their support and their incredibly detailed and insightful comments. The following individuals provided valuable comments at various stages of the revision: Abdelrahman Abdelgawad, Jim Alateras, Kasimu Aliyu, Sriharsha Annamaneni, Christopher Añorve, Amir Arzani, Rohan Baijal, Rahat Kibria Bhuiyan, Ben Caunt, Boyuan Chen, Xinyu Chen, Jon Dewitt Enriquez Dalisay, Joshua Day, Akylas Fotiadis, Joseph Free, Ján Gašper, Kalp Garg, Ritobrata Ghosh, Ben Halliwell, Erfan Hamdi, Dana Henshaw, Mark Howell, Viktor Hruska, Kevin Iverson, Lukasz Juszkiewicz, Kristian Lorenzen, Edward Mehrez, Christian Mendonidis, Khojasteh Mirza, Seth Miu, Carlos Aya-Moreno, Saad Moro, Daniel Odoemene, Brad Ofrim, Ibrahim Omer, Damien O'Rourke, Alessandro Pecile, Sabin Pokharel, Rafia Rahim, Preston Rozwood, Cooper Runyan, Ivan Rýger, Brett Schroeder, Hugo Tadashi, Milkias Teklai, Kevin Carrington Tichana, N.M.T. Vijay, Ricardo Vinuesa, Div Waghaskar, Vickie Wang, Matthias Wüstenhagen, and Efe Yarbasi. Special thanks to Ed Mehrez, Preston Rozwood, and Damien O'Rourke, who provided extremely detailed and insightful comments from start to finish. Anybody reading this book should also thank them. If I have failed to mention anyone, I am truly sorry, and this is a testament to the incredible outpouring of support from the community to make this happen.

In writing this book, I leveraged AI tools, mostly GPT-4o and GPT-5. I mainly used LLMs for co-developing code and brainstorming topics, big-picture organization, and illustrative examples. I also used it for a rough sanity check that the homework is solvable. AI tools were quite helpful in providing rough initial background sketches of topics that I could use to cross-reference against my own research. It was also useful to refine language and clarify ideas. It was not particularly helpful for creating accurate or interesting text or equations, and so the writing is my own.

I would like to express my sincerest and deepest gratitude to my Editor David Tranah at Cambridge University Press, who believed in this project from the start and provided constant support from start to finish. David saw the vision and believed in what this project could be. He gave just the right amount of feedback at key times to really improve the flow and style of the book, and he helped work around my ambitious plan to finish within a year.

And lastly, I am grateful to you, the reader. Your reading makes this book come to life, so that the ideas inside may grow and bear fruit. My sincerest hope is that you find inspiration and new perspectives and go on to change the world for the better with these tools.

Optimistically yours,
Steven L. Brunton
Seattle, WA, September 2025

1 Optimization Bootcamp

I write this as my flight approaches Charles de Gaulle airport in Paris, and I am overwhelmed by signs of optimization. The aircraft itself is a pinnacle of modern engineering optimization. Flying at nearly the speed of sound, the shape of the wings has been carefully refined for efficient flight at cruise and to provide enough lift for a heavy, fully fueled take off. The composite materials used to construct the wings and body were designed for strength and durability, and to provide higher cabin pressure and moisture to reduce hypoxia and jetlag for its passengers. Further, the aircraft has been optimized not only for flight, but for global manufacturability, as it is assembled from supply chains spanning several continents. Out the window, I see farmlands yield to city, a quiet scene of distributed optimization; fields and water management transition into planned streets and urban arterials, ensuring the city is fed. The surrounding landscape has been shaped by the scars of war, a terrible reminder of the birth of modern optimization. As we near the city, I see hills that did not exist in Roman times, signs of Napoleon's great earth-moving efforts and the birth of optimal transport[1]. The airport is likewise a city of its own: supply chains, inventories, ground crews, flight controllers, food service workers, and custodians working in shifts, all in a vast system of local optimization. And I write this on my laptop[2], which is a compact masterpiece of constrained optimization.

Our world is built on optimization. How did we get here, and what are the tools of modern, applied optimization? This chapter provides an introductory overview of key concepts in optimization, as well as several motivating applications. Although this chapter is not meant to be entirely standalone, it serves as the bulk of an initial *bootcamp* course, augmented with deeper dives into each topic in later chapters. Many of the sections here will be explored in much more detail in later chapters, and so this is mostly meant to provide a rapid overview and categorization of the major topics in optimization, important algorithms, and how the various ingredients fit together. Thus, this is also meant to be a preview of later chapters.

We begin with a categorization of different optimization problems, the so-called *anatomy* of an optimization problem. We introduce the objective function, to be maximized or minimized, and constraint equations that limit the set of feasible solutions.

[1] My favorite café in the city is on Rue Monge, named after Gaspard Monge, who developed optimal transport to efficiently reshape Paris at Napoleon's command.

[2] This book was an exercise in constrained optimization. My objective: to write the most interesting and informative book possible for scientists and engineers within the constraints of one year and 500 pages.

Next we introduce two of the main concepts in optimization: convexity and the gradient. Convexity provides a fundamental dichotomy in optimization, where convex problems are essentially solved, at least numerically, and where local optimal solutions are also globally optimal. Convexity is used throughout the book, although this is the only chapter that formally discusses the conditions for convexity and its implications for global optimization. Chapters 3 and 4 on linear programming and quadratic programming (e.g., least-squares regression) cover two broad classes of convex optimization problems. Nonconvex problems are often more challenging, and do not generically have solutions that scale computationally. However, there has been a major shift in applied optimization, where it is increasingly possible to directly solve nonconvex problems, rather than relaxing them to nearby convex problems. Although there are fewer guarantees, solving the nonconvex problem directly is often highly effective in practice. Chapter 5 explores modern approaches to nonconvex optimization problems in more detail.

The gradient is another fundamental concept in optimization, as it provides critical information about the geometry of the optimization landscape, and which direction (locally) improves the solution. This topic is explored in much greater depth in Chapter 2, including algorithms that use gradient descent, Newton's method, and modern extensions for high-dimensional systems, such as stochastic gradient descent in machine learning.

Throughout the chapter, we examine examples of unconstrained and constrained optimization, to remind the reader of basic properties and approaches. These will be simple low-dimensional optimization problems, where it is possible to solve for the optimum solution using basic calculus. We also briefly introduce the notion of Lagrange multipliers, which are an intuitive and physical way of handling constraint equations. Lagrange multipliers are closely related to the notion of a dual optimization problem, and both concepts are developed further in Chapter 6.

Modern mathematical optimization opens up a world of applications in science, engineering, and industry. We conclude with a brief look at three major application areas of modern optimization: inverse problems,, control theory, and machine learning. For example, nearly all of machine learning is built around optimization, which is used to train models to fit data. Similarly, control theory is essentially a field of applied optimization, where the problem is constrained by a differential equation that captures the underlying dynamic behavior of a system. Inverse problems form another major class of applied optimization problems, where the goal is to infer unknown quantities or parameters of a model to best match observed data. Each of these topics will be their own chapter in the later half of this book, with an additional chapter on Bayesian optimization to complement the topics of inverse problems and machine learning.

On your first read, this chapter may be skimmed, and I encourage you to peek into later topics in the book. Some of the technical results around convexity can also be skipped, especially the proofs. However, the big-picture idea that convexity leads to powerful guarantees and efficient algorithms is an important message of this chapter. But remember that convexity isn't everything, that most of the real world has horns and fangs, and that we must be ready to face nonconvexity in its many forms. The rest is meant to pique your interest and provide you with a road map for further studies. Bon voyage.

1.1 Anatomy of an Optimization Problem

In its simplest form, an optimization problem involves minimizing or maximizing an objective function $f(\mathbf{x})$ with respect to a set of decision variables $\mathbf{x} \in \mathbb{R}^n$:

$$\min_{\mathbf{x}} \quad f(\mathbf{x}). \tag{1.1}$$

This is an *unconstrained* optimization problem, as it is assumed that the decision variable may take any value. For scalar functions of a scalar variable, we have experience solving these optimization problems by finding points where the derivative of f with respect to x is equal to zero that also have a positive second derivative. See Panel 1.1–Panel 1.3.

However, many real-world optimization problems do have constraints on $\mathbf{x}$, which are written either as inequality or equality constraint equations. In Panel 1.1, we implicitly assumed that the angle α should be constrained to be between 0 and $\pi/2$. Panel 1.1–Panel 1.3 also admit natural constraints, such as restricted lake crossing zone, or positivity constraints on the coefficients of the model. In general, a *constrained optimization* optimization problem over a variable $\mathbf{x}$ may be written in the following *standard* form:

$$\begin{aligned} \min_{\mathbf{x}} \quad & f(\mathbf{x}), \\ \text{subject to} \quad & g_i(\mathbf{x}) \leq 0, \quad i = 1, \ldots, m, \\ & h_j(\mathbf{x}) = 0, \quad j = 1, \ldots, p, \end{aligned} \tag{1.2}$$

Panel 1.1: Maximizing Projectile Distance

First, let's solve for the best angle α to launch a projectile with initial speed v to maximize its range. The distance traveled is $D(\alpha) = \frac{v^2}{g}\sin(2\alpha)$.

To derive the total distance, we decompose v into the vertical velocity $v_y = v\sin(\alpha)$, and the horizontal velocity $v_x = v\cos(\alpha)$. The acceleration due to gravity is $-g$, so the vertical position is $y(t) = v\sin(\alpha)t - \frac{g}{2}t^2$. The ball lands on the ground again after time T, so $y(T) = 0$ so $T = \frac{2v}{g}\sin(\alpha)$. In this time, the projectile travels $D(\alpha) = Tv\cos(\alpha) = \frac{2v^2}{g}\sin(\alpha)\cos(\alpha) = \frac{v^2}{g}\sin(2\alpha)$ in the horizontal direction.

We now optimize this with respect to α. At a maximizing α, changing α has no affect on the distance D, so the derivative $\mathrm{d}D(\alpha)/\mathrm{d}\alpha = \frac{2v^2}{g}\cos(2\alpha)$ must be equal to zero. Thus, $2\alpha = \pi/2$, which implies an optimal angle, α^*, equal to $\pi/4$.

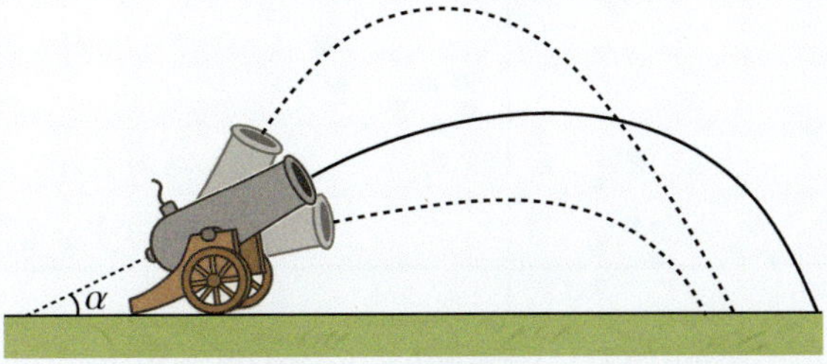

Figure 1.1: Schematic of canon trajectory for various launch angles α.

Panel 1.2: Optimal Lake Crossing

A classic optimization problem is to find the fastest path between points A and B, where some distance is traveled on land, and some is traveled over a lake. If the distance was all traveled on level ground, we know that the shortest path is a straight line. However, when the land speed is v_1 and the swimming speed is $v_2 < v_1$, the solution becomes more interesting. Specifically, what is the optimal point x to cross the lake?

This problem may be approached several ways, which we will explore in the homework. This geometric problem is a simple, discrete cousin of the famous Brachistochrone problem, which is discussed in much greater detail in Section 6.6.

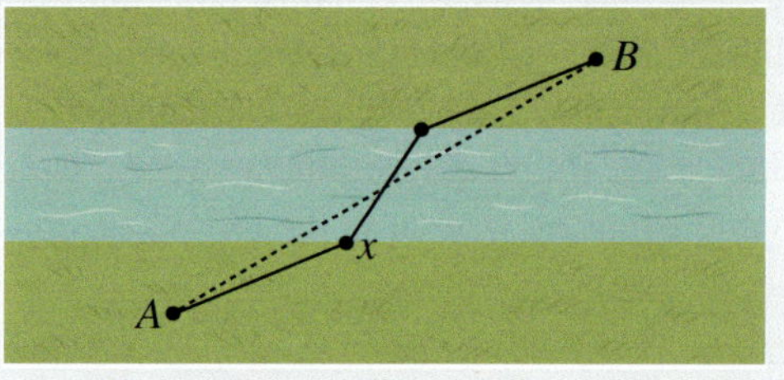

Figure 1.2: Schematic of the lake crossing problem.

Panel 1.3: Data Fitting

A more relevant example is fitting a model to describe data. In this case, we have a collection of data points $\{(x_j, y_j)\}_{j=1}^{n}$, and we optimize the parameters of either a linear or an exponential model. We explore data fitting as an optimization problem in Chapter 4 on least-squares regression and Chapter 10 on machine learning. The following code shows how simple it is to solve for a linear fit in Python.

```
import numpy as np
from scipy.optimize import minimize
x = np.linspace(0, 10, 20); y = 3*x+2 + np.random.randn(20)
def f(p):
    return np.sum((y - (p[0]*x + p[1]))**2) # least squares
res = minimize(f,[0,0]) # initial guess slope=0, intercept=0
a_opt, b_opt = res.x
```

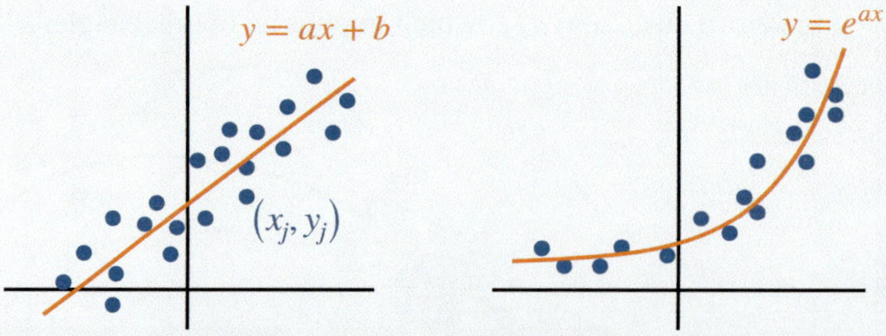

Figure 1.3: Examples of two data-fitting regression problems.

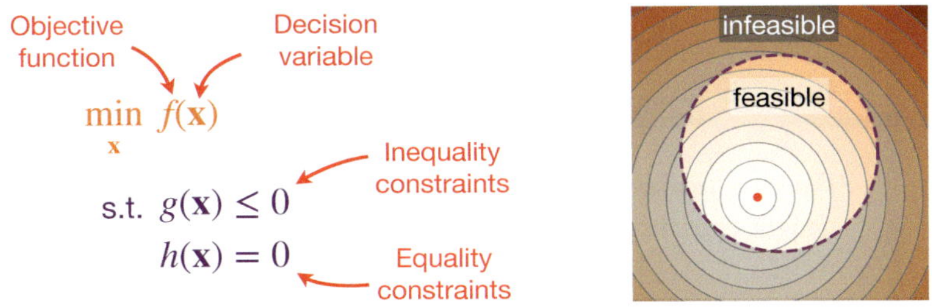

Figure 1.4 The anatomy of an optimization problem. The two main ingredients are the objective function f and the constraint equations g and h. We optimize over the decision variable $\mathbf{x}$. In the case on the right, the constraints are $g(x_1, x_2) = x_1^2 + x_2^2 \leq 1$ and the objective function is $f(x_1, x_2) = (x_1 + 0.5)^2 + (x_2 + 0.5)^2$. We visualize the objective function by logarithmically spaced contours, with the minimum shown as a red dot.

where $f(\mathbf{x})$ is the objective function to be minimized, and $g_i(\mathbf{x})$ and $h_j(\mathbf{x})$ are inequality and equality constraint equations that restrict the set of feasible $\mathbf{x}$ to search over. Typically the objective function f is a scalar function[3] over a vector input $\mathbf{x}$. Figure 1.4 shows this schematically, color coded with the colors used throughout this chapter.

We often write the constraint equations in vector form as:

$$\begin{aligned} \min_{\mathbf{x}} \quad & f(\mathbf{x}), \\ \text{s.t.} \quad & \mathbf{g}(\mathbf{x}) \leq \mathbf{0} \\ & \mathbf{h}(\mathbf{x}) = \mathbf{0}. \end{aligned} \tag{1.3}$$

The inequality and equality constraints determine a feasible set $\mathcal{S}$ of admissible values of $\mathbf{x}$ over which to optimize the objective function f. We use the symbol $\leq$ to indicate that each row of the system of equations must be an inequality; some texts use the symbol $\preccurlyeq$ for this. We also shorten *subject to* to *s.t.*. Several example constraints are considered in Panel 1.4.

When the objective function f and the feasible set $\mathcal{S}$ are both convex, then Eq. (1.2) is a *convex* optimization problem. Convex optimization problems come with strong theoretical guarantees and efficient computational algorithms, as we will discuss in Section 1.2. Linear programming and quadratic programming (e.g., least-squares regression) are two of the most important classes of convex optimization problems. The most important result is that a local minimum of a convex optimization problem is also a global minimum. If the problem is strictly convex, this minimum is unique. Convex sets and functions are discussed later in this chapter, with several examples; see Fig. 1.8 for examples of convex sets and functions.

Maximization is a trivial extension with similar theoretical guarantees when f is concave and the feasible set is convex. Such optimization problems are still considered *convex*, as maximizing a concave function f is equivalent to minimizing the convex function $-f$. For the special case of a linear program, where the linear objective function $f(\mathbf{x}) = \mathbf{c}^T\mathbf{x}$ is both convex and concave, we may pose either minimization or maximization problems.

[3] In multi-objective optimization, the objective function has multiple components and is written as a vector function $\mathbf{f}$. In this case, there is a family of optimal solutions for different weightings of the components of $\mathbf{f}$, which are known as a Pareto front. This is common in engineering applications, such as aircraft design, where there are multiple competing design objectives, (e.g., size, weight, and power). See Panel 5.2.

Panel 1.4: Simple Constraints

There are several intuitive examples of where constraints arise in optimization:

Packing problems: If we are packing a delivery truck, it may have a volume and weight capacity, given by $V_{\max}$ and $W_{\max}$, respectively. Then the total volume and weight of the cargo must be below these max values:

$$\sum_{j=1}^{n} v_j x_j \leq V_{\max} \quad \text{and} \quad \sum_{j=1}^{n} w_j x_j \leq W_{\max},$$

where x_j is the quantity of item j, v_j and w_j are the volume and weight of this item, respectively, and there are n distinct item types. These are *linear* inequality constraints of the form $\mathbf{c}^T\mathbf{x} \leq \mathbf{d}$.

Investment portfolio: Given a total budget B for an investment portfolio, with c_j the cost of asset j, and x_j the quantity purchased, we have an inequality constraint:

$$\sum_{j=1}^{n} c_j x_j \leq B.$$

Communication radius: If a drone must operate in a plane of radius R, where it is able to communicate to a base station, then we have a convex, quadratic constraint:

$$x^2 + y^2 \leq R^2.$$

Controllability ellipsoid: If we are controlling a linear dynamical system and have one unit of control input (e.g., one liter of gasoline), then the state is constrained by an ellipsoid defined by the controllability Gramian (see Chapter 9 and [77]):

$$\mathbf{x}^T\mathbf{W}\mathbf{x} \leq 1.$$

Legal and regulatory constraints: Many businesses, for example in finance, operate within the bounds of legal and regulatory constraints. And as is the nature of constraints in optimization problems, we find that optimal solutions edge right up against the boundary of these constraints.

Standard Form Optimization Problems

One of my goals is for you to learn how to recognize an optimization problem, identify the basic problem template, and choose which solution method to use. There are several standard problems in optimization that we will revisit throughout this book. Two of the most well-studied problems are those of linear programming and quadratic programming. These are the topics of Chapters 3 and 4, respectively.

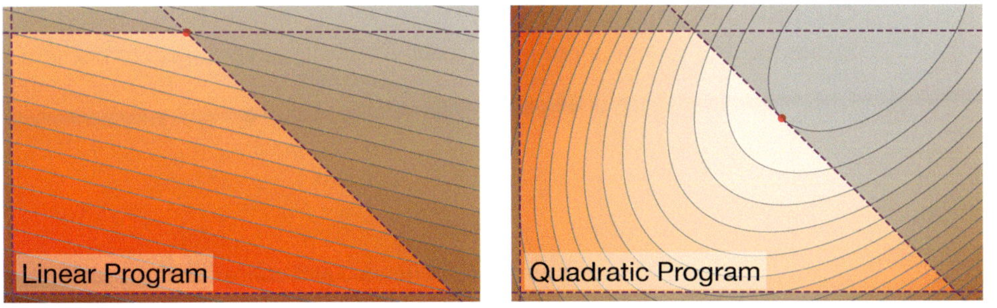

Figure 1.5 Two standard optimization problems are linear programming (left) and quadratic programming (right). These are covered in Chapters 3 and 4, respectively. Linear programs are always convex, and quadratic programs are convex if the matrix $\mathbf{Q}$ in the quadratic objective function is positive semi-definite. In linear programming, when the feasible region is a closed set, the optimal is typically at a vertex of the feasible polytope. In quadratic programming, the optimal point may be in the interior of the feasible region, or at one of the boundaries.

Linear Programming (LP)

Linear programming involves optimizing a linear objective function subject to linear equality and inequality constraints.

Let's construct a simple example. Imagine we are trying to maximize the profit associated with manufacturing two types of handbags: we produce x_1 units of a simple bag with profit $c_1 = \$100$ per unit, and x_2 units of a luxury with profit $c_2 = \$400$ per unit. The total production capacity of our factory means that the total number of units produced must be no greater than 50 and the number of units of item two must be no greater than 30. Finally, we cannot produce a negative number of units, so $\mathbf{x} \geq 0$.

This may be encoded into an optimization problem of the following form:

$$\begin{aligned} \max_{\mathbf{x}} \quad & \mathbf{c}^{\mathrm{T}}\mathbf{x}, \\ \text{s.t.} \quad & \mathbf{A}\mathbf{x} \leq \mathbf{b}, \\ & \mathbf{x} \geq \mathbf{0}. \end{aligned} \tag{1.4}$$

Here $\mathbf{c} \in \mathbb{R}^n$ is the cost vector, while $\mathbf{A} \in \mathbb{R}^{m \times n}$ and $\mathbf{b} \in \mathbb{R}^m$ define the constraints. For the example above, we have

$$\mathbf{c}^{\mathrm{T}} = \begin{bmatrix} 100 & 400 \end{bmatrix}, \quad \mathbf{A} = \begin{bmatrix} 1 & 1 \\ 0 & 1 \end{bmatrix}, \quad \text{and} \quad \mathbf{b} = \begin{bmatrix} 50 \\ 30 \end{bmatrix}.$$

The inequality constraints determine a convex feasible region $\boldsymbol{S}$, called a polytope, that generalizes the notion of a polygon to higher dimensions. The objective function $f(\mathbf{x}) = \mathbf{c}^{\mathrm{T}}\mathbf{x}$ is also convex, making this a convex optimization problem. The linear objective function and convex feasible region are shown in Fig. 1.5 (left).

Note that minimization and maximization problems are both convex, because the linear objective function is both convex and concave, a concept we will explore in a moment.

Quadratic Programming (QP)

Quadratic programming optimizes a quadratic objective subject to linear constraints:

$$\begin{aligned} \min_{\mathbf{x}} \quad & \frac{1}{2}\mathbf{x}^{\mathrm{T}}\mathbf{Q}\mathbf{x} + \mathbf{c}^{\mathrm{T}}\mathbf{x}, \\ \text{s.t.} \quad & \mathbf{A}\mathbf{x} \le \mathbf{b}, \\ & \mathbf{x} \ge \mathbf{0}. \end{aligned} \tag{1.5}$$

Quadratic programs are widely used in finance and control theory, and they are the basis of least-squares regression in data analysis and machine learning. When the matrix $\mathbf{Q} \in \mathbb{R}^{n\times n}$ is a symmetric positive semi-definite matrix, then the quadratic program is *convex*. The quadratic objective function and convex feasible region are shown in Fig. 1.5 (right).

As a concrete example, consider the least-squares regression problem, where we solve for $\mathbf{x}$ to minimize the mean-squared error (MSE)[4] of the fit $\mathbf{A}\mathbf{x} \approx \mathbf{b}$, defined as:

$$\min_{\mathbf{x}} \tfrac{1}{2}\|\mathbf{A}\mathbf{x} - \mathbf{b}\|_2^2. \tag{1.6}$$

We may expand this out as $f(\mathbf{x}) = \frac{1}{2}(\mathbf{A}\mathbf{x} - \mathbf{b})^{\mathrm{T}}(\mathbf{A}\mathbf{x} - \mathbf{b}) = \frac{1}{2}\mathbf{x}^{\mathrm{T}}\mathbf{A}^{\mathrm{T}}\mathbf{A}\mathbf{x} - \mathbf{b}^{\mathrm{T}}\mathbf{A}\mathbf{x} + \frac{1}{2}\mathbf{b}^{\mathrm{T}}\mathbf{b}$. Since the constant $\mathbf{b}^{\mathrm{T}}\mathbf{b}$ doesn't affect the optimization problem, we may remove it, and the objective function now has the form in Eq. (1.5), with $\mathbf{Q} = \mathbf{A}^{\mathrm{T}}\mathbf{A}$ and $\mathbf{c}^{\mathrm{T}} = -\mathbf{b}^{\mathrm{T}}\mathbf{A}$. The $\mathbf{Q}$ matrix in this case is symmetric and positive definite, so the problem is convex, as we will see soon. We may solve this problem with or without constraints, as explored in Chapter 4.

The data fitting problem in Panel 1.3 is a simple least squares regression problem with a quadratic objective function. In this case, the matrix $\mathbf{A}$ and vectors $\mathbf{x}$ and $\mathbf{b}$ are

$$\mathbf{A} = \begin{bmatrix} x_1 & 1 \\ x_2 & 1 \\ \vdots & \vdots \\ x_n & 1 \end{bmatrix} \quad \mathbf{x} = \begin{bmatrix} a \\ b \end{bmatrix} \quad \mathbf{b} = \begin{bmatrix} y_1 \\ y_2 \\ \vdots \\ y_n \end{bmatrix}.$$

The Objective Function FAQ

We often solve unconstrained optimization problems, in which case we are only concerned with the properties of the objective function f. In either the constrained or unconstrained cases, there are several important properties of f that determine which optimization algorithms are well suited to solve the problem.

Is f Convex or Nonconvex? This is the biggest dichotomy of optimization problems. Convex optimization problems have strong theoretical guarantees and powerful algorithms to find the globally optimal solution, even for large-scale problems. We carefully explore what makes an optimization problem convex in this chapter, and we will use this perspective throughout the remainder of the book.

[4] Here, the MSE is given by the square of the ℓ_2 norm, defined as $\|\mathbf{x}\|_2^2 = \sum_{j=1}^{n} x_j^2$. For $\mathbf{x} \in \mathbb{R}^2$ this becomes $\|\mathbf{x}\|_2^2 = x_1^2 + x_2^2$. We discuss norms in much greater detail later in this chapter.

Examples: The linear least-squares regression problem from Chapter 4 is a good example of a convex optimization problem, where there is a single, globally optimal solution. It is even possible to analytically derive an expression for this optimum. Nonlinear least-squares regression, such as training a neural network in Chapter 10 is an example of a nonconvex optimization problem. In this case, the objective function has many local minima and saddle points, making it challenging to find a global optimum.

Can We Compute the Gradient ∇f? Having access to the gradient ∇f allows us to use gradient descent and its variants to find candidate local minima where $\nabla f = \mathbf{0}$. Higher-order derivatives may be used for Newton's method, which has faster convergence rates. Both are discussed in Chapter 2. Even when f is nonsmooth, it may be possible to compute the *proximal gradient*, which is discussed in Section 5.1.

Examples: Sometimes we have an analytic expression for the gradient of f, as in least-squares regression (see Chapter 4). In other cases, we approximate the gradient numerically, for example using finite difference derivatives. When training a neural network, we use stochastic gradient descent and estimate the gradient via backpropagation.

In some cases, gradients are difficult to estimate or do not even exist. For example, some objective functions are nonsmooth and do not admit gradient descent methods. In other cases, we may be optimizing some physical process where we do not have a model of f to differentiate. For example, when optimizing the shape of a compressor blade in an experiment or simulation, we may not have access to a differentiable model of the process, and instead must resort to gradient-free iterative optimization. Gradient-free optimization techniques are typically much more expensive and inefficient; these approaches are explored further in Chapter 5.

In fact gradient-free optimization is so expensive that we often go to great lengths to come up with gradients, for example by developing smooth surrogate models f or developing adjoint simulations for complex shape optimization problems.

Is f Linear or Nonlinear? The structure of the function f plays a major role in the optimization landscape, and this is closely related to convexity above. We will consider several classes of functions f, including linear, quadratic, and more general nonlinear functions. Linear functions are convex, but there is not a clear connection between nonlinearity of f and nonconvexity. However, highly nonlinear functions, such as neural network training landscapes, are more likely to have many local minima and be nonconvex.

Examples: If f is linear, as in linear programming (see Chapter 3) then the objective function is flat, so it is both convex and concave. It is possible to compute the gradient of a linear objective function analytically, and the maximum or minimum of the function depends entirely on the geometry of the feasible set given by the constraint equations.

If f is quadratic, we may also compute the gradient of f analytically, as in Chapter 4. However, the convexity of f now depends on the Hessian matrix $\mathbf{H}$ of second partial derivatives. If $\mathbf{H}$ is positive definite, so that all its eigenvalues are positive, then f has

Panel 1.5: Minima, Maxima, and Saddle Points

An objective function $f(x)$ may have minima, maxima, and saddle points.

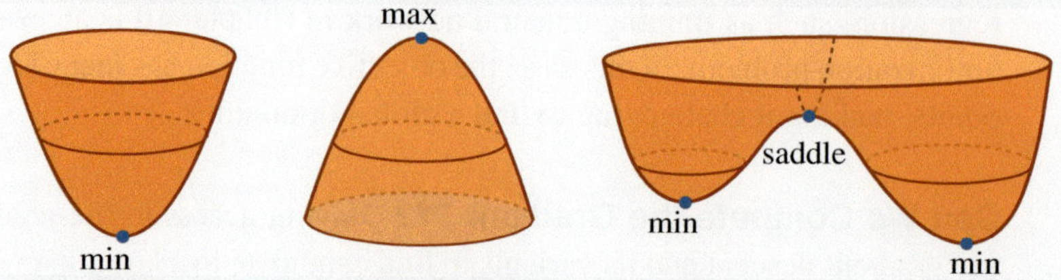

Figure 1.6: Examples of various convex, concave, and indeterminant functions.

positive curvature in every direction $\mathbf{x}$ and is convex. The linear least-squares problem discussed above is a quadratic objective function f with positive definite Hessian $\mathbf{H}$. If $\mathbf{H}$ is negative definite, then f is concave, and if it is indefinite, so that it has both negative and positive eigenvalues, then f has a saddle point and is neither convex nor concave.

For more general nonlinear f, polynomial and otherwise, the function may or may not be well behaved. For example, the function $f(x) = \exp(x)$ is convex with analytical derivative; however, the minimum of $\exp(x)$ occurs at $x \rightarrow -\infty$, so it is never attained. In contrast, the function $f(x) = \sin(x)$ has infinitely many local minima and maxima. However, both functions are relatively easy to work with and it is straightforward to compute derivatives and solve for extrema. In higher dimensions, however, enumerating and characterizing the extrema may become quite challenging or intractable.

Is f High Dimensional? Many modern problems involve very high-dimensional decision variables $\mathbf{x}$, so that gradients and Hessians of f are high-dimensional. In general, high-dimensional optimization problems are more computationally expensive and are prone to ill-conditioned solutions.

Examples: In deep learning, it is not uncommon to train a model with millions or billions of free parameters to be optimized over. This motivates a certain class of algorithm, such as stochastic gradient descent. Even for convex problems, when the dimension is large, the optimization problem may be ill-conditioned or slow to converge. For example, large-scale linear least-squares problems are often ill-conditioned and require regularization techniques to improve the quality of the optimization solution. Many modern optimization packages and algorithms, such as interior point methods (IPMs) are designed specifically to scale robustly to high-dimensional problems. These issues are discussed in Chapters 4 and 6. Further, the issues associated with nonconvex optimization tend to compound in higher dimensions, as the number of possible local minima to be checked may grow exponentially, or even combinatorially, in the dimension of $\mathbf{x}$.

Is f Known or Unknown? In many modern applications, there is not a mathematical model for f, and it must be computed by interacting with a complicated simulation or experiment. When f is not easily representable as an analytical function, or a function that can be represented with a computational graph, then the solution techniques must be modified. For example, if f is a black box simulation or experiment, we may be forced to resort to gradient-free optimization techniques in Chapter 5. However, we are increasingly able to construct data-driven surrogate models for such f using machine learning, and these surrogate models can be used to accelerate the optimization process.

Examples: In many engineering optimization problems, our model of the real world is imperfect, or too expensive to easily model or simulate in a function f. For example, when designing a Formula One car, evaluating a new test design may involve fabricating a model and testing for hours in a wind tunnel. Thus, we may need to optimize over a physical system or experiment, for which there is no well behaved function f. Similarly, we may be optimizing a simulation where some components of the simulation are legacy compiled codes and we no longer have access to the source code.

Is f Expensive to Compute or Evaluate? This is closely related to the previous issue. When the function f is expensive to evaluate, we typically must be more conscientious of how our optimization algorithm scales with function evaluations. Iterative methods, such as gradient descent and Newton's method, have well-understood scaling properties, and the computational cost of f often guides which method to use. Of course, the expense of computing f may be related to the dimension of $\mathbf{x}$ and the nonlinearity of f, compounding with the issues above.

Examples: In many engineering applications, such as aerodynamic shape optimization of a wing, it is expensive to evaluate the objective function for a single $\mathbf{x}$. This may involve a computationally intensive, high-fidelity simulation, or actually fabricating and testing the wing in a wind tunnel. When function evaluation is expensive, we require algorithms that converge in very few iterations, which often motivate surrogate models.

Does f Encode Dynamics in Time? When the system being optimized involves time dynamics, this often changes the optimization algorithm and computational cost involved. In this case, the dynamics of f must be captured via differential equations, which are typically more involved, both conceptually and computationally. Alternatively, we may average out the time dynamics and optimize over mean quantities, although this may also be expensive, requiring enough time for a converged average.

Examples: Control theory (see Chapter 9) is a good example where we optimize a control law to modify the performance of a dynamical system, which inherently varies in time. Other optimization problems, such as aerodynamic shape optimization, may involve time averaging dynamic effects, such as turbulent fluctuations. Many optimization problems we face do not explicitly involve time dynamics, such as packing or resource allocation problems in Chapter 3 or regression problems in Chapter 4.

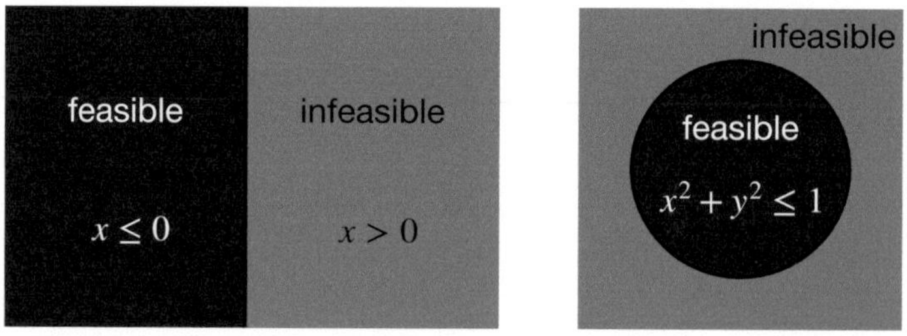

Figure 1.7 Examples of feasible and infeasible sets determined by inequalities. A single linear inequality defines a convex half-space (left); multiple linear inequalities result in intersecting half-spaces and define a polytope (see Fig. 1.11). Nonlinear inequalities define more complex feasible sets (right). In this case, the feasible region is still convex.

The Constraint Equations

Constraint equations arise naturally in optimization. Any time a decision variable involves a constrained resource, such as time, space, money, materials, labor, etc., these may be expressed as constraint equations. This is common in linear programming, which involves minimizing or maximizing an objective function under linearly constrained resources.

We will see many other types of constraints throughout this book. For example, we may want to solve a least-squares regression problem where we know that some variables must be non-negative, or within some radius (see Fig. 1.7). Control theory is another example, where the constraints are governed by the dynamics of the system.

Figure 1.7 shows two examples of feasible sets given by inequality constraints. The first example is a linear inequality constraint, and the second example is a nonlinear inequality constraint. Any linear inequality constraint, given by $g(\mathbf{x}) = \mathbf{c}^T\mathbf{x} \leq 0$, determines a feasible half-plane in the n-dimensional space. The complementary inequality $\mathbf{c}^T\mathbf{x} > 0$ determines an infeasible half plane. The nonlinear constraint $g(\mathbf{x}) = x^2 + y^2 - 1$ determines a feasible circular disk of radius equal to 1, with an infeasible region exterior to this disk. In both cases, the feasible region is convex, as we will see defined formally soon.

The Feasibility Problem

In nearly every domain using optimization – engineering design with physical constraints, finance with regulatory constraints, or machine learning with interpretability constraints – maximizing or minimizing an objective is meaningless if we cannot locate any feasible solutions. Thus, verifying feasibility is a foundational step before optimizing a solution.

For high-dimensional problems with complicated constraint equations, even finding a feasible $\mathbf{x}$ might be a challenge. The *feasibility problem* is

$$\begin{aligned} \text{find} \quad & \mathbf{x}, \\ \text{s.t.} \quad & \mathbf{g}(\mathbf{x}) \leq \mathbf{0} \\ & \mathbf{h}(\mathbf{x}) = \mathbf{0}. \end{aligned} \tag{1.7}$$

Ensuring feasibility (i.e., satisfying all constraints) can be at least as challenging as optimizing an objective. One might have to solve large sets of equations or inequalities to find a point that meets all physical, geometric, or statistical constraints, before even addressing optimality. We will see an example of a feasible set obtained via the intersection of several inequality constraints later, in Fig. 1.11, in the context of linear programming.

Feasibility is challenging because constraints may define complicated sets. High-dimensional constraints can intersect in to cause narrow or fragmented feasible sets. Sometimes it is unknown if a feasible region even exists. Many optimization problems are ill-posed, so small changes to the problem can dramatically change the feasible set. For example, in inverse problems, one must find model parameters consistent with observed data, enforcing model constraints (e.g., PDEs or boundary conditions). Feasible solutions are sensitive to these constraints. Similarly, in control theory, a controller must often satisfy closed-loop stability, performance bounds, or actuator limits. Ensuring a feasible controller even exists can be challenging, especially if the system is barely controllable.

There are several strategies for approaching the feasibility problem. For linear or convex constraints, there are specialized approaches, such as linear matrix inequality solvers, that can systematically search for feasible solutions. For more complex constraints, we may start with a simpler version of the constraints and continuously transition to the more challenging constraints, using homotopy or continuation to track the solutions through this transition. Other approaches involve penalty or barrier methods that transform constraints into large penalty terms in the objective, promoting feasible solutions. Projection and iterative methods, such as Douglas–Rachford methods [160], may also be used to project guesses onto a feasible set; these are often used in PDE-constrained problems.

Box Constraints

One common constraint is where the components of $\mathbf{x}$ must be between a lower and an upper bound, $\mathbf{l}$ and $\mathbf{u}$, respectively. These are *box* constraints:

$$\begin{aligned} \min_{\mathbf{x}} \quad & f(\mathbf{x}), \\ \text{s.t.} \quad & \mathbf{l} \leq \mathbf{x} \leq \mathbf{u}. \end{aligned} \tag{1.8}$$

It is possible to rewrite these constraints in standard form as:

$$\begin{aligned} \mathbf{l} - \mathbf{x} \leq \mathbf{0} \\ \mathbf{x} - \mathbf{u} \leq \mathbf{0}. \end{aligned}$$

Slack Variable Form

It is helpful to know how close the decision variable $\mathbf{x}$ is to the constraint boundaries. For this we introduce a *slack variable* $\mathbf{s} \in \mathbb{R}^m$ and write the constraints in slack variable form:

$$\begin{aligned} \min_{\mathbf{x},\mathbf{s}} \quad & \mathbf{c}^{\mathrm{T}}\mathbf{x}, \\ \text{s.t.} \quad & \mathbf{x}, \mathbf{s} \geq \mathbf{0}, \\ & \mathbf{A}\mathbf{x} + \mathbf{s} = \mathbf{b}. \end{aligned} \tag{1.9}$$

This is equivalent to Eq. (1.4), since $\mathbf{s} = \mathbf{b} - \mathbf{A}\mathbf{x} \geq \mathbf{0}$. Each component s_j of $\mathbf{s}$ quantifies how close $\mathbf{x}$ is to the j-th inequality boundary; $\mathbf{x}$ has as many dimensions as there are constraint equations. This is used in the simplex algorithm, which solves a linear program by iterating through vertices of the feasible region until a local optimum is found.

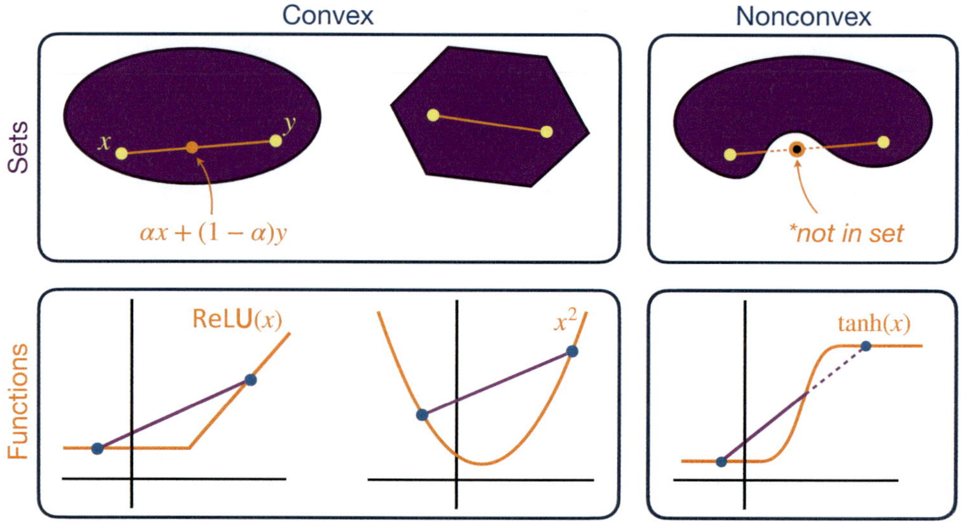

Figure 1.8 Examples of convex and nonconvex sets (top) and convex and nonconvex functions (bottom). Convex sets are defined so that the line segment connecting any two points in the set is itself entirely in the set. Convex functions are defined so that the line segment connecting any two points on the function does not dip below the function at any intermediate values of the domain x. In general, nonconvex feasible sets or nonconvex objective functions complicate the optimization problem considerably. We'll meet the ReLU (rectified linear unit) function in Panel Panel 10.1.

1.2 Convexity in Optimization

Convexity is one of the most important concepts in optimization [66], resulting in strong theoretical guarantees and powerful numerical algorithms. In a convex optimization problem, any local optimal solution is also globally optimal. This is particularly important in high-dimensional optimization problems, where searching for global solutions may be intractable. In contrast, in convex optimization, if an algorithm converges to a local minima, we have confidence that this is the global minima as well.

Convex optimization problems are problems where the objective function is convex and the feasible set determined by the constraint equations are also convex [66, 365, 366]. Occasionally these problems have analytic solutions, such as the classic least-squares regression (see Chapter 4). Other problems, such as linear programming (see Chapter 3), have mature algorithms with favorable computational scaling properties.

In this section, we introduce key notions of convex sets and functions, along with some fundamental theoretical results. Examples of convex sets and functions, as well as nonconvex sets and functions, are shown in Fig. 1.8. The categorization of optimization problems into convex and nonconvex is as fundamental as the categorization of differential equations into linear and nonlinear. Convex optimization problems are largely solved, at least algorithmically, with efficient, robust, and scalable numerical methods.

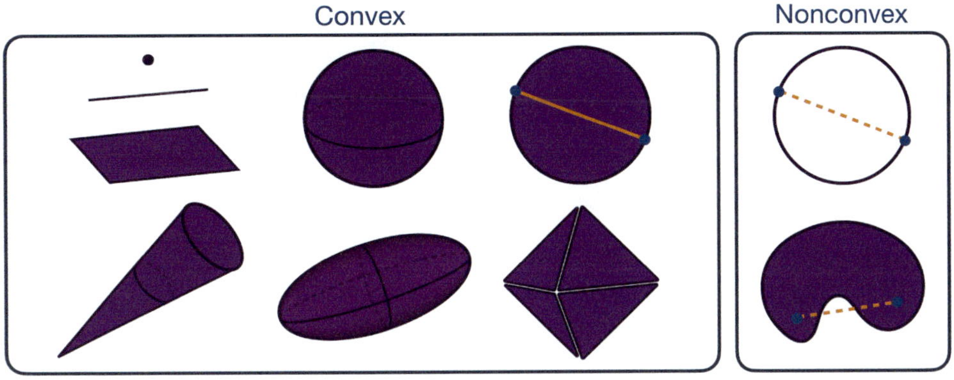

Figure 1.9 Examples of convex and nonconvex sets. Points, lines and line segments, planes and half-planes, spaces and half-spaces (called *slabs*) are all convex. A filled disk or higher-dimensional ball (e.g., a filled sphere), and filled ellipsoids are convex. Polytopes defined by the intersection of multiple inequality constraints generalize polygons to higher dimensions and are also convex. A hollow sphere or a circle without its interior is not convex, because the lines connecting points on the boundary include points that are not in the set; likewise, an object including part but not all of its boundary may be nonconvex.

Convex Sets

It is important to be able to easily assess whether or not the feasible set, defined by the constraint equations, is convex. A set $\boldsymbol{C} \subseteq \mathbb{R}^n$ is convex if and only if the line connecting any two points $\mathbf{x}, \mathbf{y} \in \boldsymbol{C}$ is also entirely in $\boldsymbol{C}$:

$$\alpha\mathbf{x} + (1-\alpha)\mathbf{y} \in \boldsymbol{C} \quad \text{for} \quad \alpha \in [0, 1]. \tag{1.10}$$

The parameter $\alpha \in [0, 1]$ defines the line segment connecting the two points $\mathbf{x}$ and $\mathbf{y}$ in the convex set $\boldsymbol{C}$; $\alpha = 0$ corresponds to the point $\mathbf{y}$, $\alpha = 1$ corresponds to the point $\mathbf{x}$, and $\alpha \in [0, 1]$ sweeps out the line segment between $\mathbf{x}$ and $\mathbf{y}$. This is shown schematically in the top row of Fig. 1.8. If any of these points $\alpha\mathbf{x} + (1-\alpha)\mathbf{y}$ are not also in $\boldsymbol{C}$, for any $\mathbf{x}, \mathbf{y} \in \boldsymbol{C}$, then the set is not convex. In this book, we assume that convex sets are defined in the context of a vector space, although there are more general notions.

Intuitively, a set is convex if there is a continuous *line of site* between any two points in the set, so that there are no disconnected regions, holes, or fingers. It is extremely difficult to search over multiple disconnected regions[5] or fingers, especially in higher dimensions, where the number of such regions may grow exponentially.

Additional examples of convex and nonconvex sets are provided in Fig. 1.9. There are broad classes of convex sets, including points, lines and line segments, planes and half-planes, spaces and half-spaces (sometimes referred to as *slabs*). In addition, several geometric shapes also define convex sets. For example, a disk (i.e., a filled in circle with its boundary) and a higher-dimensional ball

$$B_r = \{\mathbf{x} \mid \|\mathbf{x}\|_2 \leq r\}$$

[5] Disconnected regions and holes are a subject of *topology*.

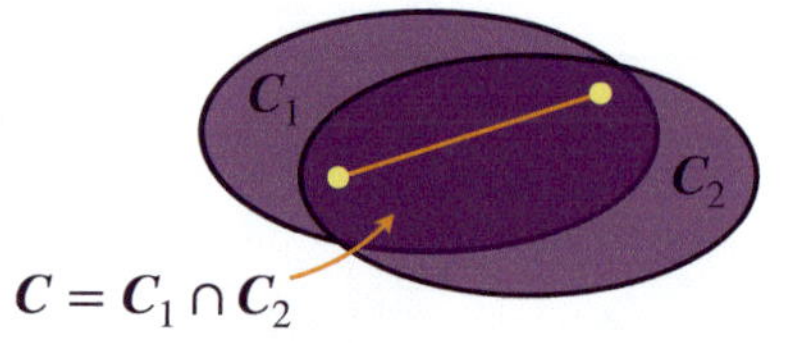

Figure 1.10 The intersection of convex sets is also convex.

(i.e., a filled in sphere) are convex sets. A hollow sphere or a circle without its interior is not a convex set, because the lines connecting points on the boundary are not in the set. Similarly, filled ellipsoids are convex sets. Filled cones, with convex cross-section, are also convex.

Intersection of Convex Sets

The intersection of two convex sets is also convex, as illustrated in Fig. 1.10. This is simple to prove mathematically. Although we won't typically prove results in this book, it is illustrative to see this simple proof.

Proof Denote the two convex sets C_1 and C_2, and let their intersection be

$$C = C_1 \cap C_2.$$

If $\mathbf{x}, \mathbf{y} \in C$, then $\mathbf{x}, \mathbf{y} \in C_1$ and $\mathbf{x}, \mathbf{y} \in C_2$. Since C_1 is convex, then

$$\alpha\mathbf{x} + (1 - \alpha)\mathbf{y} \in C_1$$

for $\alpha \in [0, 1]$. The same is true for C_2. Thus,

$$\alpha\mathbf{x} + (1 - \alpha)\mathbf{y} \in C_1 \cap C_2 = C$$

for $\alpha \in [0, 1]$ for all $\mathbf{x}, \mathbf{y} \in C$, which proves the result. □

The intersection of several linear inequality constraints, $\mathbf{Ax} \le \mathbf{b}$, defines a convex polytope, which generalizes polyhedra to higher-dimensional spaces. An example of a convex polytope is shown in Fig. 1.11. These sets are *pointy*, with the bounding surfaces, edges, and vertices being characterized by subsets of the inequality constraints. Convex polytopes will be explored extensively in the context of linear programming in Chapter 3.

Convex Hull

It is possible to *convexify* a nonconvex set C by systematically addingthe points that violate the convexity condition. The smallest convex set containing a set C is the *convex hull*, and is visualized in Fig. 1.12. This is denoted hull(C). The convex hull of a convex set C is trivially itself. Further, the convex hull of any set C is the intersection of every convex set containing C, hence making it the smallest convex set containing C.

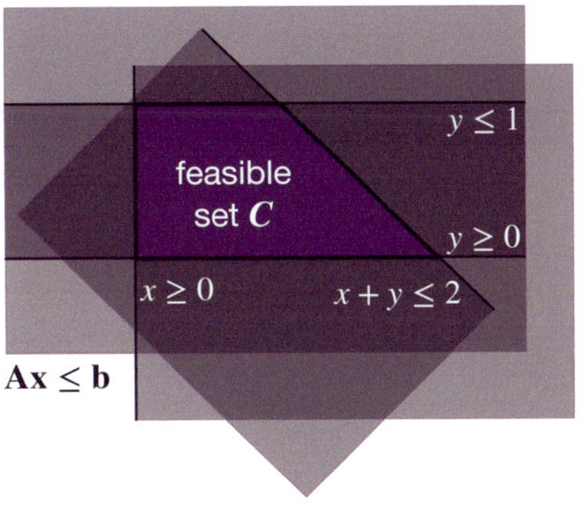

Figure 1.11 Convex polytopes are obtained by the intersection of several linear inequality constraints. Convex polytopes play an important role in several classes of optimization problem, including linear programming (see Chapter 3).

Figure 1.12 Illustration of the convex hull, hull($\mathcal{S}$), of a nonconvex set $\mathcal{S}$.

Affine Sets and Transformations

Affine sets satisfy an even more restrictive condition than convex sets. An affine set is any set in which every line passing through two points of the set lies *completely* within that set. Formally, a set $\mathcal{A}$ is affine if, for any $\mathbf{x}, \mathbf{y} \in \mathcal{A}$ the point $\alpha\mathbf{x} + (1 - \alpha)\mathbf{y}$ lies in $\mathcal{A}$ for any real α. In contrast, a *convex* set requires $\alpha\mathbf{x} + (1 - \alpha)\mathbf{y}$ to remain in the set *only* for $\alpha \in [0, 1]$. For a convex set, the line segment connecting any two points $\mathbf{x}, \mathbf{y}$ in the set must also be in the set, while for an affine set, the entire infinite line connecting them must be in the set.

Affine sets are automatically convex, because if the points $\alpha\mathbf{x} + (1 - \alpha)\mathbf{y}$ are in the set for any scalar α then they are also in the set for $\alpha \in [0, 1]$. A simple example of an affine set is a line $\{\mathbf{x}_0 + t\mathbf{v} : t \in \mathbb{R}\}$. More generally, a plane or hyperplane is affine.

An affine set can be viewed as a "translated" version of a vector subspace. This condition means that $\mathcal{A}$ behaves like a vector space *except* it need not contain the origin, although it may. Thus, an affine set is essentially a vector space that may be shifted or translated by some fixed offset. Affine sets generalize linearvector subspaces by allowing for an arbitrary translation in space. All vector spaces are affine sets, but the converse is not true.

Affine sets arise naturally in optimization when translating or shifting feasible regions. For instance, solving linearsystems $\mathbf{Ax} = \mathbf{b}$ defines an affine subspace. Many linear and affine constraints preserve affine structure, and problems that reduce to finding a point in an affine subspace are simpler than those restricted to convex but curved sets. Hence,

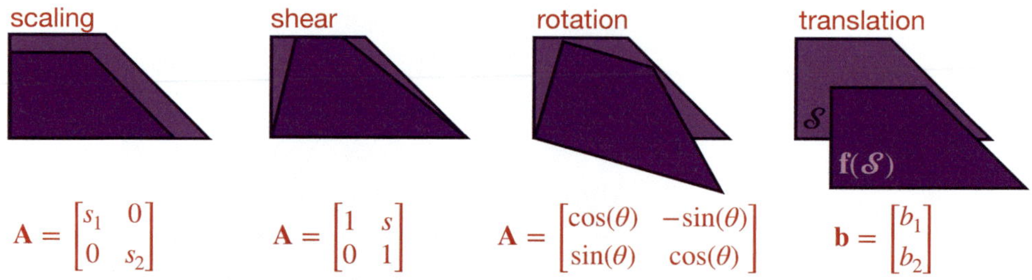

Figure 1.13 Affine transformations $\mathbf{f}(\mathbf{x}) = \mathbf{A}\mathbf{x} + \mathbf{b}$. Scaling, shear, rotation, translation, and combinations (such as reflection) thereof, are all affine transformations, which preserve convexity.

in optimization, one often solves feasibility in an affine set, such as $\mathbf{A}\mathbf{x} = \mathbf{b}$, and then superimposes convex inequalities to further restrict that region. Convex sets remain convex under affine transformations, as shown in Fig. 1.13. This is an exercise for the reader.

Panel 1.6: Dual Spaces and Supporting Hyperplanes

Dual spaces and supporting hyperplanes are another important concept in optimization. A supporting hyperplane to a set $\boldsymbol{C} \subset \mathbb{R}^n$ at a boundary point $\mathbf{x}_0$ is a hyperplane

$$\{\mathbf{x} : \mathbf{y}^\top(\mathbf{x} - \mathbf{x}_0) = 0\}$$

such that $\boldsymbol{C}$ lies entirely in one closed half-space determined by that hyperplane. Equivalently, the hyperplane touches $\boldsymbol{C}$ at $\mathbf{x}_0$ but does not cut through its interior. If $\boldsymbol{C} \subset \mathbb{R}^n$ is a convex set, one may define its *dual* set in a dual vector space. Concretely, for a bounded convex body containing the origin, the dual set $\boldsymbol{C}^*$ is given by

$$\boldsymbol{C}^* = \left\{\mathbf{y} \in \mathbb{R}^n : \mathbf{y}^\top\mathbf{x} \leq 1 \text{ for all } \mathbf{x} \in \boldsymbol{C}\right\}.$$

Geometrically, each supporting hyperplane to $\boldsymbol{C}$ is described by an equation $\mathbf{y}^\top\mathbf{x} = \alpha$, where $\mathbf{y}$ is a normal. The *dual space* is the collection of all such normal directions $\mathbf{y}$ that define half-spaces containing $\boldsymbol{C}$; this entirely characterizes the boundary of $\boldsymbol{C}$.

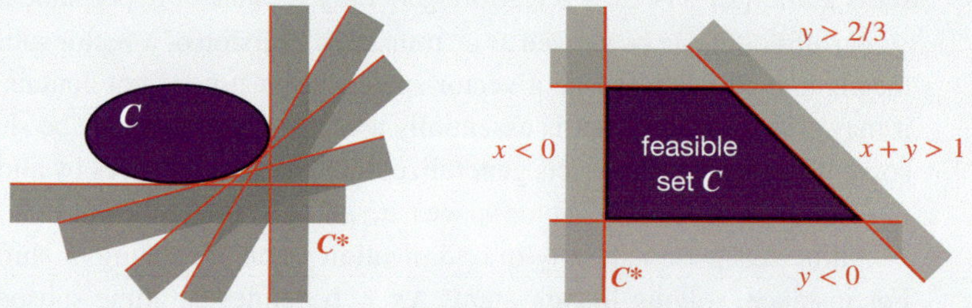

Figure 1.14: Schematic of the dual space of supporting half planes for two convex sets.

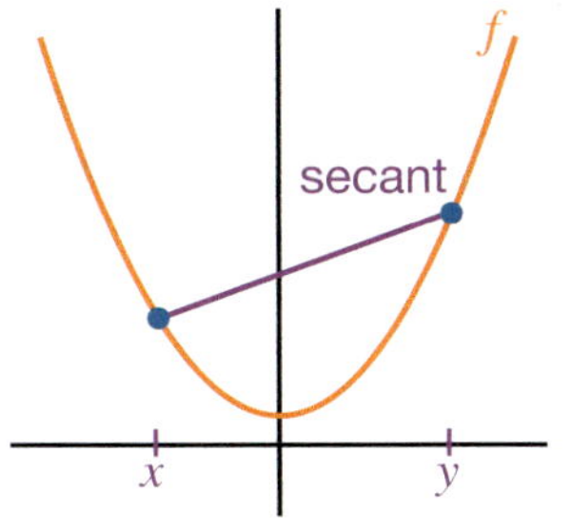

Figure 1.15 A convex function has the property that every secant line connecting any two points on the function, $(\mathbf{x}, f(\mathbf{x}))$ and $(\mathbf{y}, f(\mathbf{y}))$, does not go below the function between $\mathbf{x}$ and $\mathbf{y}$. For a strictly convex function, the secant line is strictly above the function.

Convex Functions

It is also helpful to have conditions to easily assess whether or not an objective function is convex. A function $f : \mathbb{R}^n \to \mathbb{R}$ is convex if and only if the line connecting any two points on the function lies above the function:

$$f(\alpha\mathbf{x} + (1 - \alpha)\mathbf{y}) \leq \alpha f(\mathbf{x}) + (1 - \alpha) f(\mathbf{y}). \tag{1.11}$$

This is illustrated in Fig. 1.15. The line connecting the points $\mathbf{x}$ and $\mathbf{y}$ is called the *secant*. Other examples of convex and nonconvex functions are shown in Fig. 1.16. We often use a more restrictive definition, called *strict* convexity, where the inequality in Eq. (1.11) cannot be equal. This prevents degenerate cases where the function has a flat bottom with a line of equal minima. We may also define $f(\mathbf{x})$ for values $\mathbf{x}$ on a restricted domain $\mathbf{x} \in \mathcal{D} \subseteq \mathbb{R}^n$. In this case, f may be considered convex restricted to $\mathcal{D}$ if Eq. (1.11) holds for any $\mathbf{x}, \mathbf{y} \in \mathcal{D}$.

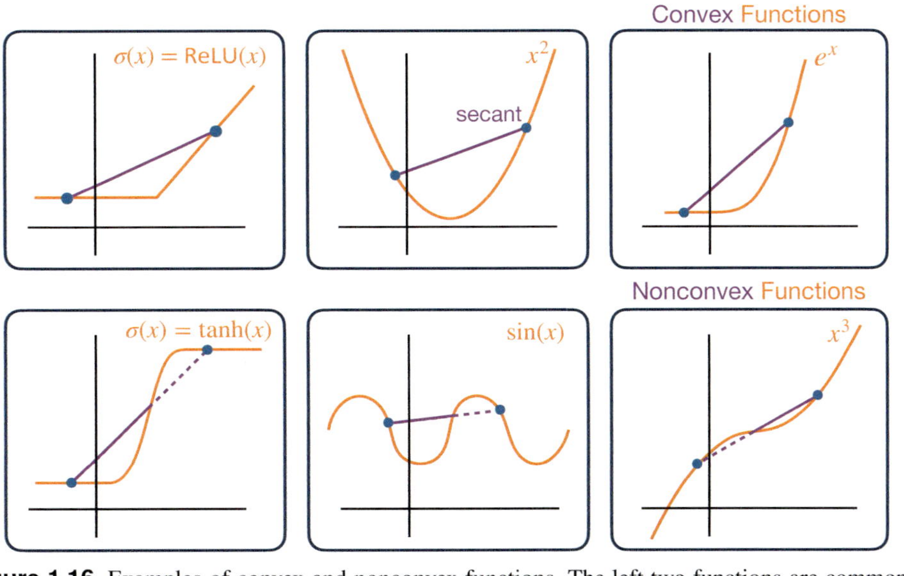

Figure 1.16 Examples of convex and nonconvex functions. The left two functions are commonly used activation functions in the nodes of neural networks, which will be discussed more in Chapter 10.

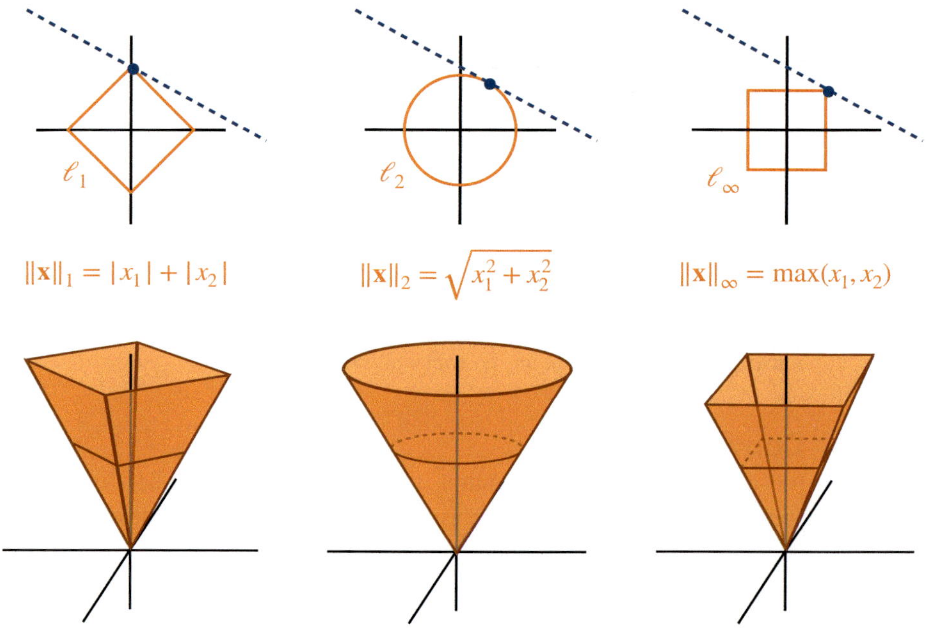

Figure 1.17 Illustration of several norms, all of which are convex functions. This is important in machine learning, where we often minimize the sum-squared error in the ℓ_2 norm, and then add additional regularization terms that are expressed in other norms. The ℓ_1 norm promotes sparsity, the ℓ_2 norm quantifies the sum-squared error, and the ℓ_∞ norm is used to penalize worst-case performance.

Norms are Convex

An important class of convex functions are *norms*, $f(\mathbf{x}) = \|\mathbf{x}\|$. Norms are widely used in optimization, as we often minimize the error of a function fit, where error is quantified in an appropriate norm, such as the ℓ_2 norm. Several norms are shown in Fig. 1.17.

A norm quantifies the length of a vector, and it must satisfy three properties:

1. $\|\alpha\mathbf{x}\| = |\alpha|\|\mathbf{x}\|$ for all scalar $\alpha \in \mathbb{R}$ (homogeneity)
2. $\|\mathbf{x} + \mathbf{y}\| \leq \|\mathbf{x}\| + \|\mathbf{y}\|$ (triangle inequality)
3. $\|\mathbf{x}\| \geq 0$ and $\|\mathbf{x}\| = 0$ only if $\mathbf{x} = \mathbf{0}$. (non-negativity)

To prove that all norms are convex, we may write the convexity condition from Eq. (1.11) when $f(\mathbf{x}) = \|\mathbf{x}\|$ as:

$$\begin{aligned} \|\alpha\mathbf{x} + (1-\alpha)\mathbf{y}\| &\leq \|\alpha\mathbf{x}\| + \|(1-\alpha)\mathbf{y}\| \\ &= \alpha\|\mathbf{x}\| + (1-\alpha)\|\mathbf{y}\| \\ &= \alpha f(\mathbf{x}) + (1-\alpha) f(\mathbf{y}). \end{aligned}$$

The most common norms we use are the ℓ_2 and ℓ_1 norms, given by

$$\|\mathbf{x}\|_2 = \left(x_1^2 + x_2^2 + \cdots + x_n^2\right)^{1/2} = \left(\sum_{k=1}^{n} x_k^2\right)^{1/2} \tag{1.12}$$

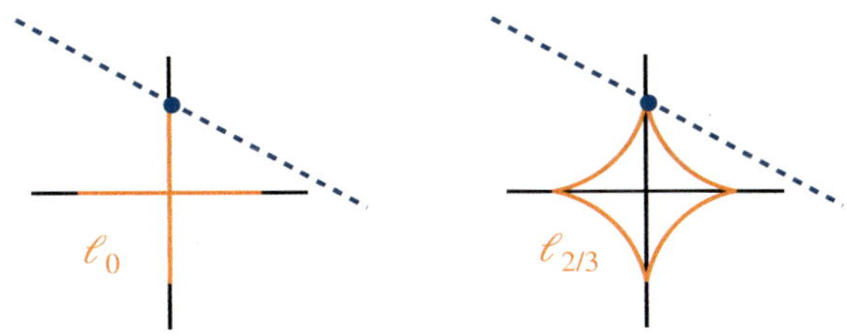

Figure 1.18 Examples of nonconvex *pseudo* norms, which are not actually norms. The ℓ_p norms for $p < 1$ are such nonconvex pseudo norms.

and

$$\|\mathbf{x}\|_1 = |x_1| + |x_2| + \cdots + |x_n| = \sum_{k=1}^{n} |x_k| \tag{1.13}$$

respectively. The ℓ_2 norm is often used to compute the sum-squared errors of a function fit to data, as in Chapters 4 and 10. The ℓ_1 norm is used to promote solutions that are sparse, meaning that many of the entries of $\mathbf{x}$ are zero.

Both the ℓ_1 and ℓ_2 norms are special cases of the ℓ_p norm for vectors $\mathbf{x}$:

$$\|\mathbf{x}\|_p = \left(|x_1|^p + |x_2|^p + \cdots + |x_n|^p\right)^{1/p} = \left(\sum_{k=1}^{n} |x_k|^p\right)^{1/p}. \tag{1.14}$$

The gradient of the ℓ_2 norm in two-dimensions, $\|\mathbf{x}\|_2^2 = x_1^2 + x_2^2$, is given by

$$\nabla\|\mathbf{x}\|_2^2 = \begin{bmatrix} 2x_1 \\ 2x_2 \end{bmatrix} = 2\mathbf{x}.$$

This is useful in the context of least-squares regression (see Chapter 4), where we must minimize the sum of square errors

$$\min_{\mathbf{x}} \|\mathbf{A}\mathbf{x} - \mathbf{b}\|_2^2 \quad \Longrightarrow \quad \nabla\|\mathbf{A}\mathbf{x} - \mathbf{b}\|_2^2 = 0. \tag{1.15}$$

The notion of a convex norm extends to infinite-dimensional L^p Hilbert spaces, where the norm of a function f is now given by

$$\|f\|_p = \left(\int |f|^p \, dx\right)^{\frac{1}{p}}. \tag{1.16}$$

This allows us to extend our optimization framework to function spaces and optimize systems governed by partial differential equations. We now use the Minkowski inequality, which is the triangle inequality for function spaces:

$$\|f + g\|_p \leq \|f\|_p + \|g\|_p.$$

There are several *pseudo* norms that are not actually norms, but are still useful. The ℓ_0 norm (see Fig. 1.18) quantifies how many non-zero entries are in a vector $\mathbf{x}$, which is a measure of its sparsity. Sparsity is essential in inverse problems and machine learning, where we promote models and solutions with the fewest nonzero terms to prevent overfitting. It is possible to minimize the ℓ_0 function, even though it is not convex, by *relaxing* the problem and minimizing the ℓ_1 norm instead. Relaxation is a fundamental technique in nonconvex optimization, and will be discussed more, for example in Chapter 4. Alternatively, it is possible to directly solve the nonconvex ℓ_0 minimization problem using the proximal gradient in Section 5.1.

Regularization and Relaxation

Relaxation and regularization are two important topics in modern optimization that appear throughout the book, starting in Chapter 4. Regularization is the process by which we add terms to the objective function to make ill-posed problems well posed, and it allows one to incorporate prior knowledge and to guide the optimization to solutions with certain properties. For example, we may want to promote sparse or low-rank solutions in many physical contexts. Many regularizing terms also arise as Bayesian priors, as discussed in Chapter 7. However, it is often the case that regularization terms are nonconvex..

It may be possible to relax nonconvex terms in the objective function to related convex approximations, and there is a large body of analysis showing how and when the optimal solutions to the relaxed problems converge to the optimal solutions of the original nonconvex

Panel 1.7: Properties of Convex Functions

There are several important properties of convex functions that we use in this book:

- If $f(\mathbf{x})$ and $g(\mathbf{x})$ are convex, then $f(\mathbf{x}) + g(\mathbf{x})$ is convex. This is useful in the context of machine learning, where we often construct sophisticated loss functions by adding several convex loss terms, as in:

$$L = \|\mathbf{x} - \hat{\mathbf{x}}\|_2^2 + \alpha\|\mathbf{x}\|_2^2 + \lambda\|\mathbf{x}\|_1.$$

- If $f(\mathbf{x})$ is convex, then $-f(\mathbf{x})$ is concave. The minimum of $f(\mathbf{x})$ becomes the maximum of $-f(\mathbf{x})$.
- If $f(\mathbf{x})$ is convex, then $f(\mathbf{A}\mathbf{x} + \mathbf{b})$ is convex. In other words, an affine transformation of $\mathbf{x}$ preserves convexity.
- If $f(\mathbf{x})$ and $g(\mathbf{x})$ are convex, then the pointwise maximum $h(\mathbf{x})$ of f and g is convex:

$$h(\mathbf{x}) = \max\{f(\mathbf{x}), g(\mathbf{x})\}.$$

- The composition of two functions $h(\mathbf{x}) = f(g(\mathbf{x}))$ is convex if either:
  f is convex and non-decreasing and g is convex;
  f is convex and non-increasing and g is concave.
  For example:
  If g is convex then $\exp(g(x))$ is convex;
  If g is concave then $-\log(-g(x))$ is convex;
  if g is concave and positive, then $1/g$ is convex.

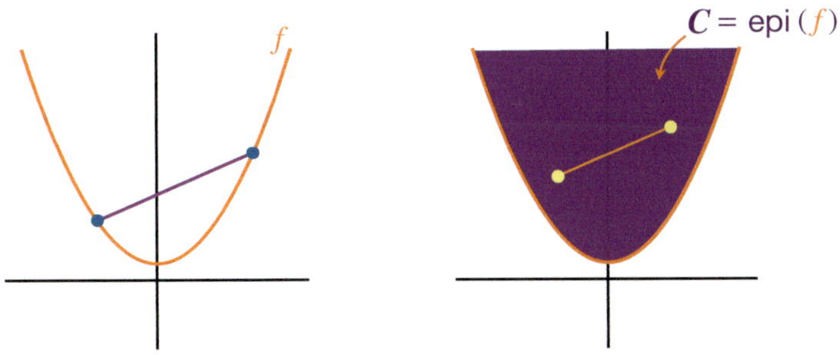

Figure 1.19 The epigraph of a function f is the set of points x, y above the graph. The function f is convex if and only if the epigraph is a convex set.

problems. Often, the addition of a nonconvex term, such as the sparsity promoting $\|\cdot\|_0$ term, leads to an NP-hard optimization problem, where it would require a brute force combinatorial search to find the global optimum. However, it is possible to relax the $\|\cdot\|_0$ pseudo norm by replacing it with the convex $\|\cdot\|_1$ norm. Similarly, if we want a low-rank matrix solution $\mathbf{X}$, we may replace the nonconvex rank $(\mathbf{X})$ with the convex nuclear norm $\|\mathbf{X}\|_*$, which is discussed more in Section 6.5.

Epigraph of a Function

The *graph* of a function is the set of all points on the function:

$$\text{graph}(f) = \{(\mathbf{x}, f(\mathbf{x})) \mid \mathbf{x} \in \mathcal{D}\}. \tag{1.17}$$

This is a useful representation of the function as a set of points in the $n+1$-dimensional space $\mathbb{R}^n \times \mathbb{R}$ given by $(\mathbf{x}, f(\mathbf{x}))$.

The *epigraph* of a function is the set of all points above the function:

$$\text{epi}(f) = \{(\mathbf{x}, y) \mid \mathbf{x} \in \mathcal{D} \text{ and } y \geq f(\mathbf{x})\}. \tag{1.18}$$

The epigraph of a convex function f is shown in Fig. 1.19. A function is convex if and only if its epigraph is a convex set, which is left as an exercise.

The *hypograph* of a function is the set of all points below the function:

$$\text{hyp}(f) = \{(\mathbf{x}, y) \mid \mathbf{x} \in \mathcal{D} \text{ and } y \leq f(\mathbf{x})\}. \tag{1.19}$$

A function is concave if and only if its hypograph is a concave set.

Jensen's Inequality

Jensen's inequality is a powerful and intuitive result for convex functions. In essence, it generalizes the convexity condition for a convex function.

Mathematically, Jensen's inequality may be stated as

$$\alpha_1 f(\mathbf{x}_1) + \alpha_2 f(\mathbf{x}_2) + \cdots + \alpha_m f(\mathbf{x}_m) \geq f(\alpha_1\mathbf{x}_1 + \alpha_2\mathbf{x}_2 + \cdots + \alpha_m\mathbf{x}_m) \tag{1.20}$$

for any $\mathbf{x}_1, \mathbf{x}_2, \ldots, \mathbf{x}_m$ and $\alpha_j \geq 0$ such that $\alpha_1 + \alpha_2 + \cdots + \alpha_m = 1$.

This inequality has an intuitive geometric interpretation. The standard condition for convexity is that the line segment connecting any two points on the function is not below

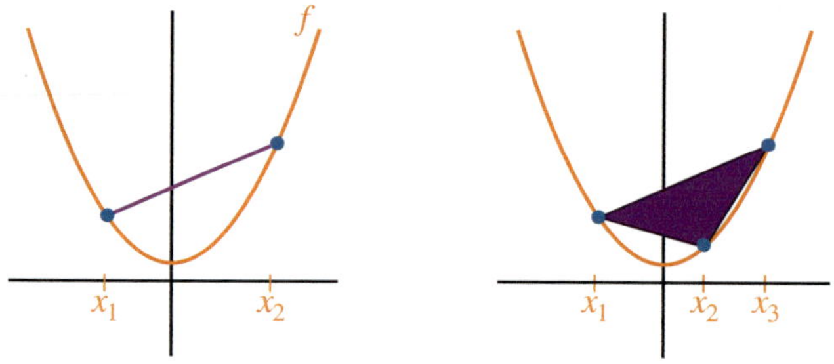

Figure 1.20 Illustration of Jensen's inequality in two-dimensions. Jensen's inequality generalizes the convexity condition that all secant lines lie above the function to state that all convex polytopes formed by m points on the function also lie above the function.

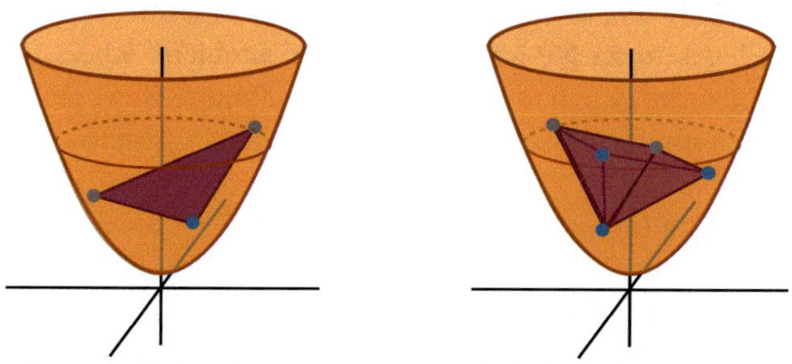

Figure 1.21 Illustration of Jensen's inequality in three-dimensions. In higher dimensions, Jensen's inequality states that any convex polytope formed from m points on the function lie entirely above the function, similar to the secant lines connecting any two points on the function.

the function. Jensen's inequality generalizes this to more than two points on the function. If we have three points (i.e., $m = 3$) on a convex function, the triangle formed with these three points as the vertices also does not lie below the function. This is shown in Fig. 1.20.

This geometric interpretation generalizes to more points and to higher dimensions. Given m points $\mathbf{x}_1, \mathbf{x}_2, \ldots, \mathbf{x}_m$, the linear combination $\alpha_1\mathbf{x}_1 + \alpha_2\mathbf{x}_2 + \cdots + \alpha_m\mathbf{x}_m$ forms a convex polytope in $\mathbb{R}^n$. A subset of the points are vertices of this polytope, and other points are interior. Similarly, the points $(\mathbf{x}_1, f(\mathbf{x}_1)), (\mathbf{x}_2, f(\mathbf{x}_2)), \ldots, (\mathbf{x}_m, f(\mathbf{x}_m))$ form a convex polytope that is entirely a subset of the epigraph of f. In other words, this convex polytope lies entirely above the function f, similar to the secant line. This is shown in Fig. 1.21.

Probability Example. A particularly useful application of Jensen's inequality is in probability. We may use Jensen's inequality to prove a fundamental property about the expected value of a function of a random variable x. If $p(x) \geq 0$ and $\int_S p(x)\,dx = 1$, then

$$f\underbrace{\left(\int_S p(x)x\,dx\right)}_{E(x)} \leq \underbrace{\int_S p(x)f(x)\,dx}_{E[f(x)]}. \tag{1.21}$$

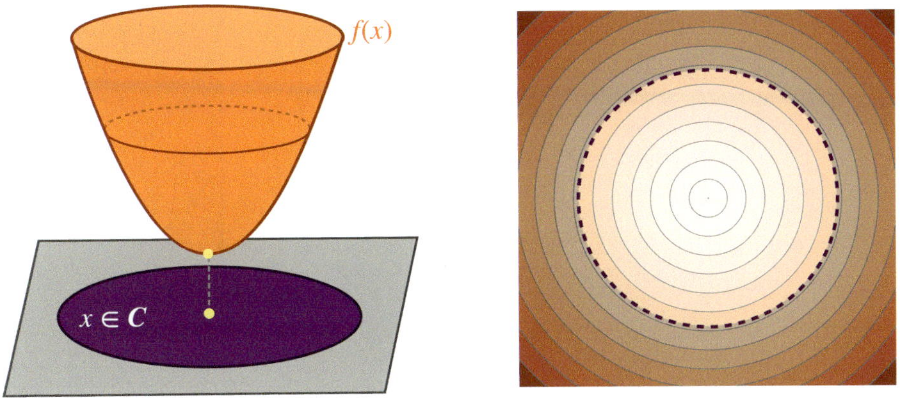

Figure 1.22 Illustration of a convex optimization problem, defined by a convex objective function over a convex feasible set. This example is strictly convex, with a unique global minima.

Thus

$$f\left(E(x)\right) \leq E\left(f(x)\right). \tag{1.22}$$

An example of this inequality is the following:

$$\left(E(x)\right)^2 \leq E\left(x^2\right). \tag{1.23}$$

To prove this using Jensen's inequality, we must extend the finite sum in Eq. (1.20) to a continuous integral. This is left as a homework exercise for the reader.

Convex Optimization Problems

Now we are able to analyze a convex optimization problem. An optimization problem with a convex function f defined over a convex feasible set $\boldsymbol{C}$ is a *convex* optimization problem, shown in Fig. 1.22. If either the function f, or the feasible set $\boldsymbol{C}$ are nonconvex, then the optimization problem is nonconvex as in Fig. 1.23. Nonconvex problems may have multiple local minima, maxima, and saddle points, and the complexity of the optimization landscape typically becomes more challenging in higher dimensions.

We now prove the central result in convex optimization: for a convex function f over a convex set $\boldsymbol{C}$, a local minimum is also a global minimum. We prove this by contradiction.

Proof We assume that $\mathbf{x}$ is a local minimum, so that there is an ϵ-ball around $\mathbf{x}$

$$B_\epsilon(\mathbf{x}) = \{\mathbf{z} \mid ||\mathbf{z} - \mathbf{x}|| < \epsilon\}$$

so that any feasible $\mathbf{z} \in B_\epsilon(\mathbf{x})$ satisfies

$$f(\mathbf{z}) \geq f(\mathbf{x}).$$

We now assume that there exists a $\mathbf{y}$ that is a *better* minima

$$f(\mathbf{y}) < f(\mathbf{x}).$$

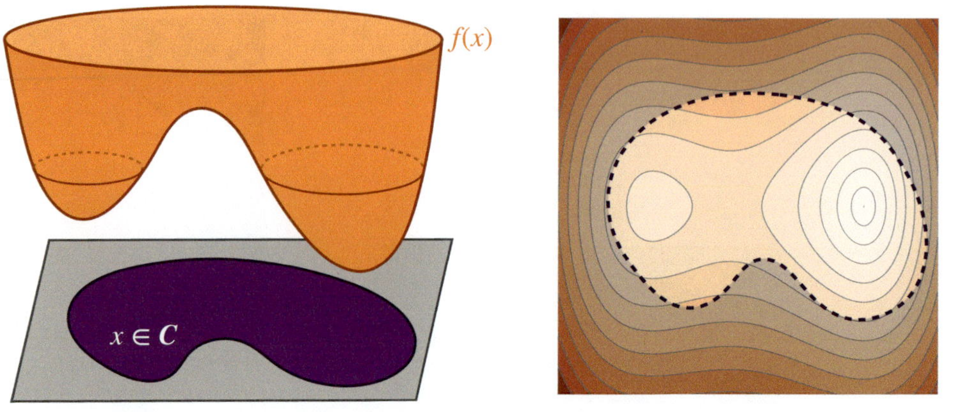

Figure 1.23 Illustration of a nonconvex optimization problem, where the objective function and feasible set are both nonconvex. This example has two local minima and a saddle point.

This is the condition that contradicts the main theorem, and so we will prove that this violates the assumption above.

Because $\mathbf{x}$ and $\mathbf{y}$ are both feasible, in the convex set $\boldsymbol{C}$, then the entire line segment connecting them is also feasible, meaning that it is also in the convex set $\boldsymbol{C}$. These points are parameterized by $\alpha \in [0, 1]$ so that $\alpha = 0$ corresponds to the point $\mathbf{y}$ and $\alpha = 1$ to the point $\mathbf{x}$.

Some of these points are inside the ϵ-Ball. Specifically, we can choose an α so that the point is exactly on the boundary of the ϵ-Ball:

$$\alpha_1 = \frac{\epsilon}{\|\mathbf{y} - \mathbf{x}\|_2}$$

and thus we may choose another α that is halfway between $\mathbf{x}$ and this boundary:

$$\alpha_2 = \frac{\epsilon}{2\|\mathbf{y} - \mathbf{x}\|_2}.$$

We denote this point by $\mathbf{z} = (1 - \alpha_2)\mathbf{x} + \alpha_2\mathbf{y}$.

However, by the convexity of f, and given that $\mathbf{y}$ is a lower minima with $f(\mathbf{y}) < f(\mathbf{x})$, then for every $\alpha \in [0, 1]$, we have

$$\alpha f(\mathbf{y}) + (1 - \alpha) f(\mathbf{x}) < f(\mathbf{x})$$

which implies that

$$f(\mathbf{z}) < f(\mathbf{x}).$$

However, since $\mathbf{z}$ is feasible and $\mathbf{z} \in B_\epsilon(\mathbf{x})$, this contradicts the assumption that $\mathbf{x}$ is a local minimum. □

This is one of the most important results in optimization, providing a powerful guarantee for a large class of convex problems. This property is especially important in Chapters 3 and 4 on linear programming and least squares regression, respectively.

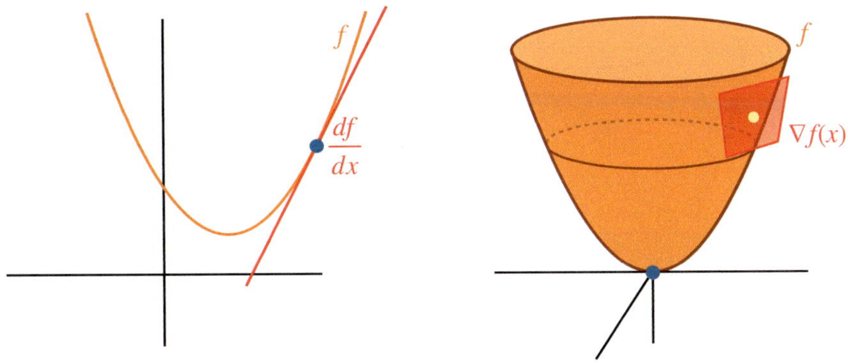

Figure 1.24 Examples of how the derivative and gradient define the tangent line and tangent space, respectively, for a one-dimensional function and a higher-dimensional function.

1.3 Gradients and the Geometry of Optimization

The gradient operator from vector calculus generalizes the derivative of a function of a single variable to the derivative of a function of multiple variables. The gradient has a deep geometric interpretation, quantifying the local direction in which the function increases most rapidly. Thus, we may use this gradient direction to iteratively walk towards a local minimum or maximum of our objective function. This idea is the basis of all gradient descent methods, including the stochastic gradient descent from machine learning, which are discussed in Chapters 2 and 10. The gradientis visualized for a scalar function and a two-dimensional function in Fig. 1.24.

The gradient of a scalar function in two dimensions, $f(x, y)$ is defined as

$$\nabla f = \begin{bmatrix} \partial f / \partial x \\ \partial f / \partial y \end{bmatrix}. \tag{1.24}$$

The components of this vector are partial derivatives that quantify how much the function f changes when x and y are varied infinitesimally. Thus, if the function increases more rapidly in the y direction, the second component will be larger. The gradient of a function $f(\mathbf{x})$ of an n-dimensional vector $\mathbf{x} \in \mathbb{R}^n$ is given by

$$\nabla f = \begin{bmatrix} \partial f / \partial x_1 \\ \partial f / \partial x_2 \\ \vdots \\ \partial f / \partial x_n \end{bmatrix}. \tag{1.25}$$

Each component of ∇f measures how much f changes when the component x_j is varied.

Just as the derivative df/dx of a scalar function $f(x)$ defines a tangent line to the function at the point x, the gradient $\nabla f(\mathbf{x})$ of a function $f(\mathbf{x})$ defines a tangent plane to the function at $\mathbf{x}$. The scalar derivative df/dx is the slope of the tangent line, and $\nabla f(\mathbf{x})$ defines the slope of the tangent plane. Specifically, the gradient $\nabla f(\mathbf{x}_0)$ defines a linear approximation

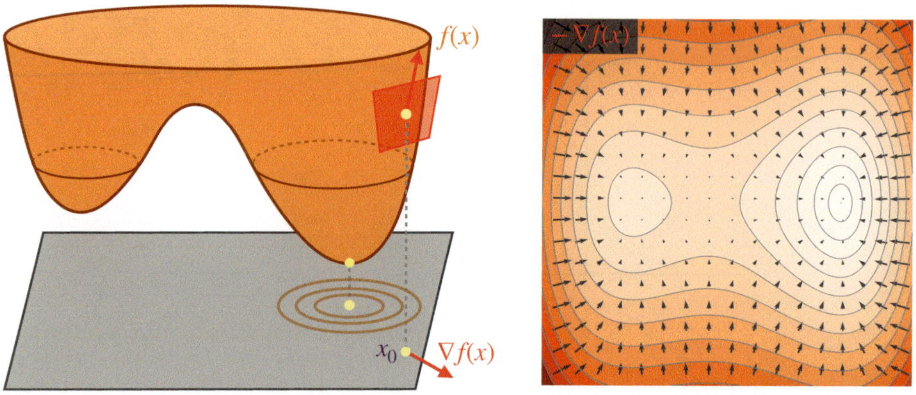

Figure 1.25 Illustration of a nonconvex function, with two local minima. The gradient ∇f is also visualized, both as a tangent plane to the objective function f and as a vector direction over the set of feasible points.

to $f(\mathbf{x})$ near $f(\mathbf{x}_0)$ via a first-order Taylor series approximation:

$$f(\mathbf{x}) \approx f(\mathbf{x}_0) + \nabla f(\mathbf{x}_0)^{\mathrm{T}} \underbrace{(\mathbf{x} - \mathbf{x}_0)}_{\Delta\mathbf{x}}. \tag{1.26}$$

The gradient tells how the function f locally changes, given a small change in $\mathbf{x}$. If we denote the small change by $\Delta\mathbf{x} = \mathbf{x} - \mathbf{x}_0$, then the change in the function is approximated by $\nabla f(\mathbf{x}_0)^{\mathrm{T}}\Delta\mathbf{x}$. In two dimensions, where $\Delta\mathbf{x} = \begin{bmatrix}\Delta x_1 & \Delta x_2\end{bmatrix}^{\mathrm{T}}$, this can be written approximately as:

$$\nabla f(\mathbf{x}_0)^{\mathrm{T}}\Delta\mathbf{x} = \frac{\partial f}{\partial x_1}\Delta x_1 + \frac{\partial f}{\partial x_2}\Delta x_2. \tag{1.27}$$

One of the most important properties of the gradient in the context of optimization is that it defines the direction in which $\Delta\mathbf{x}$ maximally increases the function f locally. This is easy to see from Eq. (1.26), since $\nabla f(\mathbf{x}_0)^{\mathrm{T}} (\mathbf{x} - \mathbf{x}_0)$ is the inner product of the vector $\nabla f(\mathbf{x}_0)$ and $\Delta\mathbf{x}$, and this inner product is largest when the two vectors point in the same direction.

Similarly, if we choose $\Delta\mathbf{x}$ in the negative gradient direction, $-\nabla f(\mathbf{x}_0)$ causes the function to locally decrease the fastest, providing the basis for gradient descent. The gradient of a nonconvex function is visualized in Fig. 1.25, where it is clear that the negative gradient may be used to locally minimize the function.

In some instances, we might not have access to a gradient, because the objective function might be nonsmooth. For example, the ℓ_1 norm does not have a continuous derivative. In this case, we use the *proximal gradient*, introduced in Section 5.1, to generalize the gradient for nonsmooth functions.

Gradients, Hessians, and Convexity

There are deep connections between the gradient of a differentiable function, $\nabla f(\mathbf{x})$, and convexity. In words, a function is convex if and only if all tangent planes defined by the gradientat every point on the function lie entirely below the function. Thus, the tangent planes are never above the function, and only bound it from below. In fact, they define the supporting hyperplanes for the convex epigraph, as discussed below.

We may write this connection mathematically. A differentiable function $f(\mathbf{x})$ is convex if and only if it lives above its first-order Taylor series approximation, given by

$$f(\mathbf{x}) \geq f(\mathbf{x}_0) + \nabla f(\mathbf{x}_0)^{\mathrm{T}} \cdot (\mathbf{x} - \mathbf{x}_0). \tag{1.28}$$

The point $\mathbf{x}_0$ is the base point of the Taylor series expansion, and $\mathbf{x}$ is any other point in the domain of f. This statement is true for all $\mathbf{x}_0$ and $\mathbf{x}$. If there is an $\mathbf{x}^*$ so that $\nabla f(\mathbf{x}^*) = \mathbf{0}$, then $f(\mathbf{x}) \geq f(\mathbf{x}^*)$ for all $\mathbf{x}$. Therefore, $\mathbf{x}^*$ is a global minimum of the function $f(\mathbf{x})$ and $f(\mathbf{x})$ is a convex function. Proving this for a vector $\mathbf{x}$ is not trivial, but it is simpler and can be illustrative to show this for a scalar x.

Second-Order Hessian Condition

Just as the gradient condition in Eq. (1.28) provides a first-order condition for convexity, there is a second-order condition in terms of the Hessian $\mathbf{H} = \nabla^2 f(\mathbf{x})$ for twice continuously

Panel 1.8: Tangent Planes are Supporting Hyperplanes

The graph of $f(\mathbf{x})$ is the pair $(\mathbf{x}, f(\mathbf{x}))$. The tangent plane may be defined by its normal vector, which is perpendicular to the gradient of this point on the graph:

$$\vec{T} = \nabla\,(\mathbf{x}, f(\mathbf{x})) = \begin{bmatrix} \mathbf{1} \\ \nabla_{\mathbf{x}} f(\mathbf{x}) \end{bmatrix},$$

where $\mathbf{1}$ is a vector of ones, obtained by taking the gradient of $\mathbf{x}$: $\nabla \mathbf{x} = \mathbf{1}$. The normal vector that is perpendicular to this tangent vector is given by:

$$\vec{n} = \begin{bmatrix} \nabla f(\mathbf{x}) \\ -\mathbf{1} \end{bmatrix}.$$

For a convex f, the set of all tangent planes defines a set of *supporting hyperplanes* that tightly surround the convex set above the function (i.e., the epigraph of f).

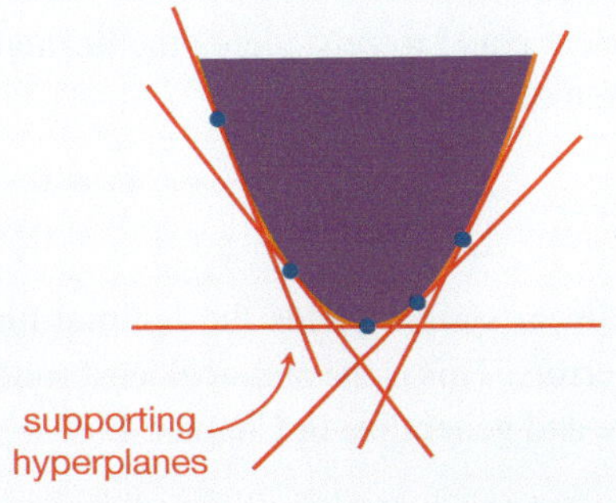

Figure 1.26: Tangents of a convex f define supporting hyperplanes to epi(f).

differentiable functions:

$$\mathbf{H} = \nabla^2 f(\mathbf{x}) = \begin{bmatrix} \frac{\partial^2 f}{\partial x_1^2} & \frac{\partial^2 f}{\partial x_1 \partial x_2} & \cdots & \frac{\partial^2 f}{\partial x_1 \partial x_n} \\ \frac{\partial^2 f}{\partial x_2 \partial x_1} & \frac{\partial^2 f}{\partial x_2^2} & \cdots & \frac{\partial^2 f}{\partial x_2 \partial x_n} \\ \vdots & \vdots & \ddots & \vdots \\ \frac{\partial^2 f}{\partial x_n \partial x_1} & \frac{\partial^2 f}{\partial x_n \partial x_2} & \cdots & \frac{\partial^2 f}{\partial x_n^2} \end{bmatrix}. \tag{1.29}$$

A function $f(\mathbf{x})$ is convex if and only if $\nabla^2 f(\mathbf{x}) \geq \mathbf{0}$ everywhere. This can be seen by expanding $f(\mathbf{x})$ in a second-order Taylor series about $\mathbf{x}_0$:

$$f(\mathbf{x}) \approx f(\mathbf{x}_0) + \nabla f(\mathbf{x}_0)^{\mathrm{T}}(\mathbf{x} - \mathbf{x}_0) + \frac{1}{2}(\mathbf{x} - \mathbf{x}_0)^{\mathrm{T}} \nabla^2 f(\mathbf{x}_0)(\mathbf{x} - \mathbf{x}_0). \tag{1.30}$$

It is clear that Eq. (1.28) is always true if $\nabla^2 f(\mathbf{x}_0)$ is positive semidefinite[6]. When $\mathbf{H} > 0$ is positive definite, then the function is *strictly* convex, meaning that it has a single global minimum; this prevents a line or region of equivalent local minima which is possible when $\mathbf{H} = 0$ and we have a *flat* portion of the objective function. If $\mathbf{H} > 0$ then the function has positive curvature everywhere, meaning it is shaped like a bowl.

Data Fitting Example

Consider the problem of fitting a line to data in Panel 1.3. Recall, the goal is to fit a data set of n points $\{(x_j, y_j)\}_{j=1}^n$ to a straight line of the form $y_j \approx ax_j + b$, where a is the slope and b is the y-intercept. We define the least squares objective as the sum of squared residuals:

$$f(a, b) = \sum_{j=1}^{n} \left(ax_j + b - y_j\right)^2.$$

This is a quadratic function in the parameters a and b, and we can minimize it by taking partial derivatives and setting them equal to zero. Let us compute the gradient:

$$\frac{\partial f}{\partial a} = \sum_{j=1}^{n} 2\left(ax_j + b - y_j\right)x_j,$$

$$\frac{\partial f}{\partial b} = \sum_{j=1}^{n} 2\left(ax_j + b - y_j\right).$$

Setting the gradient equal to zero yields the normal equations, which may also be written as a matrix system of equations:

$$\nabla f(a, b) = 0 \implies \begin{matrix} \sum_{j=1}^n \left(ax_j + b - y_j\right)x_j = 0, \\ \sum_{j=1}^n \left(ax_j + b - y_j\right) \quad = 0 \end{matrix} \implies \begin{bmatrix} \sum x_j^2 & \sum x_j \\ \sum x_j & n \end{bmatrix} \begin{bmatrix} a \\ b \end{bmatrix} = \begin{bmatrix} \sum x_j y_j \\ \sum y_j \end{bmatrix}.$$

Solving this linear system yields the optimal line parameters a^*, b^* that minimize the sum of squared errors. This is the classical least squares solution to linear regression, which we explore more and generalize in Chapter 4.

[6] A matrix $\mathbf{Q}$ is positive semidefinite if $\mathbf{x}^{\mathrm{T}}\mathbf{Q}\mathbf{x} \geq 0$ for all $\mathbf{x}$. We often denote this as $\mathbf{Q} \geq 0$. If $\mathbf{x}^{\mathrm{T}}\mathbf{Q}\mathbf{x} = 0$ only for $\mathbf{x} = 0$ then the matrix is positive definite, denoted as $\mathbf{Q} > 0$.

1.4 Constraints and Lagrange Multipliers

Constraints are a fundamental ingredient of many optimization problems, and there are several approaches to handle them. The method of Lagrange multipliers provides an elegant way to enforce constraints by embedding them into the objective function. This approach not only offers practical techniques for solving constrained problems but also introduces key ideas in duality, which are central to optimization theory. Lagrange multipliers are widely used in control theory and other engineering applications. Here we provide an intuitive overview of Lagrange multipliers, which are explored more in Chapter 6 and throughout the book.

Consider the simple constrained optimization problem with equality constraints:

$$\begin{aligned} \min_{\mathbf{x}} \quad & f(\mathbf{x}), \\ \text{s.t.} \quad & h(\mathbf{x}) = 0. \end{aligned} \tag{1.31}$$

It is possible to add the constraint equation to the loss function through the use of a penalty function $\delta_0(h(\mathbf{x}))$ which is 0 for $h(\mathbf{x}) = 0$ and ∞ for $h(\mathbf{x}) = 0$:

$$\delta_0(h(\mathbf{x})) = \begin{cases} 0, & h(\mathbf{x}) = 0 \\ \infty, & h(\mathbf{x}) \neq 0. \end{cases} \tag{1.32}$$

If we add this penalty function to the objective function

$$\min_{\mathbf{x}} \quad f(\mathbf{x}) + \delta_0(h(\mathbf{x})) \tag{1.33}$$

then the optimal solution is equivalent to that of the original constrained problem. However, the penalty function is no longer smooth or easy to optimize over. Instead we add $\lambda h(\mathbf{x})$, and as $\lambda \to \pm\infty$, this approaches $\delta_0(h(\mathbf{x}))$; specifically, if $h(\mathbf{x}) > 0$ we take $\lambda \to \infty$ and if $h(\mathbf{x}) < 0$ we take $\lambda \to -\infty$. Technically we don't need to take the limit of λ to plus or minus infinity; at some finite λ, this term becomes large enough to make the constraint active, after which the optimal solution remains the same. Mathematically, we write this as

$$\min_{\mathbf{x}} \max_{\lambda} \quad \underbrace{f(\mathbf{x}) + \lambda h(\mathbf{x})}_{\mathcal{L}(\mathbf{x},\lambda)}. \tag{1.34}$$

The variable λ is a *Lagrange multiplier*, which weights the constraint violation, and the new objective function $\mathcal{L}(\mathbf{x}, \lambda)$ is called the *Lagrangian*. We explore this more in Chapter 6.

The inner optimization operates as follows: for a given $\mathbf{x}$, we maximize the Lagrangian over λ. Again, when the constraint is violated, this maximization is unbounded[7] , since infinite λ results in an infinite Lagrangian. However, when the constraint is satisfied, this term goes to zero, and λ no longer has an affect. Then, the outer optimization searches for $\mathbf{x}$ that minimizes this maximum over λ, which is infinite when the constraint is violated, and is just $f(\mathbf{x})$ when the constraint is satisfied.

[7] Because the maximization problem is unbounded for $h(\mathbf{x}) \neq 0$, this inner $\max_\lambda$ should technically be written as $\sup_\lambda$, the supremum, which is also called the *least upper bound*. This is a technical detail we explore in Chapter 6, but for now we simply write it as max.

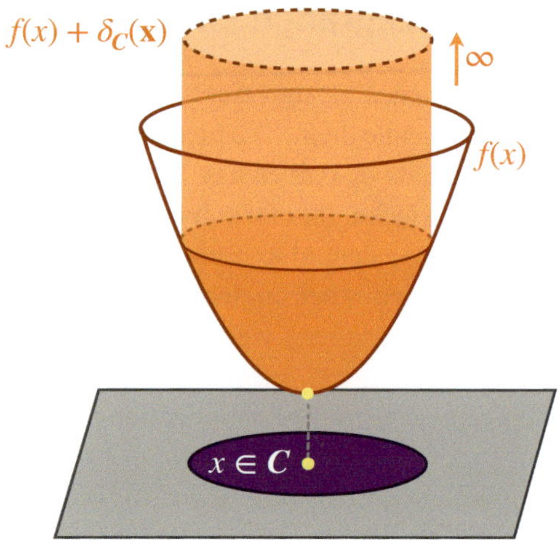

Figure 1.27 Lagrangian objective function with infinite penalty function $\delta_C(\mathbf{x})$ to enforce that $\mathbf{x} \in C$. We will replace this penalty function with Lagrange multipliers.

It is possible to reverse the order of the min and max, resulting in the *dual* optimization problem

$$\max_{\lambda} \min_{\mathbf{x}} \quad \underbrace{f(\mathbf{x}) + \lambda h(\mathbf{x})}_{\mathcal{L}}. \tag{1.35}$$

The optimization problem in Eq. (1.31) is known as the *primal* problem, and Eq. (1.34) is just a different form of the primal. Swapping the order of the $\min_{\mathbf{x}}$ and $\max_{\lambda}$ results in the *dual* problem in Eq. (1.35).

We will show in Chapter 6 that a condition for optimality of these types of constrained problem is that the gradients of the objective function and constraint equation must align:

$$\nabla_{\mathbf{x}}\mathcal{L} = \nabla f(\mathbf{x}^*) + \lambda \nabla h(\mathbf{x}^*) = 0, \quad \nabla_{\lambda}\mathcal{L} = h(\mathbf{x}^*) = 0. \tag{1.36}$$

If there are several equality constraints, so $\mathbf{h}(\mathbf{x}) = 0$, then this may be written as

$$\min_{\mathbf{x}} \quad f(\mathbf{x}) + \boldsymbol{\lambda}^{\mathrm{T}}\mathbf{h}(\mathbf{x}). \tag{1.37}$$

Here $\boldsymbol{\lambda}$ is a vector of the Lagrange multipliers.

For inequality constraints, $g(\mathbf{x}) \leq 0$, we modify the notion of the Lagrange multiplier λ to satisfy additional conditions:

$$\lambda \geq 0, \quad \lambda g(\mathbf{x}) = 0. \tag{1.38}$$

This ensures that λ enforces the constraint only when it is active, meaning that the optimal solution is at the boundary of this constraint equation, and not in the interior. This generalized theory of Lagrange multipliers is know as KKT (Karush–Kuhn–Tucker) theory, which is discussed more in Chapter 6.

For inequality constraints, we use a related penalty function

$$\delta_{\mathbb{R}^+}(g(\mathbf{x})) = \begin{cases} 0, & g(\mathbf{x}) \leq 0 \\ \infty, & g(\mathbf{x}) > 0. \end{cases} \tag{1.39}$$

Again, adding this penalty function to the objective function

$$\min_{\mathbf{x}} \quad f(\mathbf{x}) + \delta_{\mathbb{R}^+}(g(\mathbf{x})) \tag{1.40}$$

is equivalent to the constrained optimization problem. However, the penalty function is no longer smooth or easy to optimize over. Again, we add $\lambda g(\mathbf{x})$ and as $\lambda \longrightarrow \infty$ this approaches $\delta_{\mathbb{R}^+}(g(\mathbf{x}))$. Similar to above, we arrive at a minimax problem:

$$\min_{\mathbf{x}} \max_{\lambda \geq 0} \quad f(\mathbf{x}) + \lambda g(\mathbf{x}). \tag{1.41}$$

Again, it is possible to reverse the order of the min and sup, resulting in the dual problem[8]:

$$\max_{\lambda \geq 0} \min_{\mathbf{x}} \quad f(\mathbf{x}) + \lambda g(\mathbf{x}). \tag{1.42}$$

The dual provides a lower bound on the primal problem and has practical and theoretical significance, particularly in large-scale optimization and sensitivity analysis. It is not obvious that these formulations are equivalent, but under some conditions, the optimal solutions of each are in fact be equivalent. This notion of duality in optimization will be revisited several times in this book.

Examples of Lagrange Multipliers

Here we explore two examples with specific functions to see how Lagrange multipliers work. The first example involves an active constraint, so that the optimal solution is at the boundary of the constraint, and the second example shows how Lagrange multipliers work when there is a trivial constraint that doesn't change the optimum solution.

Example with Active Constraint

Consider the optimization problem

$$\begin{aligned} \min_{x} \quad & f(x), \\ \text{s.t.} \quad & g(x) \leq 0 \end{aligned}$$

with $f(x) = (x - 1)^2 + 1$ and $g(x) = x$. The minimum of the unconstrained optimization problem occurs at $x^* = 1$, but the constraint forces $x \leq 0$. It is possible to plot this objective function and constraint and verify the intuition that the constrained optimal solution is on the boundary of the constraint, so that $x^* = 0$.

Reformulating this problem using Lagrange multipliers yields the dual formulation:

$$\min_{x} \max_{\lambda \geq 0} \quad (x - 1)^2 + 1 + \lambda x.$$

[8] We discuss duality in much greater detail later, for example in Chapter 6. *Minimax* and *maximin* problems often arise in the context of dual problems, where we minimize over one variable, such as the primal variable **x**, and maximize over another variable, such as the dual variable $\boldsymbol{\lambda}$.

Because of convexity, we may swap the order of the inner and outer minimization and maximization and have guarantees that the dual optimum is also the primal optimum. We first compute the inner minimization, which results in

$$\min_x \; f + \lambda g \quad \Longrightarrow \quad \nabla f + \lambda \nabla g = 2x - 2 + \lambda = 0 \quad \Longrightarrow \quad x^* = \frac{1}{2}(2 - \lambda).$$

Next we substitute this into $f(x) + \lambda g(x)$ and maximize with respect to $\lambda \geq 0$:

$$\begin{aligned}\max_{\lambda \geq 0} f(x^*) + \lambda g(x^*) &= \max_{\lambda \geq 0} \left(1 - \frac{\lambda}{2} - 1\right)^2 + 1 + \lambda\left(1 - \frac{\lambda}{2}\right) \\ &= \max_{\lambda \geq 0} \frac{\lambda^2}{4} + 1 + \lambda - \frac{\lambda^2}{2} = \max_{\lambda \geq 0} -\frac{\lambda^2}{4} + \lambda + 1.\end{aligned}$$

The optimum λ is obtained by taking the derivative with respect to λ and setting equal to zero:

$$-\frac{\lambda}{2} + 1 = 0 \quad \Longrightarrow \quad \lambda^* = 2.$$

Plugging this back into our expression for x^* yields $x^* = 0$.

A fundamental question arises: why is $\lambda = 2$ special? Remember, we are not really trying to minimize $f(x) + \lambda g(x)$ but rather minimize $f(x)$ such that $g(x) \leq 0$. If $\lambda < 2$ the constraint is not satisfied. If $\lambda > 2$ the constraint is satisfied but $f(x)$ is not minimized. To visualize this, plot the Lagrangian function for several values of λ, along with the constraint region. This is explored more in Exercise 1.11 and in Chapter 5.

Example with Trivial Constraint

Let's now explore a case where the constraint is not active in the minimization problem. We consider the same constrained optimization problem but with $f(x) = (x + 1)^2 + 1$ and $g(x) = x$. The minimum of the unconstrained problem occurs at $x^* = -1$ which also satisfies the constraint.

Again, we reformulate this problem using the Lagrangian form:

$$\min_x \max_{\lambda \geq 0} \; (x + 1)^2 + 1 + \lambda x.$$

We swap the order of the min and max and solve the inner minimization over x:

$$\frac{\partial \mathcal{L}}{\partial x} = 2x + 2 + \lambda = 0 \quad \Longrightarrow \quad x = -1 - \frac{\lambda}{2}.$$

Substituting this into the Lagrangian yields

$$-\frac{\lambda^2}{4} - \lambda + 1.$$

Maximizing with respect to yields

$$\frac{\partial \mathcal{L}}{\partial \lambda} = -\frac{\lambda}{2} - 2 = 0 \quad \Longrightarrow \quad \lambda = -2.$$

Since this is negative, the maximum of the function for $\lambda \geq 0$ occurs when $\lambda^* = 0$, at which value $x^* = -1$. Thus, when the constraint is not active, then $\lambda^* = 0$.

1.5 Applications of Optimization

The power of optimization is particularly evident in its many modern applications. Entire books could be written about individual application areas, for example in industrial optimization or network flow optimization.

In this text, we focus on three major application areas to motivate and develop theory: inverse problems (Chapter 8), control theory (Chapter 9), and machine learning (Chapter 10). These applications are of particular modern relevance and represent three frontiers of optimization research. In addition, each of these application domains often involve interacting with or modeling the real, physical world, making them crucial for science and engineering.

Modern applications of optimization are typically high-dimensional and nonlinear, leading to nonconvex problems. There are several approaches to either relax these into convex problems or to tackle the full nonconvex optimization, which we develop in later chapters.

Inverse Problems

Inverse problems seek to infer unknown parameters or inputs that best match observed data, which is typically represented by an equation or system where the solution minimizes the discrepancy between observed and modeled data. For example, seismic inversion is a challenging inverse problem, where the goal is to estimate the geophysical properties of the Earth to best fit seismic data. Thus the goal is to find the unknown parameters $\mathbf{x}$ of a forward model $\mathbf{F}(\mathbf{x})$ that best matched observed data $\mathbf{y}_{\text{obs}}$:

$$\mathbf{F}(\mathbf{x}) \approx \mathbf{y}_{\text{obs}}. \tag{1.43}$$

We typically employ optimization to reframe the problem:

$$\min_{\mathbf{x}} ||\mathbf{F}(\mathbf{x}) - \mathbf{y}_{\text{obs}}||^2. \tag{1.44}$$

The objective function is the model mismatch $f(\mathbf{x}) = ||\mathbf{F}(\mathbf{x}) - \mathbf{y}_{\text{obs}}||^2$.

Often, the observations are partial and noisy, and thus don't contain enough information to uniquely determine the unknown parameters, making the inverse problem *ill-posed* or *ill-conditioned*. In such cases, we employ relaxation, regularization, and other prior information to make the problem well posed.

Chapter 8 provides a deep dive into applied optimization for inverse problems. Some of the main examples and case studies include:

Classical inverse problems. There is a rich history of classical inverse problems, such as finding the roots of a polynomial, solving a (potentially ill-posed) matrix system of equations, or estimating the derivative of noisy data. These all provide historical context for fundamental algorithms that are ubiquitous in machine learning and data science.

Inverse problems in medical imaging. Much of the modern work in inverse optimization has centered around inverse problems in imaging sciences, especially in medical imaging. This field is characterized by partial and indirect measurements that

Figure 1.28 Inverse problems, such as super-resolution inversion, are optimization problems, as we seek an inverse solution that minimizes some appropriate loss function. The super-resolution task of inverting a low-resolution image of Mordecai the snow dog to uncover the true high-resolution image is ill-posed, meaning that there are many high-resolution images that yield the same low-resolution input. Thus regularization and other prior information must be added to the optimization problem to promote a unique and reasonable solution.

must be used to reconstruct an image of interest. X-ray computed tomography (CT) and magnetic resonance imaging (MRI) both involve computationally intensive optimization problems, using the Radon transform and Fourier transform, respectively.

Modern computational imaging. Increasingly sophisticated tasks are becoming possible in computational imaging, that were once the subject of science fiction. Figure 1.28 provides one such example of an image-based inverse problem: the task of super resolution, or reconstructing a high-resolution image from a low-resolution image. Other examples include de-noising, de-blurring, filling in damage or occlusions, etc. These are all ill-posed or ill-conditioned inverse problems.

Image-based inverse problem with physics. If an image is obtained from a physical system, it is often possible to use physical priors to regularize the optimization problem. For example, modern computational imaging of black holes involves inverse optimization blended with physics.

PDE-constrained inversion. One of the most challenging classes of inverse problems are those that involve a physical system governed by complex partial differential equation dynamics. For example, designing the shape of a wing to optimize aerodynamic performance, is one of the computational grand challenge problems in engineering. These problems are so challenging in part because of the computational overhead associated with simulating the multiscale dynamics of these systems. Other relevant problems include seismic inversion, or localizing earthquakes and estimating geophysical parameters from seismic data.

System identification. Modeling time series data with ordinary differential equations is a broad topic with a rich history in inverse optimization. This field is closely related to control theory and has dramatically advanced in recent years with advances in applied optimization.

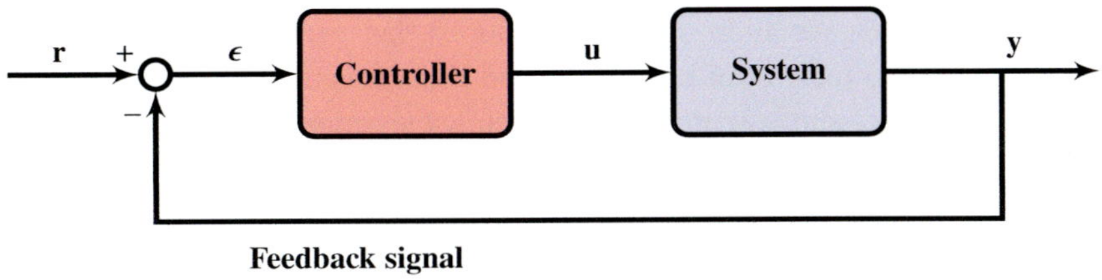

Figure 1.29 Closed-loop feedback control diagram. The sensor signal $\mathbf{y}$ is fed back and subtracted from the reference signal $\mathbf{r}$, providing information about how the system is responding to actuation and external disturbances. The controller uses the resulting error $\boldsymbol{\varepsilon}$ to determine the correct actuation signal $\mathbf{u}$ for the desired response. Feedback is often able to stabilize unstable dynamics while also effectively rejecting disturbances and attenuating sensor noise.

Control Theory

Feedback control is one of the most ubiquitous and important technologies in the modern industrial landscape. The goal of a control system is to design a control law

$$\mathbf{u} = \mathbf{k}(\mathbf{x}) \tag{1.45}$$

that minimizes a cost function

$$J \triangleq J(\mathbf{x}, \mathbf{u}), \tag{1.46}$$

subject to the constraints of a dynamical system that governs the evolution of the system:

$$\frac{\mathrm{d}}{\mathrm{d}t}\mathbf{x} = \mathbf{f}(\mathbf{x}, \mathbf{u}). \tag{1.47}$$

The cost function J is often a positive-semidefinite quadratic form, and for linear dynamics, there is an analytic closed form solution, derived using Lagrange multipliers.

In the more general case, we only have access to partial, noisy measurements $\mathbf{y} = \mathbf{g}(\mathbf{x}) + \mathbf{n}$, where $\mathbf{n}$ is the noise. Moreover, the dynamics model is nonlinear and has uncertainty, and there are disturbances to the system. All of these make the optimization problem more challenging.

Much of the technological progress in the past century can be attributed to improvements in our ability to control complex dynamical systems. There are still many challenges in control theory that will undoubtedly benefit from advanced techniques in applied optimization. Chapter 9 explores several topics in optimization for control theory, including:

Linear optimal control. The most complete theory of control has been developed for linear systems, and this topic provides an intuitive mathematical foundation for nearly every other field of control. Linear optimal control includes the linear quadratic regulator (LQR) full-state feedback control (shown in Fig. 1.29) and the Kalman filter, both of which have analytic solutions. The dynamics enters these problems as a constraint, providing an intuitive connection to Lagrange multipliers.

Nonlinear optimal control. Full nonlinear control problems are characterized by the Hamilton–Jacobi–Bellman (HJB) equations, a set of partial differential equations that must be satisfied for an optimal control solution. The HJB equations are based on variation optimal conditions that generalize the classic ideas of variational principles in Lagrangian and Hamiltonian mechanics.

Reinforcement learning. Reinforcement learning (RL) is an important class of optimization problems at the intersection of control theory and machine learning, where the optimal control actions of an agent are learned through a process of exploration and exploitation. RL is a deep topic, and a more modern treatment involving deep learning is provided in the chapter on machine learning.

Dynamic programming. Dynamic programming may be thought of as the optimization engine used to solve nonlinear optimal control problems. DP is also used extensively in reinforcement learning, although modern approaches are increasingly incorporating deep learning.

Model predictive control. Model predictive control (MPC) is perhaps the most widely-used industrial control optimization technique. MPC uses real-time optimization to determine a control action over a receding, finite-time horizon. The performance and applicability of MPC is directly related to our ability to solve challenging optimization problems in real time, and is thus one of the main applications of modern optimization.

Linear matrix inequalities. Linear matrix inequalities (LMIs) are how we solve many linear control problems in practice using modern techniques in optimization. LMIs are often not included in a course on control theory or in a course on optimization. However, they are the backbone of many control implementations.

Differentiable control. If the system model and controller are differentiable, for example represented as neural networks, then powerful gradient-based optimization techniques become available, even for high-dimensional control problems.

Machine Learning

Machine learning is one of the most ubiquitous and compelling modern application of applied optimization. Machine learning involves using optimization to tune the parameters of a model to minimize a loss function averaged over a set of training data; a schematic of a neural network is shown in Fig. 1.30. The machine learning model is a function

$$\mathbf{y} = \mathbf{f}_{\boldsymbol{\theta}}(\mathbf{x}) \tag{1.48}$$

where $\mathbf{x}$ are the inputs to the model, $\mathbf{y}$ are the outputs, and $\boldsymbol{\theta}$ are the parameters to be optimized over. The structure of the machine learning model $\mathbf{f}_{\boldsymbol{\theta}}$ defines a space of possible functions that might describe the data, parameterized by $\boldsymbol{\theta}$. For example, $\boldsymbol{\theta}$ might be the parameters of a deep neural network, making this a high-dimensional optimization problem.

In machine learning, the objective function to minimize is called a *loss function* L. The loss function typically includes several terms, with the primary term being the mean squared error (MSE)

$$L = \sum_{j=1}^{N} \|\mathbf{y}_j - \mathbf{f}_{\boldsymbol{\theta}}(\mathbf{x}_j)\|^2 \tag{1.49}$$

summed over the number of data points N.

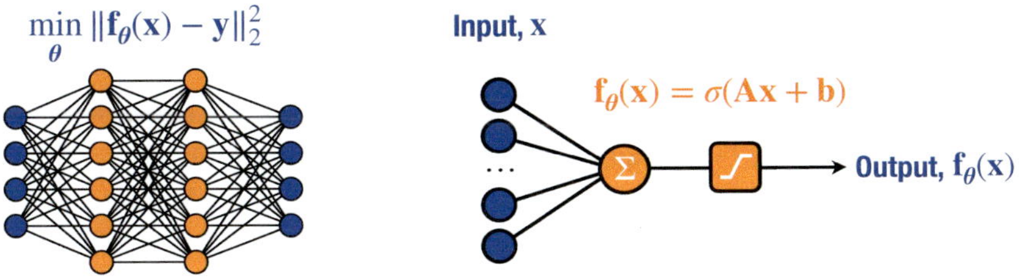

Figure 1.30 Training a neural network is inherently an optimization problem, involving tuning the weights of a neural network to minimize a loss function averaged over the training data.

Chapter 10 provides an in-depth discussion of how optimization is used in various aspects of machine learning, including:

Automatic Differentiation and Backpropagation. The algorithmic evaluation of the derivatives of complex functions is at the heart of optimization for machine learning. These so-called automatic differentiation techniques have been around since the 1960s, but have more recently been used to train neural networks using the backpropagation algorithm. Backpropagation is used to estimate the gradient of the loss function L with respect to the tunable parameters $\boldsymbol{\theta}$, so that gradient descent, and its stochastic variants, may be applied.

Deep learning. Deep learning is the fastest growing field of machine learning, where a deep neural network is used for function approximation. This field is fueled by an abundance of data, modern high-performance computing architectures, and advanced optimization techniques. Stochastic gradient descent is the optimization engine underlying deep learning. It scales to very large systems and incorporates momentum methods, such as the ADAM optimizer, to improve convergence.

Hyperparameter optimization. Machine learning algorithms involve several hyperparameters, such as the size of the neural network, weights of regularizers in the objective, etc. Tuning hyperparameters is a challenging meta-optimization problem.

Deep reinforcement learning. Deep learning has massively improved the scope and applicability of reinforcement learning, for example to improve protein folding, nuclear fusion, to dominate humans in nearly all games, and for other advanced control applications.

Generative methods. Generative AI, such as large language models and diffusion models for image generation, are becoming increasingly powerful and involve advanced optimization methods to formulate and train these models.

Physics-constrained machine learning. Incorporating prior knowledge, such as known physics, is one of the open frontiers of machine learning research. There are two main approaches that are discussed here: (1) incorporating physical priors as regularizing terms in the loss function (e.g., physics-informed neural networks), and (2) directly enforcing physics as constraints in a constrained optimization problem.

Suggested Reading

Texts

(1) **Convex Optimization**, by S. Boyd and L. Vandenberghe, 2009 [66].
(2) **Numerical Optimization**, by J. Nocedal and S. J. Wright, 1999 [370].
(3) **Practical Methods of Optimization**, by R. Fletcher, 2000 [174].
(4) **Introductory Lectures on Convex Optimization: A Basic Course**, by Y. Nesterov, 2013 [365].
(5) **Constrained Optimization and Lagrange Multiplier Methods**, by D. P. Bertsekas, 2014 [48].
(6) **Lectures on Convex Optimization**, by Y. Nesterov, 2018 [366].
(7) **Engineering Optimization: Theory and Practice**, by S. S. Rao, 2019 [408].
(8) **Practical Optimization**, by P. E. Gill, M. Walter, and M. H. Wright, 2019 [193].
(9) **Introduction to Applied Optimization**, by U. M. Diwekar, 2020 [143].
(10) **Engineering Design Optimization**, by J. R. R. A. Martins and A. Ning, 2021 [340].

Papers and reviews

(1) **CVXPY: A Python-embedded modeling language for convex optimization**, by S. Diamond and S. Boyd, *Journal of Machine Learning Research*, 2016 [140].

Homework

Exercise 1.1 An aircraft must fly 2000 miles. It costs \$1000 per hour to fly. In addition, fuel costs \$5 per gallon, and the rate of fuel consumption is $100 + .001V^2$, where V is the cruising speed, measured in miles per hour.

1. What is the minimum-cost cruising speed V, time of flight t, and trip cost C?
2. Suppose the cost of fuel doubles. What is the minimum-cost cruising speed, time of flight, and trip cost?
3. Solve this optimization problem as a function of the fuel cost and plot V, t, and C versus fuel cost.

Exercise 1.2 Consider the projectile problem in Panel 1.1. Imagine we want to hit a particular point at some distance d away. Write this as an optimization problem, and discuss how it might be solved using some of the techniques from this book.

Exercise 1.3 Consider the problem of finding the optimal point to cross a river in Panel 1.2. Let A have coordinates $(0, 0)$ and B have coordinates (L, H), where the distance H is equally spaced between the first stretch of land, the river, and the second stretch of land, so that each has width W and $H = 3W$. More formally write down the optimization problem to solve for x.
Solve for x using calculus. Now write a short computer program to solve for the best x. Compare your answers.

Exercise 1.4 Consider the following general optimization problem domains. In each, determine what are the decision variables $\mathbf{x}$, what is the objective function $f(\mathbf{x})$ and what are the constraints on the system. Try to add as much detail as possible to put these systems into a standard form.

1. Portfolio management in a hedge fund.
2. Optimizing the shape of a wing for aerodynamic performance.
3. Optimizing the pricing of concert tickets in the Sphere.

Exercise 1.5 Determine if the following functions are convex or not. What is the largest domain of convexity around the origin?

1. $f(x) = x^4$.
2. $f(x) = x^4 - .01x^2$.
3. $f(x, y) = -e^{x^2+y^2}$.
4. $f(\mathbf{x}) = -e^{\|\mathbf{x}\|}$.
5. $f(\mathbf{x}) = \sum_j |x_j|$.
6. $f(x, y) = 4\cos(x)^2 + 2\sin(y)^2$.
7. $f(x, y) = 4\cos(x)^2 - 2\sin(y)^2$.
8. $f(x, y) = xy$.

Exercise 1.6 Sketch the feasible sets for the following constraint equations:

1. $S_1 = \{(x, y) \in \mathbb{R}^2 : x \geq 0, y \geq 0\}$;

2. $S_2 = \{(x, y) \in \mathbb{R}^2 : y = 2x + 3\}$;
3. $S_3 = \{(x, y) \in \mathbb{R}^2 : y \geq x^2\}$;
4. $S_4 = \{(x, y) \in \mathbb{R}^2 : y^2 \leq x\}$;
5. $S_5 = \{(x, y) \in \mathbb{R}^2 : |x| + |y| \leq 1\}$;
6. $S_6 = \{(x, y) \in \mathbb{R}^2 : x^2 + y^2 = 1\}$;
7. $S_7 = \{(x, y) \in \mathbb{R}^2 : x^2 + y^2 \leq 1\}$;
8. $S_8 = \{(x, y) \in \mathbb{R}^2 : x^2 + y^2 \geq 1\}$;
9. $S_9 = \{(x, y) \in \mathbb{R}^2 : 1 \leq x^2 + y^2 \leq 4\}$;
10. $S_{10} = \{(x, y) \in \mathbb{R}^2 : x^2 - y^2 \leq 1\}$.

Which of these are convex or nonconvex?

Exercise 1.7 Prove the following:

1. If $f(\mathbf{x})$ is convex, then $f(\mathbf{x}) + k$ is convex for any constant $k \in \mathbb{R}$;
2. If $f(\mathbf{x})$ is convex, then $\alpha f(\mathbf{x})$ is convex for any constant $\alpha \in \mathbb{R}$;
3. If $f(\mathbf{x})$ is convex, then $f(\mathbf{Ax})$ is convex;
4. If $f(\mathbf{x})$ is convex, then $f(\mathbf{x} + \mathbf{b})$ is convex;
5. If $f(\mathbf{x})$ is convex, then $f(\mathbf{Ax} + \mathbf{b})$ is convex for any affine transformation $\mathbf{Ax} + \mathbf{b}$. If the original domain is $\mathcal{D}$, what is the new domain?

Exercise 1.8 Show that the triangle inequality holds for the following four norms:

1. $f(\mathbf{x}) = \|\mathbf{x}\|_1$;
2. $f(\mathbf{x}) = \|\mathbf{x}\|_2$;
3. $f(\mathbf{x}) = \|\mathbf{x}\|_2^2$;
4. $f(\mathbf{x}) = \|\mathbf{x}\|_\infty$.

Exercise 1.9 Compute the gradient of the four norms in the previous problem in two-dimensions and visualize in the x_1-x_2 plane.

Exercise 1.10 Compute the gradient ∇f and Hessian $\nabla^2 f$ of the following functions:

1. $f(x, y) = x^2 + y^2$;
2. $f(x, y) = x^4 - x^2y^2 + y^4$;
3. $f(x, y) = x^2y + y^2x$;
4. $f(x, y) = e^{x^2+y^2}$;
5. $f(x, y) = \sin(xy)$;
6. $f(\mathbf{x}) = \|\mathbf{x}\|^2 = \mathbf{x}^\mathrm{T}\mathbf{x}$;
7. $f(\mathbf{x}) = \|\mathbf{x}\|_\mathbf{Q}^2 = \mathbf{x}^\mathrm{T}\mathbf{Qx}$;
8. $f(\mathbf{x}) = \mathbf{a}^\mathrm{T}\mathbf{x} + b$, where $\mathbf{a} \in \mathbb{R}^n, b \in \mathbb{R}$;
9. $f(\mathbf{x}) = \ln(1 + \|\mathbf{x}\|^2)$;
10. $f(\mathbf{x}) = e^{\mathbf{a}^\mathrm{T}\mathbf{x}}$, where $\mathbf{a} \in \mathbb{R}^n$.

Exercise 1.11 Determine the extrema of the following functions, and which of these are local/global minima:

1. $f(x) = \cos(x)$;

2. $f(x) = e^{-x^2}$;
3. $f(x) = -e^{(x-5)^2}$;
4. $f(x, y) = x^2 - y^2$;
5. $f(x, y) = x^4 + y^4 - x^2y^2$;
6. $f(x, y) = x^2 + 2y^2 - 4x$;
7. $f(x) = x^6 - 3x^4 + 2x^2$;
8. $f(\mathbf{x}) = \|\mathbf{x}\|^2 + 2\mathbf{a}^\mathrm{T}\mathbf{x} + b$, where $\mathbf{x} \in \mathbb{R}^n$ and $\mathbf{a}, b$ are constants;
9. $f(\mathbf{x}) = \|\mathbf{x}\|^3$, where $\mathbf{x} \in \mathbb{R}^n$.

Exercise 1.12 Identify the following optimization problems as convex or nonconvex, and justify your answer:

1. Minimize $f(\mathbf{x}) = x_1^2 + x_2^2$ subject to $x_1 + x_2 = 1$;
2. Minimize $f(\mathbf{x}) = e^{x_1} + e^{x_2}$ subject to $\| \mathbf{x}\| \leq 2$;
3. Maximize $f(\mathbf{x}) = x_1x_2$ subject to $x_1, x_2 \geq 0$ and $x_1 + x_2 = 5$.

Exercise 1.13 Show that the L_0 "norm"

$$\|\mathbf{x}\|_0 = \#\text{nonzero entries of } \mathbf{x}$$

is not actually a norm. What property does it violate?

Exercise 1.14 Explain why the ℓ_1 norm is a suitable proxy for the ℓ_0 pseudo norm. Does this relaxation become more or less effective for higher dimensional $\mathbf{x} \in \mathbb{R}^n$?

Exercise 1.15 Prove that a function is convex if and only if its epigraph is convex.

Exercise 1.16 Prove that for three points $\mathbf{x}_1, \mathbf{x}_2, \mathbf{x}_3$ then $\alpha_1\mathbf{x}_1 + \alpha_2\mathbf{x}_2 + \alpha_3\mathbf{x}_3$ forms a convex triangle in $\mathbb{R}^n$ if $\alpha_1 + \alpha_2 + \alpha_3 = 1$. This generalizes Jensen's inequality.

Exercise 1.17 Prove that for a set of m points $\mathbf{x}_1, \mathbf{x}_2, \ldots, \mathbf{x}_m$ then $\alpha_1\mathbf{x}_1 + \alpha_2\mathbf{x}_2 + \cdots + \alpha_m\mathbf{x}_m$ forms a convex polytope in $\mathbb{R}^n$ if $\alpha_1 + \alpha_2 + \cdots + \alpha_m = 1$. Give an example where every point $\mathbf{x}_j$ is a vertex of the polytope, and give an example where some of the points are not vertices of the resulting polytope.

Exercise 1.18 Prove Jensen's inequality.

Exercise 1.19 Generalize Jensen's inequality to continuous integrals, as in Eq. (1.21).

Exercise 1.20 **[Hard]** Show that if a set $\mathcal{S} \subseteq \mathbb{R}^n$ is convex, then its projection onto any subset of of the coordinates of $\mathbb{R}^n$ is also convex.

Exercise 1.21 **[Hard]** Show that if a set $\mathcal{S} \subseteq \mathbb{R}^n$ is convex, then its affine projection onto any plane will also be convex.

Exercise 1.22 Show that if a set $\mathcal{S} \subseteq \mathbb{R}^n$ is convex, then its projection onto any affine space is convex.

Exercise 1.23 Given two points $\mathbf{a}, \mathbf{b} \in \mathbb{R}^n$, show that the set of points closer to $\mathbf{a}$ than $\mathbf{b}$ is a convex half-plane:

$$\mathcal{S} = \{\mathbf{x} \mid \|\mathbf{x} - \mathbf{a}\| \leq \|\mathbf{x} - \mathbf{b}\|\}.$$

Exercise 1.24 (Boyd and Vandenberghe 3.1) Given a convex function $f(x) : \mathbb{R} \to \mathbb{R}$, show that

$$\frac{f(x) - f(a)}{x - a} \leq \frac{f(b) - f(a)}{b - a} \leq \frac{f(b) - f(x)}{b - x}$$

for $x \in (a, b)$. Draw a geometric sketch to demonstrate this.

Exercise 1.25 This exercise will explore the convex properties of matrices.

1. Show that the maximum eigenvalue $\lambda_{\max}(\mathbf{X})$ of a matrix $\mathbf{X}$ is a convex function.
2. Show that the log of the determinant of a matrix, $\log \det \mathbf{X} = \sum_i \log \lambda_i$ is a convex function for symmetric positive definite matrices $\mathbf{X}$.
3. Show that the trace, $\mathrm{Tr}(\mathbf{X}) = \sum_i \lambda_i$ is a linear function, i.e. that $\mathrm{Tr}(\mathbf{X}+\mathbf{Y}) = \mathrm{Tr}(\mathbf{X})+\mathrm{Tr}(\mathbf{Y})$, and hence is a convex function.

Exercise 1.26 Prove that the Cartesian product of two convex sets is a convex set.

Exercise 1.27 Let C_1 and C_2 be convex. Then their Minkowski sum $C_1 + C_2 = \{x + y | x \in C_1, y \in C_2\}$ is also convex. Prove this and give some examples.

Exercise 1.28 Using Lagrange multipliers, solve the problem

$$\min_{x,y \in \mathbb{R}} \quad x^2 + y^2 \quad \text{subject to} \quad (x-1)^2 + (y-1)^2 = 2.$$

Plot the constraint region and identify the minimizing point(s).

Exercise 1.29 In this problem, we will revisit the Lagrange multiplier example from Section 1.4. Plot the Lagrangian form of the objective function

$$(x-1)^2 + 1 + \lambda x$$

for several values of $\lambda > 2$, $\lambda < 2$, and $\lambda = 2$. Include the constraint $x \leq 0$ on the plot. Which value of $\lambda \geq 0$ maximizes the Lagrangian?

Exercise 1.30 For each of the affine transformations $\mathbf{T}$ (scaling, shear, rotation, and translation), show that they preserve the convexity of a set C. In particular

$$\mathbf{T}(C) = \{\mathbf{Tx} \,|\, \mathbf{x} \in C\}$$

is convex.

Exercise 1.31 Consider two convex sets in 2D: a closed unit disk and a closed unit square, both centered at the origin. Show by hand how the affine transformations (scaling, shear, rotation, and translation) change these sets. Then write a code to apply these transformations and plot the resulting set.

Exercise 1.32 Show that all affine transformations (scaling, shear, rotation, and translation) are invertible.

2 Gradient Based Optimization

Gradient descent is one of the most essential techniques in optimization. It is simple, effective, and versatile in a wide range of scenarios; for example, stochastic gradient descent is the backbone of modern machine learning training. If you learn one method in optimization, this is it. In this chapter we introduce leading gradient-based methods and variants, and discuss scalability, stability, and performance on several numerical examples.

For a smooth, unconstrained objective function $f(\mathbf{x})$, a point $\mathbf{x}^*$ with zero gradient, $\nabla f(\mathbf{x}^*) = \mathbf{0}$, and a positive semidefinite Hessian, $\nabla^2 f(\mathbf{x}^*) \geq \mathbf{0}$, will be a local minimum. Thus, finding points $\mathbf{x}$ where $\nabla f(\mathbf{x}) = \mathbf{0}$ is a central challenge in optimization. Gradient-based approaches use information about the gradient of the objective function $\nabla f(\mathbf{x})$, to iteratively search for local minima $\mathbf{x}^*$. For example, the function f may be the loss function of a neural network, as in Chapter 10, and we will use stochastic gradient descent to find parameters $\mathbf{x}$ that locally minimize this function.

In some cases, it is possible to directly solve for points where $\nabla f(\mathbf{x}) = \mathbf{0}$, as in the least squares problem where $f(\mathbf{x})$ is a quadratic function; this will be explored thoroughly in Chapter 4. However, it is generally impractical to solve for the roots of $\nabla f(\mathbf{x})$ explicitly. Instead, we use *iterative methods*, such as gradient descent and Newton's method, to obtain a sequence of points $\mathbf{x}_k$ that eventually converge towards the local minimum $\mathbf{x}^*$. Specifically, we seek an algorithm

$$\mathbf{x}_{k+1} = \mathbf{F}(\mathbf{x}_k) \tag{2.1}$$

so that starting with an initial guess $\mathbf{x}_0$, we may iterate through $\mathbf{F}$:

$$\mathbf{x}_0 \xrightarrow{\mathbf{F}} \mathbf{x}_1 \xrightarrow{\mathbf{F}} \mathbf{x}_2 \xrightarrow{\mathbf{F}} \cdots \xrightarrow{\mathbf{F}} \mathbf{x}_k \xrightarrow{\mathbf{F}} \cdots . \tag{2.2}$$

The goal is to iteratively decrease the objective function towards a local minimum $\mathbf{x}^*$:

$$f(\mathbf{x}_0) \geq f(\mathbf{x}_1) \geq f(\mathbf{x}_2) \geq \cdots f(\mathbf{x}_k) \geq \cdots f(\mathbf{x}^*). \tag{2.3}$$

Ideally, this algorithm converges quickly, monotonically, and with low computational complexity. In gradient-based optimization, we use $\nabla f(\mathbf{x})$ to design the algorithm for the iterative update $F(\mathbf{x}_k)$.

First-order methods use first derivative information, e.g., the gradient ∇f, to build local linear approximations to f, which are then used to develop iterative algorithms. Gradient descent, and its extension to high-dimensional optimization problems in neural network training, stochastic gradient descent, are two popular first-order methods we will explore.

This chapter will develop and analyze several leading iterative, gradient-based algorithms in optimization. In each case, we will discuss convergence rate, stability, computational complexity, and scaling to higher-dimensional problems. In all iterative methods, the resulting solution is only locally optimal, unless the problem is convex. We also stress the importance of the initial guess $\mathbf{x}_0$ for nonconvex problems, as this will determine which local minima the algorithm converges to and how fast it converges. Here we mostly consider unconstrained problems, although gradient-based approaches are also quite useful for constrained optimization.

A key benefit of gradient-based optimization is that very little structure is assumed, other than the existence of derivatives. Throughout this chapter, it will help to appreciate gradient-based algorithms by imagining how we might proceed when we don't have access to the gradient, either because it doesn't exist or we can't approximate it numerically. These *nonsmooth* optimization techniques are discussed more in Chapter 5, and they usually involve some intelligent form of *perturb and observe* or *explore and exploit*, which can be quite slow to converge, especially in higher dimensions. In contrast, gradient descent scales to very large systems, and increasingly we are able to approximate derivatives of very complex functions, such as a neural network training objective function, using modern automatic differentiation and stochastic gradient descent, which will be discussed more in Chapter 10. Further, even for many nonsmooth functions, we are able to compute the proximal gradient (see Section 5.1), which may be used similarly in descent algorithms.

Higher-order derivatives of the objective function will lead to more accurate iterative algorithms with faster convergence, such as Newton's method. This is analogous to other numerical methods, where higher-order Taylor approximations yield more accurate schemes. Second-order methods use the second derivative, called the Hessian $\nabla^2 f$, to converge in fewer iterations, at the expense of added computational complexity. Newton's method, and its related Quasi-Newton methods, are second-order algorithms discussed here. Many of these algorithms are adapted for nonlinear least squares problems, discussed in Chapter 4, and for machine learning training, discussed in Chapter 10.

2.1 Gradient Descent

Gradient descent seeks a local minima of an objective function $f(\mathbf{x})$ by taking iterative steps in the negative gradient direction. Recall from Section 1.3 that the gradient of a function $\nabla f(\mathbf{x})$ determines the direction of locally steepest ascent, and so the negative gradient $-\nabla f(\mathbf{x})$ is the direction of locally steepest descent. The gradient descent algorithm is:

$$\mathbf{x}_{k+1} = \mathbf{x}_k - \gamma \nabla f(\mathbf{x}_k), \tag{2.4}$$

where γ is the step size, also referred to as the learning rate in machine learning. This is illustrated schematically in Fig. 2.1.

Gradient descent is based on a first-order Taylor series approximation of $f(\mathbf{x})$:

$$f(\mathbf{x}) \approx f(\mathbf{x}_0) + \nabla f(\mathbf{x}_0)^{\mathrm{T}} \underbrace{(\mathbf{x} - \mathbf{x}_0)}_{\Delta \mathbf{x}}. \tag{2.5}$$

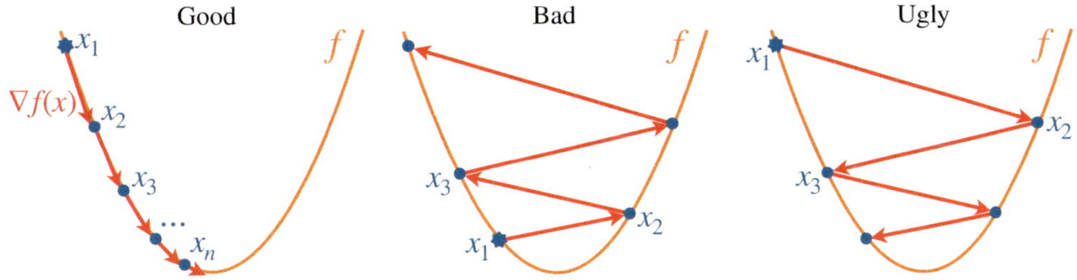

Figure 2.1 Schematic illustrating gradient descent for a simple one-dimensional, quadratic objective function. Left: As the iterate $\mathbf{x}_k$ gets closer to the minimum of f, the gradient becomes smaller (i.e., the slope of the gradient becomes closer to 0), so the gradient descent steps also become smaller. Thus, the algorithm takes larger steps when further from the minimum. Gradient descent with with an overly large step size will oscillate around the minima (right), or worse, it will diverge (middle). Here we are not plotting the actual gradient, which is always tangent to the function, but instead the actual steps $\mathbf{x}_{k+1} = \mathbf{x}_k - \gamma \nabla f(\mathbf{x}_k)$. If γ is too large, then the step will overshoot the minimum.

This is useful to determine the stability and convergence of gradient descent algorithms.

The most immediate question is how to choose a good step size γ. If γ is too large, the iteration will be unstable or barely stable (see Fig. 2.1). Specifically, the gradient step may overshoot the minimum, causing oscillations, or even divergence. Similarly, if γ is too small, the iteration will converge slowly. If we have the second derivative $\nabla^2 f$, we can use this to determine a better γ, as in Newton's method in Section 2.4.

There are several approaches to determine the step size γ:

- Fixed γ: A fixed γ may be good for a convex objective f with a well-conditioned Hessian $\nabla^2 f$. Later, we will explore a simple example with quadratic objective to see the effect of the condition number of $\nabla^2 f$ on a fixed-step gradient descent algorithm. Fixed-step schemes are simple, but may require tuning for adequate convergence.
- Decaying γ: It may help to slowly decrease the step size γ with each iteration k, so it takes larger steps initially. A simple decaying step size formula is

$$\gamma_k = \frac{\gamma_0}{1 + \beta k}, \qquad \beta > 0,$$

  where γ_0 is the initial step size and β is the decay factor. This may seem counterintuitive, since this causes the step size to shrink later in the iteration, where f is typically flatter and has a smaller gradient; we may think that larger steps later would accelerate convergence. However, for stochastic gradient descent, a decaying γ acts to *anneal* the algorithm, letting it take large exploratory steps early, and cooling to smaller steps later in the iteration. Annealing will be discussed in Chapter 5.
- Momentum methods: It is possible to use physical intuition to imbue the step size with *momentum*, so that it is less prone to getting stuck in saddle points and locally flat regions. This is used widely in machine learning and is discussed in Section 2.2.
- Adaptive methods: There are several adaptive methods that tune γ depending on the history of past iterations. This is where we will spend most of our emphasis because of their favorable properties. Root mean square propagation (RMSProp) scales the step

Panel 2.1: Line search

By far the most effective and robust strategy for smooth, deterministic objective functions f is to perform a line search for γ. A line search uses the gradient direction and then runs a second optimization over the step size γ in that direction to find the best γ according to some criteria. This secondary line-search optimization has the form:

$$\gamma_k = \underset{\gamma}{\operatorname{argmin}}\, f(\mathbf{x}_k - \gamma \nabla f(\mathbf{x}_k)).$$

We generally don't solve this optimization problem exactly, and instead iteratively sweep through γ. These methods will be discussed in Section 2.5. Line search adds computational cost and complexity per iteration, especially when it is expensive to evaluate f. However, the added robustness and faster convergence typically justify its use. We often use a line search in Newton's method, since each iteration is already expensive, justifying the additional cost of a line search to improve convergence.

size with a moving average of past squared gradients, regularizing the step. This is particularly effective for ill-conditioned problems.

Adaptive momentum estimation (ADAM) combines momentum methods and RMSProp and is the standard algorithm to train neural networks with stochastic gradient descent due to its robustness. This will be discussed in more detail in Section 2.3.

- Learning rate schedules: Various other approaches modify the learning rate in different stages of the optimization. This is closely related to annealing methods in nonconvex optimization (see Section 5.3), including exponential decay.

It is important to emphasize that gradient descent will only converge to the nearest local minima (see Fig. 2.2), so for nonconvex objective functions with multiple local minima, it is important to start with a good initialization, ideally "close" to the desired solution.

Numerical Examples

In our first example, we will apply gradient descent to find the minimum of a simple quadratic cost function

$$f(\mathbf{x}) = \tfrac{1}{2}\mathbf{x}^{\mathrm{T}}\mathbf{A}\mathbf{x} \qquad \text{where} \qquad \mathbf{A} = \begin{bmatrix} 2 & 1 \\ 1 & 3 \end{bmatrix} \tag{2.6}$$

so that $\mathbf{A}$ is positive semidefinite. This problem actually has a closed-form solution, which will be discussed more in Chapter 4. The minimum value is at the origin, $\mathbf{x}^* = \mathbf{0}$. This will provide a simple example for testing the algorithm, where we know the answer.

Figure 2.3 shows the gradient descent iteration for three step sizes: $\gamma = \{0.01, 0.1, 0.543\}$. We can clearly see the effects of γ being too small or too large,

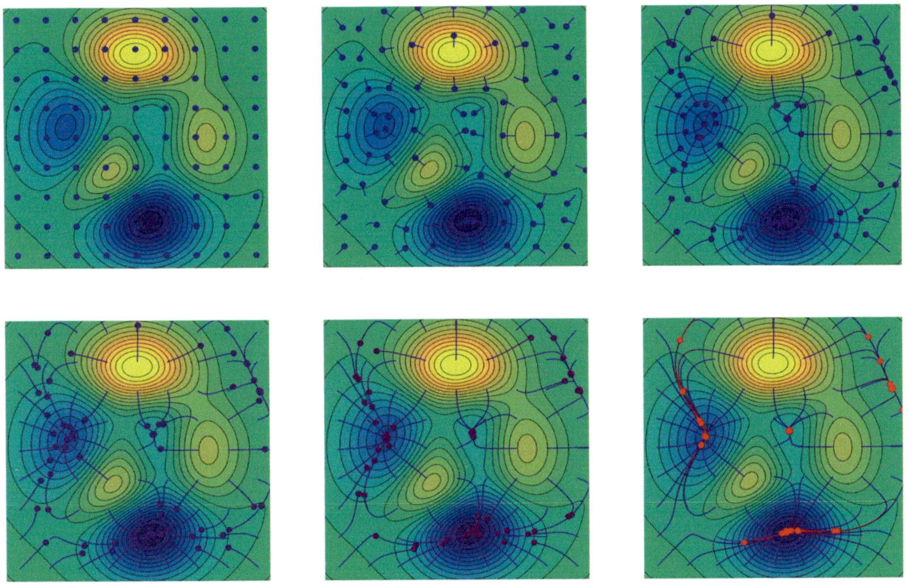

Figure 2.2 Illustration of gradient descent on a nonconvex objective function with multiple local minima. Points descend to their nearest local minimima. Reproduced from Wikipedia (https://en.wikipedia.org/wiki/File:Gradient_Descent_in_2D.webm); author Gabriel Peyré[2].

namely either slow convergence or nearly unstable oscillating convergence, respectively. A simple starter gradient descent code is provided in Code 2.1.

Code 2.1: Gradient descent code for a quadratic objective function.

```
# Define the quadratic objective function with its gradient
def quadraticObjective(X, A, b, c):
    return 0.5 * X.T @ A @ X + b.T @ X + c

def gradientDescent(A, b, c, x0, step_size, max_iter):
    xvals = [np.array(x0)] # Initialize with x0, f(x0)
    fvals = [quadraticObjective(np.array(x0), A, b, c)]
    x = np.array(x0, dtype=float)

    for _ in range(max_iter):
        grad = A @ x + b  # Gradient of objective function
        x = x - step_size * grad
        xvals.append(x.copy())
        fvals.append(quadraticObjective(x, A, b, c))

    return np.array(xvals), np.array(fvals)
```

Let's initialize the problem and run the gradient descent optimization using the following:

[2] Professor Gabriel Peyré is an amazing optimization researcher and an amazing teacher. He has provided many innovative and clear descriptions of advanced topics, and he clearly loves sharing knowledge with others. Check out his codes and tutorials on his website www.gpeyre.com and at x.com/@gabrielpeyre.

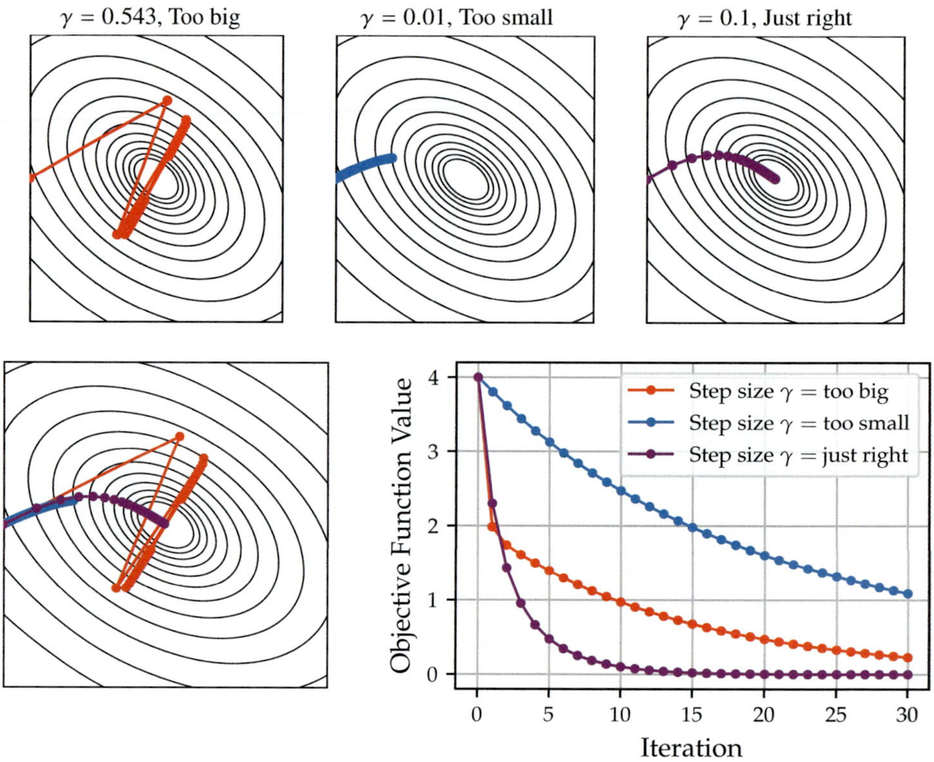

Figure 2.3 Gradient descent on a simple quadratic objective function with three step sizes: $\gamma \in \{0.01, 0.1, 0.543\}$. The step size $\gamma = 0.1$ converges rapidly and without oscillations. A smaller size $\gamma = 0.01$ converges too slowly, and a larger step size $\gamma = 0.543$ is nearly unstable, oscillating as it slowly converges. The top plot shows each trajectory individually, superimposed on the level sets of the objective function.. The bottom left plot shows all step sizes together for comparison. The bottom right plot shows the objective function versus iteration number, which is how we typically visualize convergence of an algorithm.

CodeSnip 2.a

```
A = np.array([[2, 1], [1, 3]]) # Positive-definite matrix
b = np.array([0, 0])           # Linear term
c = 0                          # Constant term
x0 = np.array([-2.0, 0.0])     # Starting point
step_size = 0.1                # Learning rate
N = 30                         # Maximum iterations

xvals, fvals = gradientDescent(A, b, c, x0, step_size, N)
```

As a more challenging example, inspired by Boyd and Vanderberghe [66], we explore gradient descent on a non-quadratic objective function:

$$f(\mathbf{x}) = e^{x_1+2x_2-.2} + e^{x_1-2x_2-.2} + e^{-x_1-.2}. \tag{2.7}$$

The gradient of this function is

$$\nabla f(\mathbf{x}) = \begin{bmatrix} e^{x_1+2x_2-.2} + e^{x_1-2x_2-.2} - e^{-x_1-.2} \\ 2e^{x_1+2x_2-.2} - 2e^{x_1-2x_2-.2} \end{bmatrix}. \tag{2.8}$$

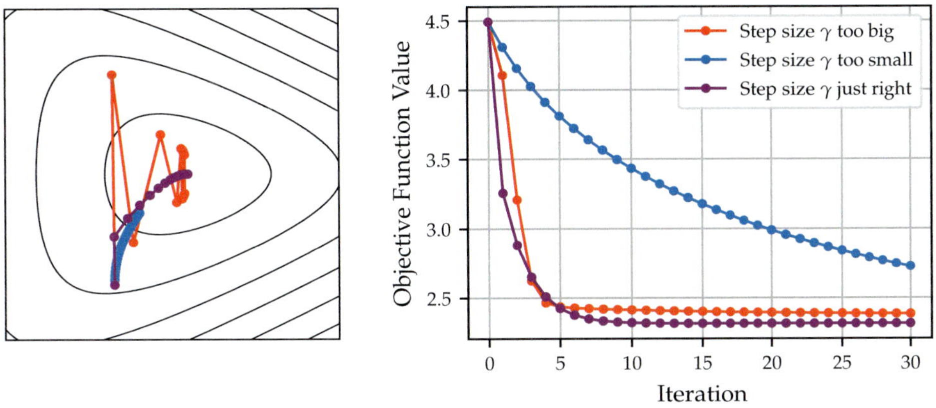

Figure 2.4 Gradient descent on a non-quadratic objective function with three step sizes.

The performance of gradient descent with various step sizes is shown in Fig. 2.4. Similar behavior is observed. When γ is too large, the gradient descent oscillates about the minimum, as depicted schematically in Fig. 2.1 (right).

Panel 2.2: Gradient Flows

The gradient descent algorithm in Eq. (2.4) may be viewed as a simple forward Euler integration scheme with timestep $\Delta t = \gamma$ for the ordinary differential equation

$$\dot{\mathbf{x}}(t) = -\nabla f(\mathbf{x}(t)).$$

This vector field $-\nabla f(\mathbf{x})$ is known as a *gradient flow*. It is a conservative, irrotational vector field, and it is incompressible if $\nabla^2 f = 0$. These properties connect gradient-based optimization to long-studied physical systems, such as mechanics, electrostatics, and fluid flows. Further, the stability of gradient descent may be related to the stability of this gradient flow.

Recall that the forward Euler integration scheme for a vector field $\dot{\mathbf{x}} = \mathbf{v}(\mathbf{x})$ is given by

$$\mathbf{x}_{k+1} = \mathbf{x}_k + \Delta t \mathbf{v}(\mathbf{x}_k).$$

It will be left as an exercise to the reader to show that the simple gradient descent algorithm is a forward Euler integration of the gradient flow equation with $\Delta t = \gamma$. A natural question is whether or not it is possible to devise a gradient descent iteration based on a better numerical integration scheme for this gradient flow, such as a higher-order Runge–Kutta integrator; this is also an exercise for the reader.

Convergence of Gradient Descent

The convergence of gradient descent has theoretical guarantees that depend on the properties of f. If f is convex and ∇f is Lipschitz continuous with Lipschitz constant[3] L, then gradient descent with a step size $\gamma = 1/L$ will have the following convergence

$$f(\mathbf{x}_k) - f(\mathbf{x}^*) \leq \frac{L}{2k} \|\mathbf{x}_0 - \mathbf{x}^*\|^2. \tag{2.9}$$

Thus the convergence rate is $O(1/k)$, i.e., sublinear in k.

If f is μ-strongly convex, this convergence rate can be improved. A function is called μ-strongly convex if

$$f(\mathbf{y}) \geq f(\mathbf{x}) + \nabla f(\mathbf{x})^{\mathrm{T}}(\mathbf{y} - \mathbf{x}) + \frac{\mu}{2}\|\mathbf{y} - \mathbf{x}\|^2 \tag{2.10}$$

for all $\mathbf{x}, \mathbf{y} \in \mathbb{R}^n$ and $\mu > 0$. In this case, the convergence rate is:

$$f(\mathbf{x}_k) - f(\mathbf{x}^*) \leq \frac{L}{2}\left(1 - \frac{\mu}{L}\right)^k \|\mathbf{x}_0 - \mathbf{x}^*\|^2. \tag{2.11}$$

As an example, the quadratic objective function in Eq. (2.6) is μ-strongly convex with $\mu = \lambda_{\min}(\mathbf{A})$. Strong convexity will also be used to develop the proximal gradient in Section 5.1 via the Moreau envelope.

Stability of Gradient Descent

Here we will explore gradient descent on a simple quadratic objective function

$$f(\mathbf{x}) = \frac{1}{2}\mathbf{x}^{\mathrm{T}}\mathbf{A}\mathbf{x} + \mathbf{b}^{\mathrm{T}}\mathbf{x} + c, \tag{2.12}$$

with symmetric $\mathbf{A}$. The gradient is given by

$$\nabla f(\mathbf{x}) = \mathbf{A}\mathbf{x} + \mathbf{b} \tag{2.13}$$

and the Hessian is

$$\mathbf{H} = \nabla^2 f(\mathbf{x}) = \mathbf{A}. \tag{2.14}$$

If $\mathbf{A}$ is not symmetric, then the gradient is $\nabla f(\mathbf{x}) = \frac{1}{2}\left(\mathbf{A} + \mathbf{A}^{\mathrm{T}}\right)\mathbf{x} + \mathbf{b}$ (see Section 2.7). In this case the Hessian is $\mathbf{H} = \frac{1}{2}\left(\mathbf{A} + \mathbf{A}^{\mathrm{T}}\right)$, so that even if $\mathbf{A}$ isn't symmetric, the Hessian is.

In many cases, we can solve for the optimal solution $\mathbf{x}^* = -\mathbf{A}^{-1}\mathbf{b}$ directly (see Chapter 4), so that $\nabla f(\mathbf{x}^*) = \mathbf{0}$. However, there are several classes of quadratic problems where we will still use gradient descent, because inverting $\mathbf{A}$ is not tractable:

- Large $\mathbf{A}$: When the dimension of $\mathbf{x} \in \mathbb{R}^n$ is large, for example n is in the millions or even billions, then $\mathbf{A}^{-1} \in \mathbb{R}^{n\times n}$ is intractable to compute and it may not even be possible to store $\mathbf{A}$. However, for large sparse $\mathbf{A}$, it may be possible to compute the matrix–vector products $\mathbf{A}\mathbf{x}$ needed for gradient descent and its variants.

[3] Lipschitz continuity of ∇f is stronger than f being continuous, but weaker than ∇f being differentiable. A function f is Lipschitz continuous with Lipschitz constant L if $|f(\mathbf{x}_1) - f(\mathbf{x}_2)| \leq L\|\mathbf{x}_1 - \mathbf{x}_2\|$ for all $\mathbf{x}_1, \mathbf{x}_2$.

- Sparse **A**: When **A** is a large, sparse matrix, it may be faster to use iterative gradient-based methods than to directly invert **A**. These matrices occur frequently in physics applications, such as fluid mechanics, where interactions are local, so terms like diffusion and convection typically lead to very sparse **A** matrices.
- Ill-conditioned **A**: In this case, it is often difficult to use standard inversion techniques, and instead we use preconditioners followed by iterative descent.

We learn a great deal by exploring gradient descent on a simple problem where we know the answer, even if it is small and we could solve it by hand.

We now examine how the stability properties of gradient descent depend on the step size γ and the eigenvalues of **A**. This is closely related to the stability of numerical integrators, such as forward Euler integration for ordinary differential equations $\dot{\mathbf{x}} = \mathbf{A}\mathbf{x}$. Even though we have the analytic solution $\mathbf{x}(t) = \exp(\mathbf{A}t)\mathbf{x}(0)$ for a linear system, we learn a great deal by analyzing the stability properties of integrators on these simple systems.

For a quadratic objective function, the gradient descent formula

$$\mathbf{x}_{k+1} = \mathbf{x}_k - \gamma \nabla f(\mathbf{x}_k) \tag{2.15}$$

becomes

$$\mathbf{x}_{k+1} = \mathbf{x}_k - \gamma\,(\mathbf{A}\mathbf{x}_k + \mathbf{b}) \tag{2.16a}$$

$$= (\mathbf{I} - \gamma\mathbf{A})\,\mathbf{x}_k - \gamma\mathbf{b}. \tag{2.16b}$$

The eigenvalues λ_j of the Hessian **A** represent the curvature of f in different directions. The maximum curvature is given by $\lambda_{\max}$ and the minimum curvature is given by $\lambda_{\min}$.

The gradient descent algorithm is stable if all eigenvalue of $\mathbf{I} - \gamma\mathbf{A}$ are stable. For the discrete-time system in Eq. (2.16b), stability means that all eigenvalues of $\mathbf{I} - \gamma\mathbf{A}$ must have magnitude less than 1. This guarantees stability of the iteration:

$$\mathbf{x}_k \to \mathbf{x}^* = -\mathbf{A}^{-1}\mathbf{b}. \tag{2.17}$$

If **A** is symmetric, the eigenvalues of $\mathbf{I} - \gamma\mathbf{A}$ are $1 - \gamma\lambda_j$ where $\lambda_j \in \text{eigs}(\mathbf{A})$. **A** is symmetric for many real quadratic optimization problems, for example least squares in Chapter 4. More generally, if **A** is diagonalizable, there exists an invertible matrix of eigenvectors **T** so that $\mathbf{T}^{-1}\mathbf{A}\mathbf{T} = \mathbf{D}$, where **D** is a diagonal matrix. In this case

$$\mathbf{T}^{-1}\,(\mathbf{I} - \gamma\mathbf{A})\,\mathbf{T} = \mathbf{I} - \gamma\mathbf{D}, \tag{2.18}$$

which is diagonal with eigenvalues $1 - \gamma\lambda_j$.

For stability we require

$$|1 - \gamma\lambda_j| < 1 \quad \forall j. \tag{2.19}$$

This may be rewritten as

$$-1 < 1 - \gamma\lambda_j < 1 \quad \Longrightarrow \quad -2 < -\gamma\lambda_j < 0 \quad \Longrightarrow \quad 0 < \gamma < \frac{2}{\lambda_{\max}}. \tag{2.20}$$

Thus, we have determined the range of stable step sizes γ.

Optimal Step Size

Within the range of stable γ, we turn to the problem of determining an optimal step size γ for fastest convergence of the gradient descent algorithm. The optimal step size for gradient descent with a quadratic objective function is given by

$$\gamma^* = \frac{2}{\lambda_{\min} + \lambda_{\max}}, \tag{2.21}$$

which arises from the spectral properties of the Hessian $\mathbf{A}$. This yields the fastest convergence rate by balancing the smallest and largest eigenvalues of $\mathbf{A}$.

The convergence rate is determined by the *spectral radius* ρ of $(\mathbf{I} - \gamma\mathbf{A})$, which is the largest absolute eigenvalue:

$$\rho(\mathbf{I} - \gamma\mathbf{A}) = \max_j |1 - \gamma\lambda_j|. \tag{2.22}$$

To minimize the convergence rate, we choose γ such that $\rho(\mathbf{I} - \gamma\mathbf{A})$ is as small as possible.

The eigenvalues of $(\mathbf{I} - \gamma\mathbf{A})$ are symmetric about the point $1 - \gamma\lambda_{\text{mid}}$, where:

$$\lambda_{\text{mid}} = \frac{\lambda_{\min} + \lambda_{\max}}{2}. \tag{2.23}$$

Showing this property will be left as an exercise. To minimize the spectral radius, the eigenvalues are distributed evenly around zero:

$$\lambda_{\max} = -\lambda_{\min}. \tag{2.24}$$

Thus we set $|1 - \gamma\lambda_{\min}| = |1 - \gamma\lambda_{\max}|$ for an optimal γ^*:

$$1 - \gamma^*\lambda_{\min} = -(1 - \gamma^*\lambda_{\max}) \tag{2.25}$$

and solve for γ^*:

$$\gamma^* = \frac{2}{\lambda_{\max} + \lambda_{\min}}. \tag{2.26}$$

This optimum step size γ^* minimizes the spectral radius of $(\mathbf{I} - \gamma\mathbf{A})$, leading to the fastest convergence for the quadratic problem. With this optimal step size, the worst-case contraction per iteration of the descent algorithm, ρ, is given by

$$\rho = \frac{\lambda_{\max} - \lambda_{\min}}{\lambda_{\max} + \lambda_{\min}}. \tag{2.27}$$

This is closely related to the condition number, which will be discussed more in Chapter 4. Specifically, a Hessian matrix $\mathbf{A}$ with a larger condition number $\kappa = \lambda_{\max}/\lambda_{\min}$ will have slower convergence.

As a simple exercise, compute the maximum step size $\gamma_{\max}$ and optimal step size γ^* for the quadratic example in Fig. 2.3. Then, test gradient descent with γ at the optimal γ^* and also slightly above and below the maximum stability boundary $\gamma^{\max}$.

2.2 Momentum Methods

Gradient descent struggles near saddle points of nonconvex objective functions, where the gradient may be negligibly small, causing the convergence to grind to a halt. This issue is especially acute when using stochastic gradient descent for high-dimensional problems, such as training a neural network, since these small gradient estimates might also be noisy.

Momentum methods, and their variants, address this issue by giving the optimization algorithm *inertia* to carry it through these flat saddle regions of small gradient. The standard gradient descent algorithm only uses the current instantaneous gradient for the next step in the iteration. In contrast, momentum methods integrate past gradient information to smooth out noisy gradient estimates and to provide inertia. This is a rather physical idea, similar to rolling a ball down the objective function landscape. If the ball gains momentum in areas of high gradient, it will have inertia to roll through flat saddle regions.

There are several approaches to momentum that improve standard gradient descent, especially in the context of machine learning, that we will discuss here. The momentum method was introduced in machine learning in the 1980s but is based on earlier work by Polyak from the 1960s [392].

Heavy Ball

The classic momentum method [431] is the simplest modification to gradient descent to include momentum:

$$\mathbf{x}_{k+1} = \mathbf{x}_k - \gamma \nabla f(\mathbf{x}_k) + \beta(\mathbf{x}_k - \mathbf{x}_{k-1}). \tag{2.28}$$

The additional term $\beta(\mathbf{x}_k - \mathbf{x}_{k-1})$ introduces a history dependence, where $\mathbf{x}_k - \mathbf{x}_{k-1}$ may be interpreted as the *velocity* of the last iterative step.

We may also formally introduce this velocity vector as

$$\mathbf{v}_{k+1} = \beta \mathbf{v}_k - \gamma \nabla f(\mathbf{x}_k) \tag{2.29}$$

so that the gradient descent update now takes the form

$$\mathbf{x}_{k+1} = \mathbf{x}_k - \gamma \nabla f(\mathbf{x}_k) + \beta \mathbf{v}_k \tag{2.30a}$$

$$\mathbf{v}_{k+1} = \beta \mathbf{v}_k - \gamma \nabla f(\mathbf{x}_k). \tag{2.30b}$$

This introduces a new hyperparameter β, the decay rate for the velocity update. The velocity $\mathbf{v}$ is an exponentially discounted sum of past gradients.

Nesterov's Accelerated Gradient Algorithm

Nesterov's accelerated gradient (NAG) algorithm [364, 525] is closely related to the classic momentum method, and it is an industry standard algorithm in machine learning. Nesterov's formulation is a relatively simple variation on Eq. (2.28):

$$\mathbf{x}_{k+1} = \mathbf{x}_k - \gamma \nabla f(\mathbf{x}_k + \beta(\mathbf{x}_k - \mathbf{x}_{k-1})) + \beta(\mathbf{x}_k - \mathbf{x}_{k-1}) \tag{2.31a}$$

$$= \mathbf{x}_k - \gamma \nabla f(\mathbf{x}_k + \beta \mathbf{v}_k) + \beta \mathbf{v}_k, \tag{2.31b}$$

where $\mathbf{v}_k = \mathbf{x}_k - \mathbf{x}_{k-1}$. The only difference is that the gradient is now evaluated at the point $\mathbf{y}_k = \mathbf{x}_k + \beta \mathbf{v}_k$ instead of $\mathbf{x}_k$. This is called a *look ahead*, and is very reminiscent of classic

Runge–Kutta integration schemes in the numerical integration of ordinary differential equations.

We may formalize this into a two-step procedure, as in other momentum methods, in terms of the look ahead vector $\mathbf{y} = \mathbf{x} + \beta\mathbf{v}$:

$$\mathbf{y}_k = \mathbf{x}_k + \beta(\mathbf{x}_k - \mathbf{x}_{k-1}) \tag{2.32a}$$

$$\mathbf{x}_{k+1} = \mathbf{y}_k - \gamma\nabla f(\mathbf{y}_k). \tag{2.32b}$$

Sutskever demonstrated how to relate NAG to classic momentum in the following way [479]. It is possible to modify the index on the look ahead variable and rewrite NAG as

$$\mathbf{y}_{k+1} = \mathbf{x}_k + \beta\left(\mathbf{x}_k - \mathbf{x}_{k-1}\right) \tag{2.33a}$$

$$\mathbf{x}_{k+1} = \mathbf{y}_{k+1} - \gamma\nabla f(\mathbf{y}_{k+1}) \tag{2.33b}$$

which can be rearranged as

$$\mathbf{y}_{k+1} = \mathbf{x}_k + \beta\mathbf{v}_k \tag{2.34a}$$

$$\mathbf{x}_{k+1} = \mathbf{x}_k + \beta\mathbf{v}_k - \gamma\nabla f(\mathbf{y}_{k+1}). \tag{2.34b}$$

Panel 2.3: Differential Equation Interpretation of Momentum Method

Recall that it is possible to recast gradient descent as a discrete forward Euler approximation to the gradient flow ODE in Panel 2.2. The heavy ball momentum method may be similarly derived from a second-order ODE:

$$m\ddot{\mathbf{x}}(t) = -\nabla f(\mathbf{x}(t)) - \delta\dot{\mathbf{x}},$$

where m is the mass of the particle and δ is the friction coefficient. Taking a simple central difference scheme approximation for $\ddot{\mathbf{x}}$ and an Euler approximation for $\dot{\mathbf{x}}$ for a time step of Δt yields

$$m\frac{\mathbf{x}_{k+1} - 2\mathbf{x}_k + \mathbf{x}_{k-1}}{\Delta t^2} \approx -\nabla f(\mathbf{x}_k) - \delta\frac{\mathbf{x}_{k+1} - \mathbf{x}_k}{\Delta t}.$$

Multiplying both sides by Δt^2 and grouping terms, this simplifies as

$$m\left(\mathbf{x}_{k+1} - 2\mathbf{x}_k + \mathbf{x}_{k-1}\right) = -\Delta t^2\nabla f(\mathbf{x}_k) - \delta\Delta t\left(\mathbf{x}_{k+1} - \mathbf{x}_k\right)$$

$$\begin{aligned}\implies \quad (m + \delta\Delta t)\mathbf{x}_{k+1} &= -\Delta t^2\nabla f(\mathbf{x}_k) + (2m + \delta\Delta t)\mathbf{x}_k - m\mathbf{x}_{k-1} \\ &= -\Delta t^2\nabla f(\mathbf{x}_k) + (m + \delta\Delta t)\mathbf{x}_k + m\left(\mathbf{x}_k - \mathbf{x}_{k-1}\right)\end{aligned}$$

$$\implies \quad \mathbf{x}_{k+1} = \mathbf{x}_k - \gamma\nabla f(\mathbf{x}_k) + \beta\left(\mathbf{x}_k - \mathbf{x}_{k-1}\right),$$

where γ and β are given by

$$\gamma = \frac{\Delta t^2}{m + \delta\Delta t} \qquad \beta = \frac{m}{m + \delta\Delta t}.$$

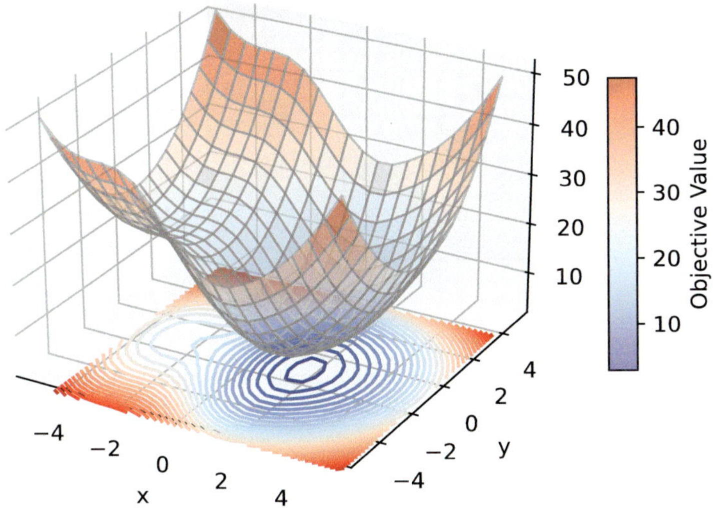

Figure 2.5 Surface plot and contour lines of objective function with saddle point and minima, given by Eq. (2.35).

Thus, comparing to Eq. (2.30), NAG essentially reverses the order of classical momentum, so that we perform the momentum update first and then proceed to gradient descent.

Numerical Example

Gradient descent performs very poorly near saddle points, as the gradient nearly vanishes, causing painfully slow convergence. This concept is illustrated for a simple objective function with a saddle, given by

$$f(\mathbf{x}) = 15e^{-.25(x+2)^2} + .5e^{-2(x-2)^2} + x^2 + y^2. \tag{2.35}$$

A three-dimensional surface plot of this objective over its contour map is shown in Fig. 2.5. Figure 2.6 shows a comparison of gradient descent with and without momentum on the objective function in Eq. (2.35) with a saddle point. It is clear that the method with momentum converges much more rapidly.

CodeSnip 2.b

```
def nesterov_accelerated_gradient(start, grad_func, step_size,
    momentum, max_iter):
    for _ in range(max_iter):
        y = x + momentum * velocity
        grad = grad_func(y[0], y[1])
        velocity = momentum * velocity - step_size * grad
        x = x + velocity
        trajectory.append(x.copy())
        values.append(double_well_potential(x[0], x[1]))

    return np.array(trajectory), np.array(values)
```

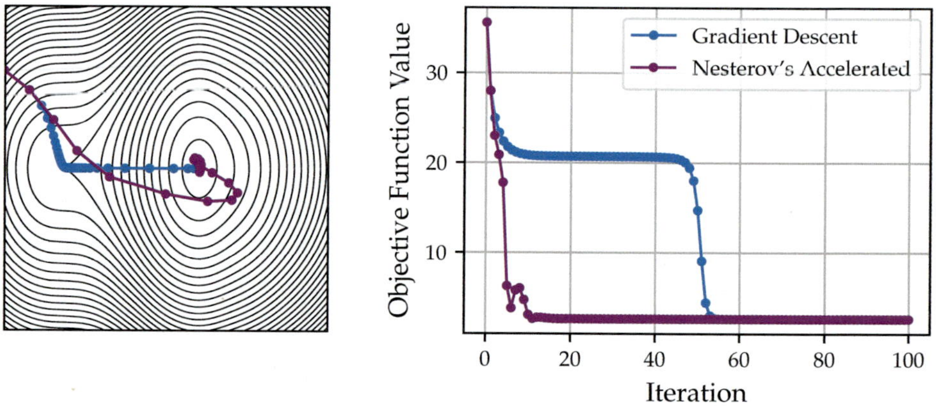

Figure 2.6 Comparison of simple gradient descent and Nesterov's accelerated method on an objective function with a saddle point.

This is then initialized with the initial condition $(-5, 3)$ and run with a step size of $\gamma = 0.1$, a momentum coefficient $\beta = 0.9$ and a maximum number of iterations of 100.

Convergence of Nesterov's Method

Nesterov's achieves faster convergence than standard gradient descent. If f is convex and ∇f is Lipschitz continuous with constant L, then the convergence of NAG is:

$$f(\mathbf{x}_k) - f(\mathbf{x}^*) \le \frac{2L}{(k+1)^2} \|\mathbf{x}_0 - \mathbf{x}^*\|^2. \tag{2.36}$$

Thus, the convergence rate is $O(1/k^2)$, which is a major improvement over the $O(1/k)$ rate of standard gradient descent.

If f is μ-strongly convex, then Nesterov's method achieves an even better linear convergence rate:

$$f(\mathbf{x}_k) - f(\mathbf{x}^*) \le 2\left(1 - \sqrt{\frac{\mu}{L}}\right)^k (f(\mathbf{x}_0) - f(\mathbf{x}^*)). \tag{2.37}$$

This shows a geometric decay rate, which is considerably faster than gradient descent.

Differential Equation Interpretation of Nesterov's Method

Just as there is an intuitive physical interpretation of the simple momentum method, there is an interpretation of Nesterov's method as the limit of a continuous second-order ordinary differential equation [478]:

$$\ddot{\mathbf{x}} + \frac{3}{t}\dot{\mathbf{x}} + \nabla f(\mathbf{x}) = 0 \tag{2.38}$$

with initial conditions $\mathbf{x}(0) = \mathbf{x}_0$ and $\dot{\mathbf{x}}(0) = \mathbf{0}$.

The time parameter is related to the step size as $t \approx k\sqrt{\gamma}$, where $\gamma \le 1/L$ is the step size and L is the Lipschitz constant of ∇f.

AdaGrad and RMSProp

AdaGrad [155] is an adaptive step method, which is related to the momentum methods above. Instead of using the discounted sum of past gradients for the velocity $\mathbf{v}$, we use the sum of past gradients squared

$$\mathbf{v}_k = \mathbf{v}_{k-1} + (\nabla f(\mathbf{x}_k))^2 \tag{2.39}$$

resulting in the following iteration

$$\mathbf{x}_{k+1} = \mathbf{x}_k - \frac{\gamma}{\sqrt{\mathbf{v}_k} + \epsilon} \nabla f(\mathbf{x}_k). \tag{2.40}$$

The small factor ϵ is added to the denominator to avoid division by zero.

The notation in Eqs. (2.40) and (2.39) is quite sloppy, although this is how it is often presented. We will be more precise here. Technically, the expression $(\nabla f(\mathbf{x}))^2$ doesn't make dimensional sense, since ∇f is a vector. Moreover, it doesn't make sense to divide by the square root of a vector. Instead we consider $(\nabla f(\mathbf{x}))^2$ to be a vector of the components of the gradient squared, so that $(\nabla f(\mathbf{x}))^2 = \begin{bmatrix}(\partial f/\partial x_1)^2 & (\partial f/\partial x_2)^2 & \cdots & (\partial f/\partial x_n)^2\end{bmatrix}^{\mathrm{T}}$. Similarly, when we divide by $\sqrt{\mathbf{v}}$, we are actually multiplying by a vector of elementwise inverses: $\mathbf{w} = \begin{bmatrix}1/\sqrt{v_1} & 1/\sqrt{v_2} & \cdots & 1/\sqrt{v_n}\end{bmatrix}^{\mathrm{T}}$, where v_j is the j-th component of $\mathbf{v}$.

The running sum $\mathbf{v}_k$ has no discount, and so it only increases, causing the iterations to slow over time. RMSProp addresses this issue by adding a decay factor to $\mathbf{v}_k$, similar to other momentum methods:

$$\mathbf{v}_{k+1} = \beta \mathbf{v}_k + (1-\beta)\left(\nabla f(\mathbf{x}_k)\right)^2. \tag{2.41}$$

Thus, we may essentially skip AdaGrad and use RMSProp, which is shorter and better.

These algorithms have been studied extensively, and there are many practical theoretical results, such as sufficient conditions for convergence [563].

2.3 Stochastic Gradient Descent and ADAM

Stochastic gradient descent (SGD) is a variant of gradient descent that scales to very high dimensional optimization problems by approximating the gradient at each step, making it suitable for large-scale neural network training. ADAM (adaptive momentum estimation) is an optimization algorithm that combines SGD with momentum and adaptive learning rates, via AdaGrad and RMSProp, making updates smoother and more responsive.

Stochastic gradient descent makes the most sense in the context of machine learning, where we are optimizing the parameters $\boldsymbol{\theta}$ of a function $\mathbf{f}_\theta(\mathbf{x})$, such as a neural network, to best fit observed data. Because this is such an important algorithm for training neural networks, it makes sense to introduce a bit more notation here. Specifically, given pairs of input–output data $\{\mathbf{x}_j, \mathbf{y}_j\}_{j=1}^N$, we seek to find the parameters $\boldsymbol{\theta}$ so that $\mathbf{f}_\theta(\mathbf{x}_j)$ best approximates $\mathbf{y}_j$, averaged over the data:

$$\min_{\theta} \sum_{j=1}^{N} \|\mathbf{f}_\theta(\mathbf{x}_j) - \mathbf{y}_j\|^2. \tag{2.42}$$

This is a *nonlinear least squares* problem, which may be approached via iterative descent methods. Nonlinear least squares will be discussed in much more detail in Chapter 4 on least squares and Chapter 10 on machine learning.

In machine learning, we call the objective a *loss function* $L(\mathbf{x}, \mathbf{y}; \boldsymbol{\theta}) = \|\mathbf{f}_{\boldsymbol{\theta}}(\mathbf{x}) - \mathbf{y}\|^2$, which we minimize over $\boldsymbol{\theta}$. It is possible to compute the gradient of the loss function with respect to $\boldsymbol{\theta}$:

$$\nabla L = 2\nabla \mathbf{f}_{\boldsymbol{\theta}}^{\mathrm{T}} \cdot (\mathbf{f}_{\boldsymbol{\theta}}(\mathbf{x}) - \mathbf{y}) . \tag{2.43}$$

The gradient of $\mathbf{f}_{\boldsymbol{\theta}}(\mathbf{x})$ may be computed analytically, or approximated using the backpropagation algorithm if it is a neural network. Backpropagation is essentially the chain rule applied to layers of a neural network using automatic differentiation [382, 31]. Automatic differentiation and backpropagation are the backbone of machine learning training and will be discussed more in Chapter 10.

Stochastic gradient descent approximates this gradient on only a subset of the data, called a batch, instead of the entire dataset. Evaluating the loss function and gradients on random batches of data adds randomness to the process. This has several key benefits, including computational speed-ups, robustness to noise and getting stuck in local minima, better exploration of the optimization landscape, and improved generalization of resulting models. Thus, stochastic gradientdescent introduces a new hyperparameter, namely the batch size.

The resulting iteration may be written as:

$$\boldsymbol{\theta}_{k+1} = \boldsymbol{\theta}_k - \gamma \nabla_{\boldsymbol{\theta}} L. \tag{2.44}$$

In principle, nothing looks to have changed from gradient descent, and all of the stochastic computations and backpropagation are hidden within the gradient $\nabla_{\boldsymbol{\theta}} L$. This is a feature, not a bug, as it means we can seamlessly extend SGD to include other gradient-based innovations like momentum and adaptive learning rates, such as ADAM.

ADAM: Adaptive Momentum Estimation

The adaptive momentum estimation (ADAM) [276] approach combines the best features of momentum methods and RMSProp, and it has become the workhorse optimization algorithm for deep learning. The original ADAM paper was the most cited paper in the world, in any field, for a decade, demonstrating its profound impact on the community.

ADAM uses a discounted running sum of squared gradients $\mathbf{v}$, as in RMSProp, and a momentum term $\mathbf{m}$ as in heavy ball momentum. The gradient update is

$$\mathbf{x}_{k+1} = \mathbf{x}_k - \gamma \frac{\hat{\mathbf{m}}_{k+1}}{\sqrt{\hat{\mathbf{v}}_{k+1}} + \epsilon} \tag{2.45}$$

where the momentum and squared gradient terms are updated according to

$$\mathbf{m}_{k+1} = \beta_1 \mathbf{m}_k + (1 - \beta_1) \nabla f(\mathbf{x}_k) \tag{2.46a}$$

$$\mathbf{v}_{k+1} = \beta_2 \mathbf{v}_k + (1 - \beta_2) \left(\nabla f(\mathbf{x}_k)\right)^2 . \tag{2.46b}$$

These terms are normalized adaptively before their use in the gradient descent update:

$$\hat{\mathbf{m}}_k = \frac{\mathbf{m}_k}{1 - \beta_1^{k-1}} \tag{2.47a}$$

$$\hat{\mathbf{v}}_k = \frac{\mathbf{v}_k}{1 - \beta_2^{k-1}}. \tag{2.47b}$$

SGD is implicitly used to compute the gradients in the update steps in ADAM. In principle, it is possible to combine other momentum and adaptive learning rate variants with SGD, such as a Nesterov accelerated SGD. However, ADAM has become the default industry standard for machine learning training because of its robustness and performance.

Example: SGD on a Linear Regression Problem

We are now ready to try out SGD on a simple linear regression problem with noisy data. We will explore linear regression in much greater depth in Chapter 4, as it is one of the foundational optimization problems. This example will also allow us to introduce the concept of batching data algorithmically. Further, we can assess all of the algorithms developed so far in a fair comparison.

First we generate data from a linear model $y = \mathbf{A}\mathbf{x} + b$ with a two-dimensional state $\mathbf{x}$ and add a relatively large amount of noise:

CodeSnip 2.c

```
N = 2000   # number of samples
A = np.random.randn(N, 2)   # design matrix
true_x = np.array([2.0, -3.0])   # true slopes
true_b = 5.0

# True y = A * x + b plus some noise
noise_std = 0.5
noise = noise_std * np.random.randn(N)
y = A.dot(true_x) + true_b + noise
```

We need code to compute the prediction $\hat{y}$ and the mean-squared error over the data set:

CodeSnip 2.d

```
def predict(A, x, b):
    # A: shape (batch_size, 2)
    # x: shape (2,)
    # b: scalar
    return A.dot(x) + b # shape (batch_size,)

def mean_squared_error(y_true, y_pred):
    return np.mean((y_true - y_pred)**2)
```

The following code computes the approximate gradient on a batch of data:

CodeSnip 2.e

```
def batch_gradients(A_batch, y_batch, x, b):
    # A_batch, x, b: same shape
    # y_batch: shape (batch_size,)
    batch_size = A_batch.shape[0]
    y_pred = predict(A_batch, x, b)
    residuals = (y_pred - y_batch)

    grad_x = (2.0 / batch_size) * A_batch.T.dot(residuals)
    grad_b = (2.0 / batch_size) * np.sum(residuals)
    return grad_x, grad_b
```

Now we define the three methods under comparison: standard (full-batch) gradient descent, mini-batch stochastic gradient descent, and stochastic gradient descent with a batch size of $n = 1$. The learning rate is γ. The standard gradient descent is:

CodeSnip 2.f

```
def full_batch_gd(A, y, gamma=0.01, n_epochs=50):
    x = np.zeros(2)  # unknown parameter vector
    b = 0.0
    mse_history = []

    for epoch in range(n_epochs):
        grad_x, grad_b = batch_gradients(A, y, x, b)
        x -= gamma * grad_x
        b -= gamma * grad_b

        y_pred = predict(A, x, b)
        mse = mean_squared_error(y, y_pred)
        mse_history.append(mse)

    return x, b, mse_history
```

the mini-batch SGD (i.e., SGD with a small batch size, $n = 50$) is:

CodeSnip 2.g

```
def mini_batch_sgd(A, y, batch_size=50, gamma=0.01, n_epochs=50)
    :
    x = np.zeros(2)
    b = 0.0
    N = A.shape[0]
    mse_history = []

    for epoch in range(n_epochs):
        indices = np.random.permutation(N)     # Shuffle data
        for start in range(0, N, batch_size):
            end = start + batch_size
            batch_idx = indices[start:end]
            A_batch = A[batch_idx]
            y_batch = y[batch_idx]

            grad_x, grad_b = batch_gradients(A_batch, y_batch, x
                , b)
            x -= gamma * grad_x
            b -= gamma * grad_b

        # Check MSE after each epoch
        y_pred = predict(A, x, b)
        mse = mean_squared_error(y, y_pred)
```

```
        mse_history.append(mse)

    return x, b, mse_history
```

and the SGD with batch size 1 is:

CodeSnip 2.h

```
def sgd_batch1(A, y, gamma=0.001, n_epochs=50):
    x = np.zeros(2)
    b = 0.0
    N = A.shape[0]
    mse_history = []

    for epoch in range(n_epochs):
        indices = np.random.permutation(N)
        for i in indices:
            A_i = A[i:i+1]          # shape (1,2)
            y_i = y[i:i+1]          # shape (1,)
            grad_x, grad_b = batch_gradients(A_i, y_i, x, b)
            x -= gamma * grad_x
            b -= gamma * grad_b

        # MSE after epoch
        y_pred = predict(A, x, b)
        mse = mean_squared_error(y, y_pred)
        mse_history.append(mse)

    return x, b, mse_history
```

Historically "true" SGD used a batch size of 1, although now in practice we almost always use mini-batch SGD. In the SGD with batch size 1, we lower gamma to reduce the sensitivity to noise and risk of divergence.

We compare the results of these three methods in Fig. 2.7. The mini-batch and SGD algorithms converge in many fewer epochs than full-batch gradient descent. However, an epoch denotes a single pass through all data, so an epoch of SGD with batch size 1 involves $n = 2000$ single steps, each using a single data point. To see the effect of batch size, you can modify the code above and increase the noise magnitude from 0.5 to 10.0. In this case, a larger batch size will be beneficial to mitigate the effect of noise.

SGD and full-batch gradient descent are the extremes of mini-batch SGD. Figures 2.8 and 2.9 compare computation time and scaling for different mini-batch sizes. A mini-batch of 10 appears to balance convergence and computation time for this example.

Example: SGD on a Nonlinear Regression Problem

We now apply SGD to a nonlinear regression problem of fitting a polynomial of high-order to noisy data. The following code generates a small amount of training and test data, $n = 50$, each with relatively large noise, for a polynomial $-10x^2 + 2x^4$. We will fit this polynomial in a space of degree 10 polynomials[4].

[4] Note that fitting a polynomial using the basis $x^0, x^1, x^2, \ldots$ is not always a great idea, as it can result in an ill-conditioned inverse problem. Instead, it is common to use orthogonal bases, such as Chebyshev polynomials. However, for this example, having a somewhat ill-conditioned inverse problem will provide a more challenging case with which to test the gradient descent methods.

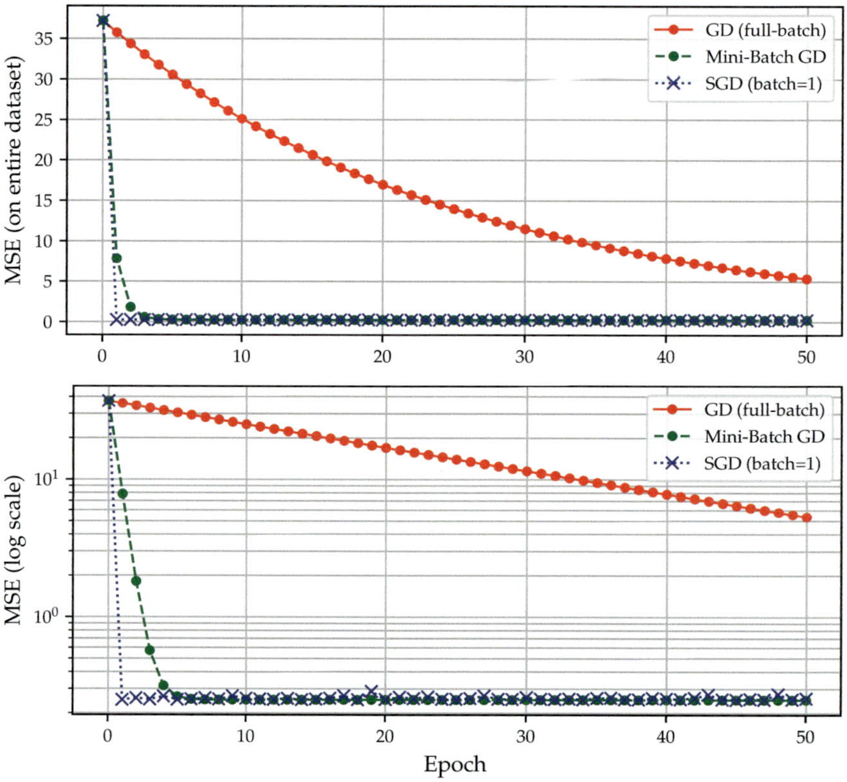

Figure 2.7 Convergence of various gradient descent methods for the linear least-squares regression problem. Methods include standard gradient descent using all data (red), mini-batch stochastic gradient descent with $n = 50$ data points (green), and stochastic gradient descent with $n = 1$ (blue). (top) Standard mean squared error on entire dataset and (bottom) MSE on log scale.

CodeSnip 2.i

```
def true_poly(x):
    return (-10.0 * x**2) + (2.0 * x**4)

num_train = 50 # Small training set makes problem challenging
num_test  = 50

x_train = np.linspace(-2, 2, num_train) # data in range [-2,2]
y_train_true = true_poly(x_train)

noise_std = 2.0 # Add relatively large noise
noise_train = noise_std * np.random.randn(num_train)
y_train = y_train_true + noise_train

np.random.seed(2023) # Test data (different seed)
x_test = np.linspace(-2, 2, num_test)
y_test_true = true_poly(x_test)
noise_test = noise_std * np.random.randn(num_test)
y_test = y_test_true + noise_test
```

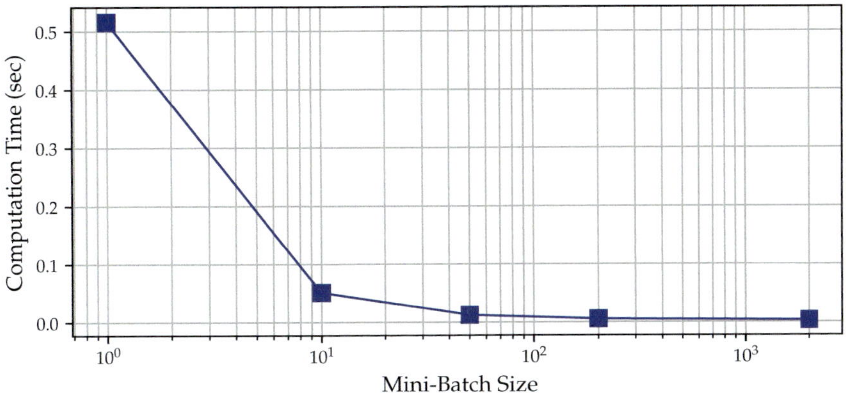

Figure 2.8 Computation time in seconds for various mini-batch sizes.

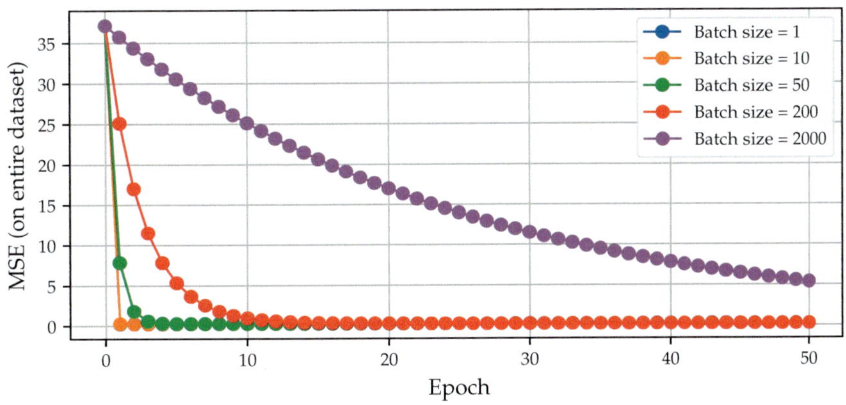

Figure 2.9 Convergence of batch gradient descent with different batch size for the linear least-squares regression problem. The batch=1 and batch=10 curves coincide.

We compute the degree 10 polynomial library:

CodeSnip 2.j

```
def create_poly_features(x, degree=10):
    # Return matrix with columns [x^0, x^1, ..., x^degree]
    return np.column_stack([x**i for i in range(degree+1)])

X_train = create_poly_features(x_train, degree=10)
X_test  = create_poly_features(x_test,  degree=10)
```

We then compute the gradient on a mini-batch of data:

CodeSnip 2.k

```
def batch_gradient(X, y, w):
    # Compute gradient of MSE w.r.t. w: -2/N * X^{\rm T} (y - Xw
        )
    N = X.shape[0]
    residuals = y - X.dot(w)
    grad = -(2.0 / N) * X.T.dot(residuals)
    return grad
```

The following code uses the ADAM implementation in PyTorch[5]. First we import torch:

CodeSnip 2.1
```
import torch
import torch.nn as nn
import torch.optim as optim
```

Next we convert our matrices and vectors to PyTorch tensors and define a linear model:

CodeSnip 2.m
```
X_train_t = torch.tensor(X_train, dtype=torch.float32)
y_train_t = torch.tensor(y_train.reshape(-1, 1), dtype=torch.
    float32)
X_test_t  = torch.tensor(X_test, dtype=torch.float32)
y_test_t  = torch.tensor(y_test.reshape(-1, 1), dtype=torch.
    float32)

# Define a linear model (# features = # polynomial terms)
model = nn.Linear(X_train.shape[1], 1)

with torch.no_grad(): # Optional: initialize weights/bias to
    zero
    model.weight.zero_()
    model.bias.zero_()
```

Finally we set up the ADAM optimizer and training loop:

CodeSnip 2.n
```
optimizer = optim.Adam(model.parameters(), gamma=1e-2)
criterion = nn.MSELoss()

n_epochs = 5000
adam_loss_history = []
for epoch in range(n_epochs):
    model.train()
    optimizer.zero_grad()
    outputs = model(X_train_t)
    loss = criterion(outputs, y_train_t)
    loss.backward()
    optimizer.step()
    adam_loss_history.append(loss.item())
    if (epoch+1) % 200 == 0:
        print(f"Epoch {epoch+1}/{n_epochs}, Loss: {loss.item()
            :.4f}")

model.eval()      # Evaluate ADAM model
with torch.no_grad():
    train_loss_adam = criterion(model(X_train_t), y_train_t).
        item()
    test_loss_adam  = criterion(model(X_test_t), y_test_t).item
        ()

y_adam_train = model(X_train_t).detach().numpy().flatten()
```

Figure 2.10 shows the training and test results for the various methods. We observe that ADAM outperforms SGD and full-batch gradient descent, both on training and testing data. Figure 2.11 shows the convergence of the methods vs. epoch, again with ADAM achieving the lowest MSE.

[5] PyTorch (`https://pytorch.org`) is an open-source machine learning library widely used for deep learning. It provides automatic differentiation and GPU acceleration, making it well suited for optimization.

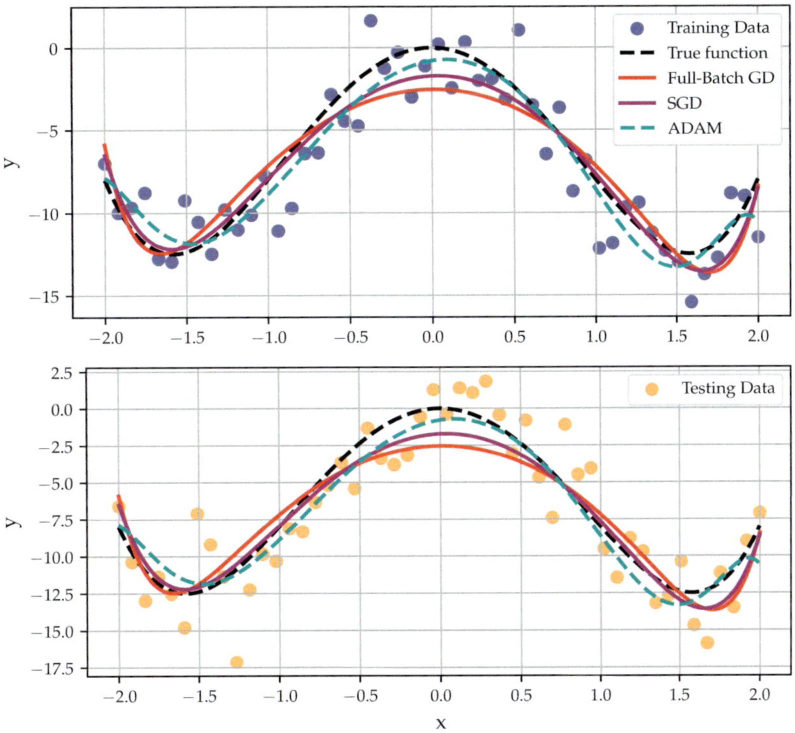

Figure 2.10 Comparison of standard gradient descent on all data, stochastic gradient descent, and SGD with ADAM on the training data (top) and testing data (bottom) for the nonlinear least-squares problem of fitting a polynomial.

2.4 Newton's Method

Originally developed to find the roots of polynomials $f(\mathbf{x})$, Newton's method may also be used to find roots of $\nabla f(\mathbf{x}) = \mathbf{0}$ in optimization. Newton's method is a second-order iterative descent scheme that uses second derivative information to improve gradient descent.

Instead of using a first-order Taylor series approximation, as in gradient descent, we will use a second-order Taylor series approximation. For a scalar variable x:

$$\tilde{f}(x) \approx f(x_0) + f'(x_0)\underbrace{(x - x_0)}_{\Delta x} + \frac{1}{2} f''(x_0)\underbrace{(x - x_0)}_{\Delta x}^2. \tag{2.48}$$

It is possible to solve for the minimum of this approximate function by setting $\nabla \tilde{f} = 0$:

$$\nabla \tilde{f} = f'(x_0) + f''(x_0)(x - x_0) = 0 \quad \Longrightarrow \quad x - x_0 = \Delta x = -\frac{f'(x_0)}{f''(x_0)}. \tag{2.49}$$

Turning this into an iterative algorithm yields Newton's method in one-dimension:

$$x_{k+1} = x_k - \frac{f'(x_k)}{f''(x_k)}. \tag{2.50}$$

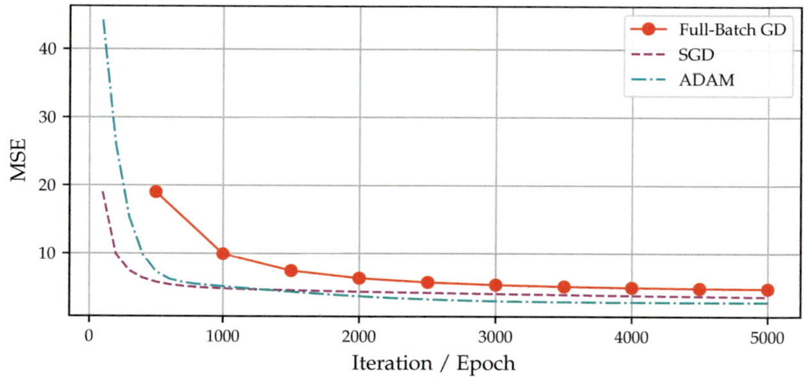

Figure 2.11 Convergence of mean square error for various methods on the nonlinear least-squares problem of fitting a polynomial.

This is very much like gradient descent in Eq. (2.4), with a step size $\gamma = 1/f''(x_k)$.

Extending to a multivariate $\mathbf{x}$, our second-order Taylor series approximation becomes

$$\tilde{f}(\mathbf{x}) \approx f(\mathbf{x}_0) + \nabla f(\mathbf{x}_0)^{\mathrm{T}}(\mathbf{x}-\mathbf{x}_0) + \frac{1}{2}(\mathbf{x}-\mathbf{x}_0)^{\mathrm{T}}\nabla^2 f(\mathbf{x}_0)(\mathbf{x}-\mathbf{x}_0). \tag{2.51}$$

Here, the Hessian $\mathbf{H} = \nabla^2 f(\mathbf{x})$ is a matrix of second partial derivatives:

$$\nabla^2 f = \begin{bmatrix} \frac{\partial^2 f}{\partial x_1^2} & \frac{\partial^2 f}{\partial x_1 x_2} & \cdots & \frac{\partial^2 f}{\partial x_1 x_n} \\ \frac{\partial^2 f}{\partial x_2 x_1} & \frac{\partial^2 f}{\partial x_2^2} & \cdots & \frac{\partial^2 f}{\partial x_2 x_n} \\ \vdots & \vdots & \ddots & \vdots \\ \frac{\partial^2 f}{\partial x_n x_1} & \frac{\partial^2 f}{\partial x_n x_2} & \cdots & \frac{\partial^2 f}{\partial x_n^2} \end{bmatrix}. \tag{2.52}$$

By the commutative property of partial derivatives, this is a symmetric matrix.

Again, it is possible to solve for the minimum of the approximate function in Eq. (2.51) by computing the gradient[6] $\nabla\tilde{f}$ and setting equal to zero:

$$\nabla\tilde{f} = \nabla f(\mathbf{x}_0) + \nabla^2 f(\mathbf{x}_0)(\mathbf{x}-\mathbf{x}_0) = \mathbf{0}$$
$$\implies \quad \mathbf{x}-\mathbf{x}_0 = \Delta\mathbf{x} = -\left(\nabla^2 f(\mathbf{x}_0)\right)^{-1} \cdot \nabla f(\mathbf{x}_0).$$

Thus, this establishes the most basic Newton's method iteration:

$$\mathbf{x}_{k+1} = \mathbf{x}_k - \mathbf{H}_k^{-1}\nabla f(\mathbf{x}_k) \tag{2.53}$$

where $\mathbf{H}_k = \nabla^2 f(\mathbf{x}_k)$ is the Hessian at $\mathbf{x}_k$. Newton's method has much faster convergence than gradient descent, but requires an expensive evaluation of the Hessian $\nabla^2 f$.

Newton's method is exact if $f(\mathbf{x})$ is a second-order polynomial. Figure 2.12 illustrates this procedure for a nonconvex objective function.

[6] Computing the gradient $\nabla\tilde{f}$ required several vector calculus identities, such as $\nabla\mathbf{c}^{\mathrm{T}}\mathbf{x} = \mathbf{c}$ and $\nabla\mathbf{x}^{\mathrm{T}}\mathbf{Q}\mathbf{x} = 2\mathbf{Q}\mathbf{x}$ for symmetric $\mathbf{Q}$. These identities are provided in Section 2.7.

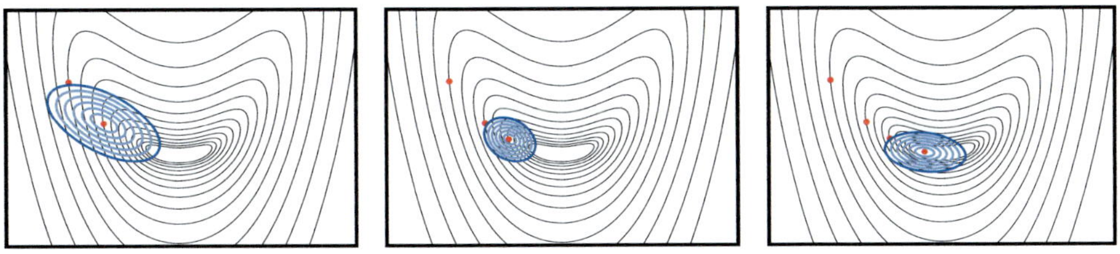

Figure 2.12 Newton's method fits a locally quadratic model at the current iteration point, and then steps to the minima of this quadratic fit for the next iteration step.

Convergence and Scaling

Newton's method has quadratic error convergence under mild assumptions. Specifically, if f is twice continuously differentiable, $\nabla^2 f$ is Lipschitz continuous near the minimizer $\mathbf{x}^*$, and $\nabla^2 f(\mathbf{x}^*)$ is positive definite, then the error $\boldsymbol{\epsilon}_k = \mathbf{x}_k - \mathbf{x}^*$ at iteration k satisfies

$$\|\boldsymbol{\epsilon}_{k+1}\| \leq \frac{1}{2}\|\boldsymbol{\epsilon}_k\|^2. \tag{2.54}$$

This quadratic convergence of the error is *much* stronger than the $O(1/k^2)$ convergence of Nesterov's momentum method and ADAM in the iteration number k. Quadratic convergence of the error means that the number of correct digits in the estimate roughly doubles at each step near the optimum. These momentum-based methods improve over standard gradient descent, but they are still much slower than Newton's method near a solution.

Lipschitz continuity of the Hessian $\nabla^2 f$ means that there exists a $L \geq 0$ such that

$$\|\nabla^2 f(\mathbf{x}) - \nabla^2 f(\mathbf{y})\| \leq L\|\mathbf{x} - \mathbf{y}\| \tag{2.55}$$

for all $\mathbf{x}, \mathbf{y} \in \mathbb{R}^n$. If $\nabla^2 f$ is continuously differentiable and the third derivatives of f are bounded, then this condition is satisfied. This is a relatively mild assumption for most smooth functions.

Of course, Newton's method comes at a significantly higher per-iteration cost. The Hessian $\mathbf{H} = \nabla^2 f$ is an $n \times n$ matrix, so its memory footprint scales as n^2 and computing its inverse scales as n^3. This leads to two key issues with Newton's method:

1. The Hessian may be unavailable or too expensive to compute.
2. The Hessian may not be positive definite, such as when the function f has a local maximum or a saddle point.

Both of these challenges motivate the development of quasi-Newton methods, such as the BFGS and LBFGS algorithms that will be discussed in Section 2.6.

The second-order, quadratic approximation in Eq. (2.51) is better than the simple first-order gradient approximation. However, away from the base point $\mathbf{x}_0$, this approximation may degrade, especially for more complex objective function landscapes. Thus, instead of using Eq. (2.53) to update $\mathbf{x}_{k+1}$, we may use it to determine a better gradient direction $\mathbf{d} = \mathbf{H}^{-1}\nabla f(\mathbf{x})$, and then do a line search in this direction. The cost of each iteration in the second-order Newton's method is expensive, so the additional cost of a line search is justified. We will explore line search in more detail in the next section.

Further, it may be necessary to add Hessian damping, which regularizes the linear system to enforce positive definiteness,

$$(\mathbf{H} + \alpha\mathbf{I})\mathbf{p} = -\nabla f, \qquad \alpha \geq 0.$$

Then, a line search is performed in the $\mathbf{p}$ direction. Levenberg–Marquardt is a classic regularized, damped Gauss–Newton method that we will discuss in Section 4.7.

Numerical Examples

We now compare gradient descent and Newton's method on the non-quadratic objective function in Eq. (2.7). Newton's method is exact for quadratic objective functions, and so it would converge to the minimum of the quadratic function in Eq. (2.6). Therefore we compare these algorithms on a more challenging nonconvex objective function.

The following short code executes Newton's method:

CodeSnip 2.o

```
def newtons_method(start, max_iter):
    trajectory = [np.array(start)]
    values = [objective_function(*start)]
    point = np.array(start, dtype=float)

    for _ in range(max_iter):
        grad = gradient(point[0], point[1])
        hess = hessian(point[0], point[1])
        hess_inv = np.linalg.inv(hess)
        point -= hess_inv @ grad
        trajectory.append(point.copy())
        values.append(objective_function(point[0], point[1]))

    return np.array(trajectory), np.array(values)

# Run updated methods
nm_trajectory, nm_values = newtons_method(start_point, max_iter)
```

Figure 2.13 shows the trajectory and objective function value versus iteration for both algorithms. We use an initial condition of $[-2, 1]$ and a max number of iterations of 30; for gradient descent, we use step size $\gamma = 0.05$. Newton's method clearly converges more rapidly than gradient descent. This plot shows the log-linear scaling of gradient descent and the incredible convergence of Newton's method. Within five steps, the Newton's method solver converges to near machine precision of the minimum.

Of course, this incredible performance of Newton's method comes at the cost of requiring Hessian information of the objective function f:

$$\mathbf{H} = \nabla^2 f(\mathbf{x}) = \begin{bmatrix} e^{x_1+2x_2-0.2} + e^{x_1-2x_2-.2} + e^{-x_1-0.2} & 2e^{x_1+2x_2-0.2} - 2e^{x_1-2x_2-0.2} \\ 2e^{x_1+2x_2-0.2} - 2e^{x_1-2x_2-0.2} & 4e^{x_1+2x_2-0.2} + 4e^{x_1-2x_2-0.2} \end{bmatrix}.$$

In code:

CodeSnip 2.p

```
# Define the objective function and its derivatives
def objective_function(x, y):
    return np.exp(x+2*y-0.2)+np.exp(x-2*y-0.2)+np.exp(-x-0.2)

def gradient(x, y):
```

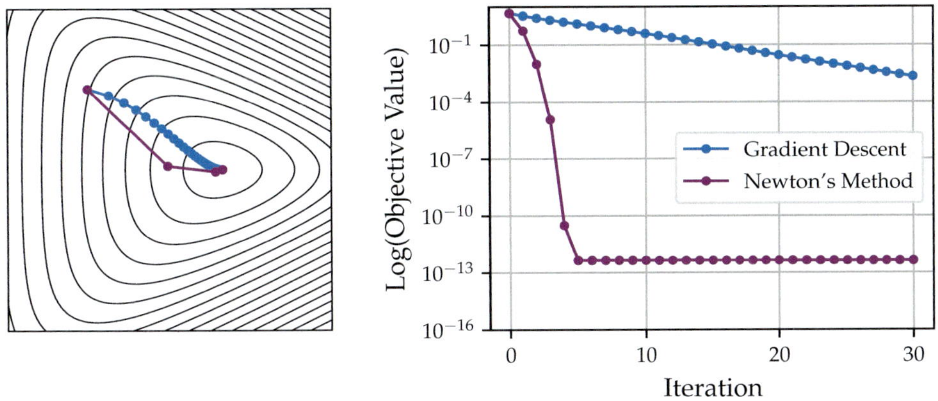

Figure 2.13 Comparison of gradient descent and Newton's method on a non-quadratic objective function, plotted on a semilog error scale to better visualize the convergence rate. It is clear that the second-order Newton's method is dramatically faster than the first-order convergence of simple gradient descent.

```
    grad_x = np.exp(x+2*y-0.2)+np.exp(x-2*y-0.2)-np.exp(-x-0.2)
    grad_y = 2*np.exp(x+2*y-0.2)-2*np.exp(x-2*y-0.2)
    return np.array([grad_x, grad_y])

def hessian(x, y):
    h_xx = np.exp(x+2*y-0.2)+np.exp(x-2*y-0.2)+np.exp(-x-0.2)
    h_xy = 2*np.exp(x+2*y-0.2)-2*np.exp(x-2*y-0.2)
    h_yy = 4*np.exp(x+2*y-0.2)+4*np.exp(x-2*y-0.2)
    return np.array([[h_xx, h_xy], [h_xy, h_yy]])
```

Comparison of All Methods

Now we compare all three methods: gradient descent, Nesterov's accelerated gradient descent, and Newton's method. We test convergence on a nonlinear objective with *banana* shaped level sets. The objective function, its gradient, and its Hessian are given by:

$$f(\mathbf{x}) = (x_2 - x_1^2)^2 + x_1^4$$
$$\nabla f(\mathbf{x}) = \begin{bmatrix} 4x_1^3 - 4x_1(x_2 - x_1^2) \\ 2(x_2 - x_1^2) \end{bmatrix}$$
$$\mathbf{H} = \nabla^2 f(\mathbf{x}) = \begin{bmatrix} 12x_1^2 - 4(x_2 - 3x_1^2) & -4x_1 \\ -4x_1 & 2 \end{bmatrix}.$$

Figure 2.14 shows the comparison of all three methods. As expected, the second-order Newton's method dramatically outperforms the first-order gradient descent methods.

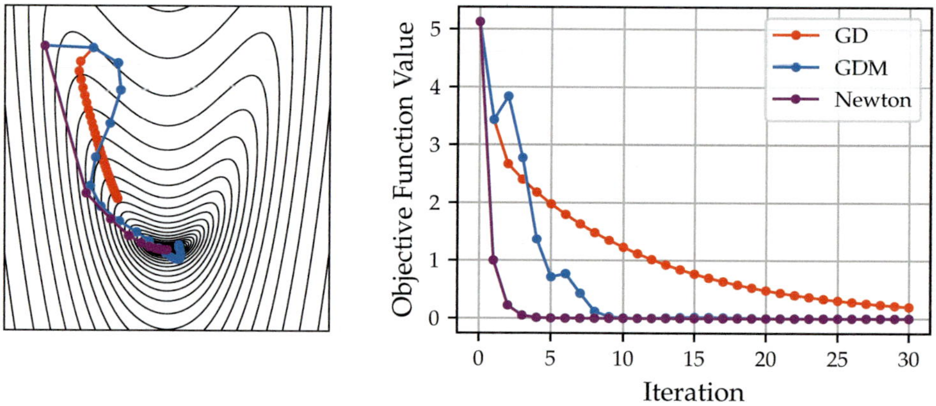

Figure 2.14 Comparison of gradient descent (GD), gradient descent with momentum (GDM), and Newton's method on a nonlinear objective function.

2.5 Line Search

Line search is an essential topic in optimization that is often glossed over in favor of flashier accelerations, such as NAG. However, in practice, it can make simple gradient descent or Newton's method vastly more effective, and competitive or superior to other methods.

At its core, line search is about choosing the step size in iterative optimization methods such as gradient descent or Newton's method. In machine learning applications, this step size is often called the learning rate. Suppose you have a current iterate $\mathbf{x}_k$ and a search direction $\mathbf{p}_k$, such as the negative gradient $-\nabla f$ for gradient descent, or the direction $-\mathbf{H}_k^{-1}\nabla f$ for Newton's method. A line search searches for a scalar $\alpha_k > 0$ such that

$$\mathbf{x}_{k+1} = \mathbf{x}_k + \alpha_k \mathbf{p}_k \tag{2.56}$$

sufficiently decreases the objective function. So the direction is determined by the algorithm, but the step size is chosen adaptively.

Line search is important for efficient and robust implementation of most descent methods. There are a few major categories of line search, and we only discuss a few here.

The *exact* line search solves the full optimization problem over α

$$\arg\min_{\alpha>0} f(\mathbf{x}_k + \alpha \mathbf{p}_k). \tag{2.57}$$

This is rarely used in practice, except for quadratic problems where there is a closed form solution. Instead, we typically perform an *inexact* line search. Three of the most common approaches are the Armijo backtracking rule, the Wolfe conditions, and the Barzilai–Borwein (BB) Step.

Armijo is both robust and easy to implement, although it is sometimes overly conservative. The Wolfe conditions combine Armijo with a curvature condition to prevent the step from being too small. Wolfe line search is widely used in quasi-Newton methods (BFGS, LBFGS). The BB step uses information from the last two iterates to approximate a quasi-Newton step size, and it is very closely related to the LBFGS discussed in the next section. We won't explore this approach here, but it is good to know that it exists, and it may be very effective in practice.

Line search requires more function evaluations per iteration. However, the number of iterations typically drops dramatically, often by an order of magnitude. In addition to the speed up, line search is considerably more robust. For first-order methods, such as gradient descent and NAG, line search adds a few function evaluations per iteration. These extra evaluations are relatively inexpensive, and have the benefit of preventing divergence and reducing iterations. For second-order methods, such as Newton and quasi-Newton methods, the overhead associated with line search is negligible compared to the Hessian computation and linear system solves.

The relative cost of computing ∇f versus f varies widely. Using finite differences, it may cost $\sim n$ evaluations of f to approximate ∇f. For regression examples, the function f and the gradient ∇f both require the same order of computations, both requiring a single pass over the data; however, the gradient may still be a constant of $2-3$ times more expensive in practice, requiring more operations and memory. For example, when using automatic differentiation in Section 10.1, the gradient has the same scaling as the objective, but may be a constant more expensive. In PDE-constrained problems, evaluating f requires a forward PDE solve, while computing ∇f requires both the forward solution and an additional adjoint PDE solve. Thus, obtaining both f and ∇f together is about twice as expensive as f alone, but the marginal cost of the gradient given f is roughly one more PDE solve.

One should always use line search for full-batch or full step gradient descent. Armijo backtracking is ideal for robustness and it makes gradient descent competitive with quasi-Newton. Line searches are particularly useful in Newton's method when the Hessian is indefinite or ill-conditioned, so that a full step may fail, and where each function evaluation is cheap compared to the Hessian calculation. For ill-conditioned problems, BB is especially effective. Line search is essential to make Newton's method globally convergent. Line search is challenging to use with SGD, and is much less common in this context.

Armijo Backtracking

The Armijo (sufficient decrease) rule is a simple, robust line search. Given a current $\mathbf{x}$ and a descent direction $\mathbf{d}$ (so $\nabla_{\mathbf{d}} f(\mathbf{x}) = \mathbf{d}^\top \nabla f(\mathbf{x}) < 0$), we seek a step size $\alpha > 0$ such that

$$f(\mathbf{x} + \alpha\mathbf{d}) \;\le\; f(\mathbf{x}) \;+\; c_1\, \alpha\, \nabla_{\mathbf{d}} f(\mathbf{x}), \qquad c_1 \in (0,1). \tag{2.58}$$

This ensures a sufficient reduction in f along $\mathbf{d}$.

Armijo line search involves the following steps:

1. Initialize $\alpha \leftarrow \alpha_0$ (often $\alpha_0 = 1$, or the size of the Newton step in Eq. (2.53)). While the Armijo inequality above is not satisfied, shrink the step geometrically: $\alpha \leftarrow \rho\,\alpha$ with $\rho \in (0,1)$ (e.g., $\rho = \frac{1}{2}$), and re-evaluate $f(\mathbf{x} + \alpha\mathbf{d})$.
2. Accept the first α that satisfies the condition and update $\mathbf{x}^+ = \mathbf{x} + \alpha\mathbf{d}$. Typical choices are $c_1 \in [10^{-4}, 10^{-2}]$ and $\rho \in [0.1, 0.8]$. (For a curvature check in addition to sufficient decrease, one can combine Armijo with a Wolfe condition.)

Armijo backtracking is shown in Fig. 2.15 for gradient descent and Newton's method. It is compared with several other approaches, including gradient descent and Newton's method

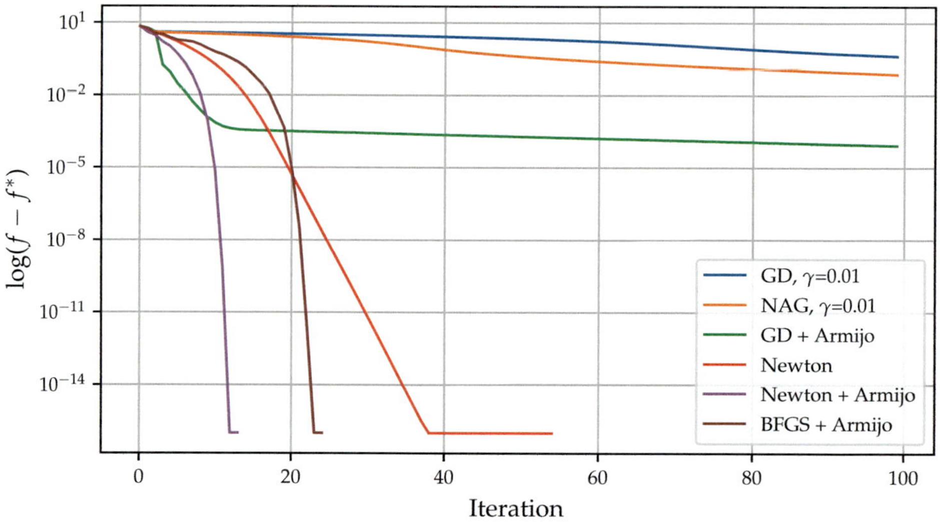

Figure 2.15 Comparison of various descent methods, with and without line search.

without line search, Nesterov's accelerated gradient, and BFGS, which will be discussed in the next section. Each method is applied to the *banana* shaped objective function from the previous section.

Wolfe Conditions

It is common to use the Wolfe conditions for the line search, which moves in the new gradient direction $\mathbf{d}$ until $\nabla_{\mathbf{d}} f = \nabla_{\mathbf{d}} f = \mathbf{d}^{\mathrm{T}} \nabla f < \tau$, where τ is a threshold that indicates when a sufficiently flat region has been found; here, $\nabla_{\mathbf{d}} f$ refers to the directional derivative of f in the direction $\mathbf{d}$. This typically involves two stages:

1. We take steps of size Δx and size $2\Delta x$, where Δx is the size of the original Newton step in Eq. (2.53), until we find a valid point where either: i) $f_2 > f_1$ or ii) $\nabla f_2 > 0$. In the first case, the objective function increased in the second step, so it must have reached a local minimum between. In the second case, the slope of the second point is positive, which also indicates a local minimum between. If these conditions are not met, then we double the step size and repeat.

2. Once a region containing a minimum has been identified, we *zoom in* to the minimum with a bisection method. This amounts to reducing the interval until a point is found. There are several fancier tricks we can play, like using spline interpolation if there are multiple points in the bisection search.

2.6 Quasi-Newton Methods and BFGS

When **H** is too expensive to compute, or when **H** is not positive definite, we may approximate this Hessian with a matrix **A**. Such approaches to approximating the Hessian are known as quasi-Newton methods [73, 138, 370]. This topic has been developed in the numerical applied mathematics community for over half a century.

The matrix **A** provides a quadratic approximation $\tilde{f}$ to our objective function $f(\mathbf{y})$ at some point $\mathbf{y}$ near our current base point $\mathbf{x}$:

$$\tilde{f}_{\mathbf{A}}(\mathbf{y}) = f(\mathbf{x}) + \nabla f(\mathbf{x})^{\mathrm{T}}(\mathbf{y}-\mathbf{x}) + \frac{1}{2}(\mathbf{y}-\mathbf{x})^{\mathrm{T}}\mathbf{A}(\mathbf{y}-\mathbf{x}). \tag{2.59}$$

To design an approximation **A** to the Hessian **H**, we need to put constraints on this new matrix:

- **A** should be symmetric, as **H** is symmetric.
- **A** should be positive definite, since we will invert it in Newton's method. Further, any positive definite **A** will yield descent.
- The quadratic approximation $\tilde{f}_{\mathbf{A}}$ should have the same gradient as f at the current point $\mathbf{x}_k$ and last point $\mathbf{x}_{k-1}$:

$$\nabla \tilde{f}(\mathbf{x}_k) = \nabla f(\mathbf{x}_k) \tag{2.60a}$$
$$\nabla \tilde{f}(\mathbf{x}_{k-1}) = \nabla f(\mathbf{x}_{k-1}). \tag{2.60b}$$

- Since **A** will be updated at each iteration step, we add a subscript $\mathbf{A}_k$. We want $\mathbf{A}_k$ to be "close" to $\mathbf{A}_{k-1}$. There are several notions of closeness we will investigate below.

Based on these conditions, there are several approaches to approximate **A**, which we discuss below and in Panel 2.4. We will build from the simplest rank-1 updates to the more commonly used, but more involved, BFGS algorithm. Since we will use the inverse of the Hessian approximation, $\mathbf{B} = \mathbf{A}^{-1}$, in the Newton iteration, some quasi-Newton methods directly construct **B**.

Now we will provide basic derivations and formulas for how to update **A** and **B** in the various schemes below. We introduce a few important points of notation:

$$\mathbf{p}_k = \mathbf{x}_{k+1} - \mathbf{x}_k \tag{2.61a}$$
$$\mathbf{q}_k = \nabla f(\mathbf{x}_{k+1}) - \nabla f(\mathbf{x}_k). \tag{2.61b}$$

These vectors quantify the change in $\mathbf{x}$ and $\nabla f(\mathbf{x})$ at each step of the iteration. We will update these vectors at every iteration, and so we drop the subscript k for ease of notation.

There is an important geometric property of the Hessian, namely that $\mathbf{H}\Delta\mathbf{x} \approx \Delta\mathbf{g}$, where $\mathbf{g} = \nabla f(\mathbf{x})$. Therefore, we should expect this to also hold for our approximate Hessian **A**:

$$\mathbf{A}\mathbf{p} = \mathbf{q}. \tag{2.62}$$

This is known as the secant equation. The inverse equation also holds for **B**:

$$\mathbf{B}\mathbf{q} = \mathbf{p}. \tag{2.63}$$

We are now ready to derive the various methods introduced briefly above. Each of these quasi-Newton methods rely on properties of the inverses of rank-1 and rank-k updates, known as the Sherman–Morrison formula and Woodbury matrix identities, respectively. These results are provided at the end of this section.

Rank-1 Update to the Hessian

We begin by illustrating quasi-Newton methods with the simplest rank-1 matrix update rule:

$$\mathbf{A}_{k+1} = \mathbf{A}_k + \mathbf{u}\mathbf{v}^{\mathrm{T}}. \tag{2.64}$$

Panel 2.4: A Brief Summary of Quasi-Newton Methods

To initialize these updates, we typically set $\mathbf{A}_0 = \mathbf{B}_0 = \mathbf{I}$, the identity matrix.

- SR1: This approach seeks a rank-1 update to $\mathbf{A}$, so that

$$\operatorname{rank}\left(\mathbf{A}_k - \mathbf{A}_{k-1}\right) = 1.$$

  Specifically, a rank-1 update will have the form $\mathbf{A}_k = \mathbf{A}_{k-1} + \mathbf{u}\mathbf{v}^{\mathrm{T}}$, where $\mathbf{u}, \mathbf{v} \in \mathbb{R}^n$ are vectors the same size as $\mathbf{x}$. This is the simplest update approach, and is often used in trust region methods. However, for general optimization, there might not be a rank-1 update that satisfies the various conditions above.
- PSB: Another approach is to find an update that has minimum Frobenius norm distance to the previous approximation:

$$\min \|\mathbf{A}_k - \mathbf{A}_{k-1}\|_F$$

  where the Frobenius norm, $\|\cdot\|_F^2$, is the sum of the squares of every element of the matrix argument. However, this approach does not guarantee that $\mathbf{A}_k$ is positive definite. We will discuss matrix norms and optimizing over matrix arguments more in Chapters 4 and 5.
- DFP: The DFP approach is similar to PSB, but incorporates a special weight matrix $\mathbf{W}$ into the Frobenius norm:

$$\min \|\mathbf{W}\left(\mathbf{A}_k - \mathbf{A}_{k-1}\right)\mathbf{W}\|_F.$$

  Depending on the choice of $\mathbf{W}$, this approach will guarantee that $\mathbf{A}_k$ is positive definite if $\mathbf{A}_{k-1}$ is positive definite. DFP is a rank-2 update.
- BFGS: This is the most stable and widely used quasi-Newton method, and it is closely related to DFP. BFGS is a rank-2 update to $\mathbf{B}$ that satisfies:

$$\min \|\mathbf{W}\left(\mathbf{B}_k - \mathbf{B}_{k-1}\right)\mathbf{W}\|_F.$$

All of these methods have analytic solutions, resulting in explicit update formulas that are efficient, requiring only a few matrix-vector products.

Using the secant equation and rearranging, we have

$$\begin{aligned} \left(\mathbf{A}_k + \mathbf{u}\mathbf{v}^{\mathrm{T}}\right)\mathbf{p} &= \mathbf{q} \\ \implies \quad \mathbf{u}\mathbf{v}^{\mathrm{T}}\mathbf{p} &= \mathbf{q} - \mathbf{A}_k\mathbf{p} \\ \implies \quad \mathbf{u} &= \frac{1}{\mathbf{v}^{\mathrm{T}}\mathbf{p}}\left(\mathbf{q} - \mathbf{A}_k\mathbf{p}\right). \end{aligned} \tag{2.65}$$

We choose $\mathbf{v} \propto \mathbf{u}$ so that $\mathbf{u}\mathbf{v}^{\mathrm{T}}$ is a symmetric rank-1 matrix, since we also want $\mathbf{A}$ to be symmetric. In this case, the update equation becomes

$$\mathbf{v} = \mathbf{q} - \mathbf{A}_k\mathbf{p} \tag{2.66a}$$

$$\mathbf{A}_{k+1} = \mathbf{A}_k + \frac{1}{\mathbf{v}^{\mathrm{T}}\mathbf{p}}\mathbf{v}\mathbf{v}^{\mathrm{T}}. \tag{2.66b}$$

Technically, we want $\mathbf{B} = \mathbf{A}^{-1}$ for Newton's iteration, which results in the following *inverse Hessian update*:

$$\mathbf{v} = \mathbf{p} - \mathbf{B}_k\mathbf{q} \tag{2.67a}$$

$$\mathbf{B}_{k+1} = \mathbf{B}_k + \frac{1}{\mathbf{v}^{\mathrm{T}}\mathbf{q}}\mathbf{v}\mathbf{v}^{\mathrm{T}}. \tag{2.67b}$$

Broyden's Update, BFGS, and DFP

There is a so-called "Broyden's family" of quasi-Newton methods, so we begin by presenting the simplest Broyden update:

$$\mathbf{A}_{k+1} = \mathbf{A}_k + \frac{1}{\mathbf{p}^{\mathrm{T}}\mathbf{A}_k\mathbf{q}}\left(\mathbf{p} - \mathbf{A}_k\mathbf{q}\right)^{\mathrm{T}}\mathbf{p}^{\mathrm{T}}\mathbf{A}_k \tag{2.68a}$$

$$\mathbf{B}_{k+1} = \mathbf{B}_k + \frac{1}{\mathbf{p}^{\mathrm{T}}\mathbf{p}}\left(\mathbf{q} - \mathbf{B}_k\mathbf{p}\right)\mathbf{p}^{\mathrm{T}}. \tag{2.68b}$$

Note that we may choose to update $\mathbf{A}$ or $\mathbf{B}$, and don't need to update both.

The Broyden–Fletcher–Goldfarb–Shanno (BFGS) [197, 173, 458, 74] algorithm is a rank-2 update, which provides more degrees of freedom to guarantee that $\mathbf{A}_{k+1}$ is symmetric, positive definite, and *close* to the previous $\mathbf{A}_k$. The updates for $\mathbf{A}$ and $\mathbf{B}$ are given by

$$\mathbf{A}_{k+1} = \left(\mathbf{I} - \frac{\mathbf{p}\mathbf{q}^{\mathrm{T}}}{\mathbf{q}^{\mathrm{T}}\mathbf{p}}\right)\mathbf{A}_k\left(\mathbf{I} - \frac{\mathbf{q}\mathbf{p}^{\mathrm{T}}}{\mathbf{q}^{\mathrm{T}}\mathbf{p}}\right) + \frac{\mathbf{p}\mathbf{p}^{\mathrm{T}}}{\mathbf{q}^{\mathrm{T}}\mathbf{p}} \tag{2.69a}$$

$$\mathbf{B}_{k+1} = \mathbf{B}_k + \frac{\mathbf{q}\mathbf{q}^{\mathrm{T}}}{\mathbf{q}^{\mathrm{T}}\mathbf{p}} - \frac{\mathbf{B}_k\mathbf{p}\left(\mathbf{B}_k\mathbf{p}\right)^{\mathrm{T}}}{\mathbf{p}^{\mathrm{T}}\mathbf{B}_k\mathbf{p}}. \tag{2.69b}$$

It is clear that this is a rank-2 update of $\mathbf{B}_{k+1}$, as each term after $\mathbf{B}_k$ is a rank-1 matrix.

The Davidon–Fletcher–Powell (DFP) [173, 126, 395, 174] algorithm is closely related to BFGS. In fact, DFP is the dual of BFGS; however, it was quickly supplanted with BFGS, which has improved stability. For completeness, we present the DFP algorithm here:

$$\mathbf{A}_{k+1} = \mathbf{A}_k + \frac{\mathbf{p}\mathbf{p}^{\mathrm{T}}}{\mathbf{p}^{\mathrm{T}}\mathbf{q}} - \frac{\mathbf{A}_k\mathbf{q}\mathbf{q}^{\mathrm{T}}\mathbf{A}_k}{\mathbf{q}^{\mathrm{T}}\mathbf{A}_k\mathbf{q}} \tag{2.70a}$$

$$\mathbf{B}_{k+1} = \left(\mathbf{I} - \frac{\mathbf{q}\mathbf{p}^{\mathrm{T}}}{\mathbf{q}^{\mathrm{T}}\mathbf{p}}\right)\mathbf{B}_k\left(\mathbf{I} - \frac{\mathbf{p}\mathbf{q}^{\mathrm{T}}}{\mathbf{q}^{\mathrm{T}}\mathbf{p}}\right) + \frac{\mathbf{q}\mathbf{q}^{\mathrm{T}}}{\mathbf{q}^{\mathrm{T}}\mathbf{p}}. \tag{2.70b}$$

Thus, DFP is the dual of BFGS, with the **p** and **q** terms swapped. However, the update for **B**, which is what we want for Newton iteration, is simpler in BFGS than in DFP.

The Broyden family of updates is obtained by interpolating between BFGS and DFP:

$$\mathbf{B}_{k+1} = (1-\alpha)\mathbf{B}_{k+1}^{\mathrm{BFGS}} + \alpha\mathbf{B}_{k+1}^{\mathrm{DFP}} \quad \text{where} \quad \alpha \in [0, 1]. \tag{2.71}$$

However, we typically set $\alpha = 0$ and use BFGS.

There are other prominent variants, such as the limited-memory BFGS (LBFGS) [342, 369, 317, 87], which is widely used in maximum entropy learning [328] and has a stochastic extension for large-scale optimization [86].

Inverse Formulas for Rank-*k* Matrix Updates

In the derivations above, we used the Sherman–Morrison formula for the inverse of a rank-1 matrix update and the Woodbury matrix identity for the inverse of a rank-k matrix update. Here, we elaborate on these formulas. In both cases, we assume that we have access to the matrix **A** and its inverse $\mathbf{A}^{-1}$. We then seek a simple formula for the inverse of the rank-k update of **A**.

The Sherman–Morrison formula [460] for the inverse of a rank-1 update is

$$\left(\mathbf{A} + \mathbf{u}\mathbf{v}^{\mathrm{T}}\right)^{-1} = \mathbf{A}^{-1} - \frac{\mathbf{A}^{-1}\mathbf{u}\mathbf{v}^{\mathrm{T}}\mathbf{A}^{-1}}{1 + \mathbf{v}^{\mathrm{T}}\mathbf{A}^{-1}\mathbf{u}}. \tag{2.72}$$

The rank-1 matrix $\mathbf{u}\mathbf{v}^{\mathrm{T}}$ is added to **A**; $\mathbf{A} + \mathbf{u}\mathbf{v}^{\mathrm{T}}$ will only have an inverse if $1 + \mathbf{v}^{\mathrm{T}}\mathbf{A}^{-1}\mathbf{u} \neq 0$.

Although this expression seems involved, it may be computed simply with just a few matrix–vector products and vector–vector inner products. More history and details are presented in the review of Hager [217]. The version presented here is from Bartlett [27].

The Woodbury matrix identity [541] is a more general result that works for a rank-k matrix update. In this case we add $\mathbf{U}\mathbf{V}^{\mathrm{T}}$ to **A**, where $\mathbf{U}, \mathbf{V} \in \mathbb{R}^{n\times k}$ are both matrices with k linearly independent columns. In this case, Woodbury's identity is

$$\left(\mathbf{A} + \mathbf{U}\mathbf{V}^{\mathrm{T}}\right)^{-1} = \mathbf{A}^{-1} - \mathbf{A}^{-1}\mathbf{U}\left(\mathbf{I} + \mathbf{V}^{\mathrm{T}}\mathbf{A}^{-1}\mathbf{U}\right)^{-1}\mathbf{V}^{\mathrm{T}}\mathbf{A}^{-1}. \tag{2.73}$$

Here **I** is a $k \times k$ identity matrix. As with the Sherman–Morrison formula, this rank-k updated matrix will only have an inverse if $\mathbf{I} + \mathbf{V}^{\mathrm{T}}\mathbf{A}^{-1}\mathbf{U}$ is invertible. This result is commonly attributed to Woodbury [541], although it was technically presented earlier in [216], as reviewed in [217].

2.7 Gradient Vector Calculus Identities

There are several important vector calculus identities we will use repeatedly when computing gradients. Here we state and derive the most common identities that we will use.

It is often the case that we must take the gradient of a scalar objective function obtained by taking an inner product of our decision vector **x** with a constant vector **c**, such as the objective function $\mathbf{c}^{\mathrm{T}}\mathbf{x}$ in linear programming in Chapter 3. This gradient is given by

$$\nabla\left(\mathbf{c}^{\mathrm{T}}\mathbf{x}\right) = \nabla\left(\mathbf{x}^{\mathrm{T}}\mathbf{c}\right) = \begin{bmatrix} c_1 \\ c_2 \\ \vdots \end{bmatrix} = \mathbf{c}. \tag{2.74}$$

This is easy to derive by expanding $\mathbf{c}^{\mathrm{T}}\mathbf{x}$ and $\mathbf{x}^{\mathrm{T}}\mathbf{c}$:

$$\mathbf{c}^{\mathrm{T}}\mathbf{x} = \begin{bmatrix} c_1 & c_2 & \cdots \end{bmatrix} \begin{bmatrix} x_1 \\ x_2 \\ \vdots \end{bmatrix} = c_1x_1 + c_2x_2 + \cdots = \begin{bmatrix} x_1 & x_2 & \cdots \end{bmatrix} \begin{bmatrix} c_1 \\ c_2 \\ \vdots \end{bmatrix} = \mathbf{x}^{\mathrm{T}}\mathbf{c}. \tag{2.75}$$

Thus, the gradient of the expanded sum in the middle, $c_1x_1 + c_2x_2 + \cdots$ is the vector **c**.

It is also common to take the gradient of a matrix–vector product, such as $\mathbf{Ax}$ or $\mathbf{x}^{\mathrm{T}}\mathbf{A}^{\mathrm{T}}$. To derive these expressions, it helps to carefully expand out the product first:

$$\mathbf{Ax} = \begin{bmatrix} a_{11} & a_{12} & \cdots & a_{1n} \\ a_{21} & a_{22} & \cdots & a_{2n} \\ \vdots & \vdots & \ddots & \vdots \\ a_{n1} & a_{n2} & \cdots & a_{nn} \end{bmatrix} \begin{bmatrix} x_1 \\ x_2 \\ \vdots \\ x_n \end{bmatrix} = \begin{bmatrix} a_{1j}x_j \\ a_{2j}x_j \\ \vdots \\ a_{nj}x_j \end{bmatrix} \tag{2.76}$$

$$\mathbf{x}^{\mathrm{T}}\mathbf{A}^{\mathrm{T}} = \begin{bmatrix} x_1 & x_2 & \cdots & x_n \end{bmatrix} \begin{bmatrix} a_{11} & a_{21} & \cdots & a_{n1} \\ a_{12} & a_{22} & \cdots & a_{n2} \\ \vdots & \vdots & \ddots & \vdots \\ a_{1n} & a_{2n} & \cdots & a_{nn} \end{bmatrix} = \begin{bmatrix} x_ja_{1j} & x_ja_{2j} & \cdots & x_ja_{nj} \end{bmatrix}. \tag{2.77}$$

The sum over j is implied, so $a_{1j}x_j = \sum_{j=1^n} a_{1j}x_j$. We will use this notation throughout. We now compute the Jacobian $\mathbf{D}(\mathbf{Ax})$ and gradient $\nabla\left(\mathbf{x}^{\mathrm{T}}\mathbf{A}^{\mathrm{T}}\right)$:

$$\mathbf{D}(\mathbf{Ax}) = \begin{bmatrix} \frac{\partial}{\partial x_1}\left(a_{1j}x_j\right) & \frac{\partial}{\partial x_2}\left(a_{1j}x_j\right) & \cdots \\ \frac{\partial}{\partial x_1}\left(a_{2j}x_j\right) & \frac{\partial}{\partial x_2}\left(a_{2j}x_j\right) & \cdots \\ \vdots & \vdots & \ddots \end{bmatrix} = \begin{bmatrix} a_{11} & a_{12} & \cdots & a_{1n} \\ a_{21} & a_{22} & \cdots & a_{2n} \\ \vdots & \vdots & \ddots & \vdots \\ a_{n1} & a_{n2} & \cdots & a_{nn} \end{bmatrix} = \mathbf{A} \tag{2.78}$$

and

$$\nabla\left(\mathbf{x}^{\mathrm{T}}\mathbf{A}^{\mathrm{T}}\right) = \begin{bmatrix} \nabla x_ja_{1j} & \nabla x_ja_{2j} & \cdots \end{bmatrix} = \begin{bmatrix} a_{11} & a_{21} & \cdots & a_{n1} \\ a_{12} & a_{22} & \cdots & a_{n2} \\ \vdots & \vdots & \ddots & \vdots \\ a_{1n} & a_{2n} & \cdots & a_{nn} \end{bmatrix} = \mathbf{A}^{\mathrm{T}}. \tag{2.79}$$

This implies that $\nabla\left(\mathbf{x}^{\mathrm{T}}\mathbf{A}\right) = \mathbf{A}$. These properties are also true for nonsquare $\mathbf{A}$ matrices, which is left as an exercise. Sometimes we use the notation $\nabla(\mathbf{A}\mathbf{x})$ for the Jacobian $\mathbf{D}(\mathbf{A}\mathbf{x})$.

Finally, we often must take the gradient of a quadratic form $\mathbf{x}^{\mathrm{T}}\mathbf{A}\mathbf{x}$, which arises in the least-squares regression problem. This gradient is given by

$$\nabla\left(\mathbf{x}^{\mathrm{T}}\mathbf{A}\mathbf{x}\right) = \left(\mathbf{A} + \mathbf{A}^{\mathrm{T}}\right)\mathbf{x}, \tag{2.80}$$

which is equal to $2\mathbf{A}\mathbf{x}$ if $\mathbf{A}$ is symmetric. To derive this, we expand the quadratic form

$$\mathbf{x}^{\mathrm{T}}\mathbf{A}\mathbf{x} = \begin{bmatrix} x_1 & x_2 & \cdots & x_n \end{bmatrix} \begin{bmatrix} a_{1j}x_j \\ a_{2j}x_j \\ \vdots \\ a_{nj}x_j \end{bmatrix} \tag{2.81a}$$

$$= x_1 \sum_j a_{1j}x_j + x_2 \sum_j a_{2j}x_j + \cdots + x_n \sum_j a_{nj}x_j. \tag{2.81b}$$

It is then possible to take the partial derivative of this expression with respect to x_i:

$$\frac{\partial \mathbf{x}^{\mathrm{T}}\mathbf{A}\mathbf{x}}{\partial x_1} = \sum_j a_{1j}x_j + [x_1 a_{11} + x_2 a_{21} + \cdots + x_n a_{n1}] \tag{2.82a}$$

$$\frac{\partial \mathbf{x}^{\mathrm{T}}\mathbf{A}\mathbf{x}}{\partial x_2} = \sum_j a_{2j}x_j + [x_1 a_{12} + x_2 a_{22} + \cdots + x_n a_{n2}] \tag{2.82b}$$

$$\vdots$$

$$\frac{\partial \mathbf{x}^{\mathrm{T}}\mathbf{A}\mathbf{x}}{\partial x_i} = \underbrace{\sum_j a_{ij}x_j}_{\mathbf{A}\mathbf{x}} + \underbrace{[x_1 a_{1i} + x_2 a_{2i} + \cdots + x_n a_{ni}]}_{\mathbf{A}^{\mathrm{T}}\mathbf{x}}, \tag{2.82c}$$

which finally yields $\nabla\left(\mathbf{x}^{\mathrm{T}}\mathbf{A}\mathbf{x}\right) = \mathbf{A}\mathbf{x} + \mathbf{A}^{\mathrm{T}}\mathbf{x}$.

We also have useful properties for the gradient and Hessian of functions $f(\mathbf{A}\mathbf{x})$:

$$\nabla\left(f(\mathbf{A}\mathbf{x})\right) = \mathbf{A}^{\mathrm{T}}\nabla f(\mathbf{A}\mathbf{x}) \tag{2.83}$$

$$\nabla^2\left(f(\mathbf{A}\mathbf{x})\right) = \mathbf{A}^{\mathrm{T}}\nabla^2 f(\mathbf{A}\mathbf{x})\mathbf{A}. \tag{2.84}$$

Panel 2.5: Summary of Gradient Vector Calculus Identities

To recap, we have the following major vector calculus identities:

$$\nabla\mathbf{c}^{\mathrm{T}}\mathbf{x} = \nabla\mathbf{x}^{\mathrm{T}}\mathbf{c} = \mathbf{c}$$

$$\mathbf{D}(\mathbf{A}\mathbf{x}) = \nabla(\mathbf{x}^{\mathrm{T}}\mathbf{A}) = \mathbf{A}$$

$$\nabla(\mathbf{x}^{\mathrm{T}}\mathbf{A}\mathbf{x}) = (\mathbf{A} + \mathbf{A}^{\mathrm{T}})\mathbf{x} = 2\mathbf{A}\mathbf{x} \text{ if } \mathbf{A} \text{ symmetric.}$$

Suggested reading

Texts

(1) Vector Calculus, by J. E. Marsden and A. Tromba, 2003 [336].

Papers and reviews

(1) Adam: A method for stochastic optimization, by D. P. Kingma and J. Ba, *arXiv:1412.6980*, 2014 [276].

(2) Learning representations by back-propagating errors, by D. E. Rumelhart, G. E. Hinton, and R. J. Williams, *Nature*, 1986 [431].

(3) Some methods of speeding up the convergence of iteration methods, by B. T. Polya, *USSR Computational Mathematics and Mathematical Physics*, 1964 [392].

(4) A method for unconstrained convex minimization problem with the rate of convergence O $(1/k^2)$, by Y. Nesterov, *Dokl. Akad. Nauk. SSSR*, 1983 [364].

(5) Nesterov's Method for Convex Optimization, by N. J. Walkington, *SIAM Review*, 2023 [525].

(6) A differential equation for modeling Nesterov's accelerated gradient method: Theory and insights, by W. Su, S. Boyd, and E. J. Candes, *Journal of Machine Learning Research*, 2016 [478].

(7) Quasi-Newton methods, motivation and theory, by J. E. Dennis Jr., and J. J. Moré, *SIAM Review*, 1977 [138].

(8) On the limited memory BFGS method for large scale optimization, by D. C. Liu and J. Nocedal, *Mathematical Programming*, 1989 [317].

Homework

Exercise 2.1 In this exercise, we will explore the effect of step size on convergence and stability of gradient descent. Consider the quadratic objective function in Eq. (2.6). Compute the step size γ where the iteration goes unstable. Simulate gradient descent with several different step sizes and plot the results. Is the stability boundary as predicted?

Exercise 2.2 Implement a simple gradient descent for the function

$$f(x) = x^2 + 2x + 5.$$

1. Show the iteration of x_k for $k = 0, \ldots, 10$ starting from $x_0 = 5$.
2. What happens if you choose a step size $\alpha = 1$? How about $\alpha = 0.01$?

Exercise 2.3 Consider the ODE $\dot{x} = -\nabla f(x)$ for $f(x) = \frac{1}{2}x^2$.

1. Solve the ODE analytically.
2. Compare the continuous gradient flow solution to the discrete updates from gradient descent with step size $\alpha = 0.1$.

Exercise 2.4 Implement both Heavy Ball *and* Nesterov's Accelerated Gradient (NAG) to minimize

$$f(x) = \sin^2(x) + 0.1x^2.$$

1. Compare convergence rates and plot the objective $f(x_k)$ vs. iteration k.
2. Discuss any overshoot or oscillations for each method.

Exercise 2.5 Use Newton's method to find the minimum of

$$f(x, y) = x^4 + x^2y^2 + y^2.$$

1. Compute the Hessian.
2. Run the method starting from $(x_0, y_0) = (2, 1)$ and use line search.
3. Compare with gradient descent for the same starting point.

Exercise 2.6 Consider the following constrained optimization problem

$$\begin{aligned} \min_{\mathbf{x}} \; & x_1^2 + x_2^2 \\ \text{s.t.} \; & x_1 \geq 2. \end{aligned}$$

Use the method of Lagrange multipliers to show that this problem has a minimum at the point $(x_1, x_2) = (2, 0)$. Compute the gradient of the objective function and the gradient of the constraint equation. What do you notice?

Now consider the modified problem

$$\begin{aligned} \min_{\mathbf{x}} \; & x_1^2 + x_2^2 \\ \text{s.t.} \; & x_1 + x_2 \geq 2. \end{aligned}$$

Show that the minimum of this problem occurs at $(1, 1)$. Compute the gradient of the objective function and the gradient of the constraint equation. What is the pattern here?

Exercise 2.7 Show that the forward Euler integration of the gradient flow ODE in Panel 2.2 is equivalent to the gradient descent iteration in Eq. (2.4) with $\Delta t = \gamma$.

Exercise 2.8 Derive a new gradient descent algorithm using a more sophisticated Runge–Kutta integration scheme for the gradient flow ODE in Panel 2.2. For example, recall that the 2nd order Runge–Kutta integration scheme is given by the following:

$$\mathbf{x}_{k+1} = \mathbf{x}_k + \Delta t \mathbf{v}_2$$

where

$$\begin{aligned}\mathbf{v}_1 &= \mathbf{v}(\mathbf{x}_k)\\ \mathbf{v}_2 &= \mathbf{v}(\mathbf{x}_k + \Delta t \mathbf{v}_1/2).\end{aligned}$$

Code this and compare with gradient descent, specifically convergence and stability.

Exercise 2.9 As a simple exercise, compute the maximum step size $\gamma_{\max}$ and optimal step size γ^* for the quadratic example in Fig. 2.3. You can then test gradient descent with a γ just above and below the maximum stability boundary and also the optimal γ^*.

Exercise 2.10 Show that the eigenvalues of $(\mathbf{I} - \gamma\mathbf{A})$ are symmetric about the point $1 - \gamma\lambda_{\text{mid}}$ from Eq. (2.23).

Exercise 2.11 Plot the convergence rate vs iteration k for Nesterov's accelerated gradient method for a Lipschitz continuous function and for a μ-strongly convex function, using Eqs. (2.36) and (2.37). Try different K and μ and discuss the trends.

Exercise 2.12 We will now explore how gradient descent scales with the dimension of an optimization problem. To make the problem more amenable to analysis, we will consider a quadratic objective function, where we know the optimal step size γ.

First, we will need to be able to generate positive semidefinite $\mathbf{Q}$ matrices of various dimensions $n \times n$. We have several choices:

1. Random: We may generate a quadratic positive semidefinite matrix $\mathbf{Q} = \mathbf{A}^{\mathrm{T}}\mathbf{A}$ by first generating a random $\mathbf{A} \in \mathbb{R}^{m\times n}$.
2. Custom Eigenvalues: It is possible to rescale the eigenvalues of $\mathbf{Q}$ as follows:
   (a) Compute the SVD of $\mathbf{A}$: $\mathbf{A} = \mathbf{U}\mathbf{\Sigma}\mathbf{V}^{\mathrm{T}}$.
   (b) Assuming that the rank of $\mathbf{A}$ is n, then the eigenvalues λ_j of $\mathbf{Q}$ will be the squares of the non-zero singular values σ_j^2.
   (c) Thus, to design custom eigenvalues $\hat{\lambda}_j > 0$, we must simply rescale (e.g., replace) the corresponding singular value with $\sqrt{\hat{\lambda}_j}$, resulting in $\hat{\mathbf{\Sigma}}$. Then we construct a new custom $\hat{\mathbf{A}}$ and $\hat{\mathbf{Q}}$

$$\hat{\mathbf{A}} = \mathbf{U}\hat{\mathbf{\Sigma}}\mathbf{V}^{\mathrm{T}} \quad \Longrightarrow \quad \hat{\mathbf{Q}} = \hat{\mathbf{A}}^{\mathrm{T}}\hat{\mathbf{A}}$$

   with the custom spectral behavior.

3. Structured: Generate a matrix with diagonal dominant structure

$$\mathbf{Q} = \begin{bmatrix} 2 & 0 & 0 & \cdots & 1 \\ 0 & 3 & 0 & \cdots & 1 \\ 0 & 0 & 4 & \cdots & 1 \\ \vdots & \vdots & \vdots & \ddots & \vdots \\ 1 & 0 & 0 & \cdots & n+1 \end{bmatrix}.$$

There are several exercises for exploring the behavior of gradient descent for different $\mathbf{Q}$:

1. Begin with the structured, diagonally-dominant $\mathbf{Q}$ from above, and a fixed step $\gamma = 0.01$. Write a code to iterate through the gradient descent algorithm, starting from an initial condition of all ones (i.e., $\mathbf{x}_0 = [1, 1, \ldots, 1]$). Plot the convergence versus iteration step for all n from $n = 1$ to $n = 1000$. Put all values of n on the same plot, and change the color of each line from blue ($n = 1$) to red ($n = 1000$). Plot the convergence in both regular and logarithmic values. Discuss the observed trends.
2. Now, repeat the exact same exercise above, but for each value of n, compute the optimal step size γ_n based on the quadratic matrix $\mathbf{Q}$.
3. Now, we will consider the case of a $\mathbf{Q}$ matrix generated from random $\mathbf{A}^T\mathbf{A}$. Write a code to iterate through the gradient descent algorithm and compute how many steps are required before we are within an absolute distance of $\epsilon = 0.01$ of the optimal value. Recall for this example the optimal $\mathbf{x}^*$ is at the origin, $\mathbf{x}^* = \mathbf{0}$. Run this code for several randomly generated $\mathbf{Q}$ with various n, say 100 random $\mathbf{Q}$ for each dimension. Generate a violin plot of the number of iteration steps as a function of dimension n.
4. Repeat for $\gamma = 0.01$ and for the optimal γ^* for each individual $\mathbf{Q}$ matrix. Also make a violin plot for the optimal γ^* as a function of n.
5. Repeat the exercise above, but now with the eigenvalues of $\mathbf{Q}$ designed to be randomly, uniformly distributed in the range $\lambda_{\min} = 1$ to $\lambda_{\max} = 2$.
6. Repeat the exercise above, but with the eigenvalues of $\mathbf{Q}$ designed to be logarithmically spaced in the range .001 to 1000.

Exercise 2.13 Run the same exact progression of exercises as above, but using SGD and mini-batch GD to see how they scale with dimension. Compare and discuss the results.

Exercise 2.14 We will now explore the method of alternating descent, which will be discussed more in Chapter 6. When gradient information is unavailable, it is still possible to design a descent algorithm as follows. We simply cycle through each coordinate x_j of the variable $\mathbf{x}$, performing a line search to minimize the objective function in that variable alone. By iterating through each coordinate several times, we will descend towards the optimal value $\mathbf{x}^*$. This method is also known as the method of coordinate descent.

Write a code to perform coordinate descent on the simple quadratic example with $\mathbf{Q} = \begin{bmatrix} 2 & 1 \\ 1 & 3 \end{bmatrix}$ and plot the results. Repeat with $\mathbf{Q} = \begin{bmatrix} 2 & 0 \\ 0 & 3 \end{bmatrix}$. In both cases, how many times do you need to alternate through each coordinate before reaching the optimal value to within 1%? Keep track of how many function evaluations $f(\mathbf{x})$ were needed for coordinate descent, and plot convergence versus function evaluations. Compare with gradient descent with $\gamma = 0.1$.

You will have several choices for your coordinate descent algorithm, including the step size, how to perform the line search, convergence criteria, etc. A good starting guideline is to use a step size $\gamma = 0.1$ and use bisection to either half or double the step at each step in the line search, based on some simple criteria. Then, end the line search in a given coordinate direction after the change in the decision variable $\mathbf{x}$ is less than some cutoff $\tau \approx 0.001$.

Exercise 2.15 Repeat the coordinate descent exercise above for larger problems with $n = 2$ to $n = 10$. Specifically, explore the following

1. Begin with the structured, diagonally-dominant $\mathbf{Q}$ from above, and a fixed step $\gamma = 0.01$. Write a code to iterate through the coordinate descent algorithm, starting from an initial condition of all ones (i.e., $\mathbf{x}_0 = [1, 1, \ldots, 1]$). Plot the convergence versus iteration step for all n from $n = 2$ to $n = 10$.
2. Now, consider a $\mathbf{Q}$ matrix generated from random $\mathbf{A}^{\mathrm{T}}\mathbf{A}$ with the eigenvalues of $\mathbf{Q}$ designed to be randomly, uniformly distributed in the range $\lambda_{\min} = 1$ to $\lambda_{\max} = 2$. Plot the convergence versus coordinate iterations, and convergence versus function evaluations, for each $n = 2$ to $n = 10$, but with a single random $\mathbf{Q}$ for each dimension.
3. Repeat the exercise above, but with the eigenvalues of $\mathbf{Q}$ designed to be logarithmically spaced in the range .001 to 1000.

Exercise 2.16 Verify that the inverse Hessian update in Eq. (2.67) is in fact the inverse of the rank-1 update in Eq. (2.66).

Exercise 2.17 Verify that the BFGS update in Eq. (2.69) satisfies the secant condition that the updated matrix remains positive definite. Compute the normed error of $\mathbf{A}_{k+1} - \mathbf{A}_k$.

Exercise 2.18 Verify that $\nabla\left(\mathbf{x}^{\mathrm{T}}\mathbf{A}\right) = \mathbf{A}$ and $\mathbf{D}(\mathbf{A}\mathbf{x}) = \mathbf{A}$ hold for non-square matrices $\mathbf{A}$.

3 Linear Programming

So far, we have mainly explored unconstrained optimization, choosing to focus on the objective function. Linear programming is an important class of convex problems with constraints, where the objective function is linear, and the problem is almost entirely determined by the structure of the linear inequality constraints (i.e., the feasible region).

Most real-world problems involve constrained resources, such as time, space, money, materials, and labor. As such, there are real tradeoffs in solving such problems, which we express as inequality constraints. Linear programming was invented by Leonid Kantorovich in 1939 [267] and developed in and after World War II, largely motivated by severe rationing and increasingly complex logistical and supply chain optimization problems. Some of the many modern applications include:

- Resource allocation and workforce scheduling;
- Factory layout, space usage, production scheduling, and inventory management;
- Project management and task scheduling;
- Crop planning and irrigation;
- Portfolio optimization, risk management, and pricing;
- Network flow: routing, bandwidth allocation, etc.;
- Power generation: scheduling, renewable mix, etc.;
- Supply chain logistics, fleet routing and delivery scheduling;
- Airline flight, crew, and maintenance scheduling;
- Healthcare: resource and staff allocation;
- Drug dosage, radiation therapy, etc.

These examples are all characterized by a relatively simple objective function that scales approximately linearly (e.g., profit or cost), and fundamental constraints on resources. Formulating these as linear programs is left as an exercise.

There is no closed-form, analytic solution to a generic linear program. However, there are several accurate and efficient algorithms that scale well to large problems. Modern linear programs may easily involve thousands of decision variables and constraints, for example in optimizing a distribution center for a large online retailer. In this chapter, we define the mathematical ingredients of a linear program, provide several examples to illustrate various applications, and discuss solution techniques and extensions.

3.1 Problem Statement and Examples

At its heart, linear programming involves a linear objective function and a set of linear constraint equations:

$$\begin{aligned} \max_{\mathbf{x}} \quad & \mathbf{c}^{\mathrm{T}}\mathbf{x}, \\ \text{s.t.} \quad & \mathbf{A}\mathbf{x} \leq \mathbf{b}, \\ & \mathbf{x} \geq \mathbf{0}, \end{aligned} \tag{3.1}$$

where $\mathbf{c} \in \mathbb{R}^n$ is the vector of linear weights for the objective function $f(\mathbf{x}) = \mathbf{c}^{\mathrm{T}}\mathbf{x}$, and $\mathbf{A} \in \mathbb{R}^{m\times n}$ and $\mathbf{b} \in \mathbb{R}^m$ define the constraint equations $\mathbf{A}\mathbf{x} \leq \mathbf{b}$. Technically we don't always require $\mathbf{x} \geq \mathbf{0}$, but it is a common constraint that variables are non-negative.

The feasible set is given by a convex polytope, which is the intersection of the half-planes defined by each of the constraint equations:

$$\mathbf{a}_j^{\mathrm{T}}\mathbf{x} \leq b_j, \quad \text{for } j = 1, \ldots, m \tag{3.2}$$

and the positive quadrant defined by

$$x_j \geq 0, \quad \text{for } j = 1, \ldots, n. \tag{3.3}$$

The intersection of two convex sets is also a convex set, so the feasible polytope is convex. A polytope is generally a *pointy* geometric object, generalizing the notion of a polygon, like a triangle or a quadrilateral, to higher dimensions. The individual constraints in Eqs. (3.2) and (3.3) capture specific properties of the constrained resources, such as total space usage, cost of goods, non-negative prices, etc. Technically, the feasible set might not be bounded, if there are not enough inequality constraints; however, we mostly focus on the case of bounded polytopes. See Panel 3.2 for an example of an unbounded polytope.

The linear objective function is both convex and concave, as it is a flat plane. Therefore, both minimization and maximization tasks are *convex* in linear programming. Although we primarily explored *minizization* tasks in Chapter 1, it is common in linear programming to *maximize* the objective function, for example when the objective function represents profit or productivity.

The gradient of the linear objective function is a constant[1]

$$\nabla f(\mathbf{x}) = \mathbf{c} \tag{3.4}$$

so that level sets of the objective function are straight lines increasing perpendicular to the gradient vector. This is visualized in Fig. 3.1.

The geometry of the linear objective function and bounded feasible polytope mean that the maximum and minimum always occur on a vertex of the feasible set. In degenerate cases, there may be a line of local maxima or minima, connecting two vertices, when the gradient is exactly perpendicular to this line. However, even in this case, the local minima still occurs on at least one vertex. See Figs. 3.1 and 3.2 for an illustration of these concepts.

Linear programs are convex optimization problems, so a local maximum (minimum) is also a global maximum (minimum). Therefore, a common strategy for optimizing a linear

[1] See the derivation of this in Section 2.7. Also see Section 1.3 for background on the gradient.

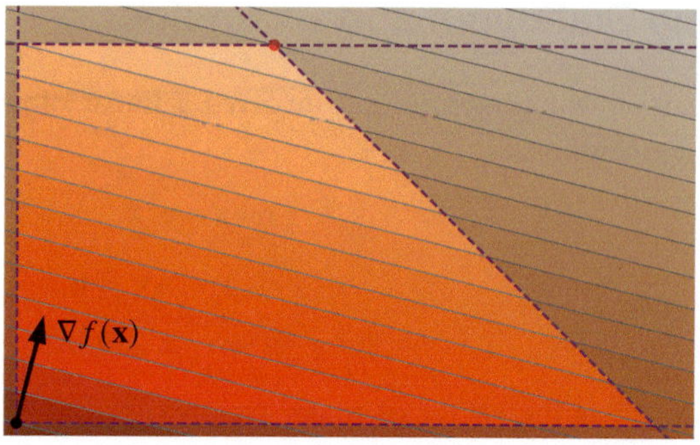

Figure 3.1 Visualization of level sets of the objective function overlaid with the convex feasible polytope. The maximum and minimum of the objective function always intersect at least one of the vertices of the polytope. Level sets of the objective are shown as gray lines, increasing from the origin. These level sets are equally spaced, since the objective is linear. This feasible polytope is consistent with the inequalities in Eq. (3.5).

program is to search across vertices until a local extremum is found, which is also a global extremum. However, the number of vertices may grow rapidly in higher dimensions[2], so exhaustively searching through all vertices is intractable. An efficient search strategy is required.

First, we will illustrate how to solve a linear program by hand on a simple low-dimensional example. We use Dantzig's simplex method [123, 124], which is a method to walk from vertex to vertex in directions that increase the objective function until no neighboring vertex increases the objective function. At this point a local maximum is attained, and by convexity this is also a global maximum. Dantzig's simplex method is the simplest *active set method*, since we identify and test vertices by making a subset of the inequality constraints active and then solve the corresponding equality-constrained subproblem. This approach is relatively straightforward and scales well to large numerical problems.

For even larger linear programs, we use *interior point methods* (IPMs), which are the other large class of solution techniques; these are discussed in Section 6.4. However, Dantzig's simplex method is simple and robust, and it is still widely used in industry.

Simple Example

Consider the following simple example of a linear program with two decision variables and two constraint equations. Say there is a factory that manufactures two types of microscopes: inexpensive microscopes for classrooms, and expensive microscopes for research labs. The profit for each teaching microscope is \$100 per unit and the profit for each research microscope is \$400 per unit. The factory can make a maximum of 50 microscopes total per month, and there is a maximum demand for research microscopes of 30 per month. How many of each type of microscope should be made each day to maximize profit?

[2] The number of vertices on an n dimensional hypercube is 2^n, so that a 10-dimensional cube has over a thousand vertices and a 20-dimensional cube has over a million vertices.

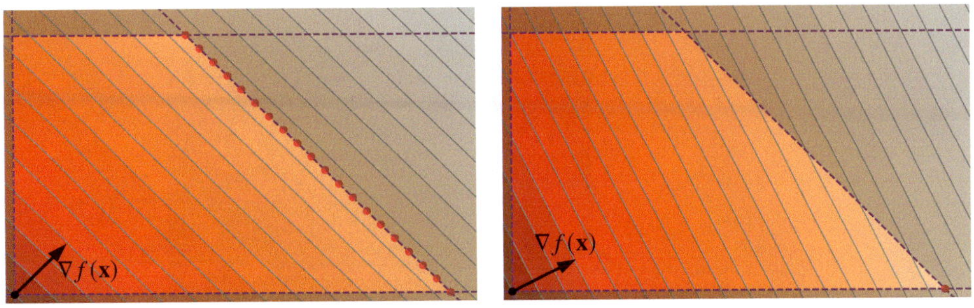

Figure 3.2 If the gradient of the objective function, given by the vector $\nabla f = \mathbf{c}$, rotates, then the maximum may occur on a different vertex. There are also cases where the gradient vector is perpendicular to one of the edges of the polytope, in which case the entire edge, including the two bounding vertices, is a local maximum.

This linear program can be formulated and solved simply, by hand. In fact, we do not need to resort to formal mathematical arguments to know that the best strategy is to make 30 of the more profitable research microscopes, up to the limit of demand, and then make the remaining 20 inexpensive teaching microscopes to fill the factory's capacity. However, stating and solving this problem formally will help illustrate how to solve more complex linear programs in a systematic way using Dantzig's simplex method.

If we let the number of teaching microscopes be x_1 and the number of research microscopes be x_2, then the linear program may be stated formally as:

$$\begin{aligned} \max_{x_1, x_2} \quad & 100x_1 + 400x_2 \\ \text{s.t.} \quad & x_1, x_2 \geq 0 \\ & x_2 \leq 30 \\ & x_1 + x_2 \leq 50. \end{aligned} \tag{3.5}$$

The inequality constraints and resulting convex feasible polytope are shown in Fig. 3.3. The maximum will occur on one of the vertices, so the simple brute force approach would be to list all of the vertices and compute the objective function on each vertex:

$$\begin{aligned} v_1 &: (0, 0) & f(v_1) &= 0 \\ v_2 &: (50, 0) & f(v_2) &= 5000 \\ v_3 &: (0, 30) & f(v_3) &= 12000 \\ v_4 &: (20, 30) & f(v_4) &= 14000. \end{aligned}$$

The maximum occurs at the vertex $(20, 30)$, which is consistent with our intuition. However, we need a more robust algorithm for solving these problems that scales to higher dimensions. To this end, we will introduce Dantzig's simplex method in a few sections to systematically solve linear programs in arbitrary dimension.

It is also straightforward to solve linear programs in Python, for example using the `pulp` package with the following code:

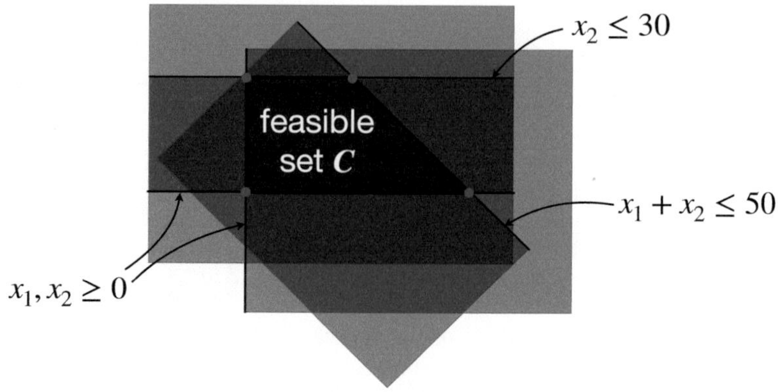

Figure 3.3 The convex feasible polytope given by the constraint equations in Eq. (3.5). Feasible solutions are limited to the positive quadrant ($x_1 \geq 0$ and $x_2 \geq 0$). The additional constraints $x_2 \leq 30$ and $x_1 + x_2 \leq 50$ form a bounded, convex polytope with four vertices.

CodeSnip 3.a

```
import pulp

# Create a linear programming problem instance
lp_problem = pulp.LpProblem("Maximize_Profit", pulp.LpMaximize)

# Define the variables
x1 = pulp.LpVariable('x1', lowBound=0, cat='Continuous')
x2 = pulp.LpVariable('x2', lowBound=0, cat='Continuous')

# Set the objective function
lp_problem += 100 * x1 + 400 * x2, "Total Profit"

# Add constraints
lp_problem += x2 <= 30, "Constraint_y_upper_bound"
lp_problem += x1 + x2 <= 50, "Constraint_combined"

lp_problem.solve()   # Solve the problem

print("Optimal value of x1:", pulp.value(x1))
print("Optimal value of x2:", pulp.value(x2))
print("Maximum Profit:", pulp.value(lp_problem.objective))
```

The output of this code is

CodeSnip 3.b

```
Optimal value of x1: 20.0
Optimal value of x2: 30.0
Maximum Profit: 14000.0
```

which agrees with our earlier vibe calculations.

3.2 Polytopes, Vertices, and High-Dimensional Geometry

Identifying vertices is challenging in higher dimensions. Before we proceed to Dantzig's simplex method, it is helpful to understand the geometry of the convex polytopes that define the feasible regions in linear programs. The essence of the simplex method involves

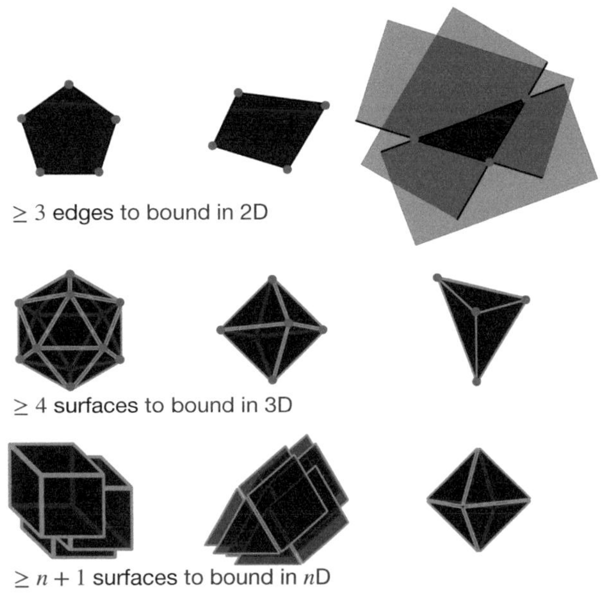

Figure 3.4 Examples of polytopes and their geometries in various dimensions.

searching from vertex to vertex in directions that locally increase the objective function until a local maximum, and thus a global maximum, is found. We need a robust strategy to relate vertices to inequality constraints, so that we can efficiently move from vertex to vertex. Fundamentally, each boundary of the polytope is determined by a particular inequality constraint being *active* or *tight*, at which point it behaves like an equality constraint. Each vertex is the intersection of several boundaries, meaning that it is defined by some subset of the inequality constraints being *active*. We will introduce the slack variable form, which allows us to flexibly relax some inequality constraints while making others active. This is an important technique in active set methods, and it will allow us to walk from vertex to vertex.

It is impossible to fully visualize geometric objects in higher-dimensional spaces[3], so it is helpful to build intuition in two-dimensions and three-dimensions, and then extrapolate rules and intuition. Figure 3.4 describes the basic notion of polytopes in two-dimensions, in three-dimensions, and in n-dimensions.

Slack Variables and Polytopes

We introduce slack variables $\mathbf{s} \in \mathbb{R}^m$ to the linear program in Eq. (3.1) to make it easier to determine which inequalities are active. This allows us to and move from vertex to vertex. The slack variables, $\mathbf{s}$, have as many dimensions as there are constraint equations in $\mathbf{Ax} \leq \mathbf{b}$.

[3] There are legends of hallucinagenic parties in the 1960s, where mathematicians would get together and try to expand their minds, to visualize higher-dimensional geometries. I am aphantasic, and can't even picture two-dimensional space with my eyes closed.

Panel 3.1: Geometric Properties and Polytopes

There are several basic properties that build our intuition:

- Every pair of adjacent vertices are connected by a one-dimensional edge.
- The (nonempty) intersection of 2 (nonparallel) edges is a vertex.
- The intersection of 2 two-dimensional surfaces is a one-dimensional edge.
- In two-dimensions, polytopes are bounded by at least 3 one-dimensional edges; picture $x_1, x_2 \geq 0$ and $x_1 + x_2 \leq 1$.
- In three-dimensions, polytopes are bounded by at least 4 two-dimensional surfaces; picture $x_1, x_2, x_3 \geq 0$ and $x_1 + x_2 + x_3 \leq 1$.
- In n-dimensions, polytopes are bounded by at least $n + 1$ distinct $(n - 1)$-dimensional surfaces; picture $\mathbf{x} \geq 0$ and $\sum_j x_j \leq 1$.
- In n-dimensions, a single inequality constraint defines a half-space bounded by an $(n - 1)$-dimensional surface; so in two-dimensions an inequality constraint defines a half-plane bounded by an edge, and in three-dimensions an inequality constraint defines a half-space bounded by a two-dimensional surface.
- Thus, the intersection of $n + 1$ inequality constraints will bound a polytope in n-dimensions. I encourage the reader to confirm that this matches the specific cases in two-dimensions and three-dimensions above.
- In n-dimensions, the vertices of a polytope are given by n active constraints.

Introducing slack variables, the linear program becomes

$$\begin{aligned} \max_{\mathbf{x}} \quad & \mathbf{c}^{\mathrm{T}}\mathbf{x}, \\ \text{s.t.} \quad & \mathbf{x}, \mathbf{s} \geq \mathbf{0}, \\ & \mathbf{A}\mathbf{x} + \mathbf{s} = \mathbf{b}. \end{aligned} \tag{3.6}$$

This is equivalent to Eq. (3.1), since $\mathbf{s} = \mathbf{b} - \mathbf{A}\mathbf{x} \geq 0$. We typically move everything except for $\mathbf{s}$ to the right of the equation, which results in

$$\begin{aligned} \max_{\mathbf{x}} \quad & \mathbf{c}^{\mathrm{T}}\mathbf{x}, \\ \text{s.t.} \quad & \mathbf{x}, \mathbf{s} \geq \mathbf{0}, \\ & \mathbf{s} = \mathbf{b} - \mathbf{A}\mathbf{x}. \end{aligned} \tag{3.7}$$

Each component s_i of $\mathbf{s}$ quantifies how close $\mathbf{x}$ is to the i-th inequality boundary. When $s_i = 0$, the i-th inequality is active, and we are on the boundary defined by this equality.

Recall that for a feasible polytope in n-dimensions, each vertex is defined by n inequality constraints being active, so they behave like equality constraints. Each corresponding slack variable is equal to zero and the decision variable is on the boundary of the constraint; i.e., it is *tight* up against the constraint boundary. The remaining constraints are inactive, and there is some *slack* between the current decision variable and those inequality boundaries[4].

[4] In some literature, the active/tight variables are called non-basic variables, and the loose/slack variables are called basic variables. We do not use this notation here, but it is useful for the reader to be aware of all of the various synonyms in the linear programming literature.

For example, at the origin, the constraints $\mathbf{x} \geq 0$ are tight, and the variables $\mathbf{s}$ are all strictly greater than zero, indicating that they are slack or loose variables.

In slack variable form, the variables on the left of equality $\mathbf{s} = \mathbf{b} - \mathbf{A}\mathbf{x}$ in Eq. (3.7) are the loose or slack variables, and those to the right are the tight variables. If we move to another vertex, where the j-th row of $\mathbf{A}\mathbf{x} \leq \mathbf{b}$ becomes an active constraint, the variable $\mathbf{s}_j$ becomes tight and moves to the right hand side of the equation, and whichever variable is now loose moves to the left hand of the equation. Writing the inequality constraints in slack variable notation makes it simple to see which constraints are active, and hence which vertex we are currently centered about. An interior point will be away from any boundary, in which case all of the variables will be loose.

Finding a new feasible vertex in higher dimensions, given by n active constraint equations, can be very computationally challenging. Solving for a vertex, given by a set of n active constraints, involves a large matrix inversion. For example, consider a problem with n decision variables $\mathbf{x} \in \mathbb{R}^n$ and m constraint equations, so that $\mathbf{A} \in \mathbb{R}^{m \times n}$. There are also generally n non-negativity constraints $\mathbf{x} \geq 0$ in addition to the m constraints in $\mathbf{A}\mathbf{x} \leq \mathbf{b}$, resulting in $n + m$ total constraints. Each vertex is determined by n of the constraints being active or tight, so vertices will be given by a subset of n out of the total $n + m$ constraints being active. For each of the n active constraints, we must solve a linear equation to move the active variables to the right hand side of the equations. For the sake of simplicity, let's say that the n active constraints are selected from the m equations in $\mathbf{A}\mathbf{x} \leq \mathbf{b}$, although in general they may be selected from these m equations and from the n non-negativity constraints. Then these constraints become equalities

$$\tilde{\mathbf{A}}\mathbf{x} = \tilde{\mathbf{b}}, \tag{3.8}$$

where $\tilde{\mathbf{A}}$ and $\tilde{\mathbf{b}}$ are obtained by selecting the given n rows of $\mathbf{A}$ and $\mathbf{b}$. We must solve this linear system for $\mathbf{x}$ via matrix inversion, typically by Gaussian elimination or a similar method. Matrix inversion becomes increasingly expensive as the system size grows, with the number of calculations being roughly proportional to the cube of the dimension, $O(n^3)$. Thus, solving for vertices becomes prohibitively expensive in higher dimensions.

If instead, we start with a vertex and walk to a new, nearby vertex that is connected via an edge, we have a much simpler set of matrix computations: we need only choose one slack variable to tighten and one tight variable to loosen. This corresponds to choosing one inactive constraint to become active, and one active constraint to become inactive. This procedure of only swapping one pair of variables is much more computationally efficient than solving for an entirely new vertex, and is the basis of the simplex method.

3.3 Dantzig's Simplex Method

Dantzig's simplex method relies on two major properties of linear programs:

- A local extremum must occur on a vertex, assuming that the feasible set is a bounded polytope.
- The local extremum is a global extremum because linear programs are convex optimization problems.

Solving a linear program thus amounts to finding a local maximum by searching across vertices. The simplest approach would be to enumerate all vertices and evaluate the objective function. However, as discussed earlier, the number of vertices scales poorly with the dimensions of the problem, and even identifying and enumerating individual vertices becomes a non-trivial linear algebra problem in higher dimensions.

An alternative approach is to start with a vertex, and intelligently walk from vertex to vertex until we find a local minimum or maximum, which will be globally optimal by convexity. The process of going from vertex to vertex is called *pivoting*, and there are several strategies. Dantzig's pivot rule is the most commonly used pivoting technique. In general, this is much more computationally efficient than solving for an entirely new vertex from a set of n active constraints, as we simply pick one active constraint to relax, and a new inactive constraint to tighten, walking us to the next vertex.

The idea is to use slack variables to walk from vertex to vertex (e.g., pivot to pivot) until the objective function cannot be improved any more in any free direction, at which point global optimality is achieved. The objective function must increase monotonically, or at least not decrease, from pivot to pivot, otherwise the function is not convex. Any loose variable may increase by a positive amount without violating any inequality constraint, as its corresponding slack variable is nonzero. Here, we first describe the procedure, and then we will work through a full example.

Dantzig's pivot rule proceeds as follows:

1. Pick one of the active constraints to relax (i.e., to loosen). This constraint corresponds to one of the inequalities that is currently *active* or *tight*, meaning that it is one of the boundaries defining the current vertex. The inequality we choose to loosen will define the boundary we walk away from, towards a new vertex.

   Dantzig's pivot rule picks the coordinate direction where the objective function increases the most, and if this variable is currently tight, then we loosen it. The direction of fastest increase of the objective function is given by the gradient ∇f. Thus, we select the slack variable corresponding to the greatest component of the gradient.

   It is convenient to define the objective function f in terms of tight variables only, which is possible using slack variable form. The gradient is then computed with respect to the tight variables, so we only look for directions of opportunity that can be loosened.
2. Once we select a tight variable to loosen (i.e., an active constraint to relax), we must choose another loose variable to tighten (i.e., an inactive constraint to make active). This will define the new vertex that we move to. Generally, we pick among the slack variables, finding the inequality that will be violated first when increasing the newly loosened variable.

   To identify which inequality will be violated first, we take the constant b_j in each candidate inequality, divided by the coefficient a_j on the newly loosened variable in that row. The inequality that will be violated first will have the smallest ratio b_j/a_j. If the newly loosened variable doesn't appear in a given row equation, then this row isn't considered for selecting a tight variable.
3. Finally, we need to recenter our problem at the new vertex. We accomplish this by taking the inequality that will be violated first, and swapping the position of the loosened and

tightened variables, so that the newly loosened variable is on the left of the equation, and the newly tightened variable is on the right of the equation. Then, we replace every instance of this newly loosened variable in the objective function and on the right hand side of every other inequality, in terms of tight variables. As we will see in the simplex tableau formulation in the next section, this is nearly identical to row reduction and pivoting in Gaussian elimination, when computing the inverse of a matrix.

Simple Example

Let's now revisit the example from Section 3.1. The linear program may be written as:

$$\begin{aligned} \max_{\mathbf{x}} \quad & 100x_1 + 400x_2 \\ \text{s.t.} \quad & x_1, x_2 \geq 0 \\ & x_1 + x_2 \leq 50 \\ & x_2 \leq 30. \end{aligned} \tag{3.9}$$

It is possible to add two slack variables (s_1, s_2) and write the constraints in slack variable notation:

$$\begin{aligned} x_1, x_2, s_1, s_2 &\geq 0 \\ x_1 + x_2 + s_1 &= 50 \\ x_2 + s_2 &= 30. \end{aligned}$$

For intuition about the geometric interpretation of s_1 and s_2, see the illustration in Fig. 3.5. Moving the slack variables to the left side of the equality constraints yields:

$$\begin{aligned} x_1, x_2, s_1, s_2 &\geq 0 \\ s_1 &= 50 - x_1 - x_2 \\ s_2 &= 30 - x_2. \end{aligned}$$

At each step of Dantzig's simplex algorithm we then walk to another vertex, choosing one variable on the right to tighten and one variable on the left to loosen, moving from one inequality constraint boundary to another. In slack variable form, we can read off s_1 and s_2 by setting x_1 and x_2 equal to zero. This means that at the current vertex we have $x_1 = x_2 = 0$ and $s_1 = 50, s_2 = 30$. Thus, this equation is written in a form that is centered at the vertex $(0, 0)$. This procedure of using slack variables to define boundaries of the feasible polytope is visualized in Fig. 3.5.

From here on, we won't keep writing $x_1, x_2, s_1, s_2 \geq 0$, as this will always be true.

Step 1. Vertex $(0, 0)$: Currently, the variables x_1, x_2 are tight and s_1, s_2 are loose. The gradient is $\nabla_{(x_1,x_2)} f = \begin{bmatrix} 100 \\ 400 \end{bmatrix}$, which is larger in the x_2 direction, so we choose to loosen the x_2 variable; note that we have explicitly written $\nabla_{(x_1,x_2)}$ to indicate that we are taking the gradient of f with respect to the variables (x_1, x_2). Since the $s_2 = 0$ constraint will be violated first when increasing x_2, we choose to tighten s_2; in the first inequality, $b_1 = 50$ and $a_1 = 1$, and in the second inequality,

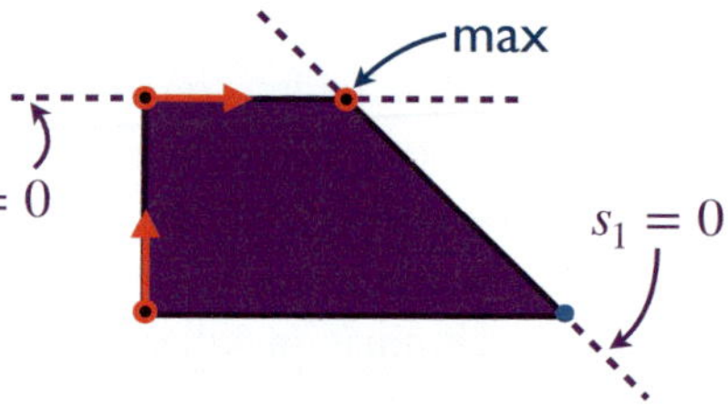

Figure 3.5 Illustration of how slack variables allow one to walk from vertex to vertex. To go from the first to the second vertex, we tighten the s_2 slack variable until $s_2 = 0$, at which point we are tight against the boundary of this inequality. Similarly, to go from the second to the third vertex, we tighten the s_1 slack variable until $s_1 = 0$, at which point we are tight against the boundary of this inequality.

$b_2 = 30$ and $a_2 = 1$, so $b_2/a_2 < b_1/a_1$ and the second inequality is violated first. This choice determines along which edge we will progress and which inequality constraint will specify the next vertex in the iteration.

Loosening x_2 and tightening s_2 changes the second slack constraint equation to[5]

$$x_2 = 30 - s_2.$$

Substituting this into the cost function and other constraint equations yields the optimization problem, centered around the new vertex $(0, 30)$.

The optimization problem now reads:

$$\begin{aligned} \max_{x_1, s_2} \quad & 100x_1 - 400s_2 + 12000 \\ \text{s.t.} \quad & s_1 = 20 - x_1 + s_2 \\ & x_2 = 30 - s_2. \end{aligned}$$

The objective function increased by 12000 at this vertex. This may be evaluated by setting the tight variables $x_1 = s_2 = 0$.

Step 2. Vertex $(0, 30)$: Now the variables x_1, s_2 are tight and s_1, x_2 are loose. The objective function now has a gradient $\nabla_{(x_1,s_2)} f = \begin{bmatrix} 100 \\ -400 \end{bmatrix}$, which only increases in the positive x_1 direction. Thus, this is the only direction we can move our vertex and increase the objective function.

We now choose x_1 to loosen, and s_1 to tighten, as the $s_1 = 0$ constraint is violated first when increasing x_1. The first slack constraint equation now becomes

$$x_1 = 20 - s_1 + s_2.$$

[5] This convention of loose variables on the left and tight variables on the right follows the mnemonic for tightening or loosening screws: "lefty loosey, righty tighty".

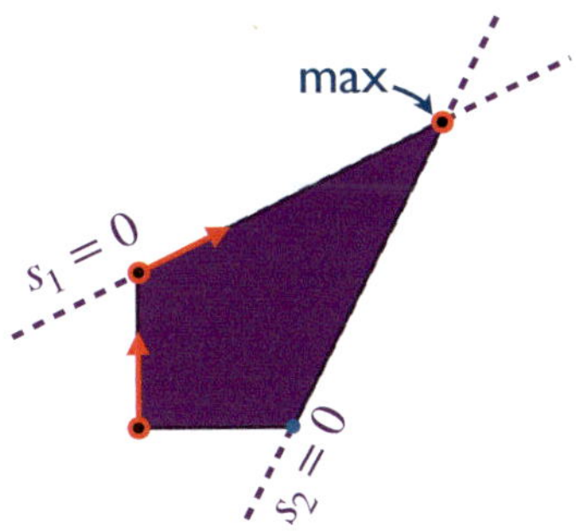

Figure 3.6 Illustration of convex feasible polytope for the linear program in Eq. (3.10). We also include the slack variable boundaries to illustrate how slack variables allow one to walk from vertex to vertex.

Substituting this into our objective function and other equations now yields a new optimization problem, centered at the vertex $(20, 30)$:

$$\begin{aligned} \max_{\mathbf{s}} \quad & -100s_1 - 300s_2 + 14000 \\ \text{s.t.} \quad & x_1 = 20 - s_1 + s_2 \\ & x_2 = 30 - s_2. \end{aligned}$$

Now the objective function value has increased to 14000 at the vertex $(x_1, x_2) = (20, 30)$, with $s_1 = s_2 = 0$. Since the gradient of the objective function decreases in both the positive s_1 and s_2 directions, this vertex is a local maximum and cannot be increased by moving in the s_1 or s_2 directions. By convexity, this local maximum is also a global maximum.

Thus, the global optimum occurs at $(x_1, x_2) = (20, 30)$ with a value of 14000.

Another Example in Two Dimensions

Let's consider another simple two-dimensional example:

$$\begin{aligned} \max_{\mathbf{x}} \quad & x_1 + 2x_2 \\ \text{s.t.} \quad & x_1, x_2 \geq 0 \\ & 2x_2 - x_1 \leq 2 \\ & 2x_1 - x_2 \leq 2. \end{aligned} \tag{3.10}$$

The feasible set is visualized in Fig. 3.6. In this example, the path from vertex to vertex does not always strictly follow a coordinate direction, as in the example above, and we will see more clearly the role of the slack variable in Dantzig's simplex method.

We begin by putting this problem in slack-variable notation:

$$\begin{aligned} \max_{\mathbf{x}} \ & x_1 + 2x_2 \\ \text{s.t.} \ & x_1, x_2, s_1, s_2 \geq 0 \\ & s_1 = 2 + x_1 - 2x_2 \\ & s_2 = 2 + x_2 - 2x_1. \end{aligned}$$

Step 1. Vertex $(0, 0)$: Currently, the variables x_1, x_2 are tight and s_1, s_2 are loose. The gradient is $\nabla_{(x_1,x_2)} f = \begin{bmatrix} 1 \\ 2 \end{bmatrix}$, which is largest in the x_2 direction, so we loosen the x_2 variable. Since the $s_1 = 0$ constraint will be violated first when increasing x_2, we tighten s_1. This choice determines which edge we walk along and which inequality constraint ($s_1 = 0$ will specify the next vertex we walk to).

Loosening x_2 and tightening s_1 changes the first slack constraint equation to

$$x_2 = 1 - s_1/2 + x_1/2.$$

Substituting this into the cost function and other constraint equations yields the optimization problem, centered around the new vertex $(0, 1)$.

The optimization problem now reads:

$$\begin{aligned} \max_{x_1,s_1} \ & 2 + 2x_1 - s_1 \\ \text{s.t.} \ & x_2 = 1 - s_1/2 + x_1/2 \\ & s_2 = 3 - 3x_1/2 - s_1/2. \end{aligned}$$

Note that the objective function increased by 2 at this vertex.

Step 2. Vertex $(0, 1)$: Now the variables x_1, s_1 are tight and x_2, s_2 are loose. The objective function has a gradient $\nabla_{(x_1,s_1)} f = \begin{bmatrix} 2 \\ -1 \end{bmatrix}$, which only increases in the positive x_1 direction. Thus, we choose x_1 to loosen and s_2 to tighten, as this is the first inequality condition that will be violated when increasing x_1. The s_2 constraint becomes

$$x_1 = 2 - s_1/3 - 2s_2/3.$$

Substituting this into our objective function and other equations now yields a new optimization problem, centered at the vertex $(2, 3)$

$$\begin{aligned} \max_{s_1,s_2} \ & 6 - 5s_1/3 - 4s_2/3 \\ \text{s.t.} \ & x_1 = 2 - s_1/3 - 2s_2/3 \\ & x_2 = 2 - 2s_1/3 - s_2/3. \end{aligned}$$

At this point the objective function is decreasing in both the positive s_1 and s_2 directions (the tight variables), so we are at a local maximum, and hence a global maximum. The maximum value is 6 at the point $(x_1, x_2) = (2, 2)$.

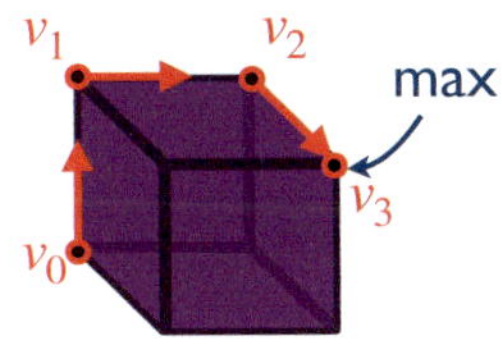

Figure 3.7 Illustration of convex feasible polytope for the linear program in Eq. (3.11). This polytope is a simple cube.

Example in Three Dimensions

We will now consider an example in three dimensions:

$$\begin{aligned} \max_{\mathbf{x}} \;& x_1 + 2x_2 + 3x_3 \\ \text{s.t. } & \mathbf{x}, \mathbf{s} \geq 0 \\ & s_1 = 1 - x_1 \\ & s_2 = 1 - x_2 \\ & s_3 = 1 - x_3. \end{aligned} \tag{3.11}$$

The feasible set is a cube, shown in Fig. 3.7.

Step 1. Vertex $(0, 0, 0)$: Currently, the variables x_1, x_2, x_3 are tight and s_1, s_2, s_3 are loose. The gradient is $\nabla f = \begin{bmatrix} 1 & 2 & 3 \end{bmatrix}^{\mathrm{T}}$, which is largest in the x_3 direction, so we loosen the x_3 variable. Since the $s_3 = 0$ constraint will be violated first when increasing x_3, we tighten s_3.

Loosening x_3 and tightening s_3 changes the third slack constraint equation to

$$x_3 = 1 - s_3.$$

Substituting this into the cost function and other constraint equations yields the optimization problem, centered around the new vertex $(0, 0, 1)$.

The optimization problem now reads:

$$\begin{aligned} \max_{x_1, x_2, s_3} \;& x_1 + 2x_2 - 3s_3 + 3 \\ \text{s.t. } & s_1 = 1 - x_1 \\ & s_2 = 1 - x_2 \\ & x_3 = 1 - s_3. \end{aligned}$$

The objective function increased by 3 at this vertex.

Step 2. Vertex $(0, 0, 1)$: Now the variables x_1, x_2, s_3 are tight and s_1, s_2, x_3 are loose. The objective function now has a gradient $\nabla f = \begin{bmatrix} 1 & 2 & -3 \end{bmatrix}^{\mathrm{T}}$, which increases most rapidly in the positive x_2 direction. Thus, we now choose x_2 to loosen, and s_2 to tighten, as the $s_2 = 0$ constraint will be the first to be violated when increasing x_2. The second slack constraint equation now becomes

$$x_2 = 1 - s_2.$$

Substituting this into our objective function and other equations now yields a new optimization problem, centered at the vertex $(0, 1, 1)$:

$$\begin{aligned}\max_{x_1,s_2,s_3} \quad & x_1 - 2s_2 - 3s_3 + 5 \\ \text{s.t.} \quad & s_1 = 1 - x_1 \\ & x_2 = 1 - s_2 \\ & x_3 = 1 - s_3.\end{aligned}$$

The objective value has increased to 5 at the vertex $(x_1, x_2, x_3) = (0, 1, 1)$.

Step 3. Vertex $(0, 1, 1)$: Now the variables x_1, s_2, s_3 are tight and s_1, x_2, x_3 are loose. The objective function now has a gradient $\nabla f = \begin{bmatrix}1 & -2 & -3\end{bmatrix}^{\mathrm{T}}$, which only increases in the positive x_1 direction. Thus, we loosen x_1 and tighten s_1, as the $s_1 = 0$ constraint will be violated first when increasing x_1. The first slack constraint equation now becomes

$$x_1 = 1 - s_1.$$

Substituting this into our objective function and other equations now yields a new optimization problem, centered at the vertex $(1, 1, 1)$:

$$\begin{aligned}\max_{s_1,s_2,s_3} \quad & - s_1 - 2s_2 - 3s_3 + 6 \\ \text{s.t.} \quad & x_1 = 1 - s_1 \\ & x_2 = 1 - s_2 \\ & x_3 = 1 - s_3.\end{aligned}$$

Since the gradient of the objective function decreases in all directions, this vertex is a local maximum and cannot be increased. By convexity, this local maximum is also a global maximum.

Thus, the global optimum point occurs at $(x_1, x_2, x_3) = (1, 1, 1)$ with a value of 6.

3.4 Simplex Tableau Method

We just saw how Dantzig's simplex algorithm conceptually operates by moving from one vertex of the feasible polytope to another, testing and improving the objective until no further improvement is possible. However, rewriting the entire linear program at each pivot step quickly becomes cumbersome in high dimensions.

The *simplex tableau method* provides a systematic matrix-based procedure for these pivot steps, making it possible to program a computer to perform a sequence of row operations similar to Gaussian elimination. The basic idea is to rewrite the linear program in a matrix format, known as the *tableau*, where it is simpler to keep track of the operations in each iteration. In each iteration, we pivot on a selected row and column entry, updating the tableau to reflect how we move from one feasible solution (vertex) to the next. The transcription of a linear program into the simplex tableau is shown in Fig. 3.8.

$$\max_{\mathbf{x}} \; f(\mathbf{x}) = \mathbf{c}^T\mathbf{x} \qquad \begin{array}{c|ccc:c} & f & \mathbf{x} & \mathbf{s} & \\ \hline f & 1 & -\mathbf{c}^T & \mathbf{0} & 0 \\ \mathbf{x} & \mathbf{0} & \mathbf{A} & \mathbf{I} & \mathbf{b} \end{array}$$
$$\text{s.t.} \quad \mathbf{x}, \mathbf{s} \geq 0$$
$$\mathbf{A}\mathbf{x} + \mathbf{s} = \mathbf{b}$$

Figure 3.8 Illustration of the mapping of a linear program into simplex tableau form.

The operations involved in the simplex tableau method will be closely related to the row operations and pivoting that we are familiar with from classic matrix inversion. Recall that to invert a matrix $\mathbf{A}$, we may form the augmented system

$$\left[\mathbf{A} \mid \mathbf{I}\right] \tag{3.12}$$

and repeat row operations and pivoting until the matrix on the left becomes the identity $\mathbf{I}$ and the matrix on the right becomes the inverse of $\mathbf{A}$:

$$\left[\mathbf{I} \mid \mathbf{A}^{-1}\right]. \tag{3.13}$$

In the simplex tableau, we follow a similar pattern of pivoting to maintain an identity-like block that represents the current slack variables at each vertex, with the remaining columns representing the active variables. This means each pivot step modifies one row (corresponding to the variable to make active) and one column (corresponding to the variable to make slack), just as we do in a single Gaussian elimination pivot.

Constructing the Initial Tableau

Initializing the tableau essentially amounts to transcribing the problem into a matrix notation, given by the following:

$$\begin{array}{c} \\ f \\ \mathbf{s} \end{array} \begin{array}{c} \begin{array}{ccc} f & \mathbf{x} & \mathbf{s} \end{array} \\ \left[\begin{array}{ccc|c} 1 & -\mathbf{c}^{\mathrm{T}} & \mathbf{0} & 0 \\ \hline \mathbf{0} & \mathbf{A} & \mathbf{I} & \mathbf{b} \end{array}\right] \end{array}. \tag{3.14}$$

The objective function and constraint equations may be read easily from this tableau. The objective function is the first row, written in terms of the active and inactive variables; as we move from vertex to vertex, different subsets of these variables will take turns being active and inactive. The constraint equations similarly appear in matrix form and can be read as

$$\mathbf{A}\mathbf{x} + \mathbf{s} = \mathbf{b}. \tag{3.15}$$

The first column corresponds to the objective function f, the next n columns to the active variables (initially the variables $\mathbf{x}$), the next m columns to the inactive slack variables (initially the variables $\mathbf{s}$), and the last column to the right hand side, which will include information about the current value of f at the current vertex. The first row encodes the objective function equation and the second row encodes the m constraint equations. We typically label the left of the second row with the names of the slack variables for that row equation (initially $\mathbf{s}$).

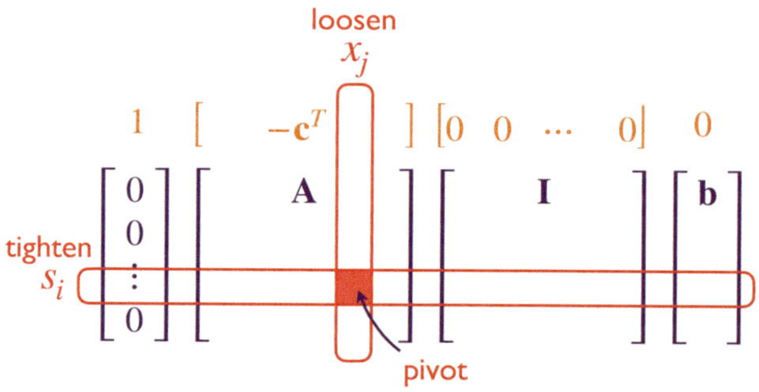

Figure 3.9 Illustration of Simplex Tableau method. Similar to row-reduction when computing the inverse of a matrix, we iteratively choose pivot locations and eliminate all terms above and below the pivot with row operations.

Pivot Step and Iteration

We now define an iterative procedure that is equivalent to how we walked from vertex to vertex in the simplex method above. This is shown schematically in Fig. 3.9. Each iteration picks:

1. *Tight Variable to Loosen (column):* Among the active variables, choose one whose coefficient in the objective row suggests an increase in objective if we increase it in the positive direction. A simple rule is Dantzig's pivot rule: pick the variable with the most negative coefficient (in a maximization problem) in the objective row. This variable will be selected to be loosened.
2. *Loose Variable to Tighten (row):* Among the slack variables, choose the one that becomes zero first if we increase the newly loosened variable, to avoid violating constraints. This is determined by the minimum ratio test: pick the row with the smallest ratio RHS/pivot coefficient.
3. *Row Reduction about Pivot:* Perform row operations so that the pivot entry becomes 1, and all other entries in that pivot column become 0. This is exactly like a Gaussian elimination pivot. The newly chosen column variable becomes loose and the row variable becomes tight. Sometimes we say that the row variable leaves the basis.

After the pivot, we obtain an updated tableau reflecting a new feasible vertex with a higher objective. We repeat until no negative coefficients remain in the objective row, or until we meet a stopping criterion in other pivot rules. At that point, we have found the optimum.

The simplex tableau method is the matrix-level realization of Dantzig's simplex algorithm. Each pivot corresponds to row operations akin to Gaussian elimination, systematically replacing one loose variable with another and updating the objective row. Despite alternative polynomial-time interior-point methods, the tableau approach remains a cornerstone of linear programming and is still used in many modern solvers, thanks to its interpretability and strong performance on typical industrial linear programs.

Simple Example, Revisited

We now revisit the example from Eq. (3.9), which we solved by hand using Dantzig's simplex method. Here we will solve this using the simplex tableau method. Comparing each step side-by-side, it will become clear that these two methods are equivalent representations of the same basic steps.

We begin by initializing the tableau, as described in Eq. (3.14) above, by first transcribing the objective function into the first row and the inequality constraint equations into the remaining rows:

$$\begin{array}{c} \\ f \\ s_1 \\ s_2 \end{array} \begin{array}{c} \begin{array}{cccccc} f & x_1 & x_2 & s_1 & s_2 & \end{array} \\ \left[\begin{array}{ccccc|c} 1 & -100 & -400 & 0 & 0 & 0 \\ \hline 0 & 1 & 1 & 1 & 0 & 50 \\ 0 & 0 & 1 & 0 & 1 & 30 \end{array}\right] \end{array}. \tag{3.16}$$

Now we proceed by identifying a pivot and performing row operations to eliminate entries in the same column. The pivot in this step will correspond to the x_2 variable, as the gradient has the largest magnitude in this direction. After row operations, the simplex tableau becomes:

$$\begin{array}{c} \\ f \\ s_1 \\ x_2 \end{array} \begin{array}{c} \begin{array}{cccccc} f & x_1 & x_2 & s_1 & s_2 & \end{array} \\ \left[\begin{array}{ccccc|c} 1 & -100 & 0 & 0 & 400 & 12000 \\ \hline 0 & 1 & 0 & 1 & -1 & 20 \\ 0 & 0 & 1 & 0 & 1 & 30 \end{array}\right] \end{array}. \tag{3.17}$$

It is possible to read off key information about the optimization problem from the tableau at this step. For example, the objective function value is 12000. Similarly, based on the pivot locations and the values in the right column, we see that we are at the vertex given by $(s_1, x_2) = (20, 30)$, or by $(x_1, x_2) = (0, 30)$.

We now identify the next pivot, x_1, as there is still room to optimize in the objective function by loosening this variable. Performing the given row operations with this column yields:

$$\begin{array}{c} \\ f \\ x_1 \\ x_2 \end{array} \begin{array}{c} \begin{array}{cccccc} f & x_1 & x_2 & s_1 & s_2 & \end{array} \\ \left[\begin{array}{ccccc|c} 1 & 0 & 0 & 100 & 300 & 14000 \\ \hline 0 & 1 & 0 & 1 & -1 & 20 \\ 0 & 0 & 1 & 0 & 1 & 30 \end{array}\right] \end{array}. \tag{3.18}$$

Now we see that there is no more room to optimize, so we have found a local, and hence global optimum solution. The objective function at this pivot is 14000. Similarly, based on the pivot locations and the values in the right column, we see that we are at the vertex $(x_1, x_2) = (20, 30)$.

Exceptions and Different Problem Types

The simplex tableau method is robust and flexible, but there are some exceptions and edge cases that require special attention. Specifically, we will consider cases when the feasible polytope is unbounded, when constraints become degenerate, and large-scale problems where the tableau method becomes intractable. We will also discuss cases where the linear program doesn't fit easily into slack variable form.

Unbounded Problems. It may be the case that we don't have enough inequality constraints to properly bound the feasible polytope. If during the minimum ratio test all pivot coefficients are less than zero, or no feasible pivot is found, the objective can increase indefinitely, indicating an unbounded linear program (see Panel 3.2).

Degeneracy. Sometimes multiple constraints become active at once, or the ratio test yields a tie. This can cause *cycling*, where we pivot among the same vertices. Solvers use anti-cycling rules, like Bland's rule, to break these ties.

Panel 3.2: Unbounded Polytopes

Until now, we have assumed that the inequality constraints defined a closed and bounded polytope, which guarantees the existence of an optimal solution (e.g., a minimum or a maximum) that is attainable, meaning that the solution is finite. However, some inequality constraints result in unbounded polytopes, as in the following

$$\begin{aligned} \max_{x_1,x_2} \quad & x_1 + x_2 \\ \text{s.t.} \quad & x_1, x_2 \geq 0 \\ & 2x_1 - x_2 \geq 2 \\ & 2x_2 - x_1 \geq 2. \end{aligned}$$

In this example, $x_1 = x_2 \to \infty$ is a feasible solution, which is not attainable. You may confirm that the dual problem is infeasible, meaning that the feasible set is empty. This will require duality theory, which we will discuss in Section 3.5.

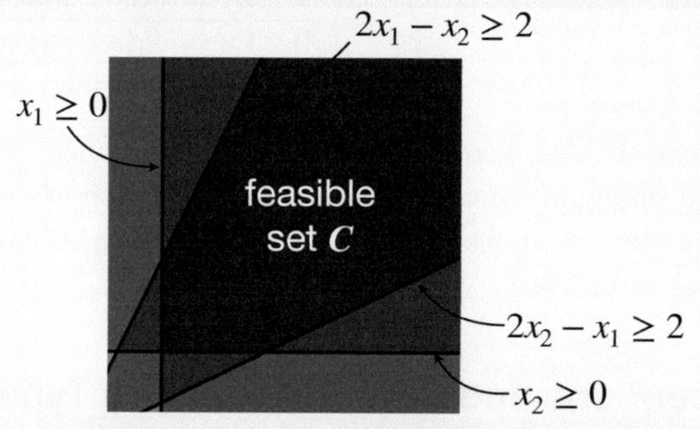

Figure 3.10: Unbounded polytope.

Large-Scale Problems. For extremely large problems, the tableau method can become memory-intensive or slow. Modern linear program solvers often switch to interior point methods, which are discussed in Section 6.4. Still, the simplex tableau remains conceptually simple, robust, and widely used for mid-sized linear programs.

Nonstandard Problems: Phase I and Artificial Variables

Sometimes a linear program does not naturally fit into slack form with $\mathbf{x} \geq 0$ and $\mathbf{Ax}+\mathbf{s} = \mathbf{b}$, for example if we have equality constraints $\mathbf{Ax} = \mathbf{b}$, or if our original constraints included terms like $\mathbf{Ax} \geq \mathbf{b}$ instead of $\mathbf{Ax} \leq \mathbf{b}$. These are still convex and linear, hence they fit into the framework of linear programming. However, they are not in the *standard form* $\mathbf{Ax} \leq \mathbf{b}$ with $\mathbf{x} \geq 0$.

The simplex method relies on starting with a basic feasible solution that satisfies all of the constraints with nonnegative values. In standard form, this is easy, as we simply set $\mathbf{x} = 0$ and $\mathbf{s} = \mathbf{b}$. However, for non-standard linear programs, with equality constraints or greater-than-or-equal constraints, we may not be able to simply start at $\mathbf{x} = 0$, as this might violate some of the constraints if $\mathbf{b} \neq \mathbf{0}$. In this case, we run a "Phase I" simplex to find a feasible solution about which to initialize.

The Phase I simplex involves introducing artificial variables $r_j \geq 0$ to each problematic constraint that does not admit a slack variable. For this discussion, consider a set of equality constraints $\mathbf{Ax} = \mathbf{b}$, which we augment to $\mathbf{Ax} + \mathbf{r} = \mathbf{b}$. For any feasible solution, we require $\mathbf{r}$ to equal zero. Thus, our initial simplex solves the following minimization

$$
\begin{aligned}
\min_{\mathbf{x},\mathbf{r}} \quad & \mathbf{1}^{\mathrm{T}}\mathbf{r}, \\
\text{s.t.} \quad & \mathbf{Ax} + \mathbf{r} = \mathbf{b}, \\
& \mathbf{x}, \mathbf{r} \geq \mathbf{0}.
\end{aligned}
\tag{3.19}
$$

If the minimum value is 0, corresponding to $\mathbf{r} = \mathbf{0}$, then we may drop the $\mathbf{r}$ and we have a feasible initial vertex. If the minimum value is greater than 0, then there is no set of $\mathbf{x}$ that satisfies the initial equalities, meaning that the problem is infeasible.

Once we find a feasible solution about which to initialize, we remove the artificial variables and proceed with a standard tableau method in "Phase II". Hence, Phase I is about finding a feasible initialization for nonstandard problems, while Phase II solves the original linear program from that feasible starting point. Once $r_j = 0$ for each artificial variable, they no longer appear in the constraints or objective, effectively removing them from the problem.

3.5 Duality in Linear Programming

For every linear program (called the *primal*), there is an associated *dual* linear program that provides an upper bound on the optimal value of the primal. The objective and constraints of the primal and are closely related to those of the dual through a systematic transposing procedure that we briefly introduce here. Duality is one of the most elegant and useful results in optimization, and it is discussed more in Chapter 6. It is the foundation of interior point methods, which are used to solve a broad class of convex optimization problems, including linear programs. Duality is particularly intuitive and straightforward in the case of a linear program. It may be helpful to cross reference against Panel 6.3 in Chapter 6.

Consider the following primal problem with decision variables $x_1, x_2 \geq 0$:

$$\begin{aligned} \max_{\mathbf{x}} \quad & x_1 + 4x_2, \\ \text{s.t.} \quad & x_1 + x_2 \leq 5, \\ & x_1 \leq 4, \\ & x_2 \leq 3, \\ & x_1, x_2 \geq 0. \end{aligned}$$

The dual problem seeks an upper bound for this maximum. A good upper bound is larger than the objective for any feasible point. Thus, we may begin with the constraint equations. Specifically, we multiply each of the first three constraint equations by y_1, y_2, and y_3:

$$\begin{aligned} & y_1(x_1 + x_2) \leq 5y_1, \\ & y_2(x_1) \leq 4y_2, \\ & y_3(x_2) \leq 3y_3 \end{aligned}$$

and then we add up all inequalities and regroup terms:

$$\begin{aligned} & y_1(x_1 + x_2) + y_2(x_1) + y_3(x_2) \leq 5y_1 + 4y_2 + 3y_3 \\ \implies \quad & x_1(y_1 + y_2) + x_2(y_1 + y_3) \leq 5y_1 + 4y_2 + 3y_3. \end{aligned}$$

Recall that we want to bound the maximum of $x_1 + 4x_2$. Thus, as long as $y_1 + y_2 \geq 1$ and $y_1 + y_3 \geq 4$, then $5y_1 + 4y_2 + 3y_3$ provides an upper bound. To make this bound as tight as possible, we may minimize this expression $5y_1 + 4y_2 + 3y_3$ subject to the constraints $y_1 + y_2 \geq 1$ and $y_1 + y_3 \geq 4$:

$$\begin{aligned} \min_{\mathbf{y}} \quad & 5y_1 + 4y_2 + 3y_3, \\ \text{s.t.} \quad & y_1 + y_2 \geq 1, \\ & y_1 + y_3 \geq 4, \\ & y_1, y_2, y_3 \geq 0. \end{aligned}$$

Thus we have arrived at the dual problem. The weightings $\mathbf{y}$ of the primal constraints have become the dual variables. The $\mathbf{c}$ vector from the primal objective becomes the $\mathbf{b}$ vector in the dual, the $\mathbf{b}$ vector from the primal becomes the $\mathbf{c}$ vector in the dual, and the $\mathbf{A}$ matrix in the primal becomes the $\mathbf{A}^{\mathrm{T}}$ matrix in the dual.

General Form: Primal and Dual

More generally, for the primal linear program

$$\begin{aligned} \max \quad & \mathbf{c}^{\mathrm{T}}\mathbf{x}, \\ \text{s.t.} \quad & \mathbf{x} \geq 0, \\ & \mathbf{A}\mathbf{x} \leq \mathbf{b}, \end{aligned} \tag{3.20}$$

the dual is given by

$$\begin{aligned} \min \quad & \mathbf{b}^{\mathrm{T}}\mathbf{y}, \\ \text{s.t.} \quad & \mathbf{y} \geq 0, \\ & \mathbf{A}^{\mathrm{T}}\mathbf{y} \geq \mathbf{c}. \end{aligned} \tag{3.21}$$

There is an important concept of weak versus strong duality, which we explore in Chapter 6. *Weak duality* means that for any feasible primal $\mathbf{x} \in \mathcal{X}$ and any feasible dual $\mathbf{y} \in \mathcal{Y}$

$$\mathbf{c}^{\mathrm{T}}\mathbf{x} \leq \mathbf{b}^{\mathrm{T}}\mathbf{y}. \tag{3.22}$$

Specifically, this means that the maximum feasible value of the primal (or the supremum, if the max it doesn't exist) is less than or equal to the minimum feasible value of the dual (or the infimum, if the min doesn't exist):

$$\max_{\mathbf{x}\in\mathcal{X}} \mathbf{c}^{\mathrm{T}}\mathbf{x} \leq \min_{\mathbf{y}\in\mathcal{Y}} \mathbf{b}^{\mathrm{T}}\mathbf{y}. \tag{3.23}$$

The gap between these optima is called the *duality gap*. Under some conditions, such as the existence of an optimal basic feasible solution in the primal, we have *strong duality*, which implies that the maximum primal objective and minimum dual objective coincide at optimality. In this case, we say that there is no duality gap.

Many algorithms, such as interior point methods, solve the primal and dual problems simultaneously, since the dual is often simpler to handle or interpret. If the primal and dual solutions $\mathbf{x}$ and $\mathbf{y}$ differ, we continue updating one or both until the duality gap closes, indicating a globally optimal solution. Once strong duality holds, each feasible solution in one problem certifies optimality in the other. Although this may seem to complicate the linear program, the dual perspective offers key advantages. It provides optimality certificates, bounds on error, and estimates of the sensitivity of solutions to parameters. The dual lets us interpret Lagrange multipliers directly as these sensitivities. We will explore duality more in Section 6.3 and with an economics interpretation in Panel 6.3.

Duality, Supporting Hyperplanes, and Lagrange Multipliers. Another way to think about duality is via *supporting hyperplanes*: each dual variable y_j can be thought of as a Lagrange multiplier enforcing the j-th constraint in the primal. The dual objective $\mathbf{b}^{\mathrm{T}}\mathbf{y}$ measures the total "cost" of satisfying these constraints. Meanwhile, the condition $\mathbf{A}^{\mathrm{T}}\mathbf{y} \geq \mathbf{c}$ ensures that any attempted linear combination of constraints is sufficient to bound the primal objective $\mathbf{c}^{\mathrm{T}}\mathbf{x}$ from above. Again, these also have a further interpretation as the sensitivity of the optimal objective value to changes in $\mathbf{b}$ and $\mathbf{c}$. For example, these are *shadow prices* or *marginal values* in economics, as we will see in Panel 6.3.

3.6 Integer Programming and NP-Hard Problems

An important extension of linear programming is the case where some or all of the decision variables must be integer-valued, often $\{0, 1\}$ in the case of a yes/no decision. For example, in packing problems such as the knapsack problem (see page 109), the choice to include an item in the pack is a binary yes or no decision. This transforms a linear program into an integer linear program (ILP), which is typically much harder to solve. In fact, many ILPs turn out to be *NP-hard* problems, meaning that no known polynomial-time algorithm can solve them in the worst case. Fortunately, we may often relax ILPs to nearby linear programs, which are efficient to solve. The optimal solution of the linear program may closely match the optimal solution of the ILP. In our example of factory production earlier, the decision variables x and y must be integer valued, as it doesn't make sense to make a fraction of a microscope. In this example, the constraint equations and profit model all involved integer values and resulted in the vertices of the feasible polytope also having integer values, so the solutions of the ILP and linear program are identical.

An ILP is typically formulated as

$$\begin{aligned} \max_{\mathbf{x}} \quad & \mathbf{c}^{\mathrm{T}}\mathbf{x} \\ \text{s.t.} \quad & \mathbf{A}\,\mathbf{x} \le \mathbf{b}, \\ & \mathbf{x} \in \{0, 1\}^n \ (\text{or } \mathbb{N}^n). \end{aligned} \tag{3.24}$$

It is possible to relax this to a linear program by allowing $\mathbf{x}$ to be continuous, satisfying $\mathbf{x} \ge 0$. However, the solution to the relaxed problem may not satisfy the integrality constraints. Nevertheless, the linear program relaxation provides a bound on the best possible integer solution, since any feasible integer solution is also feasible for the relaxation. It also helps guide branch-and-bound and cutting-plane strategies.

In branch-and-bound, the relaxed linear program is solved, ignoring integrality constraints. Then for each x_j that is non-integer, we create two subproblems with additional constraints, one where $x_j \le \lfloor x_j \rfloor$ and another where $x_j \ge \lceil x_j \rceil$. These branches are each solved using standard linear programming techniques, and if the bound of a subproblem is worse than a known feasible integer solution, this branch is pruned, since it cannot improve the optimum. This creates a search tree structure for exploring integer solutions that are near the relaxed linear programming solution.

Cutting planes are another approach where we start from the relaxed linear program and iteratively add linear constraints, or cuts, that remove the non-integer solutions but retain all feasible integer points. Each cut tightens the feasible region, bringing the LP relaxation closer to the true integer hull. By repeatedly solving the tightened LP and adding cuts, we converge to an integer-optimal solution once no fractional points remain. There are a number of types of cuts, including Gomory cuts, cover cuts, and clique inequalities.

Most modern integer programming solvers use a hybrid branch-and-cut approach that combines branch-and-bound to create a global search structure with cutting planes to locally tighten relaxations and reduce branching.

The Knapsack Problem

In the knapsack problem, the task is to select several items to fill a knapsack in order to maximize the value of the accumulated items while not exceeding the knapsack's weight capacity W. The knapsack formulation is important as a prototype for many resource allocation problems. Whether we are packing boxes into a truck, scheduling tasks on a supercomputer cluster finite processors, or allocating some limited budget among competing investments, the structure of "items with value and resource consumption" is the same. Hence, the knapsack problem and its variants have driven much research in branch-and-bound methods and heuristics for large-scale combinatorial optimization.

Consider a set of n items, each with a value (profit) p_j and a weight w_j. If item j is selected then $x_j = 1$, and $x_j = 0$ otherwise. The optimization problem of maximizing the total profit without exceeding capacity can be written as:

$$\begin{aligned} \max_{x_j \in \{0,1\}} \quad & \sum_{j=1}^{n} p_j x_j \\ \text{s.t.} \quad & \sum_{j=1}^{n} w_j x_j \leq W. \end{aligned} \tag{3.25}$$

The objective function is written in vector form as $\mathbf{p}^{\mathrm{T}}\mathbf{x}$, and the constraints are written as $\mathbf{w}^{\mathrm{T}}\mathbf{x} \leq W$. This integer linear program is famously NP-hard: enumerating all subsets of items grows exponentially in n (i.e., there are 2^n possible subsets of n items). Similarly, selecting m objects from n is given by $\binom{n}{m} = \frac{n!}{m!(n-m)!}$. Nonetheless, it is central to many industrial and scheduling applications where a resource limit, like weight or volume, bounds the number of items or tasks that can be selected. For example, many airlines use excess cargo space to pack freight, and this packing problem can be formulated as an ILP.

Traveling Salesman Problem (TSP)

Given a list of cities or nodes and the pairwise distances or costs of traveling between them, the traveling salesman problem (TSP) requires finding a single route that visits each city exactly once, returning to the start, while minimizing total travel distance:

$$\begin{aligned} \min_{x_{ij} \in \{0,1\}} \quad & \sum_{(i,j) \in E} c_{ij} x_{ij} \\ \text{s.t.} \quad & \sum_{j} x_{ij} = 1 \ \forall i, \\ & \sum_{i} x_{ij} = 1 \ \forall j. \end{aligned} \tag{3.26}$$

Here, E is the set of edges connecting cities and $x_{ij} = 1$ if the path connects city i to city j; otherwise, $x_{ij} = 0$. Eliminating smaller subtours (i.e., cycles not visiting all nodes) requires additional constraints, making the TSP a complicated integer linear program. TSP is among the most famous NP-hard problems, and its structure is well studied, so that modern heuristics can handle extremely large instances effectively in practice.

Panel 3.3: Solving the Knapsack Problem with `pulp`

Consider a knapsack problem where we have ten items, with weights and values $(4, 15)$, $(2, 11)$, $(5, 20)$, $(4, 12)$, $(8, 26)$, $(5, 21)$, $(8, 29)$, $(5, 15)$, $(10, 30)$, $(4, 12)$, and a weight limit of 20. Finding the optimal packing to maximize value is NP-hard. However, for moderate sized problems, it is possible to solve an approximation efficiently by relaxing to a linear program.

The following code solves the knapsack problem very efficiently using `pulp`:

CodeSnip 3.c

```
from pulp import *

n = 10
weights = [4,2,5,4,8,5,8,5,10,4]
values = [15,11,20,12,26,21,29,15,30,12]
C = 20

# initialize problem
model = LpProblem(sense=LpMaximize)
vars = [LpVariable(name=f"x_{i}", cat=LpBinary) for
    i in range(n)]
model += lpDot(weights, vars) <= C
model += lpDot(values, vars)

# solve
status = model.solve(PULP_CBC_CMD(msg=False))
print("print:", model.objective.value())
print("b_j:", *[vars[i].value() for i in range(n)])
```

The optimal integer linear programming solution is to select items 1, 2, 5, and 6, for a total weight of 20 and a value of 81.

Because this problem is small, with $n = 10$, there are $2^{10} = 1024$ combinations to search over. Thus, we may also solve it by brute force:

CodeSnip 3.d

```
import itertools
best_value = 0
best_subset = None

# Iterate over all subsets
for mask in itertools.product([0, 1], repeat=n):
    total_w = sum(w * m for w, m in zip(weights,
        mask))
    if total_w <= C:
        total_p = sum(p * m for p, m in zip(values,
            mask))
        if total_p > best_value:
            best_value = total_p
            best_subset = mask
```

The output of this combinatorial search also selects items 1, 2, 5, and 6, for a maximum value of 81, confirming that the result above is globally optimal.

Geometric Packing and Floor Arrangement

There are also geometric packing and layout tasks that combine integer decisions with geometric constraints. For instance, arranging equipment on a factory floor may demand discrete variables indicating whether or not to include certain machines, akin to a knapsack choice, while also incorporating geometric feasibility constraints: no overlapping pieces, no exceeding the factory boundary, and possibly minimum aisle spaces for safety. Similarly, we may want to optimally pack cargo into a truck or a train car; in this case, weight may be less important than optimal use of space.

Consider a scenario where we are trying to pack items into a two-dimensional space. Each item i has a value v_i and a binary variable $x_i \in \{0, 1\}$ that represents whether we decide to place that item at all. There will also be continuous variables (X_i, Y_i) for the position of item i. We thus obtain a *mixed-integer* geometric model:

$$\begin{aligned} \max \quad & \sum_{i=1}^{n} v_i x_i \\ \text{s.t.} \quad & x_i \in \{0, 1\}, \quad (X_i, Y_i) \in \mathbb{R}^2 \end{aligned} \tag{3.27}$$

plus geometric packing constraints such as no objects overlapping:

$$\begin{aligned} & |X_i - X_j| \geq L_i + L_j + \delta \\ & |Y_i - Y_j| \geq H_i + H_j + \delta \quad \text{for any } i \neq j \text{ with } x_i = x_j = 1, \end{aligned}$$

and boundary constraints ensuring that each item stays within the container. In the above, L_i is the length of object i and H_i is the height, assuming it fits inside of a simple rectangle. In reality, preventing overlap requires a disjunctive constraint, either separating in X or in Y, often handled by additional binary variables. The above inequalities are a simplification to illustrate the idea of geometric constraints. Real-world geometric constraints may be considerably more sophisticated.

This mixture of integer and geometric constraints tends to be even more complex than standard integer linear programs, because overlap constraints are often nonlinear. One approach is to approximate or discretize the floor into a grid, leading to a large 0–1 layout model. These geometric ILPs or nonlinear IPs are used in several contexts, such as: factory floor planning to decide on which machines to include (integer decisions) and where to place them (continuous geometry); in warehouse layout to determine item placements on a 2D or 3D grid subject to collision constraints; and VLSI chip design which involves a 2D or 3D packing of logic blocks with integrality and adjacency constraints. In all cases, the geometry combined with integrality can make the problem extremely challenging, and practitioners rely on approximation schemes, heuristics, or specialized constraint solvers that handle both the discrete choices and the continuous layout constraints. Thus, while these are arguably "packing" variants of integer programming, they require more sophisticated geometric reasoning and typically are not purely linear.

Suggested Reading

Texts

(1) **Linear Programming and Extensions**, by G. B. Dantzig, 1963 [124].
(2) **Introduction to Linear Optimization**, by D. Bertsimas and J. N. Tsitsiklis, 1997 [50].
(3) **Integer and Combinatorial Optimization**, by L. A. Wolsey and G. L. Nemhauser, 2014 [540].
(4) **Theory of Linear and Integer Programming**, by A. Schrijver, 1998 [450].
(5) **Introduction to Operations Research**, by F. S. Hillier, and G. J. Lieberman, 2021 [228].
(6) **Combinatorial Optimization: Theory and Algorithms**, by B. Korte, and J. Vygen, 2018 [284].
(7) **Knapsack Problems: Algorithms and Computer Implementations**, by S. Martello and P. Toth, 1990 [338].

Papers and reviews

(1) **Maximization of a linear function of variables subject to linear inequalities**, by G. B. Dantzig, *in T.C. Koopmans (Ed.) Activity Analysis of Production and Allocation*, 1951 [123].
(2) **A new polynomial-time algorithm for linear programming**, by N. Karmarkar, *Combinatorica*, 1984 [270].
(3) **Outline of an algorithm for integer solutions to linear programs**, by R. E. Gomory, *Bulletin of the American Mathematical Society*, 1960 [203].

Homework

For the next several problems, consider the following linear program:

$$
\begin{aligned}
\max_{\mathbf{x}} \ & x_1 + 2x_2 \\
\text{s.t. } & x_1, x_2 \geq 0 \\
& x_1 \leq 300 \\
& x_2 \leq 400 \\
& x_1 + x_2 \leq 500.
\end{aligned}
$$

Exercise 3.1 Draw the feasible region and identify all of the vertices. Which vertex has the maximum value?

Exercise 3.2 For each vertex, write down the inequality relationships that are active at that vertex. Also write them in slack variable notation.

Exercise 3.3 Solve this problem using Dantzig's simplex method.

Exercise 3.4 Solve this problem using the simplex tableau method, and compare with the answer from above.

Exercise 3.5 Write this linear program in dual form, interpret the dual variables, and solve using any method you'd like. Do we have weak or strong duality?

Exercise 3.6 Write the dual of the dual form of this problem from the last exercise and show that it is the same as the original linear program.

Exercise 3.7 Repeat all of the problems above for the following linear program:

$$
\begin{aligned}
\max_{\mathbf{x}} \ & 7x_1 + 8x_2 + 10x_3 \\
\text{s.t. } & x_1, x_2, x_3 \geq 0 \\
& 2x_1 + 3x_2 + 2x_3 \leq 1000 \\
& x_1 + x_2 + 2x_3 \leq 800.
\end{aligned}
$$

Exercise 3.8 Suppose you are a farmer choosing how to allocate land to farm corn and agave. You have a total of 50 acres. Corn yields a profit of \$100 per acre, while agave yields a profit of \$400 per acre. However, there is a maximum demand of 30 acres worth of agave. How much of each crop should you produce?

Formulate this as a linear program and solve using any method you like. Draw a picture of the feasible region with the vertices.

Exercise 3.9 Go through several of the examples from the first page of this chapter and formulate a linear program in standard form: what is the objective function to maximize or minimize, what are the constrained resources, and what are the decision variables?

Exercise 3.10 (Space allocation) A factory has a total floor area of 100,000 square feet that can be allocated among different production lines (e.g., producing *Product A* vs. *Product B*) to maximize total profit. Each product requires a certain minimum area (or at most some maximum area) and has a certain profit (per square foot or per unit produced). Specifically, suppose product A yields \$50 per 100 sq. ft. (i.e. \$0.50 per sq. ft.) and product B yields \$1.00 per sq. ft. Moreover, we require at least 20,000 sq. ft. for product A to keep it viable.

Write this as a linear program and solve using any method you like.

Exercise 3.11 (Portfolio management) An investor chooses how much capital to allocate among different assets (e.g., stocks, bonds, etc.) to maximize expected return. Each asset has an expected return (per dollar invested), but also a maximum allowable fraction or total risk limit.

Suppose the investor has \$100,000 total to distribute among three assets.

- Asset 1 has expected return 8% (0.08) per year, max \$60,000,
- Asset 2 has expected return 6% per year, no explicit max, and
- Asset 3 has expected return 11% but at most \$20,000 can be placed to keep risk moderate.

Formulate a linear program to maximize total annual expected return subject to $\sum x_i \leq 100000$ and the given max constraints on x_1 and x_3.

Exercise 3.12 Construct a 2D polytope in the (x_1, x_2)-plane by intersecting 4 or more half-planes.

1. Sketch this polytope.
2. Formulate a linear objective you wish to minimize or maximize.
3. Identify all vertices and check which one(s) optimize your objective.

Exercise 3.13 Solve the following linear program graphically:

$$\begin{aligned} \min \quad & 3x + 4y \\ \text{s.t.} \quad & x + 2y \geq 6, \\ & x, y \geq 0. \end{aligned}$$

(a) Sketch the feasible region. (b) Identify all vertices. (c) Determine which point(s) minimize the objective. (d) Verify if the solution is bounded and feasible.

Exercise 3.14 Solve by the simplex method:

$$\begin{aligned} \max \quad & 2x_1 + 3x_2 \\ \text{s.t.} \quad & x_1 + x_2 \leq 8, \\ & x_1 + 3x_2 \leq 15, \\ & x_1, x_2 \geq 0. \end{aligned}$$

(a) Convert to standard form. (b) Set up the initial simplex tableau. (c) Perform pivot steps until optimal. (d) Report the optimal point (x_1^*, x_2^*) and the maximum value obtained.

Exercise 3.15 Solve by the simplex method:

$$\begin{aligned}\max \quad & 2x_1 + 3x_2 \\ \text{s.t.} \quad & x_1 + x_2 \le 8, \\ & x_1 + 3x_2 \le 15, \\ & x_1, x_2 \ge 0.\end{aligned}$$

(a) Convert to standard form. (b) Set up the initial simplex tableau. (c) Perform pivot steps until optimal. (d) Report the optimal vertex (x_1^*, x_2^*) and the max value.

Exercise 3.16 Consider the following linear program with 10 constraint equations and 3 variables:

$$\begin{aligned}\max_{\mathbf{x}} \quad & x_1 + x_2 + x_3 \\ \text{s.t.} \quad & \mathbf{x} \ge 0 \\ & x_1 \le 3 \\ & x_2 \le 4 \\ & x_3 \le 5 \\ & x_1 + x_2 \le 5 \\ & x_1 + x_3 \le 5 \\ & x_2 + x_3 \le 5 \\ & x_1 + x_2 + x_3 \le 6.\end{aligned}$$

Write down all vertices, and please indicate which constraint equations are active for each vertex.

Exercise 3.17 Write the dual of the linear program in the previous example. How many vertices are they, and which constraint equations are active for each vertex?

Exercise 3.18 Solve the linear program and its dual in the exercises above. Which is simpler to solve?

Exercise 3.19 Solve the following linear program using the simplex method by hand:

$$\begin{aligned}\min \quad & 3x_1 + 2x_2 \\ \text{s.t.} \quad & x_1 + x_2 \ge 4, \\ & 2x_1 + x_2 \ge 6, \\ & x_1,\ x_2 \ge 0.\end{aligned}$$

Show each step of your tableau.

Exercise 3.20 Write the dual of the above linear program. Interpret the dual variables.

Exercise 3.21 Suppose we are given the primal problem:

$$\begin{aligned} \max \quad & 4x_1 + x_2 \\ \text{s.t.} \quad & 2x_1 + 3x_2 \le 12, \\ & x_1 + x_2 \ge 5, \\ & x_1, x_2 \ge 0. \end{aligned}$$

(a) Convert to a standard form suitable for taking the dual. (b) Write the dual problem explicitly. (c) Identify which variables in the dual are unrestricted vs. nonnegative if any. (d) State the dual objectives and constraints clearly.

Exercise 3.22 Suppose you have the primal:

$$\begin{aligned} \max \quad & 5x_1 + 3x_2 \\ \text{s.t.} \quad & 2x_1 + x_2 \le 20, \\ & 3x_1 + 4x_2 \le 48, \\ & x_1, x_2 \ge 0. \end{aligned}$$

Assume that x_1 and x_2 are two products that require constrained resources to produce (see Panel 6.3). (a) Write the dual. (b) Solve the primal and dual. (c) Interpret the dual variables as *shadow prices* or *resource valuations*. (d) Suppose the right-hand side in the first constraint changes from 20 to 21. Predict the new optimal value using dual prices.

Exercise 3.23 Consider the linear program:

$$\begin{aligned} \max \quad & 2x + 5y \\ \text{s.t.} \quad & x + 2y \le b_1, \\ & 3x + 4y \le b_2, \\ & x, y \ge 0. \end{aligned}$$

(a) Solve it for $(b_1, b_2) = (8, 20)$. (b) Perform sensitivity analysis to find allowable ranges of b_1, b_2 for which the current optimal vertex remains optimal. (c) Illustrate how the optimal solution or basis might change if $b_1 > 10$, or $b_2 < 16$.

Exercise 3.24 (Fleet scheduling) An airline must decide how to allocate planes across different routes or time blocks from a single airport. Each route has a certain revenue per plane, and there's a limited number of planes available. The objective is to maximize total revenue, subject to demand or flight constraints.

Let the airline have 10 planes total. There are 3 routes:

- Route 1: revenue \$3000 per plane, max 6 planes,
- Route 2: revenue \$4000 per plane, max 4 planes, and
- Route 3: revenue \$2500 per plane, no explicit max but the total is 10.

Formulate a linear program (or ILP, for integer planes) to assign planes to each route for maximum total revenue.

Exercise 3.25 (Knapsack problem) Consider the following collection of items:

Item k	1	2	3	4	5	6	7	8	9	10
Weight (kg)	5	2	7	3	1	2	5	6	8	3
Value ($)	2	3	2	4	1	7	3	5	1	6

Table for Exercise 3.25

Solve this knapsack problem to find the most valuable collection that fits with a weight of 18 kg or less. Solve this using a linear program relaxation and also using the combinatorial brute force search.

Exercise 3.26 (ILP to linear program relaxation) Consider the integer program:

$$\begin{aligned} \min \quad & 4x_1 + 5x_2 \\ \text{s.t.} \quad & 2x_1 + x_2 \geq 8, \\ & x_1 + 3x_2 \geq 12, \\ & x_1, x_2 \in \{0, 1, 2, 3, \dots\}. \end{aligned}$$

(a) Formulate the linear program relaxation by allowing $x_1, x_2 \geq 0$ real. (b) Solve the relaxation. (c) Show how a naive rounding might yield an infeasible or suboptimal integer solution. (d) Find an approach that tries to fix this via a better *rounding* or *branch-and-bound* approach.

Exercise 3.27 (ILP $\rightarrow$ LP relaxation and rounding) Consider the integer covering problem

$$\begin{aligned} & \min 4x_1 + 5x_2 \\ & \text{s.t.;} 2x_1 + x_2 \geq 8, \\ & \quad x_1 + 3x_2 \geq 12, \\ & \quad x_1, x_2 \in \mathbb{Z}_+. \end{aligned}$$

1. Write the LP relaxation by replacing $x_1, x_2 \in \mathbb{Z}_+$ with $x_1, x_2 \geq 0$.
2. Solve the relaxation and report the optimal (x_1^*, x_2^*) and objective value.
3. Show how naive component-wise rounding can fail: *rounding down* can violate constraints; *rounding up* can be feasible but costly. Give concrete rounded solutions and check feasibility and cost.
4. Propose a better approach that increase variables until all constraints are satisfied.

4 Least-Squares Regression

Least-squares regression is one of the most foundational concepts in applied mathematics and statistics, with applications in data analysis, inverse problems, control theory, and machine learning. The linear least-squares regression problem involves solving for the unknown parameters $\mathbf{x}$ of a model that is written as a linear system of equations

$$\mathbf{A}\mathbf{x} = \mathbf{b}. \tag{4.1}$$

We have already seen an example of how data fitting can be recast into a least-squares problem in Panel 1.3 and Eq. (1.6) from Chapter 1.

Linear least-squares regression is a convex optimization problem, where the model parameters $\mathbf{x}$ are chosen to minimize the sum-squared error of the fit over the data

$$\min_{\mathbf{x}} \|\mathbf{A}\mathbf{x} - \mathbf{b}\|_2^2. \tag{4.2}$$

When the matrix $\mathbf{A}$ is square and invertible (see Fig. 4.1), then we may solve for $\mathbf{x} = \mathbf{A}^{-1}\mathbf{b}$ by inverting $\mathbf{A}$. However, when $\mathbf{A}$ is non-square or singular, this requires an optimization solution to find the *best* fit (see Fig. 4.2). Typically the problem will either be *overdetermined*, meaning that no $\mathbf{x}$ will exactly solve this equation, or *underdetermined*, meaning that infinitely many $\mathbf{x}$ vectors solve the equation. In both cases, we formulate an optimization problem to solve the least-squares regression problem. We also typically add regularization to the least-squares objective to improve the numerical stability of this inverse problem and to promote certain types of solutions.

As a more concrete example, consider a bridge with many beams and joints. Each joint experiences forces from the connected beams and their applied loads, and the sum of these forces must balance in equilibrium. We can write one equation per joint, so the sum of applied loads plus the unknown internal forces in the beams equals zero. Collecting these equations gives us the rows of a matrix $\mathbf{A}$, describing the connection geometry, a vector $\mathbf{x}$ for the unknown beam forces, and a right-hand side vector $\mathbf{b}$ for the external loads. Solving for the unknown $\mathbf{x}$ is an *inverse* problem, and when there is noise on the measurements of $\mathbf{b}$ or imperfections in our formulation of $\mathbf{A}$, this becomes a least-squares regression problem: to find the forces $\mathbf{x}$ that minimize $\mathbf{A}\mathbf{x} - \mathbf{b}$ to make the residual error as small as possible in a sum of squares sense.

Linear least-squares regression is a rare optimization problem that has an analytic, closed-form solution. It is possible to compute the gradient of the objective function and

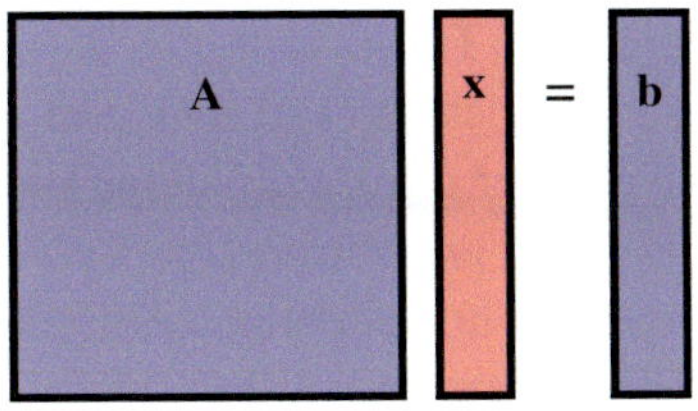

Figure 4.1 Solving a matrix system of equations, $\mathbf{Ax} = \mathbf{b}$, with the same number of equations as unknowns involves matrix inversion.

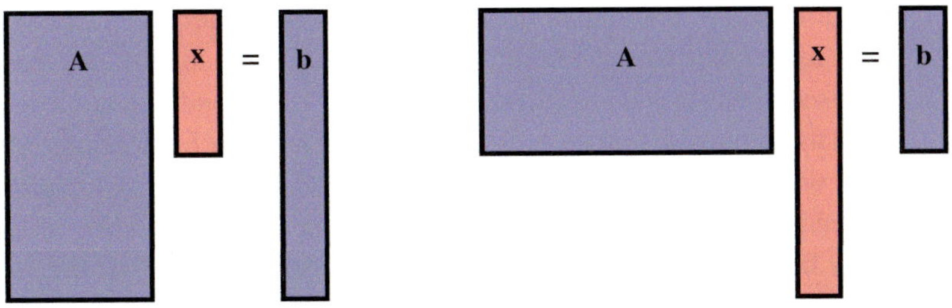

Figure 4.2 The least-squares regression problem may be overdetermined (left) so that there are more equations than unknowns, or underdetermined (right) so that there are fewer equations than unknowns.

set it equal to zero, even for non-square matrices **A**, which results in the solution

$$\mathbf{x} = \mathbf{A}^{\dagger}\mathbf{b} \tag{4.3}$$

where $\mathbf{A}^{\dagger}$ is the Moore–Penrose pseudo-inverse. In practice, we compute $\mathbf{A}^{\dagger}$ for large systems of equations using the singular value decomposition (SVD), which is itself an optimal matrix decomposition. Specifically, the SVD yields the best low-rank approximation to a matrix, which we will see in Section 4.2.

Nonlinear least squares generalizes this problem, where we now optimize over the parameters $\boldsymbol{\theta}$ of a nonlinear function fit $\mathcal{N}_{\theta}(\mathbf{x})$

$$\min_{\theta} \quad ||\mathcal{N}_{\theta}(\mathbf{x}) - \mathbf{y}||_2^2. \tag{4.4}$$

For example, a neural network is such a function $\mathcal{N}_{\theta}$. In general this is a nonconvex optimization problem that requires iterative methods, such as stochastic gradient descent from Chapter 2 or Gauss–Newton methods that will be discussed later in this chapter.

In this chapter, we begin with the simple linear least-squares regression problem in Eq. (4.2) for the overdetermined case, where there are more equations than unknowns and we must minimize the sum squared errors. This is a typical scenario in regression, since it corresponds to the case when there is more data than the number of free parameters in the linear mode. We then introduce the singular value decomposition, which is particularly useful for solving least-squares problems, especially for large, low-rank matrices.

At this point we revisit the underdetermined regression case (see Fig. 4.2, right). Since there are typically infinitely many **x** that solve a system equations with more unknowns than equations, we solve for the **x** with the minimum norm $||\mathbf{x}||_2$ in the so-called *minimum-*

norm regression problem. This is formulated as a constrained least-squares problem using Lagrange multipliers, which we will revisit in more depth in Chapter 6. We also briefly introduce a special class of inverse problems known as *compressed sensing*, where a high-dimensional sparse signal is reconstructed from surprisingly few measurements. This may be viewed as a minimum-norm regression problem where we minimize the $\|\mathbf{x}\|_1$ norm to promote sparsity. This example will be revisited and developed in homework problems in Chapters 5 and 8.

The SVD naturally leads to a discussion on the condition number of the matrix $\mathbf{A}$, which determines how much small errors on the vector $\mathbf{b}$ get amplified when (pseudo)inverting $\mathbf{A}$ to solve for $\mathbf{x}$. Once we have introduced the notion of ill-conditioned matrix inverse problems, we discuss how to add regularization to improve the condition number. We discuss Tikhonov regularization, also known as ridge regression, and the LASSO regression, which add either $\|\mathbf{x}\|_2^2$ or $\|\mathbf{x}\|_1$ to the objective function to promote regularity or sparsity in the solution $\mathbf{x}$, respectively. These regularization terms decrease the variance in the solution $\mathbf{x}$ given a noisy $\mathbf{b}$, but introduce a bias, in one of our first examples of the *bias–variance tradeoff*. Regularization is a key topic in optimization and these norms will be discussed more in Chapters 5 and 7.

After dealing extensively with unconstrained least squares, we will next add constraints using Lagrange multipliers and the KKT conditions. This is a topic of central importance, as we often have constraints on the vector $\mathbf{x}$ that must be enforced. We will discuss constraints in much more detail in Chapter 6. In linear control theory (Chapter 9) the dynamics enter as constraints to a quadratic optimization problem, and in machine learning (Chapter 10) we often use KKT least-squares to enforce symmetries and other prior information on the coefficients of a model to make it more physical.

The least-squares problems introduced so far fall into a broad class of optimization called *quadratic programming*. We discuss this large category of optimization problems, how least squares fits in, and some of the leading computational approaches to solving these problems.

Finally, we explore nonlinear least-squares problems introduced in Eq. (4.4), which encompasses a broad class of nonlinear inverse optimization problems. Most of machine learning, such as training a neural network, falls into this class of nonlinear least-squares regression, since we almost always use the sum-squared error (or mean-squared error) as the first and primary term in our loss function. In this chapter, we will introduce more classical approaches to solve these problems, including Gauss–Newton methods and the conjugate gradient method, which is used to solve very large scale least-squares problems. More modern approaches, such as stochastic gradient descent, were introduced in Chapter 2 and will be discussed more in Chapter 10 on machine learning.

Throughout this chapter, we assume that the reader is comfortable with how to compute the gradients of linear and quadratic functions over a vector $\mathbf{x}$. These relationships and derivations are provided in Section 2.7 if a refresher is needed.

4.1 Least-Squares Regression

Given a dataset of predictor variables and their associated outcomes, arranged as rows of a matrix $\mathbf{A} \in \mathbb{R}^{m \times n}$ and a vector $\mathbf{b} \in \mathbb{R}^m$, respectively, regression seeks to find the relationship between the columns of $\mathbf{A}$ that is most consistent with the outcomes in $\mathbf{b}$. This relationship is quantified by the linear system of equations

$$\mathbf{Ax} \approx \mathbf{b}, \tag{4.5}$$

which indicates that the vector $\mathbf{b}$ may be approximated as a linear combination of the columns of $\mathbf{A}$. This linear combination is given by the vector $\mathbf{x} \in \mathbb{R}^n$, which is to be determined. The least-squares objective function in Eq. (4.2) is visualized in two-dimensions in Fig. 4.3. The vector of outcomes, $\mathbf{b}$, is often contaminated with measurement noise, which we typically model as the addition of a vector $\boldsymbol{\epsilon}$ of independent, identically distributed Gaussian white noise: $\mathbf{b} = \mathbf{b}_{\text{true}} + \boldsymbol{\epsilon}$.

It is possible to solve the linear least-squares problem in Eq. (4.2) analytically by computing the gradient of the objective function and setting it equal to zero. We may expand the objective function as

$$\begin{aligned}
\|\mathbf{Ax} - \mathbf{b}\|_2^2 &= (\mathbf{Ax} - \mathbf{b})^{\mathrm{T}} (\mathbf{Ax} - \mathbf{b}) && (4.6a)\\
&= \left(\mathbf{x}^{\mathrm{T}}\mathbf{A}^{\mathrm{T}} - \mathbf{b}^{\mathrm{T}}\right)(\mathbf{Ax} - \mathbf{b}) && (4.6b)\\
&= \mathbf{x}^{\mathrm{T}}\mathbf{A}^{\mathrm{T}}\mathbf{Ax} - \mathbf{x}^{\mathrm{T}}\mathbf{A}^{\mathrm{T}}\mathbf{b} - \mathbf{b}^{\mathrm{T}}\mathbf{Ax} + \mathbf{b}^{\mathrm{T}}\mathbf{b}. && (4.6c)
\end{aligned}$$

Computing the gradient and setting equal to zero yields

$$\begin{aligned}
\nabla\|\mathbf{Ax} - \mathbf{b}\|_2^2 &= 2\mathbf{A}^{\mathrm{T}}\mathbf{Ax} - \mathbf{A}^{\mathrm{T}}\mathbf{b} - \mathbf{A}^{\mathrm{T}}\mathbf{b} = \mathbf{0} && (4.7a)\\
\implies \quad & \mathbf{A}^{\mathrm{T}}\mathbf{Ax} = \mathbf{A}^{\mathrm{T}}\mathbf{b} && (4.7b)\\
\implies \quad & \mathbf{x} = \left(\mathbf{A}^{\mathrm{T}}\mathbf{A}\right)^{-1}\mathbf{A}^{\mathrm{T}}\mathbf{b}. && (4.7c)
\end{aligned}$$

The matrix $\mathbf{A}^{\dagger} = \left(\mathbf{A}^{\mathrm{T}}\mathbf{A}\right)^{-1}\mathbf{A}^{\mathrm{T}}$ is known as the Moore–Penrose pseudo-inverse, and it is defined even for non-square matrices $\mathbf{A}$. However, this is only valid when $\left(\mathbf{A}^{\mathrm{T}}\mathbf{A}\right)$ is invertible. The matrix $\mathbf{A}^{\mathrm{T}}\mathbf{A}$ is invertible when $\mathbf{A}$ has n linearly independent columns and $m \geq n$. Thus, this formula is only applicable for the overdetermined case in Fig. 4.2. A more technical discussion of exactly when a matrix is overdetermined or underdetermined is presented at the end of this section.

In practice, we compute this pseudo-inverse for large matrices using the singular value decomposition, which is discussed extensively in the next section. In brief, given the SVD of a real matrix $\mathbf{A}$:

$$\mathbf{A} = \mathbf{U\Sigma V}^{\mathrm{T}} \tag{4.8}$$

then the Moore–Penrose pseudo-inverse may be written as

$$\mathbf{A}^{\dagger} = \mathbf{V\Sigma}^{-1}\mathbf{U}^{\mathrm{T}} \quad \implies \quad \mathbf{x} = \mathbf{V\Sigma}^{-1}\mathbf{U}^{\mathrm{T}}\mathbf{b}. \tag{4.9}$$

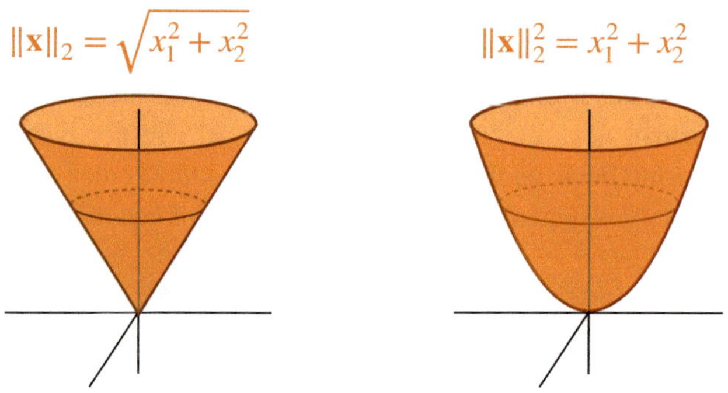

Figure 4.3 The root-mean-square error norm $\|\mathbf{x}\|_2 = \sqrt{x_1^2 + x_2^2}$ (left) and the sum-squared error norm $\|\mathbf{x}\|_2^2 = x_1^2 + x_2^2$ (right) visualized in two dimensions.

The matrices $\mathbf{U}$ and $\mathbf{V}$ are unitary, so it is simple and efficient to invert them by taking their transpose: $\mathbf{U}^T\mathbf{U} = \mathbf{I}$ and $\mathbf{V}^T\mathbf{V} = \mathbf{I}$. Similarly, $\mathbf{\Sigma}$ is a diagonal matrix, which is also easy to invert.

Examples of Least-Squares Regression

Regression is an important statistical tool for relating relate variables to one another based on data [329]. At the beginning of this chapter we already discussed a motivating example where the matrix $\mathbf{A}$ represents demographics and behavioral data for a number of individual patients, the $\mathbf{b}$ represents health outcomes, and $\mathbf{x}$ represents the prognostic model. This is visualized in Fig. 4.4.

We will now explore several simple computational regression examples in Python. All of these examples are unconstrained. We will revisit more complex examples with constraints in Section 4.6.

Example: Random Data. As a first example, we will generate random data $\mathbf{A}$ and $\mathbf{b}$ and will solve for $\mathbf{x}$ using the pseudo-inverse, computed via the SVD.

CodeSnip 4.a

```
import numpy as np

# Initialize A (m x n) and b (m x 1).
np.random.seed(42)
m, n = 20, 5
A = np.random.randn(m,n)
b = np.random.randn(m)

# Solve Least Squares Ax=b using SVD
U, S, VT = np.linalg.svd(A,full_matrices=0)
x = VT.T @ np.linalg.inv(np.diag(S)) @ U.T @ b

x = np.linalg.pinv(A) @ b    # Alternative
```

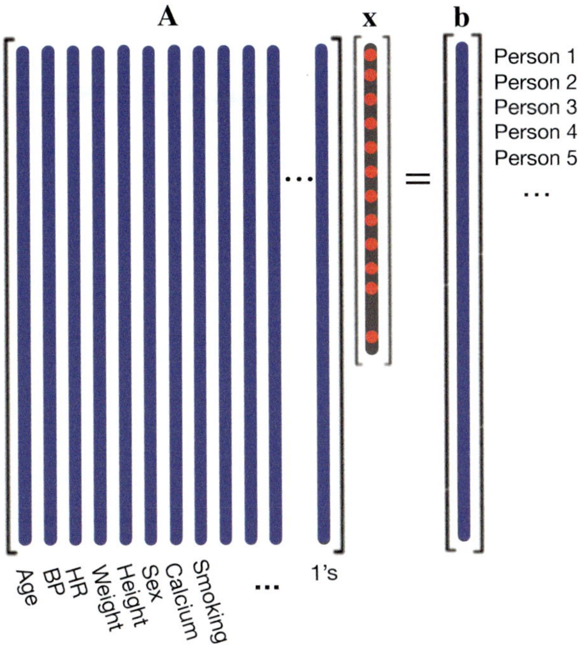

Figure 4.4 Illustrative schematic of how data from a regression problem is formed into matrices and vectors. In this example, alluded to in the introduction to this chapter, we have health data (**A**) and outcomes (**b**) for several individual patients. The unknown vector **x** represents the model coefficients for how we predict **b** given the information in **A**. In later sections, we will use sparse regularization to find the fewest columns of **A** that best correlate with **b**. Note that we may pad the matrix with an extra column of ones, as shown. Most regression fits have an explicit intercept term in the underlying model, which is captured with this column of ones.

Example: Cement Heat Generation. Now, we explore a simple engineering data set that describes the heat generation for various cement mixtures that comprise four basic ingredients. In this problem, we are solving $\mathbf{A}\mathbf{x} = \mathbf{b}$, where $\mathbf{A} \in \mathbb{R}^{13\times 4}$, since there are four ingredients and heat measurements for 13 unique mixtures. The goal is to determine the weighting $\mathbf{x}$ that relates the proportions of the four ingredients to the heat generation. It is possible to find the minimum error solution using the SVD, as shown in the code below.

CodeSnip 4.b

```
A = np.loadtxt('hald_ingredients.csv')
b = np.loadtxt('hald_heat.csv')

# Solve Ax=b using SVD
U, S, VT = np.linalg.svd(A,full_matrices=0)
x = VT.T @ np.linalg.inv(np.diag(S)) @ U.T @ b

x = np.linalg.pinv(A) @ b    # Alternative
```

The results of the regression fit are shown in Fig. 4.5.

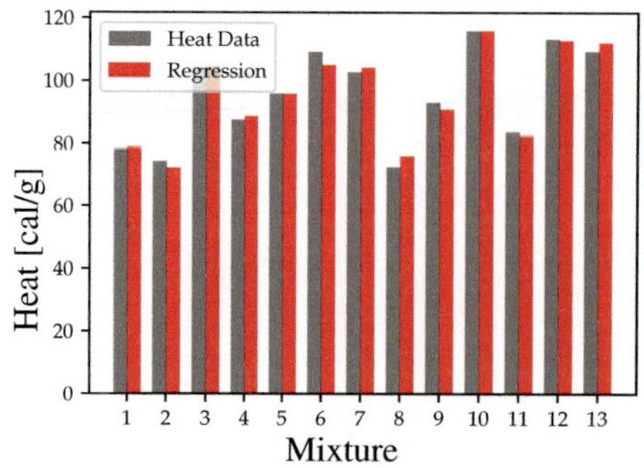

	Ingredient	x_j
1	Tricalcium aluminate	2.1930
2	Tricalcium silicate	1.1533
3	Tetracalcium alumiferrite	0.7585
4	Beta-dicalcium silicate	0.4863

Figure 4.5 Heat data for cement mixtures containing four basic ingredients.

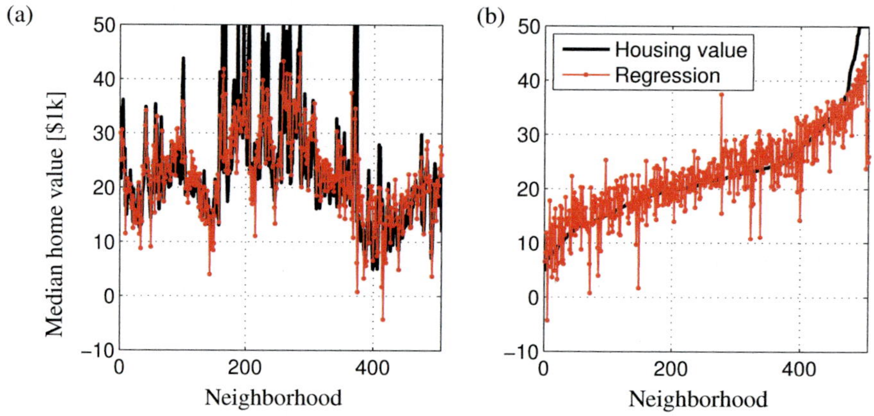

Figure 4.6 Multi-linear regression of home prices using various factors: (a) unsorted data and (b) data sorted by home value.

Example: Boston Housing Data. In this example, we explore a larger data set to determine which factors best predict prices in the Boston housing market [219]. This data is available from the UCI Machine Learning Repository [18].

There are 13 attributes that are correlated with house price, such as per-capita crime rate and property-tax rate. These features are regressed onto the price data, the best-fit price prediction is plotted against the true house value in Fig. 4.6. Although the house value is not perfectly predicted, the trend agrees quite well. It is often the case that the highest value outliers are not well captured by simple linear fits, as in this example.

This data contains prices and attributes for 506 homes, so the attribute matrix is of size 506×13. It is important to pad this matrix with an additional column of ones, to take into account the possibility of a non-zero constant offset in the regression formula. This corresponds to the "y intercept" in a simple one-dimensional linear regression. Note that padding the $\mathbf{A}$ matrix with a column of ones is the same as normalizing by subtracting its column mean. The code for this example is nearly identical to the example above, and is available on the book's GitHub.

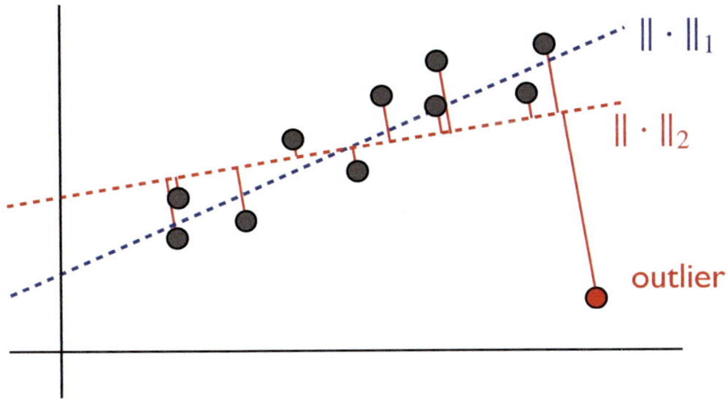

Figure 4.7 Least-squares regression, i.e. minimizing the $\|\cdot\|_2^2$ norm, is sensitive to outliers, as the square of the distance to the outlier skews the entire regression fit. Least absolute deviation regression, i.e., minimizing the $\|\cdot\|_1$ norm, is more robust to outliers.

Outliers and Other Norms

Least-squares regression is sensitive to large outliers in the data, which skew the entire regression fit. This is because the error associated with outliers are squared, which weights them disproportionally compared with data points having small error. This skewing of the regression fit is shown schematically in Fig. 4.7.

To address the robustness of least squares to outliers, it may be possible to instead minimize the sum of absolute values of the errors,

$$\|\mathbf{Ax} - \mathbf{b}\|_1, \tag{4.10}$$

known as the least absolute deviation (LAD) regression. Because errors are not squared, large outliers do not disproportionally skew the regression. The robust $\|\cdot\|_1$ norm regression is shown in Fig. 4.7 and Code 4.1. This is a theme in optimization, that using the $\|\cdot\|_1$ norm of the error as the loss function is typically more robust to outliers.

Code 4.1: Use of ℓ_1 norm for robust statistical regression.

```
x = np.sort(4*(np.random.rand(25,1)-0.5),axis=0) # data
b = 0.9*x + 0.1*np.random.randn(len(x),1) # Noisy line y=ax
atrue = np.linalg.lstsq(x,b,rcond=None)[0] # Least-squares a

b[-1] = -5.5   # Introduce outlier
acorrupt = np.linalg.lstsq(x,b,rcond=None)[0] # New slope

## L1 optimization to reject outlier
def L1_norm(a):
    return np.linalg.norm(a*x-b,ord=1)

a0 = acorrupt    # initialize to L2 solution
res = minimize(L1_norm, a0)
aL1 = res.x[0]   # aL1 is robust
```

When is a Matrix Overdetermined or Underdetermined?

So far we have only considered the case where $\mathbf{A}$ is overdetermined. Now we will zoom out and look at both overdetermined and underdetermined systems of equations. A non-square matrix $\mathbf{A}$ may either be *underdetermined* or *overdetermined*:

Underdetermined: In this case $\mathbf{A} \in \mathbb{R}^{m\times n}$ and $m \ll n$ (i.e., $\mathbf{A}$ is a short-fat matrix), so that there are fewer equations than unknowns. This type of system is likely to have columns that span all of $\mathbb{R}^m$, since it has many more columns than are required for a linearly independent basis.[1] Generically, if a short-fat $\mathbf{A}$ has m linearly independent columns (i.e., its column space spans $\mathbb{R}^m$), then there are infinitely many solutions $\mathbf{x}$ for every $\mathbf{b}$. The system is called *underdetermined* because there are not enough values in $\mathbf{b}$ to uniquely determine the higher-dimensional $\mathbf{x}$.

Overdetermined: Similarly, consider the *overdetermined system*, where $\mathbf{A}$ is a tall-skinny matrix with $m \gg n$, so that there are more equations than unknowns. This matrix cannot have m linearly independent columns, so it is guaranteed that there are vectors $\mathbf{b}$ that have no solution $\mathbf{x}$. In fact, there will only be a solution $\mathbf{x}$ if $\mathbf{b}$ is in the column space of $\mathbf{A}$, i.e., $\mathbf{b} \in \text{col}(\mathbf{A})$.

Panel 4.1: Least Absolute Deviation or Least Squares?

Interestingly, there has been a major debate about the use of the $\|\cdot\|_2$ norm versus the $\|\cdot\|_1$ norm in regression for over two hundred years. Gauss and Legendre independently introduced the least-squares regression in the early 1800s. However, fifty years earlier, in 1757, Boscovich introduced the LAD regression to solve problems related to estimating the shape of the earth. There have been ongoing debates for the past two centuries about which norm is appropriate for which applications.

Today, we have firm mathematical foundations for what prior information these norms quantify, related to Bayesian statistics and inverse problems, which will be discussed more in Chapters 7 and 8, respectively. Roughly speaking, the least-squares regression is an estimator for the mean of a distribution, whereas the least-absolute-deviation regression estimates the median; the median is known to be more robust to outliers[a]. In fact, the least-squares regression is an *unbiased* estimator for the mean of a distribution assuming Gaussian noise on the measurement vector $\mathbf{b}$.

[a] As an example, consider the mean versus median wealth in the United States of America, as of September 2024. The median wealth is $\sim$ \$200k, while the mean wealth is $\sim$ \$1000k because of a relatively small number of extremely wealthy individuals, aka outliers.

[1] It is easy to construct degenerate examples where the columns of a short-fat matrix do not form a full basis for $\mathbb{R}^m$, such as

$$\mathbf{A} = \begin{bmatrix} 1 & 1 & 1 & 1 \\ 1 & 1 & 1 & 1 \end{bmatrix}.$$

Technically, there may be some choices of $\mathbf{b}$ that admit infinitely many solutions $\mathbf{x}$ for a tall–skinny matrix $\mathbf{A}$ and other choices of $\mathbf{b}$ that admit zero solutions even for a short-fat matrix. The solution space to the system in (4.1) is determined by the following four fundamental subspaces of $\mathbf{A}$:

- The column space, $\text{col}(\mathbf{A})$, is the span of the columns of $\mathbf{A}$, also known as the *range*.
- The row space, $\text{row}(\mathbf{A})$, is the span of the rows of $\mathbf{A}$. Note that $\text{row}(\mathbf{A}) = \text{col}(\mathbf{A}^{\text{T}})$.
- The kernel space, $\ker(\mathbf{A})$, is the orthogonal complement to $\text{row}(\mathbf{A})$, and is also known as the *null space*. The null space is the subspace of vectors that map through $\mathbf{A}$ to zero, i.e., $\mathbf{Ax} = \mathbf{0}$.
- The orthogonal complement to $\text{col}(\mathbf{A})$ is $\ker(\mathbf{A}^{\text{T}})$.

More precisely, if $\mathbf{b} \in \text{col}(\mathbf{A})$ and if $\dim(\ker(\mathbf{A})) \neq 0$, then there are infinitely many solutions $\mathbf{x}$. Note that the condition $\dim(\ker(\mathbf{A})) \neq 0$ is guaranteed for a short–fat matrix. Similarly, if $\mathbf{b} \notin \text{col}(\mathbf{A})$, then there are no solutions, and the system of equations in (4.1) is called *inconsistent*.

The fundamental subspaces above satisfy the following properties:

$$\begin{aligned} \text{col}(\mathbf{A}) \oplus \ker(\mathbf{A}^{\text{T}}) &= \mathbb{R}^n, \\ \text{col}(\mathbf{A}^{\text{T}}) \oplus \ker(\mathbf{A}) &= \mathbb{R}^n. \end{aligned}$$

Here, the symbol $\oplus$ means a direct sum of subspaces, $\mathbf{U} \oplus \mathbf{V} = \{\mathbf{u} + \mathbf{v} : \mathbf{u} \in \mathbf{U}, \mathbf{v} \in \mathbf{V}\}$, with the additional property that the decomposition is unique, $\mathbf{U} \cap \mathbf{V} = \{\mathbf{0}\}$. Thus $\text{col}(\mathbf{A}) \oplus \ker(\mathbf{A}^{\text{T}}) = \mathbb{R}^n$ means that every vector in $\mathbb{R}^n$ can be uniquely written as a sum of a vector in the column space and a vector in the null space, and the two subspaces only intersect at the zero vector.

Remark 4.1 *There is an extensive literature on random matrix theory, where the above stereotypes are almost certainly true, meaning that they are true with high probability. For example, a system $\mathbf{Ax} = \mathbf{b}$ is extremely unlikely to have an exact solution for a random matrix $\mathbf{A} \in \mathbb{R}^{m\times n}$ and random vector $\mathbf{b} \in \mathbb{R}^m$ with $m \gg n$, since there is little chance that $\mathbf{b}$ is exactly in the column space of $\mathbf{A}$. In this case, we must settle for solutions that minimize the sum of squared errors (i.e., the least squares solution). These properties of random matrices play an important role in sparse inverse optimization problems, which we discuss throughout the book.*

In the overdetermined case when no solution exists, we would often like to find the solution $\mathbf{x}$ that minimizes the sum-squared error $\|\mathbf{Ax} - \mathbf{b}\|_2^2$, the so-called *least-squares* solution. Note that the least-squares solution also minimizes $\|\mathbf{Ax} - \mathbf{b}\|_2$. In the underdetermined case when infinitely many solutions exist, we may like to find the solution $\mathbf{x}$ with minimum norm $\|\mathbf{x}\|_2$ so that $\mathbf{Ax} = \mathbf{b}$, the so-called *minimum-norm* solution. We will explore both cases extensively in this chapter, and we will also add regularization terms, such as $\|\mathbf{x}\|_2^2$ or $\|\mathbf{x}\|_1$ to promote solutions with certain properties.

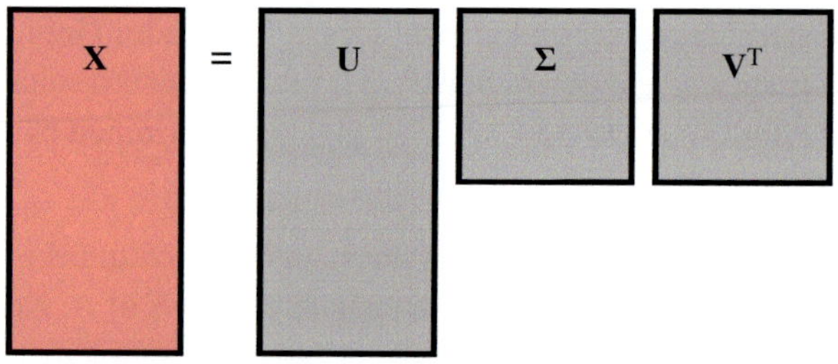

Figure 4.8 The singular value decomposition of a matrix **X** provides a hierarchical decomposition into a set of orthogonal rows and columns that best capture the variance in the data.

4.2 Singular Value Decomposition

The singular value decomposition (SVD) is the technique of choice for large matrix inverse problems. The SVD provides a numerically stable matrix decomposition that can be used for a variety of purposes and is guaranteed to exist. We will use the SVD to obtain optimal low-rank approximations to matrices and to perform pseudo-inverses of non-square matrices to find the solution of a system of equations $\mathbf{Ax} = \mathbf{b}$. The SVD is also the algorithm underlying principal component analysis (PCA), where high-dimensional data is decomposed into its most statistically descriptive factors.

We denote the full SVD as

$$\mathbf{A} = \mathbf{U}\mathbf{\Sigma}\mathbf{V}^{\mathrm{T}} \tag{4.11}$$

where $\mathbf{U} \in \mathbb{R}^{m\times m}$ and $\mathbf{V} \in \mathbb{R}^{n\times n}$ are orthogonal matrices[2] with orthonormal columns, and $\mathbf{\Sigma} \in \mathbb{R}^{m\times n}$ is a matrix with real, non-negative entries on the diagonal and zeros off the diagonal. The columns of $\mathbf{U}$ are called *left singular vectors* of $\mathbf{A}$ and the columns of $\mathbf{V}$ are *right singular vectors*. The diagonal elements of $\mathbf{\Sigma}$ are called *singular values* and they are ordered from largest to smallest: $\sigma_1 \geq \sigma_2 \geq \cdots \geq \sigma_n \geq 0$. The SVD is visualized in Fig. 4.8.

The above can easily be generalized for complex-valued matrices $\mathbf{A} \in \mathbb{C}^{m\times n}$, in which case we use the complex conjugate transpose * instead of the real transpose ^{T}. For real-valued matrices, the complex conjugate transpose is the same as the regular transpose $\mathbf{A}^* = \mathbf{A}^{\mathrm{T}}$.

When $m \geq n$, the matrix $\mathbf{\Sigma}$ has at most n non-zero elements on the diagonal. If there are linearly dependent columns of $\mathbf{A}$, then there will be even fewer non-zero singular values. If we denote the diagonal matrix of non-zero singular values as $\hat{\mathbf{\Sigma}}$, then it is possible to *exactly* represent $\mathbf{A}$ using the *economy* SVD:

$$\mathbf{A} = \mathbf{U}\mathbf{\Sigma}\mathbf{V}^{\mathrm{T}} = \begin{bmatrix}\hat{\mathbf{U}} & \hat{\mathbf{U}}^{\perp}\end{bmatrix}\begin{bmatrix}\hat{\mathbf{\Sigma}} \\ \mathbf{0}\end{bmatrix}\mathbf{V}^{\mathrm{T}} = \hat{\mathbf{U}}\hat{\mathbf{\Sigma}}\mathbf{V}^{\mathrm{T}}. \tag{4.12}$$

[2] A square matrix $\mathbf{U}$ is orthogonal if $\mathbf{U}\mathbf{U}^{\mathrm{T}} = \mathbf{U}^{\mathrm{T}}\mathbf{U} = \mathbf{I}$. If $\mathbf{A}$ is complex valued, then $\mathbf{U}$ and $\mathbf{V}$ become *unitary* matrices, where $\mathbf{U}^*\mathbf{U} = \mathbf{U}\mathbf{U}^* = \mathbf{I}$ and $\mathbf{V}^*\mathbf{V} = \mathbf{V}\mathbf{V}^* = \mathbf{I}$, where * denotes the complex conjugate transpose. Orthogonal matrices are always unitary.

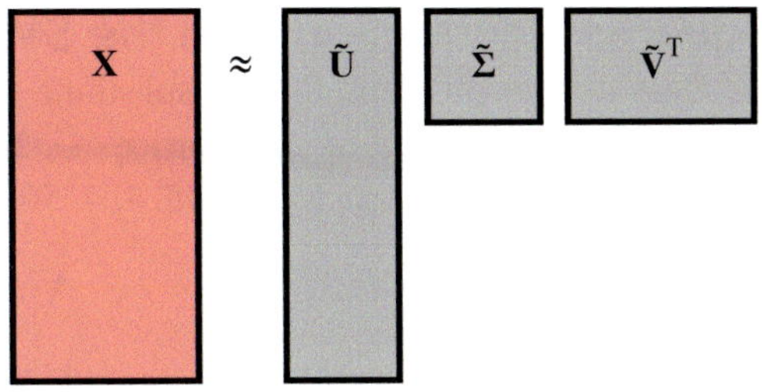

Figure 4.9 The truncated singular value decomposition approximates a matrix $\mathbf{X}$ using the first r singular values and singular vectors.

The matrix $\hat{\mathbf{U}}$ are the columns corresponding to nonzero singular values in $\hat{\mathbf{\Sigma}}$, and $\hat{\mathbf{U}}^{\perp}$ is the orthogonal complement to this. The rank of $\mathbf{A}$ is equal to the number of non-zero singular values. The SVD can also be used to obtain an optimal rank-r approximation of $\mathbf{A} = \tilde{\mathbf{U}}\tilde{\mathbf{\Sigma}}\tilde{\mathbf{V}}^{\mathrm{T}}$, where the rank $r < n$ is chosen to retain the largest singular values. In this case, the matrices $\tilde{\mathbf{U}}$ and $\tilde{\mathbf{V}}$ contain the first r columns of $\mathbf{U}$ and $\mathbf{V}$, and $\tilde{\mathbf{\Sigma}}$ contains the first $r \times r$ block of $\mathbf{\Sigma}$. This is shown schematically in Fig. (4.9).

The four fundamental subspaces of $\mathbf{A}$ may be related to the $\mathbf{U}$ and $\mathbf{V}$ matrices in the SVD as follows:

- The column space of $\mathbf{A}$ is the same as the column space of $\hat{\mathbf{U}}$.
- The orthogonal complement to the column space of $\mathbf{A}$ is given by the column space of $\hat{\mathbf{U}}^{\perp}$.
- The row space of $\mathbf{A}$ is spanned by the columns of $\mathbf{V}$.
- The kernel space of $\mathbf{A}$ is given by $\mathrm{col}(\mathbf{V}^{\perp})$.

Panel 4.2: The SVD as a Sum of Rank-1 Matrices

In essence, because $\mathbf{\Sigma}$ is diagonal, the SVD may be thought of as a sum of rank-1 matrices, given by $\sigma_j \mathbf{u}_j \mathbf{v}_j^{\mathrm{T}}$. This will be useful when computing error bounds of the SVD approximation, for example in the Eckart–Young theorem.

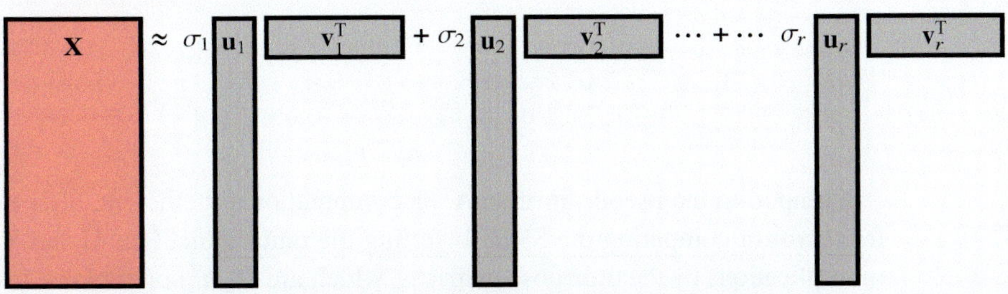

Figure 4.10: The SVD as a sum of rank-1 matrices.

Least-Squares Regression with the SVD

To solve the least-squares problem, first we substitute an exact truncated SVD $\mathbf{A} = \hat{\mathbf{U}}\hat{\mathbf{\Sigma}}\mathbf{V}^{\mathrm{T}}$ in for $\mathbf{A}$. We then "invert" each of the matrices $\hat{\mathbf{U}}$, $\hat{\mathbf{\Sigma}}$, and $\mathbf{V}^{\mathrm{T}}$ in turn, resulting in the Moore–Penrose *left pseudo-inverse* [387, 388, 424, 561] $\mathbf{A}^{\dagger}$ of $\mathbf{A}$:

$$\mathbf{A}^{\dagger} \triangleq \mathbf{V}\hat{\mathbf{\Sigma}}^{-1}\hat{\mathbf{U}}^{\mathrm{T}} \quad \Longrightarrow \quad \mathbf{A}^{\dagger}\mathbf{A} = \mathbf{V}\mathbf{V}^{\mathrm{T}}. \tag{4.13}$$

Note that $\mathbf{A}^{\dagger}\mathbf{A}$ will only equal the identity $\mathbf{I}_{n\times n}$ if the truncated SVD captures all non-zero singular values; otherwise if we use a rank-r approximation, then $\tilde{\mathbf{V}}\tilde{\mathbf{V}}^{\mathrm{T}} \neq \mathbf{I}_{n\times n}$, and it will only approximate the identity. This may be used to find both the minimum-norm and least-squares solutions to (4.1):

$$\mathbf{A}^{\dagger}\mathbf{A}\mathbf{x}^{*} = \mathbf{A}^{\dagger}\mathbf{b} \quad \Longrightarrow \quad \mathbf{x}^{*} = \mathbf{V}\hat{\mathbf{\Sigma}}^{-1}\hat{\mathbf{U}}^{\mathrm{T}}\mathbf{b}. \tag{4.14}$$

Plugging the solution $\mathbf{x}^{*}$ back in to (4.1) results in

$$\mathbf{A}\mathbf{x}^{*} = \hat{\mathbf{U}}\hat{\mathbf{\Sigma}}\mathbf{V}^{\mathrm{T}}\mathbf{V}\hat{\mathbf{\Sigma}}^{-1}\hat{\mathbf{U}}^{\mathrm{T}}\mathbf{b} \tag{4.15a}$$

$$= \hat{\mathbf{U}}\hat{\mathbf{U}}^{\mathrm{T}}\mathbf{b}. \tag{4.15b}$$

In other words, this solution is a projection of the vector $\mathbf{b}$ onto the space spanned by the columns of $\hat{\mathbf{U}}$. If $\mathbf{b}$ is not in the span of $\mathbf{A}$, and hence in the span of $\hat{\mathbf{U}}$, then $\mathbf{x}^{*}$ will only approximately solve this equation and there will be some residual error from the components of $\mathbf{b}$ that are not in the span of $\mathbf{A}$. Note that $\hat{\mathbf{U}}\hat{\mathbf{U}}^{\mathrm{T}}$ is not the identity when $n < m$, even though $\mathbf{V}^{\mathrm{T}}\mathbf{V}$ is the identity since $\mathbf{V}$ is a full rank $n \times n$ orthogonal matrix.

Computing the SVD

Algorithmically, the SVD is computed by first reducing the matrix $\mathbf{A}$ to a bidiagonal matrix and then using an iterative algorithm to compute the SVD of the bidiagonal matrix. For matrices with high aspect ratio (i.e., $m \gg n$), the first step may be achieved by first computing a QR factorization to reduce $\mathbf{A}$ to an upper triangular matrix, followed by Householder reflections to reduce this upper triangular matrix into a bidiagonal form [77, 500]. The second step may be performed using a modified QR algorithm developed by Golub & Kahan [199]. There are numerous variations, alternatives, and important results on the computation of the SVD; for a more thorough discussion, see [199, 85, 201].

The SVD is computed in Python using the following code:

CodeSnip 4.c

```
>>> import numpy as np
>>> X = np.random.rand(5, 3)  # create random data matrix
>>> U, S, VT = np.linalg.svd(X,full_matrices=True) #full SVD
>>> Uhat, Shat, VThat = np.linalg.svd(X,full_matrices=False)  #
    economy SVD
```

Computing the pseudo-inverse $\mathbf{A}^{\dagger}$ is computationally efficient, after the expensive up-front cost of computing the SVD. Inverting the unitary matrices $\tilde{\mathbf{U}}$ and $\tilde{\mathbf{V}}$ involves matrix multiplications by the transpose matrices, which are $O(m^2)$ operations. Inverting $\tilde{\mathbf{\Sigma}}$ is even more efficient, since it is a diagonal matrix, requiring $O(m)$ operations. In contrast, inverting a dense square matrix would require an $O(m^3)$ operation.

Optimal Approximation and Error Bounds

The SVD is itself the globally optimal solution of an optimization problem. For a given rank r, there is no better approximation for $\mathbf{X}$, in the ℓ_2 sense, than the truncated SVD approximation $\tilde{\mathbf{X}}$. The Eckart–Young theorem below will state this precisely and provide expressions for the error of the truncated approximation. Thus, high-dimensional data may be well described by a few dominant patterns given by the columns of $\tilde{\mathbf{U}}$ and $\tilde{\mathbf{V}}$.

Schmidt (of Gram–Schmidt) generalized the SVD to function spaces and developed an approximation theorem, establishing the truncated SVD $\tilde{\mathbf{X}}$ as the optimal low-rank approximation of the matrix $\mathbf{X}$ [446]. Schmidt's approximation theorem was rediscovered by Eckart and Young [159], and is sometimes referred to as the Eckart–Young theorem.

Theorem 4.2 (Eckart–Young [159]) *The optimal rank-r approximation to $\mathbf{X}$, in a least-squares sense, is given by the rank-r SVD truncation $\tilde{\mathbf{X}}$:*

$$\underset{\tilde{\mathbf{X}},\text{s.t. rank}(\tilde{\mathbf{X}})=r}{\operatorname{argmin}} \|\mathbf{X} - \tilde{\mathbf{X}}\|_F = \tilde{\mathbf{U}}\tilde{\mathbf{\Sigma}}\tilde{\mathbf{V}}^{\mathrm{T}}. \tag{4.16}$$

Again, $\tilde{\mathbf{U}}$ and $\tilde{\mathbf{V}}$ contain the first r columns of $\mathbf{U}$ and $\mathbf{V}$, and $\tilde{\mathbf{\Sigma}}$ contains the leading $r \times r$ sub-block of $\mathbf{\Sigma}$. The Frobenius norm is defined as $\|\mathbf{X}\|_F = \sqrt{\sum_{i=1}^{n} \sum_{j=1}^{m} |X_{ij}|^2}$, which is equivalent to the two-norm of the vectorized $\mathbf{X}(:)$. The Eckart–Young theorem says that the truncated SVD provides the best matrix approximation of a given rank, in the Frobenius norm. It is also possible to exactly quantify the error of the rank-r SVD approximation:

$$\|\mathbf{X} - \tilde{\mathbf{X}}\|_F^2 = \sum_{k=r+1}^{m} \sigma_k^2. \tag{4.17}$$

Thus, all other rank-r matrices $\tilde{\mathbf{X}}$ will have at least this much error. Because the error scales with the size and magnitude of $\mathbf{X}$, it is often more useful to consider the relative error

$$\frac{\|\mathbf{X} - \tilde{\mathbf{X}}\|_F^2}{\|\mathbf{X}\|_F^2}. \tag{4.18}$$

This expression for the relative error in the Frobenius norm has the interpretation of the amount of missing variance in the approximation $\tilde{\mathbf{X}}$, a property that is used extensively in the principal component analysis (PCA). Remarkably, the SVD also provides an optimal rank-r approximation in the matrix 2-norm, also known as the spectral norm:

$$\underset{\tilde{\mathbf{X}},\text{s.t. rank}(\tilde{\mathbf{X}})=r}{\operatorname{argmin}} \|\mathbf{X} - \tilde{\mathbf{X}}\|_2 = \tilde{\mathbf{U}}\tilde{\mathbf{\Sigma}}\tilde{\mathbf{V}}^{\mathrm{T}}. \tag{4.19}$$

The 2-norm of a matrix $\mathbf{X}$ given by $\|\mathbf{X}\|_2 = \max_{\mathbf{v}\neq\mathbf{0}} \frac{\|\mathbf{X}\mathbf{v}\|_2}{\|\mathbf{v}\|_2}$. The error expression for the rank-r SVD approximation is even simpler in the 2-norm:

$$\|\mathbf{X} - \tilde{\mathbf{X}}\|_2 = \sigma_{r+1}. \tag{4.20}$$

This error expression is rather simple to derive by expanding

$$\mathbf{X} - \tilde{\mathbf{X}} = \sum_{k=r+1}^{m} \sigma_k \mathbf{u}_k \mathbf{v}_k^{\mathrm{T}}. \tag{4.21}$$

The $\mathbf{v}_k$ vectors are orthonormal so $\mathbf{v} = \mathbf{v}_{r+1}$ maximizes $\|(\mathbf{X} - \tilde{\mathbf{X}})\mathbf{v}\|_2 = \sigma_{r+1}$.

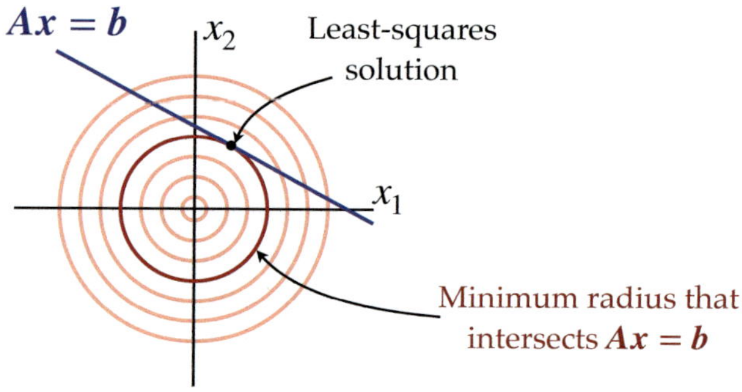

Figure 4.11 Level sets of the ℓ_2 norm of $\mathbf{x}$ are circles, and the least-squares solution gives the point $\mathbf{x}$ on $\mathbf{Ax} = \mathbf{b}$ that intersects the minimum radius level set of this norm.

4.3 Minimum-Norm Regression

The minimum-norm regression problem is similar to the least-squares problem above, except for the case of an underdetermined system of equations

$$\mathbf{Ax} = \mathbf{b} \tag{4.22}$$

where $\mathbf{A}$ is a short–fat matrix, so there will be fewer equations than unknowns. In this case, under mild conditions, there will be infinitely many vectors $\mathbf{x}$ that solve this system of equations.

The minimum-norm regression problem seeks the $\mathbf{x}$ that satisfies the system of equations and has the minimum ℓ_2 norm:

$$\begin{aligned} \min_{\mathbf{x}} \quad & \|\mathbf{x}\|_2^2 \\ \text{s.t.} \quad & \mathbf{Ax} = \mathbf{b}. \end{aligned} \tag{4.23}$$

Importantly, here we are minimizing the norm $\|\mathbf{x}\|_2^2$ and not $\|\mathbf{Ax} - \mathbf{b}\|_2$, and instead the system of equations now enters the optimization problem as a constraint. The minimum-norm solution is visualized in Fig. 4.11 for the ℓ_2 norm. This idea may be generalized to other norms, as shown in Fig. 4.12.

We may use the Lagrange-multiplier form, introduced in Section 1.4 and discussed more in Chapter 6, and expand the objective function $\|\mathbf{x}\|_2^2 = \mathbf{x}^{\mathrm{T}}\mathbf{x}$, resulting in:

$$L(\mathbf{x}, \boldsymbol{\lambda}) = \mathbf{x}^{\mathrm{T}}\mathbf{x} + \boldsymbol{\lambda}^{\mathrm{T}}(\mathbf{Ax} - \mathbf{b}). \tag{4.24}$$

The gradient condition on $\mathbf{x}$ becomes

$$\nabla_{\mathbf{x}} L = 2\mathbf{x} + \mathbf{A}^{\mathrm{T}}\boldsymbol{\lambda} = 0 \quad \Longrightarrow \quad \mathbf{x} = -\frac{1}{2}\mathbf{A}^{\mathrm{T}}\boldsymbol{\lambda}, \tag{4.25}$$

and the gradient condition on $\boldsymbol{\lambda}$ becomes

$$\nabla_{\boldsymbol{\lambda}} L = \mathbf{Ax} - \mathbf{b} = 0 \quad \Longrightarrow \quad -\frac{1}{2}\mathbf{AA}^{\mathrm{T}}\boldsymbol{\lambda} = \mathbf{b} \tag{4.26a}$$

$$\Longrightarrow \quad \boldsymbol{\lambda} = -2\left(\mathbf{AA}^{\mathrm{T}}\right)^{-1}\mathbf{b}, \tag{4.26b}$$

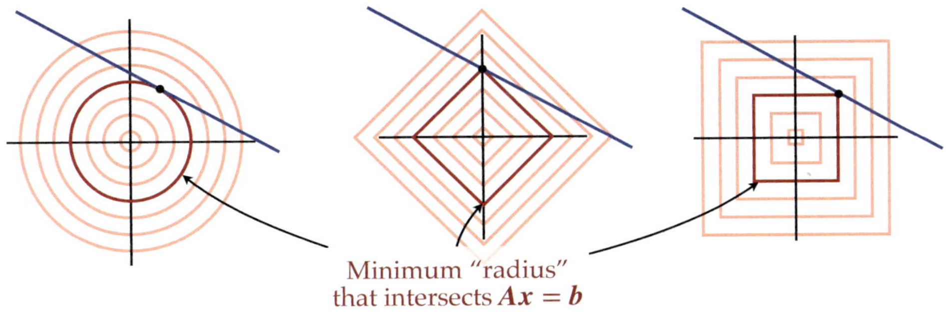

Figure 4.12 The ℓ_2, ℓ_1, and ℓ_∞ norms of $\mathbf{x}$ visualized with the line $\mathbf{Ax} = \mathbf{b}$ in two dimensions.

which leads to

$$\mathbf{x} = -\frac{1}{2}\mathbf{A}^{\mathrm{T}}\lambda = \mathbf{A}^{\mathrm{T}}\left(\mathbf{A}\mathbf{A}^{\mathrm{T}}\right)^{-1}\mathbf{b}. \tag{4.27}$$

This is the Moore–Penrose *right* pseudo-inverse:

$$\mathbf{x} = \mathbf{A}^{\mathrm{T}}\left(\mathbf{A}\mathbf{A}^{\mathrm{T}}\right)^{-1}\mathbf{b}. \tag{4.28}$$

This may be computed using the singular value decomposition, as in the least-squares regression. If $\mathbf{A} = \mathbf{U\Sigma V}^{\mathrm{T}}$ then $\mathbf{x} = \mathbf{V\Sigma}^{-1}\mathbf{U}^{\mathrm{T}}\mathbf{b}$.

Compressed Sensing

We will now go deeper into a particularly interesting recent application of underdetermined regression. Most natural signals, such as images and audio, are highly compressible [77]. When the signal is written in an appropriate basis (e.g., Fourier or wavelet), only a few modes are active, having large coefficients. The remaining modes have negligible effect on the signal, and their small coefficients may be truncated to zero, reducing the number of coefficients that must be stored and transmitted for an accurate representation. This compressibility is the basis of modern image and audio compression, such as JPEG and MP3.

However, until recently, images and audio signals were typically collected in full resolution, only to be subsequently compressed for storage and transfer. Recently, within the past twenty years, this paradigm has been turned upside down in what is now known as *compressed sensing*. In compressed sensing, the idea is to collect a small but random subsampling of the original high-dimensional data and then infer which Fourier or wavelet coefficients are large and active in the signal. For example, instead of collecting all million pixels in a megapixel image, it may be possible to collect 10% of the pixels randomly and infer the large Fourier modes in the compressed JPEG image.

Compressed sensing is a very challenging underdetermined regression problem, where instead of solving for the minimum $\|\cdot\|_2$-norm solution, we search for the *sparsest* solution to a system of equations. Here we will set up the basic compressed sensing regression problem and state the associated optimization problem. Then we will have several homework exercises to explore various aspects of this problem in more detail.

A compressible signal $\mathbf{x} \in \mathbb{R}^n$ may be written as a sparse vector $\mathbf{s} \in \mathbb{R}^n$ (containing mostly zeros) in a transform basis $\boldsymbol{\Psi} \in \mathbb{R}^{n \times n}$:

$$\mathbf{x} = \boldsymbol{\Psi}\mathbf{s}. \tag{4.29}$$

Specifically, the vector $\mathbf{s}$ is called K-sparse in $\boldsymbol{\Psi}$ if there are exactly K nonzero elements. If the basis $\boldsymbol{\Psi}$ is generic, such as the Fourier or wavelet basis, then only the few active terms in $\mathbf{s}$ are required to reconstruct the original signal $\mathbf{x}$, reducing the data required to store or transmit the signal. Images and audio signals are both compressible in Fourier or wavelet bases, so after transforming, most coefficients are small and may be set exactly equal to zero with negligible loss of quality. With knowledge of the sparse vector $\mathbf{s}$ it is possible to reconstruct the signal $\mathbf{x}$ from (4.29).

The goal of compressed sensing is to measure a small subset of the original signal in a measurement vector $\mathbf{y} \in \mathbb{R}^p$ and infer the sparsest vector $\mathbf{s}$ that is consistent with these measurements. The measurements are related to the full state $\mathbf{x}$ via $\mathbf{y} = \mathbf{C}\mathbf{x}$, where $\mathbf{C} \in \mathbb{R}^{p \times n}$ is a measurement matrix with $p \ll n$:

$$\mathbf{y} = \mathbf{C}\mathbf{x} = \mathbf{C}\boldsymbol{\Psi}\mathbf{s} = \boldsymbol{\Theta}\mathbf{s}. \tag{4.30}$$

The system of equations in (4.30) is underdetermined since there are infinitely many consistent solutions $\mathbf{s}$. The *sparsest* solution $\hat{\mathbf{s}}$ satisfies the following optimization problem:

$$\hat{\mathbf{s}} = \underset{\mathbf{s}}{\operatorname{argmin}} \, \|\mathbf{s}\|_0 \quad \text{s.t.} \quad \mathbf{y} = \mathbf{C}\boldsymbol{\Psi}\mathbf{s}, \tag{4.31}$$

where $\|\cdot\|_0$ denotes the ℓ_0-pseudo-norm, given by the number of nonzero entries; this is also referred to as the cardinality of $\mathbf{s}$.

The optimization in (4.31) is nonconvex, and in general, the global solution can only be found with a brute-force search that is combinatorial in n and K. In particular, all possible K-sparse vectors in $\mathbb{R}^n$ must be checked; if the exact level of sparsity K is unknown, the search is even broader. Because this search is combinatorial, solving (4.31) is intractable for even moderately large n and K, and the prospect of solving larger problems does not improve with Moore's law of exponentially increasing computational power.

Fortunately, under certain conditions on the measurement matrix $\mathbf{C}$, it is possible to relax the optimization in (4.31) to a convex ℓ_1-minimization [93, 146]:

$$\hat{\mathbf{s}} = \underset{\mathbf{s}}{\operatorname{argmin}} \, \|\mathbf{s}\|_1 \quad \text{s.t.} \quad \mathbf{y} = \mathbf{C}\boldsymbol{\Psi}\mathbf{s}. \tag{4.32}$$

The ℓ_1-norm is also known as the taxicab or Manhattan norm because it represents the distance a taxi would take between two points on a rectangular grid. The ℓ_1-minimum-norm solution is sparse, while the ℓ_2-minimum-norm solution is not.

We will also see in Section 5.1 that it is possible to use the proximal gradient and splitting methods to accurately and efficiently solve for local solutions to this nonconvex problem. Further, it is possible to use the global ℓ_1 solution as an initialization to the proximal gradient descent on the original ℓ_0 problem.

4.4 Condition Number and Regularization

Very often the least-squares regression problem is ill-conditioned, meaning that the columns of $\mathbf{A}$ are nearly linearly dependent. In this case, the pseudo-inverse amplifies noise on the vector $\mathbf{b}$, introducing high variance and a large degree of uncertainty into the estimate $\mathbf{x}$.

In this section, we will formally define the condition number, which is a fundamental concept in regression and optimization that will be used throughout the remainder of this book. We then introduce the Tikhonov regularization, also known as ridge regression, that improves the conditioning of ill-conditioned least-squares problems, decreasing variance but introducing bias. We also introduce the related LASSO regression, which promotes sparsity of the solution vector $\mathbf{x}$.

Condition Number

The condition number of a matrix $\mathbf{A}$ is a measure of how sensitive matrix multiplication and inversion are to errors in the input. Larger condition number indicates higher sensitivity and worse performance. The condition number $\kappa(\mathbf{A})$ is directly related to the singular values of the matrix:

$$\kappa(\mathbf{A}) = \frac{\sigma_{\text{max}}(\mathbf{A})}{\sigma_{\text{min}}(\mathbf{A})}. \tag{4.33}$$

Panel 4.3: Alternative Formulation of Compressed Sensing

In addition to the ℓ_1-minimization in (4.32), there are alternative optimization approaches based on *greedy algorithms* [501, 502, 504, 503, 506, 226, 505, 190, 507, 188, 362, 189] that determine the sparse solution of (4.30) through an iterative matching pursuit problem. For instance, the compressed sensing matching pursuit (CoSaMP) [362] is computationally efficient and easy to implement.

When the measurements $\mathbf{y}$ have additive noise, say white noise of magnitude ε, then the following variant of (4.32) may be more robust [513]:

$$\hat{\mathbf{s}} = \underset{\mathbf{s}}{\text{argmin}}\, ||\mathbf{s}||_1, \quad \text{s.t.} \quad ||\mathbf{C}\boldsymbol{\Psi}\mathbf{s} - \mathbf{y}||_2 < \varepsilon.$$

A related convex optimization is the following [11]:

$$\hat{\mathbf{s}} = \underset{\mathbf{s}}{\text{argmin}}\, ||\mathbf{C}\boldsymbol{\Psi}\mathbf{s} - \mathbf{y}||_2 + \lambda||\mathbf{s}||_1,$$

where $\lambda \geq 0$ is a parameter that weights the importance of sparsity. These two equations are closely related [504]. This is very similar to the LASSO regression we will introduce in Eq. (4.41), but for the underdetermined compressed sensing problem. Comparing the various approaches to solve compressed sensing will be an important exercise for the reader.

The condition number is a central concept in all of numerical linear algebra and applied computation. It is easiest to understand the effect of a large condition number when considering the linear system of equations $\mathbf{Ax} = \mathbf{b}$. If the vector $\mathbf{b}$ is not specified perfectly, but instead has some error $\boldsymbol{\epsilon}_\mathbf{b}$, then the system becomes

$$\mathbf{A}(\mathbf{x} + \boldsymbol{\epsilon}_\mathbf{x}) = \mathbf{b} + \boldsymbol{\epsilon}_\mathbf{b}, \tag{4.34}$$

where $\boldsymbol{\epsilon}_\mathbf{x}$ is the corresponding error in $\mathbf{x}$.

Assuming a worst-case scenario, where $\boldsymbol{\epsilon}_\mathbf{b}$ is aligned with the singular vector corresponding to the minimum singular value σ_{min}, and $\mathbf{b}$ is aligned with the singular vector corresponding to the maximum singular value σ_{max}, the estimated solution for $\mathbf{x} + \boldsymbol{\epsilon}_\mathbf{x}$ is

$$\mathbf{x} + \boldsymbol{\epsilon}_\mathbf{x} \approx \mathbf{A}^\dagger(\mathbf{b} + \boldsymbol{\epsilon}_\mathbf{b}) = \frac{1}{\sigma_{\text{max}}}\mathbf{b} + \frac{1}{\sigma_{\text{min}}}\boldsymbol{\epsilon}_\mathbf{b}. \tag{4.35}$$

The signal-to-noise of the estimated $\mathbf{x}$ is given by

$$\|\mathbf{x}\|/\|\boldsymbol{\epsilon}_\mathbf{x}\| = \frac{\sigma_{\text{min}}}{\sigma_{\text{max}}}\|\mathbf{b}\|/\|\boldsymbol{\epsilon}_\mathbf{b}\|, \tag{4.36}$$

which is degraded by a factor equal to the condition number $\kappa(\mathbf{A})$. Thus, noise on $\mathbf{b}$ is amplified by the condition number in the inversion process.

Even if the vectors $\mathbf{b}$ and $\boldsymbol{\epsilon}_\mathbf{b}$ are not perfectly aligned with the worst-case directions, they are likely to have some component of all singular vectors. In this case, components of the error will still experience large amplification relative to other components of the desired output. One approach to mitigate a large condition number is to truncate the SVD more aggressively, increasing the effective minimum singular value σ_{min}. However, this comes at the cost of decreasing the size of the subspace $\tilde{\mathbf{U}}$ used to approximate the output. The following code truncates the SVD based on a desired condition number:

CodeSnip 4.d

```
>> kappanew = 1e-5 # Desired condition number
>> U, S, Vt = np.linalg.svd(A, full_matrices=False)
>> r = np.sum(S > np.max(S) * kappanew)
>> invA = Vt[:r,:].T @ np.linalg.inv(np.diag(S[:r])) @ U[:,:r].T
     # Rank-r approximation
```

Tikhonov Regularization

Tikhonov regularization [496], also known as ridge regression [232], addresses the poor condition number of the least-squares regression when the columns of $\mathbf{A}$ are very nearly parallel and the condition number of the pseudo-inverse is large.

Recall that when $\mathbf{A}$ is non-square we cannot invert it to solve for $\mathbf{x}$ and so instead we compute the pseudo-inverse via the singular value decomposition:

$$\mathbf{A} = \mathbf{U\Sigma V}^{\mathrm{T}} \quad \Longrightarrow \quad \mathbf{A}^\dagger = \mathbf{V\Sigma}^{-1}\mathbf{U}^{\mathrm{T}} = \cdots = \left(\mathbf{A}^{\mathrm{T}}\mathbf{A}\right)^{-1}\mathbf{A}^{\mathrm{T}}. \tag{4.37}$$

When the columns of $\mathbf{A}$ are nearly parallel, the condition number of $\mathbf{A}$ is large, and hence the condition number of $\mathbf{A}^{\mathrm{T}}\mathbf{A}$ is even larger. This manifests as the diagonal matrix $\boldsymbol{\Sigma}$ having nearly degenerate singular values, so that inverting $\boldsymbol{\Sigma}^{-1}$ involves dividing by very small numbers.

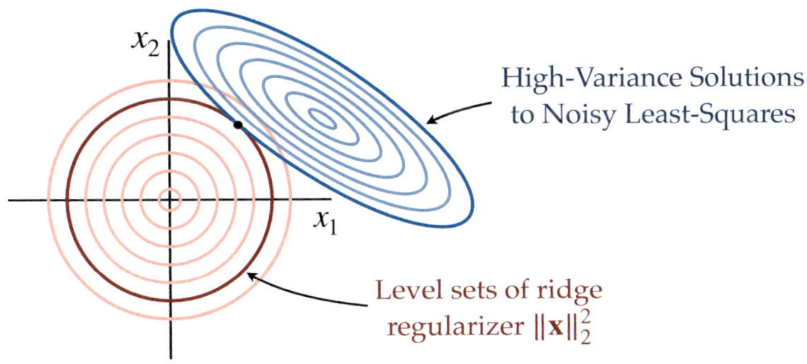

Figure 4.13 Schematic of Tikhonov regularization, aka ridge regression. The original ill-conditioned least-squares problem has high variance, as visualized by the eccentric blue ellipses, which represent the solutions given Gaussian noise on the vector **b**. Tikhonov regularization pulls these solutions towards level sets of $\|\mathbf{x}\|_2^2$, shown as red circles.

Any noise on **b** will be amplified by this ill-conditioned pseudo-inverse. This has the effect of some entries of **x** being large and opposite, leading to high variance in the resulting solution. Said another way, if we sample from many instances of small noise on **b**, we will get a large, highly eccentric ellipsoid of solutions **x** centered around the true solution, as shown in Fig. 4.13.

One solution is to regularize the least-squares problem with an additional loss term $\|\mathbf{x}\|_2^2$:

$$\mathbf{x} = \underset{\mathbf{x}'}{\operatorname{argmin}} \|\mathbf{A}\mathbf{x}' - \mathbf{b}\|_2^2 + \alpha\|\mathbf{x}'\|_2^2, \tag{4.38}$$

where α is a hyperparameter that weights this regularizing term.

It is again possible to compute the gradient of this augmented objective function and set it equal to zero, resulting in the analytic solution to the ridge regression problem

$$\mathbf{x} = \left(\mathbf{A}^{\mathrm{T}}\mathbf{A} + \alpha\mathbf{I}\right)^{-1} \mathbf{A}^{\mathrm{T}}\mathbf{b}. \tag{4.39}$$

We may view the additional term $\alpha\|\mathbf{x}\|_2^2$ as arising from a Lagrange multiplier enforcing the condition that $\|\mathbf{x}\|_2^2 = c$, where c is a constant (i.e., $\alpha(\|\mathbf{x}\|_2^2 - c)$). This will work for any constant c, as it essentially promotes the individual entries of **x** being similar in magnitude, as opposed to having very large and opposite entries canceling each other out, as is the case when **A** is ill conditioned.

This leads to a fundamental property in many regularized optimization problems: the bias–variance tradeoff. The original problem has high variance, so we add regularization. However, this regularization term introduces a bias towards a solution where the entries of **x** are on the unit circle, as in Fig. 4.14.

There is also a Bayesian perspective to ridge regression, where the ridge regularizer arises from a Bayesian prior that the entries of **x** are Gaussian random variables with zero mean and the same variance. We will discuss this more in Chapter 7.

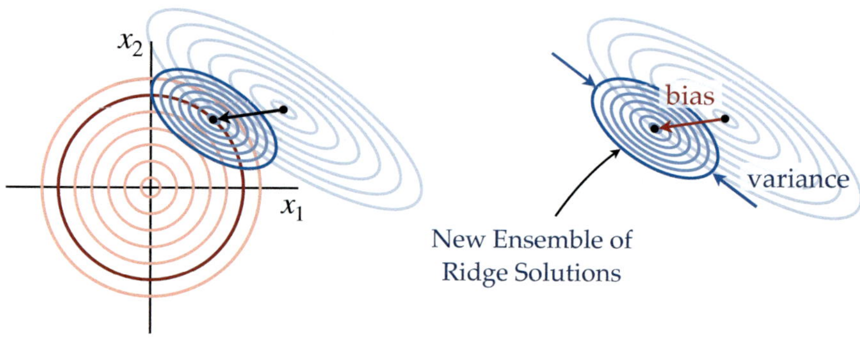

Figure 4.14 Schematic of bias–variance tradeoff in Tikhonov regularization. The regularized solution is represented by the dark blue ellipses, which have lower variance than the original solution, given by the light blue ellipses. However, Tikhonov regularization introduces a bias, shifting the mean of the new distribution.

Ridge Regression via SVD. At this point, you won't be surprised that the formula in Eq. (4.39) can be computed using the SVD, without inverting the square matrix $(\mathbf{A}^T\mathbf{A}+\alpha\mathbf{I})$. Substituting $\mathbf{A} = \mathbf{U}\mathbf{\Sigma}\mathbf{V}^T$, we have

$$(\mathbf{A}^T\mathbf{A} + \alpha\mathbf{I})^{-1} = (\mathbf{V}\mathbf{\Sigma}^2\mathbf{V}^T + \alpha\mathbf{I})^{-1} \tag{4.40a}$$

$$= (\mathbf{V}\mathbf{\Sigma}^2\mathbf{V}^T + \alpha\mathbf{V}\mathbf{I}\mathbf{V}^T)^{-1} \tag{4.40b}$$

$$= (\mathbf{V}(\mathbf{\Sigma}^2 + \alpha\mathbf{I})\mathbf{V}^T)^{-1}. \tag{4.40c}$$

This may be inverted simply, as $\mathbf{V}$ is unitary and $\mathbf{\Sigma}^2 + \alpha\mathbf{I}$ is diagonal. So essentially, we add α times the identity matrix to the squared singular values in the matrix inversion step.

Sparse Regression

The least absolute shrinkage and selection operator (LASSO) is an ℓ_1-penalized regression technique that balances model complexity with descriptive capability [495]. LASSO is quite similar to ridge regression, but with an ℓ_1 norm penalty on $\mathbf{x}$ instead of a ℓ_2 norm penalty. The ℓ_1-norm was introduced earlier as a penalty in 1969 [389], with specialized algorithms in 1973 [111], and as a robust loss in geophysics in 1973 [107]. Other related sparsity-promoting techniques have subsequently been explored [17, 95, 423, 23, 548].

This principle of *parsimony* in a model is also a reflection of Occam's razor, stating that among all possible descriptions, the simplest correct model is probably the true one. Since its inception by Tibshirani in 1996 [495], the LASSO has become a cornerstone of statistical modeling, with many modern variants and related techniques [220, 539, 247]. The LASSO is closely related to the earlier nonnegative garrote of Breimen [68], and is also related to earlier work on soft-thresholding by Donoho and Johnstone [147, 148].

Least-squares regression will tend to result in a vector $\mathbf{x}$ that has nonzero coefficients for all entries, indicating that *all* columns of $\mathbf{A}$ must be used to predict $\mathbf{b}$. However, we often believe that the statistical model should be *simpler*, so that $\mathbf{x}$ is sparse. The LASSO adds an ℓ_1 penalty term to *regularize* the least-squares regression problem; i.e., to prevent

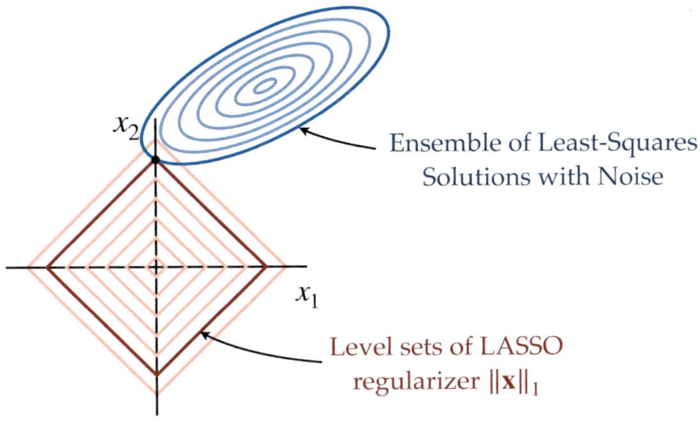

Figure 4.15 Schematic of LASSO regularization, to promote sparsity.

overfitting:

$$\mathbf{x} = \underset{\mathbf{x}'}{\operatorname{argmin}} \|\mathbf{A}\mathbf{x}' - \mathbf{b}\|_2^2 + \lambda\|\mathbf{x}'\|_1. \tag{4.41}$$

Typically, the parameter λ is varied through a range of values and the fit is *validated* against a test set of holdout data. If there is not enough data to have a sufficiently large training and test set, it is common to repeatedly train and test the model on random selection of the data (often 80 % for training and 20 % for testing), resulting in a *cross-validated* performance. This cross-validation procedure enables the selection of a parsimonious model that has relatively few terms and avoids overfitting.

LASSO regression is frequently used to build statistical models for disease, such as cancer and heart failure, since there are many possible predictors, including demographics, lifestyle, biometrics and genetic information. Interpretability is important in statistical models, as these models are often communicated to a non-technical audience, including business leaders and policy makers. Generally, a regression model is more interpretable if it has fewer terms that bear on the outcome, motivating yet another perspective on sparsity.

Many statistical systems are overdetermined, as there are more observations than candidate predictors. Fortunately, the LASSO regression works well with overdetermined problems. Note that an early version of the geometric picture in Fig. 4.15 to explain the sparsity-promoting nature of the ℓ_1-norm was presented in Tibshirani's 1996 paper [495].

Elastic Net. The elastic net [564] combines the ridge and LASSO regularization terms:

$$\mathbf{x} = \underset{\mathbf{x}'}{\operatorname{argmin}} \|\mathbf{A}\mathbf{x}' - \mathbf{b}\|_2^2 + \lambda\|\mathbf{x}'\|_1 + \alpha\|\mathbf{x}'\|_2^2. \tag{4.42}$$

Elastic net regression is widely used in various machine learning applications. Level sets of the elastic net regularization $\lambda\|\mathbf{x}\|_1 + \alpha\|\mathbf{x}\|_2^2$ are shown in Fig. 4.16.

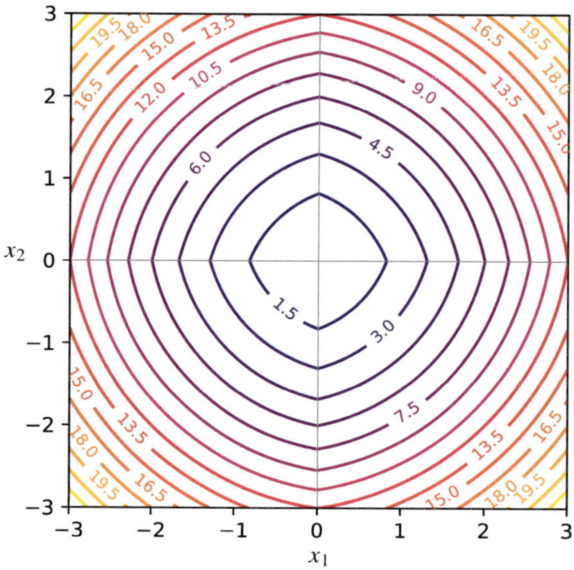

Figure 4.16 Level sets of the elastic net regularization in Eq. (4.42) for $\lambda = \alpha = 1$. This is analogous to the level sets of the LASSO regularizer in Fig. 4.15. Note that the elastic net contours look similar to those of the $\ell_{3/2}$ norm.

4.5 Constrained Least Squares

Up until now we have only considered unconstrained least-squares regression. However, many systems have constraints on the admissible values of $\mathbf{x}$. For example, we might know that some values of $\mathbf{x}$ must be non-negative, or we may have other prior knowledge, such as a certain symmetry in the coefficients of $\mathbf{x}$ or that the entries of $\mathbf{x}$ must be within some lower and upper bounds (i.e., box constraints). Other common constraints include some equality constraints on the coefficients $\mathbf{Cx} = \mathbf{d}$. In Section 10.5 we will use constrained least squares to enforce physical constraints in machine learning models, for example to enforce energy conservation in quadratic models of fluids [320]. We will discuss constraints in much more generality and detail in the next chapter. Here we briefly introduce the simplest equality constraints in least-squares.

We may derive the constrained least-squares solution for linear equality constraints:

$$\begin{aligned} \min_{\mathbf{x}} \quad & \|\mathbf{Ax} - \mathbf{b}\|_2^2 \\ \text{s.t.} \quad & \mathbf{Cx} = \mathbf{d}. \end{aligned} \tag{4.43}$$

Expanding the objective function, we may write this as a quadratic program with linear equality constraints as

$$\begin{aligned} \min_{\mathbf{x}} \quad & \frac{1}{2}\mathbf{x}^T\mathbf{Q}\mathbf{x} + \mathbf{c}^T\mathbf{x} \\ \text{s.t.} \quad & \mathbf{Cx} = \mathbf{d} \end{aligned} \tag{4.44}$$

with the following matrices:

$$\mathbf{Q} = 2\mathbf{A}^{\mathrm{T}}\mathbf{A} \qquad \mathbf{c}^{\mathrm{T}} = -2\mathbf{b}^{\mathrm{T}}\mathbf{A}. \tag{4.45}$$

We may then use Lagrange multipliers to augment the objective function with the constraint equation, resulting in the following augmented Lagrangian:

$$\mathcal{L}(\mathbf{x}, \lambda) = \frac{1}{2}\mathbf{x}^{\mathrm{T}}\mathbf{Q}\mathbf{x} + \mathbf{c}^{\mathrm{T}}\mathbf{x} + \lambda^{\mathrm{T}}\left(\mathbf{C}\mathbf{x} - \mathbf{d}\right). \tag{4.46}$$

The conditions for optimality are that $\nabla_{\mathbf{x}}\mathcal{L} = 0$ and $\nabla_{\lambda}\mathcal{L} = 0$, which result in the stationarity condition and equality constraint condition, respectively:

$$\nabla_{\mathbf{x}}\mathcal{L} = \mathbf{Q}\mathbf{x} + \mathbf{c} + \mathbf{C}^{\mathrm{T}}\lambda = 0 \tag{4.47a}$$

$$\nabla_{\lambda}\mathcal{L} = \mathbf{C}\mathbf{x} - \mathbf{d} = 0. \tag{4.47b}$$

Rewriting these equations in matrix form:

$$\left.\begin{aligned} \mathbf{Q}\mathbf{x} + \mathbf{C}^{\mathrm{T}}\lambda &= -\mathbf{c} \\ \mathbf{C}\mathbf{x} &= \mathbf{d} \end{aligned}\right\} \implies \begin{bmatrix} \mathbf{Q} & \mathbf{C}^{\mathrm{T}} \\ \mathbf{C} & \mathbf{0} \end{bmatrix} \begin{bmatrix} \mathbf{x} \\ \lambda \end{bmatrix} = \begin{bmatrix} -\mathbf{c} \\ \mathbf{d} \end{bmatrix}. \tag{4.48}$$

Writing this in terms of the original matrices and vectors from the constrained least-squares problem yields:

$$\begin{bmatrix} 2\mathbf{A}^{\mathrm{T}}\mathbf{A} & \mathbf{C}^{\mathrm{T}} \\ \mathbf{C} & \mathbf{0} \end{bmatrix} \begin{bmatrix} \mathbf{x} \\ \lambda \end{bmatrix} = \begin{bmatrix} 2\mathbf{A}^{\mathrm{T}}\mathbf{b} \\ \mathbf{d} \end{bmatrix}. \tag{4.49}$$

It is possible to solve for $\mathbf{x}$ and λ by inverting this matrix system of equations, again using SVD or QR factorization. It is clear in the above that if we have no constraints, this is equivalent to the standard least-squares problem.

Minimum Norm as a Constrained Problem

If $\mathbf{A} = \mathbf{I}$ and $\mathbf{b} = \mathbf{0}$, then the constrained least squares reduces to the minimum-norm problem above.

Specifically, if we solve

$$\begin{aligned} \min_{\mathbf{x}} \quad & \|\mathbf{x}\|_2^2 \\ \text{s.t.} \quad & \mathbf{C}\mathbf{x} = \mathbf{d} \end{aligned} \tag{4.50}$$

then the constrained problem becomes

$$\begin{bmatrix} \mathbf{I} & \mathbf{C}^{\mathrm{T}} \\ \mathbf{C} & \mathbf{0} \end{bmatrix} \begin{bmatrix} \mathbf{x} \\ \lambda \end{bmatrix} = \begin{bmatrix} \mathbf{0} \\ \mathbf{d} \end{bmatrix} \tag{4.51}$$

which has optimal solution $\mathbf{x}^* = \mathbf{C}^{\mathrm{T}}\left(\mathbf{C}\mathbf{C}^{\mathrm{T}}\right)^{-1}\mathbf{d}$.

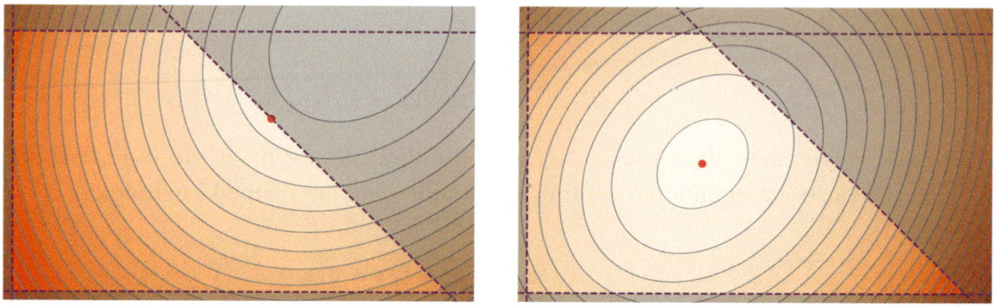

Figure 4.17 A quadratic program may have *inactive* constraints (left) so that the minimum lies in an interior point of the feasible region. Alternatively, a quadratic program may have *active* constraints (right) so that the minimum lies on the boundary of the feasible region. Here we visualize the convex feasible polytope in purple dashed lines with level sets of the quadratic objective function in red.

4.6 Quadratic Programming

Linear least squares may be written as a *quadratic programming* problem, which generalizes linear programming to quadratic objective functions.

A quadratic program is written in standard form as:

$$\begin{aligned} \min_{\mathbf{x}} \quad & \frac{1}{2}\mathbf{x}^{\mathrm{T}}\mathbf{Q}\mathbf{x} + \mathbf{c}^{\mathrm{T}}\mathbf{x}, \\ \text{Subject to:} \quad & \mathbf{A}\mathbf{x} \leq \mathbf{b}, \end{aligned} \tag{4.52}$$

where $\mathbf{Q}$ is a symmetric matrix.

If $\mathbf{Q}$ is positive semi-definite, so that $\mathbf{x}^{\mathrm{T}}\mathbf{Q}\mathbf{x} \geq 0$ for all $\mathbf{x} \in \mathbb{R}^n$, then this quadratic program is convex and any local minimum is a global minimum. If $\mathbf{Q}$ is indefinite, then the quadratic program is nonconvex, and the approach to solving it may be much more involved.

In the case that $\mathbf{Q}$ is positive semidefinite, there are several cases based on whether or not the constraints are *active* or not. If the constraints are inactive, then the optimum value may occur at an interior point of the feasible set. If the constraints are active, then the optimum may live anywhere on the boundary of the feasible set. In linear programming, the solution occurred on vertices, but for quadratic programming with active constraints, the optimum value can occur on an edge, a face, or a vertex. Thus, it may be harder to search for solutions because there is no obvious *simplex* search method. This is illustrated schematically in Fig. 4.17.

Instead, there are several approaches to solve quadratic programs, and we will focus on two of these approaches:

- Active set methods: The idea behind active set methods is to guess which constraints are active and then solve a smaller problem with only those active constraints. Roughly speaking, if $\mathbf{a}_j^{\mathrm{T}}\mathbf{x} \leq \mathbf{b}_j$ is active, then $\mathbf{a}_j^{\mathrm{T}}\mathbf{x} = \mathbf{b}_j$, and we may solve the constrained least-squares problem with this active constraint using KKT methods. Dantzig's simplex method from Chapter 3 is an active set method.

- Interior point methods: IPMs are among the leading approaches to solve both linear and quadratic programming problems in high-dimensions. They scale well to large systems when we have active linear constraints like $\mathbf{Ax} \leq \mathbf{b}$. These will be discussed in much more detail in Chapter 6.

Sequential quadratic programming (SQP) is a powerful approach for solving nonlinear constrained optimization problems by iteratively solving a sequence of quadratic programming subproblems. At each iteration, SQP approximates the original nonlinear objective and constraints using a second-order Taylor expansion and linearization, respectively. The resulting QP captures both curvature and constraint structure, and is solved using active-set or interior-point methods. The original problems don't need to be convex, although there is no guarantee of global optimality in this case. SQP methods offer fast local convergence, often superlinear, and are widely used in engineering and control applications due to their robustness and compatibility with constraint handling.

Least Squares as a Quadratic Program

It is simple to see that the linear least-squares problem is a quadratic program:

$$\min_{\mathbf{x}} \quad \|\mathbf{Ax} - \mathbf{b}\|_2^2 = (\mathbf{Ax} - \mathbf{b})^{\mathrm{T}} (\mathbf{Ax} - \mathbf{b}) \tag{4.53a}$$

$$= \mathbf{x}^{\mathrm{T}}\mathbf{A}^{\mathrm{T}}\mathbf{Ax} - 2\mathbf{b}^{\mathrm{T}}\mathbf{Ax} + \mathbf{b}^{\mathrm{T}}\mathbf{b}. \tag{4.53b}$$

Thus, this loss function is a quadratic form with

$$\mathbf{Q} = \mathbf{A}^{\mathrm{T}}\mathbf{A} \qquad \mathbf{c}^{\mathrm{T}} = -2\mathbf{b}^{\mathrm{T}}\mathbf{A}. \tag{4.54}$$

We see that $\mathbf{Q}$ is symmetric and positive semi-definite for any $\mathbf{A}$ matrix in the original least-squares problem, so the resulting quadratic program is convex.

Quadratic Programming Examples in Python

In Python, one can solve convex QPs conveniently with `cvxpy`, similar to how linear programs are solved. Other libraries include `qpsolvers` and `OSQP`, and there are several other libraries with specialized solvers. The package `cvxpy` automatically analyzes whether your problem is a QP, even with additional constraints. If the problem is a QP, a specialized solver (e.g. `OSQP`) is chosen if available. For large QPs, direct factorization or warm starts with `OSQP` or `GUROBI` might be beneficial. The user can also fine-tune performance, for example by setting `prob.solve(solver=cp.OSQP, warm_start=True)`.

Any least-squares regression can be solved as a quadratic program: we minimize a quadratic objective (i.e., the squared residuals) subject to linear or affine constraints on the parameters. We will explore constrained least-squares more in Chapter 6 on constraints and duality.

Below are some code examples illustrating different constraint types.

Python Example: Least-Squares as a Quadratic Program. Here we use `cvxpy` to solve a least-squares regression problem, first *without* any constraints. Again, `cvxpy` recognizes squared norms automatically, so we can write $\|\mathbf{Ax}-\mathbf{b}\|_2^2$ as a quadratic objective.

CodeSnip 4.e

```
import cvxpy as cp
import numpy as np

# Initialize A (m x n) and b (m x 1).
np.random.seed(0)
m, n = 20, 5
A = np.random.randn(m,n)
b = np.random.randn(m)

# Unconstrained Least Squares
x = cp.Variable(n)  # Regression coefficients
objective = cp.Minimize(cp.sum_squares(A@x - b))

prob = cp.Problem(objective)
result = prob.solve()

print("Optimal value:", result)
print("Optimal x:", x.value)
```

We now include the *linear constraint* $x_j \geq 0.5$ and $\sum_j x_j \leq 3$:

CodeSnip 4.f

```
# Constrained Least Squares
x_constrained = cp.Variable(n)
constraints = [ x_constrained >= 0.5,
                cp.sum(x_constrained) <= 3.0 ]

obj_constrained = cp.Minimize(cp.sum_squares(A@x_constrained-b))
prob_constrained = cp.Problem(obj_constrained, constraints)
result_constrained = prob_constrained.solve()
```

Portfolio Optimization (Markowitz Model). A canonical QP arises in *portfolio optimization*: we choose $\mathbf{x}$, the fraction of total budget in each asset, to minimize risk subject to a target return and other constraints:

$$\begin{aligned} \min_{\mathbf{x}} \quad & \tfrac{1}{2}\mathbf{x}^{\mathrm{T}}\boldsymbol{\Sigma}\,\mathbf{x} \\ \text{s.t.} \quad & \mathbf{r}^{\mathrm{T}}\mathbf{x} \geq R_0, \\ & \mathbf{1}^{\mathrm{T}}\mathbf{x} = 1, \\ & \mathbf{x} \geq 0. \end{aligned} \tag{4.55}$$

Here, $\boldsymbol{\Sigma}$ is the covariance matrix of returns, indicating some spread of risk, $\mathbf{r}$ is the vector of expected returns per asset, R_0 is a minimal acceptable return, and $\mathbf{x} \geq 0$ ensures no short-selling.

The following code solves this portfolio optimization problem:

CodeSnip 4.g

```
n = 5 # Example data has n=5 points
Sigma = np.array([[0.05,0.01,0.02,0.0,0.0],
                  [0.01,0.06,0.0,0.01,0.0],
                  [0.02,0.0,0.07,0.01,0.0],
                  [0.0,0.01,0.01,0.05,0.01],
                  [0.0,0.0,0.0,0.01,0.04]]) # 5x5 covariance
r = np.array([0.10,0.12,0.14,0.09,0.11])  # expected returns
R0 = 0.11                                 # target return
```

```
x = cp.Variable(n, nonneg=True)  # decision variable

# Objective: minimize x^{\rm T} Sigma x
objective = cp.Minimize(0.5 * cp.quad_form(x, Sigma))

constraints = [
    cp.sum(x) == 1,       # sum of allocations = 1
    r @ x >= R0           # minimum return
]

prob = cp.Problem(objective, constraints)
result = prob.solve(solver=cp.SCS)
print("Optimal risk:", result)
print("Optimal x:", x.value)
print("Return of portfolio:", r @ x.value)
```

Here, each constraint $x_j \leq x_{j+1}$ is linear, so the feasible set is a polyhedron defined by the entries of the vector $\mathbf{x}$ monotonically increasing. The sum of squares objective is strictly convex. The solver can be a direct QP solver like `OSQP`, or the default `SCS`.

Regularized Least-Squares with Box Constraints. Here we implement a Tikhonov-regularized least-squares with additional box constraints on the state: $l_j \leq x_j \leq u_j$. We will revisit box constraints in Section 6.2 using Lagrange multipliers and the KKT conditions.

CodeSnip 4.h

```
np.random.seed(42)
m, n = 8, 4
A = np.random.randn(m,n)
b = np.random.randn(m)
alpha = 0.2

x = cp.Variable(n)  # Decision variable x in R^n

# Objective: (1/2)*||A x - b||^2 + alpha/2 * ||x||^2
objective = cp.Minimize( 0.5*cp.sum_squares(A@x - b) + 0.5*alpha
    *cp.sum_squares(x) )

# Example box constraints: -1 <= x_i <= 2
l, u = -1.0, 2.0
constraints = [ x >= l, x <= u ]

prob = cp.Problem(objective, constraints)
res = prob.solve(solver=cp.OSQP)
print("Optimal value:", res)
print("x =", x.value)
```

This problem is a quadratic objective with linear constraints, solvable by standard QP solvers like `OSQP`, `ECOS`, `SCS`, or others in `cvxpy`.

4.7 Nonlinear Least-Squares: Gauss–Newton Methods

When we move from simple least-squares regression in Eq. (4.2) to the nonlinear least squares in Eq. (4.4), the problem is no longer convex and there is no closed-form solution.

Gauss–Newton methods are a powerful class of iterative techniques to solve these nonlinear least-squares problems. We will use the following notation for this nonlinear least-squares problem:

$$L = \frac{1}{2}\sum_{j=1}^{m} \|\mathbf{y}_j - \mathcal{N}(\mathbf{x}_j;\boldsymbol{\theta})\|_2^2 = \frac{1}{2}\sum_{j=1}^{m} f_j(\boldsymbol{\theta})^2 = \frac{1}{2}\|\mathbf{f}(\boldsymbol{\theta})\|_2^2. \tag{4.56}$$

The sum from $j = 1$ to m is typically a sum over m data points $\{(\mathbf{x}_j, \mathbf{y}_j)\}_{j=1}^{m}$, so that we are minimizing the sum-squared error of a function fit over all data. The function $\mathbf{f}(\boldsymbol{\theta})$ is a vector of normed *residuals*, so that each $f_j(\boldsymbol{\theta}) = \|\mathbf{r}_j\|_2$ where the residual error is $\mathbf{r}_j = \mathbf{y}_j - \mathcal{N}(\mathbf{x}_j)$ evaluated at the j-th data point. The variables $\boldsymbol{\theta}$ are the parameters of the nonlinear regression model and are the variables we will optimize over. This nonlinear least-squares formulation is very general, and encompasses much of neural network training, among other nonlinear regression problems. In the case of a neural network, discussed more in Chapter 10, $\mathcal{N}$ represents the neural network and $\boldsymbol{\theta}$ is the vector of weights and biases that we train to minimize the loss function.

Solving for $\min_{\boldsymbol{\theta}} \|\mathbf{f}(\boldsymbol{\theta})\|^2$ amounts to finding points where $\nabla_{\boldsymbol{\theta}} L = \mathbf{0}$ with positive curvature. This is a nonlinear system of equations, leading to a nonconvex optimization problem, so we solve with iterative methods, such as gradient descent or Newton's method.

Similar to other Newton methods for large-scale problems, here we will approximate the Hessian. In this way, Gauss–Newton is similar to the quasi-Newton method from Chapter 2, in that both approximate the Hessian with a positive definite construction. However, they are often considered distinct classes of methods, as they have diverged in implementations and use cases.

Here we compute the gradient and Hessian of the loss function L, with respect to the decision variable $\boldsymbol{\theta}$. The gradient is

$$\nabla_{\boldsymbol{\theta}} L = \sum_{j=1}^{m} f_j(\boldsymbol{\theta}) \nabla f_j(\boldsymbol{\theta}) = \mathbf{J}^{\mathrm{T}}\mathbf{f} \tag{4.57}$$

where $\mathbf{J}$ is the Jacobian matrix of derivatives

$$\mathbf{J} = \mathbf{Df} = \begin{bmatrix} \frac{\partial f_1}{\partial \theta_1} & \frac{\partial f_1}{\partial \theta_2} & \cdots \\ \frac{\partial f_2}{\partial \theta_1} & \frac{\partial f_2}{\partial \theta_2} & \cdots \\ \vdots & \vdots & \ddots \end{bmatrix} = \begin{bmatrix} — & \nabla f_1^{\mathrm{T}} & — \\ — & \nabla f_2^{\mathrm{T}} & — \\ & \vdots & \end{bmatrix} \tag{4.58}$$

and its transpose is given by

$$\mathbf{J}^{\mathrm{T}} = \begin{bmatrix} | & | & \\ \nabla f_1 & \nabla f_2 & \cdots \\ | & | & \end{bmatrix}. \tag{4.59}$$

Thus $\mathbf{J}^{\mathrm{T}}\mathbf{f}$ yields the sum in Eq. (4.57).

We now write an expression for the Hessian $\mathbf{H} = \nabla_{\theta}^{2} L$, which we will derive in the next subsection below:

$$\mathbf{H} = \nabla_{\theta}^{2} L = \mathbf{J}^{\mathrm{T}}\mathbf{J} + \sum_{j=1}^{m} f_j(\boldsymbol{\theta})\nabla^2 f_j(\boldsymbol{\theta}) = \mathbf{J}^{\mathrm{T}}\mathbf{J} + \mathbf{Q}. \tag{4.60}$$

The first term only includes first derivatives, and the second term contains second derivatives.

For many problems, it is too expensive to compute $\mathbf{Q}$. Fortunately, because $f_j(\boldsymbol{\theta})$ are the residual errors, it is common to assume that these are small, and approximate the Hessian just using the first order terms:

$$\mathbf{H} \approx \mathbf{J}^{\mathrm{T}}\mathbf{J}. \tag{4.61}$$

This is the primary assumption used in Gauss–Newton methods.

Newton's method then becomes:

$$\boldsymbol{\theta}_{k+1} = \boldsymbol{\theta}_k - \mathbf{H}^{-1}\nabla_{\theta} L. \tag{4.62}$$

However, instead of storing and computing $\mathbf{H}^{-1}$, which would involve $O(n^3)$ operations, we instead solve the linear system of equations

$$\mathbf{H}\mathbf{s}_k = -\nabla_{\theta} L \tag{4.63}$$

and directly use $\mathbf{s}_k$, which should equal $\mathbf{H}^{-1}\nabla_{\theta} L$, for the update in Newton's method:

$$\boldsymbol{\theta}_{k+1} = \boldsymbol{\theta}_k + \mathbf{s}_k. \tag{4.64}$$

Using the Gauss–Newton approximation to the Hessian in Eq. (4.61), this becomes

$$\mathbf{J}_k^{\mathrm{T}}\mathbf{J}_k\mathbf{s}_k \approx -\mathbf{J}_k^{\mathrm{T}}\mathbf{f}(\boldsymbol{\theta}_k) \tag{4.65}$$

where $\mathbf{J}_k = \mathbf{J}(\boldsymbol{\theta}_k) = \mathbf{Df}(\boldsymbol{\theta}_k)$.

Therefore, the full Gauss–Newton iteration is

$$\text{Solve} \quad \mathbf{J}_k^{\mathrm{T}}\mathbf{J}_k\mathbf{s}_k = -\mathbf{J}_k^{\mathrm{T}}\mathbf{f}(\boldsymbol{\theta}_k) \tag{4.66a}$$

$$\text{Iterate} \quad \boldsymbol{\theta}_{k+1} = \boldsymbol{\theta}_k + \mathbf{s}_k. \tag{4.66b}$$

Thus, Gauss–Newton methods reduce the nonlinear least-squares problem to a sequence of linearized regular least-squares problems[3]. This can be interpreted several ways. In each iteration step, solving

$$\mathbf{J}^{\mathrm{T}}\mathbf{J}\mathbf{s} = -\mathbf{J}^{\mathrm{T}}\mathbf{f} \tag{4.67}$$

is equivalent to a least-squares problem

$$\min_{\mathbf{s}} \frac{1}{2}\|\mathbf{J}\mathbf{s} + \mathbf{f}\|^2. \tag{4.68}$$

[3] Solving these linearized regular least-squares problems may be challenging for large-scale problems, which motivates the conjugate gradient method in the next section.

Specifically, we may expand this function as

$$\frac{1}{2}\left[(\mathbf{Js}+\mathbf{f})^{\mathrm{T}}(\mathbf{Js}+\mathbf{f})\right]=\frac{1}{2}\left[\mathbf{s}^{\mathrm{T}}\mathbf{J}^{\mathrm{T}}\mathbf{Js}+2\mathbf{s}^{\mathrm{T}}\mathbf{J}^{\mathrm{T}}\mathbf{f}+\mathbf{f}^{\mathrm{T}}\mathbf{f}\right]. \tag{4.69}$$

Minimizing this with respect to **s** amounts to computing the gradient with respect to **s** and setting equal to **0**, which yields:

$$\mathbf{J}^{\mathrm{T}}\mathbf{Js}=\mathbf{J}^{\mathrm{T}}\mathbf{f}. \tag{4.70}$$

Deriving the Hessian

To derive the Hessian expression in Eq. (4.60), we first introduce a helpful bit of notation. If $\nabla_{\boldsymbol{\theta}}L=\mathbf{J}^{\mathrm{T}}\mathbf{f}(\boldsymbol{\theta})$, then we denote the j-th component as:

$$\nabla_{\theta_j}L=\sum_{i=1}^{m}f_i(\boldsymbol{\theta})\frac{\partial f_i}{\partial\theta_j}. \tag{4.71}$$

Similarly, we denote the (j,k)-th component of the Hessian as $H_{jk}=\nabla^2_{\theta_j\theta_k}L$, which is given by:

$$H_{jk}=\nabla^2_{\theta_j\theta_k}L=\sum_i\left(\frac{\partial f_i}{\partial\theta_k}\frac{\partial f_i}{\partial\theta_j}+f_i\frac{\partial^2 f_i}{\partial\theta_j\partial\theta_k}\right)=\mathbf{J}^{\mathrm{T}}\mathbf{J}+\mathbf{Q}. \tag{4.72}$$

Again, we often drop the quadratic terms, since they are expensive to compute and assumed to be small. To see that $\mathbf{J}^{\mathrm{T}}\mathbf{J}=\sum_i\frac{\partial f_i}{\partial\theta_k}\frac{\partial f_i}{\partial\theta_j}$, we expand as

$$\mathbf{J}^{\mathrm{T}}\mathbf{J}=\begin{bmatrix}— & \nabla f_1^{\mathrm{T}} & —\\ — & \nabla f_2^{\mathrm{T}} & —\\ & \vdots & \end{bmatrix}\begin{bmatrix}| & | & \\ \nabla f_1 & \nabla f_2 & \cdots\\ | & | & \end{bmatrix} \tag{4.73}$$

so the (j,k)-th element of $\mathbf{J}^{\mathrm{T}}\mathbf{J}$ is given by the j-th row of $\mathbf{J}^{\mathrm{T}}$ times the k-th column of $\mathbf{J}$:

$$\begin{bmatrix}\partial f_1/\partial\theta_j & \partial f_2/\partial\theta_j & \cdots\end{bmatrix}\begin{bmatrix}\partial f_1/\partial\theta_k\\ \partial f_2/\partial\theta_k\\ \vdots\end{bmatrix}=\sum_i\frac{\partial f_i}{\partial\theta_k}\frac{\partial f_i}{\partial\theta_j}. \tag{4.74}$$

Challenges with Gauss–Newton

There are a number of well-known issues with Gauss–Newton methods, and there is a vast literature on how to address these challenges. The Gauss–Newton approximation to the Hessian matrix as the Jacobian of the residuals is efficient, but it assumes the residuals are small and ignores second-order terms. This can lead to divergence of the iteration when residuals are large, so that Gauss–Newton performs poorly in regions far from the solution. Moreover, there are several failure modes when $\mathbf{J}^{\mathrm{T}}\mathbf{J}$ is singular or nearly singular. Fundamentally, Gauss–Newton methods have first-order convergence, unless the residuals are very small.

There are several issues that arise when $\mathbf{J}^{\mathrm{T}}\mathbf{J}$ is nearly singular. First,the update $\mathbf{J}^{\mathrm{T}}\mathbf{J}\Delta\mathbf{x}=-\nabla L(\mathbf{x})$ may result in a large step $\Delta\mathbf{x}$, which can cause the iteration to overshoot the solution, leading to divergence. Second, when $\mathbf{J}^T\mathbf{J}$ is nearly singular, then the linear system in Eq. (4.66) is ill conditioned, and solving it becomes numerically sensitive to errors. Small errors in the residuals or the Jacobian can cause large changes in the solution $\Delta\mathbf{x}$. This may lead to erratic or unstable behavior of the iteration. Finally, if $\mathbf{J}^T\mathbf{J}$ is nearly singular, the update $\Delta\mathbf{x}$ might point in a direction that does not improve the objective function. This can cause the optimization to stall.

The Levenberg–Marquardt algorithm improves the robustness of Gauss–Newton methods for nonlinear least-squares problems with ill-conditioned Hessian approximations, and we discuss it in the next subsection.

Levenberg–Marquardt Algorithm

Gauss–Newton methods are efficient for well-conditioned problems, but have known failure modes when the Hessian approximation is ill-conditioned. The Levenberg–Marquardt (LM) algorithm [307, 334, 185] combines Gauss–Newton with gradient descent to improve the robustness for poorly conditioned or ill-posed problems. LM also helps when the initial guess is far from the solution and residuals are large, making the Hessian approximation in Gauss–Newton inaccurate. This hybrid approach is highly effective and is used in a wide range of application domains that involve solving nonlinear systems, including computer vision, curve fitting, machine learning, parameter estimation in control, and robotics.

The fundamental idea behind Levenberg–Marquardt is to regularize the Hessian approximation with a weighted identity matrix, so that

$$\mathbf{H} \approx \left(\mathbf{J}^T\mathbf{J} + \alpha\mathbf{I}\right). \tag{4.75}$$

Thus the update equation becomes

$$\left(\mathbf{J}^T\mathbf{J} + \alpha\mathbf{I}\right)\Delta\boldsymbol{\theta} = -\mathbf{J}^T\mathbf{f}. \tag{4.76}$$

The damping parameter α blends between Gauss–Newton ($\alpha = 0$) and gradient descent ($\alpha \to \infty$). The transition to gradient descent as $\alpha \to \infty$ is simple to see, as the update becomes

$$\Delta\boldsymbol{\theta} = -\tfrac{1}{\alpha}\mathbf{J}^T\mathbf{f} \tag{4.77}$$

when the $\alpha\mathbf{I}$ term dominates the Hessian approximation.

The LM modification to the Hessian approximation in Eqs. (4.75) and (4.76) may also be viewed as adding Tikhonov regularization (see Section 4.4), which is known to improve the robustness of ill-conditioned least-squares regression problems.

Typically, the parameter α is adapted dynamically, based on the residuals. When the initial guess is far from the solution, a large α is used, and the gradient-descent behavior stabilizes the updates. As the iteration approaches the optimal solution and residuals decrease, LM automatically shifts to the faster Gauss–Newton method. Thus, LM achieves fast quadratic convergence near the solution. In a sense, LM may be viewed as transitioning between exploration (i.e., gradient descent) and exploitation (i.e., Gauss–Newton) in different phases of the iteration.

There are several approaches for modifying α dynamically. The simplest approach is to simply increase α if the loss function does not decrease sufficiently, and to decrease α if the change in the loss function is sufficient. This is achieved by multiplying or dividing α by a constant factor $\nu > 1$. Since the loss function naturally decreases toward the solution, the changes to the loss function become smaller with successful iterations, and so we use a relative change in the loss function with respect to the quadratic approximation to determine if the change in loss is sufficient. This is quantified by the relative ratio ρ:

$$\rho = \frac{\Delta L_{\text{actual}}}{\Delta L_{\text{predicted}}}, \tag{4.78}$$

where

$$\Delta L_{\text{actual}} = L(\boldsymbol{\theta}_k) - L(\boldsymbol{\theta}_{k+1}) \tag{4.79}$$

is the actual reduction, and

$$\Delta L_{\text{predicted}} = -\mathbf{J}^{\mathrm{T}}\mathbf{f}\Delta\boldsymbol{\theta} - \frac{1}{2}\Delta\boldsymbol{\theta}^{\mathrm{T}}(\mathbf{J}^{\mathrm{T}}\mathbf{J} + \alpha\mathbf{I})\Delta\boldsymbol{\theta} \tag{4.80}$$

is the predicted reduction from the quadratic model.

Even if ΔL_{actual} becomes small near the minimum, ρ ensures that the reduction is evaluated relative to the quadratic Hessian prediction. This approach avoids issues with absolute magnitudes of reductions. The parameter α is usually initialized with a small positive value (e.g., 10^{-3}).

This simple approach of increasing or decreasing α based on the relative ratio ρ has a *trust region* interpretation [396, 357, 139, 370, 112], where α controls the size of the region in which the quadratic model is trusted. For example, we may use a condition such as:

- If $\rho > 0.75$, reduce α (the model is trustworthy).
- If $\rho < 0.25$, increase α (the model is unreliable).

LM explicitly adjusts α based on the norm of the step $\Delta\boldsymbol{\theta}$, so that

$$\alpha = \max(\alpha\nu, \epsilon\|\Delta\boldsymbol{\theta}\|), \tag{4.81}$$

where $\epsilon > 0$ ensures sufficient regularization.

It is also possible to introduce more sophisticated regularizations, beyond Tikhonov regularization. For example, instead of adding $\alpha\mathbf{I}$ to the Hessian approximation, it is possible to adaptively scale each diagonal entry individually:

$$\mathbf{H} = \mathbf{J}^{\mathrm{T}}\mathbf{J} + \alpha\,\mathrm{diag}(\mathbf{J}^{\mathrm{T}}\mathbf{J}). \tag{4.82}$$

This approach dynamically scales damping based on the local curvature.

4.8 The Conjugate Gradient Method

The conjugate[4] gradient (CG) method is an iterative algorithm for solving large, sparse systems of linear equations [500, 10],

$$\mathbf{A}\mathbf{x} = \mathbf{b}, \tag{4.83}$$

where $\mathbf{A} \in \mathbb{R}^{n\times n}$ is symmetric positive definite, $\mathbf{x} \in \mathbb{R}^n$ is the unknown vector, and $\mathbf{b} \in \mathbb{R}^n$ is the right-hand side. For example, $\mathbf{A}$ might be the Hessian, or the Hessian approximation from the Gauss–Newton method above. The CG method is particularly efficient for large, sparse systems because it avoids explicitly storing or factorizing $\mathbf{A}$. Instead, it builds an approximate solution iteratively using Krylov subspaces and ensures convergence in at most n iterations. It is often used in scientific computing, for example to solve the Poisson equation in computational fluid dynamics.

The conjugate gradient algorithm is also frequently used to accelerate Gauss–Newton methods for very large, sparse systems. Specifically, the conjugate gradient method is used to solve the linear system[5]:

$$\mathbf{J}^{\mathrm{T}}\mathbf{J}\Delta\mathbf{x} = -\mathbf{J}^{\mathrm{T}}\mathbf{f}, \tag{4.84}$$

Instead of directly inverting the approximate Hessian $\mathbf{J}^{\mathrm{T}}\mathbf{J} \approx \mathbf{H}$, the conjugate gradient method solves the system iteratively, avoiding the explicit computation of the Hessian matrix. Instead, it requires only matrix–vector products, which dramatically improves the memory efficiency of the algorithm for large-scale problems, where it might be prohibitive to construct and store the Hessian approximation. The improvement in memory and speed is even more pronounced when the Hessian is sparse.

In addition to improving the scalability of Gauss–Newton methods to large-scale problems, the conjugate gradient method also improves numerical stability by avoiding explicit inversion of the Hessian, which may be ill-conditioned. Similarly, it is possible to add regularization and preconditioners flexibly in the conjugate gradient formulation. However, convergence still depends on the condition number of $\mathbf{J}^{\mathrm{T}}\mathbf{J}$ and preconditioning is often necessary.

The idea behind the conjugate gradient method is to expand the solution $\mathbf{x}$ in Eq. (4.83) in a basis of conjugate vectors $\mathbf{p}_k$, that are orthogonal with respect to the inner product defined by $\mathbf{A}$:

$$\mathbf{p}_i^{\mathrm{T}}\mathbf{A}\mathbf{p}_j = 0 \quad \text{for all } i \neq j. \tag{4.85}$$

If $\mathbf{A}$ is a symmetric, positive definite matrix, the set of $\{\mathbf{p}_j\}_{j=1}^n$ form a basis for $\mathbb{R}^n$, and any vector $\mathbf{x}$ may be expanded in this basis. If these vectors are chosen judiciously, then we may be able to approximate $\mathbf{x}$ quite accurately without all n of these vectors.

[4] Here, conjugate just means orthogonal with respect to the inner product $\langle \mathbf{u}, \mathbf{v}\rangle_{\mathbf{A}} = \mathbf{u}^{\mathrm{T}}\mathbf{A}\mathbf{v} = 0$. We will iteratively construct and re-orthogonalize a basis in this method.

[5] Note that here we have returned to using $\mathbf{x}$ for the decision variable, while in the discussion on Gauss–Newton we used $\boldsymbol{\theta}$.

Given a set of conjugate vectors $\{\mathbf{p}_j\}_{j=1}^{n}$, we approximate the solution $\mathbf{x}$ as

$$\mathbf{x} = \sum_{j=1}^{n} \alpha_j \mathbf{p}_j. \tag{4.86}$$

To approximate the k-th coefficient α_k, we use $\mathbf{Ax} = \mathbf{b}$

$$\mathbf{Ax} = \sum_{j=1}^{n} \alpha_j \mathbf{A}\mathbf{p}_j = \mathbf{b} \tag{4.87}$$

and take the inner product of both sides with $\mathbf{p}_k$:

$$\mathbf{p}_k^{\mathrm{T}}\mathbf{b} = \sum_{j=1}^{n} \alpha_j \mathbf{p}_k^{\mathrm{T}}\mathbf{A}\mathbf{p}_j. \tag{4.88}$$

Since $\mathbf{p}$ are conjugate vectors, the only nonzero term in the sum is $\alpha_k \mathbf{p}_k^{\mathrm{T}}\mathbf{A}\mathbf{p}_k$:

$$\mathbf{p}_k^{\mathrm{T}}\mathbf{b} = \alpha_k \mathbf{p}_k^{\mathrm{T}}\mathbf{A}\mathbf{p}_k \tag{4.89}$$

so that

$$\alpha_k = \frac{\mathbf{p}_k^{\mathrm{T}}\mathbf{b}}{\mathbf{p}_k^{\mathrm{T}}\mathbf{A}\mathbf{p}_k}. \tag{4.90}$$

Thus, the conjugate gradient method iteratively builds up this basis $\mathbf{p}_k$ and the approximation to $\mathbf{x}$, which is reconstructed by the coefficients α_k in the basis $\mathbf{p}_k$.

Gradient Descent Formulation. The CG method can be viewed as an enhancement of the classical gradient descent algorithm for minimizing the quadratic objective function:

$$f(\mathbf{x}) = \frac{1}{2}\mathbf{x}^{\mathrm{T}}\mathbf{A}\mathbf{x} - \mathbf{b}^{\mathrm{T}}\mathbf{x}. \tag{4.91}$$

The gradient of $f(\mathbf{x})$ is:

$$\nabla f(\mathbf{x}) = \mathbf{Ax} - \mathbf{b}. \tag{4.92}$$

Since we are trying to set $\nabla f(\mathbf{x}) = \mathbf{0}$, we use the negative gradient as our first vector $\mathbf{p}_1 = \mathbf{b} - \mathbf{Ax}$. Thus, all future vectors will be conjugate to the gradient direction, hence the name.

Iterative Construction of Conjugate Vectors. We first introduce the simplest way to understand the construction of the conjugate vectors $\mathbf{p}$, but it turns out that there is a more efficient algorithm. We will present the intuitive construction first here, followed by the streamlined algorithm.

The residual at each step is $\mathbf{r}_k = \mathbf{b} - \mathbf{A}\mathbf{x}_k$, so the initial conjugate vector $\mathbf{p}_1$ is also the first residual. Classic gradient descent would take the next step in this residual direction, but the conjugate gradient method requires that the next step is in a direction that is conjugate to the previous $\mathbf{p}$ vectors. Thus, the direction of the next step is obtained by taking the residual $\mathbf{r}_k$

and removing all of its components in each of the previous $\mathbf{p}_j$ directions. This amounts to a Gram–Schmidt orthonormalization in the weighted inner product space determined by $\mathbf{A}$:

$$\mathbf{p}_k = \mathbf{r}_k - \sum_{j=1}^{k-1} \frac{\mathbf{p}_j^{\mathrm{T}} \mathbf{A} \mathbf{r}_k}{\mathbf{p}_j^{\mathrm{T}} \mathbf{A} \mathbf{p}_j} \mathbf{p}_j. \tag{4.93}$$

Each term in the sum can be understood in the following way: the numerator is the projection of $\mathbf{r}_k$ in the $\mathbf{p}_j$ direction, the denominator is a normalization factor, and the $\mathbf{p}_j$ to the right of the fraction indicates that this vector is being subtracted from $\mathbf{r}_k$ in exactly the amount that is contained in $\mathbf{r}_k$. Thus, this sum removes all components from all previous $\mathbf{p}_j$ conjugate vectors, so that $\mathbf{p}_k$ is guaranteed to be conjugate to all prior vectors.

Once the new vector $\mathbf{p}_k$ is constructed, we may update $\mathbf{x}$ as

$$\mathbf{x}_{k+1} = \mathbf{x}_k + \alpha_k \mathbf{p}_k = \mathbf{x}_k + \frac{\mathbf{p}_k^{\mathrm{T}} \mathbf{b}}{\mathbf{p}_k^{\mathrm{T}} \mathbf{A} \mathbf{p}_k} \mathbf{p}_k \tag{4.94}$$

where α_k was previously derived in Eq. (4.90).

In actuality, there are ways of compressing and streamlining several of these steps, which we will explore below.

Krylov Subspaces and Conjugate Directions. At each iteration, CG minimizes $f(\mathbf{x})$ within the k-dimensional Krylov subspace:

$$\mathcal{K}_k = \operatorname{span}\{\mathbf{r}_0, \mathbf{A}\mathbf{r}_0, \mathbf{A}^2\mathbf{r}_0, \ldots, \mathbf{A}^{k-1}\mathbf{r}_0\}, \tag{4.95}$$

where $\mathbf{r}_0 = \mathbf{b} - \mathbf{A}\mathbf{x}_0$ is the initial residual. Unlike gradient descent, the conjugate gradient method constructs a sequence of search directions $\mathbf{p}_k$ that are $\mathbf{A}$-*conjugate*, as defined before. Conjugate directions ensure that the minimization along each direction does not interfere with progress along previously chosen directions, leading to faster convergence.

The CG method iteratively updates $\mathbf{x}_k$, the residual $\mathbf{r}_k$, and the search direction $\mathbf{p}_k$ as follows.

We begin with an initial guess $\mathbf{x}_0$:

$$\mathbf{x}_0, \quad \mathbf{r}_0 = \mathbf{b} - \mathbf{A}\mathbf{x}_0, \quad \mathbf{p}_0 = \mathbf{r}_0. \tag{4.96}$$

Then, for each iteration k we compute:

$$\alpha_k = \frac{\mathbf{r}_k^{\mathrm{T}} \mathbf{r}_k}{\mathbf{p}_k^{\mathrm{T}} \mathbf{A} \mathbf{p}_k}, \tag{4.97a}$$

$$\mathbf{x}_{k+1} = \mathbf{x}_k + \alpha_k \mathbf{p}_k, \tag{4.97b}$$

$$\mathbf{r}_{k+1} = \mathbf{r}_k - \alpha_k \mathbf{A} \mathbf{p}_k, \tag{4.97c}$$

and update the next conjugate direction:

$$\beta_k = \frac{\mathbf{r}_{k+1}^{\mathrm{T}} \mathbf{r}_{k+1}}{\mathbf{r}_k^{\mathrm{T}} \mathbf{r}_k}, \tag{4.98a}$$

$$\mathbf{p}_{k+1} = \mathbf{r}_{k+1} + \beta_k \mathbf{p}_k. \tag{4.98b}$$

Notice here that we modified our expression for α_k to only include the current residual $\mathbf{r}_k$ and the current conjugate vector $\mathbf{p}_k$. It will also be helpful for the reader to verify that the new expression for the residual is equivalent to $\mathbf{r}_k = \mathbf{A}\mathbf{x}_k - \mathbf{b}$.

The iterations terminate when $\|\mathbf{r}_k\|$ is below a predefined tolerance or after n iterations. The CG method converges in at most n iterations because the Krylov subspace expands to include the entire space $\mathbb{R}^n$. However, practical performance depends on the condition number $\kappa(\mathbf{A}) = \lambda_{\max}/\lambda_{\min}$, where $\lambda_{\max}$ and $\lambda_{\min}$ are the largest and smallest eigenvalues of $\mathbf{A}$. The convergence rate is approximately:

$$\|\mathbf{x}_k - \mathbf{x}^*\|_{\mathbf{A}} \leq C\left(\frac{\sqrt{\kappa(\mathbf{A})} - 1}{\sqrt{\kappa(\mathbf{A})} + 1}\right)^k, \tag{4.99}$$

where $\mathbf{x}^*$ is the true solution and C is a constant depending on the initial error. Thus, the performance of CG deteriorates for poorly conditioned systems (large $\kappa(\mathbf{A})$), which slows convergence. Preconditioning is commonly used to improve the condition number of $\mathbf{A}$, transforming the system into:

$$\mathbf{M}^{-1}\mathbf{A}\mathbf{x} = \mathbf{M}^{-1}\mathbf{b}, \tag{4.100}$$

where $\mathbf{M}$ is a preconditioner chosen to approximate $\mathbf{A}$ while being easy to invert.

CG is also memory efficient, requiring storage only for a few vectors ($\mathbf{x}_k, \mathbf{r}_k, \mathbf{p}_k$), making it ideal for large-scale problems. It avoids explicit matrix factorization and is particularly suited for sparse systems.

Panel 4.4: Generalized Minimal Residual Method (GMRES)

The generalized minimal residual method (GMRES) [432] is a closely related Krylov subspace algorithm for solving linear systems $\mathbf{A}\mathbf{x} = \mathbf{b}$ when $\mathbf{A}$ is not symmetric positive definite. Whereas CG minimizes the energy norm $\|\mathbf{x} - \mathbf{x}^*\|_{\mathbf{A}}$, GMRES directly minimizes the two-norm of the residual $\|\mathbf{r}\|_2$ over the Krylov subspace. It does so using the Arnoldi iteration to build an orthonormal basis, which leads to increasing storage and orthogonalization costs as iterations grow.

In practice, GMRES is often used in a restarted form, GMRES(m), where the basis is truncated after m steps to control cost. Like other Krylov methods, GMRES is highly dependent on effective preconditioning, which can dramatically accelerate convergence. Despite higher per-iteration cost than CG, GMRES is robust and broadly applicable, making it a workhorse iterative solver in scientific computing, particularly for nonsymmetric and indefinite systems arising in fluid mechanics, electromagnetics, and optimization.

Suggested Reading

Texts

(1) **Data-Driven Science and Engineering: Machine Learning, Dynamical Systems, and Control**, by S. L. Brunton and J. N. Kutz, 2022 [77].
(2) **An Introduction to Statistical Learning**, by G. James, D. Witten, T. Hastie, and R. Tibshirani, 2013 [247].
(3) **Approximation of Large-Scale Dynamical Systema**, by A.C. Antoulas, 2005 [10].
(4) **Numerical Linear Algebra**, by L.N. Trefethen, 2022 [500].

Papers and reviews

(1) **On the stability of inverse problems**, by A. N. Tikhonov *Dokl. Akad. Nauk SSSR*, 1943 [496].
(2) **Tikhonov regularization and total least squares**, by G. H. Golub, P. C. Hansen and D. P. O'Leary, *SIAM Journal on Matrix Analysis and Applications*, 1999 [200].
(3) **Regression shrinkage and selection via the Lasso**, by R. Tibshirani, *Journal of the Royal Statistical Society B*, 1996 [495].
(4) **Regularization and variable selection via the elastic net**, by H. Zou and T. Hastie, *Journal of the Royal Statistical Society: Series B* , 2005 [564].
(5) **Compressed sensing**, by D. L. Donoho, *IEEE Transactions on Information Theory*, 2006 [146].
(6) **Compressive sensing**, by R. G. Baraniuk, *IEEE Signal Processing Magazine*, 2007 [25].
(7) **Stable signal recovery from incomplete and inaccurate measurements**, by E. J. Candès, J. Romberg, and T. Tao, *Communications in Pure and Applied Mathematics*, 2006 [93].

Homework

Exercise 4.1 Recall that the minimum norm solution may be formulated in the KKT framework with the following system of equations:

$$\begin{bmatrix} \mathbf{I} & \mathbf{A}^{\mathrm{T}} \\ \mathbf{A} & \mathbf{0} \end{bmatrix} \begin{bmatrix} \mathbf{x} \\ \boldsymbol{\lambda} \end{bmatrix} = \begin{bmatrix} \mathbf{0} \\ \mathbf{b} \end{bmatrix}. \tag{4.101}$$

Show that the solution of this system is $\mathbf{x}^* = \mathbf{A}^{\mathrm{T}} \left(\mathbf{A}\mathbf{A}^{\mathrm{T}}\right)^{-1} \mathbf{b}$.

Exercise 4.2 Show that $\nabla(\mathbf{A}\mathbf{x}) = \mathbf{A}$ and $\nabla(\mathbf{x}^{\mathrm{T}}\mathbf{A}) = \mathbf{A}$ even for non-square matrices $\mathbf{A}$.

Exercise 4.3 For each of the following matrices, indicate if it is positive definite, semi-definite, indefinite, etc.

1. $\mathbf{Q} = \begin{bmatrix} 1 & 2 \\ -2 & 1 \end{bmatrix}$
2. $\mathbf{Q} = \begin{bmatrix} 1 & 2 \\ -2 & -1 \end{bmatrix}$
3. $\mathbf{Q} = \begin{bmatrix} 2 & 0 \\ 2 & 2 \end{bmatrix}$
4. $\mathbf{Q} = \begin{bmatrix} 2 & -1 \\ -1 & 2 \end{bmatrix}$
5. $\mathbf{Q} = \begin{bmatrix} 1 & 2 & 2 \\ 2 & 2 & 2 \\ 2 & 2 & 1 \end{bmatrix}.$

Exercise 4.4 For each of the matrices above, compute the eigenvalues, the trace, and the determinant.

Exercise 4.5 Derive the gradient result in Eq. (4.7) using the result $\nabla\left(f(\mathbf{A}\mathbf{x})\right) = \mathbf{A}^{\mathrm{T}}\nabla f(\mathbf{A}\mathbf{x})$.

Exercise 4.6 Generate examples of matrices $\mathbf{A} \in \mathbb{R}^{3\times 2}$ that are overdetermined and underdetermined. Repeat for matrices $\mathbf{A} \in \mathbb{R}^{2\times 3}$.

Exercise 4.7 Consider the following matrix

$$\mathbf{Q} = \begin{bmatrix} a & b \\ c & d \end{bmatrix}.$$

What are conditions on a, b, c, d for this matrix to be positive definite? Positive semi-definite?

Exercise 4.8 Show that the trace of a 2×2 matrix is the sum of its eigenvalues. Show that the determinant of this matrix is the product of its eigenvalues.

Exercise 4.9 Show that the trace of a general $n \times n$ matrix is the sum of its eigenvalues.

Exercise 4.10 Show that the determinant of a general $n \times n$ matrix is the product of its eigenvalues.

Exercise 4.11 Show that a matrix $\mathbf{Q}$ is positive semidefinite if $\mathbf{Q} = \mathbf{A}^{\mathrm{T}}\mathbf{A}$. Show that this is also a necessary condition for $\mathbf{Q}$ to be positive semidefinite.

Exercise 4.12 Give an example of a real-valued 2×2 matrix that is positive definite and not symmetric. How does this relate to the requirement that $\mathbf{Q} = \mathbf{A}^{\mathrm{T}}\mathbf{A}$ above?

Exercise 4.13 Show that if $\mathbf{Q} = \mathbf{A}^{\mathrm{T}}\mathbf{A}$ for a non-singular $\mathbf{A}$, then $\mathbf{Q}$ is positive definite.

Exercise 4.14 Consider a symmetric matrix $\mathbf{A}$ with maximum eigenvalue λ. Show that $\mathbf{Q} = \alpha\mathbf{I} - \mathbf{A}$ is positive semidefinite for $\alpha \geq \lambda$.

Exercise 4.15 A matrix is called diagonally dominant if the magnitude of each diagonal entry is larger than the sum of magnitudes of all other entries in the same row:

$$|a_{jj}| > \sum_{i \neq j} |a_{ji}|.$$

Show that a diagonally dominant matrix with positive diagonal entries is positive semi-definite.

Exercise 4.16 The matrices $\mathbf{X}^{\mathrm{T}}\mathbf{X}$ and $\mathbf{X}\mathbf{X}^{\mathrm{T}}$ are *correlation* matrices for the columns and rows of $\mathbf{X}$, respectively. Specifically, the ij-th entry of $\mathbf{X}^{\mathrm{T}}\mathbf{X}$ is the inner product of the i-th column with the j-th column of $\mathbf{X}$. Expand $\mathbf{X}^{\mathrm{T}}\mathbf{X}$ and $\mathbf{X}\mathbf{X}^{\mathrm{T}}$ in terms of the SVD, $\mathbf{X} = \mathbf{U}\boldsymbol{\Sigma}\mathbf{V}^{\mathrm{T}}$, and simplify.

Exercise 4.17 Show that the matrices $\mathbf{X}^{\mathrm{T}}\mathbf{X}$ and $\mathbf{X}\mathbf{X}^{\mathrm{T}}$ have the same nonzero eigenvalues.

Exercise 4.18 Show that $\mathbf{U}$ are eigenvectors of the symmetric matrix $\mathbf{X}\mathbf{X}^{\mathrm{T}}$. Similarly show that $\mathbf{V}$ are eigenvectors of $\mathbf{X}^{\mathrm{T}}\mathbf{X}$. (Hint: remember that $\mathbf{U}$ and $\mathbf{V}$ are unitary matrices, so $\mathbf{U}^{\mathrm{T}}\mathbf{U} = \mathbf{I}$ and $\mathbf{V}^{\mathrm{T}}\mathbf{V} = \mathbf{I}$.)

Exercise 4.19 Show that if $\mathbf{X}$ is self-adjoint (i.e., $\mathbf{X} = \mathbf{X}^{\mathrm{T}}$) then the singular values of $\mathbf{X}$ are equal to the absolute value of the eigenvalues of $\mathbf{X}$.

Exercise 4.20 Load a high resolution image, like the image of Mordecai the snow dog from Fig. 4.18. This image has 2000×1500 pixels.

1. Compute the SVD of this image and plot the singular values on a logarithmic plot.
2. For various values of r from 1 to 100, compute the rank-r SVD approximation and compute the the error between the original matrix image $\mathbf{X}$ and the reconstruction $\hat{\mathbf{X}}$; use the Frobenius norm. Plot this error versus the rank r.
3. Now compute the error of the rank-r approximation using Eq. (4.17) and compare with the actual error.
4. What rank r is needed to reconstruct the image to 90% accuracy? How about for 99%?

Exercise 4.21 This exercise will test the concept of condition number. We will test the accuracy of solving $\mathbf{A}\mathbf{x} = \mathbf{b}$ when noise is added to $\mathbf{b}$ for matrices $\mathbf{A}$ with different condition numbers.

Figure 4.18 High resolution image to demonstrate SVD compression.

1. To build the two matrices, generate a random $\mathbf{U} \in \mathbb{R}^{100\times100}$ and $\mathbf{V} \in \mathbb{R}^{100\times100}$ and then create two $\boldsymbol{\Sigma}$ matrices: the first $\boldsymbol{\Sigma}$ will have singular values spaced logarithmically from 100 to 1, and the second $\boldsymbol{\Sigma}$ will have singular values space logarithmically from 100 to 1e-6. Use these matrices to create two $\mathbf{A}$ matrices, one with a condition number of 100 and the other with a condition number of 100 million. Now create a random $\mathbf{b}$ vector, solve for $\mathbf{x}$ using the two A matrices and compare the results. Add a small $\boldsymbol{\epsilon}$ to $\mathbf{b}$, with norm 1e-6 smaller than the norm of $\mathbf{b}$. Now solve for $\mathbf{x}$ using this new $\mathbf{b} + \boldsymbol{\epsilon}$ and compare the results.
2. Repeat the experiment above with many different noise vectors $\boldsymbol{\epsilon}$ and compute the distribution of the error; plot this error as a histogram and explain the shape.
3. Repeat the above experiment comparing two $\mathbf{A}$ matrices with different singular value distributions: the first $\boldsymbol{\Sigma}$ will have values spaced linearly from 100 to 1 and the second $\boldsymbol{\Sigma}$ will have value spaced logarithmically from 100 to 1. Does anything change? Please explain why yes or no.
4. Repeat the above experiment, but now with a $\mathbf{A}$ matrix that has size 100×10. Explain any changes.

Exercise 4.22 There are very specific conditions that must be met for the ℓ_1-minimization in (4.32) to converge with high probability to the sparsest solution in (4.31) [90, 92, 25]:

1. The measurement matrix $\mathbf{C}$ must be *incoherent* with respect to the sparsifying basis $\boldsymbol{\Psi}$, meaning that the rows of $\mathbf{C}$ are not correlated with the columns of $\boldsymbol{\Psi}$,

2. The number of measurements p must be sufficiently large, on the order of

$$p \approx \mathcal{O}(K \log(n/K)) \approx k_1 K \log(n/K). \tag{4.102}$$

The constant multiplier k_1 depends on how incoherent $\mathbf{C}$ and $\boldsymbol{\Psi}$ are.

Roughly speaking, these two conditions guarantee that the matrix $\mathbf{C}\boldsymbol{\Psi}$ acts as a unitary transformation on K sparse vectors $\mathbf{s}$, preserving relative distances between vectors and enabling almost certain signal reconstruction with ℓ_1 convex minimization. This is often referred to as the restricted isometry property.

1. Construct a measurement matrix $\mathbf{C}$ with p rows and n columns. To start, use $p = 100$ and $n = 1000$. The entries of $\mathbf{C}$ should be independent, identically distributed Gaussian elements with mean 0 and standard deviation 1.
   Similarly construct the basis $\boldsymbol{\Psi}$ given by the inverse discrete cosine transform.
2. Compute $\mathbf{C}\boldsymbol{\Psi} = \boldsymbol{\Theta}$. What is its condition number?
3. Compute random pairs of vectors $\mathbf{x}$ and $\mathbf{y}$ with given amounts of sparsity in the basis $\boldsymbol{\Psi}$ (e.g., $K = 1, 2, \cdots 100$).
4. Compute the normalized inner product of these vectors in the high-dimensional ambient space, $\langle \mathbf{x}, \mathbf{y} \rangle$, and compare with the normalized inner product in compressed space. Specifically $\langle \boldsymbol{\Theta}\mathbf{x}, \boldsymbol{\Theta}\mathbf{y} \rangle$.
5. Plot the error between these inner products as a function of the sparsity K. To get good statistics, with error bounds, you will likely need to test several pairs of $\mathbf{x}$ and $\mathbf{y}$ for each sparsity K (maybe 100 pairs).
6. Repeat the above error plots with more or less rows of $\mathbf{C}$ (i.e., vary p from 10 to 1000). Plot the error in a two-dimensional `pcolor` plot of K vs p. What trends do you notice?
7. Now, for the vectors $\mathbf{x}$ generated above, solve the compressed sensing problem for $\mathbf{s}$, the sparse vector, and compute the ℓ_0 error between the estimated $\mathbf{s}$ and the true $\mathbf{s}$. Plot the average error in a `pcolor`. How does this compare to the above?

Exercise 4.23 Repeat the above problem with different measurement matrices $\mathbf{C}$.

- Try $\mathbf{C}$ with each entry uniformly distributed between 0 and 1.
- Try $\mathbf{C}$ with each entry a Bernoulli random variable with equal probability of being a 0 or a 1.
- Try $\mathbf{C}$ with each entry a Bernoulli random variable with 95% probability of being a 0.
- Try $\mathbf{C}$ with each row a single nonzero entry of 1, which corresponds to measuring a single spatial location in $\mathbf{x}$.

How do the trends on the error versus sparsity and number of measurements p change?

Exercise 4.24 Construct a $\mathbf{C}$ and $\boldsymbol{\Psi}$ matrix from above with $p = 100$ and $n = 1000$ and $\mathbf{C}$ a Gaussian random matrix. Construct a sparse vector $\mathbf{s}$ with $K = 10$ and $\mathbf{x} = \boldsymbol{\Psi}\mathbf{s}$. Solve this sparse, underdetermined system using the three methods discussed in this chapter:

- Use the standard compressed sensing formulation from Eq. (4.32);
- Use the constrained ℓ_1 minimization with constraints $||\mathbf{C}\boldsymbol{\Psi}\mathbf{s} - \mathbf{y}|| < \epsilon$ from Panel 4.3;
- Use the Dantzig selector from Panel 4.3, which is more akin to the classic LASSO.

Add noise of varying magnitude to the measurement in $\mathbf{y} = \mathbf{C\Theta s}$ and repeat the methods above. Try different ϵ and λ and discuss the various trends and findings.

Exercise 4.25 To see the sparsity promoting effects of the ℓ_1 norm, we consider a generic underdetermined system of equations. First construct a random matrix system of equations $\mathbf{y} = \mathbf{\Theta s}$ with $p = 200$ rows (measurements) and $n = 1000$ columns (unknowns). In general, there are infinitely many solutions $\mathbf{s}$ that are consistent with these equations, unless we are very unfortunate and the row equations are linearly dependent while the measurements are inconsistent in these rows. In fact, this is an excellent example of the probabilistic thinking used more generally in compressed sensing: if we generate a linear system of equations at random, that has sufficiently many more unknowns than knowns, then the resulting equations will have infinitely many solutions with *high probability*.

Solve this system of equations using both the minimum ℓ_1 norm and minimum ℓ_2 norm solutions. The minimum ℓ_2 norm solution is obtained using the pseudo-inverse, related to the SVD. The minimum ℓ_1 norm solution is obtained via the **cvx** package.

Plot the histogram of the coefficients of both solutions $\mathbf{x}$ and comment on the results.

Exercise 4.26 Consider the following one-dimensional linear regression problem, where we are simply trying to find the constant x that best scales a vector $\mathbf{a}$ into a vector $\mathbf{b}$:

$$\begin{bmatrix} | \\ \mathbf{b} \\ | \end{bmatrix} = \begin{bmatrix} | \\ \mathbf{a} \\ | \end{bmatrix} x = \mathbf{U\Sigma V}^{\mathrm{T}} x.$$

The solution is given by

$$x = \mathbf{V\Sigma}^{-1}\mathbf{U}^{\mathrm{T}}\mathbf{b}.$$

Please write down $\mathbf{U}$, $\mathbf{\Sigma}$, and $\mathbf{V}$ in terms of $\mathbf{a}$, $\mathbf{b}$, and x.

Write down the solution x in terms of the vectors $\mathbf{a}$ and $\mathbf{b}$.

Exercise 4.27 As a simple example, construct an artificial data set consisting of 100 observations of an outcome, arranged in a vector $\mathbf{b} \in \mathbb{R}^{100}$. Make it so that each outcome in $\mathbf{b}$ is given by a combination of exactly 2 out of 10 candidate predictors, whose observations are arranged in the rows of a matrix $\mathbf{A} \in \mathbb{R}^{100\times 10}$.

Add a small amount of noise (5%) to the vector $\mathbf{b}$ and solve for $\mathbf{x}$ using ordinary least squares, Tikhonov regularized least squares, and LASSO regression. Please comment on the results.

Compare the LASSO prediction before and after de-biasing, which involves taking the sparse pattern returned by LASSO and performing a final least squares regression just onto these terms.

Exercise 4.28 In this exercise, we will reproduce the Tikhonov regularization schematic in Figs. 4.13 and Fig. 4.14.

First, generate a matrix **A** with condition number κ; start with $\kappa = 100$. To do this, you can use the SVD with

$$\mathbf{U} = \begin{bmatrix} 1 & 1 \\ -1 & 1 \end{bmatrix} \quad \boldsymbol{\Sigma} = \begin{bmatrix} \kappa & 0 \\ 0 & 1 \end{bmatrix} \quad \mathbf{V} = \begin{bmatrix} 1 & 1 \\ -1 & 1 \end{bmatrix}.$$

Generate an ensemble of **b** vectors centered around $\mathbf{b} = \begin{bmatrix} 1 \\ 1 \end{bmatrix}$ with additive Gaussian noise with zero mean and standard deviation of 0.1.

Solve for **x** using least squares and plot the ellipsoid of solutions.

Now, solve for **x** using Tikhonov regularized least squares and plot the ellipsoid of solutions. Try different regularization strengths α.

Estimate the bias and the decrease in variance with various α.

Exercise 4.29 Repeat the exercise above, but with the LASSO regression.

5 Nonsmooth and Global Optimization

In this chapter, we explore several methods for nonsmooth (i.e., gradient-free) and nonconvex optimization. This chapter will have a bit less structure than the previous chapters, because the topic of nonconvex optimization is itself less structured. This is a vast field, and we will only cover some of the major highlights.

First, we will develop the proximal gradient method, which is an incredibly useful extension of gradient descent for nonsmooth problems where the gradient may not exist. Many optimization problems have a smooth objective, such as the least squares error function, and a nonsmooth regularizer, such as the ℓ_1 regularized LASSO regression from Chapter 4. The proximal gradient approach is a powerful and intuitive algorithmic approach to handle these situations. The proximal operator, used in the proximal gradient method, will also be useful in the next chapter, when we introduce the alternating direction method of multipliers (ADMM). Proximal algorithms are essential when, in our objective function, we have nonsmooth terms such as $\|\cdot\|_1$, the nuclear norm $\|\cdot\|_*$, the total variation used in image processing in Chapter 8, and other nonsmooth regularizers used in neural network training (see Chapter 10).

Next, we will explore two workhorse algorithms that assume very little structure in the optimization problem: the Nelder–Mead downhill simplex method and simulated annealing. The downhill simplex method, not to be confused with Dantzig's simplex method from linear programming, uses a simplex (i.e., a polytope) to sample the objective function landscape and then take steps that deform the simplex towards better objectives. Simulated annealing replicates the natural process of annealing, or cooling, when creating strong metal alloys, where a specific cooling schedule is used to promote strong crystal structures in the material. In simulated annealing, the optimization algorithm begins "hot", exploring the objective function landscape, and gradually "cools" to begin exploiting favorable local minima.

We then introduce evolutionary algorithms, including genetic algorithms and genetic programming, the last being particularly interesting in the context of modern explainable and interpretable AI.

Finally, we will cover distributed and swarm optimization approaches, such as particle swarm optimization and ant colony optimization. These approaches leverage distributed optimization to mimic the collective behavior of natural biological systems, such as fish schooling, birds flocking, and insects swarming.

5.1 Proximal Gradient Algorithms

Standard gradient descent from Chapter 2 no longer applies to optimization problem with a nonsmooth regularizer, such as the ℓ_1 norm in LASSO or total variation penalties, due to the non-differentiability of the objective. The *proximal gradient method* extends gradient descent to handle such composite optimization problems of the form

$$\min_{\mathbf{x}} \quad f(\mathbf{x}) + g(\mathbf{x}), \tag{5.1}$$

where $f(\mathbf{x})$ is smooth, and $g(\mathbf{x})$ is possibly nonsmooth. Typically, we assume additional structure on f and g, such as f being convex with Lipschitz-continuous gradient, and g being convex. However, these are not strictly necessary for much of the discussion and practical implementation. A Lipschitz-continuous gradient of f means that we can bound the function from above by a simple quadratic, which will be used in the derivation below.

Panel 5.1: The Convex Hegemony

The centrality of convexity in optimization emerged in the mid-20th century. After Dantzig's simplex method [123] made linear programming practical, researchers realized that convexity was the underlying property that guaranteed tractability. Local minima of convex problems are also global, and convex functions admit powerful duality theory. Rockafellar's 1970 book *Convex Analysis* [419] codified this view, giving mathematicians and engineers a common language for convex sets, cones, and subgradients. In the 1970s and 1980s, convex quadratic programming and conic formulations became essential for operations research, control, and engineering design. Interior point methods, in particular, reinforced the idea that convex optimization was not only elegant but also scalable to real-world problems [270, 367, 542]. By the 1990s, the *convex hegemony* was firmly established: researchers habitually reshaped models to fit convex templates because these were the only problems they could solve reliably at scale.

In the past two decades, this dominance has softened. The incredible growth of computing power and available data, together with algorithmic advances in splitting and proximal methods, showed that many nonconvex problems could be attacked directly. Forward–backward splitting and proximal gradient methods [110, 109], ADMM [65], and more general convergence frameworks [16, 59] demonstrated practical ways to handle structured nonconvex problems. Meanwhile, the success of deep learning [204] made nonconvexity mainstream: billions of parameters trained by first-order methods challenged the old intuition that nonconvex meant intractable. Today convex optimization is still foundational [66], guiding algorithms and serving as a benchmark. However, the field now is less dogmatic, with nonconvex approaches now seen as legitimate, effective tools rather than mathematical dead ends.

As a concrete example to motivate this discussion, we let $f(\mathbf{x})$ be the least-squares loss function and $g(\mathbf{x})$ be the sparsity-promoting ℓ_1 regularizer, discussed in Chapter 4:

$$\min_{\mathbf{x}} \quad \|\mathbf{A}\mathbf{x} - \mathbf{b}\|_2^2 + \lambda\|\mathbf{x}\|_1. \tag{5.2}$$

We will proceed for now with general f and g functions, but if the discussion becomes too abstract, just go back to this simple example, which we will revisit and solve in a moment.

The idea behind the proximal gradient method is to replace a gradient step

$$\mathbf{x}_{k+1} = \mathbf{x}_k - \gamma\nabla f(\mathbf{x}_k) \tag{5.3}$$

Panel 5.2: Multi-Objective Optimization and the Pareto Front

We often have competing objectives, such as the goodness of fit $\|\mathbf{A}\mathbf{x} - \mathbf{b}\|_2^2$ and the sparsity of the solution $\|\mathbf{x}\|_0$, which form the objective function

$$f(\mathbf{x}) = \|\mathbf{A}\mathbf{x} - \mathbf{b}\|_2^2 + \lambda\|\mathbf{x}\|_0.$$

For a fixed λ, this is a scalar objective function, but often we are interested in the performance tradeoffs for different λ weighting these multiple objectives. In this case, we form the Pareto family of optimal solutions by sweeping through λ and solving for the various optima. The resulting *Pareto front* denotes the family of solutions so that any decrease in one objective will result in a corresponding increase in another. This can be very useful in design optimization, where the particular weighting of different terms is not decided, and instead it is more useful to understand the various design tradeoffs to look for a good balance of objectives.

Jumps in the Pareto curve can help identify good values for λ. For example, in model fitting, a jump in the Pareto curve often corresponds to the point where overfitting begins. If test data is available, we validate the error on this test data, but if only training data is available, then jumps in the Pareto curve can be useful.

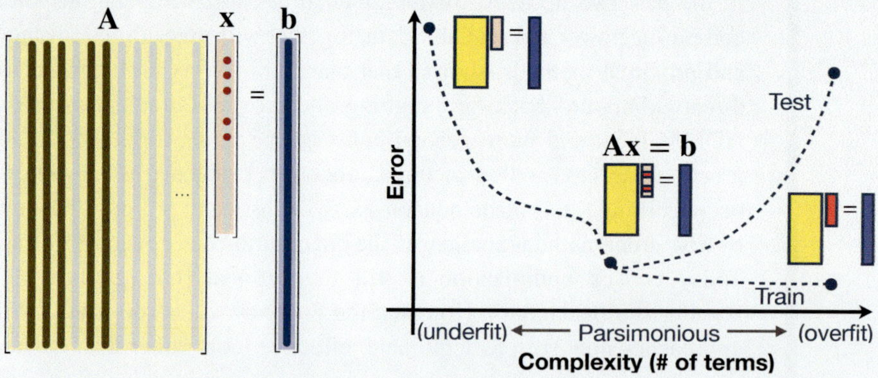

Figure 5.1: Linear regression problem (left) and Pareto curve (right).

with a *proximal* gradient step

$$\mathbf{x}_{k+1} = \text{prox}_{\gamma,g}\left(\mathbf{x}_k - \gamma \nabla f(\mathbf{x}_k)\right) \tag{5.4}$$

that takes into account the nonsmooth term g. This notation $\text{prox}_{\gamma,g}(\mathbf{z})$ indicates that the proximal operator explicitly depends on the step size γ of the unregularized problem (i.e., min f without g) and the nonsmooth function g. Its argument $\mathbf{z}$ in Eq. (5.4) is the unregularized gradient step $\mathbf{z} = \mathbf{x}_k - \gamma \nabla f(\mathbf{x}_k)$. This algorithm alternates between a gradient step on the smooth part f and a proximal step on the nonsmooth regularizer g.

The proximal operator of g with step size γ is defined as

$$\text{prox}_{\gamma,g}(\mathbf{z}) := \arg\min_{\mathbf{x}} \left(\frac{1}{2}\|\mathbf{x} - \mathbf{z}\|^2 + \gamma g(\mathbf{x}) \right). \tag{5.5}$$

In the case of $g(\mathbf{x}) = \lambda\|\mathbf{x}\|_1$, the proximal step corresponds to *soft thresholding*, and the method reduces to the classical iterative shrinkage-thresholding algorithm (ISTA). We will derive this next. Accelerated variants, such as FISTA [32], improve convergence from $O(1/k)$ to $O(1/k^2)$ for convex problems, by combining ISTA with Nesterov's accelerated gradient method from Section 2.2. The proximal gradient is particularly useful for sparse or low-rank models, such as ℓ_1-regularized least squares (LASSO) and nuclear-norm minimization for low-rank matrix recovery. These will be explored throughout this section.

We will derive the proximal operator for several regularizers $g(\mathbf{x})$ in a moment. For now, the main takeaway is that the proximal operator has been computed for dozens of commonly used regularizers [109, 105] $g(\mathbf{x})$, so we substitute it directly in for the gradient step in gradient descent from Chapter 2. However, for some problems, computing the proximal gradient may be quite challenging, and is not guaranteed to simplify the optimization problem. Fortunately, for most cases we care about, there are simple and efficient implementations of the prox operator. The proximal operator will also be useful in several other algorithms that are not based on gradient descent, for example the SR3 algorithm later in this section, and the ADMM algorithm in Section 6.5.

Deriving the Proximal Operator

We now derive a mathematical expression for the proximal operator. To motivate this, we take a small digression to discuss properties of the unregularized problem. We typically assume that f is convex with Lipschitz-continuous gradient, so that we may write it as

$$f(\mathbf{x}) \leq f(\mathbf{x}_0) + (x - \mathbf{x}_0)^{\mathrm{T}} \nabla f(\mathbf{x}_0) + \tfrac{\beta}{2}\|\mathbf{x} - \mathbf{x}_0\|_2^2 \tag{5.6a}$$

$$= f(\mathbf{x}_0) + \frac{\beta}{2}\left\|\mathbf{x} - \left(\mathbf{x}_0 - \tfrac{1}{\beta}\nabla f(\mathbf{x}_0)\right)\right\|_2^2 - \tfrac{1}{2\beta}\|\nabla f(\mathbf{x}_0)\|_2^2 \equiv m_0(\mathbf{x}). \tag{5.6b}$$

The second equation is obtained from the first by completing the square, and this is known as the Moreau envelope, denoted $m_0(\mathbf{x})$. Since only the middle term depends on $\mathbf{x}$, minimizing $m_0(\mathbf{x})$ with respect to $\mathbf{x}$ simply amounts to setting $\mathbf{x} = \mathbf{x}_0 - \frac{1}{\beta}\nabla f(\mathbf{x}_0)$, which is exactly gradient descent with a step size $\gamma = \frac{1}{\beta}$ obtained via Newton's method on the quadratic Moreau envelope with curvature $\frac{\beta}{2}$. More formally, we may say that

$$\mathbf{x}_1 = \arg\min_{\mathbf{x}} m_0(\mathbf{x}) \quad \Longrightarrow \quad \mathbf{x}_1 = \mathbf{x}_0 - \tfrac{1}{\beta}\nabla f(\mathbf{x}_0). \tag{5.7}$$

Importantly, this establishes some very useful bounds. First,

$$f(\mathbf{x}_1) \leq m_0(\mathbf{x}_1) \leq m_0(\mathbf{x}_0) = f(\mathbf{x}_0) \tag{5.8}$$

by the definition of m_0. We may also write down

$$f(\mathbf{x}_1) \leq f(\mathbf{x}_0) - \tfrac{1}{2\beta}\|\nabla f(\mathbf{x}_0)\|_2^2 \tag{5.9}$$

which provides a lower bound for the descent. This is equal to 0 if $\mathbf{x}_0$ is stationary.

We are now ready to derive the proximal operator, by adding $g(\mathbf{x})$ to the equation above. We follow the same procedure, adding $g(\mathbf{x})$ to both sides of the equation:

$$f(\mathbf{x}) + g(\mathbf{x}) \leq f(\mathbf{x}_0) + (x - \mathbf{x}_0)^{\mathrm{T}}\nabla f(\mathbf{x}_0) + \tfrac{\beta}{2}\|\mathbf{x} - \mathbf{x}_0\|_2^2 + g(\mathbf{x}) \tag{5.10a}$$

$$= f(\mathbf{x}_0) + \frac{\beta}{2}\left\|\mathbf{x} - \left(\mathbf{x}_0 - \tfrac{1}{\beta}\nabla f(\mathbf{x}_0)\right)\right\|_2^2 - \tfrac{1}{2\beta}\|\nabla f(\mathbf{x}_0)\|_2^2 + g(\mathbf{x}). \tag{5.10b}$$

Now, there are two terms in the last equation that depend on $\mathbf{x}$, and we solve for the $\mathbf{x}$ that minimizes this expression:

$$\mathbf{x}_1 = \arg\min_{\mathbf{x}}\ (m_0(\mathbf{x}) + g(\mathbf{x})) = \arg\min_{\mathbf{x}}\ \left(g(\mathbf{x}) + \tfrac{\beta}{2}\left\|\mathbf{x} - \left(\mathbf{x}_0 - \tfrac{1}{\beta}\nabla f(\mathbf{x}_0)\right)\right\|^2\right) \tag{5.11a}$$

$$\equiv \operatorname{prox}_{\frac{1}{\beta},g}\left(\mathbf{x}_0 - \tfrac{1}{\beta}\nabla f(\mathbf{x}_0)\right) = \operatorname{prox}_{\frac{1}{\beta},g}(\mathbf{z}) \tag{5.11b}$$

where $\mathbf{z} = \mathbf{x}_0 - \frac{1}{\beta}\nabla f(\mathbf{x}_0)$.

Thus, the proximal gradient is a simple gradient step on f to obtain $\mathbf{z}$, followed by a "prox" step on the function $g(\mathbf{x})$. Note that we have not explicitly computed the arg min of $m_0(\mathbf{x}) + g(\mathbf{x})$, and in practice this may be very challenging. But for many common regularizers $g(\mathbf{x})$ there are known prox gradient solutions. In a sense, by writing the optimization problem like this, we abstract away this challenging step, relying on the optimization community to come up with efficient solutions for many common problems[1].

As before, we have a bound on the descent:

$$(f + g)(\mathbf{x}_1) \leq (m_0 + g)(\mathbf{x}_1) \leq (m_0 + g)(\mathbf{x}_0) = (f + g)(\mathbf{x}_0). \tag{5.12}$$

For a least-squares regression problem with $f(\mathbf{x}) = \|\mathbf{Ax} - \mathbf{b}\|_2^2$, the bounding curvature $\frac{\beta}{2}$ is rather pessimistic, and may result in slow descent when β is large. Recall that descent algorithms also have poor convergence for ill-conditioned $\mathbf{A}$, since the number of steps is related to the condition number. This motivates the SR3 and ADMM algorithms, where we split the objective function and solve the least-squares portion with a better conditioned algorithm, such as a pre-conditioned least-squares from Chapter 4.

Notice that we have assumed that $f(\mathbf{x})$ has a particular structure, namely that it has a Lipschitz-continuous gradient, so that the smooth step involves minimizing a quadratic objective function. This is true for the case of a least-squares regression objective, and it also holds for many other objective functions. Even if this is not true, we may still proceed

[1] Sasha Aravkin likes to think of abstracting away the proximal operator as a *side quest* on the path to solving the main quest of minimizing $f(\mathbf{x}) + g(\mathbf{x})$. With the side quest in hand, the method is just as simple as gradient descent in Chapter 2. The proximity operator repository (`https://proximity-operator.net/`) is an excellent source [105], with Matlab and Python implementations in the `proxop` library.

with proximal gradient descent, although we may descend to a local, rather than a global minimum. Also, similar to Newton's method, the gradient step $\gamma = \frac{1}{\beta}$ might be rather inefficient, and so instead we may use a line search to determine this step size.

Example: Solving the LASSO Problem with Prox Gradient

Let's examine the proximal gradient operator on the motivating example above. It is possible to explicitly compute the proximal gradient for the LASSO regularizer, $g(\mathbf{x}) = \lambda \|\mathbf{x}\|_1$:

$$\text{prox}_{\frac{1}{\beta}, \lambda\|\cdot\|_1}(\mathbf{z}) = \text{argmin}_{\mathbf{x}} \frac{\beta}{2}\|\mathbf{x} - \mathbf{z}\|_2^2 + \lambda\|\mathbf{x}\|_1 \tag{5.13a}$$

$$= \text{argmin}_{\mathbf{x}} \sum_{j=1}^{n} \left(\frac{\beta}{2}(x_j - z_j)^2 + \lambda|x_j|\right). \tag{5.13b}$$

Because both the quadratic objective and the ℓ_1 norm can each be decomposed into a sum of expressions that depend on the individual coordinates x_j and z_j, we may explicitly compute the solution for each index j separately. Specifically,

$$\mathbf{x}^* = \text{argmin}_{\mathbf{x}} \sum_{j=1}^{n} \left(\frac{\beta}{2}(x_j - z_j)^2 + \lambda|x_j|\right) \implies x_j^* = \text{argmin}_{x_j} \left(\frac{\beta}{2}(x_j - z_j)^2 + \lambda|x_j|\right).$$

To solve for the prox operator, we solve the piecewise linear optimization problem coordinate-wise for each x_j. Specifically, we compute the gradient of the expression above and set it equal to zero. Because $|x_j|_1$ is nonsmooth, its gradient depends on the sign of x_j:

$$\beta(x_j - z_j) + \lambda\, \text{sign}(x_j) = 0.$$

Thus, we have the following relationships for $x_j > 0$ and $x_j < 0$:

$$x_j > 0 \implies \beta(x_j - z_j) + \lambda = 0 \implies x_j - z_j = -\frac{\lambda}{\beta}$$
$$x_j < 0 \implies \beta(x_j - z_j) - \lambda = 0 \implies x_j - z_j = \frac{\lambda}{\beta}$$

However, we don't want a condition on x_j, but rather on the argument z_j. When $x_j > 0$, we know that $z_j > \frac{\lambda}{\beta}$, and when $x_j < 0$, we know that $z_j < -\frac{\lambda}{\beta}$, so we have:

$$z_j > \frac{\lambda}{\beta} \implies x_j^* = z_j - \frac{\lambda}{\beta}$$
$$z_j < -\frac{\lambda}{\beta} \implies x_j^* = z_j + \frac{\lambda}{\beta}.$$

Figure 5.2 Proximal gradient operator for ℓ_1 (left) and ℓ_0 (right) norms. The x-axis is the input z_j and the y-axis is the output x_j^*. The threshold τ is determined by the problem parameters β and λ.

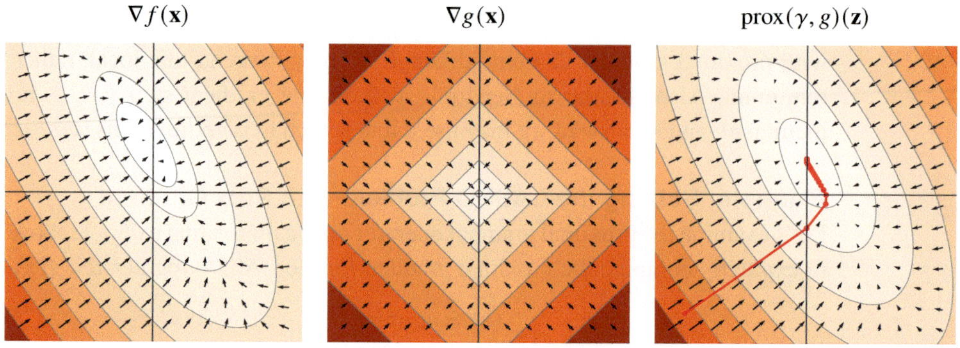

Figure 5.3 Gradient vector fields and loss level sets for the LASSO regression problem: (left) least squares loss $f(\mathbf{x})$; (middle) ℓ_1 loss $g(\mathbf{x})$; (right) proximal gradient vector field $\text{prox}_{\gamma,g}(\mathbf{x} - \gamma\nabla f(\mathbf{x}))$ and proximal gradient descent.

When $|z_j| < \frac{\lambda}{\beta}$ we set $x_j^* = 0$. This is exactly the iterative *soft* threshold algorithm (ISTA) with threshold $\tau = \frac{\lambda}{\beta}$, shown in Fig. 5.2. This is also known as the iterative *shrinkage*-thresholding algorithm. Further, we show the proximal gradient vector field and prox gradient descent for this ℓ_1 regularized least-squares in Fig. 5.3.

Example: Solving the ℓ_0 Problem with Prox Gradient

Similarly, we may solve for the proximal gradient operator for the ℓ_0 regularized regression problem with $g(\mathbf{x}) = \lambda\|\mathbf{x}\|_0$:

$$\text{prox}_{\frac{1}{\beta},\lambda\|\cdot\|_0}(\mathbf{z}) = \text{argmin}_{\mathbf{x}} \frac{\beta}{2}\|\mathbf{x} - \mathbf{z}\|_2^2 + \lambda\|\mathbf{x}\|_0 \tag{5.14a}$$

$$= \text{argmin}_{\mathbf{x}} \sum_{j=1}^{n} \left(\frac{\beta}{2}(x_j - z_j)^2 + \lambda\mathbf{1}(x_j \neq 0)\right). \tag{5.14b}$$

Here the indicator function $\mathbf{1}(x_j \neq 0)$ is equal to 1 if $x_j \neq 0$ and is equal to 0 if $x_j = 0$. Instead of computing the gradient and setting equal to zero, as with the ℓ_1 regularized problem above, we directly solve for the x_j^* that minimizes the objective above. Specifically, we are balancing the two terms in the objective function, so we choose x_j^* to minimize the sum. Either we will choose $x_j^* = 0$ to remove the second term, in which case our error from the j-th term is $\frac{\beta}{2}z_j^2$, or we will choose $x_j^* \neq 0$, in which case the second term is equal to λ. In this case, we might as well choose $x_j = z_j$ to eliminate the first term. This makes the choice relatively simple, since we have the following cases:

$$\begin{aligned} x_j^* = 0 &\implies \frac{\beta}{2}z_j^2 \\ x_j^* = z_j \neq 0 &\implies \lambda. \end{aligned}$$

Depending on z_j, we choose the x_j^* that yields the smaller of $\frac{\beta}{2}z_j^2$ and λ. When $z_j^2 < \frac{2\lambda}{\beta}$ we choose $x_j^* = 0$, and when $z_j^2 > \frac{2\lambda}{\beta}$ we choose $x_j^* = z_j$. This is exactly the iterative *hard* threshold, shown in Fig. 5.2, with threshold $\tau = \sqrt{\frac{2\lambda}{\beta}}$. We often use the iterative hard threshold to promote sparsity in optimization problems. The ℓ_0 problem is neither differentiable nor is it continuous, and it is still possible to compute the proximal gradient step.

Constrained Optimization and the Projected Gradient

It is possible to handle constraints with the proximal gradient, by adding them directly into the objective function, via an infinite penalty for $\mathbf{x}$ not in a feasible set C (see Fig. 1.27 for an example of the infinite penalty function):

$$g(\mathbf{x}) = \delta_C(\mathbf{x}) = \begin{cases} 0, & \mathbf{x} \in C \\ \infty, & \mathbf{x} \notin C. \end{cases} \tag{5.15}$$

This function is nonsmooth, discontinuous, and unbounded. However, we can still compute the proximal gradient of this function, which will allow us to develop a simple gradient-based descent algorithm for constrained optimization problems.

Specifically, we may write the proximal gradient operator explicitly

$$\begin{aligned} \text{prox}_{\frac{1}{\beta},\delta_C} &= \text{argmin}_{\mathbf{x}} \ \tfrac{\beta}{2}\|\mathbf{x}-\mathbf{z}\|_2^2 + \delta_C(\mathbf{x}) && \text{(5.16a)} \\ &= \text{argmin}_{\mathbf{x}\in C} \ \tfrac{\beta}{2}\|\mathbf{x}-\mathbf{z}\|_2^2 && \text{(5.16b)} \\ &\equiv \text{proj}_C(\mathbf{z}). && \text{(5.16c)} \end{aligned}$$

The proximal gradient operator takes a gradient step on $f(\mathbf{x})$ and then projects onto the feasible set C. This projection operator may or may not be simple to compute. For example, if C is a sphere or is a box, then it is simple to project $\mathbf{z}$ onto this set.

The basic idea of the projected gradient algorithm is to start with an initial $\mathbf{x}_0$ in the feasible set, take an unconstrained gradient iteration step, and project the solution back into the feasible set. We then iterate until convergence.

Projected Gradient and Box Constraints

Here we will analyze the constrained least-squares regression problem with box constraints on the regression variable $\mathbf{x}$:

$$\begin{aligned} \min_{\mathbf{x}} \quad & \tfrac{1}{2}\|\mathbf{A}\mathbf{x}-\mathbf{b}\|_2^2 \\ \text{s.t.} \quad & \boldsymbol{\ell} \le \mathbf{x} \le \mathbf{u}. \end{aligned} \tag{5.17}$$

We may also write the box constraints on each coordinate x_j as $\ell_j \le x_j \le u_j$.

We may enforce box constraints with the projected gradient algorithm. We begin with an initial $\mathbf{x}_0$ in the box $[\boldsymbol{\ell}, \mathbf{u}]$, take an unconstrained step:

$$\mathbf{z}_{k+1} = \mathbf{x}_k - \gamma \nabla f(\mathbf{x}_k) \tag{5.18}$$

and project it back into the feasible box:

$$\mathbf{x}_{k+1} = \mathcal{P}_{[\boldsymbol{\ell},\mathbf{u}]}(\mathbf{z}_{k+1}) \quad \text{where} \quad \left[\mathcal{P}_{[\boldsymbol{\ell},\mathbf{u}]}(\mathbf{z})\right]_j = \min\left\{u_j, \max\{\ell_j, z_j\}\right\}. \tag{5.19}$$

We denote the unconstrained least-squares objective function by $f(\mathbf{x}) = \frac{1}{2}\|\mathbf{A}\mathbf{x}-\mathbf{b}\|_2^2$, so $\nabla f(\mathbf{x}) = \mathbf{A}^{\mathrm{T}}(\mathbf{A}\mathbf{x}-\mathbf{b})$. We will revisit box-constrained least-squares in the context of Lagrange multipliers in Chapter 6.

Splitting Methods

The proximal operator, in the proximal gradient method above, generalizes to a broad family of operator splitting methods in which one decomposes a problem with multiple terms in the objective function, introducing extra variables to handle each term separately. These separate variables are then coupled with an additional term in the objective. This has several benefits that we will explore in this section via the sparse, relaxed, regularized regression (SR3) method, and via ADMM[2] in Section 6.5.

The general idea behind splitting methods is to transform a problem of the form

$$\min_{\mathbf{x}} \quad f(\mathbf{x}) + g(\mathbf{x}) \tag{5.20}$$

into the form

$$\min_{\mathbf{x},\mathbf{z}} \quad f(\mathbf{x}) + g(\mathbf{z}) + \frac{\rho}{2}\|\mathbf{x} - \mathbf{z}\|^2. \tag{5.21}$$

Again, the regularization term $g(\mathbf{x})$ could be used to promote specific properties, such as sparsity, or enforce constraints, as discussed in Chapters 4 and 6. These terms each have different decision variables $\mathbf{x}$ and $\mathbf{z}$, which are coupled through the quadratic $\frac{\rho}{2}\|\mathbf{x} - \mathbf{z}\|^2$. The parameter ρ determine the tightness of this coupling. Technically, Eq. (5.21) solves a different problem that approximates Eq. (5.20); to exactly solve Eq. (5.20), we add the constraint $\mathbf{x} = \mathbf{z}$ via Lagrange multipliers, which is the approach taken in ADMM.

This approach has several benefits. We break a complex problem into simpler subproblems, which are each better conditioned.

We may also use different algorithms for each term, for example using a preconditioned least squares solver for the $f(\mathbf{x})$, while using a proximal gradient descent algorithm for the nonsmooth $g(\mathbf{x})$.

Many problems can be written in this form. For example, the least-squares regression problem with objective function $f(\mathbf{x}) = \frac{1}{2}\|\mathbf{A}\mathbf{x} - \mathbf{b}\|^2$ can be written

$$\min_{\mathbf{x},\mathbf{z}} \frac{1}{2}\|\mathbf{A}\mathbf{x} - \mathbf{b}\|^2 + g(\mathbf{z}) + \frac{\rho}{2}\|\mathbf{x} - \mathbf{z}\|^2. \tag{5.22}$$

It is possible to incorporate nonsmooth and nonconvex norms via proximal gradients, as above. LASSO regression uses $g(\mathbf{z}) = \lambda\|\mathbf{z}\|_1$, and ridge regression uses $g(\mathbf{z}) = \alpha\|\mathbf{z}\|_2^2$. Occasionally, algorithms will add an additional matrix $\mathbf{C}$ in the last term, $\frac{\rho}{2}\|\mathbf{C}\mathbf{x} - \mathbf{z}\|^2$.

Sparse, Relaxed, Regularized Regression (SR3)

Sparse, relaxed, regularized regression (SR3) [557] is a powerful framework for handling nonsmooth and nonconvex sparse optimization problems in a fast and accurate way. Real-world optimization problems are often ill-posed or numerically unstable. Regularization and relaxation are two common strategies to mitigate these challenges. *Regularization* modifies an objective function to improve stability, interpretability, or generalization.

[2] ADMM is a *primal-dual* method, meaning that it relies on both the primal and dual formulation of the optimization problem, while SR3 and proximal gradient descent are *primal only* algorithms, meaning that they only depend on the original primal form of the optimization problem. Primal methods are often easier to formulate and work with. We will discuss dual methods more in Chapter 6.

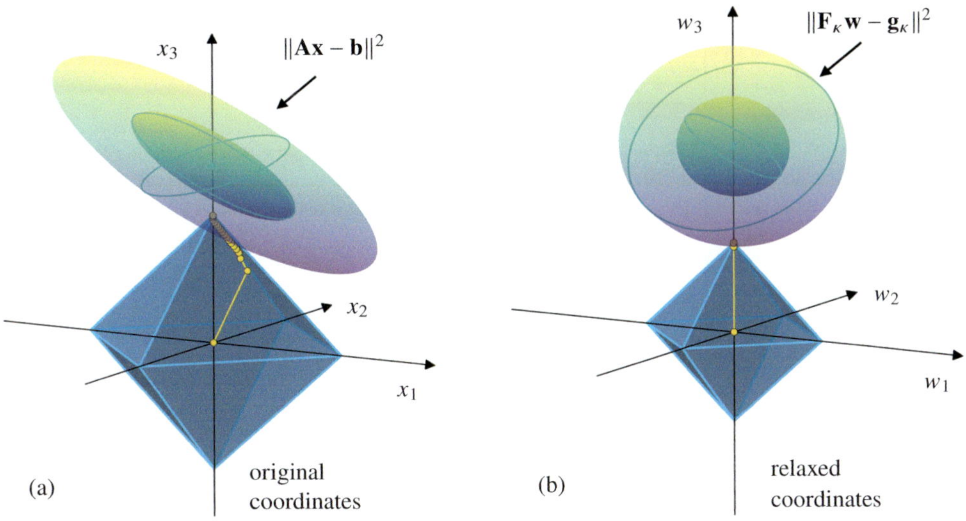

Figure 5.4 (a) Level sets (shaded ellipsoids) of the quadratic part of LASSO and corresponding path of prox-gradient to the solution (40 iterations) in $\mathbf{x}$-coordinates. (b) Level sets (shaded spheres) of the quadratic part of the SR3 loss and corresponding SR3 solution path (2 iterations) in relaxed coordinates $\mathbf{w}$. Shaded octahedra show the ℓ_1 ball in each set of coordinates. SR3 decreases the singular values of $\mathbf{F}_\kappa$ relative to those of $\mathbf{A}$ with a weaker effect on the small ones, 'squashing' the level sets into approximate spheres, accelerating convergence, and improving performance. Reproduced from Zheng et al. [557].

Relaxation replaces nonconvex terms in the constraints or the objective function with convex approximations that are amenable to computations. When combined, regularization and relaxation provide a powerful framework for systematically handling large-scale, ill-conditioned, or combinatorially complex problems. Splitting methods are effective ways of solving regularized, relaxed optimization problems. Instead of satisfying all constraints and objectives with the same decision variable, splitting introduces new decision variables for each term, adding a final regularization term to tightly couple these independent optimization problems. This makes stiff optimization problems more tractable. For convex problems, the SR3 formulation is part of a larger class of problems [30], where it is possible to use alternating minimization algorithms [2]. The original alternating projection methods go back to the 1950s by Von Neumann [522] and Cheney and Goldstein [103].

Sparse optimization is ubiquitous, and SR3 splits the sparsity and accuracy requirements in Eq. (5.22) between the $\mathbf{z}$ and $\mathbf{x}$ variables. The algorithm iterates between solving the subproblems in $\mathbf{x}$ and $\mathbf{z}$ individually, combining $\frac{\rho}{2}||\mathbf{x} - \mathbf{z}||_2^2$ and $g(\mathbf{z})$ in a proximal gradient step. The $\mathbf{x}$ term can be solved for using preconditioned least squares. It is possible to partially minimize $\mathbf{x}$ to obtain a function entirely in $\mathbf{z}$, which has been shown to have nearly spherical level sets, shown schematically in Fig. 5.4 from Zheng et al. [557]. This is in contrast to the highly eccentric elliptical level sets of $||\mathbf{Ax} - \mathbf{b}||^2$.

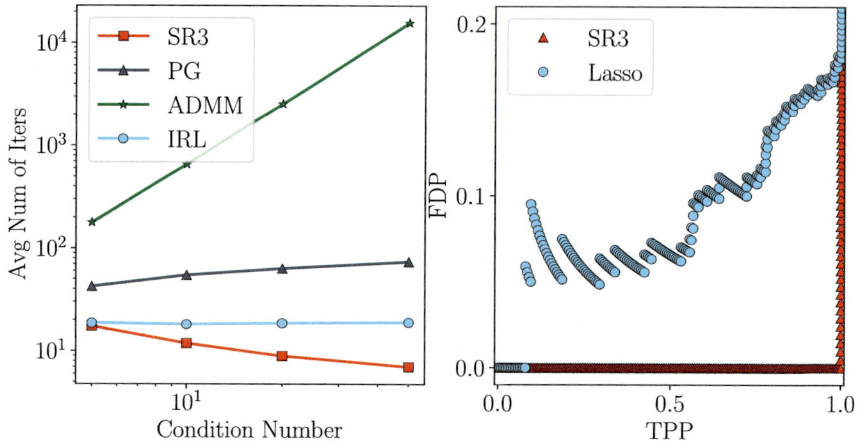

Figure 5.5 Top Left: SR3 approach (red) is orders of magnitude faster than ADMM (green) or other first-order methods such as prox-gradient (gray). While iteratively reweighted least squares (IRL) (blue) requires a comparable number of iterations, its cost per iteration is more expensive than SR3. Top Right: True Positives vs. False Positives along the LASSO path (blue) and along the SR3 path (red). Reproduced from Zheng et al. [557].

These concepts will be used in the later chapters on inverse problems[3], control theory, and machine learning, where relaxation and regularization are essential. Machine learning examples include sparse regression (LASSO), ridge regression, robust fitting, and training neural networks under constraints, such as derived from physics. In signal processing and image analysis, we use total variation (TV) regularization to denoise, wavelet-based sparsity constraints and sparse regularization for image reconstruction given sparse measurements, and nuclear-norm regularization for matrix completion. We have already seen examples of relaxing integer constraints to a continuous linear program. Finally, in robust control and PDE-constrained optimization, small changes in model parameters can produce unstable or ill-posed results, necessitating regularization and relaxation.

As a concrete example of the improvement using SR3, we consider the LASSO problem, investigated in Zheng et al. [557]. It was shown in Su et al. [477] that when performing LASSO, false discoveries occur early in the optimization path that cannot be reversed, leading to many false positives compared to true positives. This is largely attributed to the eccentricity of the geometry of the optimization problem, which is largely mitigated with SR3. In SR3, there are nearly no false positives, leading to a more accurate and more efficient algorithm. This is shown in Fig. 5.5.

[3] In fact, SR3 was originally developed in a discussion with Sasha Aravkin about the surprising effectiveness of our sequentially thresholded least-squares (STLS) algorithm used in the sparse identification of nonlinear dynamics (SINDy) algorithm [81]. Peng Zheng and Sasha showed that our algorithm for solving the ℓ_0 sparse optimization problem was a proximal gradient descent on a particular value function, and they went on to generalize this approach for several other sparse optimization problems. This will be discussed more in Section 8.7.

5.2 Nelder–Mead Downhill Simplex

The Nelder–Mead downhill simplex method [363] is a popular gradient-free, local search technique for unconstrained optimization problems, introduced in 1965. Unlike gradient descent, which requires the gradient of an objective function, Nelder–Mead works purely with function value evaluations. It is particularly useful when the objective function is non-differentiable or discontinuous in certain regions, or when numerical or analytical gradients are difficult or expensive to obtain. This approach works well when the dimension of the problem is relatively small, commonly up to a few dozen variables.

Despite its popularity in practice and its ease of implementation, the Nelder–Mead method does not have strong theoretical convergence guarantees. In fact, there are known counterexamples, even for strictly convex functions [346], where the method fails to converge or its performance degrades. Nevertheless, it remains widely used in engineering and applied mathematics for small to medium-scale problems [475], particularly when function evaluations are expensive and derivative approximations or adjoint-based approaches are complicated or infeasible.

The idea behind the downhill simplex method is quite simple. Consider a minimization problem. Given a decision variable $\mathbf{x} \in \mathbb{R}^n$, we initialize a simplex in $\mathbb{R}^n$ with $n+1$ vertices $\{\mathbf{x}_j\}_{j=0}^{n}$, evaluate the objective function at all vertices, and modify the vertex with the highest objective function value to ideally move to a lower objective function value. In this way, we iteratively move the centroid of the simplex to a lower objective function value, one vertex at a time, without needing to compute a gradient at any step. In a sense, this simplex "rolls down" the objective function landscape; see Fig. 5.6 for an illustration. There are several details and options for how to modify the vertex with the highest value, depending on the dimension of the problem, the structure of the objective function landscape, among other considerations. In addition, we may shrink or expand the size of the simplex depending on properties of the objective function landscape.

This may be mathematically formalized as follows. At each step, the current simplex is defined by the points $\{\mathbf{x}_0, \mathbf{x}_1, \ldots, \mathbf{x}_n\}$. We assume that the points are sorted so that

$$f(\mathbf{x}_0) \leq f(\mathbf{x}_1) \leq \cdots \leq f(\mathbf{x}_n), \tag{5.23}$$

where f is the objective function being minimized. We define the centroid of the n best points in the simplex, omitting the $(n+1)$th, as

$$\bar{\mathbf{x}} = \frac{1}{n} \sum_{j=0}^{n-1} \mathbf{x}_j. \tag{5.24}$$

Then, at each step of the Nelder–Mead iteration, we modify the simplex according to the following procedure:

Reflection We reflect the point $\mathbf{x}_n$ about the centroid $\bar{\mathbf{x}}$

$$\mathbf{x}' = \bar{\mathbf{x}} + \alpha(\bar{\mathbf{x}} - \mathbf{x}_n) \tag{5.25}$$

where α is a coefficient that either increases the distance from the centroid ($\alpha > 1$), decreases it ($\alpha < 1$) or keeps it the same ($\alpha = 1$). Typically $\alpha \approx 1$ is a good choice.

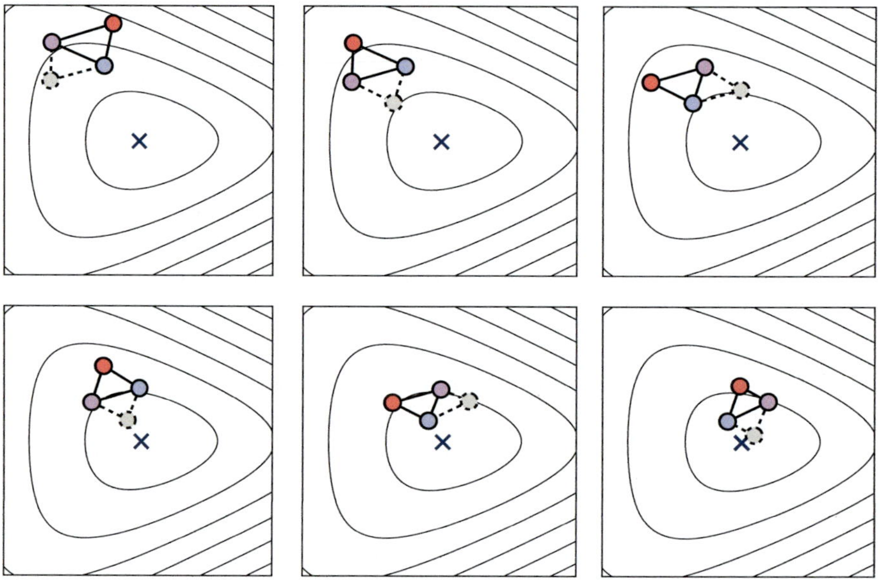

Figure 5.6 Schematic of the downhill simplex method on a simple two-dimensional objective function with a single global minima. The simplex is illustrated by the three colorful vertices, with blue having the lowest objective function value, red having the highest, with purple in the middle. In each reflection, the red vertex becomes the gray dashed vertex. In this example, simple reflections suffice to drive the simplex towards the minima.

We now compare $f(\mathbf{x}')$ against the objective value at the other vertices. If the new reflected vertex $f(\mathbf{x}')$ falls between the best and second best vertex,

$$f(\mathbf{x}_0) \leq f(\mathbf{x}') < f(\mathbf{x}_{n-1}) \tag{5.26}$$

then replace $\mathbf{x}_n$ with $\mathbf{x}'$, finish the iteration, and proceed back to step 1 for the next iteration. Otherwise, proceed to the next steps, depending on the ranking of $f(\mathbf{x}')$.

Expansion If $\mathbf{x}'$ is the new best vertex, so that

$$f(\mathbf{x}') < f(\mathbf{x}_0) \tag{5.27}$$

then expand the point further in this new reflected direction:

$$\mathbf{x}'' = \bar{\mathbf{x}} + \gamma(\bar{\mathbf{x}} - \mathbf{x}_n) \tag{5.28}$$

where $\gamma \approx 2\alpha > 1$ is the expansion coefficient. If the new value is still the best, i.e. $f(\mathbf{x}'') < f(\mathbf{x}')$, then replace $\mathbf{x}_n$ with $\mathbf{x}''$, otherwise replace with $\mathbf{x}'$. The iteration ends and we proceed with step 1 in the next iteration.

Contraction If $\mathbf{x}'$ is still the worst vertex, so that

$$f(\mathbf{x}') \geq f(\mathbf{x}_{n-1}) \tag{5.29}$$

then we perform a contraction

$$\mathbf{x}'' = \bar{\mathbf{x}} + \rho(\bar{\mathbf{x}} - \mathbf{x}_n) \tag{5.30}$$

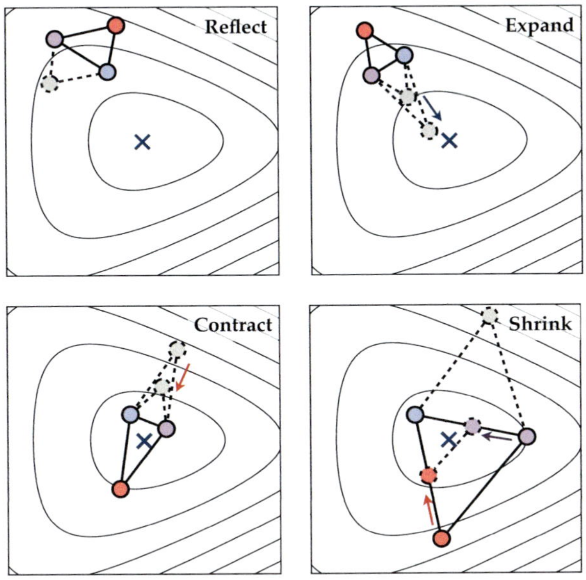

Figure 5.7 Illustration of the various downhill simplex operations. The first operation is a simple reflection. If the reflected point is the new optimum in the simplex, then we move to the expand operation, moving more in this favorable direction. If the new point is still the worst, then we contract, moving closer to the centroid. If all else fails, we move all points closer to the current optimum point.

where $\rho \approx 0.5\alpha < 1$ is the contraction coefficient. If the new vertex $\mathbf{x}''$ is better than $\mathbf{x}'$, i.e. $f(\mathbf{x}'') < f(\mathbf{x}')$, then replace $\mathbf{x}_n$ with $\mathbf{x}''$ and end this iteration. Otherwise, when all else fails, as a dousing of cold water, we proceed to the shrinkage operation.

Shrink Here we contract all other vertices $j = 1, \ldots, n$ towards the best vertex $\mathbf{x}_0$:

$$\mathbf{x}_i \leftarrow \mathbf{x}_0 + \sigma\big(\mathbf{x}_i - \mathbf{x}_0\big), \tag{5.31}$$

where $0 < \sigma < 1$ is the *shrinkage* coefficient. Typically $\sigma \approx 0.5$ is reasonable.

These operations are shown schematically in Fig. 5.7. The above parameters are the defaults originally proposed by Nelder and Mead. Adjusting them can change the aggressiveness of exploration and the method's stability.

There are several common stopping criteria: if the function values at all simplex vertices are sufficiently close to one another (e.g., $\max_j |f(\mathbf{x}_j) - f(\mathbf{x}_0)| < \epsilon$); if the simplex size (e.g., the maximal distance between vertices) is below a tolerance; or if the number of function evaluations or iterations reaches a specified maximum.

Now that we have seen the algorithm, and before exploring a coding example, it is important to reiterate the strengths and limitations of this approach. The downhill simplex method is simple to implement and requires no gradient or higher-order derivative information, making it a reasonable choice for many situations. It is also possible to code up from scratch and implement special or custom features. Thus, it is often effective for

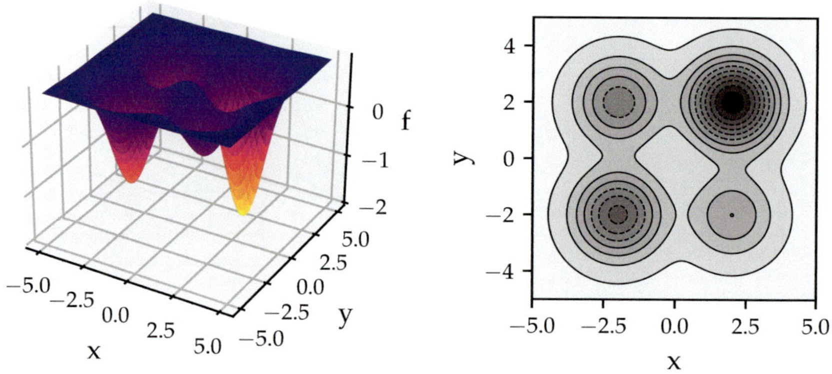

Figure 5.8 The objective function with four local minima, visualized in three-dimensions (left) and in contours (right).

small-scale problems where function evaluations are costly, or gradient approximations are infeasible. However, there are also several drawbacks. There are relatively few convergence results for nonconvex problems, and convergence to a local minima is not guaranteed to be global. However, the algorithm can often escape local minima depending on the size of the initial simplex and optimization hyperparameters. In higher dimensions, the downhill simplex may perform poorly, often requiring many function evaluations. Finally, without careful tuning, the method might stagnate or wander slowly around an optimum.

However, in practice, the Nelder–Mead downhill simplex method remains one of the most widely used derivative-free local optimization techniques [113]. Its simple conceptual framework and relatively straightforward hyperparameter setup make it attractive for problems in low to moderate dimensions, particularly when function evaluations are expensive or noisy but one still prefers a local method to refine near a suspected optimum.

Example in Python

Here we will demonstrate the ability of the downhill simplex method to find a global optimum of a nonconvex landscape with four local minima.

The objective function is given by

$$f(\mathbf{x}) = 1 + f_1(\mathbf{x}) + f_2(\mathbf{x}) + f_3(\mathbf{x}) + f_4(\mathbf{x}) \tag{5.32}$$

where each $f_j(\mathbf{x})$ is defined as

$$f_j(\mathbf{x}) = -\alpha_j e^{-\frac{1}{2}\|\mathbf{x}-\mathbf{x}_j\|^2} \tag{5.33}$$

and $\mathbf{x}_j$ is the center of the j-th minima and α_j is the depth. For this example, we use $(\alpha_1, \alpha_2, \alpha_3, \alpha_4) = (3, 1.5, 1, 2)$ at centers $\mathbf{x}_1 = (2, 2), \mathbf{x}_2 = (-2, 2), \mathbf{x}_3 = (2, -2)$, and $\mathbf{x}_4 = (-2, -2)$. This function is visualized in Fig. 5.8.

The downhill simplex algorithm is simple to implement, and there are many established libraries (e.g., `scipy.optimize` with `method='Nelder--Mead'`). The following code performs a single iteration of the Nelder–Mead downhill simplex method, where α, γ, ρ, and σ are the reflection, expansion, contraction, and shrinkage coefficients.

CodeSnip 5.a

```
%\lstinputlisting[caption={Nelder--Mead downhill simplex method
    .}, label={Code:CH02:neldermead}]
%{\begin{lstlisting}[style=customP]

def nelder_mead_step(vertices, alpha, gamma, rho, sigma):
    vals = np.array([f(v) for v in vertices])
    order = np.argsort(vals)
    vertices = vertices[order]
    vals = vals[order]

    x1, x2, x3 = vertices # vertices : np.ndarray of shape (3,2)
    f1, f2, f3 = vals     # best, middle, worst
    center = (x1 + x2) / 2.0 # center of best and middle points

    # Reflection
    x_reflect = center + alpha * (center - x3)
    f_reflect = f(x_reflect)

    if f_reflect < f1:          # Expansion
        x_expand = center + gamma * (x_reflect - center)
        f_expand = f(x_expand)
        if f_expand < f_reflect:
            vertices[2] = x_expand
        else:
            vertices[2] = x_reflect
    elif f_reflect < f2:
        vertices[2] = x_reflect
    else        # Reflection is worse than mid
        if f_reflect < f3:      # Outside contraction
            x_contr = center + rho * (x_reflect - center)
            f_contr = f(x_contr)
            if f_contr <= f_reflect:
                vertices[2] = x_contr
            else:               # Shrink
                vertices[1] = x1 + sigma*(x2 - x1)
                vertices[2] = x1 + sigma*(x3 - x1)
        else:                   # Inside contraction
            x_contr = center - rho * (center - x3)
            f_contr = f(x_contr)
            if f_contr < f3:
                vertices[2] = x_contr
            else:               # Shrink
                vertices[1] = x1 + sigma*(x2 - x1)
                vertices[2] = x1 + sigma*(x3 - x1)
    return vertices
```

This code would be called in the following way:

CodeSnip 5.b

```
    # Initialize a large simplex in 2D
    vertices = np.array([[4.0, 4.0],
        [-4.0, -3.0],
        [-2.0,  4.0],])

    # Perform Nelder--Mead updates
```

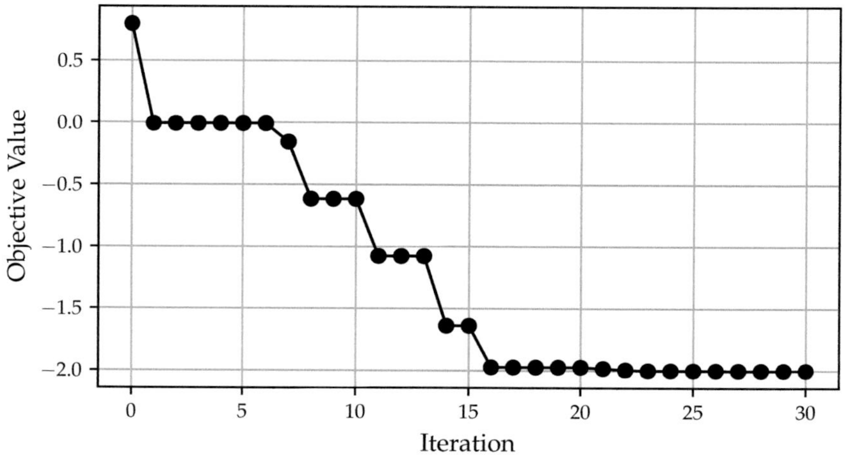

Figure 5.9 Objective function versus iteration for the downhill simplex method on the nonconvex system with multiple local minima.

```
num_iterations = 30
for it in range(1, num_iterations+1):
    vertices = nelder_mead_step(vertices)
```

The objective function versus iteration is shown in Fig. 5.9. Figure 5.10 shows the evolution of the vertex in the first sixteen iterations. After an initial search, it finds the global minima and then descends to the best value.

5.3 Simulated Annealing

Simulated annealing is a stochastic global optimization technique inspired by the metallurgical process of annealing, in which a material is heated and then slowly cooled to decrease defects and obtain a strong crystalline structure [279, 521]; see Panel 5.3. In metallurgical annealing, the specific cooling schedule makes a tremendous difference in the final properties of the metal. In optimization, simulated annealing uses a similar idea, leveraging an initially higher "temperature" to explore the solution space and escape local minima, then gradually lowering the temperature to focus on refining solutions near good local minima. In simulated annealing, the temperature refers to the step size of random perturbations that are made to the current solution at each iteration step, so that a higher temperature leads to larger, more exploratory, random fluctuations.

Simulated annealing is a robust, global search algorithm that is suitable for nonconvex optimization problems. Moreover, it does not require gradient information and is able to handle stochastic or noisy optimization problems, as well as problems with discrete, continuous, or mixed variables. Simulated annealing is simple to implement, as it only involves evaluations of the objective function f and some basic logic. However, the cooling schedule and other hyperparameters play an essential role, and must be tuned carefully

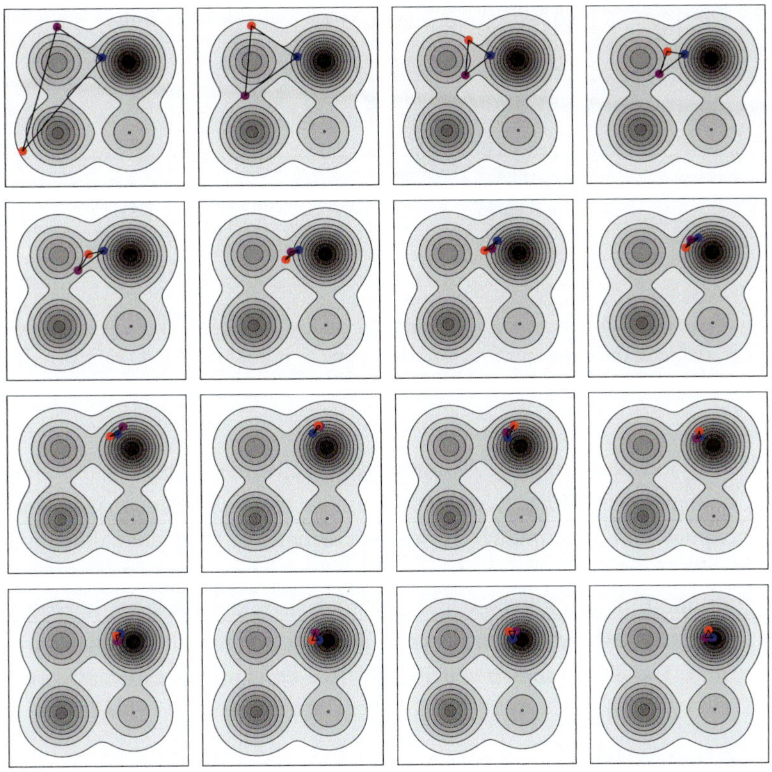

Figure 5.10 Illustration of the downhill simplex method on the nonconvex system with multiple local minima. After an initial exploration phase, the simplex eventually finds the global minima and descends.

Panel 5.3: Annealing Metal Alloys

Annealing is an ancient practice of heating and cooling metal alloys to promote certain material properties, such as strength and durability.

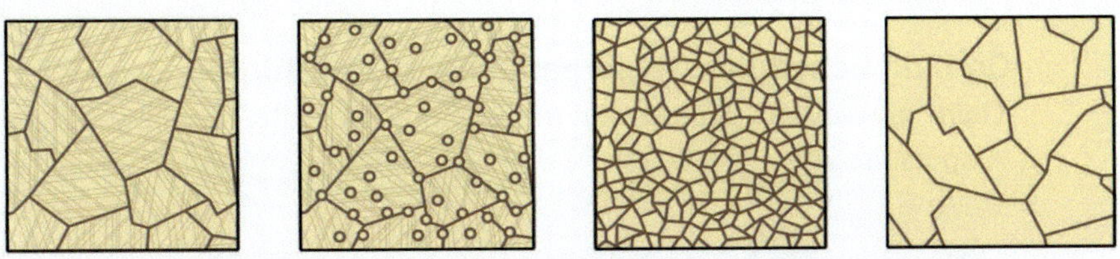

Figure 5.11: Schematic of nucleation and recrystallization in annealing, from Wikimedia by Daniele Pugliesi.

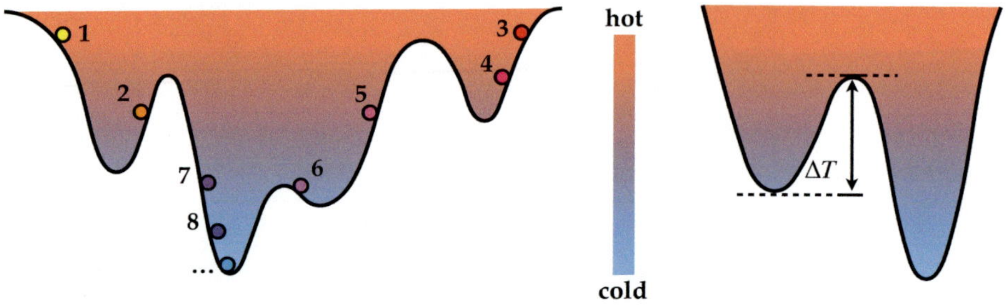

Figure 5.12 Schematic of simulated annealing. Left: We see an example path of the decision variable in the objective function landscape as it goes from hot (yellow/red) to cold (blue). Right: The current temperature determines how likely it is to jump from one local minimum to another.

for good performance. As the cooling schedule is itself a function, with infinitely many degrees of freedom, we typically choose from a parameterized class of functions, such as exponential or geometric decay. A schematic of simulated annealing is shown in Fig. 5.12.

There are a number of interesting convergence results for simulated annealing in the infinite-time limit, for specific cooling schedules. However, in practice, while it is designed for global exploration, there is no guarantee that a global minimum will be found in finite time. The global search can also be time-consuming, especially for large problems.

Algorithm

We aim to minimize an objective function $f(\mathbf{x})$. The decision variable $\mathbf{x}$ may be continuous (i.e., $\mathbf{x} \in \mathbb{R}^n$) or discrete. Simulated annealing iteratively samples a new candidate solution $\mathbf{x}_{k+1}$ in the neighborhood of the current solution $\mathbf{x}_k$. The difference in objective values $\Delta = f(\mathbf{x}_{k+1}) - f(\mathbf{x}_k)$ guides the acceptance of $\mathbf{x}_{k+1}$:

1. If $\Delta \leq 0$ then the new solution is better, and we accept it unconditionally.
2. If $\Delta > 0$, the new solution is worse, and we accept it with probability $p = \exp\left(-\frac{\Delta}{T}\right)$, where T is the current temperature.

Over time, T is decreased according to the cooling schedule, reducing the acceptance of worse solutions. The algorithm then focuses more on local refinement near good solutions. Typical stopping criteria include a minimum temperature threshold, a maximum number of iterations, or detecting convergence.

Cooling Schedules, Hyperparameters, and Tuning

There are several choices for the cooling schedule. A classical geometric decay for temperature is

$$T_{k+1} = \alpha T_k, \tag{5.34}$$

where $0 < \alpha < 1$ is a constant (e.g., $0.9 \leq \alpha \leq 0.99$). More sophisticated schedules adjust temperature dynamically based on recent acceptance rates. There are also some theoretical guarantees for cooling rates T_k that lie between the curves $1/k$ and $1/k^2$.

In addition, there are several hyperparameters that must be tuned:

- Initial temperature T_0: This should be high enough to encourage global exploration early, but too large a T_0 will increase the number of iterations required before the temperature is low enough for exploitation of a local minimum.
- Cooling factor α: For geometric decay, this determines how fast T is reduced in each iteration. Too fast, and the algorithm won't have time to explore, and too slow and the algorithm may not converge quickly or to a good minimum.
- Neighborhood function or "move operator": This determines how new candidate solutions are generated from the current solution. In continuous domains, one might add small random Gaussian or uniform perturbations. In discrete domains (e.g., traveling salesman), it might be a random permutation swap.
- Stopping criteria: Several choices include reaching a minimum temperature $T_{\min}$, a maximum number of iterations, or no improvement for a number of steps.

Code Example

Here we explore the objective function in Eq. (5.32) with simulated annealing. Figure 5.13 shows the progression of the decision variable, starting from a "hot" exploratory condition in red to the final "cold" nearly optimal condition in blue. It is simple to implement a basic simulated annealing code. The code in CodeSnip5.c tracks the best objective value at each iteration and returns the path of visited points.

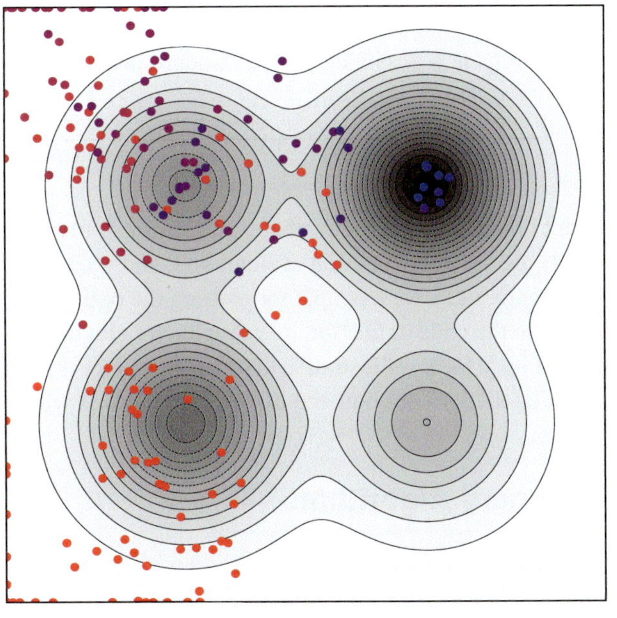

Figure 5.13 Illustration of the simulated annealing process, where initial "hot" points are red, and as they "cool" they become more blue. Eventually the global minimum is found.

CodeSnip 5.c

```
def simulated_annealing(init_point, n_iter=200, T0=5.0, alpha
    =0.98, step_size=1.0, bounds=(-5,5)):
    current = np.array(init_point, dtype=float)
    current_val = f(current)
    path = [current.copy()]
    temp = T0 # Initial temperature

    for i in range(n_iter):
        # Propose random step
        candidate = current + np.random.randn(2) * step_size
        candidate = np.clip(candidate, bounds[0], bounds[1])
        candidate_val = f(candidate)
        delta = candidate_val - current_val

        # Accept if better or with probability e^(-delta/T)
        if delta < 0:
            current, current_val = candidate, candidate_val
        else:
            if np.random.rand() < np.exp(-delta / temp):
                current, current_val = candidate, candidate_val

        temp *= alpha # Lower the temperature
        path.append(current.copy()) # Store the path
    return np.array(path)
```

We would then call this function as follows:

CodeSnip 5.d

```
xIC = np.array([-4.0, -4.0])  # Initial guess (bad)
path = simulated_annealing(
    init_point=xIC,n_iter=200,T0=10.0,alpha=0.98,step_size=1.)
```

Alternatively, it is possible to use a standard package.

CodeSnip 5.e

```
from scipy.optimize import dual_annealing
bounds = [(-5, 5), (-5, 5)] # Domain for x,y in [-5, 5]
result = dual_annealing(objective_2d, bounds=bounds, maxiter
    =1000, seed=0)
```

The result object has several attributes, including `x` (the best point), `fun` (the objective at that point), `nfev` (# of function evaluations), and `nit` (# of iterations). To access internal data at each iteration, for example for plotting the optimization path, we may use a callback function. See the code online for an example. We plot the cost per iteration of the annealing process in Fig. 5.14. It is illustrative to repeat this process for an ensemble of many annealing paths, in Fig. 5.15, to see how the distribution evolves with iteration number.

5.4 Evolutionary Algorithms

Evolutionary algorithms form a broad class of methods for nonconvex optimization, based on the principles of biological evolution and survival of the fittest. Evolutionary operations, such as replication, mutation, and exchange of genetic information, are used to iteratively evolve populations of candidate solutions. With a judicious choice of hyperparameters, the performance distribution of this population will improve and individuals in the population may approach globally optimal solutions. Evolutionary algorithms broadly comprise two

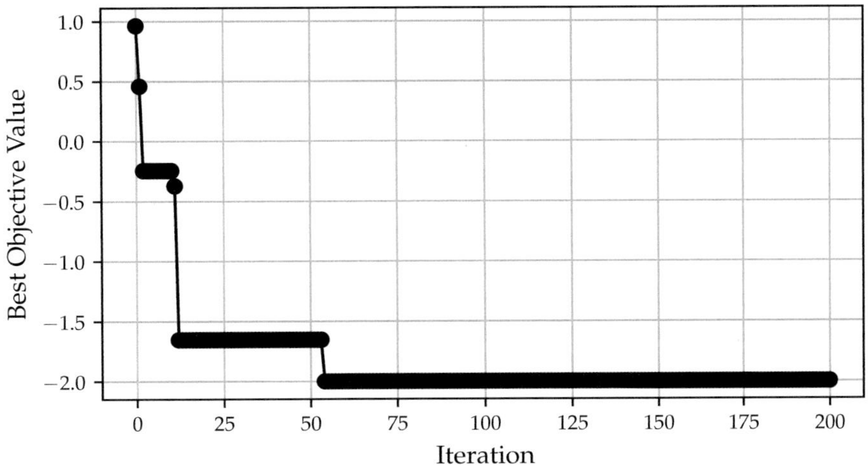

Figure 5.14 The objective function versus iteration for simulated annealing is shown.

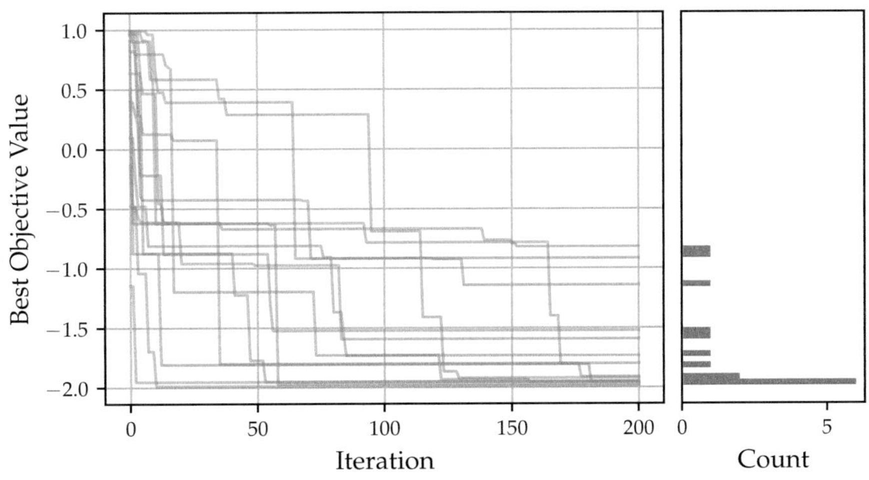

Figure 5.15 Ensemble of multiple simulated annealing paths (left), with a histogram of the final optimized value (right)

major categories of methods: genetic algorithms for parameter optimization and genetic programming for function optimization.

Parameter optimization is a simpler problem, where we learn the parameters of a function with known structure. The goal is to find the optimal parameters given some objective function, for example training a neural network fitting $\hat{y} = f_\theta(\mathbf{x})$ to fit data y, or trying to learn the parameters of a control law with optimal performance.

Function optimization is more challenging, because we are learning both the structure and parameters of a function that optimizes a given objective function. Again, the task may be to fit data with a learned approximation $f(\mathbf{x})$ or learn a control law $u = f(\mathbf{x})$. However, in this case, we will also learn the structure of the function f.

We begin with genetic algorithms, because of their simplicity and broad applicability, and then proceed to the more challenging, but more flexible, class of genetic programming which is often referred to as symbolic regression, because it yields analytic representations of the function being optimized. This is useful for explainable AI [116].

Genetic algorithms

Genetic algorithms (GAs) are used for parameter optimization [235, 127, 196]. In GA, a population of many realizations with different parameter values compete to minimize a cost function, and successful parameter values are propagated to future generations through a set of *genetic* rules. The parameters of a system are generally represented by a binary sequence, as in Fig. 5.16. For example, these parameters may represent the three control gains K_P, K_I, and K_D for a proportional-integral-derivative (PID) controller [77]. A number of realizations with different parameter values, called *individuals*, are initialized in a population and their performance is evaluated and compared on a well-defined task. Successful individuals with a lower cost have a higher probability of being selected to advance to the next generation, according to the following genetic operations:

Replication: An individual is selected to advance to the next generation.

Crossover: Two individuals are selected to exchange a portion of their code and advance to the next generation; crossover exploits and enhances successful strategies.

Mutation: An individual is selected to have some of its code modified with new values; mutation promotes diversity and increases the exploration of parameter space.

Elitism (optional): A set number of the most fit individuals are advanced directly to the next generation. Elitism is closely related to replication, but it involves a fixed number of the best individuals, rather than a random selection.

For replication, crossover, and mutation, individuals are randomly selected to advance to the next generation with the probability of selection increasing with fitness. The genetic operations are illustrated for the PID control example in Fig. 5.17. These generations are evolved until the fitness of the top individuals converges or other stopping criteria are met.

Genetic algorithms are generally used to find nearly globally optimal parameter values, as they are capable of exploring and exploiting local wells in the cost function. They provide a middle ground between a brute-force search and a convex optimization, and are an alternative to expensive Monte Carlo sampling, which does not scale to high-dimensional parameter spaces. However, there is no guarantee that genetic algorithms will converge to a globally optimal solution. There are also a number of hyperparameters that may affect performance, including the size of the populations, number of generations, and relative selection rates of the various genetic operations.

Genetic Programming

Genetic programming (GP) [287, 286] is a powerful generalization of genetic algorithms that simultaneously optimizes both the structure and parameters of an input–output map

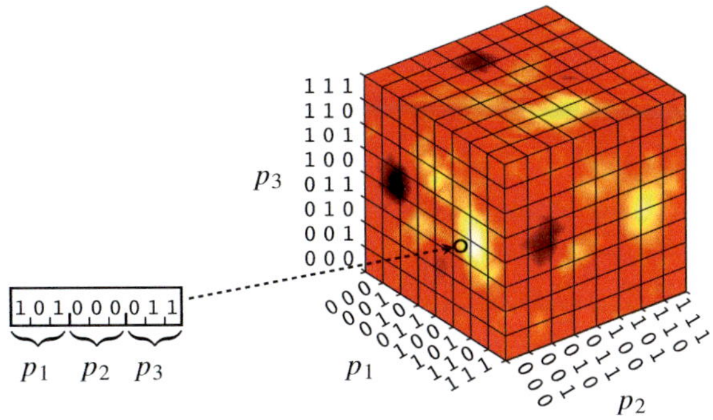

Figure 5.16 Depiction of a parameter cube with three parameters p_1, p_2, and p_3. For example, these three parameters may represent the proportional, integral, and derivative gains in a PID controller. The genetic algorithm represents a given parameter value as a *genetic sequence* that concatenates the various parameters. In this example, the parameters are expressed in binary representation that is scaled so that **000** is the minimum bound and **111** is the upper bound. Color indicates the cost associated with each parameter value.

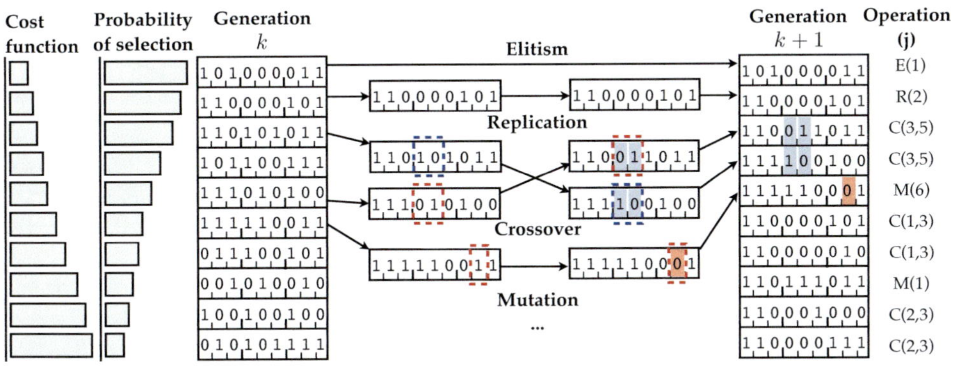

Figure 5.17 Schematic illustrating evolution in a genetic algorithm. Individuals in generation k are each evaluated and ranked based on their cost function, which is inversely proportional to their probability of selection for genetic operations. Individuals are chosen based on this weighted probability to advance to generation $k + 1$ using the four operations: elitism, replication, crossover, and mutation. Generation iterations are repeated until the population statistics converges or another stopping criterion is reached. Note that there are extensions to handle floating point numbers as well.

(i.e., a function). More recently, genetic programming has been used extensively to learn dynamical systems [61, 447, 116] or control laws [158].

The function tree representation in genetic programming is quite flexible, enabling the encoding of complex functions of the input data $\mathbf{x}$ through a recursive tree structure. Each branch is a signal, and the merging points are mathematical operations. Inputs and constants are the leaves, and the overall function output $f(\mathbf{x})$ is the root. The genetic operations of crossover, mutation, and replication are shown schematically in Fig. 5.18. Each of these operations has a corresponding probability, $p(C)$, $p(M)$, and $p(R)$.

In general, although evolutionary algorithms are extremely general and powerful, there are very few theoretical guarantees on convergence and stability. That said, they have

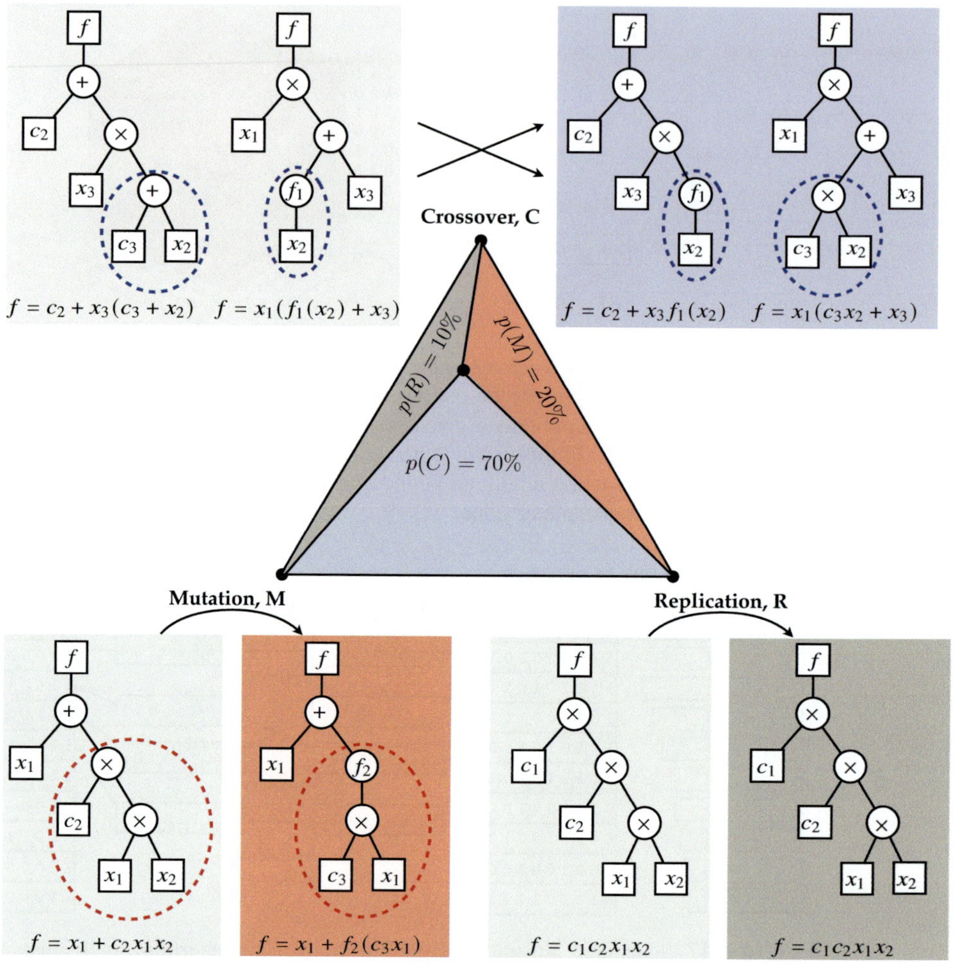

Figure 5.18 Genetic operations used to advance function trees in genetic programming. In this example, the probability of replication is 10%, the probability of mutation is 20%, and the probability of crossover is 70%.

had a resurgence of interest recently as a technique for learning human-interpretable and explainable representations of machine learning models [119].

Genetic Programming in Python with PySR

Here we explore a simple example of symbolic regression using the PySR package [116]. The following code generates data from the Lorenz system and learns a differential equation $\dot{\mathbf{x}} = \mathbf{f}(\mathbf{x})$, shown in Fig. 5.19. We will revisit this as an inverse problem in Chapter 8.

CodeSnip 5.f

```
from pysr import PySRRegressor
from sympy import lambdify, symbols

# Generate data from Lorenz system
x0 = [5, 5, 5] # Initial state
```

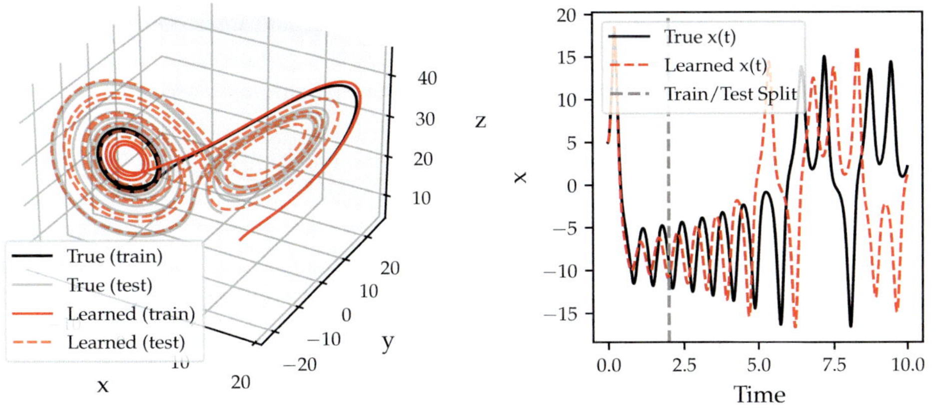

Figure 5.19 Output of PySR on Lorenz system. The model is trained on the first 2 time units and tested on the next 8. The model performance is in excellent agreement.

```
sol_train = solve_ivp(lorenz, (0, 2), x0, t_eval=t_train)
X_train = sol_train.y.T
dXdt_train = np.gradient(X_train, dt, axis=0) # compute dx/dt

# --- Train PySR models for dx/dt, dy/dt, dz/dt ---
models = []
rhs_exprs = []
variables = symbols("x0 x1 x2")
for i, label in enumerate(["dx/dt", "dy/dt", "dz/dt"]):
    model = PySRRegressor(
        model_selection="best",
        binary_operators=["+", "-", "*", "/"],
        unary_operators=["sin", "cos", "exp"],
        loss="loss(x, y) = (x - y)^2",
        niterations=100, maxsize=10, populations=10)
    model.fit(X_train, dXdt_train[:, i])
    models.append(model)
    rhs_exprs.append(lambdify(variables, model.get_best().
        sympy_format, modules=["numpy"]))

# --- Define learned system using all three expressions ---
def learned_lorenz(t, state):
    x, y, z = state
    dx = rhs_exprs[0](x, y, z)
    dy = rhs_exprs[1](x, y, z)
    dz = rhs_exprs[2](x, y, z)
    return [dx, dy, dz]

# Simulate full learned model over [0, 2]
sol_learned = solve_ivp(learned_lorenz,(0,2),x0,t_eval=t_full)
```

5.5 Distributed and Swarm Optimization

With the rise of big data, parallel computing, and interconnected systems, optimization strategies that leverage multiple agents to solve sub-problems in a distributed manner have become essential. Here we discuss two main paradigms: distributed optimization and swarm optimization.

Distributed optimization aims to solve large-scale problems by decomposing them into subproblems that run in parallel, synchronizing partial solutions through communication protocols [65, 361]. By contrast, *swarm optimization* takes inspiration from natural phenomena, such as birds flocking, fish schooling, and ants foraging, to design optimization algorithms. Particle swarm optimization (PSO) [274], ant colony optimization (ACO) [149], and related *swarm intelligence* methods embody this approach.

Distributed optimization can be traced back to early studies on decentralized control and parallel processing in the 1970s and 1980s [509]. Over time, as networking technologies matured, applications of distributed optimization proliferated in sensor networks, communication systems, and large-scale machine learning (e.g., federated learning). Meanwhile, swarm intelligence took shape in the 1990s with foundational algorithms such as *particle swarm optimization* by Kennedy and Eberhart [274] and *ant colony optimization* by Dorigo and co-authors [149]. These bio-inspired algorithms emphasize simplicity, adaptability, and the emergent behavior arising from the coordination of many simple agents.

Both distributed and swarm approaches can be viewed as leveraging the idea of *collective decision-making* to handle large, complex problems. Distributed optimization typically focuses on rigorous mathematical decomposition, while swarm methods favor iterative search processes guided by heuristics and emergent coordination.

Distributed Optimization

A common scenario in distributed optimization arises when one has a large-scale objective

$$\min_{\mathbf{x}} \quad f(\mathbf{x}) = \sum_{j=1}^{m} f_j(\mathbf{x}), \tag{5.35}$$

where each $f_j(\mathbf{x})$ is associated with a different agent (or node) $j \in \{1, \ldots, m\}$ in a network [361]. Each agent knows only its local function $f_j(\mathbf{x})$ and is responsible for processing or storing part of the data. The agents must collaboratively find a global minimizer $\mathbf{x}^*$.

Agents communicate over a network, exchanging local information, such as gradients, subgradients, or partial solutions. A fundamental concept is *consensus*, where all agents agree on a common solution vector, or feasible point. Methods like the *consensus algorithm* iteratively average local solutions to converge to a consistent global solution. Although sparse network communications are not necessary, this assumption is often included to make the scenario more realistic and robust.

The alternating direction method of multipliers (ADMM) is one popular choice for decomposing a large problem into subproblems, each solved locally, with updates to ensure global consistency [65]. We will discuss ADMM in much more detail in Section 6.5.

Distributed optimization has several attractive aspects, including scalability, privacy, and efficient parallelism. Because this approach decomposes large problems across multiple machines or agents, it may, if well designed, leverage parallel computation and reduce communication overhead. Another key advantage is that data can remain local to each node, which is important for privacy and ownership in sensitive applications (e.g., federated learning).

However, these algorithms are relatively complex to implement, requiring careful handling of asynchrony, message passing, and partial updates. Convergence may also be slower than centralized solvers if the communication or coordination overhead is large. Finally convergence depends on communication reliability and network topology.

Particle Swarm Optimization

Swarm optimization (PSO) methods view a population of agents (e.g., particles, ants, bees) that each maintaining a candidate solution. The overall solution quality emerges from how these agents *interact* and *update* themselves based on shared experience or local heuristics [274, 58]. Swarm optimization, in contrast to distributed optimization, is relatively easy to implement and is versatile for many black-box or nonsmooth problems. It relies on global search heuristics (e.g., swarm intelligence) to escape local minima in nonconvex problems. Like distributed optimization, it is parallelizable, as swarm updates can be computed in parallel for each agent.

There are no exact guarantees for swarm optimization, as the heuristics can fail to find global optima or converge slowly. The performance is also sensitive to hyperparameter tuning. Finally, the algorithm scaling is not always favorable, as it may require large swarms and many iterations to converge in high-dimensional space.

In PSO, each particle has a position $\mathbf{x}_j$ (candidate solution) and velocity $\mathbf{v}_j$ in the search space. The algorithm updates these based on the best known position the particle has individually found and the best known position found by any particle:

$$\mathbf{v}_j^{(k+1)} = \omega \mathbf{v}_j^{(k)} + c_1 r_1 \big(\mathbf{x}_j^{(k)*} - \mathbf{x}_j^{(k)}\big) + c_2 r_2 \big(\mathbf{x}^{(k)*} - \mathbf{x}_j^{(k)}\big), \tag{5.36a}$$

$$\mathbf{x}_j^{(k+1)} = \mathbf{x}_j^{(k)} + \mathbf{v}_j^{(k+1)}, \tag{5.36b}$$

where $\mathbf{x}_j^{(k)*}$ is particle j's best historical position, $\mathbf{x}^{(k)*}$ is the global best across all particles, and r_1, r_2 are random scalars in $[0, 1]$. The parameters ω, c_1, c_2 control the balance between exploration and exploitation.

Code Example

Here we revisit the example objective function in Eq. (5.32) with PSO. The code below will initialize a swarm of particles and iterate through the PSO algorithm. Figure 5.20 shows the progression of the swarm versus iteration, with early iterations being plotted in red and later iterations plotted in blue. The swarm converges toward the optimum value in relatively few iterations, shown in Fig. 5.21.

CodeSnip 5.g

```
def particle_swarm_optimization(n_particles=30, n_iter=20, w
    =0.5, c1=1.5, c2=1.5, bounds=(-5, 5)):
    positions = np.random.uniform(bounds[0], bounds[1], size=(
        n_particles, 2))
    velocities = np.zeros_like(positions)
    p_best = positions.copy()
```

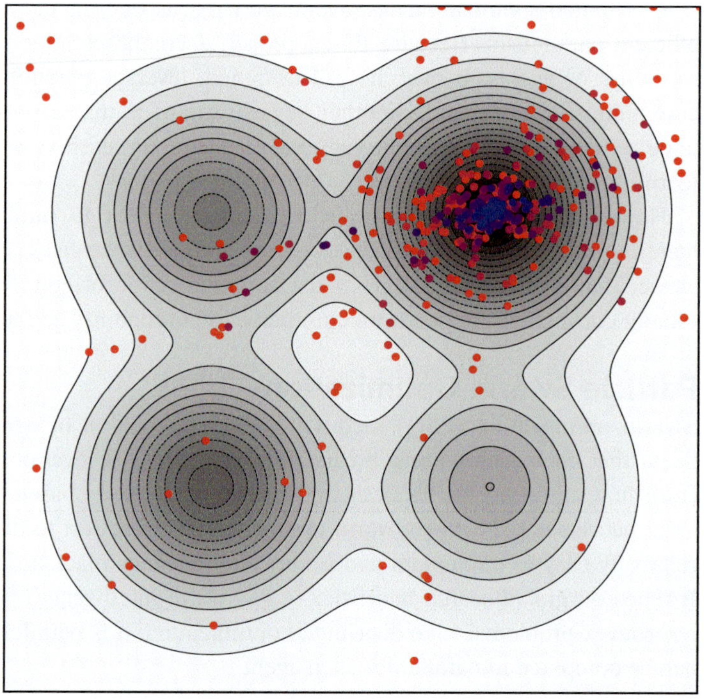

Figure 5.20 Illustration of the particle swarm optimization, where the initial swarm locations are plotted in red, and as they optimize they become more blue. Eventually the global minimum is found.

```
    p_best_vals = np.apply_along_axis(lambda p: f(*p),1,p_best)
    g_best = p_best[np.argmin(p_best_vals)]
    history = [positions.copy()]

    for _ in range(n_iter):
        r1 = np.random.rand(n_particles, 1)
        r2 = np.random.rand(n_particles, 1)
        velocities = w * velocities + c1 * r1 * (p_best -
            positions) + c2 * r2 * (g_best - positions)
        positions = np.clip(positions + velocities, bounds[0],
            bounds[1])
        vals = np.apply_along_axis(lambda p: f(*p),1,positions)

        improved = vals < p_best_vals
        p_best[improved] = positions[improved]
        p_best_vals[improved] = vals[improved]
        g_best = p_best[np.argmin(p_best_vals)]
        history.append(positions.copy())

    return np.array(history), g_best
```

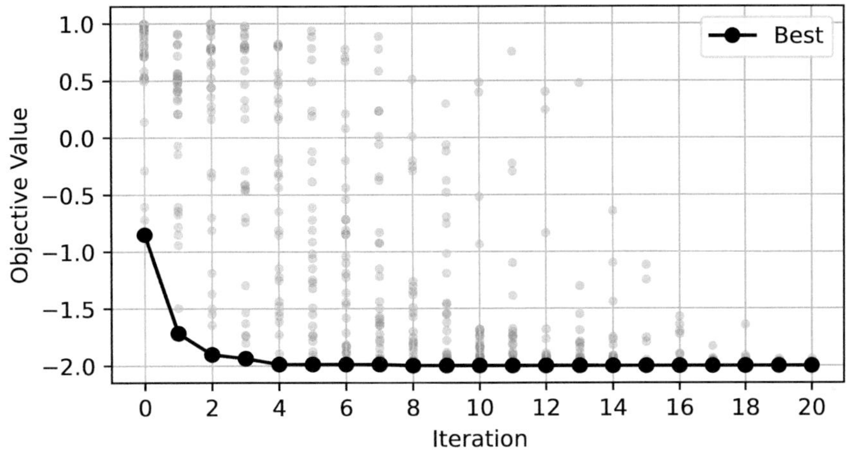

Figure 5.21 Objective function versus iteration for particle swarm optimization.

Ant Colony Optimization (ACO)

Ant Colony Optimization (ACO) [149], inspired by how ants deposit and follow pheromone trails, represents solutions as paths in a graph. At each step, an ant chooses its next edge with probability of moving from node i to node j given by

$$P_{i,j} = \frac{\tau_{i,j}^{\alpha}\,\eta_{i,j}^{\beta}}{\sum_{k\in\mathcal{N}_i}\tau_{i,k}^{\alpha}\,\eta_{i,k}^{\beta}}, \tag{5.37}$$

where $\tau_{i,j}$ is the pheromone level on edge (i, j), $\eta_{i,j}$ is a heuristic measure (e.g., inverse of distance), and α, β tune the relative influence of pheromone vs. heuristic. The denominator normalizes over all feasible neighbors $\mathcal{N}_i$ so that probabilities sum to one. After constructing candidate solutions, ants update the pheromone trails, reinforcing promising edges while allowing others to evaporate:

$$\tau_{i,j} \leftarrow (1-\rho)\,\tau_{i,j} + \sum_{\text{ants } k} \Delta\tau_{i,j}^{k}, \tag{5.38}$$

with evaporation rate $\rho \in (0, 1]$ and reinforcement $\Delta\tau_{i,j}^{k} = Q/L_k$ if ant k used edge (i, j) in a solution of cost L_k, and zero otherwise.

Both PSO and ACO rely on simple local rules at the agent level. Remarkably, the collective system can converge toward high-quality solutions, even in complex, high-dimensional landscapes. Convergence proofs typically rely on assumptions about problem structure and parameter tuning.

Suggested Reading

Texts

(1) Numerical Optimization, by J. Nocedal and S. J. Wright, 1999 [370].
(2) Proximal Splitting Methods in Signal Processing, P. L. Combettes and J.-Ch. Pesquet, 2011 [109].
(3) Simulated Annealing, by V. Laarhove, J. M. Peter, and E. H. L. Aarts, 1987 [516].
(4) Real-Time Optimization by Extremum-Seeking Control, by K. B. Ariyur and M. Krstić, 2003 [12].
(5) Genetic Algorithms, by D. E. Goldberg, 2006 [196].
(6) Genetic Programming as a Darwinian Invention Machine, by J. R. Koza, F. H. Bennett III, and O. Stiffelman, 1999 [287].

Papers and reviews

(1) Proximal algorithms, by N. Parikh and S. Boyd, *Foundations and Trends® in Machine Learning*, 2014 [380].
(2) Sparse relaxed regularized regression: SR3, by P. Zheng, T. Askham, S. L. Brunton, J. N. Kutz, and A. Y. Aravkin, *IEEE Access*, 2019 [557].
(3) Just relax: Convex programming methods for identifying sparse signals in noise, by J. A. Tropp, *IEEE Transactions on Information Theory*, 2006 [504].
(4) Convergence properties of the Nelder–Mead simplex method in low dimensions, by J. C. Lagarias, J. A. Reeds, M. H. Wright, and P. E. Wright, *SIAM Journal on optimization*, 1998 [294].
(5) Simulated annealing, by D. Bertsimas and J. Tsitsiklis, *Statistical Science*, 1993 [49].
(6) Interpretable machine learning for science with PySR and SymbolicRegression.jl, by M. Cranmer, *arXiv:2305.01582*, 2023 [118].
(7) Particle swarm optimization, by J. Kennedy and R. Eberhart, *Proceedings of the IEEE International Conference on Neural Networks*, 1995 [274].
(8) Distributed subgradient methods for multi-agent optimization, by A. Nedić and A. Ozdaglar, *IEEE Transactions on Automatic Control*, 2009 [361].

Homework

Exercise 5.1 This exercise will compare several sparse regression algorithms, including the LASSO, proximal gradient descent with ℓ_1 and ℓ_0 regularization, and SR3. We will consider the system $\mathbf{Ax} \approx \mathbf{b}$, where $\mathbf{x} \in \mathbb{R}^{10}$ has exactly 2 nonzero elements. Generate a random $\mathbf{A} \in \mathbb{R}^{100\times 10}$, so that our observations $\mathbf{b} \in \mathbb{R}^{100}$ are a combination of exactly 2 columns of $\mathbf{A}$. Add a small amount of noise (say 1%) to the vector $\mathbf{b}$ and solve for $\mathbf{x}$ using the following methods:

- LASSO regression for various λ;
- Proximal gradient descent on the ℓ_1-regularized regression problem, for various λ;
- Proximal gradient descent on the ℓ_0-regularized regression problem, for various λ;
- Sequentially thresholded least squares (STLS), consisting of iteratively alternating between computing least squares and thresholding small coefficients below a threshold τ;
- SR3.

For a good value of λ, plot the proximal gradient descent versus iteration. How many iterations does SR3 take to converge? LASSO? Discuss the results.

Exercise 5.2 Repeat the exercise above for increasing amount of noise, from 1% to 25%. Discuss the trends and results.

Exercise 5.3 We will reconsider the same dataset from the first exercise, but now we will create a second test dataset, $\mathbf{b}_2 = \mathbf{A}_2\mathbf{x}$, where $\mathbf{A}_2$ is another random matrix of the same size as $\mathbf{A}$. Solve the regression problem on the first training data set using STLS and LASSO for various thresholds τ and regularization parameters λ. Plot the error of the training data fit versus the number of nonzero terms in $\mathbf{x}$. Now, on top of these plots, add the testing data error of the learned model. Do you notice any trends or correlations between the training error and testing error?

Exercise 5.4 Derive the projected gradient algorithm for a feasible set $C = \{(x, y) \mid x^2 + y^2 \leq r^2\}$ given by a disk of radius r. This derivation will be similar to the projected gradient with box constraints. Implement this algorithm to minimize the objective function $f(x, y) = (x - 0.5)^2 + (y - 0.2)^2$ starting from an initial condition $(1, 1)$.

Exercise 5.5 Derive the proximal gradient of a ridge regularization term $\alpha\|\mathbf{x}\|_2^2$. Does the proximal gradient descent on the ridge regression problem $\min_{\mathbf{x}} \|\mathbf{Ax} - \mathbf{b}\|_2^2 + \alpha\|\mathbf{x}\|_2^2$ make sense?

Exercise 5.6 Derive the proximal gradient of the nuclear norm.

Exercise 5.7 Reproduce Figure 5.5, taken from Zheng et al. [557].

Exercise 5.8 Explore the downhill simplex method on the four-minima problem from Fig. 5.9. Start with several different random initial simplexes, of different sizes. Plot the ensemble of objective function values versus iteration number, as in Fig. 5.15.

Exercise 5.9 Explore the simulated annealing method on the four-minima problem from Fig. 5.9. Try several different cooling schedules and compare results.

Exercise 5.10 Now we will explore the downhill simplex method on a convex problem with a single global minimum: a quadratic objective function $f(\mathbf{x}) = \|\mathbf{x}\|_2^2$. In this problem, you will vary the different parameters of the algorithm, as well as the dimension of $\mathbf{x} \in \mathbb{R}^n$. Plot the ensemble of objective function values versus iterations, for the various choices of downhill simplex parameters and system dimension. Explore the convergence rate, success rate, etc. What are the optimal parameters?

Now repeat the problem above with the objective function $f(\mathbf{x}) = \frac{1}{2}\mathbf{x}^T\mathbf{Q}\mathbf{x}$. Try a few different $\mathbf{Q}$ with better and worse condition number. For example, for $\mathbf{x} \in \mathbb{R}^2$, try $\mathbf{Q}$ diagonal with entries 1 and 100. Try a non-normal $\mathbf{Q} = \begin{bmatrix} 10 & 1 \\ 0 & 10 \end{bmatrix}$.

Exercise 5.11 Repeat the exercise above with simulated annealing.

Exercise 5.12 Try the downhill simplex method on the objective function with a saddle point in Eq. (2.35) from Chapter 2. Explore the convergence rate for different parameters.

Now, attempt to add momentum to the algorithm, similar to how we added momentum in Chapter 2. How does this change the convergence rate and stability?

Exercise 5.13 Try the simulated annealing method on the objective function with a saddle point in Eq. (2.35) from Chapter 2. Explore the convergence rate for different parameters.

Exercise 5.14 Repeat the exercise above with particle swarm optimization.

Exercise 5.15 Use `PySR` to learn the following functions from data:

- $f(x) = \sin(\exp(x^2 + 3)))$
- $f(x, y) = \cos(x^2 + y^2) - \sin(x^2 - y^2)$
- $f(x, y) = x - \frac{1}{6}x^3 + \frac{1}{120}x^5$.

Start with 100 data points uniformly distributed on the interval $[-1, 1]$. Try with 100 randomly distributed points on this interval. Add noise until the algorithm breaks.

Exercise 5.16 Now we will explore the symbolic regression example where we learned the Lorenz system in Fig. 5.19. Explore how much noise the algorithm can tolerate, and how little data is required to learn the correct model.

Exercise 5.17 (Extremum Seeking) Even when it is not possible to compute the gradient of a function, it may be possible to estimate the gradient by injecting a sinusoidal input perturbation and cross-correlating with the resulting output perturbation. The method of *extremum seeking control* (ESC) [291, 12] uses this approach to track the maximum of a function, even when the objective function varies in time. In the simplest form, when the objective function is static, this may be viewed as a form of gradient-free optimization.

Given a convex objective function $J(u)$, extremum-seeking adds a sinusoidal perturbation to the estimate of the input that maximizes the objective, $\hat{u}$: $u = \hat{u} + a\sin(\omega t)$. This perturbation results in an objective function J that varies sinusoidally about a mean value. The output J is high-pass filtered to remove the mean, resulting in the oscillating ρ. The

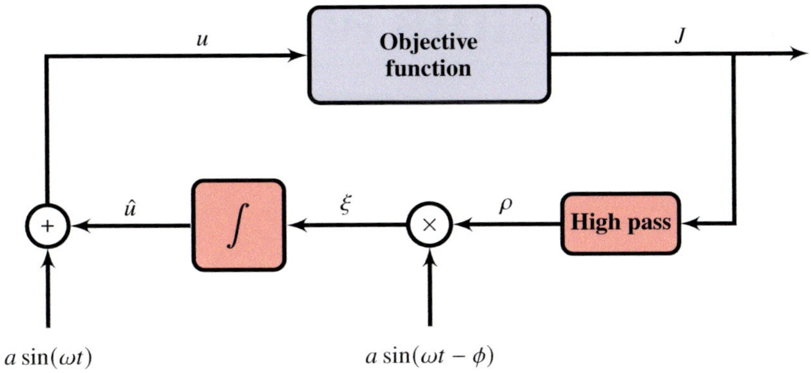

Figure 5.22 Schematic of extremum seeking. A sinusoidal perturbation is added to the input $\hat{u}$, which results in an output perturbation. A high-pass filter removes the mean. The oscillating ρ is multiplied by the input perturbation. The resulting ξ estimates the gradient of the objective and is integrated into the the optimizing input $\hat{u}$.

high-pass filter is chosen to pass the perturbation frequency ω. The high-pass filtered ρ is multiplied by the input sinusoid, possibly with a phase shift ϕ, resulting in ξ: $\xi = a\sin(\omega t - \phi)\rho$, which estimates the gradient of J. This signal ξ is mostly positive if $u < u^*$ and is mostly negative if $u > u^*$. Thus, ξ is integrated into $\hat{u}$, the best estimate of the optimizing value, $\frac{\mathrm{d}}{\mathrm{d}t}\hat{u} = k\,\xi$, so that the system estimate $\hat{u}$ is steered towards the optimal input u^*. Here, k is an integral gain, which determines how aggressively the algorithm climbs gradients in J. A schematic of extremum seeking is shown in Fig. 5.22.

- Show that the average ξ_{avg} is proportional to the gradient of the objective J with respect to u. First, write $J(u) = J(\hat{u} + a\sin(\omega t))$ and expand $J(u)$ in the perturbation amplitude a, which is small. Then average $\xi = a\sin(\omega t - \phi)\rho$ over one period.
- Use extremum seeking with perturbation frequency $\omega = 10\,\text{Hz}$ and amplitude $a = 0.2$ to maximize a quadratic objective $J(u) = 25 - (5 - u)^2$. Plot the tracking of the optimal value and discuss. Hint: use a simple Butterworth high-pass filter.
- Use extremum seeking to find the peak of a quartic cost function $J(u) = 16 - (2 - u)^4$. Do you need to modify the extremum seeking control parameters? Plot the response and discuss the performance.
- Use extremum seeking to track the maximum of a time-varying quadratic objective $J(u, t) = 25 - (\sin(.1t) - u)^2$.

6 Constraints and Duality

Constraints are an essential component of many optimization problems. Most real-world problems are constrained by resources, time, material properties, cost, etc. Until now we have mostly either considered unconstrained problems, such as least-squares regression and gradient descent, or problems with the simplest constraints and cost functions, such as linear programming.

Constraints shape the feasible set of an optimization problem. In Chapter 3 on linear programming, we saw that the constraints almost entirely determine the problem, as the optimum always lives on the vertex of the feasible polytope defined by the constraints. We also saw constraints arise in Chapter 4 in the context of constrained linear regression. The importance of constraints in optimization extends far beyond these simple linear and quadratic problems, becoming critical for modeling a wide range of real-world systems.

We begin this chapter with an in-depth discussion about Lagrange multipliers for enforcing equality constraints. We will derive the conditions for optimality of a constrained problem, with several examples. Then, we formally introduce the more general Karush–Kuhn–Tucker (KKT) conditions, which hold for inequality constraints as well. Lagrange multipliers and the KKT conditions have an intuitive geometric interpretation in terms of directional derivatives of the objective and constraint functions. Specifically, locally optimal solutions will be those such that the gradient of the objective function has no direction tangent to the constraint equations. Thus, the objective normal must be in the span of the constraint normal directions, and the linear combination of these constraint normals will be the Lagrange multipliers.

The interactions between constraints and the objective function are surprisingly rich and interesting. The concept of *duality* provides crucial insight into this connection. At its core, duality provides a deeper perspective on constrained optimization, often by reformulating the original, primal problem into a complementary dual problem. Analyzing this dual problem can reveal tight bounds on the original objective and provide methods for efficient computation. In some cases, the dual formulation may be simpler or more computationally tractable. We will explore duality and constraints, expanding and developing the earlier perspective in Section 1.4 based on infinite penalty functions. Dual variables also have the useful interpretation as the sensitivities of the objective function with respect to active constraints.

Duality is most easily understood through linear programming (via the primal–dual pair of linear programs) and quadratic programming (via the classical use of

Lagrange multipliers and KKT conditions). However, the concept of duality is much more widely applicable. In mechanics and physics, for instance, dual principles connect variational formulations of equilibrium conditions with complementary energy or stress-based descriptions. In the calculus of variations, dual arguments can simplify or solve certain boundary value problems by re-expressing constraints or re-labelling the roles of independent and dependent variables.

Duality then leads us to interior point methods (IPMs), which are a powerful class of algorithms for constrained optimization. There are several variants of IPMs that leverage primal and dual formulations to efficiently find optimal solutions to convex problems. These methods may also be used for nonconvex problems, as were discussed in Chapter 5, although there are typically fewer guarantees in the nonconvex case, and extra care must be taken.

We then proceed to introduce one of the most widely used splitting algorithms, the alternating direction method of multipliers (ADMM), which is a powerful primal–dual method that is used to handle complex problems with nonsmooth regularizers. ADMM relies on the proximal operator from Section 5.1, and is closely related to the sparse, relaxed, regularized regression from the same section. Both methods involve operator splitting and proximal steps; however, SR3 is primal only, while ADMM is primal–dual. This introduces subtle differences that we will discuss throughout this section.

Finally, we will introduce infinite-dimensional optimization problems and the calculus of variations. This is a rich field with an equally rich history. Although it may seem beyond the scope of this chapter, the concepts of constraints and duality were first introduced in these infinite-dimensional problems in the context of classical mechanics. The Euler–Lagrange equations, which are used to derive the equations of motion in physics, also provide optimality conditions for infinite-dimensional optimization problems. Further, the Legendre transformation is an important transformation of the Lagrangian into a dual Hamiltonian formulation.

Variational optimization is extremely relevant for broad and useful classes of infinite-dimensional problems, including control theory, machine learning, and shape optimization in fluid applications. We will investigate several key infinite-dimensional optimization problems, such as the classic brachistochrone and tautochrone, which have broad implications for modern optimization problems. We already saw a very simplified cousin of the brachistochrone problem in Example Panel 1.2 in Chapter 1, and we will also explore how to solve the full brachistochrone problem using modern techniques in automatic differentiation in Chapter 10.

To summarize, this chapter will accomplish three major goals:

1. Describe how to handle constraints, and how they may be tractably optimized through Lagrange multipliers and KKT conditions;
2. Introduce the related concept of duality, which is important in advanced algorithms, such as IPMs, discussed in Section 6.4, and ADMM, discussed in Section 6.5; and
3. Discuss infinite-dimensional problems where the objective is a functional of a function to be optimized.

6.1 Lagrange Multipliers

We begin by discussing equality constraints $\mathbf{h}(\mathbf{x}) = \mathbf{0}$, which are the simplest class of constraints to handle. The constrained optimization problem is written as

$$\begin{aligned} \min_{\mathbf{x}} \quad & f(\mathbf{x}), \\ \text{s.t.} \quad & \mathbf{h}(\mathbf{x}) = \mathbf{0}. \end{aligned} \tag{6.1}$$

The method of Lagrange multipliers reformulates the constrained optimization problem, moving the constraints into the objective function. This introduces additional decision variables, the Lagrange multipliers $\boldsymbol{\lambda}$, that act as weights to enforce the constraint:

$$\min_{\mathbf{x}} \sup_{\boldsymbol{\lambda}} \quad f(\mathbf{x}) + \boldsymbol{\lambda}^{\mathrm{T}}\mathbf{h}(\mathbf{x}). \tag{6.2}$$

This new objective $\mathcal{L}(\mathbf{x}, \boldsymbol{\lambda}) = f(\mathbf{x}) + \boldsymbol{\lambda}^{\mathrm{T}}\mathbf{h}(\mathbf{x})$ is called the Lagrangian, and it is closely related to the Lagrangian from classical mechanics (Section 6.6) and control theory (Chapter 9). Now, in addition to minimizing over $\mathbf{x}$, we also extremize over $\boldsymbol{\lambda}$ in the supremum[1].

We derived a simplified form of this in Section 1.4 by adding the infinite penalty function $\delta_0(\mathbf{h}(\mathbf{x}))$ to the objective; this function is zero when the argument is zero and infinite when the argument is not. If we add this penalty function to the objective function

$$\min_{\mathbf{x}} \quad f(\mathbf{x}) + \delta_0(\mathbf{h}(\mathbf{x})) \tag{6.3}$$

then the optimal solution is equivalent to that of the original constrained problem. We then replace this nonsmooth term with $\boldsymbol{\lambda}^{\mathrm{T}}\mathbf{h}(\mathbf{x})$, and as λ_j goes to infinity, this term is large enough to enforce the $h_j(\mathbf{x}) = 0$ constraint. Writing $\min_{\mathbf{x}} \sup_{\boldsymbol{\lambda}}$ takes care of this detail, as it ensures that when the constraint is violated the inner supremum is infinite, and when the constraint is satisfied, it reduces to the original constrained problem in Eq. (6.1).

Equation (6.1) is referred to as the *primal* optimization problem, and Eq. (6.2) is the Lagrangian form of the primal. In Section 6.3 we will see that swapping the order of the min and max in Eq. (6.2) and partially minimizing over $\mathbf{x}$ will result in the *dual* optimization problem, which provides a lower bound on the original primal optimization problem. Specifically, we have the following relationship

$$\min_{\mathbf{x}} \max_{\boldsymbol{\lambda}} \quad f(\mathbf{x}) + \boldsymbol{\lambda}^{\mathrm{T}}\mathbf{h}(\mathbf{x}) \quad \geq \quad \max_{\boldsymbol{\lambda}} \min_{\mathbf{x}} \quad f(\mathbf{x}) + \boldsymbol{\lambda}^{\mathrm{T}}\mathbf{h}(\mathbf{x}). \tag{6.4}$$

[1] The *supremum* (sup) of a function is its least upper bound. It generalizes the maximum when the maximum might not be attained. We write $\sup_{\mathbf{x}} f(\mathbf{x})$ to refer to the greatest value that $f(\mathbf{x})$ *approaches* over the feasible domain $\mathbf{x} \in \mathcal{S}$, even if no $\mathbf{x}$ actually achieves it. We use sup to ensure well-defined expressions, even when the Lagrangian does not attain its maximum. For example, when the constraint is not satisfied, and $\mathbf{h}(\mathbf{x}) \neq \mathbf{0}$, then it is possible to drive this term $\boldsymbol{\lambda}^{\mathrm{T}}\mathbf{h}(\mathbf{x})$ to infinity by choosing increasingly large $\boldsymbol{\lambda}$; thus the supremum is infinity, even though the maximum is not attained. This is a technical point that we won't dwell on much here. The supremum is always defined, even when max is not. The sup can be replaced with max when it exists, and we will often simply use max for simplicity.
Example: Let $f(x) = 1/x$ over the open interval $(0, 1)$. The supremum of f is equal to infinity, $\sup_{x\in(0,1)} f(x) = \infty$, even though $\max_{x\in(0,1)} f(x)$ does not exist because $f(x) \to \infty$ as $x \to 0^+$, but $0 \notin (0, 1)$. Similarly, $f(x) = 1 - 1/x$ over the open interval $(0, \infty)$ has a supremum of 1; the function never attains this upper bound for any finite x, but the values get arbitrarily close to this supremum as $x \to \infty$. Related to above, $\sup_{\lambda} \lambda x$ is ∞ for $x \neq 0$ and 0 for $x = 0$; thus $\min_x \sup_{\lambda} \lambda x = 0$.

We will show in Section 6.3 that under certain conditions of convexity and regularity, this bound will be tight, and the solutions to the primal and dual problems have the same value. However, even when these conditions do not hold, the dual problem provides a useful lower bound on the primal. Minimax inequalities are discussed more in Panel 6.1.

Rewriting Eq. (6.1) as Eq. (6.2), we have converted a lower-dimensional constrained problem into a higher-dimensional unconstrained problem with a more challenging minimax structure. In the next section, we will show that optimality comes in the form of the KKT equations, which we solve with standard techniques, such as Newton's method.

Geometry of Lagrange Multipliers

Geometrically, for $\mathbf{x}^*$ to be an optimal solution, the gradient of $f(\mathbf{x}^*)$ can have no component tangent to $\mathbf{h}(\mathbf{x}) = \mathbf{0}$ at $\mathbf{x}^*$. A tangent direction $\mathbf{d}$ to $\mathbf{h}(\mathbf{x}) = \mathbf{0}$ satisfies $\nabla\mathbf{h}(\mathbf{x}) \cdot \mathbf{d} = 0$. These tangent directions are *feasible* directions in which we can move the variable $\mathbf{x}$ to improve the objective function while still maintaining the constraint equation. If ∇f improves in one of these feasible tangent directions, then the current point is not optimal. Thus, ∇f must be normal to $\mathbf{h}(\mathbf{x})$, meaning that it is in the span of the normals $\nabla h_j(\mathbf{x})$.

This is simple to visualize for a single constraint $h(\mathbf{x}) = 0$ in two-dimensions, $\mathbf{x} \in \mathbb{R}^2$, as in Fig. 6.1. If $f(\mathbf{x})$ decreases in a direction that is tangent to h, meaning that $\nabla f(\mathbf{x})$ has a component tangent to h, then we can achieve a solution that is closer to optimal by moving in this direction. Thus, a necessary condition for optimality is that $\nabla f(\mathbf{x})$ is orthogonal to the tangent space at h, which means it is in the direction of ∇h. We will revisit and generalize this geometric intuition to higher dimensions and more constraints soon.

Panel 6.1: Minimax Inequality

To derive the general form of the minimax inequality used in Eq. (6.4), we let $f : X \times Y \to \mathbb{R}$ be any real-valued function. Then we have

$$\min_{x\in X}\max_{y\in Y} \quad f(x,y) \quad \geq \quad \max_{y\in Y}\min_{x\in X} \quad f(x,y).$$

This inequality holds for all $x \in X$ and $y \in Y$, regardless of whether f is continuous or convex. This may be derived using the following pointwise inequality:

$$\min_{x'\in X} f(x',y) \leq f(x,y) \leq \max_{y'\in Y} f(x,y').$$

Taking $\max_y$ on the left and $\min_x$ on the right yields the desired minimax inequality:

$$\max_{y\in Y}\min_{x\in X} \quad f(x,y) \quad \leq \quad \min_{x\in X}\max_{y\in Y} \quad f(x,y).$$

In zero-sum games, $f(x, y)$ represents the payoff for Player A choosing x and Player B choosing y. Then $\min_x \max_y f(x, y)$ is the best worst-case outcome Player A can guarantee, while $\max_y \min_x f(x, y)$ is the best guaranteed outcome Player B can ensure. The minimax inequality reflects that, in general, the first player is at a disadvantage, since they commit to a strategy without knowing the other's move.

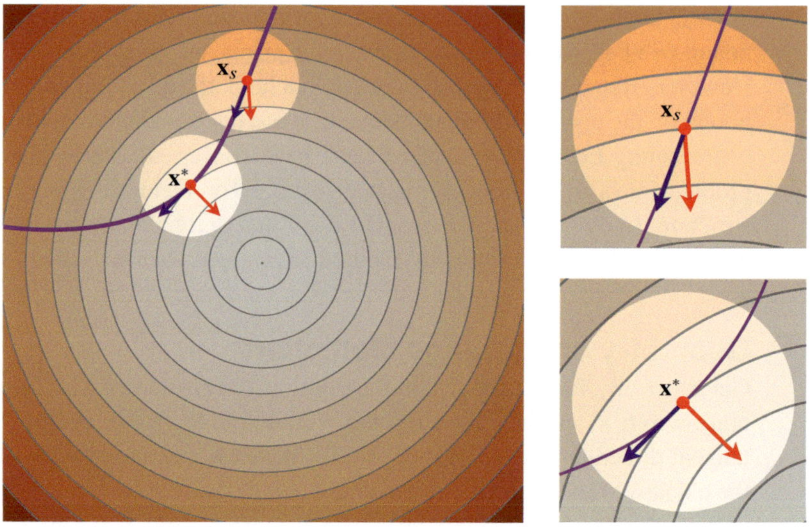

Figure 6.1 Illustration of the geometry of an optimal solution $\mathbf{x}^*$ in terms of the gradient of the objective function (red arrow) and tangent of the constraint equation (purple arrow); the constraint is shown in purple. We zoom into two points on the right, indicated by shaded regions. In the top, we see a sub-optimal $\mathbf{x}_s$, where the constraint tangent and objective normal are not orthogonal, so we may improve the objective function by moving in the feasible constraint tangent direction. In the bottom, we see the optimal $\mathbf{x}^*$, where the constraint tangent and objective normal directions are orthogonal.

The conditions for optimality of a solution $(\mathbf{x}^*, \lambda^*)$ are the following:

$$\nabla_{\mathbf{x}} \mathcal{L}(\mathbf{x}^*, \lambda^*) = \mathbf{0}, \tag{6.5a}$$

$$\nabla_{\lambda} \mathcal{L}(\mathbf{x}^*, \lambda^*) = \mathbf{0}. \tag{6.5b}$$

These are the *first-order* optimality conditions or the stationarity conditions, and are a subset of the Karush–Kuhn–Tucker (KKT) conditions, to be discussed in Section 6.2.

The second condition amounts to the constraint being enforced at $(\mathbf{x}^*, \lambda^*)$, as

$$\nabla_{\lambda} \mathcal{L}(\mathbf{x}^*, \lambda^*) = \mathbf{0} \quad \Longrightarrow \quad \mathbf{h}(\mathbf{x}^*) = \mathbf{0}. \tag{6.6}$$

The first condition yields the following expression

$$\nabla_{\mathbf{x}} \mathcal{L}(\mathbf{x}^*, \lambda^*) = \mathbf{0} \quad \Longrightarrow \quad \nabla_{\mathbf{x}} f(\mathbf{x}^*) + \lambda^{*T} \nabla_{\mathbf{x}} \mathbf{h}(\mathbf{x}^*) = \mathbf{0}. \tag{6.7}$$

This implies the following

$$\nabla_{\mathbf{x}} f(\mathbf{x}^*) = -\lambda^{*T} \nabla_{\mathbf{x}} \mathbf{h}(\mathbf{x}^*). \tag{6.8}$$

For a single equality constraint $h(\mathbf{x}) = 0$, this simplifies to

$$\nabla_{\mathbf{x}} f(\mathbf{x}^*) = -\lambda^* \nabla_{\mathbf{x}} h(\mathbf{x}^*), \tag{6.9}$$

which is the same as the geometric interpretation that ∇f must have no component in the tangent direction to $h(\mathbf{x}) = 0$ at an optimal $\mathbf{x}^*$. Said another way, ∇f must point in a direction normal to the constraint surface defined by $h(\mathbf{x}) = 0$, given by ∇h. If this is not true, then we can locally improve the objective by moving in a feasible direction.

In this example, *feasible directions* $\mathbf{d}$ at a point $\mathbf{x}$ satisfy

$$\nabla h(\mathbf{x}) \cdot \mathbf{d} = 0. \tag{6.10}$$

Optimality of $\mathbf{x}^*$ requires that the directional derivative of the objective function vanishes in every *feasible* direction $\mathbf{d}$:

$$\nabla f(\mathbf{x}^*) \cdot \mathbf{d} = 0 \quad \forall \mathbf{d} \text{ with } \nabla h(\mathbf{x}^*) \cdot \mathbf{d} = 0. \tag{6.11}$$

Hence $\nabla f(\mathbf{x}^*)$ must be orthogonal to the tangent space and therefore lies in the one-dimensional normal space spanned by $\nabla h(\mathbf{x}^*)$:

$$\nabla f(\mathbf{x}^*) = -\lambda \nabla h(\mathbf{x}^*). \tag{6.12}$$

The scalar λ is the *Lagrange multiplier*, providing the exact rescaling of the normals.

When we have more constraint equations, we will see that there may be several normal directions to the constraint equations, and the gradient of the objective function must be in this constraint-normal subspace. The values of these constraint normals that add up to equal $-\nabla f$ are the Lagrange multipliers in higher dimensions.

Simple Example

Here we introduce a concrete example to illustrate the method of Lagrange multipliers to solve a constrained optimization problem. Consider $f(\mathbf{x})$ to be a linear objective function $\mathbf{c}^T\mathbf{x} = x + 2y$ with the constraint that our solution remains on the unit circle $x^2 + y^2 = 1$:

$$\begin{aligned} \min_{x,y} \quad & x + 2y \\ \text{s.t.} \quad & x^2 + y^2 = 1. \end{aligned}$$

We may solve this problem by hand and analyze the behavior of the Lagrangian for various values of λ. Note, this optimization problem is nonconvex because of the constraint equation.

First, we move the constraints to the objective function via the Lagrangian $\mathcal{L}(\mathbf{x}, \lambda)$:

$$\min_{x,y} \max_{\lambda} \quad x + 2y + \lambda(x^2 + y^2 - 1).$$

Notice here that we rearranged the constraint $x^2 + y^2 = 1$ to be in the standard form $h(x, y) = x^2 + y^2 - 1 = 0$.

Based on the inequality in Eq. (6.4), we may swap the order of the min and max and solve the dual problem, which provides a lower bound on the primal. We first minimize over (x, y) and then maximize over λ by taking the various partial derivatives and setting equal to zero. This yields the optimal (x^*, y^*, λ^*) of the dual:

$$\begin{aligned} \min_{x} \quad x + 2y + \lambda(x^2 + y^2 - 1) \quad &\Longrightarrow \quad 1 + 2\lambda x = 0 \quad \Longrightarrow \quad x^* = -\frac{1}{2\lambda} \\ \min_{y} \quad x + 2y + \lambda(x^2 + y^2 - 1) \quad &\Longrightarrow \quad 2 + 2\lambda y = 0 \quad \Longrightarrow \quad y^* = -\frac{1}{\lambda}. \end{aligned}$$

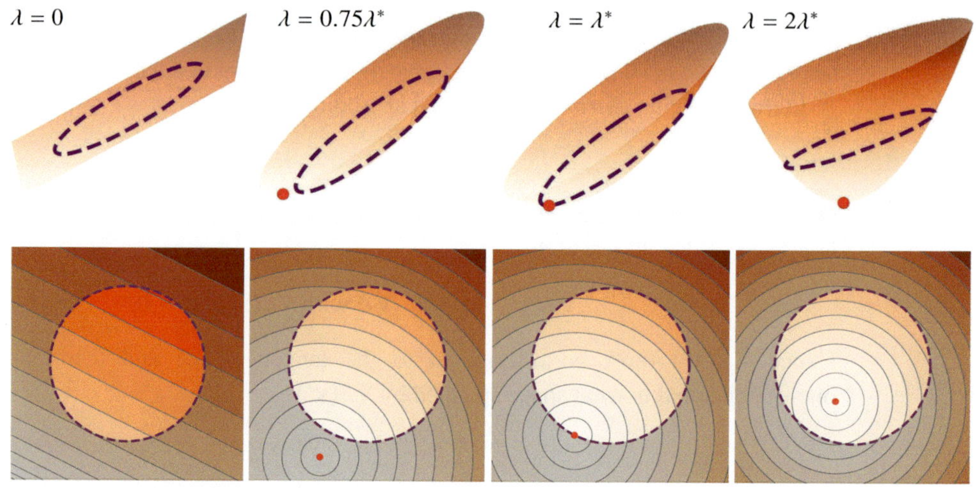

Figure 6.2 Illustration of the geometry of the Lagrangian as we vary λ. The top row shows the three-dimensional surface plots of the Lagrangian, and the bottom row shows the two-dimensional contour plots of the same; the purple circle denotes the constraint $x^2 + y^2 = 1$. At λ^*, the optimum value of the Lagrangian is on the constraint boundary. For smaller $\lambda < \lambda^*$, the optimum is outside of the boundary, and for larger $\lambda > \lambda^*$ the optimum is inside the boundary. Therefore, the optimum λ^* is the Lagrange multiplier required so that the minimum is on the boundary of the constraint and the objective function has the same value as the Lagrangian at that point. Note that even though we shade the exterior of the constraint circle to emphasize the boundary, we are considering an equality constraint, so solutions are restricted to be *on* the circle.

Substituting these back into the Lagrangian and optimizing over λ we have:

$$\max_{\lambda} -\frac{1}{2\lambda} - \frac{2}{\lambda} + \lambda\left(\frac{1}{4\lambda^2} + \frac{1}{\lambda^2} - 1\right) = \max_{\lambda} \frac{-5}{4\lambda} - \lambda \quad \Longrightarrow \quad \lambda^* = \frac{\sqrt{5}}{2}. \tag{6.13}$$

We may plug the optimal solution into the dual problem in Eq. (6.13) to obtain a maximum of $-\sqrt{5}$ and into the primal problem to obtain a minimum of $-\sqrt{5}$. These are equal, so the duality gap is zero and this solution is a global optimum. As an exercise, confirm that the gradient of the objective function is normal to the constraint equation at this point.

Note that we explicitly derived the dual problem in Eq. (6.13), which is a maximization over λ, after the inner minimization over (x, y) has been performed. We will discuss duality more in Section 6.3. The dual optimal technically only provides a lower bound on the primal. However, since we evaluated the primal at this optimum point and the value was equal to the value of the dual, then the inequality becomes equal, which shows that we have actually found the global optimum, even of this nonconvex problem[2]!

It is interesting to visualize the Lagrangian $L(x, y, \lambda)$ as a function of λ, as in Fig. 6.2. Here we plot the Lagrangian, both as a three-dimensional surface, and as a two-dimensional contour plot. For $\lambda = 0$, we recover the simple linear objective function. For small $\lambda < \lambda^*$, the optimum of the Lagrangian moves toward the constraint boundary. At λ^* the optimum

[2] The dual is concave on the branch where $\lambda > 0$, and x and y are unique for a given λ. So even though the overall problem is nonconvex, the circle constraint is well-behaved enough to yield a unique solution.

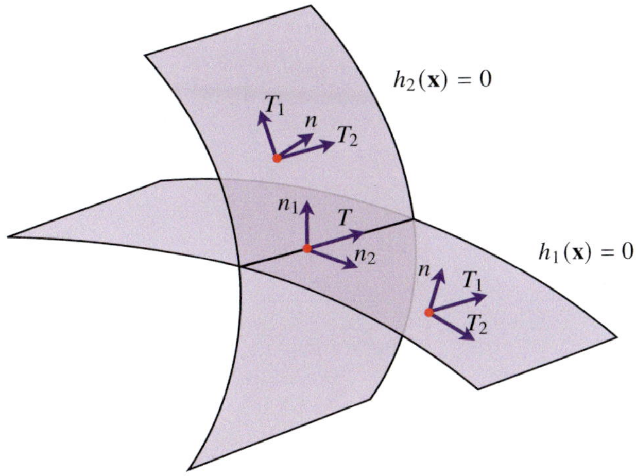

Figure 6.3 Geometric interpretation of constraint equations in higher dimensions as defining feasible submanifolds.

of the Lagrangian is exactly on the constraint boundary, so $\lambda^* h(x) = 0$ and the constraint is satisfied. This also means the value of the Lagrangian is the same as the original objective at this point. For larger $\lambda > \lambda^*$, we see the optimal value of the Lagrangian moves inside the feasible region. Therefore, the Lagrange multiplier λ weights the constraint equation so that the optimum of the Lagrangian is *exactly* on the constraint boundary.

More Constraints

When we have more constraints, then $\mathbf{h}(\mathbf{x}) = \mathbf{0}$ defines a *constraint manifold* in $\mathbb{R}^n$. If the constraints are convex, they define a constraint subspace in $\mathbb{R}^n$; recall that convex equality constraints are affine, $\mathbf{h}(\mathbf{x}) = \mathbf{A}\mathbf{x} - \mathbf{b}$. In either case, a necessary condition for optimality of a point $\mathbf{x}^*$ is that the gradient of f must have no component in the tangent directions to $\mathbf{h}(\mathbf{x}) = \mathbf{0}$ at $\mathbf{x}^*$. Otherwise, we could improve the solution by moving in this direction.

The tangent space $T_{\mathbf{x}}$ to the constraint manifold at a point $\mathbf{x}$ is the linear subspace:

$$T_{\mathbf{x}} = \{\mathbf{d} \mid \nabla h_j(\mathbf{x}) \cdot \mathbf{d} = 0 \, \forall j\}. \tag{6.14}$$

The tangent space $T_{\mathbf{x}}$ and the space of constraint normals $N_{\mathbf{x}} = \text{span}\{\nabla h_j(\mathbf{x})\}$ are complementary subspaces, so $\mathbb{R}^n = T_{\mathbf{x}} \oplus N_{\mathbf{x}}$. Thus, for the gradient of the objective function ∇f to not have any component in the tangent space, it must be a linear combination of the constraint normals $\{\nabla h_j(\mathbf{x})\}$. The necessary condition in Eq. (6.8) establishes exactly this linear combination, where the Lagrange multipliers λ_j are exactly the coordinates of $\nabla f(\mathbf{x})$ in the basis of constraint normals. If $\nabla f = -\lambda^{\mathrm{T}} \nabla \mathbf{h}$, then we may write the following:

$$\nabla f(\mathbf{x}^*) = -\lambda_j \nabla h_j(\mathbf{x}^*), \tag{6.15}$$

which implies that

$$\nabla f \cdot \nabla h_j \neq 0, \tag{6.16}$$

so that ∇f is not in the tangent space $T_{\mathbf{x}^*}$. This can be visualized geometrically in Fig. 6.3.

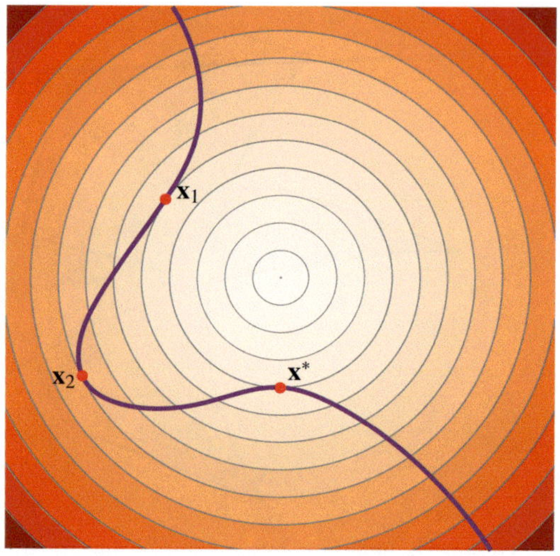

Figure 6.4 For a nonconvex optimization problem, having the constraint tangent and objective normal directions be orthogonal is a necessary but not sufficient condition for the solution to be globally optimal. Here we see three points that satisfy this condition: $\mathbf{x}_1$ is a local minimum, $\mathbf{x}_2$ is a local maximum, and $\mathbf{x}^*$ is a global minimum.

Tilt and Match Interpretation. There is an intuitive geometric interpretation of Lagrange multipliers, building on the intuition from Fig. 6.2. We think of the Lagrange multiplier as defining a family of tilted objective functions

$$f_\lambda(\mathbf{x}) = f(\mathbf{x}) + \lambda^{\mathrm{T}}\mathbf{h}(\mathbf{x}). \tag{6.17}$$

We then vary λ until a level set $f_{\lambda=\lambda^*}$ becomes *tangent* to the constraint surface $\mathbf{h}(\mathbf{x}) = \mathbf{0}$. At that critical λ^*, the first-order condition $\nabla f + \lambda^{\mathrm{T}}\nabla\mathbf{h} = \mathbf{0}$ is satisfied.

Convexity and Sufficient Conditions

The optimality conditions above are *necessary* conditions for a state $\mathbf{x}^*$ to be optimal. In the case of a convex problem, then they are also *sufficient*. Again, for the problem to be convex, $\mathbf{h}(\mathbf{x}) = \mathbf{0}$ must be an affine constraint $\mathbf{A}\mathbf{x} = \mathbf{b}$. However, if either f or the feasible set defined by $\mathbf{h}(\mathbf{x}) = \mathbf{0}$ are nonconvex, then the optimality conditions may not be sufficient. Essentially, these conditions just indicate a stationary point, but it may be a local minimum, maximum, or saddle point. We must then introduce second-order conditions to test for optimality. Again, we let $T_{\mathbf{x}}$ be the tangent space to the constraint manifold at a point $\mathbf{x}$:

$$T_{\mathbf{x}} = \{\mathbf{d} \mid \nabla\mathbf{h}(\mathbf{x}) \cdot \mathbf{d} = 0\}. \tag{6.18}$$

We conclude that $\mathbf{x}^*$ is a strict local minimum if

$$\mathbf{d}^{\mathrm{T}}\nabla^2_{\mathbf{xx}}\mathcal{L}(\mathbf{x}^*, \lambda^*)\mathbf{d} > 0 \quad \forall \mathbf{d} \in T_{\mathbf{x}} \tag{6.19}$$

as long as $\mathbf{d} \neq \mathbf{0}$. Figure 6.4 shows an example of a nonconvex constraint equation that gives rise to several locally optimal points that satisfy the necessary conditions.

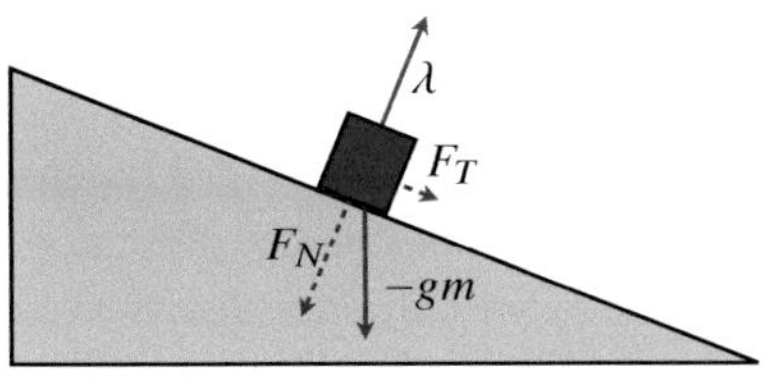

Figure 6.5 Here a simple physical constraint arises from a block sliding down an incline without friction. Gravity pulls down, and the ramp applies a normal force so that the block stays on top.

Regularity Conditions

The use of Lagrange multipliers requires a mild set of regularity conditions, called constraint qualifications, to ensure the KKT conditions hold at a solution. These conditions rule out pathological cases, such as when the active constraints at a point are not linearly independent or when the feasible region has no interior.

For equality-constrained problems, a key regularity condition is that gradients of the active constraints must be linearly independent at $\mathbf{x}^*$, so the Jacobian $\nabla \mathbf{h}(\mathbf{x}^*)$ has full rank:

$$\text{rank}\left[\nabla h_1(\mathbf{x}^*) \quad \cdots \quad \nabla h_m(\mathbf{x}^*)\right] = m.$$

This ensures that the set of feasible directions (i.e., the tangent space to the constraint surface) has the expected dimension. If the constraint gradients are linearly dependent, then instead of intersecting transversely to form a well-behaved surface, the constraints may define a degenerate feasible region. Thus, there may not be Lagrange multipliers that satisfy the stationarity condition, even at a point that minimizes the objective and satisfies the constraints. For example, two equality constraints in 2D whose gradients point in the same direction define parallel lines so the feasible region may be empty or ill-defined. Similar regularity conditions exist for the more general KKT theory of inequality constraints.

History

Lagrange first introduced the method of multipliers to eliminate constraint forces in holonomic[3] mechanics [296, 335]. Figure 6.5 shows a simple mechanical system of a block sliding down an inclined plane, where the plane exerts a force to keep the block exactly on the boundary; this force is a *Lagrange multiplier* enforcing the constraint that the block remains exactly on the boundary of the inclined plane. We will discuss the broader connections of constraints, Lagrange multipliers, and classical mechanics in Section 6.6.

Nearly two hundred years later, this approach was generalized to handle inequality constraints by Karush [272] and popularized by Kuhn and Tucker [292], marking the birth of modern nonlinear programming. Since then, the field has developed rapidly, with the theory of duality and the primal–dual algorithms that followed.

[3] A *holonomic constraint* is a constraint on a mechanical system that can be expressed as an explicit equation relating the generalized coordinates, typically in the form $h(q_1, q_2, \ldots, q_n, t) = 0$. Such constraints restrict the configuration space but not the velocities independently. In contrast, *nonholonomic* constraints involve the velocities and cannot be integrated to purely positional constraints. For example, a coin rolling without slipping is a non-holonomic system, while a block sliding down an incline is a holonomic system.

6.2 Karush–Kuhn–Tucker (KKT) Conditions

In the previous section, we saw how Lagrange multipliers allow us to enforce equality constraints by augmenting the objective function. Here we introduce the KKT conditions for optimality [272, 292], which generalize the Lagrange multiplier approach to also handle inequality constraints of the form $\mathbf{g}(\mathbf{x}) \leq \mathbf{0}$:

$$\begin{aligned} \min_{\mathbf{x}} \quad & f(\mathbf{x}), \\ \text{s.t.} \quad & \mathbf{h}(\mathbf{x}) = \mathbf{0} \\ & \mathbf{g}(\mathbf{x}) \leq \mathbf{0}. \end{aligned} \tag{6.20}$$

As in Section 6.1, the constrained minimization is recast in Lagrangian form,

$$\mathcal{L}(\mathbf{x}, \boldsymbol{\lambda}, \boldsymbol{\mu}) = f(\mathbf{x}) + \boldsymbol{\lambda}^{\mathrm{T}}\mathbf{h}(\mathbf{x}) + \boldsymbol{\mu}^{\mathrm{T}}\mathbf{g}(\mathbf{x}) \tag{6.21a}$$

$$= f(\mathbf{x}) + \sum_{j=1}^{m} \lambda_j h_j(\mathbf{x}) + \sum_{j=1}^{p} \mu_j g_j(\mathbf{x}) \tag{6.21b}$$

where $\boldsymbol{\lambda} \in \mathbb{R}^m$ and $\boldsymbol{\mu} \in \mathbb{R}_+^p$ are the Lagrange multipliers for equality and inequality constraints, respectively. Here $\mathbb{R}_+$ indicates the non-negative reals, so $\boldsymbol{\mu} \geq \mathbf{0}$.

The primal optimization problem in Eq. (6.20) may be recast in the following form

$$\min_{\mathbf{x}} \sup_{\boldsymbol{\lambda};\boldsymbol{\mu}\geq 0} \quad f(\mathbf{x}) + \boldsymbol{\lambda}^{\mathrm{T}}\mathbf{h}(\mathbf{x}) + \boldsymbol{\mu}^{\mathrm{T}}\mathbf{g}(\mathbf{x}), \tag{6.22}$$

where the optimal solution is the global minimum over $\mathbf{x}$ and a global maximum over the Lagrange multipliers $\boldsymbol{\lambda}$ and $\boldsymbol{\mu}$. This is derived in the same way as before, by adding the infinite penalty functions $\delta_0(\mathbf{h}(\mathbf{x}))$ and $\delta_{\mathbb{R}_-}(\mathbf{g}(\mathbf{x}))$ to the objective

$$\min_{\mathbf{x}} \quad f(\mathbf{x}) + \delta_0(\mathbf{h}(\mathbf{x})) + \delta_{\mathbb{R}_-}(\mathbf{g}(\mathbf{x})). \tag{6.23}$$

We then replace the nonsmooth terms with $\boldsymbol{\lambda}^{\mathrm{T}}\mathbf{h}(\mathbf{x})$ and $\boldsymbol{\mu}^{\mathrm{T}}\mathbf{g}(\mathbf{x})$ and take the supremum over $\boldsymbol{\lambda}$ and $\boldsymbol{\mu} \geq 0$. Note that $\boldsymbol{\mu}$ must be non-negative, else this supremum will be infinite, even when the constraint $\mathbf{g}(\mathbf{x}) \leq 0$ is satisfied.

Panel 6.2: Infinite Penalty Function δ_C

We already saw a sketch of how to derive Lagrange multipliers to enforce constraints in Section 1.4 using the infinite penalty function δ. We generalize the notation here, introducing $\delta_C(\mathbf{x})$ to denote an infinite penalty if $\mathbf{x}$ is not in the feasible set $\boldsymbol{C}$:

$$\delta_C(\mathbf{x}) = \begin{cases} 0, & \mathbf{x} \in \boldsymbol{C} \\ \infty, & \mathbf{x} \neq \boldsymbol{C}. \end{cases}$$

Thus, for equality constraints, we may use $\delta_0(\mathbf{h}(\mathbf{x}))$ and for inequality constraints we may use $\delta_{\mathbb{R}_-}(\mathbf{g}(\mathbf{x}))$, where $\mathbb{R}_-$ denotes the set $\mathbf{x} \leq 0$.

The first-order KKT conditions provide a set of necessary conditions for the point $(\mathbf{x}^*, \boldsymbol{\lambda}^*, \boldsymbol{\mu}^*)$ to be optimal:

$$\text{(stationarity)} \quad \nabla f(\mathbf{x}^*) + \boldsymbol{\lambda}^{*T} \nabla \mathbf{h}(\mathbf{x}^*) + \boldsymbol{\mu}^{*T} \nabla \mathbf{g}(\mathbf{x}^*) = \mathbf{0} \tag{6.24a}$$

$$\text{(primal feasibility)} \quad \mathbf{h}(\mathbf{x}^*) = \mathbf{0}, \ \mathbf{g}(\mathbf{x}^*) \leq \mathbf{0} \tag{6.24b}$$

$$\text{(dual feasibility)} \quad \boldsymbol{\mu}^* \geq \mathbf{0}, \tag{6.24c}$$

$$\text{(complementary slackness)} \quad \boldsymbol{\mu}^{*T} \mathbf{g}(\mathbf{x}^*) = 0. \tag{6.24d}$$

If $f(\mathbf{x})$ and $\mathbf{g}(\mathbf{x})$ are convex and $\mathbf{h}(\mathbf{x})$ are affine, so the optimization problem is convex, then the KKT conditions are necessary and sufficient for a point to be globally optimal[4]. For non-convex problems, the KKT system gives necessary first-order conditions. First order means the conditions only involve first derivatives of the objective and constraints.

The stationarity condition comes from $\nabla_{\mathbf{x}} \mathcal{L}(\mathbf{x}^*, \boldsymbol{\lambda}^*, \boldsymbol{\mu}^*) = \mathbf{0}$, and this generalizes the unconstrained optimality condition that $\nabla_{\mathbf{x}} f(\mathbf{x}^*) = \mathbf{0}$. We see that in the absence of inequality constraints, we recover the stationarity condition for equality constraints in Eq. (6.7). The stationarity equation also states that the gradient of the objective function is in the span of the active constraint normals. Recall that some of the inequality constraints may be inactive, with $\mu_j = 0$, so only the active constraint normals are considered.

The primal feasibility equation states that $\mathbf{x}^*$ must be in the feasible set. Dual feasibility states that each μ_j must be non-negative, since if the Lagrange multipliers can be negative, the supremum in Eq. (6.22) would be infinite even when the constraints are satisfied. In Section 6.3, we will rewrite the primal problem in Eq. (6.20) in terms of the maximization of a dual objective $\phi(\boldsymbol{\lambda}, \boldsymbol{\mu})$ in $\boldsymbol{\lambda}$ and $\boldsymbol{\mu}$. This dual objective will have a term $\boldsymbol{\mu}^{\mathrm{T}} \mathbf{g}(\mathbf{x})$, so if any μ_j is negative, then multiplying with any inactive negative constraints $g_j(\mathbf{x}) < 0$, we can drive this maximization to infinity. Thus, $\boldsymbol{\mu} \geq \mathbf{0}$ is required to prevent this.

The fourth complementary slackness condition, Eq. (6.24d), states that $\mu_j^* g_j(\mathbf{x}^*) = 0$ for each j, so either $\mu_j = 0$ or $g_j(\mathbf{x}^*) = 0$. Either $g_j(\mathbf{x}^*) = 0$ is *active*, in which case it acts like a standard equality constraint and $\mu_j > 0$ is a standard Lagrange multiplier, or $g_j(\mathbf{x}^*) < 0$ is *inactive*, in which case $\mu_j = 0$. This is closely related to the slack variable form used in Chapter 3 for solving linear programs with Dantzig's simplex method. When the constraint is inactive, so $\mu_j = 0$, this constraint cannot push on the optimum solution via the gradient in the stationarity condition in Eq. (6.24a). These two cases are shown in Fig. 6.6.

Geometry of Complementary Slackness and Feasible Directions

Consider a single inequality constraint $g_j(\mathbf{x}) \leq 0$. The boundary $g_j(\mathbf{x}) = 0$ defines a surface with outward-pointing normal $\nabla g_j(\mathbf{x})$. At a point on the boundary, motion is permitted only in directions $\mathbf{d}$ that satisfy $\nabla g_j(\mathbf{x}) \cdot \mathbf{d} \leq 0$, thereby maintaining feasibility. If the constraint is active, the KKT stationarity condition includes the term $\mu_j \nabla g_j(\mathbf{x})$, where $\mu_j \geq 0$ acts as a Lagrange multiplier *pushing* against the boundary of the constraint. If the constraint is inactive, so $g_j(\mathbf{x}) < 0$, then stationarity omits $\nabla g_j(\mathbf{x})$, since $\mu_j = 0$. Thus, the complementary slackness condition $\mu_j g_j(\mathbf{x}) = 0$ ensures that the gradient contribution $\mu_j \nabla g_j(\mathbf{x})$ appears in the stationarity condition if and only if the constraint is active.

[4] There are mild additional conditions that must also be satisfied, and we will discuss these soon.

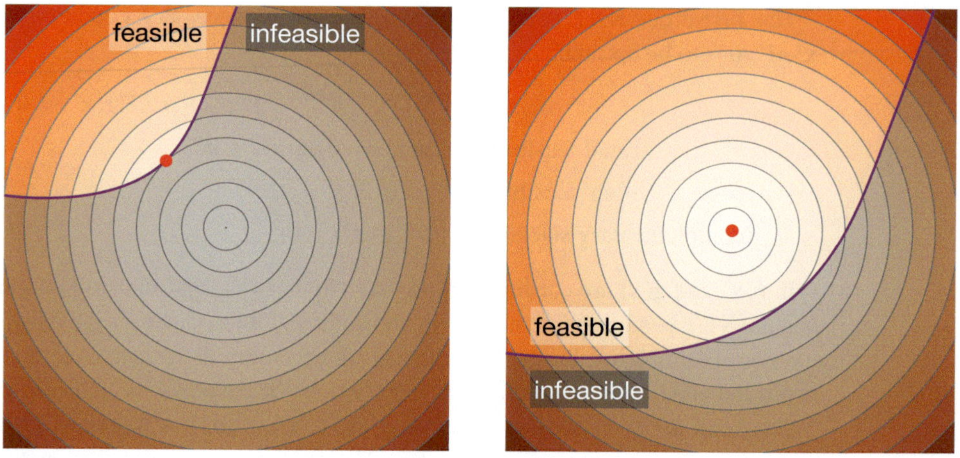

Figure 6.6 Here are two examples of an inequality-constrained problem. On the left, the optimal point is on the edge of the feasible set, so the constraint is active. On the right, the optimal point is on the interior of the feasible set, so the constraint is inactive. In both cases, the complementary slackness KKT condition $\boldsymbol{\mu}^{\mathrm{T}}\mathbf{g}(\mathbf{x}) = 0$ is satisfied.

Algorithmic Perspectives

The KKT system of equations underpin many of the most powerful modern algorithms for constrained optimization, including active-set, interior-point, and augmented Lagrangian methods. These algorithms are closely related to the notion of duality, and we will discuss them more in this chapter and in Chapter 5. We provide a brief overview here:

Active-Set Methods (ASM) and Sequential Quadratic Programming Active set methods involve guessing which inequality constraints are active and become equality constraints. Then the unconstrained Lagrangian in Eq. (6.2) is solved, for example with Newton's method. Finally, constraints are tested to see if they enter or exit after this iteration. In sequential quadratic programming, we approximate $f(\mathbf{x})$ with a quadratic.

Interior Point Methods (IPM) IPMs iteratively solve for an optimal point $(\mathbf{x}^*, \boldsymbol{\lambda}^*, \boldsymbol{\mu}^*)$ that satisfy the KKT conditions. We replace each $g_j(\mathbf{x}) \leq 0$ in the Lagrangian objective with a log-barrier function $-t\log(-g_j(\mathbf{x}))$. A Newton iteration on the barrier-augmented objective function will enforce $g_j(\mathbf{x}) \leq 0$. As $t \to 0$, the solution converges to a KKT point. We explore IPMs in Section 6.4.

Augmented Lagrangian Methods (ALM) Here we rewrite our inequality constraints $\mathbf{g}(\mathbf{x}) \leq \mathbf{0}$ as equality constraints by introducing slack variables: $\mathbf{g}(\mathbf{x}) + \mathbf{s} = \mathbf{0}$ with $\mathbf{s} \geq \mathbf{0}$. We then write the augmented Lagrangian to include penalty terms $\frac{\rho}{2}\|\mathbf{g}(\mathbf{x}) + \mathbf{s}\|^2$ to enforce feasibility. An outer-loop optimization updates the multipliers $(\boldsymbol{\lambda}, \boldsymbol{\mu})$, while the inner-loop optimization minimizes the augmented Lagrangian over $(\mathbf{x}, \mathbf{s})$, often using unconstrained solvers. We discuss augmented Lagrangian methods in Section 6.5.

Regularity Conditions

There are regularity conditions, known as constraint qualifications, for when a constrained optimizer $\mathbf{x}^*$ will satisfy the KKT conditions. The regularity conditions are more complicated in the case of mixed equality and inequality constraints than those for simple equality constraints. However, they typically are used to ensure that an optimal solution even exists. In this case, several constraint qualifications can be used:

Slater's Condition. For convex problems, Slater's condition is a go-to regularity test that is both powerful and practical. It simply requires the existence of a strictly feasible point $\mathbf{x}_0$:

$$\mathbf{h}(\mathbf{x}_0) = \mathbf{0} \quad \text{and} \quad \mathbf{g}(\mathbf{x}_0) < \mathbf{0}$$

so that all inequality and equality constraints are satisfied strictly at $\mathbf{x}_0$. This must only be true at a single feasible $\mathbf{x}_0$, and gives a global result. Slater's condition guarantees strong duality and ensures the KKT conditions are necessary and sufficient for optimality in convex problems. For linear programs, Slater's condition automatically holds if the feasible set has a non-empty interior. For nonconvex problems, Slater's condition does not guarantee strong duality or the sufficiency of the KKT conditions. It may still ensure that the KKT conditions are necessary for optimality, but only if other constraint qualifications, such as LICQ or MFCQ, are also satisfied.

Linear Independence Constraint Qualification (LICQ). This is one of the most commonly used regularity conditions. As before, the gradients of all active constraints (i.e., equality constraints $\mathbf{h}(\mathbf{x}) = \mathbf{0}$ and inequality constraints $g_j(\mathbf{x}^*) = 0$ active at $\mathbf{x}^*$) must be linearly independent. LICQ ensures that if $\mathbf{x}^*$ is a local minimizer, then there exist unique Lagrange multipliers $(\boldsymbol{\lambda}, \boldsymbol{\mu})$ that satisfy the KKT conditions. LICQ alone does not ensure existence of an optimizer or sufficiency of KKT. In the convex case, LICQ is stronger than Slater for existence and isn't needed for sufficiency.

Mangasarian–Fromovitz Constraint Qualification (MFCQ). This is a bit weaker than LICQ and states that the equality constraint gradients $\nabla\mathbf{h}(\mathbf{x}^*)$ are linearly independent, and that there exists a direction $\mathbf{d} \in \mathbb{R}^n$ such that:

$$\nabla h_i(\mathbf{x}^*) \cdot \mathbf{d} = 0 \quad \text{for all } i, \qquad \nabla g_j(\mathbf{x}^*) \cdot \mathbf{d} < 0 \quad \text{for all active } j.$$

This guarantees the existence of a feasible direction that moves away from the active inequality constraints. Unlike LICQ, MFCQ allows for linear dependence among active inequalities, as long as there is a direction of descent that maintains feasibility.

Example: Least Squares with Box Constraints

Consider the constrained least-squares regression with box constraints on $\mathbf{x}$

$$\begin{aligned} \min_{\mathbf{x}} \quad & \tfrac{1}{2}\|\mathbf{A}\mathbf{x} - \mathbf{b}\|_2^2 \\ \text{s.t.} \quad & \boldsymbol{\ell} \le \mathbf{x} \le \mathbf{u}. \end{aligned} \tag{6.25}$$

We solved this problem with Python in Section 4.6, and using the projected gradient algorithm in Section 5.1. To solve this using the KKT conditions, we introduce dual variables $\boldsymbol{\mu}$ and $\boldsymbol{\nu}$ for the lower and upper bounds $\boldsymbol{\ell}$ and $\mathbf{u}$, respectively. The Lagrangian is

$$\mathcal{L}(\mathbf{x}, \boldsymbol{\mu}, \boldsymbol{\nu}) = \tfrac{1}{2}\|\mathbf{A}\mathbf{x} - \mathbf{b}\|_2^2 + \boldsymbol{\mu}^{\mathrm{T}}(\boldsymbol{\ell} - \mathbf{x}) - \boldsymbol{\nu}^{\mathrm{T}}(\mathbf{x} - \mathbf{u}).$$

The KKT conditions then become

$$\begin{aligned} &\mathbf{A}^{\mathrm{T}}(\mathbf{A}\mathbf{x} - \mathbf{b}) - \boldsymbol{\mu} + \boldsymbol{\nu} = \mathbf{0} \\ &\boldsymbol{\ell} \le \mathbf{x} \le \mathbf{u}, \\ &\boldsymbol{\mu}, \boldsymbol{\nu} \ge \mathbf{0} \\ &\boldsymbol{\mu}^{\mathrm{T}}(\boldsymbol{\ell} - \mathbf{x}) = 0, \quad \boldsymbol{\nu}^{\mathrm{T}}(\mathbf{x} - \mathbf{u}) = 0. \end{aligned}$$

We may solve for an optimal point that satisfies the KKT conditions using a variety of algorithms. Since the problem is convex, the KKT conditions are both necessary and sufficient[5]. Here we sketch the interior point method approach. We will revisit IPMs in Section 6.4. We will also briefly discuss connections to the projected gradient approach from Section 5.1.

Interior Point Approach

The idea is to replace each constraint in the Lagrangian with a logarithmic barrier function

$$\phi_t(\mathbf{x}) = \tfrac{1}{2}\|\mathbf{A}\mathbf{x} - \mathbf{b}\|^2 - t\sum_j \left(\log(x_j - \ell_j) + \log(u_j - x_j)\right).$$

This results in an unconstrained problem that may be solved via Netwon's method. We will show in Section 6.4 that as $t \longrightarrow 0$, this will converge to a solution of the KKT conditions, since the function $-t\log(\mathbf{x})$ provides a steep barrier that rapidly goes to infinity as $t \longrightarrow \infty$. The associated dual Lagrange multipliers are $\mu_j = t/(x_j - \ell_j)$ and $\nu_j = t/(u_j - x_j)$.

Projected Gradient Descent

We enforce feasibility at each step by projecting the unconstrained gradient onto the box $[\boldsymbol{\ell}, \mathbf{u}]$. This is a special case where each bound constraint is treated as a separate inequality, and the associated multipliers are recovered by a soft-thresholding operation on the gradient.

The stationarity condition $\nabla f(\mathbf{x}^*) + \boldsymbol{\mu} - \boldsymbol{\nu} = \mathbf{0}$ for box constraints is $\frac{\partial f}{\partial x_j}(\mathbf{x}^*) = \nu_j - \mu_j$ in coordinates. Thus, we may explicitly write down the KKT conditions for a minimizer $\mathbf{x}^*$:

- If $\ell_j < x_j^* < u_j$, then both constraints are inactive, $\mu_j = \nu_j = 0$ and $\frac{\partial f}{\partial x_j}(\mathbf{x}^*) = 0$.
- If $x_j^* = \ell_j$ then the lower bound is active, $\mu_j \ge 0$, $\nu_j = 0$ and $\frac{\partial f}{\partial x_j}(\mathbf{x}^*) = -\mu_j \le 0$.

[5] We will leave it as an exercise to verify the regularity conditions are satisfied for this problem.

- If $x_j^* = u_j$ then the upper bound is active, $\mu_j = 0,\ \nu_j \geq 0$ and $\frac{\partial f}{\partial x_j}(\mathbf{x}^*) = \nu_j \geq 0$.

Thus, the signs of the gradient components tell us which bound is active and correspond directly to the Lagrange multipliers μ_j and ν_j. This is a special case of the KKT conditions for general inequality-constrained optimization.

There is another, more technical way to state this. At a minimizer $\mathbf{x}^*$, the KKT conditions imply $-\nabla f(\mathbf{x}^*) \in N_{[\boldsymbol{\ell},\mathbf{u}]}(\mathbf{x}^*)$, where $N_{[\boldsymbol{\ell},\mathbf{u}]}(\mathbf{x}^*)$ is defined as the *normal cone* of the box constraint. The normal cone is a set of vectors that *push outward* from the constraint set at $\mathbf{x}^*$, meaning a set of vectors that are perpendicular to the boundary of the feasible set.

Saddle Point Property of KKT Optimal Point

A KKT optimal point $(\mathbf{x}^*, \boldsymbol{\lambda}^*, \boldsymbol{\mu}^*)$ will be a saddle of the constrained Lagrangian $\mathcal{L}(\mathbf{x}, \boldsymbol{\lambda}, \boldsymbol{\mu})$:

$$\mathcal{L}(\mathbf{x}^*, \boldsymbol{\lambda}, \boldsymbol{\mu}) \leq \mathcal{L}(\mathbf{x}^*, \boldsymbol{\lambda}^*, \boldsymbol{\mu}^*) \leq \mathcal{L}(\mathbf{x}, \boldsymbol{\lambda}^*, \boldsymbol{\mu}^*) \tag{6.26}$$

for all $\mathbf{x}$ in the primal feasible set and for all $\boldsymbol{\lambda}$ and $\boldsymbol{\mu}$ in the dual feasible set. The right inequality says that $\mathbf{x}^*$ is a primal minimizer of the Lagrangian over feasible $\mathbf{x}$ for fixed $\boldsymbol{\lambda}^*$ and $\boldsymbol{\mu}^*$. Similarly, the left inequality says that $\boldsymbol{\lambda}^*$ and $\boldsymbol{\mu}^*$ are dual maximizers of the Lagrangian over feasible $\boldsymbol{\lambda}, \boldsymbol{\mu}$ for a fixed $\mathbf{x}^*$. In general, for nonconvex problems, this saddle point condition holds only locally, as KKT points may be local but not global optima.

This saddle-point property is necessary for strong duality, but sufficiency requires more conditions. Strong duality means that the primal and dual optimal values coincide[6]:

$$\min_{\mathbf{x} \in \mathcal{S}} f(\mathbf{x}) = \min_{\mathbf{x}} \sup_{\boldsymbol{\lambda}; \boldsymbol{\mu} \geq 0} \mathcal{L}(\mathbf{x}, \boldsymbol{\lambda}, \boldsymbol{\mu}) = \max_{\boldsymbol{\lambda}; \boldsymbol{\mu} \geq 0} \min_{\mathbf{x}} \mathcal{L}(\mathbf{x}, \boldsymbol{\lambda}, \boldsymbol{\mu}). \tag{6.27}$$

Let $\mathbf{x}^*$ be a primal minimizer, and $(\boldsymbol{\lambda}^*, \boldsymbol{\mu}^*)$ be a dual maximizer. Then:

$$\mathbf{x}^* \in \arg\min_{\mathbf{x}} \mathcal{L}(\mathbf{x}, \boldsymbol{\lambda}^*, \boldsymbol{\mu}^*),$$

and

$$(\boldsymbol{\lambda}^*, \boldsymbol{\mu}^*) \in \arg\max_{\boldsymbol{\lambda}; \boldsymbol{\mu} \geq 0} \mathcal{L}(\mathbf{x}^*, \boldsymbol{\lambda}, \boldsymbol{\mu}).$$

Hence, the pair $(\mathbf{x}^*, \boldsymbol{\lambda}^*, \boldsymbol{\mu}^*)$ satisfies the saddle point inequality:

$$\mathcal{L}(\mathbf{x}^*, \boldsymbol{\lambda}, \boldsymbol{\mu}) \leq \mathcal{L}(\mathbf{x}^*, \boldsymbol{\lambda}^*, \boldsymbol{\mu}^*) \leq \mathcal{L}(\mathbf{x}, \boldsymbol{\lambda}^*, \boldsymbol{\mu}^*) \tag{6.28}$$

for all $\mathbf{x} \in \mathcal{S}$, $\boldsymbol{\mu} \geq 0$, and unconstrained $\boldsymbol{\lambda} \in \mathbb{R}^m$.

For a convex optimization problem with differentiable functions f, $\mathbf{h}$, and $\mathbf{g}$, and when Slater's condition is satisfied, then the following statements are all equivalent:

1. The KKT conditions hold at $(\mathbf{x}^*, \boldsymbol{\lambda}^*, \boldsymbol{\mu}^*)$;
2. The point $(\mathbf{x}^*, \boldsymbol{\lambda}^*, \boldsymbol{\mu}^*)$ is a saddle point of $\mathcal{L}(\mathbf{x}, \boldsymbol{\lambda}, \boldsymbol{\mu})$; and

[6] Technically the $\min_{\mathbf{x}}$ will be replaced with the *infimum* $\inf_{\mathbf{x}}$ which we will introduce in the next section.

3. Strong duality holds and optima are attained[7]:

$$\min_{\mathbf{x}\in\mathcal{S}} f(\mathbf{x}) = \max_{\boldsymbol{\lambda};\boldsymbol{\mu}\geq 0} \min_{\mathbf{x}} \mathcal{L}(\mathbf{x},\boldsymbol{\lambda},\boldsymbol{\mu}) = \min_{\mathbf{x}} \max_{\boldsymbol{\lambda};\boldsymbol{\mu}\geq 0} \mathcal{L}(\mathbf{x},\boldsymbol{\lambda},\boldsymbol{\mu}).$$

6.3 Duality

Duality is one of the most profound and useful concepts in optimization. For every constrained optimization problem, there is a *dual* optimization problem that provides a lower bound to the optimum of the original *primal* problem. We have already seen a simple example of a dual optimization problem in Chapter 3 on linear programming. These ideas generalize naturally to other convex and nonconvex optimization problems. Regardless of the convexity of the primal problem, the dual problem is always concave.

The dual optimization problem is closely related to Lagrange multipliers and the KKT conditions. We will show that the Lagrange multipliers $\boldsymbol{\lambda}$ and $\boldsymbol{\mu}$ are actually decision variables for the dual problem. When only equality constraints are present, this dual problem is unconstrained. For a mix of equality and inequality constraints, the dual has positivity constraints on the Lagrange multipliers $\boldsymbol{\mu}$ corresponding to the inequality constraints.

Primal–dual interior point methods are a class of algorithms for modern high-dimensional constrained optimization, discussed in Section 6.4 Duality is also important in the alternating direction method of multipliers (ADMM), discussed in Section 6.5, and branch-and-bound algorithms for combinatorial and nonconvex problems. These approaches are central for many problems we have already seen, such as box-constrained least squares, LASSO, and ridge regression. When the dual equals the primal, then no further improvement is possible, providing a *certificate* of optimality.

Given a constrained optimization problem

$$\begin{aligned} \min_{\mathbf{x}} \quad & f(\mathbf{x}), \\ \text{s.t.} \quad & \mathbf{h}(\mathbf{x}) = \mathbf{0} \\ \text{and} \quad & \mathbf{g}(\mathbf{x}) \leq \mathbf{0}, \end{aligned} \tag{6.29}$$

the dual objective function is a partial minimization[8] of the Lagrangian with respect to $\mathbf{x}$:

$$\phi(\boldsymbol{\lambda},\boldsymbol{\mu}) = \inf_{\mathbf{x}} \mathcal{L}(\mathbf{x},\boldsymbol{\lambda},\boldsymbol{\mu}) = \inf_{\mathbf{x}} \left(f(\mathbf{x}) + \boldsymbol{\lambda}^{\mathrm{T}}\mathbf{h}(\mathbf{x}) + \boldsymbol{\mu}^{\mathrm{T}}\mathbf{g}(\mathbf{x}) \right). \tag{6.30}$$

[7] Note that for strong duality without qualification, we should write inf and sup instead of min and max. Slater's condition specifically guarantees that the optima is attained, so we may write min and max here.

[8] The *infimum* (inf) of a function is its greatest lower bound. It generalizes the minimum when the minimum might not be attained. We write $\inf_{\mathbf{x}} f(\mathbf{x})$ to refer to the lowest value that $f(\mathbf{x})$ *approaches* over the feasible domain $\mathbf{x} \in \mathcal{S}$, even if no $\mathbf{x}$ actually achieves it. We use *inf* to ensure well-defined expressions, even when the primal function does not attain its minimum. It can be replaced with min when it exists.
Example: Let $f(x) = x$ over the open interval $(0,1)$. The infimum of f is equal to zero, $\inf_{x\in(0,1)} f(x) = 0$, even though $\min_{x\in(0,1)} f(x)$ does not exist because $f(x) \to 0$ as $x \to 0^+$, but $0 \notin (0,1)$. Similarly, $f(x) = 1/x$ over the open interval $(0,1)$ has an infimum of 0.

We don't restrict the infimum to the feasible set $\mathbf{x} \in \mathcal{S}$, and instead the infimum is taken over all $\mathbf{x}$. The dual function is always concave, even if the primal problem is nonconvex, as Eq. (6.30) is the pointwise infimum of functions that are affine in $(\boldsymbol{\lambda}, \boldsymbol{\mu})$.

The dual problem

$$\max_{\boldsymbol{\lambda};\boldsymbol{\mu}\geq\mathbf{0}} \quad \phi(\boldsymbol{\lambda}, \boldsymbol{\mu}) \tag{6.31}$$

provides a lower bound on the primal optimum.

We have the following fundamental relationship between the primal and dual optimization problems, called weak duality, which states that the optimum value of the dual problem is always a lower bound to the optimum value of the primal problem:

$$\max_{\boldsymbol{\lambda};\boldsymbol{\mu}\geq\mathbf{0}} \phi(\boldsymbol{\lambda}, \boldsymbol{\mu}) \leq \min_{\mathbf{x}\in\mathcal{S}} f(\mathbf{x}). \tag{6.32}$$

This is an extremely useful property for algorithms and optimality certificates. Specifically, the dual variables (i.e., the Lagrange multipliers $(\boldsymbol{\lambda}, \boldsymbol{\mu})$) allow us to apply pressure to the constraints, so that we may tune these dual variables to maximize the guaranteed lower bound. The dual function measures how low the objective function can be pushed down if constraint violations are allowed but penalized. Thus, dual solutions give safe lower bounds on the primal problem, even if infeasible points are allowed.

Weak duality is a more fundamental primal–dual relationship that always holds, even for non-optimal points. Any feasible primal $\mathbf{x} \in \mathcal{S}$ and feasible dual $(\boldsymbol{\lambda}, \boldsymbol{\mu})$ satisfy

$$\phi(\boldsymbol{\lambda}, \boldsymbol{\mu}) \leq f(\mathbf{x}). \tag{6.33}$$

Thus, the value of any dual feasible point is a lower bound on the value of any primal feasible point. This is always true, even for nonconvex problems.

Strong duality is a special case when the dual and primal optima are equal. Strong duality holds for convex problems under mild regularity conditions, such as Slater's condition, namely the existence of a strictly feasible point. Under these mild assumptions, strong duality states that the primal and dual optimum values are equal to each other[9]:

$$\phi(\boldsymbol{\lambda}^*, \boldsymbol{\mu}^*) = f(\mathbf{x}^*). \tag{6.34}$$

For nonconvex problems strong duality might not hold and there may be a duality gap.

Dual variables are also important to quantify the sensitivity of the optimum to perturbations in the constraint right-hand sides. When the necessary conditions are met, dual variables are the partial derivatives of the optimal value with respect to constraint right-hand sides. This connection bridges optimization theory with sensitivity analysis and supports the practical use of duality in engineering, economics, and control.

[9] Technically, this equality is true only if the primal optimum is attained, which also may require compactness of the feasible set (e.g., the convex polytope in linear programming being closed and compact). If the primal is not attained, then we must write $\inf_{\mathbf{x}\in\mathcal{S}} f(\mathbf{x}) = \sup_{\boldsymbol{\lambda};\boldsymbol{\mu}\geq 0} \phi(\boldsymbol{\lambda}, \boldsymbol{\mu})$.

Duality for Equality Constraints Only

If the primal has equality constraints and no inequality constraints, the dual becomes

$$\max_{\lambda} \ \phi(\lambda) \tag{6.35}$$

in terms of a dual function

$$\phi(\lambda) = \inf_{\mathbf{x}} \mathcal{L}(\mathbf{x}, \lambda) = \inf_{\mathbf{x}} \left(f(\mathbf{x}) + \lambda^{\mathrm{T}} \mathbf{h}(\mathbf{x}) \right). \tag{6.36}$$

There are no constraints on the dual variables λ. We may then solve the unconstrained dual problem using simpler techniques. For example, if $\phi(\lambda)$ is differentiable or twice differentiable, we may use gradient descent or Newton's method, respectively.

Dual Linear Programs

We now illustrate the key concepts of duality for the linear program in Chapter 3. We will begin with a simple and straightforward derivation of the dual problem, and then will provide some intuition for what this dual linear program means.

Deriving the Dual Linear Program

We begin with the basic linear program, where we write the positivity constraint $\mathbf{x} \geq \mathbf{0}$ in the maximization statement:

$$\begin{aligned} \max_{\mathbf{x} \geq \mathbf{0}} \quad & \mathbf{c}^{\mathrm{T}} \mathbf{x}, \\ \text{s.t.} \quad & \mathbf{A}\mathbf{x} \leq \mathbf{b}. \end{aligned} \tag{6.37}$$

To be explicit, we will rearrange the inequality constraints so that $\mathbf{b} - \mathbf{A}\mathbf{x} \geq \mathbf{0}$:

$$\begin{aligned} \max_{\mathbf{x} \geq \mathbf{0}} \quad & \mathbf{c}^{\mathrm{T}} \mathbf{x}, \\ \text{s.t.} \quad & \mathbf{b} - \mathbf{A}\mathbf{x} \geq \mathbf{0}. \end{aligned} \tag{6.38}$$

Now we are able to move the constraints into the Lagrangian objective function using Lagrange multipliers $\mathbf{y}$

$$\max_{\mathbf{x} \geq \mathbf{0}} \min_{\mathbf{y} \geq \mathbf{0}} \quad \underbrace{\mathbf{c}^{\mathrm{T}} \mathbf{x} + \mathbf{y}^{\mathrm{T}} (\mathbf{b} - \mathbf{A}\mathbf{x})}_{\mathcal{L}(\mathbf{x},\mathbf{y})}. \tag{6.39}$$

This is max min instead of min max because we began with a maximization problem. We also assumed that the feasible set is closed, bounded, and non-empty so there exists a primal maximizer. The Lagrange multiplier $\mathbf{y}$ is non-negative as it corresponds to inequality constraints, playing the role of $\boldsymbol{\mu}$ in Eq. (6.21). We use $\mathbf{y}$ instead of $\boldsymbol{\mu}$, since $\mathbf{y}$ will be the decision variable of the dual linear program in Eq. (6.44) below.

Now we rearrange the terms in the Lagrangian, collecting terms with $\mathbf{x}$, which will become our Lagrange multipliers for the dual problem:

$$\max_{\mathbf{x} \geq \mathbf{0}} \min_{\mathbf{y} \geq \mathbf{0}} \quad \mathbf{y}^{\mathrm{T}} \mathbf{b} + (\mathbf{c}^{\mathrm{T}} - \mathbf{y}^{\mathrm{T}} \mathbf{A})\mathbf{x}. \tag{6.40}$$

We may also transpose terms in the objective function, because it is a scalar:

$$\max_{\mathbf{x}\geq\mathbf{0}} \min_{\mathbf{y}\geq\mathbf{0}} \quad \mathbf{b}^{\mathrm{T}}\mathbf{y} + \mathbf{x}^{\mathrm{T}}(\mathbf{c} - \mathbf{A}^{\mathrm{T}}\mathbf{y}). \tag{6.41}$$

This is still a formulation of the primal in Eq. (6.37). Because we have strong duality[10], we may equate this with the dual problem, swapping the order of the $\max_{\mathbf{x}}$ and $\min_{\mathbf{y}}$:

$$\min_{\mathbf{y}\geq\mathbf{0}} \max_{\mathbf{x}\geq\mathbf{0}} \quad \mathbf{b}^{\mathrm{T}}\mathbf{y} + \mathbf{x}^{\mathrm{T}}(\mathbf{c} - \mathbf{A}^{\mathrm{T}}\mathbf{y}). \tag{6.42}$$

The term $\mathbf{x}^{\mathrm{T}}$ is now the Lagrange multiplier for the inequality constraints $\mathbf{c} - \mathbf{A}^{\mathrm{T}}\mathbf{y} \leq \mathbf{0}$ in the dual problem. We move these terms back to inequality constraints

$$\begin{aligned} \min_{\mathbf{y}\geq\mathbf{0}} \quad & \mathbf{b}^{\mathrm{T}}\mathbf{y}, \\ \text{s.t.} \quad & \mathbf{A}^{\mathrm{T}}\mathbf{y} - \mathbf{c} \geq \mathbf{0} \end{aligned} \tag{6.43}$$

so that we arrive at the dual problem in standard form

$$\begin{aligned} \min_{\mathbf{y}\geq\mathbf{0}} \quad & \mathbf{b}^{\mathrm{T}}\mathbf{y}, \\ \text{s.t.} \quad & \mathbf{A}^{\mathrm{T}}\mathbf{y} \geq \mathbf{c}. \end{aligned} \tag{6.44}$$

The relationship between the primal and dual problems is shown schematically in Fig. 6.7 for this special case where we have strong duality and both primal and dual optima are attained.

Intuition Behind the Dual Linear Program

It is helpful to discuss the intuition behind the dual form of a linear program, as this is the simplest case to understand, and it will provide a basis for more complex primal–dual problems and algorithms.

Given a maximization $\max_{\mathbf{x}\geq\mathbf{0}} \mathbf{c}^{\mathrm{T}}\mathbf{x}$ subject to the constraints $\mathbf{A}\mathbf{x} \leq \mathbf{b}$, the idea is to try and provide the tightest upper bound on $\mathbf{c}^{\mathrm{T}}\mathbf{x}$ by taking linear combinations of the constraint equations $\mathbf{a}_j\mathbf{x} \leq b_j$. The coefficients of the linear combinations of these equations will be denoted by $\mathbf{y}$, which we saw above are the dual variables and the Lagrange multipliers enforcing the constraints. We know that any feasible solution will satisfy all of these constraints, so any linear combination will also be satisfied for any feasible point. This linear combination of the constraints yields the following equation:

$$\mathbf{y}^{\mathrm{T}}\mathbf{A}\mathbf{x} \leq \mathbf{y}^{\mathrm{T}}\mathbf{b}. \tag{6.45}$$

Thus, the product $\mathbf{y}^{\mathrm{T}}\mathbf{A}$ provides the coefficients of $\mathbf{x}$. If these coefficients are greater than or equal to $\mathbf{c}^{\mathrm{T}}$, then $\mathbf{y}^{\mathrm{T}}\mathbf{b}$ provides a direct upper bound on $\mathbf{c}^{\mathrm{T}}\mathbf{x}$; note that this relies on $\mathbf{x} \geq \mathbf{0}$, although this condition may be relaxed. This means that for any $\mathbf{y}^{\mathrm{T}}\mathbf{A} \geq \mathbf{c}^{\mathrm{T}}$, then $\mathbf{y}^{\mathrm{T}}\mathbf{b}$ will

[10] For convex linear programs we have strong duality and the expression

$$\max_{\mathbf{x}\geq\mathbf{0}} \min_{\mathbf{y}\geq\mathbf{0}} \quad \mathbf{c}^{\mathrm{T}}\mathbf{x} + \mathbf{y}^{\mathrm{T}}(\mathbf{b} - \mathbf{A}\mathbf{x}) \qquad \leq \qquad \min_{\mathbf{y}\geq\mathbf{0}} \max_{\mathbf{x}\geq\mathbf{0}} \quad \mathbf{b}^{\mathrm{T}}\mathbf{y} + \mathbf{x}^{\mathrm{T}}(\mathbf{c} - \mathbf{A}^{\mathrm{T}}\mathbf{y})$$

becomes an equality under mild assumptions, such as the feasible polytope being closed and bounded. Thus, we may also swap their order in Eq. (6.41) to obtain Eq. (6.42).

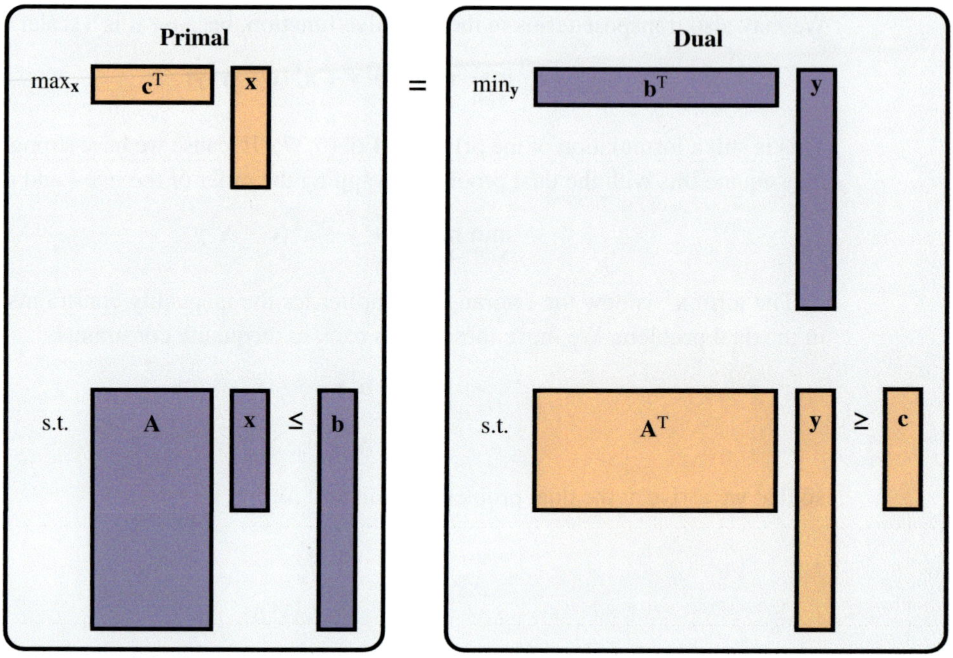

Figure 6.7 Visualization of primal (left) and dual (right) linear programs. In this case, the primal has more constraints than decision variables, while the dual has more decision variables than constraints. The dual minimization always provides an upper bound for the primal maximization. Since linear programs are convex, this duality becomes strong, and the bound is tight under mild assumptions, such as a closed and bounded feasible polytope. In this case, the dual solution is equal to the primal solution.

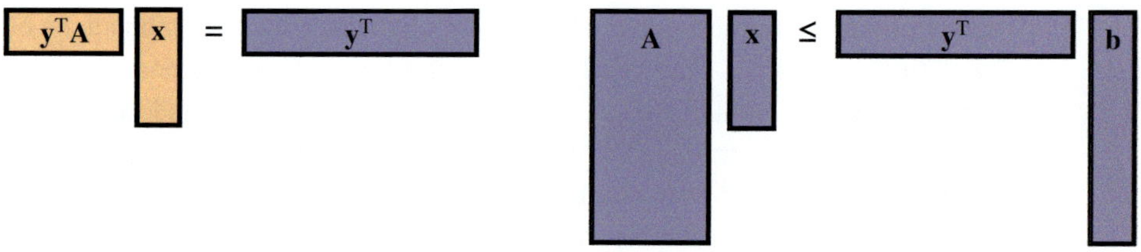

Figure 6.8 The idea behind a dual linear program is to find a linear combination of constraint equations that provide a tight bound for the primal problem. In this case, the combination of constraint equations is given by the vector $\mathbf{y}^T$, so that $\mathbf{y}^T\mathbf{A}$ are the coefficients multiplying $\mathbf{x}$, and $\mathbf{y}^T\mathbf{A}\mathbf{x}$ is bounded from above by $\mathbf{y}^T\mathbf{b}$. We will then minimize $\mathbf{y}^T\mathbf{b}$ subject to $\mathbf{y}^T\mathbf{A} \geq \mathbf{c}^T$, where the original objective function was $\mathbf{c}^T\mathbf{x}$. Notice that this is the transpose of the typical dual problem formulation, although they are equivalent because they are both scalar functions and are therefore equal to their transposes.

provide an upper bound on the primal. Since we seek the tightest bound, we minimize $\mathbf{y}^T\mathbf{b}$ over $\mathbf{y} \geq \mathbf{0}$ such that $\mathbf{y}^T\mathbf{A} \geq \mathbf{c}^T$. This is shown schematically in Fig. 6.8.

Simple Example. The following example illustrates the idea of duality quite simply.

$$\begin{array}{lll} \text{(Primal)} & & \text{(Dual)} \\ \min_{x_1,x_2} \; 3x_1 + 5x_2 & \Longleftrightarrow & \max_{y} \; 6y \\ \text{s.t.} \; 2x_1 + x_2 = 6 & & \text{s.t.} \; 2y \le 3 \\ \quad x_1, x_2 \ge 0 & & \quad y \le 5 \end{array}$$

The dual optimum $y^* = \frac{3}{2}$ gives a lower bound $6y^* = 9$, matching the primal optimum $(x_1^*, x_2^*) = (3, 0)$, which gives a cost of 9. Strong duality holds in this case.

Duality for Constrained Least Squares

We will revisit the constrained-least squares problem we analyzed in Chapter 4, from the perspective of duality. Consider the constrained least squares problem

$$\begin{aligned} \min_{\mathbf{x}} \quad & \tfrac{1}{2}\|\mathbf{A}\mathbf{x} - \mathbf{b}\|_2^2 \\ \text{s.t.} \quad & \mathbf{C}\mathbf{x} = \mathbf{d}. \end{aligned} \tag{6.46}$$

We may formulate the dual problem by first forming the Lagrangian

$$\mathcal{L}(\mathbf{x}, \boldsymbol{\lambda}) = \tfrac{1}{2}\|\mathbf{A}\mathbf{x} - \mathbf{b}\|_2^2 + \boldsymbol{\lambda}^{\mathrm{T}}(\mathbf{C}\mathbf{x} - \mathbf{d}). \tag{6.47}$$

The dual function is $\min_{\mathbf{x}} \mathcal{L}$, which amounts to finding a stationary $\mathbf{x}$ that satisfies

$$\underbrace{(\mathbf{A}^{\mathrm{T}}\mathbf{A})}_{\mathbf{Q}}\,\mathbf{x} = \underbrace{\mathbf{A}^{\mathrm{T}}\mathbf{b} - \mathbf{C}^{\mathrm{T}}\boldsymbol{\lambda}}_{\mathbf{r}(\boldsymbol{\lambda})} \quad \Longrightarrow \quad \mathbf{x}^* = \mathbf{Q}^{\dagger}\mathbf{r}(\boldsymbol{\lambda}).$$

Expanding the Lagrangian yields:

$$\mathcal{L}(\mathbf{x}, \boldsymbol{\lambda}) = \tfrac{1}{2}\mathbf{x}^{\mathrm{T}}\mathbf{Q}\mathbf{x} - \mathbf{r}(\boldsymbol{\lambda})^{\mathrm{T}}\mathbf{x} + \tfrac{1}{2}\mathbf{b}^{\mathrm{T}}\mathbf{b} - \boldsymbol{\lambda}^{\mathrm{T}}\mathbf{d}. \tag{6.48}$$

We could plug in $\mathbf{x}^*$, which is a mess. Instead, we use the standard quadratic identity

$$\min_{\mathbf{x}} \left\{\tfrac{1}{2}\mathbf{x}^{\mathrm{T}}\mathbf{Q}\mathbf{x} - \mathbf{r}^{\mathrm{T}}\mathbf{x}\right\} = \begin{cases} -\tfrac{1}{2}\langle \mathbf{r}, \mathbf{Q}^{\dagger}\mathbf{r}\rangle, & \mathbf{r} \in \text{range}(\mathbf{Q}), \\ -\infty, & \text{otherwise.} \end{cases}$$

Thus, we obtain the dual function

$$\phi(\boldsymbol{\lambda}) = \min_{\mathbf{x}} \mathcal{L}(\mathbf{x}, \boldsymbol{\lambda}) = -\tfrac{1}{2}\big\langle \mathbf{r}(\boldsymbol{\lambda}), \mathbf{Q}^{\dagger}\mathbf{r}(\boldsymbol{\lambda})\big\rangle + \tfrac{1}{2}\|\mathbf{b}\| - \boldsymbol{\lambda}^{\mathrm{T}}\mathbf{d}, \tag{6.49}$$

with the constraint that $\mathbf{r}(\boldsymbol{\lambda})$ is in the range of $\mathbf{Q}$, which is equivalent to being in the range of $\mathbf{A}^{\mathrm{T}}$. If $\mathbf{A}^{\mathrm{T}}\mathbf{A}$ is invertible, then we replace $\mathbf{Q}^{\dagger}$ with $\mathbf{Q}^{-1}$.

We may write the dual using norms instead of inner products as

$$\phi(\boldsymbol{\lambda}) = -\tfrac{1}{2}\Big\| \underbrace{((\mathbf{A}^{\mathrm{T}}\mathbf{A})^{\dagger})^{-1/2}(\mathbf{A}^{\mathrm{T}}\mathbf{b} - \mathbf{C}^{\mathrm{T}}\boldsymbol{\lambda})}_{(\mathbf{Q}^{\dagger})^{-1/2}\mathbf{r}(\boldsymbol{\lambda})} \Big\|_2^2 + \tfrac{1}{2}\|\mathbf{b}\|_2^2 - \boldsymbol{\lambda}^{\mathrm{T}}\mathbf{d}. \tag{6.50}$$

Now we see the benefit of writing constrained least squares in dual form. The primal KKT system in Eq. (4.49) requires an $(n + m) \times (n \times m)$ indefinite system of equations,

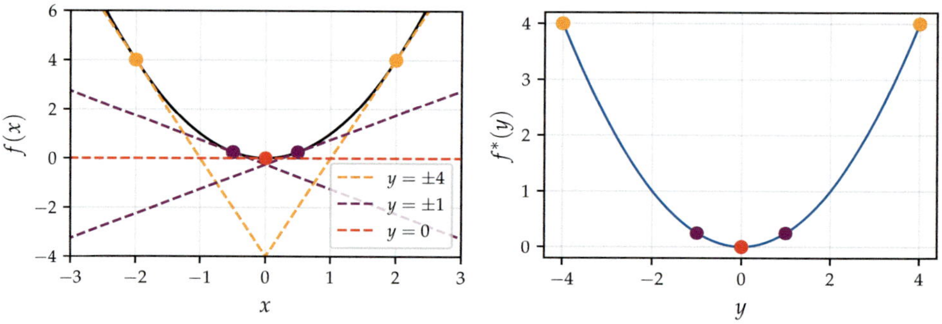

Figure 6.9 The Fenchel conjugate of the function $f(x) = x^2$ is $f^*(y) = \frac{1}{4}y^2$. We derive this by taking a point x and looking at the tightest affine lower bound, which is the line with slope given by the derivative $f'(x) = 2x$. The conjugate at a slope y is the y-intercept of this tangent line to $f(x)$ at x. For example, at $y = 1$ the line with slope 1 on the left figure intercepts the vertical axis at $\frac{1}{4}$.

while Eq. (6.50) solve the $m \times m$ definite Schur complement system:

$$\mathbf{C}\mathbf{Q}^{\dagger}\mathbf{C}^{\mathrm{T}}\boldsymbol{\lambda} = \mathbf{C}(\mathbf{A}^{\mathrm{T}}\mathbf{A})^{-1}\mathbf{A}^{\mathrm{T}}\mathbf{b} - \mathbf{d}. \tag{6.51}$$

Once we solve for $\boldsymbol{\lambda}^*$ we can recover $\mathbf{x}^* = (\mathbf{A}^{\mathrm{T}}\mathbf{A})^{-1}(\mathbf{A}^{\mathrm{T}}\mathbf{b} - \mathbf{C}^{\mathrm{T}}\boldsymbol{\lambda}^*)$. Solving for the matrix $\mathbf{Q}^{\dagger}$ or $(\mathbf{A}^{\mathrm{T}}\mathbf{A})^{-1}$ still involves factoring $\mathbf{A}$, usually with QR or SVD. However, the KKT matrices are indefinite, while the dual Schur complement is symmetric positive semidefinite, which is preferable for numerical linear algebra. When the number of constraints m is smaller than the number of variables n, this dual viewpoint is computationally appealing. Many numerical linear algebra solvers exploit this dual structure in their implementation by reducing the primal KKT system to the Schur complement form. Specifically, they factorize $\mathbf{A}$, form the Schur complement $\mathbf{C}(\mathbf{A}^{\mathrm{T}}\mathbf{A})^{-1}\mathbf{C}^{\mathrm{T}}$, solve for $\boldsymbol{\lambda}$, and then back-solve for $\mathbf{x}$. So in practice, solving the KKT system already reduces to the dual step internally.

Conjugate Functions and Fenchel Duality

To understand duality in optimization, it helps to examine the more general Fenchel–Rockafellar duality [419, 281], which arises from convex conjugate functions. The Fenchel conjugate, also known as the convex conjugate, bridges convex analysis and Lagrangian duality [419, 281]. The Fenchel conjugate[11] $f^*(\mathbf{y})$, of a function $f : \mathbb{R}^n \to \mathbb{R}$ is given by

$$f^*(\mathbf{y}) = \sup_{\mathbf{x}} \left(\mathbf{y}^{\mathrm{T}}\mathbf{x} - f(\mathbf{x}) \right). \tag{6.52}$$

Intuitively, $f^*(\mathbf{y})$ captures the tightest affine lower bound of $f(\mathbf{x})$ with slope $\mathbf{y}$. Thus, we construct f^* as the lower affine supporting hyperplanes of f, as shown in Fig. 6.9.

[11] The Fenchel conjugate is a generalization of the classical Legendre transform (see Section 6.6). Both transformations relate a function to a dual representation based on its linear supporting hyperplanes.

The Biconjugate and the Fenchel–Moreau Theorem

The biconjugate of f is the conjugate of the conjugate:

$$f^{**}(\mathbf{x}) = \sup_{\mathbf{y}} \left(\mathbf{x}^{\mathrm{T}}\mathbf{y} - f^{*}(\mathbf{y}) \right). \tag{6.53}$$

The Fenchel–Moreau Theorem says for f proper[12] convex, and lower semicontinuous[13]:

$$f(\mathbf{x}) = f^{**}(\mathbf{x}), \quad \text{for all } \mathbf{x} \in \mathrm{dom}(f). \tag{6.54}$$

Moreover, if f is nonconvex, then f^{**} is the closed convex hull of f.

Example: Convex Envelope of a Nonconvex Function

Consider the nonconvex function:

$$f(x) = \min(x^2, |x|), \tag{6.55}$$

which is formed by taking the pointwise minimum of two convex functions. Figure 6.10 shows the nonconvex function $f(x)$ and its convex envelope $f^{**}(x)$. The function $f(x)$ is nonconvex, and no single supporting hyperplane bounds the function from below.

The convex envelope of a function f is the largest convex function that lies below f everywhere. In this case, the convex envelope of f is:

$$f^{**}(x) = \begin{cases} x^2 & \text{if } |x| \le 0.5, \\ |x| - 0.25 & \text{if } |x| > 0.5, \end{cases} \tag{6.56}$$

which is the tightest convex function lying entirely beneath $f(x)$. This convex envelope is the *Huber* function, which is widely used in robust statistics.

This is precisely what arises when we compute the conjugate of the conjugate. If f is not convex, its conjugate f^* is still convex, and the biconjugate f^{**} satisfies:

$$f^{**}(x) \le f(x), \tag{6.57}$$

with equality if and only if f is convex and lower semicontinuous. It is left as a homework problem to compute the conjugate f^* and biconjugate f^{**} for this function and confirm that they are both convex.

Fenchel–Rockafellar Duality

Now, consider the dual function

$$\phi(\boldsymbol{\lambda}, \boldsymbol{\mu}) = \inf_{\mathbf{x}} \left(f(\mathbf{x}) + \boldsymbol{\lambda}^{\mathrm{T}}\mathbf{h}(\mathbf{x}) + \boldsymbol{\mu}^{\mathrm{T}}\mathbf{g}(\mathbf{x}) \right), \tag{6.58}$$

[12] Proper means that $f(\mathbf{x}) < +\infty$ for at least one $\mathbf{x}$, and $f(\mathbf{x}) > -\infty$ for all $\mathbf{x}$; see Rockafeller [419] for details.

[13] In convex optimization, lower semicontinouous means that the function $f : \mathbb{R}^n \to \mathbb{R} \cup \{+\infty\}$ satisfies:

$$\liminf_{\mathbf{x} \to \mathbf{x}_0} f(\mathbf{x}) \ge f(\mathbf{x}_0) \quad \text{for all } \mathbf{x}_0 \in \mathbb{R}^n.$$

This property guarantees that the function does not jump downward, i.e., it doesn't have discontinuities where it suddenly takes a lower value than nearby points. This also means that epi(f) is a closed set. A continuous function is always lower semicontinuous.

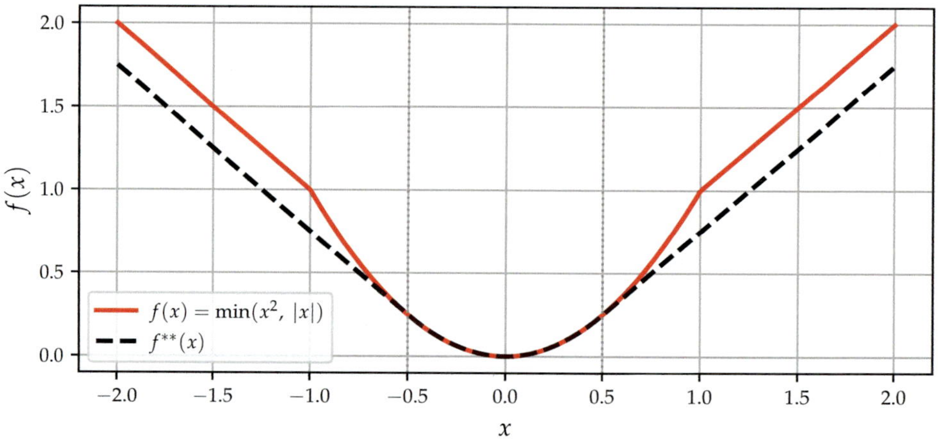

Figure 6.10 Illustration that the biconjugate of a nonconvex function is the convex envelope of the function. In this example, the nonconvex function $f(x) = \min(x^2, |x|)$ relaxes to the convex envelope $f^{**}(x)$.

whichis a concave function, being a pointwise infimum of affine functions in $(\boldsymbol{\lambda}, \boldsymbol{\mu})$. If f is convex and the constraints are convex, this construction fits naturally into the Fenchel duality framework.

Equality Constraints. In the special case where $\mathbf{h}(\mathbf{x}) = \mathbf{A}\mathbf{x} - \mathbf{b}$ and $\mathbf{g}(\mathbf{x}) = \mathbf{0}$, then:

$$\phi(\boldsymbol{\lambda}) = \inf_{\mathbf{x}} \left(f(\mathbf{x}) + \boldsymbol{\lambda}^{\mathrm{T}}(\mathbf{A}\mathbf{x} - \mathbf{b}) \right) = -\boldsymbol{\lambda}^{\mathrm{T}}\mathbf{b} + \inf_{\mathbf{x}} \left(f(\mathbf{x}) + \boldsymbol{\lambda}^{\mathrm{T}}\mathbf{A}\mathbf{x} \right), \tag{6.59}$$

which can be rewritten as[14]:

$$\phi(\boldsymbol{\lambda}) = -\boldsymbol{\lambda}^{\mathrm{T}}\mathbf{b} - f^*(-\mathbf{A}^{\mathrm{T}}\boldsymbol{\lambda}), \tag{6.60}$$

where f^* is the Fenchel conjugate of f. This gives an explicit form of the dual function in terms of the conjugate of the objective function.

Inequality Constraints. Now suppose the problem also includes convex inequality constraints $\mathbf{g}(\mathbf{x}) \leq \mathbf{0}$, where each g_j is convex. Then the dual function becomes

$$\phi(\boldsymbol{\lambda}, \boldsymbol{\mu}) = \inf_{\mathbf{x}} \left(f(\mathbf{x}) + \boldsymbol{\lambda}^{\mathrm{T}}(\mathbf{A}\mathbf{x} - \mathbf{b}) + \boldsymbol{\mu}^{\mathrm{T}}\mathbf{g}(\mathbf{x}) \right), \quad \text{with } \boldsymbol{\mu} \geq \mathbf{0}. \tag{6.61}$$

[14] We use the identity that $-\sup_{\mathbf{x}} f(\mathbf{x}) = \inf_{\mathbf{x}} -f(\mathbf{x})$ to rewrite

$$\inf_{\mathbf{x}} \left(f(\mathbf{x}) + \boldsymbol{\lambda}^{\mathrm{T}}\mathbf{A}\mathbf{x} \right) = -\sup_{\mathbf{x}} \left(\mathbf{x}^{\mathrm{T}} \underbrace{(-\mathbf{A}^{\mathrm{T}}\boldsymbol{\lambda})}_{\mathbf{y}} - f(\mathbf{x}) \right) = -f^*(-\mathbf{A}^{\mathrm{T}}\boldsymbol{\lambda}).$$

This uses the definition of the Fenchel conjugate $f^*(\mathbf{y}) = \sup_{\mathbf{x}} (\mathbf{x}^{\mathrm{T}}\mathbf{y} - f(\mathbf{x}))$ with $\mathbf{y} = -\mathbf{A}^{\mathrm{T}}\boldsymbol{\lambda}$.

We can write this as

$$\phi(\boldsymbol{\lambda}, \boldsymbol{\mu}) = -\boldsymbol{\lambda}^{\mathrm{T}}\mathbf{b} + \inf_{\mathbf{x}} \left(f(\mathbf{x}) + \boldsymbol{\lambda}^{\mathrm{T}}\mathbf{A}\mathbf{x} + \boldsymbol{\mu}^{\mathrm{T}}\mathbf{g}(\mathbf{x}) \right). \tag{6.62}$$

Define the function

$$\tilde{f}(\mathbf{x}) := f(\mathbf{x}) + \boldsymbol{\mu}^{\mathrm{T}}\mathbf{g}(\mathbf{x}), \tag{6.63}$$

which remains convex. Then the dual function becomes

$$\phi(\boldsymbol{\lambda}, \boldsymbol{\mu}) = -\boldsymbol{\lambda}^{\mathrm{T}}\mathbf{b} - \tilde{f}^*(-\mathbf{A}^{\mathrm{T}}\boldsymbol{\lambda}), \tag{6.64}$$

where $\tilde{f}^*$ is the Fenchel conjugate of the function $\tilde{f}(\mathbf{x}) = f(\mathbf{x}) + \boldsymbol{\mu}^{\mathrm{T}}\mathbf{g}(\mathbf{x})$. This provides an explicit form of the dual in terms of conjugates.

Geometric Interpretation of Duality

A key geometric insight behind Lagrange duality is the relationship between *supporting hyperplanes*[15] and the construction of the dual function. In convex optimization, duality is not just an algebraic formula, but it is also a geometric process of constructing the best possible linear lower bound on a convex function, subject to constraints.

For simplicity, we consider equality constraints $\mathbf{h}(\mathbf{x}) = \mathbf{0}$. In the primal, we search for the lowest level set of f that intersects the feasible region defined by $\mathbf{h}(\mathbf{x}) = \mathbf{0}$. In the dual, we search over $\boldsymbol{\lambda}$ for the best affine lower bound (i.e., supporting hyperplane) on $f(\mathbf{x})$ that respects the constraint. The optimal $\boldsymbol{\lambda}^*$ is the slope of this best supporting hyperplane.

Duality and Supporting Hyperplanes

Recall that the dual function

$$\phi(\boldsymbol{\lambda}) = \inf_{\mathbf{x}} \left(f(\mathbf{x}) + \boldsymbol{\lambda}^{\mathrm{T}}\mathbf{h}(\mathbf{x}) \right), \tag{6.65}$$

is concave in $\boldsymbol{\lambda}$ and gives a family of lower bounds on the primal objective.

In the case of convex equality constraints $\mathbf{h}(\mathbf{x}) = \mathbf{0}$ (i.e., affine constraints $\mathbf{A}\mathbf{x} = \mathbf{b}$), then each choice of $\boldsymbol{\lambda}$ defines a linear perturbation of the objective function

$$f(\mathbf{x}) + \boldsymbol{\lambda}^{\mathrm{T}}(\mathbf{A}\mathbf{x} - \mathbf{b}). \tag{6.66}$$

This can be interpreted as tilting the function f in the direction $\boldsymbol{\lambda}^{\mathrm{T}}\mathbf{A}$. Taking its infimum provides a lower bound on the primal optimal value. The dual function is then:

$$\phi(\boldsymbol{\lambda}) = -\boldsymbol{\lambda}^{\mathrm{T}}\mathbf{b} + \inf_{\mathbf{x}} \left(f(\mathbf{x}) + \boldsymbol{\lambda}^{\mathrm{T}}\mathbf{A}\mathbf{x} \right). \tag{6.67}$$

Geometrically, the pair $(\boldsymbol{\lambda}, \phi(\boldsymbol{\lambda}))$ defines a supporting hyperplane to the epigraph of f. Specifically, the inequality

$$f(\mathbf{x}) \geq -\boldsymbol{\lambda}^{\mathrm{T}}\mathbf{A}\mathbf{x} + \left(\phi(\boldsymbol{\lambda}) + \boldsymbol{\lambda}^{\mathrm{T}}\mathbf{b} \right) \tag{6.68}$$

[15] Recall that a supporting hyperplane to a set $\mathcal{S}$ at $\mathbf{x}_0$ touches the set at exactly $\mathbf{x}_0$, and lies entirely on one side of $\mathcal{S}$. Mathematically, this means there exists a vector $\mathbf{a} \in \mathbb{R}^n$ so that $\mathbf{a}^{\mathrm{T}}(\mathbf{x} - \mathbf{x}_0) \geq 0$ for all $\mathbf{x} \in C$.

defines an affine lower bound on f that is tight at the minimizing $\mathbf{x}$ for that value of λ, assuming strong duality. Thus, the dual variables λ parameterize a family of affine underestimators of f, one for each direction of perturbation, forming the set of supporting hyperplanes to $\text{epi}(f)$. The *optimal* dual variable λ^* corresponds to the slope of the tightest such supporting hyperplane that lies below the graph of f, touches the graph at an optimal point $\mathbf{x}^*$, and respects the constraint $\mathbf{h}(\mathbf{x}) = \mathbf{0}$. We will generalize this geometric picture to handle inequality constraints in the homework.

Example: Scalar Convex Problem

To build geometric intuition for Lagrangian duality, we study a scalar convex problem:

$$\begin{aligned} \min_x \quad & f(x) = (x-2)^2 + 1 \\ \text{s.t.} \quad & h(x) = x - 1 = 0. \end{aligned}$$

This problem is simple, since there is a single feasible point, $x = 1$, and hence this must also be the optimal value of x; see Fig. 6.11. However, it is still helpful as an example.

We form the Lagrangian, which are linear *tilts* of $f(x)$:

$$\mathcal{L}(x, \lambda) = f(x) + \lambda h(x) = (x-2)^2 + 1 + \lambda(x-1),$$

If we minimize over x, we obtain the dual function:

$$\phi(\lambda) = \inf_x \mathcal{L}(x, \lambda) = \inf_x \left((x-2)^2 + 1 + \lambda(x-1)\right) = -\frac{1}{4}\lambda^2 + \lambda + 1, \tag{6.69}$$

which is affine in λ and always satisfies $\phi(\lambda) \leq f(x)$ for any feasible x. In geometric terms, $\phi(\lambda)$ corresponds to the vertical intercept of the tightest line with slope λ that lies below the epigraph of f. The dual achieves its maximum at $\lambda^* = 2$, corresponding to the primal optimal point $x^* = 1$ and value $f(x^*) = 2$. This value of λ^* defines the best affine lower bound that respects the constraint.

The optimal multiplier λ^* is determined by the KKT condition $f'(x^*) + \lambda^* = 0$, so that $\lambda^* = -f'(x^*)$. In this example the feasible point is $x^* = 1$, where $f'(1) = -2$, hence $\lambda^* = 2$. Geometrically, the tilt $\lambda(x-1)$ exactly cancels the slope of f at $x = 1$, so that the tilted function has a minimum at the feasible point. The optimal value λ^* corresponds to the negative of the slope of the tightest such supporting line that touches the graph of f at the feasible point $x^* = 1$. The corresponding supporting line to the epigraph of f at $x^* = 1$ is $t = f(1) + f'(1)(x-1) = 2 - 2(x-1)$, with slope $f'(1) = -2$.

Figure 6.11 illustrates this geometry: the parabola $f(x)$, along with several tilted functions $f(x) + \lambda(x-1)$, and the supporting line at $x^* = 1$. The dual function $\phi(\lambda)$ records the lowest value of each tilted function, yielding a concave curve in λ whose maximum matches the primal optimum $f(x^*) = 2$. This demonstrates weak and strong duality in a concrete geometric setting.

Dual Variables as Sensitivities

Dual variables, i.e. Lagrange multipliers, represent the sensitivity of the optimal value of the objective function, denoted $v = f(\mathbf{x}^*)$, to perturbations in the constraint right-hand

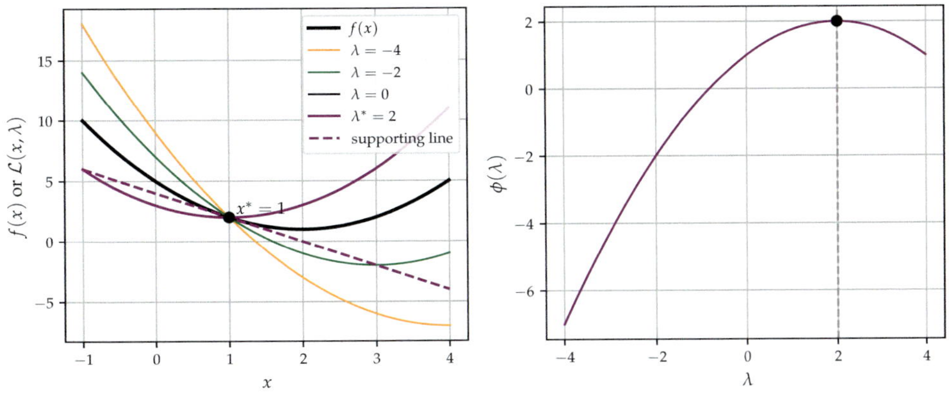

Figure 6.11 Left: Illustration of original function $f(x)$ along with the *tilted* Lagrangian $\mathcal{L}(x, \lambda)$ for several values of λ, including $\lambda^* = 2$. The optimal dual variable $\lambda^* = 2$ is the negative of the slope of the tightest supporting hyperplane at the optimal/feasible point $x^* = 1$. Right: Illustration of the dual function $\phi(\lambda)$.

sides. If we write the constraints as $\mathbf{h}(\mathbf{x}) = \mathbf{b}$ and $\mathbf{g}(\mathbf{x}) \leq \mathbf{c}$, then we may write $v(\mathbf{b}, \mathbf{c})$ as a function of the right hand sides $\mathbf{b}$ and $\mathbf{c}$. Under conditions[16] provided in Rockafellar and Wets [422], then the optimal multipliers $(\boldsymbol{\lambda}^*, \boldsymbol{\mu}^*)$ satisfy

$$\lambda_j^* = -\frac{\partial v}{\partial b_j}, \qquad \mu_j^* = -\frac{\partial v}{\partial c_j}. \tag{6.70}$$

In vector form

$$-\nabla v(\mathbf{b}, \mathbf{c}) = \begin{bmatrix} \boldsymbol{\lambda}^* \\ \boldsymbol{\mu}^* \end{bmatrix}. \tag{6.71}$$

Each dual variable λ_j^* quantifies how much the optimum f^* increases or decreases per unit increase in the right-hand side b_j of an equality constraint. Similarly, if an inequality constraint $g_j(\mathbf{x}) \leq c_j$ is *active* at $\mathbf{x}^*$, then μ_j^* measures the marginal cost or benefit of loosening or tightening that constraint. If the inequality constraint is *inactive*, then $\mu_j^* = 0$ and has no local effect on the optimum. Equivalently, for small perturbations ϵ, we may write

$$v(\mathbf{b} + \epsilon \mathbf{e}_j, \mathbf{c}) = v(\mathbf{b}, \mathbf{c}) - \epsilon \lambda_j^* + o(\epsilon), \qquad v(\mathbf{b}, \mathbf{c} + \epsilon \mathbf{e}_j) = v(\mathbf{b}, \mathbf{c}) - \epsilon \mu_j^* + o(\epsilon),$$

where $\mathbf{e}_j$ is a unit vector with a one in the j-th component, perturbing the j-th boundary.

Shadow Prices and Marginal Value in Economics

Consider a linear program for profit maximization of a firm producing n goods. Each x_j represents the units of good j produced, with profit p_j per unit, so the total profit is $\mathbf{p}^T\mathbf{x}$.

[16] Specifically, we require that the problem is convex and Slater's condition holds at $(\mathbf{b}, \mathbf{c})$, so strong duality and dual attainment hold. Further, we assume v is differentiable at $(\mathbf{b}, \mathbf{c})$. If v is not differentiable, then we say that the multipliers are in the sub-gradient of v: $(-\boldsymbol{\lambda}^*, -\boldsymbol{\mu}^*) \in \partial v(\mathbf{b}, \mathbf{c})$.

Panel 6.3: A Concrete Example Illustrating Marginal Value

Consider a cement factory producing two types of cement: regular cement (x_1) and special cement (x_2). The profit of the regular cement is \$1 per kg and the profit of the special cement is \$2 per kg. Each requires labor and material:

	x_1	x_2
Labor (hours)	1	1
Material (kg)	0	1

so that one unit of x_1 uses 1 hour of labor, one unit of x_2 uses 1 hour of labor and 1 kg of material, with 4 hours of labor (r_1) and 2 kg of material (r_2) available.

The optimization problem is

$$\max_{x_1,x_2\geq 0} \quad x_1 + 2x_2 \quad \text{s.t.} \quad \begin{cases} x_1 + x_2 \leq 4 & \text{(labor)}, \\ x_2 \leq 2 & \text{(material)}. \end{cases}$$

The feasible region is a polygon with corners $(0,0)$, $(4,0)$, $(2,2)$, $(0,2)$. The optimal solution is $(x_1^*, x_2^*) = (2,2)$ with maximum profit $v = 6$.

Value Function. In this example, we may construct the value function $v(r_1, r_2)$ explicitly. Because product x_2 yields higher profit per labor unit than x_1, we use all available material on x_2 first, up to the labor cap, then fill remaining labor with x_1:

$$x_2^*(r_1, r_2) = \min\{r_2,\ r_1\}, \qquad x_1^*(r_1, r_2) = r_1 - x_2^*(r_1, r_2) \ (\geq 0).$$

Thus, the maximum profit is the piecewise–linear, concave function

$$v(r_1, r_2) = \begin{cases} r_1 + r_2, & \text{if } 0 \leq r_2 \leq r_1, \\ 2r_1, & \text{if } r_2 \geq r_1 \geq 0. \end{cases}$$

Shadow Prices of Resources. Away from $r_1 = r_2$ the value function is differentiable. Since $r_2 < r_1$, we have $\nabla v(r_1, r_2) = \begin{bmatrix} 1 & 1 \end{bmatrix}^{\mathrm{T}}$. Thus, the dual multipliers are $\mu^*_{\text{labor}} = 1$ and $\mu^*_{\text{material}} = 2$. Thus, if the firm had one more hour of labor ($r_1 = 5$), the new optimum would be $(3,2)$ with profit 7, so each extra unit of labor increases profit by \$1. If the firm had one more kg of material ($r_2 = 3$), the new optimum would be $(1,3)$ with profit 7, so each extra unit of material also increases profit by \$1.

When $r_2 < r_1$, both resources are scarce: one extra hour of labor or one extra kg of material raises profit by \$1. Since both resources are scarce and valuable, then both shadow prices are positive. When $r_2 > r_1$, labor is the bottleneck (value \$2 per hour) while extra material has no marginal value. This illustrates the economic principle that only binding constraints correspond to resources with positive marginal value.

Production consumes m resources, such as labor, raw materials, or machine hours. Each unit of good j requires A_{ij} units of resource i, and the available supply of each resource is r_i. The firm wishes to maximize profit subject to resource constraints:

$$\begin{aligned} \max_{\mathbf{x}\geq\mathbf{0}} \quad & \mathbf{p}^{\mathrm{T}}\mathbf{x} \\ \text{s.t.} \quad & \mathbf{Ax} \leq \mathbf{r}. \end{aligned} \tag{6.72}$$

The value function $v(\mathbf{r})$ gives the maximum profit for a given resource availability $\mathbf{r}$. The dual multipliers $\boldsymbol{\mu}^*$ satisfy

$$\mu_j^* = -\frac{\partial v}{\partial r_j},$$

so that μ_j^* measure the firm's maximum profit increase if resource j is increased by one unit. If $\mu_j^* = 0$, the resource is not scarce; if $\mu_j^* > 0$, it is a bottleneck and has positive value. We call μ_j^* the shadow price, which is the marginal profit per extra unit of resource j.

6.4 Interior Point Methods

Interior point methods[17] IPMs are a family of iterative algorithms for solving constrained optimization problems by traversing the *interior* of the feasible region rather than exploring the boundary [66, 370]. We've seen the simplex method solve linear programs in Chapter 3, which moves from boundary vertex to vertex until an optimal solution is found. IPMs are an alternative approach that scale to very high-dimensional problems with complex boundaries and generalize to more sophisticated convex and nonconvex problems.

The interior point method finds solutions that satisfy the KKT conditions, and hence are optimal. For convex problems this optimality is global, with polynomial-time complexity. The basic principle of IPMs is to replace the original constrained optimization problem with a series of unconstrained subproblems. A common example is the barrier approach, which enforces strict interior feasibility by introducing a barrier, or penalty, that becomes infinite at the boundary of the feasible region. This is the approach we take here. Specifically, we present a primal approach[18] that utilizes a logarithmic barrier function.

It is often difficult to solve the KKT conditions in Eq. (6.24) directly, so instead we solve a family of simpler related problems that converge to the KKT solution. We explain this for a single constraint $g(\mathbf{x}) \leq 0$, though it generalizes to more constraints. The first three KKT conditions stay the same, and the final condition changes from $\mu g(\mathbf{x}) = 0$ to $\mu g(\mathbf{x}) = -t$, where t is an auxiliary variable that we slowly relax to $t = 0$, in which case we recover

[17] Interior point methods were developed several times independently. They were originally discovered in the 1960s by Dikin [142] and by Fiacco and McCormick in the context of penalty-barrier methods [169, 170]. However, they were considered impractical for large-scale problems until Karmarkar's primal–dual IPM algorithm for linear programming [270] in 1984, which was competitive with the simplex method [123]. For linear programming, the convergence is provably linear in the dimension of the problem. In practice, the iteration count is often even independent of dimension.

[18] In practice, primal–dual IPMs are more widely used, as they have faster convergence and better numerical stability. They also measure the duality gap, which makes it easier to monitor progress and stopping criteria.

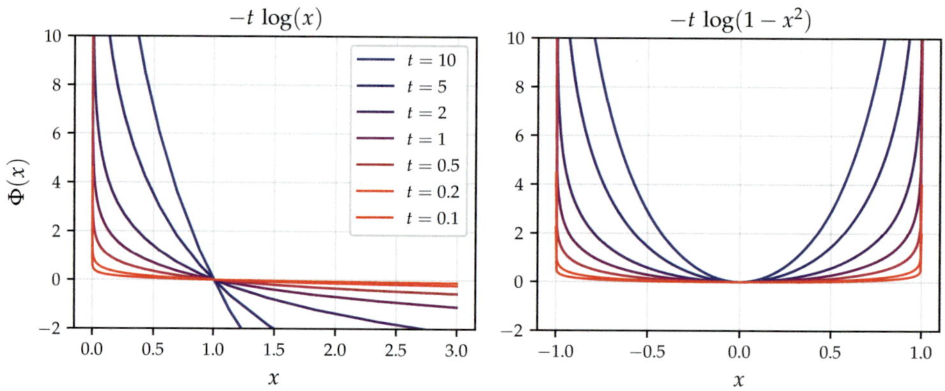

Figure 6.12 Visualization of the logarithmic barrier function $\Phi(x) = -t\log(g(x))$ for various t. When $g(x) = x \geq 0$ (left), the log barrier function becomes $\Phi(x) = -t\log(x)$, which rapidly provides a steep barrier preventing x from moving into the left half plane. For a nonlinear $g(x) = x^2 - 1 \leq 0$ (right), the log barrier function becomes $\Phi(x) = -t\log(1-x^2)$. Similarly, this provides a steep barrier preventing x from leaving the interval $(-1, 1)$.

the original KKT system. It is simpler to solve these perturbed equations for $\mathbf{x}^*(t)$, and we may then vary t to 0, using the solution $\mathbf{x}^*(t_{k-1})$ as the initial condition to solve for $\mathbf{x}^*(t_k)$. In the limit, we satisfy the KKT conditions and the inequality constraint. This is known as continuation, or homotopy, and it is a very useful technique beyond optimization.

When $t \neq 0$, it is possible to solve for μ directly from $\mathbf{x}$ via

$$\mu = \frac{-t}{g(\mathbf{x})}. \tag{6.73}$$

We then substitute this back into the gradient equation and solve for $\mathbf{x}$:

$$\nabla f(\mathbf{x}) + \frac{-t}{g(\mathbf{x})}\nabla g(\mathbf{x}) = 0. \tag{6.74}$$

A little calculus shows that this is equivalent to the following:

$$\nabla\left(f(\mathbf{x}) - t\log(-g(\mathbf{x}))\right) = 0, \tag{6.75}$$

so we are looking for the solution to the following minimization problem:

$$\min_{\mathbf{x}} f(\mathbf{x}) - t\log(-g(\mathbf{x})). \tag{6.76}$$

A solution to this problem will satisfy the first KKT condition and will also ensure that $g(\mathbf{x}) \leq 0$, since $-g(\mathbf{x})$ must be positive for $\log(-g(\mathbf{x}))$ to be finite. Similarly, from Eq. (6.73), $\mu \geq 0$. Thus, all modified KKT conditions will be satisfied. The function $-t\log(-g(\mathbf{x}))$ is known as a *barrier function*, since it is a smooth convex function, defined in the interior of the feasible set, that becomes very large near the boundaries. As $t \longrightarrow 0$, this barrier becomes infinite, as shown in Fig. 6.12 for two inequality constraints.

Since this is an unconstrained optimization problem, we may use simpler techniques, such as Newton's method, which converges rapidly with a good initial condition. When t is small, the penalty function $t\log(-g(\mathbf{x}))$ will be very steep at the boundaries, and shallow elsewhere, so Newton's method may converge slowly. However, using the optimal solution $\mathbf{x}^*(t_k)$ as the initial condition for Newton's method at t_{k+1} accelerates the algorithm.

If we start with a large t, the objective function appears to have no $f(\mathbf{x})$ term, and instead we minimize $-\log(-g(\mathbf{x}))$, which is equivalent to minimizing $g(\mathbf{x})$. We can solve this using Newton's method to find $\mathbf{x}^*(\infty)$, called the *analytic center* of the feasible set, which will be an initialization for our first value of t. The sequence of intermediate solutions $\mathbf{x}^*(t_k)$ is called the *central path*, and it is always in the interior of the feasible set.

Extensions to More Constraints. It is simple to incorporate additional constraints, $g_j(\mathbf{x}) \leq 0$ for $j = 0, \ldots, m$. In this case, we design a composite barrier function that is the sum of the various individual barrier functions corresponding to each individual constraint:

$$\Phi(\mathbf{x}) = -\sum_{j=0}^{m} \log\left(-g_j(\mathbf{x})\right). \tag{6.77}$$

Barrier Method for Linear Programming

It is possible to solve linear programming using the logarithmic barrier method, as described above. We begin with a standard linear program

$$\begin{aligned} \min_{\mathbf{x}} \quad & \mathbf{c}^{\mathrm{T}}\mathbf{x} \\ \text{subject to} \quad & \mathbf{A}\mathbf{x} \leq \mathbf{b} \\ & \mathbf{x} \geq 0. \end{aligned} \tag{6.78}$$

The barrier method addresses the inequality constraints by introducing logarithmic barrier functions to penalize solutions that approach the constraint boundaries. Specifically, the barrier formulation replaces the original constrained problem with a sequence of unconstrained subproblems:

$$\min_{\mathbf{x}} f_t(\mathbf{x}) = \min_{\mathbf{x}} \left(\mathbf{c}^{\mathrm{T}}\mathbf{x} - t\left(\sum_{j=1}^{m} \log\left(b_j - \mathbf{a}_j^{\mathrm{T}}\mathbf{x}\right) + \sum_{j=1}^{n} \log\left(x_j\right)\right)\right), \tag{6.79}$$

where $t \geq 0$ is the parameter that we decrease to zero, and $\mathbf{a}_j^{\mathrm{T}}\mathbf{x} = b_j$ denotes the jth row of the matrix system of equations $\mathbf{A}\mathbf{x} = \mathbf{b}$. As $t \to 0$, the solution to this unconstrained problem converges to the optimal solution of the original problem.

With the barrier scaling, the gradient and Hessian become:

$$\nabla f_t(\mathbf{x}) = \mathbf{c} + t\left(\sum_{j=1}^{m} \frac{\mathbf{a}_j}{b_j - \mathbf{a}_j^{\mathrm{T}}\mathbf{x}} - \sum_{j=1}^{n} \frac{\mathbf{e}_j}{x_j}\right), \tag{6.80a}$$

$$\nabla^2 f_t(\mathbf{x}) = t\left(\sum_{j=1}^{m} \frac{\mathbf{a}_j\mathbf{a}_j^{\mathrm{T}}}{(b_j - \mathbf{a}_j^{\mathrm{T}}\mathbf{x})^2} + \sum_{j=1}^{n} \frac{\mathbf{e}_j\mathbf{e}_j^{\mathrm{T}}}{x_j^2}\right). \tag{6.80b}$$

Here, $\mathbf{e}_j$ is a unit vector in the x_j direction. Newton's method yields the direction $\Delta\mathbf{x}$:

$$\nabla^2 f_t(\mathbf{x})\Delta\mathbf{x} = -\nabla f_t(\mathbf{x}).$$

Typically we would select the step size α via line search to ensure feasibility (i.e., $\mathbf{Ax} < \mathbf{b}$ remains satisfied), then update:

$$\mathbf{x} \leftarrow \mathbf{x} + \alpha\Delta\mathbf{x}.$$

We then update the barrier parameter t, for example $t_{k+1} = 0.9t_k$. We then continue until the duality gap and constraint residuals fall below pre-chosen tolerances.

Linear Programming Example in Python

Modern optimization packages implement efficient and robust primal–dual interior point methods that exploit sparsity, primal–dual relationships, and predictor–corrector steps to efficiently solve large-scale linear programming problems[19]. For example, we solve the linear program from Section 3.1 using SciPy's built-in solver, as follows:

CodeSnip 6.a

```
import numpy as np
from scipy.optimize import linprog

# Define constraints
A = np.array([ [-1, 0],     # -x <= 0   => x >= 0
    [0, -1],                # -y <= 0   => y >= 0
    [0, 1],                 # y <= 30
    [1, 1] ])               # x + y <= 50
b = np.array([0, 0, 30, 50])
c = np.array([-100, -400]) # Max c^T x = 100x+400y => min -c^T x

# Solve using a modern interior-point method (HiGHS-IPM)
res = linprog(c, A_ub=A, b_ub=b, method='highs-ipm')
```

Note that here we have embedded the non-negativity constraints $\mathbf{x} \geq 0$ directly into the first two rows of the matrix system $\mathbf{Ax} \leq \mathbf{b}$.

It is simple to implement a barrier-method IPM ourselves. Note that the following example is primal and not primal–dual. Initializing the same matrix $\mathbf{A}$ and vectors $\mathbf{c}$ and $\mathbf{b}$ as above, the code is straightforward. First, we must define the logarithmic barrier function $\Phi(\mathbf{x})$, its gradient $\nabla\Phi(\mathbf{x})$, and its Hessian $\nabla^2\Phi(\mathbf{x})$:

CodeSnip 6.b

```
def phi(x):
    return -np.sum(np.log(b - A @ x))

def grad_phi(x):
    return A.T @ (1 / (b - A @ x))

def hess_phi(x):
    vals = 1 / (b - A @ x)
    return A.T @ np.diag(vals**2) @ A
```

We may then implement the barrier-method IPM as in Code 6.1.

19 In practice, modern implementations for large-scale LP or convex problems often outperform the simplex method for certain classes of large or sparse problems. However, for small or medium-scale LPs, or those with special structure, an optimized simplex method can still be competitive and may outperform IPMs in practice.

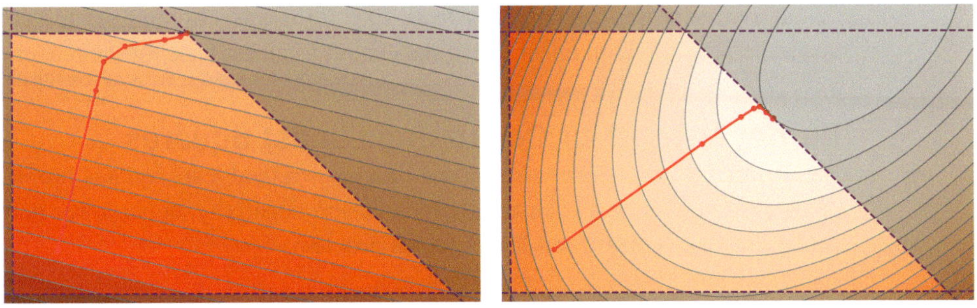

Figure 6.13 Central path of the interior point method for a convex linear program (left) and quadratic program (right). In both cases, the trajectory reaches the optimum value in very few iterations. On the right, the trajectory approaches the boundary quickly, and then slowly hugs it until the optimum. This illustrates the shallowness of the log-barrier objective landscape for some problems, making Newton's method slow to converge.

Code 6.1: Interior point method for a linear program. Here we multiply the barrier function by $1/t$ instead of t and let $t \longrightarrow \infty$ instead of $t \longrightarrow 0$, to be consistent with the literature.

```
def barrier_method_lp(A, b, c, x0, mu=10, tol=1e-8, max_iter=50)
    :
    m, n = A.shape
    x = x0.copy()
    t = 1.0
    path = [x.copy()]

    for _ in range(max_iter):
        for _ in range(20):  # Newton steps
            grad = t * c + grad_phi(x)
            hess = hess_phi(x)
            dx = -np.linalg.solve(hess, grad)

            s = 1.0
            while np.any(A @ (x + s * dx) >= b):
                s *= 0.5
                if s < 1e-16:
                    break

            x += s * dx
            path.append(x.copy())

            if np.linalg.norm(grad) < tol:
                break

        if n / t < tol:
            break
        t *= mu
    return np.array(path)
```

It is easy to call this code with the following two lines:

CodeSnip 6.c

```
x0 = np.array([5.0, 5.0])  # Feasible initial condition
path = barrier_method_lp(A, b, c, x0)
```

The output of this code is shown in Fig. 6.13 (left). Modifying this code to solve a quadratic program with linear constraints is left as an exercise for the reader.

Barrier Method for Constrained Least Squares

Barrier-method IPMs may also be used for constrained least-squares problems and other quadratic programs with constraints such as non-negativity, upper/lower bounds, or linear constraints. The constrained-least squares problem

$$\begin{aligned} \min_{\mathbf{x}} \quad & \tfrac{1}{2}\|\mathbf{A}\mathbf{x}-\mathbf{b}\|^2 \\ \text{s.t.} \quad & \mathbf{C}\mathbf{x} \leq \mathbf{d} \end{aligned}$$

may be solved efficiently with IPMs [66]. In this case, the Hessian is $\mathbf{A}^{\mathrm{T}}\mathbf{A} + t\nabla^2\Phi(\mathbf{x})$, so each iteration involves solving a Newton system that includes both the quadratic term $\mathbf{A}^{\mathrm{T}}\mathbf{A}$ and the barrier term's diagonal contributions from $\Phi(\mathbf{x})$. This is particularly powerful in high-dimensional settings when combined with sparse linear algebra techniques.

6.5 Alternating Direction Method of Multipliers (ADMM)

The alternating direction method of multipliers (ADMM) [65, 380] is a splitting method. It allows us to solve convex optimization problems by decomposing them into smaller, more manageable sub-problems. It combines the decomposability of dual ascent methods with the robustness of the method of multipliers. ADMM is widely used in applications, including signal processing, machine learning, and control. Several specific applications include robust PCA [91], MAP inference [339, 348], LP decoding [26], graphical models [122], and sparse feedback control [311]. We will explore robust PCA in the next subsection. ADMM is related to earlier operator splitting methods and penalty-duality schemes [151, 195, 183, 316, 160]. Because of its widespread applicability, ADMM has inspired many efficient software implementations [65, 374]. There are also a number of works to develop theoretical results concerning guarantees and rates of convergence [222, 198, 161].

There are two equivalent formulations, the consensus form and the composite form.

Consensus Form of ADMM. A standard form of ADMM may be written as

$$\begin{aligned} \min_{\mathbf{x},\mathbf{z}} \quad & f(\mathbf{x}) + g(\mathbf{z}) \\ \text{s.t.} \quad & \mathbf{A}\mathbf{x} + \mathbf{B}\mathbf{z} = \mathbf{c}. \end{aligned} \tag{6.81}$$

This form of ADMM is suitable for splitting across $\mathbf{x}$ and $\mathbf{z}$. It allows for distributed computing when f and g are separable and makes use of the augmented Lagrangian:

$$\mathcal{L}_\rho(\mathbf{x},\mathbf{z},\lambda) = f(\mathbf{x}) + g(\mathbf{z}) + \lambda^{\mathrm{T}}(\mathbf{A}\mathbf{x}+\mathbf{B}\mathbf{z}-\mathbf{c}) + \frac{\rho}{2}\|\mathbf{A}\mathbf{x}+\mathbf{B}\mathbf{z}-\mathbf{c}\|^2. \tag{6.82}$$

Composite Form of ADMM. A special form of consensus ADMM may be written as

$$\min_{\mathbf{x}} \ f(\mathbf{x}) + g(\mathbf{A}\mathbf{x}). \tag{6.83}$$

using a variable substitution:

$$\mathbf{z} = \mathbf{A}\mathbf{x}. \tag{6.84}$$

Then the problem becomes:

$$\min_{\mathbf{x},\mathbf{z}} \quad f(\mathbf{x}) + g(\mathbf{z}) \quad \text{s.t.} \quad \mathbf{A}\mathbf{x} - \mathbf{z} = \mathbf{0}. \tag{6.85}$$

This matches the consensus form with:

$$\mathbf{B} = -\mathbf{I}, \quad \mathbf{c} = \mathbf{0}. \tag{6.86}$$

The two forms are equivalent; the second is a more structured special case that is easier to apply to problems like LASSO, total variation denoising, or compressed sensing.

We introduce the new variables $\mathbf{z} = \mathbf{A}\mathbf{x}$ to decouple the problem, making it less stiff and more easily distributed. If g is nonsmooth or difficult to minimize in composition, then minimizing $f(\mathbf{x}) + g(\mathbf{A}\mathbf{x})$ directly might be hard. Introducing $\mathbf{z}$ allows one to alternatively minimize $f(\mathbf{x})$ with a quadratic penalty, then minimize $g(\mathbf{z})$ separately, for example using known proximal operators; this process is repeated until convergence. The constraint $A\mathbf{x} = \mathbf{z}$ is then enforced softly via the dual variable and the quadratic penalty.

Primal–Dual Interpretation. ADMM a primal–dual algorithm, and it is a special case of the *augmented Lagrangian method*, which is also a primal–dual method. In this more general approach, we form the augmented Lagrangian with respect to the constraint (e.g., $\mathbf{A}\mathbf{x} + \mathbf{B}\mathbf{z} = \mathbf{c}$), and then alternate the updates, first minimizing with respect to the primal variables ($\mathbf{x}$, $\mathbf{z}$), and then performing a gradient ascent on the dual variable (λ). This differs from standard Lagrangian dual ascent, which may suffer from slow convergence or instability. ADMM can be viewed as a dual ascent method with proximal regularization, which dampens the dual updates with a penalty to encourage stability and convergence. This is an application of the Douglas-Rachford algorithm [151] to the dual problem [281].

Specifically, given the optimization problem

$$\begin{aligned} \min_{\mathbf{x},\mathbf{z}} \quad & f(\mathbf{x}) + g(\mathbf{z}) \\ \text{s.t.} \quad & \mathbf{x} = \mathbf{z}. \end{aligned} \tag{6.87}$$

Then ADMM iterates through the following

$$\mathbf{x}_{k+1} = \arg\min_{\mathbf{x}} \Big(f(\mathbf{x}) + \frac{\rho}{2} \|\mathbf{x} - \mathbf{z}_k + \mathbf{u}_k\|^2 \Big), \tag{6.88a}$$

$$\mathbf{z}_{k+1} = \arg\min_{\mathbf{z}} \Big(g(\mathbf{z}) + \frac{\rho}{2} \|\mathbf{x}_{k+1} - \mathbf{z} + \mathbf{u}_k\|^2 \Big), \tag{6.88b}$$

$$\mathbf{u}_{k+1} = \mathbf{u}_k + \mathbf{x}_{k+1} - \mathbf{z}_{k+1}, \tag{6.88c}$$

where $\mathbf{u}$ is a dual variable (Lagrange multiplier), and $\rho > 0$ is a penalty parameter. This framework generalizes to multiple constraints. This may be written in pseudo-code as

CodeSnip 6.d

```
initialize x, z, u  # primal variables and scaled dual variable
repeat until convergence:
    x := prox_f( -z - u , 1/rho )
    # Solves x := argmin_x [ f(x) + (rho/2) * ||x + z  + u||^2 ]
```

```
z := prox_g( -x - u , 1/rho )
# Solves z := argmin_z [ g(z) + (rho/2) * ||x + z  + u||^2 ]
u := u + x + z    # dual variable update
```

The functions prox_f and prox_g are often simple to derive and execute. For example, the proximal gradient for $f(\mathbf{x}) = \|\mathbf{x}\|_1$ is given by the soft thresholding operator:

CodeSnip 6.e

```
def prox_l1(x, lam, rho):
    return np.sign(x) * np.maximum(np.abs(x) - lam/rho, 0)
```

Code Example: LASSO

Here we implement ADMM to solve the LASSO regression problem

$$\min_{\mathbf{x}} \quad \|\mathbf{A}\mathbf{x} - \mathbf{b}\|_2^2 + \lambda\|\mathbf{x}\|_1 \tag{6.89}$$

by introducing a new variable $\mathbf{z}$ to handle the $\lambda\|\cdot\|_1$ term, and a Lagrange multiplier $\mathbf{u}$ to enforce $\mathbf{z} = \mathbf{x}$. Specifically, we solve the new problem

$$\min_{\mathbf{x},\mathbf{z}} \quad \|\mathbf{A}\mathbf{x} - \mathbf{b}\|_2^2 + \lambda\|\mathbf{z}\|_1 + \frac{\rho}{2}\|\mathbf{x} - \mathbf{z}\|_2^2 \tag{6.90}$$

using the proximal gradient for the nonsmooth term $\|\mathbf{z}\|_1$ (e.g., the soft thresholding operator). The $f(\mathbf{x})$ and $g(\mathbf{z})$ functions are

$$f(\mathbf{x}) = \tfrac{1}{2}\|\mathbf{A}\mathbf{x} - \mathbf{b}\|_2^2 = \tfrac{1}{2}\mathbf{x}^{\mathrm{T}}\mathbf{A}\mathbf{x} - \left(\mathbf{A}^{\mathrm{T}}\mathbf{b}\right)^{\mathrm{T}}\mathbf{x} + \mathbf{b}^{\mathrm{T}}\mathbf{b} \tag{6.91a}$$

$$g(\mathbf{z}) = \|\mathbf{z}\|_1. \tag{6.91b}$$

Based on these equations, the ADMM updates are given by

$$\mathbf{x}_{k+1} = \left(\mathbf{A}^{\mathrm{T}}\mathbf{A} + \rho\mathbf{I}\right)^{-1}\left(\mathbf{A}^{\mathrm{T}}\mathbf{b} + \rho\left(\mathbf{z}_k - \mathbf{u}_k\right)\right) \tag{6.92a}$$

$$\mathbf{z}_{k+1} = \mathcal{S}_{\lambda/\rho}\left(\mathbf{x}_{k+1} + \mathbf{u}_k\right) \tag{6.92b}$$

$$\mathbf{u}_{k+1} = \mathbf{u}_k + \mathbf{x}_{k+1} - \mathbf{z}_{k+1}. \tag{6.92c}$$

Here $\mathcal{S}$ is a shrinkage operator $\mathcal{S}_\tau(x) = \text{sign}(x)\max(|x| - \tau, 0)$. The updates for $\mathbf{x}$ and $\mathbf{z}$ correspond to the following proximal updates:

$$\mathbf{x}_{k+1} = \arg\min_{\mathbf{x}} \quad f(\mathbf{x}) + \frac{\rho}{2}\|\mathbf{x} - \mathbf{z}_k + \mathbf{u}_k\|_2^2 = \text{prox}_{f,\rho}(\mathbf{z}_k - \mathbf{u}_k) \tag{6.93a}$$

$$\mathbf{z}_{k+1} = \arg\min_{\mathbf{z}} \quad \lambda g(\mathbf{z}) + \frac{\rho}{2}\|\mathbf{x}_{k+1} - \mathbf{z} + \mathbf{u}_k\|_2^2 = \text{prox}_{\lambda g,\rho}(\mathbf{x}_{k+1} + \mathbf{u}_k). \tag{6.93b}$$

CodeSnip 6.f

```
def ADMM_LASSO(A, b, lam, rho=1.0, max_iter=1000, tol=1e-4):
    # min (1/2)||Ax - b||^2 + lam * ||z||_1    s.t. x = z
    m, n = A.shape
    x = np.zeros(n)          # primal variable for f
    z = np.zeros(n)          # primal variable for g
    u = np.zeros(n)          # scaled dual variable

    # Precompute static matrices
    AtA = A.T @ A
    Atb = A.T @ b
    LHS = AtA + rho * np.eye(n)
    linsolve = lambda rhs: np.linalg.solve(LHS, rhs)
```

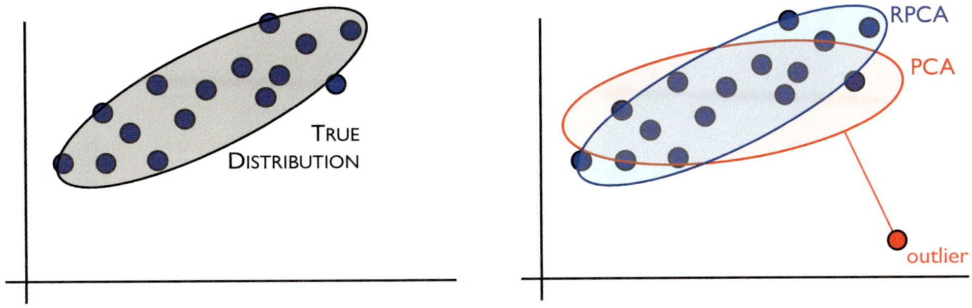

Figure 6.14 Principal component analysis seeks a Gaussian distribution that best captures the variance in a set of observation data (left). Outlier data can skew the entire estimated distribution, making standard PCA fragile to outliers (right, red). Robust PCA (right, blue) is more robust to outlier data.

```
    for k in range(max_iter):
        # (1) x-update: minimize f(x) + (rho/2)||x - z + u||^2
        q = Atb + rho * (z - u)
        x = linsolve(q)

        # (2) z-update: proximal operator for lambda * ||z||_1
        v = x + u
        z_old = z.copy()
        z = np.sign(v) * np.maximum(np.abs(v) - lam / rho, 0.0)

        # (3) dual update
        u += x - z

        # (4) convergence check
        r_norm = np.linalg.norm(x - z)          # primal residual
        s_norm = rho * np.linalg.norm(z - z_old) # dual residual
        if r_norm < tol and s_norm < tol:
            break
    return z  # final solution: z approx x
```

The keyword `lambda` defines an anonymous function `linsolve` with argument `rhs`.

Application: Robust Principal Component Analysis (RPCA)

As mentioned in Chapter 4, least-squares regression is highly susceptible to outliers and corrupted data. Principal component analysis (PCA) suffers from the same weakness, making it *fragile* so that outliers may shift the entire estimated distribution; see Fig. 6.14. To address this, Candès et al. [91] developed a robust principal component analysis (RPCA) that decomposes a data matrix $\mathbf{X}$ into a low-rank matrix $\mathbf{L}$ and a sparse matrix $\mathbf{S}$ containing outliers and corrupt data:

$$\mathbf{X} = \mathbf{L} + \mathbf{S}. \tag{6.94}$$

The principal components of $\mathbf{L}$ are *robust* to outliers and corrupt data in $\mathbf{S}$. This decomposition has implications for many modern problems of interest, including video

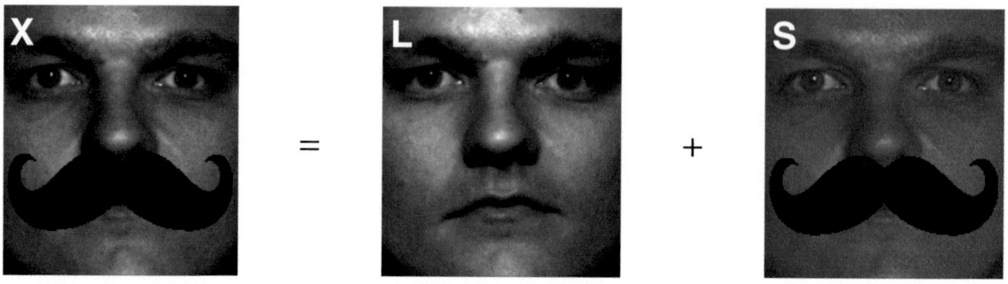

Figure 6.15 A classic example application of RPCA is in image processing, where it is possible to use a large library of human faces to build a robust statistical distribution of faces, making it possible to remove or "subtract" occlusions, such as a fake mustache.

surveillance (where the background objects appear in $\mathbf{L}$ and foreground objects appear in $\mathbf{S}$), face recognition (eigenfaces are in $\mathbf{L}$ and shadows, occlusions, etc. are in $\mathbf{S}$), natural language processing and latent semantic indexing, and ranking problems[20]. Figure 6.15 shows an example in facial recognition, where a fake mustache is removed from a face, revealing the most statistically likely features underneath.

Mathematically, the goal is to find $\mathbf{L}$ and $\mathbf{S}$ that satisfy the following:

$$\min_{\mathbf{L},\mathbf{S}} \operatorname{rank}(\mathbf{L}) + \|\mathbf{S}\|_0 \quad \text{subject to} \quad \mathbf{L} + \mathbf{S} = \mathbf{X}. \tag{6.95}$$

Neither the $\operatorname{rank}(\mathbf{L})$ nor $\|\mathbf{S}\|_0$ are convex. As in compressed sensing, it is possible to solve for the optimal $\mathbf{L}$ and $\mathbf{S}$ with *high probability* using a convex relaxation of (6.95):

$$\min_{\mathbf{L},\mathbf{S}} \|\mathbf{L}\|_* + \lambda\|\mathbf{S}\|_1 \quad \text{subject to} \quad \mathbf{L} + \mathbf{S} = \mathbf{X}. \tag{6.96}$$

Here, $\|\cdot\|_*$ is the nuclear norm, given by the sum of singular values, which is a proxy for rank. The solution to (6.96) converges to the solution of (6.95) with high probability if $\lambda = 1/\sqrt{\max(n, m)}$, where $\mathbf{X} \in \mathbb{R}^{n\times m}$, if $\mathbf{L}$ is not sparse and $\mathbf{S}$ is not low-rank.

The convex problem[21] in (6.96) is known as *principal component pursuit* (PCP), and may be solved using the augmented Lagrange multiplier (ALM) algorithm. Specifically, an augmented Lagrangian may be constructed:

$$\mathcal{L}(\mathbf{L}, \mathbf{S}, \mathbf{Y}) = \|\mathbf{L}\|_* + \lambda\|\mathbf{S}\|_1 + \langle \mathbf{Y}, \mathbf{X} - \mathbf{L} - \mathbf{S}\rangle + \frac{\mu}{2}\|\mathbf{X} - \mathbf{L} - \mathbf{S}\|_F^2. \tag{6.97}$$

A general solution would solve for the $\mathbf{L}_k$ and $\mathbf{S}_k$ that minimize $\mathcal{L}$, update the Lagrange multipliers $\mathbf{Y}_{k+1} = \mathbf{Y}_k + \mu(\mathbf{X} - \mathbf{L}_k - \mathbf{S}_k)$, and iterate until the solution converges. However, for this specific system, the alternating directions method of multipliers (ADMM) approach provides a simple procedure to find $\mathbf{L}$ and $\mathbf{S}$ [312, 549].

First, a shrinkage operator $\mathcal{S}_\tau(x) = \operatorname{sign}(x)\max(|x| - \tau, 0)$ is constructed.

[20] The ranking problem may be thought of in terms of the Netflix prize for matrix completion. In the Netflix prize, a large matrix of preferences is constructed, with rows corresponding to users and columns corresponding to movies. This matrix is sparse, as most users only rate a handful of movies. The Netflix prize seeks to fill in the missing entries of the matrix, revealing the likely user rating for movies the user has not seen.

[21] For better or worse, the community often chooses convex relaxation because of guarantees on convex models. However, it is also possible to directly solve the nonconvex problem in Eq. (6.95) using SR3 from Section 5.1.

CodeSnip 6.g

```
def shrink(X,tau):
    Y = np.abs(X)-tau
    return np.sign(X) * np.maximum(Y, 0)
```

Next, the singular value threshold (SVT) operator $\mathrm{SVT}_\tau(\mathbf{X}) = \mathbf{U}\mathcal{S}_\tau(\mathbf{\Sigma})\mathbf{V}^*$ is constructed.

CodeSnip 6.h

```
def SVT(X,tau):
    U,S,VT = np.linalg.svd(X,full_matrices=False)
    out = U @ np.diag(shrink(S,tau)) @ VT
    return out
```

Finally, we use $\mathcal{S}_\tau$ and SVT operators iteratively to solve for $\mathbf{L}$ and $\mathbf{S}$ in ADMM Code:

CodeSnip 6.i

```
def RPCA(X):
    n1,n2 = X.shape
    mu = n1*n2/(4*np.sum(np.abs(X)))
    lambd = 1/np.sqrt(np.maximum(n1,n2))
    thresh = 10**(-7) * np.linalg.norm(X)

    S = np.zeros_like(X)
    Y = np.zeros_like(X)
    L = np.zeros_like(X)
    count = 0
    while (np.linalg.norm(X-L-S) > thresh) and (count < 1000):
        L = SVT(X-S+(1/mu)*Y,1/mu)
        S = shrink(X-L+(1/mu)*Y,lambd/mu)
        Y = Y + mu*(X-L-S)
        count += 1
    return L,S
```

RPCA is demonstrated on the eigenface data with the following code

CodeSnip 6.j

```
X = faces[:,:nfaces[0]]
L,S = RPCA(X)
```

6.6 Calculus of Variations and the Euler–Lagrange Equations

The calculus of variations is a field of mathematical analysis that deals with maximizing or minimizing functionals, which are mappings from a space of functions to the real numbers [187, 163]. For example, when optimizing the shape of a wing to maximize lift and minimize drag, the shape may be represented as a function to be optimized, given some constraints (structural, size, weight, etc.). In contrast to conventional optimization, where the unknowns are finite-dimensional vectors, variational problems involve optimizing over infinite-dimensional spaces of functions. This broader setting underlies many topics, including optimal control, PDE-constrained optimization, and machine learning.

The calculus of variations dates back to Johann Bernoulli's *brachistochrone problem* in the late 17th century. Euler and Lagrange developed the field in the 18th century, including the Euler–Lagrange equation and functional derivatives. Over time, the field evolved to include more general functional spaces, such as Sobolev and Banach spaces.

Today, variational calculus is vital in many areas:

- Physics and Mechanics: Action principles [1], based on variational principles, are central in classical mechanics, general relativity, and quantum field theory.

- PDE-Constrained Optimization (Chapter 8): Many engineering problems involve optimizing a geometric shape given the constraints of some physics, typically expressed as a partial differential equation. For example, designing the shape of cooling fins in a heat exchanger [299], or the shape of a wing on a modern commercial aircraft [248] are both constrained variational optimization problems.
- Optimal Control (Chapter 9): Most optimal control problems involve solving for a function to minimize or maximize a given objective function, constrained by dynamics. The Pontryagin maximum principle, the Kalman filter, and dynamic programming are three concrete examples that have roots in variational arguments [60, 473, 37].
- Machine Learning (Chapter 10): Many problems in machine learning may be viewed through a variational lens [552, 250, 277]. Deep neural networks often approximate functionals, such as deep operator networks [323] and neural operators [310, 309].

Panel 6.4: Robust Processing of Scientific Data

Robust methods may also be used to process scientific data, for example to remove outliers and fill inclusions in experimental fluid velocity fields [440]. PCA, based on the SVD, is already widely used to extract coherent structures from fluid datasets.

Typically, fluid flow field images are acquired experimentally using a process called particle image velocimetry (PIV), where a laser sheet illuminates neutrally buoyant particles moving with the flow, and high speed camera images are used to compute velocity fields via optical flow. These velocity fields are then reshaped into vectors $\mathbf{x}_j$ at time $t_k = k\Delta t$. The dimension of these flow field vectors, called *snapshots* are often very large, sometimes in the millions or billions of components, corresponding to pixels in the original image. RPCA may be used to disambiguate the low-rank coherent structures in $\mathbf{L}$ with the sparse outliers and occlusions in $\mathbf{S}$:

$$\underbrace{\begin{bmatrix} | & | & & | \\ \mathbf{x}_1 & \mathbf{x}_2 & \cdots & \mathbf{x}_m \\ | & | & & | \end{bmatrix}}_{\mathbf{X}} = \underbrace{\begin{bmatrix} | & | & & | \\ \tilde{\mathbf{x}}_1 & \tilde{\mathbf{x}}_2 & \cdots & \tilde{\mathbf{x}}_m \\ | & | & & | \end{bmatrix}}_{\mathbf{L}} + \underbrace{\begin{bmatrix} | & | & & | \\ \mathbf{s}_1 & \mathbf{s}_2 & \cdots & \mathbf{s}_m \\ | & | & & | \end{bmatrix}}_{\mathbf{S}}$$

The result when applied to scientific data, such as flow fields, is a robust de-noising and cleaning without removing key coherent structures.

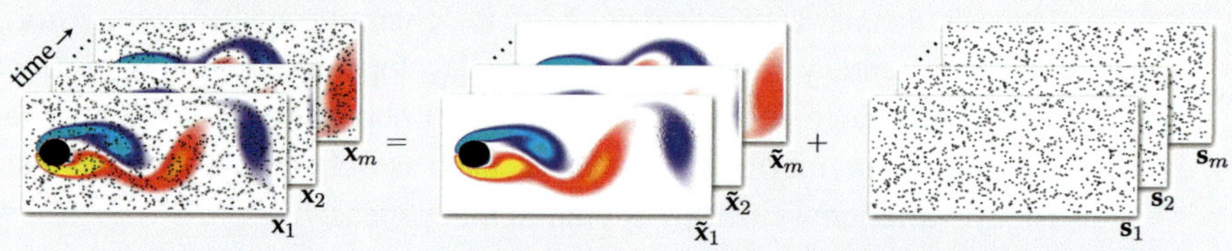

Figure 6.16: RPCA on corrupted fluid flow data, from Scherl et al. [440].

Figure 6.17 The equations for the pendulum (left) are easier to derive and solve in the intrinsic coordinates $\mathbf{q} = \begin{bmatrix}\theta & \dot{\theta}\end{bmatrix}^{\mathrm{T}}$, which naturally handle the constraint, rather than in (x, y) coordinates. These benefits only increase for more complex systems, such as the double pendulum (right).

A functional $\mathcal{J}$ takes a function $y(x)$ as input and returns a real number:

$$\mathcal{J}[y] = \int_a^b F\left(y(x), y'(x), x\right)\, dx, \tag{6.98}$$

where F is typically a scalar-valued function that depends on $y(x)$, its derivative $y'(x)$, and the independent variable x. A *variational problem* is to find a function $y^*(x)$ that extremizes (minimizes or maximizes) $\mathcal{J}[y]$ subject to boundary conditions or other constraints.

In the simplest unconstrained setting, the condition for $y(x)$ to optimize $\mathcal{J}$ is given by the Euler–Lagrange equation, which we will derive shortly:

$$\frac{\partial F}{\partial y} - \frac{\mathrm{d}}{\mathrm{d}x}\left(\frac{\partial F}{\partial y'}\right) = 0. \tag{6.99}$$

This is the continuous analog of setting the gradient to zero [187].

Mechanics Interpretation. In classical mechanics and physics, the Euler–Lagrange equations represent the principle of least action, which is perhaps the most fundamental and unifying principle in all of physics. The Euler–Lagrange equations were originally derived to describe the motion of mechanical systems in arbitrary coordinate systems (see Fig. 6.17), although they provide a much more general set of optimality conditions for variational problems where we want to optimize a functional over a function. To be consistent with this literature in physics and engineering, we will adopt physical notation here.

In the context of a physical system, we replace F with the Lagrangian $L(\mathbf{q}(t), \dot{\mathbf{q}}(t), t)$, which is written in terms of generalized coordinates $\mathbf{q}(t) \in \mathbb{R}^n$ that describe the system. For example, when analyzing a simple pendulum, the generalized coordinates may be $\mathbf{q} = \begin{bmatrix}\theta & \dot{\theta}\end{bmatrix}^{\mathrm{T}}$; see Fig. 6.17 for a schematic. Now the independent variable is time, t, and $\dot{\mathbf{q}}(t) = \frac{d}{dt}\mathbf{q}(t)$. We now call the functional $\mathcal{J}$ the action integral and denote it by $\mathcal{A}(\mathbf{q})$:

$$\mathcal{A}[\mathbf{q}] = \int_{t_0}^{t_1} L(\mathbf{q}(t), \dot{\mathbf{q}}(t), t)\, dt. \tag{6.100}$$

The Lagrangian is often formulated as the difference between kinetic energy $T(\mathbf{q}, \dot{\mathbf{q}}, t)$ and potential energy $V(\mathbf{q}, \dot{\mathbf{q}}, t)$: $L = T - V$. Energy transfer between kinetic and potential energy is called the *action*. The principle of least action states that among all possible trajectories between two fixed endpoints in configuration space, the physical path is the one that makes the action integral stationary, meaning it extremizes the functional $\mathcal{A}$; we will formalize this in a moment. More informally, the physical path $\mathbf{q}(t)$ will typically minimize, or at least be a stationary point, of the action integral in Eq. (6.100).

The Euler–Lagrange equations may now be written as

$$\frac{\partial L}{\partial q_j} - \frac{\mathrm{d}}{\mathrm{d}t}\left(\frac{\partial L}{\partial \dot{q}_j}\right) = 0, \tag{6.101}$$

where q_j is the jth component of $\mathbf{q}$. The Euler–Lagrange equations are the *necessary* conditions for a solution of the variational problem, much as $\nabla_{\mathbf{x}} f(\mathbf{x}) = \mathbf{0}$ is a necessary condition for a minimum of an unconstrained objective over a discrete decision variable $\mathbf{x}$.

Next, we will derive the Euler–Lagrange equations, analyze several classical examples, and explore duality in the context of infinite-dimensional optimization problems. Specifically, we introduce the Legendre transformation and establish that the Hamiltonian is the dual of the Lagrangian. Finally, we discuss how constraints fit into this framework.

Deriving the Euler–Lagrange Equations

We will now derive the Euler–Lagrange equations as a necessary condition for optimality of variational optimization problems. The idea is that if a function $\mathbf{q}(t)$ extremizes the functional $\mathcal{A}$, then $\mathcal{A}$ must be insensitive to infinitesimal perturbations to $\mathbf{q}$, similar to how an optimal $\mathbf{x}^*$ of an objective must satisfy $\nabla f(\mathbf{x}^*) = \mathbf{0}$, meaning that infinitesimal variations in $\mathbf{x}^*$ won't change the objective. The derivative of $\mathcal{A}$ with respect to infinitesimal variations in $\mathbf{q}(t)$ is called the *first variation* $\delta\mathcal{A}$. We will vary the trajectory $\mathbf{q}(t)$ in Eq. (6.100) and require that the first variation of the action vanishes. This will result in the Euler–Lagrange equations as a necessary condition for optimality. The principle of least action says a function $\mathbf{q}(t)$ that extremizes the functional $\mathcal{A}[\mathbf{q}]$ must satisfy

$$\delta\mathcal{A} = 0, \tag{6.102}$$

where $\delta\mathcal{A}$ is the first variation of $\mathcal{A}$ with respect to variations of $\mathbf{q}(t)$ that fix the endpoints $\mathbf{q}(t_0)$ and $\mathbf{q}(t_1)$. To carry out this variation explicitly, we consider a family of curves

$$\mathbf{q}_\varepsilon(t) = \mathbf{q}(t) + \varepsilon\boldsymbol{\eta}(t), \tag{6.103}$$

where $\boldsymbol{\eta}(t)$ is a smooth variation with $\boldsymbol{\eta}(t_0) = \boldsymbol{\eta}(t_1) = \mathbf{0}$ to ensure fixed endpoints. The functional $\mathcal{A}$ may then be evaluated at $\mathbf{q}_\varepsilon$ and we may take the derivative with respect to ε

$$\delta\mathcal{A} = \frac{\mathrm{d}}{\mathrm{d}\varepsilon}\mathcal{A}[\mathbf{q}_\varepsilon]|_{\varepsilon=0} = \frac{\mathrm{d}}{\mathrm{d}\varepsilon}\left[\int_{t_0}^{t_1} L(\mathbf{q}_\varepsilon, \dot{\mathbf{q}}_\varepsilon, t)\,dt\right]_{\varepsilon=0}. \tag{6.104}$$

We may now compute $\frac{d}{d\varepsilon}\mathcal{A}[\mathbf{q}_\varepsilon]$ explicitly:

$$\frac{\mathrm{d}}{\mathrm{d}\varepsilon}\int_{t_0}^{t_1} L(\mathbf{q} + \varepsilon\boldsymbol{\eta}, \dot{\mathbf{q}} + \varepsilon\dot{\boldsymbol{\eta}}, t)\,dt = \int_{t_0}^{t_1}\left[\frac{\partial L}{\partial \mathbf{q}}\boldsymbol{\eta} + \frac{\partial L}{\partial \dot{\mathbf{q}}}\dot{\boldsymbol{\eta}}\right]dt \tag{6.105}$$

where $\frac{\partial L}{\partial \mathbf{q}}$ and $\frac{\partial L}{\partial \dot{\mathbf{q}}}$ are evaluated at $(\mathbf{q} + \varepsilon\boldsymbol{\eta}, \dot{\mathbf{q}} + \varepsilon\dot{\boldsymbol{\eta}}, t)$. Note that there is no term from the time dependence because $\frac{dt}{d\varepsilon} = 0$.

We now evaluate this derivative at $\varepsilon = 0$, which yields

$$\delta\mathcal{A} = \frac{\mathrm{d}}{\mathrm{d}\varepsilon}\mathcal{A}[\mathbf{q}_\varepsilon]|_{\varepsilon=0} = \int_{t_0}^{t_1}\left[\frac{\partial L}{\partial \mathbf{q}}(\mathbf{q},\dot{\mathbf{q}},t)\boldsymbol{\eta}(t) + \frac{\partial L}{\partial \dot{\mathbf{q}}}(\mathbf{q},\dot{\mathbf{q}},t)\dot{\boldsymbol{\eta}}(t)\right] dt. \tag{6.106}$$

This may be written in coordinate form as

$$\delta\mathcal{A} = \frac{\mathrm{d}}{\mathrm{d}\varepsilon}\mathcal{A}[\mathbf{q}_\varepsilon]|_{\varepsilon=0} = \int_{t_0}^{t_1}\sum_{j=1}^{n}\left[\frac{\partial L}{\partial q_j}(\mathbf{q},\dot{\mathbf{q}},t)\eta_j(t) + \frac{\partial L}{\partial \dot{q}_j}(\mathbf{q},\dot{\mathbf{q}},t)\dot{\eta}_j(t)\right] dt. \tag{6.107}$$

Finally, we apply integration by parts to the second set of terms including $\dot{\eta}_j$:

$$\int_{t_0}^{t_1}\frac{\partial L}{\partial \dot{q}_j}\dot{\eta}_j\, dt = \left[\frac{\partial L}{\partial \dot{q}_j}\eta_j\right]_{t_0}^{t_1} - \int_{t_0}^{t_1}\frac{\mathrm{d}}{\mathrm{d}t}\left(\frac{\partial L}{\partial \dot{q}_j}\right)\eta_j\, dt. \tag{6.108}$$

The boundary term vanishes because $\boldsymbol{\eta}(t_0) = \boldsymbol{\eta}(t_1) = \mathbf{0}$. Substituting this back into the first variation of the action from Eq. (6.107) we arrive at

$$\delta\mathcal{A} = \int_{t_0}^{t_1}\left[\frac{\partial L}{\partial q_j} - \frac{\mathrm{d}}{\mathrm{d}t}\frac{\partial L}{\partial \dot{q}_j}\right]\eta_j\, dt. \tag{6.109}$$

The summation over j is implied. This integral must vanish for all variations $\boldsymbol{\eta}(t)$, so the integrand must be zero pointwise, resulting in the Euler–Lagrange equations:

$$\frac{\partial L}{\partial q_j} - \frac{\mathrm{d}}{\mathrm{d}t}\left(\frac{\partial L}{\partial \dot{q}_j}\right) = 0. \tag{6.110}$$

They are a necessary condition for $\mathbf{q}$ to extremize the functional $\mathcal{A}[\mathbf{q}]$, given fixed endpoints. Modifying this for non-fixed endpoints will be left as an exercise.

This viewpoint links mechanics to a broader class of variational problems in mathematical optimization. The presence of constraints is addressed either by embedding them into the Lagrangian via multipliers or by restricting the set of admissible curves, and we will discuss this more at the end of this section.

We will explore several examples that use the Euler–Lagrange equations, including the brachistochrone problem and finding geodesics on surfaces.

Brachistochrone Problem

The brachistochrone problem is one of the earliest and most classical variational problems that motivated much of the modern theory presented so far. The objective of the brachistochrone problem is to find the curve between two points so that a particle constrained to this path will reach the end point as quickly as possible. Although simple to state, this problem defied several of the greatest minds at the time[22].

A simple guess for the optimal path may be a straight line, as this is the shortest path between two points. However, adding a slight curvature below the line will cause the particle to travel faster initially and gain momentum to arrive sooner. Several curves are shown in

[22] Bernoulli posed this problem as an open challenge to the mathematical community. As the story goes, Newton, long retired from mathematics and arriving home after a day at work jotted down the solution that evening. Although submitted anonymously, Bernoulli, when reading the solution knew it was Newton, saying "I could tell the lion by his paw print".

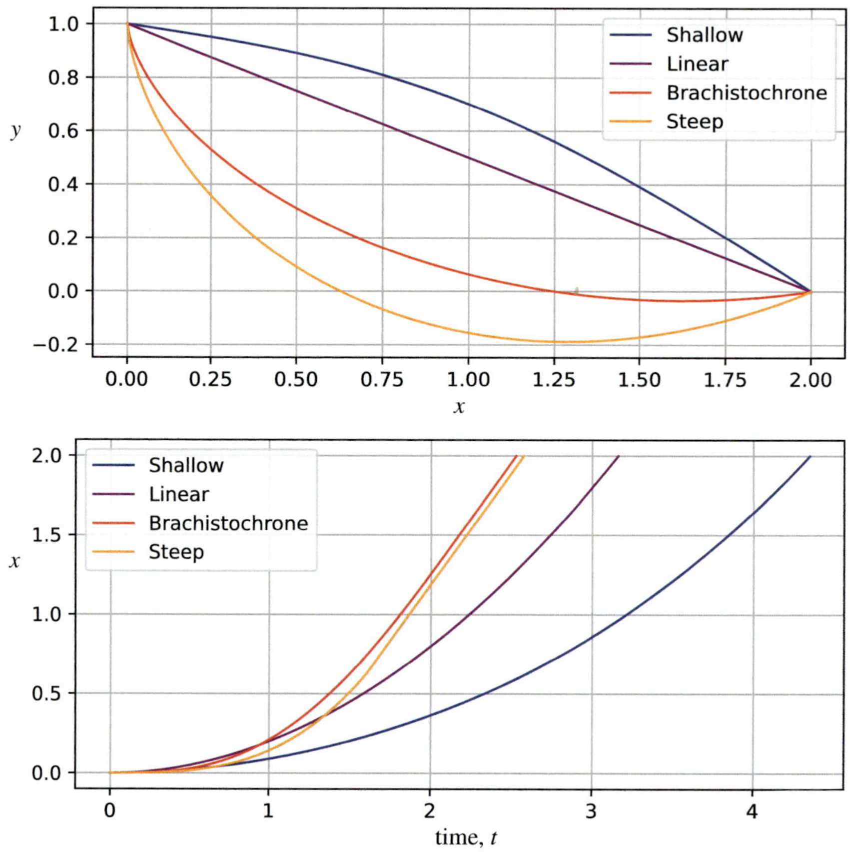

Figure 6.18 The solution to the brachistochrone problem is the cycloid (red) which would take a particle the minimum time to slide from the start to the end. Upper: Several candidate curves, including the optimal cycloid. Lower: The x position versus t. We see that the cycloid reaches the end point $x = 2$ in the shortest time.

Fig. 6.18, along with their travel times. Freeze frames of the progress along these various paths are shown in Fig. 6.19. The optimal solution to the brachistochrone problem is called a cycloid, and we will derive it now.

We want a path $y(x)$, where y is measured vertically down from the top, that minimizes the total time T it takes for a frictionless bead to slide under uniform gravity from $(x = 0, y = 0)$ to $(x = x_1, y = y_1)$.

Assuming the bead starts at $y = 0$ with zero speed, then its speed v at any point depends on the drop in height:

$$v = \sqrt{2gy}.$$

An infinitesimal path element ds along the curve satisfies

$$ds = \sqrt{1 + (y'(x))^2}\, dx,$$

since $ds^2 = dx^2 + dy^2$ and $y'(x) = dy/dx$.

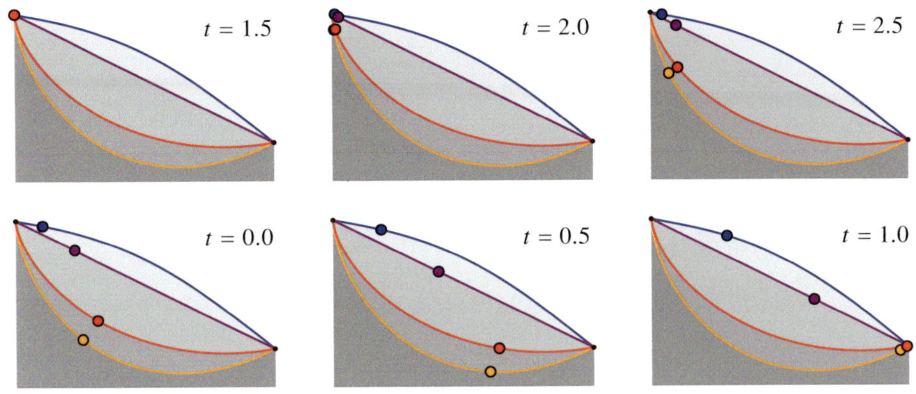

Figure 6.19 Snapshots of the various particles rolling down each of the four ramps from Fig. 6.18. The red particle on the optimal cycloid reaches the end first.

The time to traverse an infinitesimal arc is

$$dT = \frac{\mathrm{d}s}{v} = \frac{\sqrt{1+y'^2}\,dx}{\sqrt{2gy}}.$$

Hence the total time is

$$T[y(\cdot)] = \int_0^{x_1} \frac{\sqrt{1+y'^2}}{\sqrt{2gy}}\,dx.$$

We can pull out the constant $\sqrt{1/(2g)}$, so the functional to be minimized becomes

$$T[y(\cdot)] = \sqrt{\frac{1}{2g}}\int_0^{x_1} \frac{\sqrt{1+y'^2}}{\sqrt{y}}\,dx = \sqrt{\frac{1}{2g}}\int_0^{x_1} F(y,y')\,dx, \text{ where } F(y,y') = \frac{\sqrt{1+y'^2}}{\sqrt{y}}.$$

In standard variational–calculus notation, we take

$$L(x,y,y') = \frac{\sqrt{1+y'^2}}{\sqrt{y}}.$$

A function $y(x)$ that extremizes T must then satisfy

$$\frac{\mathrm{d}}{\mathrm{d}x}\left(\frac{\partial L}{\partial y'}\right) - \frac{\partial L}{\partial y} = 0.$$

Notice that L does not explicitly depend on x. This lets us use a simpler form, the Beltrami identity, which we will show next.

Beltrami Identity. Since $L = L(y, y')$ has no explicit x dependence, we have

$$L(y,y') - y'\frac{\partial L}{\partial y'} = \text{constant}.$$

In our notation,

$$L(y, y') = \frac{\sqrt{1 + y'^2}}{\sqrt{y}}, \qquad \frac{\partial L}{\partial y'} = \frac{1}{\sqrt{y}} \frac{y'}{\sqrt{1 + y'^2}}.$$

Hence

$$L - y'\frac{\partial L}{\partial y'} = \frac{\sqrt{1 + y'^2}}{\sqrt{y}} - y'\left(\frac{1}{\sqrt{y}} \frac{y'}{\sqrt{1 + y'^2}}\right) = \frac{1}{\sqrt{y}\sqrt{1 + y'^2}} = \text{constant}.$$

We will let the constant be $1/(2a)$, so that

$$\frac{1}{\sqrt{y}\sqrt{1 + y'^2}} = \frac{1}{2a} \implies \sqrt{1 + y'^2} = 2a\sqrt{y}.$$

Solution: A Cycloid. We may now solve for $y'(x)$

$$1 + y'^2 = 4a^2 y \implies y'^2 = 4a^2 y - 1, \qquad y' = \sqrt{4a^2 y - 1}.$$

This is a first-order differential equation that we can solve by separation of variables. A standard approach is to parametrize by the angle of a rolling circle:

$$y(t) = a(1 - \cos t), \quad x(t) = \frac{a}{2}(t - \sin t).$$

The constants a and t_{end} are fixed by the boundary conditions. The curve $(x(t), y(t))$ is a cycloid, which is the extremal, minimal-time curve, matching the classical solution.

It is possible to numerically verify that this solution is optimal by trying small variations $q_\varepsilon(t) = q(t) + \varepsilon\eta$. We visualize the ensemble of nearby trajectories and their travel times in Fig. 6.20. We plot the arrival times for the ensemble as a histogram in Fig. 6.21, where we see that the cycloid is in fact the curve with shortest arrival time.

Tautochrone Problem

The tautochrone problem, first posed by Huygens in the 17th century[23], asks for the curve along which a bead, sliding without friction under uniform gravity, reaches the lowest point in the same amount of time regardless of its starting point on the curve. As in the brachistochrone problem, we assume the particle starts from rest at height $y = 0$ and slides down a curve $y(x)$, where y is measured vertically downward. We will show that the solution is also a cycloid, visualized in Fig. 6.22.

The particle's speed at height y is determined by conservation of energy:

$$v = \sqrt{2gy}.$$

The arc length of the curve element is

$$ds = \sqrt{1 + y'^2}\, dx.$$

[23] Huygens even constructed a pendulum with a cycloidal constraint to ensure isochronous motion, a major step in the design of accurate timekeeping devices.

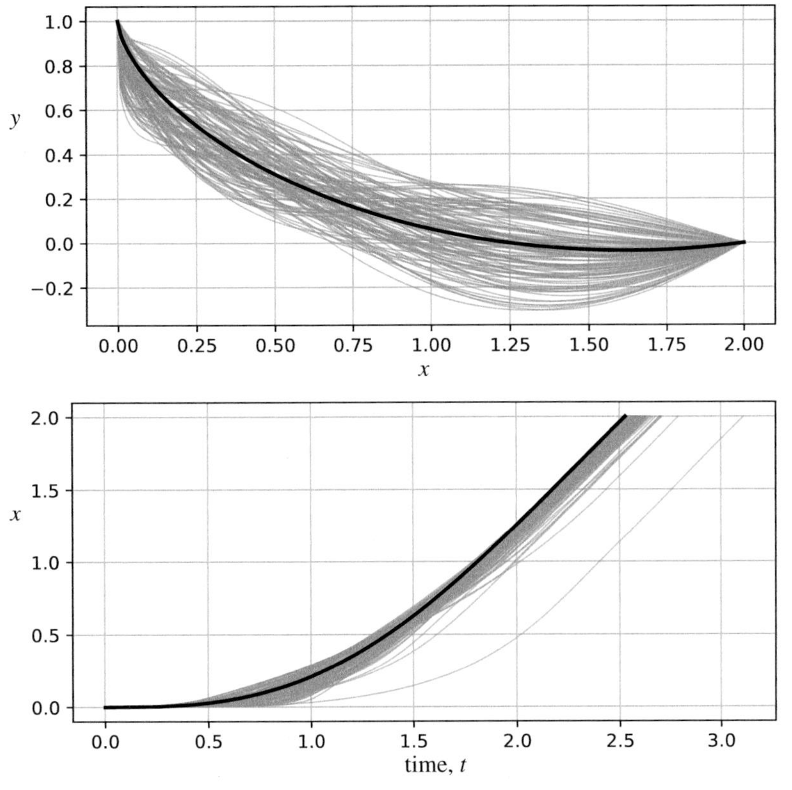

Figure 6.20 Here we compute the trajectories for several variations on the optimal cycloid. Each variation is visualized in gray. (top) The shape of the curves and (bottom) the position x versus time t. Among all variations, the cycloid has the minimum time to the end point $(2, 0)$. A histogram of the final time to reach the goal is shown in Fig. 6.21.

So the time to traverse an arc is

$$dT = \frac{ds}{v} = \frac{\sqrt{1 + y'^2}}{\sqrt{2gy}}\, dx.$$

Let x be the horizontal coordinate corresponding to a starting point and x^* the endpoint at the bottom of the curve. The total time to reach the bottom from point x is

$$T[x] = \int_x^{x^*} \frac{\sqrt{1 + y'^2}}{\sqrt{2gy}}\, dx.$$

Tautochrone Condition. We now impose the tautochrone condition: the time $T[x]$ must be constant for all x, regardless of where the particle is released along the curve. This can only be true if the integrand is an exact differential of a function that increases linearly with arc length or with a parameter t such that $T = t^* - t$.

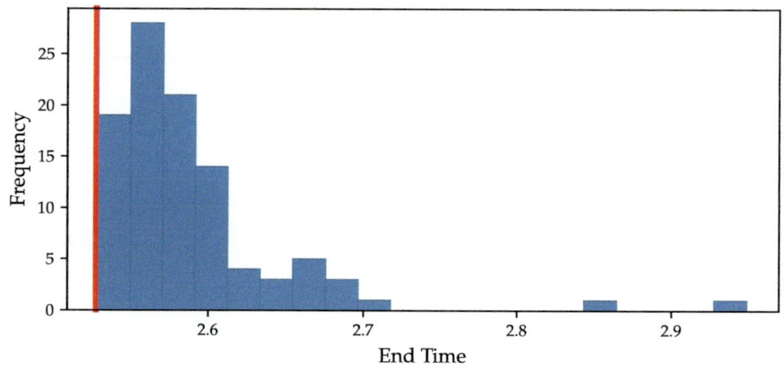

Figure 6.21 Histogram of the final time to reach the goal state for the optimal cycloid (red) and the ensemble of variations around the cycloid (blue).

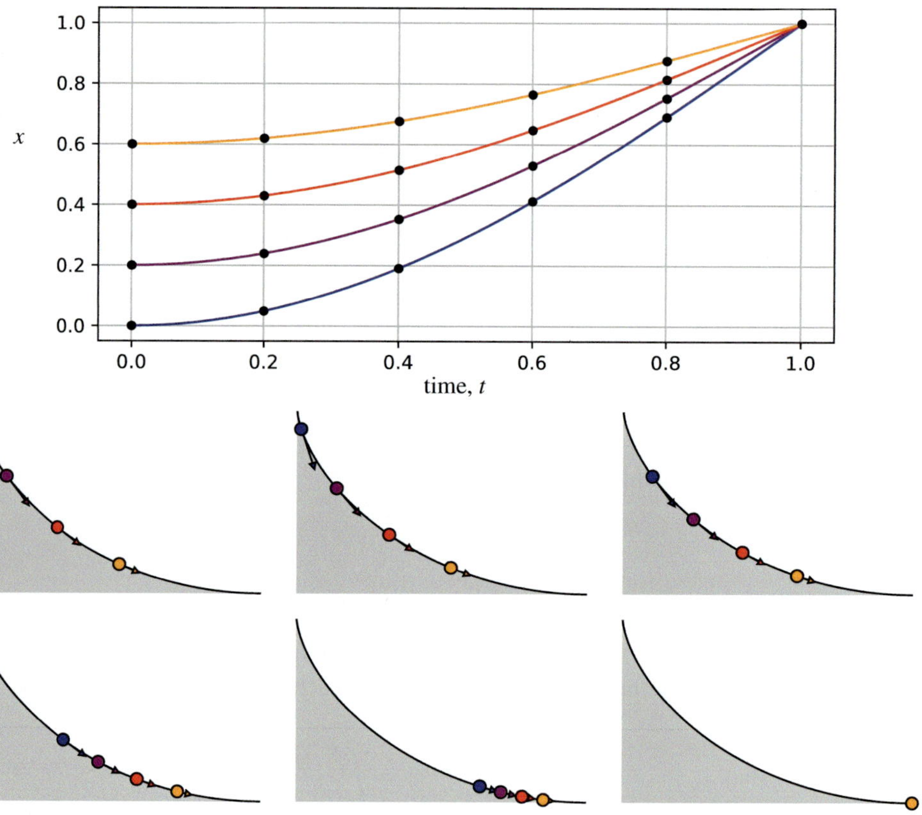

Figure 6.22 (top) Plots of four different solutions of the tautochrone, starting from different points on the curve. Each trajectory reaching the bottom ($x = 1$) at the same time ($t = 1$). (bottom) Several snapshots of the different trajectories at various times. The times are indicated in the top plot as black dots. Modified from `https://en.wikipedia.org/wiki/File:Tautochrone_curve.gif`; original code by Claudio Rocchini.

A classical way to enforce this is to change variables and parametrize the curve with respect to time t. Let

$$x = x(t), \quad y = y(t),$$

and assume the bead reaches the bottom at time $t = t^*$. Then, as before,

$$ds = \sqrt{\dot{x}^2 + \dot{y}^2}\, dt, \quad v = \sqrt{2gy},$$

so the total time becomes

$$T = \int_0^{t^*} dt.$$

But we can also write

$$dt = \frac{ds}{v} = \frac{\sqrt{\dot{x}^2 + \dot{y}^2}}{\sqrt{2gy}}\, dt.$$

This implies that

$$\frac{\sqrt{\dot{x}^2 + \dot{y}^2}}{\sqrt{2gy}} = 1 \quad \Longrightarrow \quad \dot{x}^2 + \dot{y}^2 = 2gy.$$

We may now solve this differential equation, which describes a particle moving in a central gravitational field. Recognizing the solution as a cycloid, we may now choose a standard cycloidal parametrization:

$$x(t) = a(t - \sin t), \quad y(t) = a(1 - \cos t).$$

Then,

$$\dot{x} = a(1 - \cos t), \quad \dot{y} = a \sin t,$$

so

$$\dot{x}^2 + \dot{y}^2 = a^2(1 - \cos t)^2 + a^2 \sin^2 t = 2a^2(1 - \cos t),$$

and since

$$y(t) = a(1 - \cos t),$$

we get:

$$\dot{x}^2 + \dot{y}^2 = 2a^2 y(t).$$

Therefore, the condition $\dot{x}^2 + \dot{y}^2 = 2gy$ is satisfied if $a^2 = g$.

The solution to the tautochrone problem is again a cycloid

$$x(t) = a(t - \sin t), \quad y(t) = a(1 - \cos t),$$

where a is a constant related to the acceleration due to gravity. This is exactly the same curve that solves the brachistochrone problem. Thus, the cycloid is the unique curve that both minimizes the time of descent and equalizes descent time from all starting points.

Example: Circle Maximizes Area for a Fixed Perimeter

Below is a classical Euler–Lagrange style derivation of the isoperimetric problem – finding the plane curve that maximizes the enclosed area A among all curves of fixed perimeter L. We show that the solution must be a circle.

Unlike the simpler brachistochrone problem, or other geodesic problems, here we have both a functional to be maximized (the area) and a constraint (the curve's perimeter is fixed). In the calculus of variations, one introduces a Lagrange multiplier to handle the constraint, then applies the Euler–Lagrange equation to the resulting combined functional.

First we parameterize the curve. A convenient choice is polar coordinates $[r(\theta), \theta]$, with θ going from 0 to 2π. The curve is

$$(x(\theta), y(\theta)) = (r(\theta)\cos\theta,\ r(\theta)\sin\theta), \qquad \theta \in [0, 2\pi].$$

The perimeter is given by

$$L[r] = \int_0^{2\pi} \sqrt{r(\theta)^2 + \dot{r}(\theta)^2}\, d\theta,$$

where $\dot{r} = dr/d\theta$, and the area is given by

$$A[r] = \frac{1}{2}\int_0^{2\pi} r(\theta)^2\, d\theta.$$

To enforce the constraints, we introduce a Lagrange multiplier λ and consider

$$\Phi[r] = \int_0^{2\pi} \left[\tfrac{1}{2} r^2 - \lambda\sqrt{r^2 + \dot{r}^2}\right] d\theta + \lambda L_0,$$

with L_0 the prescribed perimeter. The integrand is

$$\mathcal{L}(r, \dot{r}) = \tfrac{1}{2} r^2 - \lambda\sqrt{r^2 + \dot{r}^2}.$$

The Euler–Lagrange equation is then

$$\frac{\mathrm{d}}{\mathrm{d}\theta}\left(-\lambda \frac{\dot{r}}{\sqrt{r^2 + \dot{r}^2}}\right) - \left[r - \lambda \frac{r}{\sqrt{r^2 + \dot{r}^2}}\right] = 0.$$

A constant radius $r(\theta) \equiv R$ satisfies this (choose $\lambda = R$). The length constraint gives $R = L_0/(2\pi)$, and the enclosed area is πR^2, maximal among all curves of perimeter L_0. Thus, the stationary solution is a circle, which uniquely maximizes area for a given perimeter.

Legendre Transformation and the Dual Hamiltonian Form

To connect to the Hamiltonian picture, one performs the Legendre transform of the Lagrangian. For a single coordinate q_j, define the momentum

$$p_j = \frac{\partial L}{\partial \dot{q}_j}. \tag{6.111}$$

Then the Hamiltonian is

$$H(\mathbf{q}, \mathbf{p}, t) = \mathbf{p}\,\dot{\mathbf{q}} - L(\mathbf{q}, \dot{\mathbf{q}}, t), \tag{6.112}$$

with $\dot{\mathbf{q}}$ viewed as a function of $(\mathbf{q}, \mathbf{p})$ by inverting $\mathbf{p} = \frac{\partial L}{\partial \dot{\mathbf{q}}}$. In coordinates[24], $p_j = \frac{\partial L}{\partial \dot{q}_j}$, and

$$H(q_1, \ldots, q_n, p_1, \ldots, p_n, t) = \sum_{j=1}^{n} p_j \dot{q}_j - L(\mathbf{q}, \dot{\mathbf{q}}, t). \tag{6.113}$$

The condition that the action is stationary is equivalent to Hamilton's canonical system:

$$\dot{q}_j = \frac{\partial H}{\partial p_j}, \qquad \dot{p}_j = -\frac{\partial H}{\partial q_j}. \tag{6.114}$$

From a variational perspective, one can treat $(\mathbf{q}, \mathbf{p})$ as independent variables and impose the symplectic structure[25] $\omega = \sum dq_j \wedge dp_j$ and read off the Hamilton equations as a statement that H is constant along solutions if H has no explicit time dependence. The Hamiltonian is thus a dual form of the Lagrangian, in the sense that the energy function arises from a Legendre transform, turning velocities into momenta.

Constraints in Hamiltonian Formalism. In the presence of constraints $\mathbf{h}(\mathbf{q}) = 0$ (and possibly momenta constraints $\mathbf{h}(\mathbf{q}, \mathbf{p}) = 0$), one must carry them forward into the Hamiltonian framework. The consistent procedure is to identify the primary constraints in phase space, then incorporate them using Dirac–Bergmann or simpler Lagrange multiplier methods. The final effect is that the Hamiltonian is augmented by $\boldsymbol{\mu}^{\mathrm{T}}\mathbf{h}(\mathbf{q}, \mathbf{p})$, or we restrict the dynamics to a constraint submanifold of phase space. Either way, the principle of stationary action is preserved, but now in the space of $(\mathbf{q}, \mathbf{p})$ subject to these constraints.

Relationship to Fenchel Conjugate. The Legendre transform applies to smooth, convex functions and maps $f(x)$ to $f^*(y) = \sup_x(xy - f(x))$ where $y = \nabla f(x)$. The Fenchel conjugate extends this to any proper convex function, including those that are non-differentiable or take $+\infty$ values. Both describe the tightest linear underestimator of f at a given slope and are involutive under regularity conditions: $f^{**} = f$. In optimization, the Fenchel conjugate underlies duality theory. In physics, the Legendre transform connects dual variables (e.g., energy vs. entropy, Lagrangian vs. Hamiltonian).

The formula for the Legendre transform is identical to that of the Fenchel conjugate, but the Legendre transform assumes differentiability and full-domain convexity. In particular, the Fenchel conjugate allows functions that take the value $+\infty$ to encode constraints, and it applies even to nondifferentiable functions such as norms or indicator functions.

[24] Often the momentum variables are denoted with a subscript p_j while the configuration coordinates are denoted with a superscript q^j. This notation is more classical in physics, representing that $\mathbf{p}$ is a dual vector to the velocity $\dot{\mathbf{q}}$, and is often used for shorthand, such as the Einstein summation convention. However, here we choose to use subscripts for all coordinates to conform to simpler engineering vector notation.

[25] The term *symplectic* refers to the special geometric structure underlying Hamiltonian mechanics: it encodes how positions and momenta are paired. The wedge product $\wedge$ is a compact symbol from differential forms to indicate an oriented area element $dq_j \wedge dp_j$ in the (q_j, p_j) plane. In practice, you can think of the symplectic form ω as the bookkeeping device that ensures Hamilton's equations preserve volume in phase space.

Quadratic Function. Let $f(x) = \frac{1}{2}x^2$. Then

$$f^*(y) = \sup_{x \in \mathbb{R}} \left(xy - \frac{1}{2}x^2 \right).$$

Setting the derivative to zero yields the maximizer $x = y$, so

$$f^*(y) = y^2 - \frac{1}{2}y^2 = \frac{1}{2}y^2.$$

Here, the Legendre transform and the Fenchel conjugate coincide, because f is convex and differentiable everywhere.

KKT Conditions and Duality in Infinite Dimensions

One can state the infinite-dimensional optimization problem as:

$$\begin{aligned} &\text{minimize} && \mathcal{J}[y] \\ &\text{subject to} && y \in \mathcal{Y}, \\ &&& g_j[y] \geq 0, \quad j = 1, \ldots, m, \end{aligned} \tag{6.115}$$

where $\mathcal{Y}$ is an admissible function space, and $c_i[y]$ are (possibly integral) constraints. The KKT conditions in the infinite-dimensional setting are similar to those in the finite-dimensional case. They still involve stationarity, via a functional derivative condition, akin to the Euler–Lagrange equations; primal feasibility, meaning that the constraints are satisfied; dual feasibility, meaning that the Lagrange multipliers satisfty $\mu_j \geq 0$; complementary slackness, $\mu_j g_j[y] = 0$ for each i; and possibly boundary or transversality conditions. These conditions are infinite-dimensional analogs of those in finite-dimensional convex optimization [230].

In indirect methods, one formally applies the Euler–Lagrange equation, or KKT conditions, to obtain differential equations and solves them analytically or numerically. In contrast, direct methods discretize the function space, via finite elements or finite differences, and then solve the resulting large, finite-dimensional problem using standard optimization tools such as interior point methods or sequential quadratic programming [370].

A common workflow in PDE-constrained optimization involves representing $y(x)$ by a set of finite parameters (e.g., piecewise polynomials), approximating $\mathcal{J}[y]$ and any constraints by summations or quadrature, and solving the resulting finite-dimensional optimization problem using gradient or Newton iterations. This approach is popular due to existing solvers that can handle large-scale, sparse systems efficiently.

Alternatively, one may solve the variational KKT system at the continuous level first and then discretize the continuous solution. The advantage is a more precise expression of conditions such as adjoint equations. However, it can be more complex to implement and analyze.

Constraints as Additional Conditions. In many mechanical systems, one has holonomic constraints expressible as $\phi(q) = 0$. Such a constraint can be incorporated into the Lagrangian through the method of Lagrange multipliers or by reduction [57]. When one sets up $L' = L + \lambda\,\phi(q)$, the variation with respect to λ yields $\phi(q) = 0$, and variation with respect to q yields the Euler–Lagrange equations with an additional $\lambda\,\partial\phi/\partial q$ term.

An alternative is to interpret constraints as restricting the space of admissible curves $q(t)$. The variation is then taken only among curves satisfying the constraints, $\phi(q(t)) \equiv 0$, for all t. The final outcome is that the constraint is satisfied at every time instant, and the Euler–Lagrange equations incorporate the constraint forces, typically quantified by a Lagrange multiplier.

Types of Derivatives

Here, we outline various types of derivatives encountered in variational and infinite-dimensional optimization. These include total derivatives from finite-dimensional calculus, as well as their infinite-dimensional analogues, the Gâteaux and Fréchet derivatives. Each are used to derive and understand optimality conditions such as the Euler–Lagrange equations.

Total Derivative. The total derivative $\mathbf{D}f$ of a function $f : \mathbb{R}^n \to \mathbb{R}$, defined on a finite-dimensional $\mathbf{x}$, measures the rate of change of f in all directions in $\mathbb{R}^n$. It is defined by the gradient vector of partial derivatives:

$$\mathbf{D}f(\mathbf{x}) = \nabla f(\mathbf{x}) = \begin{bmatrix} \frac{\partial f}{\partial x_1} \\ \vdots \\ \frac{\partial f}{\partial x_n} \end{bmatrix}, \tag{6.116}$$

and it satisfies the approximation[26]

$$f(\mathbf{x} + \boldsymbol{\delta}) = f(\mathbf{x}) + \nabla f(\mathbf{x})^{\mathrm{T}}\boldsymbol{\delta} + o(\|\boldsymbol{\delta}\|). \tag{6.117}$$

This concept generalizes to Banach spaces through directional and functional derivatives.

Functional Derivative. Let $\mathcal{A}[\mathbf{q}]$ be a functional that maps a function $\mathbf{q}(t)$ to a scalar, as in Eq. (6.98). The functional derivative $\delta\mathcal{A}/\delta\mathbf{q}(t)$ generalizes the notion of a gradient to infinite-dimensional function spaces. It satisfies the identity:

$$\delta\mathcal{A}[\mathbf{q};\boldsymbol{\eta}] := \frac{\mathrm{d}}{\mathrm{d}\varepsilon}\mathcal{A}[\mathbf{q} + \varepsilon\boldsymbol{\eta}]\bigg|_{\varepsilon=0}. \tag{6.118}$$

When $\mathcal{A}$ is an integral, as in Eq. (6.98), it is possible to show that this may be written as

$$\delta\mathcal{A}[\mathbf{q};\boldsymbol{\eta}] = \int_{t_0}^{t_1} \frac{\delta\mathcal{A}}{\delta\mathbf{q}(t)} \cdot \boldsymbol{\eta}(t)\,dt. \tag{6.119}$$

[26] The term $o(\|\boldsymbol{\delta}\|)$ means that the remainder vanishes faster than linearly as $\|\boldsymbol{\delta}\| \to 0$, which is the defining condition for differentiability. In contrast, $O(\|\boldsymbol{\delta}\|)$ only guarantees that the error is bounded by a constant multiple of $\|\boldsymbol{\delta}\|$ near zero. For precise first-order expansions, little-o is the appropriate notation.

This defines $\delta\mathcal{A}/\delta\mathbf{q}(t)$ as the function, or distribution, that plays the role of the gradient in the inner product with any admissible variation $\boldsymbol{\eta}(t)$. Showing this last step will be an exercise for the reader. The Euler–Lagrange equations result from setting the functional derivative to zero under admissible variations.

Here we will discuss two particulary useful types of functional derivatives, the Gâteaux and Fréchet derivatives. The Gâteaux derivative is more common and requires less stringent conditions than the Fréchet derivative. In practice, the Gâteaux derivative is easiest to compute but weakest, and the Fréchet derivative ensures stronger properties like continuity. It is also possible to use these definitions to define directional derivatives, extending the finite-dimensional notion to infinite dimensions.

Gâteaux Derivative. Let $\mathcal{A} : \mathcal{X} \to \mathbb{R}$ be a functional defined on a normed vector space $\mathcal{X}$. The Gâteaux derivative at $\mathbf{q} \in \mathcal{X}$ in the direction $\boldsymbol{\eta} \in \mathcal{X}$ is defined as:

$$\mathbf{D}f(\mathbf{q};\boldsymbol{\eta}) := \lim_{\varepsilon\to 0} \frac{\mathcal{A}(\mathbf{q}+\varepsilon\boldsymbol{\eta}) - \mathcal{A}(\mathbf{q})}{\varepsilon}. \tag{6.120}$$

It generalizes the directional derivative and may exist even if $\mathcal{A}$ is not differentiable in the stronger Fréchet sense. If this limit exists for all $\boldsymbol{\eta}$, then $\mathcal{A}$ is Gâteaux differentiable at $\mathbf{q}$.

Fréchet Derivative. The Fréchet derivative is a stronger and more rigorous version of the Gâteaux derivative. A functional $\mathcal{A} : \mathcal{X} \to \mathbb{R}$ is Fréchet differentiable at $\mathbf{q}$ if there exists a bounded linear operator $L : \mathcal{X} \to \mathbb{R}$ such that

$$\lim_{\|\boldsymbol{\eta}\|\to 0} \frac{|\mathcal{A}(\mathbf{q}+\boldsymbol{\eta}) - \mathcal{A}(\mathbf{q}) - L(\boldsymbol{\eta})|}{\|\boldsymbol{\eta}\|} = 0. \tag{6.121}$$

We write this as

$$\mathbf{D}\mathcal{A}(\mathbf{q})[\boldsymbol{\eta}] = L(\boldsymbol{\eta}), \tag{6.122}$$

and call $\mathbf{D}\mathcal{A}(\mathbf{q})$ the Fréchet derivative. If a functional is Fréchet differentiable, then it is also Gâteaux differentiable, but the converse is not necessarily true. Formally, one often uses Fréchet derivatives in Banach spaces (e.g., $W^{1,2}$ Sobolev spaces) to handle more complex problems [163]. In the exercises, we will explore examples where the Gâteaux and Fréchet derivatives agree, and examples where they are different to solidify these concepts.

Suggested Reading

Texts

(1) **Constrained Optimization and Lagrange Multiplier Methods**, by D. P. Bertsekas, 2014 [48].
(2) **Conjugate Duality and Optimization**, by R. T. Rockafellar, 1974 [420].
(3) **Conjugate Duality in Convex Optimization**, by R. I. Bot, 2009 [62].
(4) **Primal-Dual Interior-Point Methods**, by S. J. Wright, 1997 [542].
(5) **Introduction to Mechanics and Symmetry**, by J. E. Marsden and T. S. Ratiu, 2009 [335].
(6) **Variational Analysis**, by R. T. Rockafellar and R. J. B. Wets, 1998 [422].

Papers and reviews

(1) **Lagrange multipliers and optimality**, by R. T. Rockafellar, *SIAM Review*, 1993 [421].
(2) **Duality in nonconvex optimization**, by J. F. Toland, *Journal of Mathematical Analysis and Applications*, 1978 [498].
(3) **Playing with duality: An overview of recent primal-dual approaches for solving large-scale optimization problems**, by N. Komodakis, J.-Ch. Pesquet, *IEEE Signal Processing Magazine*, 2015 [281].
(4) **Interior-point methods**, by F. A. Potra and S. J. Wright, *Journal of Computational and Applied Mathematics*, 2005 [394].
(5) **Distributed optimization and statistical learning via the alternating direction method of multipliers**, by S. Boyd, N. Parikh, E. Chu, B. Peleato, and J. Eckstein, *Foundations and Trends® in Machine Learning*, 2011 [65].
(6) **Robust principal component analysis**, by E. J. Candès, X. Li, Y. Ma, and J. Wright, *Journal of the ACM*, 2011 [91].
(7) **Discrete mechanics and optimal control**, by O. Junge, J. E. Marsden, and S. Ober-Blöbaum, *IFAC Proceedings Volumes*, 2005 [258].

Homework

Exercise 6.1 Solve the constrained problem:

$$\min_{x,y\in\mathbb{R}} (x-2)^2 + (y+1)^2 \quad \text{subject to} \quad x^2 + y^2 = 4.$$

Find the Lagrange multiplier and interpret the geometric meaning of this solution.

Exercise 6.2 Verify that Eq. (6.13) yields a solution where the gradient of the objective function is normal to the constraint equation.

Exercise 6.3 Consider

$$\min_{x\in\mathbb{R}^2} f(\mathbf{x}) \quad \text{subject to} \quad g_i(\mathbf{x}) \geq 0,\ i = 1, 2.$$

List the KKT conditions in terms of the Lagrange multipliers μ_1, μ_2. Discuss complementary slackness and stationarity in your own words.

Exercise 6.4 Verify that the regularity conditions are satisfied for the box-constrained least squares problem in Section 6.2.

Exercise 6.5 We will now explore the *saddle point* property of the KKT conditions for the case of equality constraints. Consider the following constrained optimization problem:

$$\begin{aligned} \min_{\mathbf{x}} \quad & f(\mathbf{x}) = x_1^2 + x_2^2 \\ \text{s.t.} \quad & x_1 + x_2 = 1. \end{aligned}$$

1. Use the KKT conditions to solve for the optimal point.
   (a) Form the Lagrangian $\mathcal{L}(\mathbf{x}, \lambda)$.
   (b) Compute the gradient with respect to $\mathbf{x}$ and λ.
   (c) Set $\nabla_{\mathbf{x}}\mathcal{L} = \mathbf{0}$ and $\nabla_{\lambda}\mathcal{L} = 0$, and solve for the optimal point $\mathbf{x}^* = (x_1^*, x_2^*)$ and the multiplier λ^*.
2. Verify the saddle point property. We have just shown that the KKT point $(\mathbf{x}^*, \lambda^*)$ is a stationary point of the Lagrangian, i.e., $\nabla_{\mathbf{x}}\mathcal{L} = \mathbf{0}$ and $\nabla_{\lambda}\mathcal{L} = 0$.
   Specifically, show that $\mathcal{L}(\mathbf{x}^*, \lambda) \leq \mathcal{L}(\mathbf{x}^*, \lambda^*)$ for all λ and that $\mathcal{L}(\mathbf{x}, \lambda^*) \geq \mathcal{L}(\mathbf{x}^*, \lambda^*)$ for all $\mathbf{x}$ satisfying the constraint $x_1 + x_2 = 1$. Thus, you may conclude that $(\mathbf{x}^*, \lambda^*)$ is a true saddle point of the Lagrangian.
3. Plot the Lagrangian $\mathcal{L}(\mathbf{x}, \lambda^*)$ as a function of (x_1, x_2) on the constraint line $x_1 + x_2 = 1$ and plot $\mathcal{L}(\mathbf{x}^*, \lambda)$ as a function of λ. Verify the saddle behavior through these plots.

Exercise 6.6 We now explore an example where the saddle point property is a bit more subtle. Consider the following constrained optimization problem:

$$\begin{aligned} \min_{x} \quad & f(x) = 3x \\ \text{s.t.} \quad & x \leq 3 \\ & x \geq -1. \end{aligned}$$

1. Use the KKT conditions to solve for the optimal point.

(a) Form the Lagrangian $\mathcal{L}(x, \boldsymbol{\mu})$.
(b) Compute the gradient with respect to x and $\boldsymbol{\mu}$.
(c) Derive the KKT conditions and solve for the optimal point x^* and $\boldsymbol{\mu}^*$.

2. Verify the saddle point property.
   (a) At the KKT point (x^*, μ_1^*, μ_2^*) evaluate the full gradient $\nabla_{x,\mu_1,\mu_2}\mathcal{L}$. Show that it is not equal to zero at the optimum point, and explain why this does not contradict the saddle point interpretation.
   (b) Explain what the saddle point condition means here. Specifically, show

$$\mathcal{L}(x^*, \mu_1, \mu_2) \leq \mathcal{L}(x^*, \mu_1^*, \mu_2^*) \leq \mathcal{L}(x, \mu_1^*, \mu_2^*)$$

   holds for all $\mu_1, \mu_2 \geq 0$ and feasible x. This is the saddle point property of the constrained Lagrangian.
3. Plot the Lagrangian $\mathcal{L}(x, \mu_1^*, \mu_2^*)$ as a function of x, and describe how the minimum over x is achieved. Then plot $\mathcal{L}(x^*, \mu_1, \mu_2)$ as a function of μ_1, $\mu_2 \geq 0$, and verify that the KKT point is a saddle point over the feasible region.
4. Explain why the constrained Lagrangian does not generally have a zero gradient at the saddle point when inequality constraints are present. Summarize how the saddle-point interpretation of the KKT conditions differs from the usual unconstrained interpretation of saddle points as critical points with zero gradient.

Exercise 6.7 Derive the Lagrangian and the dual function for

$$\begin{aligned}
\min_{\mathbf{x}} \quad & x_1 + x_2 \\
\text{s.t.} \quad & x_1^2 + x_2^2 \leq 25, \\
& x_1 + x_2 \geq 2, \\
& x_1, x_2 \geq 0.
\end{aligned}$$

Write the dual problem explicitly and discuss strong vs. weak duality here.

Solve the primal and the dual problem explicitly and discuss the results.

Exercise 6.8 Derive the dual program for the simple linear program with equality constraints:

$$\begin{aligned}
\min_{\mathbf{x}} \quad & \mathbf{c}^{\mathrm{T}}\mathbf{x} \\
\text{s.t.} \quad & \mathbf{A}\mathbf{x} = \mathbf{b}.
\end{aligned}$$

Write down the Lagrangian, the dual function, and the dual program. Solve both the primal and dual problem for a specific example of your choice. Solve this optimization problem using an interior point method.

Exercise 6.9 Show that

$$f(x) = \begin{cases} 0 & \text{if } x = 0, \\ 1 & \text{if } x \neq 0, \end{cases}$$

is not lower semicontinuous.

Exercise 6.10 Show that dual function in Eq. (6.30) is always concave, even if the primal problem is non-convex, because it is the pointwise infimum of functions that are affine in (λ, μ). First, explain how it is the pointwise infimum of functions that are affine in (λ, μ). Specifically, show that the augmented Lagrangian is affine in λ and μ. Then show that the pointwise infimum of affine functions is always concave.

Exercise 6.11 Prove that the dual function $\phi(\lambda, \mu)$ in Eq. (6.62) is always concave.

Exercise 6.12 Fill in the details of Eq. (6.69). Specifically, compute the infimum over x to solve for $x^*(\lambda)$, and substitute back in to arrive at $\phi(\lambda) = -\frac{1}{4}\lambda^2 + \lambda + 1$.

Exercise 6.13 Here we will extend the geometric interpretation of duality to include inequality constraints $\mathbf{g}(\mathbf{x}) \leq \mathbf{0}$. In Section 6.3, we saw that for the case of convex equality constraints $\mathbf{h}(\mathbf{x}) = \mathbf{A}\mathbf{x} - \mathbf{b} = \mathbf{0}$, the pair $(\lambda, \phi(\lambda))$ defined a supporting hyperplane to the epigraph of f, defining a tight lower bound on f. The slope of this hyperplane is λ.

In this exercise, we extend this geometric picture to convex programs with inequality constraints. The supporting hyperplane now arises not from the epigraph of f but from the geometry of the feasible region. Consider the convex optimization problem:

$$\min_{\mathbf{x}} \quad f(\mathbf{x}) \quad \text{s.t.} \quad \mathbf{g}(\mathbf{x}) \leq \mathbf{0},$$

where $f : \mathbb{R}^n \to \mathbb{R}$ is convex and lower semicontinuous, and $\mathbf{g} : \mathbb{R}^n \to \mathbb{R}^m$ is a vector of convex constraint functions. Let $\mathbf{x}^*$ be an optimal solution, and let $\boldsymbol{\mu}^* \geq 0$ be the associated vector of Lagrange multipliers.

1. The feasible set $\boldsymbol{C}$ defined by the constraints $\mathbf{g}(\mathbf{x}) \leq \mathbf{0}$ is convex. Show that the set of active constraints at $\mathbf{x}^*$ determines a face of the boundary $\partial \boldsymbol{C}$. Prove that for each active constraint, its gradient $\nabla g_j(\mathbf{x}^*)$ defines an inward-pointing normal to $\boldsymbol{C}$ at $\mathbf{x}^*$.
2. Show that the vector

$$\sum_{\text{active}} \mu_j^* \nabla g_j(\mathbf{x}^*)$$

is a *conic combination* of the active constraint normals, meaning it is a linear combination with nonnegative coefficients. A conic combination of $\{\mathbf{v}_1, \ldots, \mathbf{v}_k\}$ is

$$\sum_{j=1}^{k} \alpha_j \mathbf{v}_j, \quad \text{where } \alpha_j \geq 0 \text{ for all } j.$$

Geometrically, the sum lies in the *conic hull* (i.e., the positive span) of the vectors.
3. (Optional) Show that the conic hull of a set of vectors is always convex.
4. Show that the vector $\sum_{\text{active}} \mu_j^* \nabla g_j(\mathbf{x}^*)$ defines the normal to a supporting hyperplane of $\boldsymbol{C}$ at $\mathbf{x}^*$, i.e., it touches $\boldsymbol{C}$ at a point and lies entirely on one side of it.
5. Use the KKT stationarity condition to conclude that the gradient $\nabla f(\mathbf{x}^*)$ is in the span of the constraint normals. What does this imply geometrically about feasible directions of descent?
6. Show that no feasible descent direction exists at $\mathbf{x}^*$, and that any direction $\mathbf{d}$ with $\nabla f(\mathbf{x}^*)^{\mathrm{T}}\mathbf{d} < 0$ must violate at least one constraint, thus establishing local optimality.

Exercise 6.14 We will now explore the previous geometric interpretation of duality on a specific problem:

$$\begin{aligned} \min_{\mathbf{x}} \quad & x_1^2 + x_2^2 \\ \text{s.t.} \quad & x_1 + x_2 - 1 \le 0, \\ & -x_1, -x_2 \le 0. \end{aligned}$$

1. Determine and plot the feasible region.
2. Identify the optimal point $\mathbf{x}^*$ and the active constraints.
3. Show that the gradients of the active constraints form inward normals to the boundary.
4. Verify the KKT stationarity condition and visualize the balance of forces.
5. Visualize the supporting hyperplane determined by the active constraints.

Exercise 6.15 Verify that the convex envelope of

$$f(x) = \min(x^2, \ |x|),$$

is equal to $f^{**}(x)$ from Eq. (6.56), as in Fig. 6.10. Compute the conjugate f^* and the biconjugate f^{**} and confirm that both are convex.

Exercise 6.16 Show that the solution to the optimization problem in Eq. (6.76) will satisfy the first KKT condition.

Exercise 6.17 Show that the gradient and Hessian of the log-barrier-augmented objective function in Eq. (6.79) for linear programming is given by the expressions in Eq. (6.80).

Exercise 6.18 Modify the interior point method code 6.1 to solve a quadratic program with the same linear constraints as in the example. The quadratic objective is given by

$$f(\mathbf{x}) = \tfrac{1}{2}\mathbf{x}^{\mathrm{T}}\mathbf{Q}\mathbf{x} + \mathbf{c}^{\mathrm{T}}\mathbf{x}, \quad \text{where} \quad \mathbf{Q} = \begin{bmatrix} 2 & -1 \\ -1 & 2 \end{bmatrix} \quad \text{and} \quad \mathbf{c}^{\mathrm{T}} = \mathbf{x}_0\mathbf{Q} \quad \text{for} \quad \mathbf{x}_0 = \begin{bmatrix} 40 \\ 30 \end{bmatrix}.$$

The solution path is shown in Fig. 6.13 (right).

Also show by hand that the analytic minimum of $f(\mathbf{x})$ is at $\mathbf{x} = \mathbf{x}_0$.

Exercise 6.19 Solve the box-constrained least-squares problem in Eq. (5.17) using the interior point method. Plot the central path for different IPM parameters.

Exercise 6.20 For the box-constrained least-squares problem in Eq. (5.17), derive the following expressions for the Lagrange multipliers: $\mu_j = t/(x_j - l_j)$ and $\nu_j = t/(u_j - x_j)$.

Exercise 6.21 Solve the box-constrained least-squares problem in Eq. (5.17) using the projected gradient algorithm for various step sizes γ. Compare with the IPM solution above.

Exercise 6.22 Solve the LASSO regression problem in Exercise 5.1 using the ADMM LASSO code from Section 6.5.

Exercise 6.23 Consider the system from Exercise 5-1, given by $\mathbf{Ax} \approx \mathbf{b}$, where $\mathbf{x} \in \mathbb{R}^{10}$ has exactly 2 nonzero elements. Generate a random $\mathbf{A} \in \mathbb{R}^{100\times10}$, so that our observations $\mathbf{b} \in \mathbb{R}^{100}$ are a combination of exactly 2 columns of $\mathbf{A}$. Add a small amount of noise (1%) to the vector $\mathbf{b}$ and solve for $\mathbf{x}$ using the ADMM LASSO code from Section 6.5.

Exercise 6.24 Compare the ADMM LASSO results in the previous exercise with SR3, ISTA, and FISTA. Plot the objective versus iterations and the resulting solution sparsity. Comment on the relative strengths and weaknesses of each approach.

Exercise 6.25 Here we will solve the total-variation denoising problem with ADMM. Consider a one-dimensional TV denoising problem

$$\min_{\mathbf{x}\in\mathbb{R}^n} \tfrac{1}{2}\|\mathbf{x}-\mathbf{y}\|_2^2 + \lambda\|\mathbf{Dx}\|_1,$$

where $\mathbf{D} \in \mathbb{R}^{(n-1)\times n}$ is the first-difference operator. Derive ADMM with splitting $\mathbf{Dx} = \mathbf{z}$ and show the linear system $(\mathbf{I} + \rho\mathbf{D}^{\mathrm{T}}\mathbf{D})\mathbf{x} = \mathbf{y} + \rho\mathbf{D}^{\mathrm{T}}(\mathbf{z}-\mathbf{u})$ and $\mathbf{z} = \mathrm{soft}(\mathbf{Dx}+\mathbf{u}, \lambda/\rho)$. Implement 1D denoising with FFT-based or banded solvers for $(\mathbf{I} + \rho\mathbf{D}^{\mathrm{T}}\mathbf{D})$. Explore λ.

Exercise 6.26 Here we will solve the box-constrained QP using ADMM and compare to projected gradient. Specifically, we will solve

$$\min_{\mathbf{x}} \tfrac{1}{2}\mathbf{x}^{\mathrm{T}}\mathbf{Qx} + \mathbf{c}^{\mathrm{T}}\mathbf{x} \quad \text{s.t.} \quad \mathbf{l} \le \mathbf{x} \le \mathbf{u}, \;\; \mathbf{Q} > 0$$

by deriving ADMM with splitting $\mathbf{x} = \mathbf{z}$ and projection $\Pi_{[\mathbf{l},\mathbf{u}]}(\cdot)$. Implement this algorithm and compare to projected-gradient descent with optimal fixed step $1/L$ for $L = \|\mathbf{Q}\|_2$.

Exercise 6.27 A mass on a frictionless plane is constrained to move on a curve given by $h(x, y) = 0$. Write the equations of motion using Lagrange multipliers to enforce the constraint. Compare this to the standard Newton's laws for constrained motion. Write down the specific equations of motion for a simple parabolic $h(x, y) = y - x^2 = 0$.

Exercise 6.28 Modify the Euler–Lagrange equations to handle non-fixed endpoints.

Exercise 6.29 Derive the expression in Eq. (6.105) by expanding the Lagrangian inside the integral in powers of ε and then taking the derivative with respect to ε at 0.

Exercise 6.30 Show that a straight line is the shortest between two points A and B by defining a functional and verifying that the Euler–Lagrange equations are satisfied.

Exercise 6.31 Derive the Beltrami identity, that if $L = L(y, y')$ has no x dependence, then

$$L(y, y') - y'\frac{\partial L}{\partial y'} = \text{constant}.$$

Exercise 6.32 Plug the cycloid solution for the Brachistichrone problem back into the Euler–Lagrange equations to verify that it is locally extremal.

Exercise 6.33 Similarly, show that the cycloid solution is also a solution to the tautochrone problem by showing that it solves the Euler–Lagrange equations.

Exercise 6.34 Derive the formula for the great arc as a geodesic (e.g., shortest distance path between two points on a curved surface) on the surface of a sphere.

Exercise 6.35 Here we will explore Lagrange multipliers for the simple mechanical pendulum. Derive the equations of motion with the three canonical approaches:

1. Newton's laws in (x, y) coordinates with constraints as Lagrange multipliers;
2. The Euler–Lagrange equations in angular coordinates $(\theta, \dot{\theta})$; and
3. Hamilton's equations in conjugate coordinates (q, p), with $q = \theta$.

Compare and discuss these various equations and how the constraints are being satisfied.

Exercise 6.36 Show that

$$\delta\mathcal{A}[\mathbf{q};\boldsymbol{\eta}] := \frac{\mathrm{d}}{\mathrm{d}\varepsilon}\mathcal{A}[\mathbf{q}+\varepsilon\boldsymbol{\eta}]\bigg|_{\varepsilon=0} = \int_{t_0}^{t_1} \frac{\delta\mathcal{A}}{\delta\mathbf{q}(t)} \cdot \boldsymbol{\eta}(t)\, dt$$

when $\mathcal{A}$ is an integral, as in Eq. (6.98). First, expand the perturbed functional to first order:

$$\mathcal{A}[\mathbf{q}+\varepsilon\boldsymbol{\eta}] = \mathcal{A}[\mathbf{q}] + \varepsilon \int \mathcal{B}(\boldsymbol{\eta})\, dt + O(\varepsilon^2),$$

and extract the integrand of the linear term as $\delta\mathcal{A}/\delta\mathbf{q}(t)$.

Exercise 6.37 Compute the functional derivative of

$$\mathcal{A}[\mathbf{q}] = \int_{t_0}^{t_1} \left(\tfrac{1}{2}\|\dot{\mathbf{q}}(t)\|^2 - V(\mathbf{q}(t))\right) dt.$$

Specifically, consider $\mathbf{q}_\varepsilon(t) = \mathbf{q}(t) + \varepsilon\boldsymbol{\eta}(t)$ and compute:

$$\delta\mathcal{A} = \int_{t_0}^{t_1} \left(-\nabla V(\mathbf{q}(t)) \cdot \boldsymbol{\eta}(t) + \dot{\mathbf{q}}(t) \cdot \dot{\boldsymbol{\eta}}(t)\right) dt.$$

Integrate the second term by parts with fixed endpoints, and set equal to zero to yield a simple form of the Euler–Lagrange equations for a particle in a potential well $V(\mathbf{q})$.

Exercise 6.38 Here we will explore an example where the Gâteaux and Fréchet derivatives are the same. Let

$$F[q] = \int_0^1 q(t)^2\, dt$$

where $q \in L^2[0,1]$. Define a perturbation $q_\varepsilon(t) = q(t) + \varepsilon\eta(t)$, where $\eta(t)$ is an admissible variation in $L^2[0,1]$.

1. Show that the Gâteaux derivative of F in the direction η is:

$$\delta F[q;\eta] = \lim_{\varepsilon\to 0} \frac{F[q+\varepsilon\eta] - F[q]}{\varepsilon} = 2\int_0^1 q(t)\eta(t)\, dt.$$

2. Verify that this directional derivative is linear and continuous in η.
3. Conclude that the Fréchet derivative exists and equals the Gâteaux derivative.
4. Show that the functional derivative exists in the distributional sense and satisfies $\frac{\delta F}{\delta q(t)} = 2q(t)$, meaning that for any admissible $\eta(t)$,

$$\delta F = \int_0^1 \frac{\delta F}{\delta q(t)}\eta(t)\, dt.$$

Exercise 6.39 Here we will explore an example where the Gâteaux and Fréchet derivatives are different. Now, let

$$F[q] = \left(\int_0^1 |q(t)|\, dt \right)^2$$

with $q \in L^1[0,1]$. Define $q_\varepsilon(t) = q(t) + \varepsilon\eta(t)$ for any $\eta \in L^1[0,1]$.

1. Show that F is well-defined and finite for all $q \in L^1[0,1]$.
2. Compute the Gâteaux derivative of F at $q = 0$ in the direction η.
3. Show that this derivative equals zero for all directions η, i.e., $\delta F[0;\eta] = 0$.
4. Argue why this Gâteaux derivative is not linear or continuous in η in any neighborhood of $q = 0$.
5. Conclude that the Fréchet derivative does not exist at $q = 0$.
6. Discuss whether a functional derivative exists in this case, and what it would mean.

7 Bayesian Modeling and Estimation

Bayesian statistics provides a flexible framework for balancing prior knowledge about a system with new evidence, in the form of observation data. Many methods in optimization are either rooted in Bayesian statistics, or have a Bayesian interpretation. For example, the Tikhonov and LASSO regularized least-squares from Chapter 4 come from specific prior distributions on the unknown parameters. Much of modern machine learning also has foundations in statistics and benefit from a Bayesian interpretation.

Prior information comes in many forms, such as the mild assumption that unknown variables are Gaussian, or the strong belief in conservation laws from physics. Bayesian methods are particularly useful when we have small data sets or complex models. They are used extensively for hyperparameter tuning of machine learning models, for experimental design, and for applications in engineering that involve complex partial differential equations. Much of Chapter 8 on inverse problems also benefits from Bayesian methods.

The problem of estimating a statistical distribution from data is fundamentally an optimization problem. It is also the basis of much of machine learning. Density estimation is a vast topic, and the simplest approach is to assume a known probability distribution and estimate the parameters of the distribution using a best-fit optimization approach, such as least squares regression. If we assume there is Gaussian noise on the observations, then there is a statistical interpretation of least-squares as a *maximum likelihood estimation*. If we include Bayesian prior information, this becomes a *maximum a posteriori* estimation problem. These are all statistical estimation problems involving optimization.

This chapter provides a high-level view of Bayesian methods in optimization, starting with a brief recap of statistical estimation problems. Key concepts, such as density estimation, regularization, and expectation maximization will be explored here. We provide simple examples with code and more extensive examples in the homework.

It is assumed that the reader has a basic knowledge of probability and statistics, including the common probability distributions (e.g., a Bernoulli distribution describes a single coin flip with probability of heads p, a binomial distribution describes the outcome of n repeated trials of a Bernoulli event, this becomes a normal distribution in the large n limit or a Poisson distribution for the large n but small np limit, etc.). For a quick recap, see the *Probability Bootcamp*[1] and the first several homework exercises.

[1] https://www.youtube.com/playlist?list=PLMrJAkhIeNNR3sNYvfgiKgcStwuPSts9V
https://www.youtube.com/playlist?list=PLMrJAkhIeNNT14qn1c5qdL29A1UaHamjx

7.1 Parameter Estimation in Statistics

Parameter estimation and fitting distributions is a central topic in data analysis and statistics, and it is the basis of much of modern machine learning. Typically, we will have data from a system and will try to find the probability distribution that best fits the data, resulting in an optimization problem.

The simplest approach is to assume a known probability distribution, such as a Gaussian or a Poisson distribution, and estimate the parameters of the model from data. For example, if we are modeling the radioactive decay of Carbon-14, we know from physical principles that a Poisson distribution will be appropriate, and we need only infer the unknown decay rate λ. Similarly, if we are modeling the height of a population of college students, it is reasonable to assume a Gaussian distribution, with unknown mean μ and standard deviation σ. Statistical methods for estimating the parameters of a known distribution are known as *parametric* methods.

More complex data will likely not be described perfectly by a named probability distribution, and instead we fit a more general parameterized probability density using machine learning. Natural images are a good example of data that *does* follow some probability distribution, but that is not well modeled by classical, named distributions. Similarly, problems in physics, such as turbulence, follow probability distributions that do not yet have names. These data are often high-dimensional and have complex joint probability density functions (PDFs) between the variables. Statistical approaches for learning distributions that do not have known parameterizations are called *nonparametric* methods.

We will begin with the simplest parametric estimation problem, based on maximum likelihood estimation. In the absence of any prior information about the parameters of the distribution, we simply use the data to find the most likely parameters. We then extend this approach to incorporate prior information in Bayesian methods. Finally, we introduce more general, nonparametric methods in later sections.

Parametric Probability Distributions

A random variable X is an observable quantity of a system that depends on random events. For example, if we flip a faircoin, we might say that a random variable X takes a value of 1 if the coin lands heads and 0 if it lands tails, and each event has an equal probability, 0.5. If the set of values X can take is *discrete* (e.g., heads or tails for a coin, or a nonnegative integer number of events recorded in a given time) then it is a *discrete random variable*. If X can take on a *continuous* range of values (e.g., the height of a person, or the time between emails) then it is a *continuous random variable*.

For a discrete random variable X, we define a parameterized probability distribution[2]

$$p(x \mid \theta) = P(X = x \mid \theta), \tag{7.1}$$

[2] We use capital P to denote the probability of an event and lower case $p(x)$ to denote the probability distribution. Some notation includes the random variable X as a subscript $p_X(x)$, which can be helpful to remember that this is the distribution of X. However, we drop this most of the time for simplicity.

which is the probability of X taking a specific value x given the parameters θ. For example, given a radioactive element with decay rate λ, the number of alpha particle emissions per unit time follows a discrete Poisson[3] distribution

$$P(X = x \mid \lambda) = \frac{\lambda^x e^{-\lambda}}{x!}. \tag{7.2}$$

Thus, the decay rate $\theta = \lambda$ parametrizes the distribution. The probability distribution for a discrete random variable X is called a *probability mass function*. Thus, we say that $p(x \mid \theta)$ is the probability of measuring a data point x in a random trial, given the parameters θ. This is known as the *likelihood function*.

For a continuous random variable X, there are infinitely many values it can take, and so the probability of it taking any specific value x becomes infinitesimal, and we must integrate over x to obtain a finite probability. For continuous random variables we define the *probability density function*

$$p(x \mid \theta) \tag{7.3}$$

so that the probability of X being in the range $[a, b]$ is given by

$$P(a \leq X \leq b) = \int_a^b p(x \mid \theta)\, \mathrm{d}x. \tag{7.4}$$

Thus, we integrate the "density" to arrive at a "mass". For example, the time between discrete alpha particle emissions in radioactive decay is a continuous random variable that follows an exponential distribution:

$$p(x \mid \lambda) = \lambda e^{-\lambda x} \quad \text{for} \quad x \geq 0$$

and $p(x \mid \lambda) = 0$ for $x < 0$, as we don't consider negative times.

Perhaps a more common example is the normal distribution, given by

$$p(x \mid \mu, \sigma^2) = \frac{1}{\sqrt{2\pi\sigma^2}} e^{-\frac{(x-\mu)^2}{2\sigma^2}}, \tag{7.5}$$

where the parameters are given by the mean and variance, $\boldsymbol{\theta} \equiv [\mu, \sigma^2]$. If we know these parameters, we can determine the probability of X taking a particular value x.

Probability is focused on constructing a model of the random process governing the variable X to arrive at the probability distribution $p(x \mid \theta)$ of observing future data x. In statistics, we are concerned with the inverse problem of determining the model parameters θ that are most consistent with the observed data x. This may be posed as an optimization problem by finding θ that maximizes the likelihood $p(x \mid \theta)$. This is the approach taken in the method of maximum likelihood estimation (MLE), which will be discussed after the following example. More generally, it would be convenient to directly have the probability of a given set of parameters θ given the data x, $p(\theta \mid x)$, which is the focus of Bayesian methods, discussed in the next section.

[3] The Poisson distribution is derived from the binomial distribution in the limit of small np, as it was recognized that a normal distribution will have a significant probability of negative events, which is fishy.

Table 7.1 The first row is the number of clicks measured in a 10-second interval, and the second row is the number of 10-second intervals observed with that many clicks.

0	1	2	3	4	5	6	7	8	9	10	11	12	13	14	15	16	17
1	6	11	28	56	105	126	146	164	161	123	101	74	53	23	15	9	5

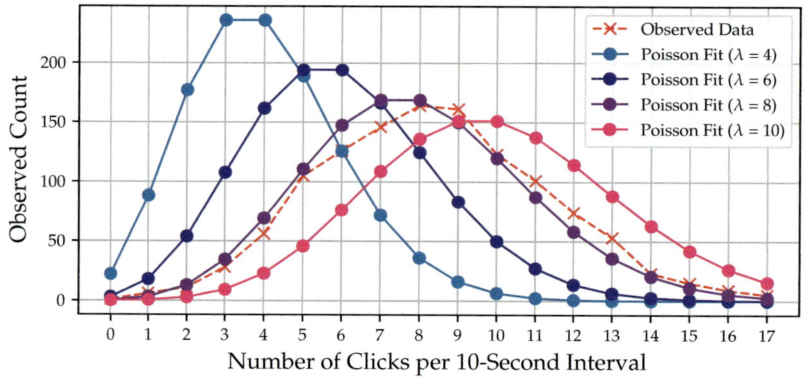

Figure 7.1 Alpha particle emission data (red) and several Poisson fits. The distribution with $\lambda = 8$ agrees best with the data, although we can find an optimal λ.

Example: Fitting Radioactive Decay with a Poisson Distribution

We begin with a simple and intuitive example of fitting the radioactive decay of alpha particles emitted by Americium-241. This is based on an example from Rice [414], taken from Berkson 1966 [46]. As we just discussed, the number of alpha particle emissions per unit time follows a Poisson distribution. The precise timing of 10,094 Geiger counter clicks (i.e., detected alpha particles) was recorded, and the data was binned into 10-second intervals. The number of clicks per 10-second interval are in Table 7.1.

The goal is to find the parameters of a Poisson distribution that best fit this data, which will involve optimization. Figure 7.1 shows several Poisson distributions for various values of λ, along with the data. The Poisson distribution with $\lambda = 8$ is closest to the data, but we can do better with optimization. Figure 7.2 shows the optimal Poisson distribution, where we fit the parameter λ using the sample mean $\bar{x} = \frac{1}{N}\sum_{j=1}^{N} x_j$ of the N samples of data x_j. For a Poisson distribution, the optimal estimate for λ is given by $\bar{x}$, as we will show using the maximum likelihood estimate. The fitting is achieved via the following code.

CodeSnip 7.a

```
from scipy.stats import poisson

ClicksPerTenSec = np.arange(18)
ObservedCounts = [1, 6, 11, 28, 56, 105, 126, 146, 164, 161,
    123, 101, 74, 53, 23, 15, 9, 5]

sample_mean = np.average(ClicksPerTenSec,weights=ObservedCounts)

# Fit a Poisson distribution with lambda = sample_mean
poisson_dist = poisson.pmf(ClicksPerTenSec, mu=sample_mean) * np
    .sum(ObservedCounts)  # Scale to match observed data
```

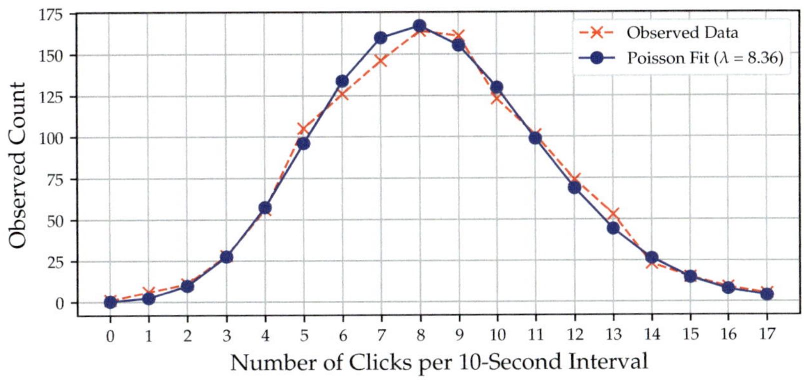

Figure 7.2 Alpha particle emission data (red) and Poisson fit (blue) using maximum likelihood estimate $\lambda = \bar{x}$.

It is useful to know the variance of the estimated parameter λ. Specifically, if we resample the system, how much do we expect the estimate λ to vary? We may estimate the distribution of λ without collecting new data, using a parametric *bootstrap* [162]. This involves using the estimated λ to sample new data, and then computing the best-fit λ to this synthetic data. Figure 7.3 shows the bootstrap distribution, which is approximately normally distributed, by the central limit theorem. The following code computes the bootstrap distribution.

CodeSnip 7.b

```
num_bootstrap = 1000  # Number of bootstrap samples
bootstrap_lambdas = []

# Bootstrap procedure
for _ in range(num_bootstrap):
    # Generate a synthetic dataset using the estimated lambda
    synthetic_data = np.random.poisson(lam=sample_mean, size=np.
        sum(Observed))

    # Bin synthetic data (# of counts per 10-second interval)
    synthetic_hist, _ = np.histogram(synthetic_data, bins=range
        (19))

    # Calculate new lambda estimate from synthetic data
    new_lambda = np.average(n, weights=synthetic_hist)

    # Store new lambda estimate
    bootstrap_lambdas.append(new_lambda)

# Calculate sample standard deviation of observed data
obsv_std = np.sqrt(np.average((n-sample_mean)**2,weights=
    Observed))

# Calculate CLT-based standard deviation for lambda estimates
clt_std = obsv_std / np.sqrt(sum(Observed))
```

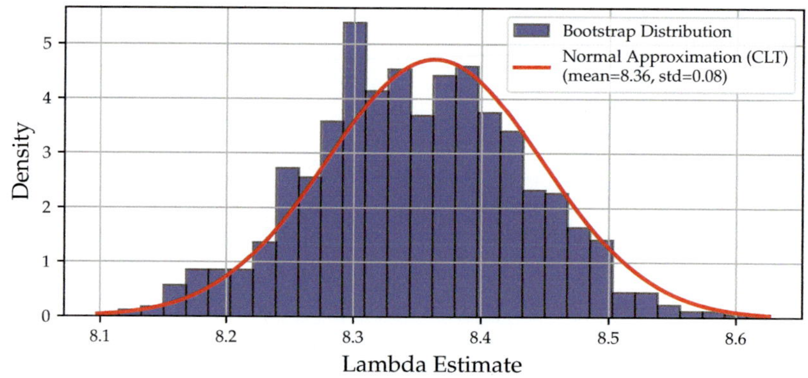

Figure 7.3 Bootstrap estimate of the distribution of the estimated λ parameter, obtained by sampling an ensemble of synthetic data from estimated Poisson distribution, and then computing the optimal λ values for each synthetic data set. The bootstrap estimate λ follows a normal distribution, via the central limit theorem, which is also computed and shown here.

Maximum Likelihood Estimation (MLE)

MLE is a method for estimating the parameters of a probability distribution from data. It frames statistical estimation as an optimization problem. This is the basis of many algorithms in machine learning, and it is the foundation for Bayesian parameter estimation.

Let the random variables $X_1, X_2, \ldots, X_n$ have a joint probability density function

$$p(x_1, x_2, \ldots, x_n \mid \theta). \tag{7.6}$$

If the random variables X_j are independent and identically distributed (i.i.d) then

$$p(x_1, x_2, \cdots, x_n \mid \theta) = \prod_{j=1}^{n} p(x_j \mid \theta). \tag{7.7}$$

The assumption that X_j is i.i.d. is common for repeated trials or observations of a system that is not changing, such as sequential coin flips or measurements of radioactive decay.

Given a realization of these n random variables in the form of data $x_1, x_2, \ldots, x_n$, we substitute this data into the joint distribution and arrive at the likelihood function of θ:

$$\text{lik}(\theta) = p(x_1, x_2, \cdots, x_n \mid \theta) = \prod_{j=1}^{n} p(x_j \mid \theta). \tag{7.8}$$

Thus, we evaluate the probability distribution[4] on the data x_j. This function tells us how likely the data is given the parameters, and it is only a function of θ. Thus, we can explicitly

[4] Some texts denote data with a capital X_j to make it more clear when the probability density is written in terms of a generic x versus being conditioned on actual data. However, this is also confusing because X_j is a random variable and not its outcome; once data is measured it is no longer a random variable. We will choose the lower-case x for observation data.

maximize the likelihood over θ:

$$\hat{\theta} = \underset{\theta}{\operatorname{argmax}} \ \text{lik}(\theta). \tag{7.9}$$

The optimal parameter estimate is $\hat{\theta}$. In MLE, we usually maximize the logarithm of the likelihood function $\ell(\theta) = \log(\text{lik}(\theta))$. Because the logarithm is a smooth, monotonic function, the maximum of the log likelihood is the same as the maximum of the likelihood. The log-likelihood is easier to optimize, as the product of PDFs becomes a sum of log-PDFs:

$$\log(\text{lik}(\theta)) = \log\left(\prod_{j=1}^{n} p(x_j \mid \theta)\right) = \sum_{j=1}^{n} \log\left(p(x_j \mid \theta)\right). \tag{7.10}$$

Note that until data is actually collected, we may think of $\hat{\theta}$ as a random variable that depends on $X_1, X_2, \ldots, X_n$. Thus $\hat{\theta}$ has its own distribution, including a mean and standard deviation. This can be useful to estimate the expected variation of estimated $\hat{\theta}$ across different data realizations.

Example: Normal Distribution

Consider n random variables that are distributed according to a normal distribution, $X_j \sim \text{normal}(\mu, \sigma^2)$, so that the joint PDF is given by

$$p(x_1, x_2, \ldots, x_n \mid \mu, \sigma^2) = \prod_{j=1}^{n} \frac{1}{\sigma\sqrt{2\pi}} e^{-\frac{1}{2}\left(\frac{x_j-\mu}{\sigma}\right)^2}.$$

The log-likelihood function is given by the log of $p(\boldsymbol{\theta}) := p(\mu, \sigma)$:

$$\begin{aligned}\ell(\mu, \sigma) := \log(p) &= \sum_{j=1}^{n}\left[-\log(\sigma) - \frac{1}{2}\log(2\pi) - \frac{1}{2}\left(\frac{x_j - \mu}{\sigma}\right)^2\right] \\ &= -n\log(\sigma) - \frac{n}{2}\log(2\pi) - \frac{1}{2\sigma^2}\sum_{j=1}^{n}\left(x_j - \mu\right)^2.\end{aligned}$$

To maximize the log-likelihood, we compute the partial of ℓ with respect to μ and σ and set both equal to zero:

$$\begin{aligned}\frac{\partial \ell}{\partial \mu} &= \frac{1}{\sigma^2}\sum_{j=1}^{n}(x_j - \mu) = \frac{1}{\sigma^2}(n\bar{x} - n\mu) \\ \frac{\partial \ell}{\partial \sigma} &= -\frac{n}{\sigma} + \frac{1}{\sigma^3}\sum_{j=1}^{n}(x_j - \mu)^2,\end{aligned}$$

where $\bar{x}$ is the sample mean. Setting both expressions equal to zero yields the maximum likelihood estimates $\hat{\mu}$ and $\hat{\sigma}^2$ for the parameters μ and σ^2:

$$\begin{aligned}\frac{\partial \ell}{\partial \mu} = 0 \quad &\Longrightarrow \quad \hat{\mu} = \bar{x} \\ \frac{\partial \ell}{\partial \sigma} = 0 \quad &\Longrightarrow \quad \hat{\sigma}^2 = \frac{1}{n}\sum_{j=1}^{n}\left(x_j - \bar{x}\right)^2.\end{aligned}$$

Thus $\hat{\mu}$ is the sample mean and $\hat{\sigma}^2$ is the sample variance. Note that the log-likelihood is concave in μ and σ^2, so these stationary points must be global maximizers. Prove that these values maximize the log-likelihood by explicitly computing the second derivative to show that the curvature is positive, is left as an exercise.

Example: Poisson Distribution

Consider random variables that are distributed according to a Poisson distribution, $X_j \sim \text{Poisson}(\lambda)$, so that the joint PDF is given by

$$p(x_1, x_2, \ldots, x_n \mid \theta) = \prod_{j=1}^{n} p(x_j \mid \theta) = \prod_{j=1}^{n} \frac{\lambda^{x_j} e^{-\lambda}}{x_j!}.$$

The log-likelihood function for this joint distribution is given by

$$\begin{aligned} \ell(\theta) = \log(\text{lik}(\theta)) &= \log\left(p(x_1, x_2, \ldots, x_n \mid \theta)\right) \\ &= \log \prod_{j=1}^{n} \frac{\lambda^{x_j} e^{-\lambda}}{x_j!} \\ &= \sum_{j=1}^{n} \log\left(\frac{\lambda^{x_j} e^{-\lambda}}{x_j!}\right) \\ &= \sum_{j=1}^{n} \left(x_j \log(\lambda) - \lambda - \log(x_j!)\right). \end{aligned}$$

To maximize this log-likelihood, we now compute the derivative of this with respect to $\theta = \lambda$ and set it equal to zero:

$$\frac{\mathrm{d}}{\mathrm{d}\lambda}\ell(\lambda) = \frac{1}{\lambda}\sum_{j=1}^{n} x_j - n = 0 \quad \Longrightarrow \quad \lambda = \frac{1}{n}\sum_{j=1}^{n} x_j,$$

which is the sample mean $\bar{x}$. This was how we fit λ in the previous example.

Properties of MLE

The MLE has several useful statistical properties in the large data limit.

Consistency: The estimate $\hat{\theta}$ is consistent, meaning that $\hat{\theta} \longrightarrow \theta$ in the large data limit, or as $n \longrightarrow \infty$.

Asymptotic normality: Further, the estimate tends towards a normally distributed random variable for large n, with mean given by the true θ and whose variance decreases as n increases:

$$\hat{\theta} \sim \text{normal}\left(\theta_{\text{true}}, \frac{1}{nI(\theta_{\text{true}})}\right).$$

This is closely related to the central limit theorem, although it is subtly different. This is a very useful expression, as we can use it to construct confidence intervals on $\hat{\theta}$. Showing that $\hat{\theta}$ is asymptotically normal is quite challenging.

The function $I(\theta)$ is the Fisher information, and it is given by

$$I(\theta) = \underbrace{\mathbb{E}\left[\frac{\partial}{\partial\theta}\log p(x \mid \theta)\right]^2}_{\mathbb{E}[\ell'(\theta)]^2} = \underbrace{-\mathbb{E}\left[\frac{\partial^2}{\partial\theta^2}\log p(x \mid \theta)\right]}_{-\mathbb{E}[\ell''(\theta)]}. \tag{7.11}$$

The first equation corresponds to the variance form, and the second equation is the negative expected Hessian form.

This distribution for $\hat{\theta}$ means that the MLE is asymptotically efficient, so that this estimate converges to the true value of θ as efficiently as possible, in the large n limit. This uses something important, called the Cramer–Rao inequality: Given data $x_1, \ldots, x_n$ from an i.i.d. process and an unbiased estimate of the parameters of the distribution, $\hat{\theta} = \hat{\theta}(x_1, \ldots, x_n)$[5], then

$$\mathrm{Var}(\hat{\theta}) \geq \frac{1}{nI(\theta)} \tag{7.12}$$

for all smooth $p(x \mid \theta)$.

Beyond MLE, there are other approaches for distribution fitting, such as the method of moments, which matches the moments of the distribution to the moments of the data.

One major drawback to MLE is that it may be fragile to bad data (e.g., outliers, etc.) and doesn't allow one to include prior knowledge or belief into the estimates. As a concrete example, imagine that we are estimating the probability θ of a coin being heads or tails. For a fair coin the probability of heads is 0.5, so $\theta = 0.5$. If we flip the coin three times and get three heads in a row, {H,H,H}, then the MLE yields $\hat{\theta} = 1$. In this case, the MLE will overfit to this data and suggest that the coin will always flip heads for all future flips. This is a major limitation, as we have additional prior knowledge that most coins are fair and three heads in a row is not particularly unlikely. These issues will be directly addressed in the Bayesian extension, maximum a posteriori (MAP) estimation, in the next section.

7.2 Bayesian Inference

In the following, we will now take the Bayesian perspective on statistical estimation. Before, we began with a parameterized probability distribution $p(x \mid \theta)$, which gives the probability of finding data x given parameters θ. However, in many cases we actually want the probability of a given set of parameters θ given the data x:

$$p(\theta \mid x).$$

This inverse model $p(\theta \mid x)$ is known as the *posterior* distribution, and the forward model $p(x \mid \theta)$ is known as the *likelihood*. They are related by Bayes' theorem as

$$p(\theta \mid x) = \frac{p(x \mid \theta)p(\theta)}{p(x)}. \tag{7.13}$$

[5] Recall that the estimated parameters are in fact functions of the data $\{x_j\}_{j=1}^n$.

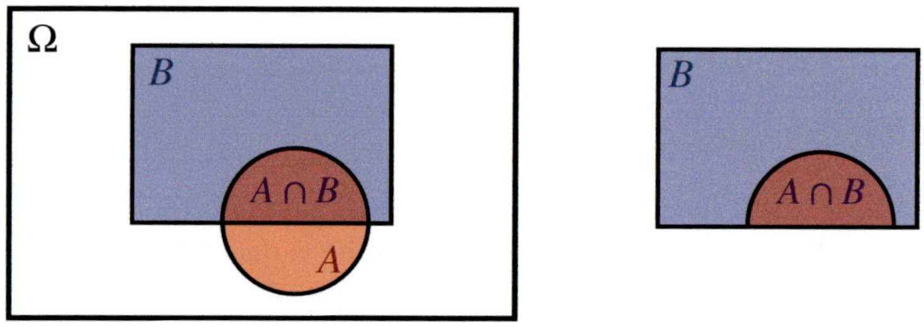

Figure 7.4 Illustration of conditional probabilities. Left: Given a sample space Ω and two events A and B, we denote $A \cap B$ as the event that both A and B occur. If we know that B occurs, then the probability of A given B, $P(A \mid B)$ is the ratio of $P(A \cap B)$ and $P(B)$, as shown in the right, where we have restricted ourselves to the set B, since we assume that B is known to have occurred.

The distribution $p(\theta)$ is called the *prior* distribution, and it encodes our prior knowledge about θ. For example, we may have prior knowledge that a coin is usually fair.

We can read off the optimal θ directly from $p(\theta \mid x)$, but typically we only have the forward model $p(x \mid \theta)$. Thus, this is an inverse problem, which is the subject of Chapter 8. In fact, the posterior distribution is called the Bayesian inverse of the likelihood function.

Bayes' Theorem

We are now ready to derive Bayes' theorem, which is a clever and useful way to invert conditional probability distributions of the data given parameters, $p(x \mid \theta)$, to obtain the distribution of the parameters given the data, $p(\theta \mid x)$.

We first state Bayes' theorem in simplified notation. Let Ω be the sample space of all possible outcomes of a system. For example, given a standard deck of 52 cards with four suits and thirteen ranks, then there are 52 unique outcomes if we deal a single card. Let A and B be specific events; for example, event A may be that a heart is drawn, or event B may be that a seven is drawn. These are independent events, since the probability of a card being a heart does not change if we know that it is a seven, and vice versa.

The conditional probability of event A occurring, given that event B has occurred, is

$$P(A \mid B) = \frac{P(A \cap B)}{P(B)} \tag{7.14}$$

where $P(A \cap B)$ is the probability that both A and B occur. This is shown in Fig. 7.4. We may multiply both sides by $P(B)$, yielding a formula for the probability $P(A \cap B)$:

$$\begin{aligned} P(A \cap B) &= P(A \mid B)P(B) \\ &= P(B \mid A)P(A) \end{aligned} \tag{7.15}$$

where the second line is true since

$$P(B \mid A) = \frac{P(B \cap A)}{P(A)} \tag{7.16}$$

and $P(A \cap B) = P(B \cap A)$ by symmetry. If A and B are independent, then $P(A \cap B) = P(A)P(B)$; for example, a card being a heart (A) or a seven (B) are independent.

Rearranging the terms in Eq. (7.15) results in Bayes' theorem

$$P(B \mid A) = \frac{P(A \mid B)P(B)}{P(A)}, \tag{7.17}$$

which allows us to solve for $P(B \mid A)$ given $P(A \mid B)$, $P(A)$, and $P(B)$.

When we extend this idea to continuous random variables, for example the parameters θ and data x, by replacing the probabilities above with probability distributions. Thus, the probability of the unknown parameters θ given the observed data x is

$$p(\theta \mid x) = \frac{p(x \mid \theta)p(\theta)}{p(x)}. \tag{7.18}$$

More General Forms of Bayes' Theorem

We may not know $P(A)$ in the denominator of Eq. (7.17), so instead we write it in terms of B and its complement $B^C = \Omega \setminus B$:

$$P(B \mid A) = \frac{P(A \mid B)P(B)}{P(A \mid B)P(B) + P(A \mid B^C)P(B^C)}. \tag{7.19}$$

This uses the law of total probability that either B or *not* B (i.e., B^C) must occur, so $P(B) + P(B^C) = 1$. Thus, the probability of A occurring, $P(A)$, must be equal to the sum of $P(A \mid B)P(B)$ and $P(A \mid B^C)P(B^C)$.

More generally, if we partition the sample space Ω into n disjoint sets

$$\Omega = \bigcup_{j=1}^{n} B_j \quad \text{with} \quad P(B_i \cap B_j) = 0 \quad \text{for } i \neq j \tag{7.20}$$

then we may write the law of total probability as

$$P(A) = \sum_j P(A \mid B_j)P(B_j). \tag{7.21}$$

This gives a more general form of Bayes' theorem

$$P(B_j \mid A) = \frac{P(A \mid B_j)P(B_j)}{\sum_{j=1}^{n} P(A \mid B_j)P(B_j)}. \tag{7.22}$$

Panel 7.1: Laplace: The Father of Modern Statistics

Bayes' theorem is a rule for inverting conditional probabilities, based on the law of total probability. Bayes only explored the binomial distribution, and Laplace independently developed an extended and more general version of this theory in his 1812 *Théorie Analytique des Probabilités* [298]. The modern theory was mainly developed by Laplace [121], in what he called the *probability of causes*, so perhaps the result should be called the Bayes–Laplace theorem. Laplace may be thought of as the father of modern statistical thinking, being the first scientist to rigorously think of measurement data as a random variable with some associated distribution of measurement error.

Maximum a Posteriori (MAP) Estimation

We now use Bayes' theorem to extend the maximum likelihood estimator to incorporate prior knowledge about θ, for example that a coin flip is usually fair.

In MLE we only use the likelihood $p(x \mid \theta)$. But if we have prior knowledge about θ, $p(\theta)$, then we can multiply these and get:

$$p(\theta \mid x) = \frac{p(x \mid \theta)p(\theta)}{p(x)}, \tag{7.23}$$

which is a statement of Bayes' theorem. In practice, we want the posterior $p(\theta \mid x)$, which is the probability of the parameters θ given the data x. The factor $p(x)$ in the denominator is a normalization constant, since it won't affect any optimization over θ. Often we write

$$p(\theta \mid x) \propto p(x \mid \theta)p(\theta) \tag{7.24}$$

to remove the denominator.

Instead of computing the maximum of $\log p(x \mid \theta)$ we compute the maximum of $\log p(\theta \mid x)$, which we call the Bayesian maximum a posteriori (MAP) estimate:

$$\max_{\theta} \log p(\theta \mid x) = \max_{\theta} \log(p(x \mid \theta)p(\theta)). \tag{7.25}$$

Thus, MAP estimation is the Bayesian extension of the MLE, and it allows us to incorporate prior knowledge about the distribution on θ. We have replaced the log-likelihood function from the MLE with the joint likelihood of the data and prior information about θ. We will show in Section 7.5 that the LASSO and Tikhonov regularized least-squares solutions may be viewed as the MAP estimates for particular choices of priors.

Iterative Bayesian Updates

In an iterative Bayesian updating scheme, we update our prior as new information becomes available. This Bayesian update may be written as:

$$p(\theta \mid x_k) = \frac{p(x_k \mid \theta)p(\theta)}{p(x_k)}, \tag{7.26}$$

so the posterior may be updated with new data. At each iteration, the posterior from the previous step becomes the prior distribution for the next step:

$$p(\theta) = p(\theta \mid x_k). \tag{7.27}$$

In this way, our prior beliefs update with new information. This scheme balances data with prior beliefs via the model $p(\theta)$. This is closely related to the Kalman filter from Section 9.1, which is also concerned with balancing measurement data and model predictions.

There are also well-known downsides to Bayesian statistics. Results are highly sensitive to the prior, which may be subjective or hard to quantify mathematically. A bad prior will significantly degrade a Bayesian model. Updates are also computationally complex, often involving Monte Carlo sampling and bootstrapping to estimate distributions from data; when the model is complex with a high-dimensional parameter $\boldsymbol{\theta}$, these methods suffer from the curse of dimensionality. Machine learning is helping to alleviate these scaling issues, though they are still challenging.

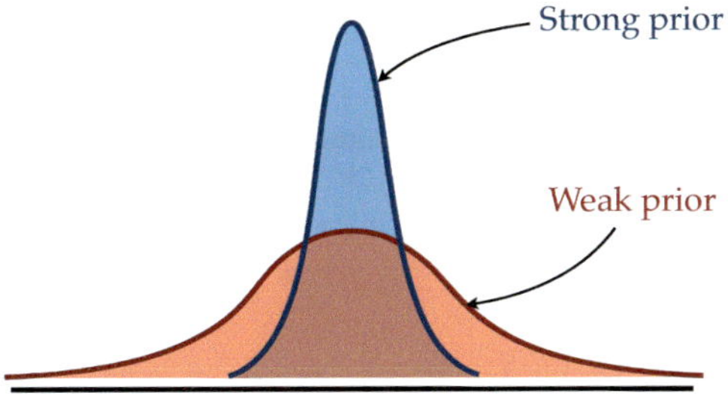

Figure 7.5 Example of a strong Gaussian prior (blue) and a weak Gaussian prior (red).

Figure 7.5 shows an example of a weak prior and a strong prior, both from a Gaussian distribution. Strong priors are harder to change with evidence, providing firmer belief in a given model. In many cases, we want *sticky* priors; for example when flipping a coin, we assume it is exactly fair until we have considerable counter evidence.

The posterior and prior distributions must be parametrized in some way. Using a known distribution, such as a Poisson or a normal distribution, is the simplest choice. However, the data may not follow a named distribution. Several approaches exist for parameterizing and estimating more general probability distributions, such as Gaussian mixture models (GMMs) and density estimation for unknown $p(\theta)$ and $p(x \mid \theta)$. Kernel density estimation (KDE) and GMMs will be explored in the next two subsections. In addition, it is increasingly common to parameterize the distribution with a neural network, sometimes in combination with a Gaussian prior on some latent state, as in the variational autoencoder [277].

Conjugate Priors and Updates

A natural issue arises in the Bayesian update. We want the posterior, $p(\theta \mid x)$ to follow the same distribution as the prior, $p(\theta)$, since the updated posterior becomes the prior for the next iteration. However, multiplying the prior by the likelihood function, $p(x \mid \theta)$, gives

$$p(\theta \mid x) \propto p(x \mid \theta) p(\theta), \tag{7.28}$$

which may change the posterior distribution.

There is a clever solution. We often know the likelihood function, which is our model of the probability density $p(x \mid \theta)$. We then choose the prior from a distribution that, when multiplied by the likelihood function, remains in the same distribution. These are called *conjugate* priors, where conjugate refers to its relationship to the likelihood function.

It is helpful to demonstrate this idea with examples. First, Gaussian distributions are conjugate with Gaussians, since their product remains Gaussian; this is a straightforward exercise for the reader. Another common example is related to coin flips. Consider a random variable X that follows a binomial distribution:

$$X \sim \text{bin}(n, \theta) = \binom{n}{h} \theta^h (1 - \theta)^{n-h},$$

where θ is the probability of a single flip landing heads and h is the total number of heads given n coin flips. The conjugate prior distribution to a Bernoulli random variable is the beta distribution, so that $\theta \sim \text{beta}(\alpha, \beta)$:

$$p(\theta \mid \alpha, \beta) = k\, \theta^{\alpha-1}(1-\theta)^{\beta-1},$$

where k is a normalization constant related to the gamma functions[6] of α and β:

$$k = \frac{\Gamma(\alpha+\beta)}{\Gamma(\alpha)\Gamma(\beta)}.$$

Panel 7.2: Exponential Family of Distributions

A major theoretical advance in statistics in the 1930s established that many common distributions belong to a large family of distributions with known conjugate prior distributions [125, 282, 390]. The *exponential* family of distributions is given by

$$p(x \mid \theta) = h(x)e^{\theta^{\mathrm{T}}\phi(x)-A(\theta)}.$$

where $h(x)$ may be absorbed into the exponential and $A(\theta)$ is a normalization constant.

All exponential family members have a conjugate prior distribution given by

$$p(\theta \mid \gamma) = h_c(\theta)\exp\left(\begin{bmatrix}\gamma_1 & \gamma_2\end{bmatrix}\begin{bmatrix}\theta \\ -A(\theta)\end{bmatrix} - A_c(\gamma)\right)$$

where γ_1 and θ are vectors of dimension n and γ_2 and $-A(\theta)$ are scalars.

Example: Gaussian Distribution. At first, it is not obvious how a common distribution, such as the normal distribution, may be written in this form. Here we will show how to represent a Gaussian distribution in the exponential family of distributions. For $X \sim \text{normal}(\mu, \sigma^2)$, we have the following:

$$\boldsymbol{\theta} = \begin{bmatrix}\frac{\mu}{\sigma^2} \\ -\frac{1}{2\sigma^2}\end{bmatrix} \qquad \boldsymbol{\phi}(x) = \begin{bmatrix}x \\ x^2\end{bmatrix}$$

so that $\boldsymbol{\theta}^{\mathrm{T}}\boldsymbol{\phi}(x) = \frac{\mu x}{\sigma^2} - \frac{x^2}{2\sigma^2}$. If we take the exponential, we recover our Gaussian:

$$\begin{aligned}
p(x \mid \theta) \propto e^{\boldsymbol{\theta}^{\mathrm{T}}\boldsymbol{\phi}(x)} &= e^{-\frac{1}{2\sigma^2}(x^2-2\mu x+\mu^2)+\frac{\mu^2}{2\sigma^2}} \\
&= e^{-\frac{1}{2\sigma^2}(x-\mu)^2}\, e^{\frac{\mu^2}{2\sigma^2}} \\
&\propto e^{-\frac{1}{2\sigma^2}(x-\mu)^2} \sim \text{normal}(\mu, \sigma^2).
\end{aligned}$$

The Gaussian is in the exponential family so it has a conjugate prior. For the univariate case with unknown mean and variance, the conjugate prior is the normal–inverse-gamma distribution. For the multivariate Gaussian with unknown mean and covariance, the conjugate prior is the normal–inverse-Wishart distribution.

[6] Recall that the Γ function is related to the factorial, so that $\Gamma(n) = (n-1)!$ for positive integers n. It generalizes the factorial for non-integer n.

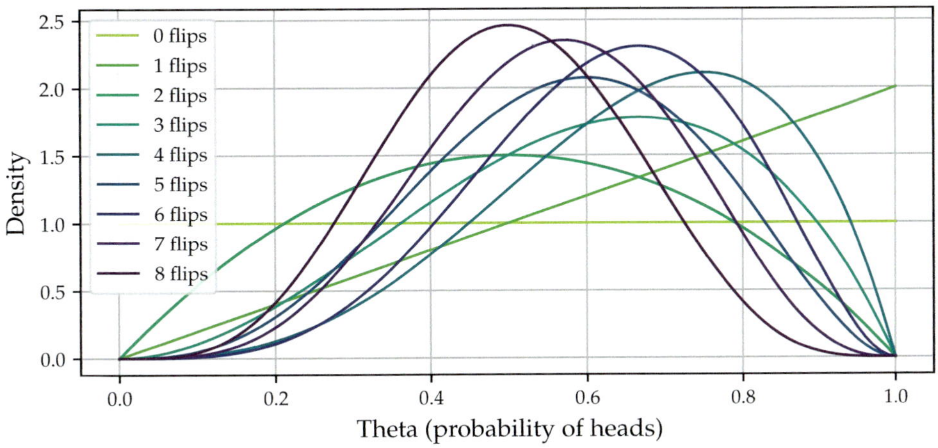

Figure 7.6 Evolution of the posterior for the coin flip example with increasing information. The posterior is approximated using the binomial distribution with the conjugate prior beta distribution.

Example: Coin Flips

A simple example of Bayesian inference is to iteratively estimate the behavior of a coin by accumulating data from sequential coin flips. We will investigate several Bayesian approaches based on parametric and nonparametric estimation. First, we assume that a single coin flip follows a Bernoulli distribution with an unknown probability of heads, θ. The probability of h heads in n flips follows a binomial distribution with a beta distribution as conjugate prior, making it simple to accumulate information from new coin flips.

Figure 7.6 shows the posterior distribution evolving for several coin flips, starting with a uniform prior. Within a relatively small number of coin flips the posterior distribution approaches a binomial distribution centered around $\theta = 0.5$, which represents a fair coin.

As additional data is accumulated, we may follow the evolution of the posterior, along with its mean, in Fig. 7.7. We also show the ± 1 standard deviation regions about the mean in orange. We can see that there is considerable wander in the mean θ, and even after 100 coin flips, the estimated probability of heads is off of the true value by about 10%.

Bayesian methods are sensitive to a bad initial prior. Figure 7.8 shows the evolving posterior distribution for a good and a bad initial prior. The bad prior (dashed) converges much more slowly to the true value of $\theta = 0.5$, while the good prior stays closer to $\theta = 0.5$.

The Bayesian update code with a binomial prior and a beta conjugate prior is given by:

CodeSnip 7.c

```
from scipy.stats import beta

def likelihood(theta, data):
    # Binomial likelihood for coin flips (Bernoulli trials)
    heads = np.sum(data == 'H')
    tails = np.sum(data == 'T')
    return theta**heads * (1 - theta)**tails

# Compute posterior distribution using Bayesian update
def bayesian_update(prior_alpha, prior_beta, data):
    # Calculate the number of heads and tails in the data
```

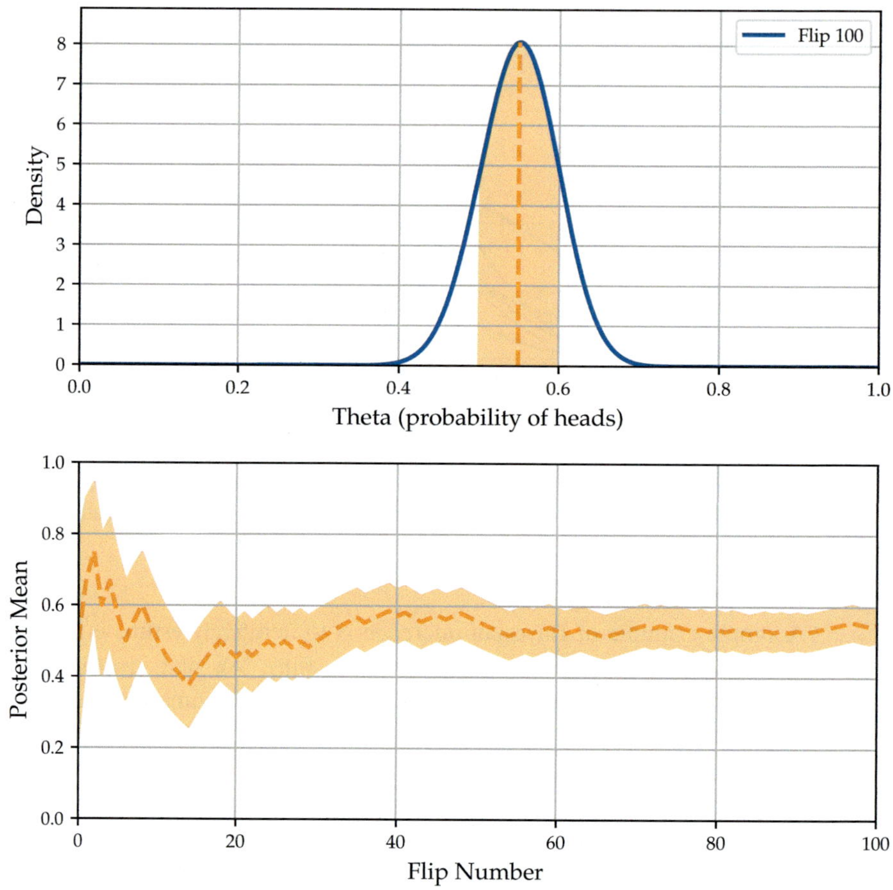

Figure 7.7 Posterior distribution (top) and posterior mean, for the standard binomial distribution with the conjugate prior beta distribution. The orange shaded regions denote ±1 standard deviation.

```
    heads = np.sum(data == 'H')
    tails = np.sum(data == 'T')
    # Update the prior (beta distribution) parameters
    posterior_alpha = prior_alpha + heads
    posterior_beta = prior_beta + tails
    return posterior_alpha, posterior_beta

# Prior: Uniform distribution (Beta(1,1))
prior_alpha = 1
prior_beta = 1

# Simulate a sequence of coin flips (H for heads, T for tails)
coin_flips = np.array(['H', 'T', 'H', 'H', 'T', 'H', 'T', 'T'])
# Perform Bayesian updates of posterior after each flip
for i, flip in enumerate(coin_flips):
    prior_alpha, prior_beta = bayesian_update(prior_alpha,
        prior_beta, np.array([flip]))
```

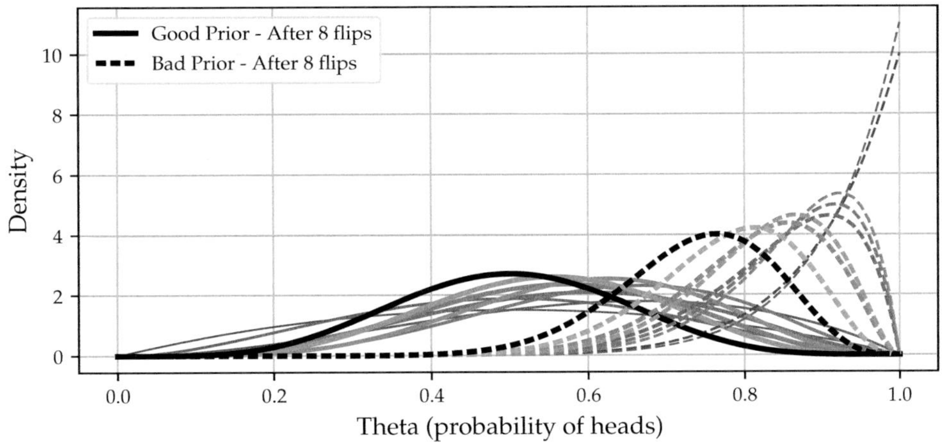

Figure 7.8 Comparison of Bayesian updates for the coin flip example with a good (solid) and a bad (dashed) prior. The lines get thicker as the number of flips increases.

7.3 Density Estimation and Empirical Priors

We have investigated several examples where the likelihood function is a well known distribution, like a binomial or normal, and we can choose a good conjugate prior distribution.

However, for most real-world scenarios, the density functions will not be nice, well-behaved functions, such as a single Gaussian or Poisson distribution. Density estimation is used to estimate these more complex probability densities from data using a sum of simple distributions. Density estimation is a direct predecessor of many modern machine learning methods. The learned distributions will be known as the empirical prior and posterior distributions. The simple building-block distributions will usually be Gaussians, for example in the Gaussian mixture model.

Two of the most common techniques are called kernel density estimation (KDE) and Gaussian mixture models (GMMs):

KDE This technique treats every data point x like its own collocation point that will have a local Gaussian, and we estimate the PDF at other points by smoothing or interpolating across these functions.

$$p(x) = \frac{1}{Nh} \sum_{j=1}^{N} \mathcal{K}\left(\frac{x - x_j}{h}\right). \tag{7.29}$$

The kernel $\mathcal{K}$ is often chosen to be a Gaussian, N is the number of data points, and h is a smoothing width, which is a hyperparameter to smooth out the fit. Each x_j is a data sample. In essence, every time we see a data point x_j, this makes it more likely to see a new point near x_j in subsequent trials. In this way, KDE has a *frequentist* flavor. However, it is possible to encode smoothness priors in more general kernel regression techniques, which would be a Bayesian approach. A schematic of kernel density estimation is shown in Fig. 7.9. Kernel density

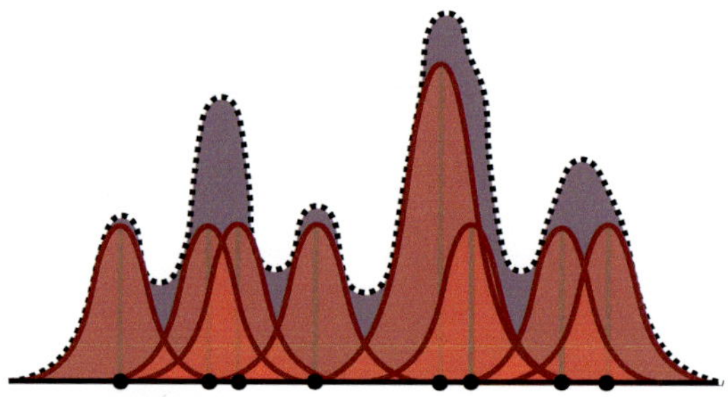

Figure 7.9 Kernel density estimation (KDE) may be thought of as a special case of a Gaussian mixture model, where there are as many Gaussians as there are data points.

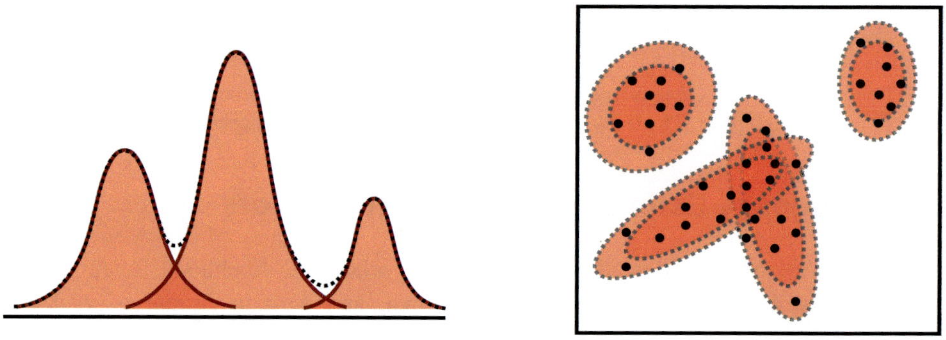

Figure 7.10 Gaussian mixture models (GMMs) visualized in one dimension (left) and two dimensions (right). The PDF $p(\theta)$ is approximated by the sum of a relatively small number of Gaussians. The fit is typically obtained through an optimization algorithm, typically expectation maximization.

estimation is closely related to manifold learning, such as the diffusion maps embedding.

GMM These also also estimate the distribution using a sum of Gaussians. However, in KDE, there was a Gaussian for every single data point x_j, while for GMMs, we try to estimate the distribution with many fewer Gaussians. Finding the specific number of Gaussians and parameters is an optimization problem, typically using an expectation maximization (EM) algorithm, discussed in Section 7.4. A schematic of GMMs is shown in Fig. 7.10.

Empirical distributions are *generative models*. If we have one of these empirical distributions, we can generate new data that follows these distributions. Generative models are the basis of modern methods in *generative AI*, such as large-language models and stable diffusion for image generation from text prompts.

Kernel Density Estimation

Here we demonstrate the kernel density estimation method to learn the empirical posterior distribution for a coin. Figure 7.11 shows the evolution of the posterior as new information is accumulated. Figure 7.12 shows the distribution and the evolution of the mean and quantiles with the number of coin flips. The following code implements the KDE method:

CodeSnip 7.d

```
from sklearn.neighbors import KernelDensity

# Compute empirical binomial likelihood of data given samples
def likelihood(samples, data):
    heads = np.sum(data == 'H')
    tails = np.sum(data == 'T')
    return samples**heads * (1 - samples)**tails

# Function to update the posterior distribution using KDE
def bayesian_update_kde(prior_samples, data):
    # Compute the empirical likelihood of the data
    likelihoods = likelihood(prior_samples, data)

    # Weight the prior samples by the likelihood
    posterior_weights = likelihoods / np.sum(likelihoods)

    # Resample prior based on likelihoods for posterior samples
    posterior_samples = np.random.choice(prior_samples, size
        =1000, p=posterior_weights)

    # Kernel Density Estimation to estimate posterior
    kde = KernelDensity(kernel='gaussian', bandwidth=0.05).fit(
        posterior_samples[:, np.newaxis])
    return posterior_samples, kde

# Prior: Uniform distribution (between 0 and 1)
prior_samples = np.random.uniform(0, 1, 1000)

# Simulate a sequence of coin flips (H for heads, T for tails)
coin_flips = np.array(['H', 'T', 'H', 'H', 'T', 'H', 'T', 'T'])

# Perform Bayesian updates after each flip
samples_list = [prior_samples] # Store prior samples to plot
for i, flip in enumerate(coin_flips):
    prior_samples, kde = bayesian_update_kde(prior_samples, np.
        array([flip]))    # Update posterior using KDE method
    samples_list.append(prior_samples) # Store new posterior

# Create histogram (plotting omitted)
theta_range = np.linspace(0, 1, 1000)
for i, samples in enumerate(samples_list):
    kde = KernelDensity(kernel='gaussian', bandwidth=0.05).fit(
        samples[:, np.newaxis])
    log_density = kde.score_samples(theta_range[:, np.newaxis])
    density = np.exp(log_density)
```

Monte Carlo Sampling

When the posterior distribution is particularly complex, it may not be well described by a named distribution. In most cases when we don't know the posterior, we resort to numerical methods, such as Monte Carlo, to obtain a bootstrap estimate of the posterior distribution. If the likelihood function and the prior are not conjugates, so the posterior is a different

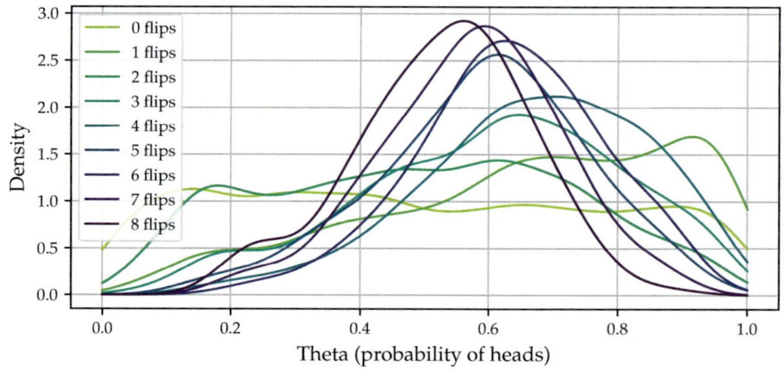

Figure 7.11 Evolution of the posterior for the coin flip example with increasing information. The posterior is approximated using kernel density estimation.

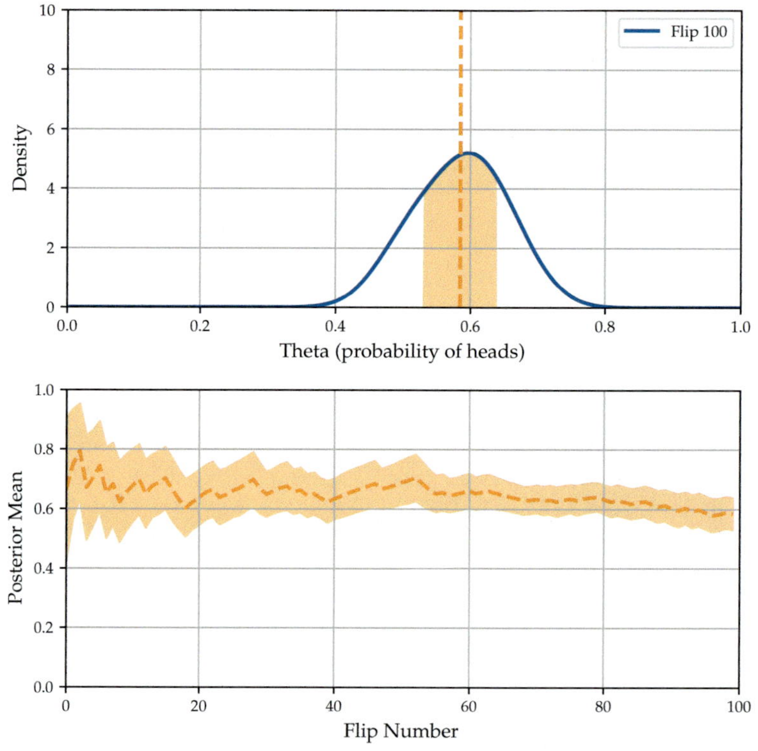

Figure 7.12 Posterior distribution (top) and posterior mean (bottom), with quantiles, for the kernel density estimation. The orange shaded regions denote ±1 standard deviation.

distribution than the prior, then we may use Monte Carlo sampling. Alternatively, if the parameter space is high dimensional, we might not want to compute the posterior directly, and instead resort to sampling. There are many extensions to Monte Carlo sampling, such as Markov chain Monte Carlo (MCMC) [191] and Hamiltonian Monte Carlo [360].

We used Monte Carlo sampling earlier to generate ensembles of data to obtain a bootstrap estimate of the unknown parameter $\hat{\theta}$ in the Poisson distributed alpha particle emissions. Here we extend this to the more general Bayesian inference framework. We assume we have access to the prior distribution $p(\theta)$ and the likelihood $p(x \mid \theta)$. There are several steps:

- Sample: We begin by sampling multiple θ_i from the prior distribution $p(\theta)$. This collection $\{\theta_i\}_{i=1}^n$ is an ensemble of model parameters.
- Weight: We weight each of these samples θ_i by the likelihood of the data x given this parameter: $w_i = p(x \mid \theta_i)$.
- Resampling/Bootstrapping: We are trying to approximate the posterior, which is the product of the likelihood and the prior, but with the ensemble of parameters $\{\theta_i\}$ and weights $\{w_i\}$. To do this, we resample the collection of parameters $\{\theta_i\}$ but weighted by $\{w_i\}$. This means creating a new collection $\{\hat{\theta}_j\}$ by sampling, with replacement, from $\{\theta_i\}$. This will shift the distribution $p(\theta)$ based on the weights obtained from the likelihood function $p(x \mid \theta)$. The histogram of the weighted, resampled distribution $\{\hat{\theta}_j\}$ will follow the unknown posterior distribution.

We repeat this process, so the posterior $\{\hat{\theta}_j\}$ becomes the prior sample in the next step of the iteration. Figure 7.13 shows the posterior distribution and mean using batched Monte Carlo estimation. The following code computes this Monte Carlo sample-based estimation.

CodeSnip 7.e

```
from scipy.stats import binom, uniform

theta_true = 0.5    # True value of theta (probability of heads)
batch_size = 10     # Number of coin flips per batch
total_flips = 250   # Total coin flips
num_samples = 5000 # Number of Monte Carlo samples

# Generate the true coin flip outcomes based on the true theta
np.random.seed(42)  # For reproducibility
coin_flips=np.random.binomial(1,theta_true,total_flips) # 1=head

# Prior: Uniform distribution for theta (= Beta(1, 1))
prior_samples = np.random.uniform(0, 1, num_samples)

def update(batch_number):
    global prior_samples, std_shade1, std_shade2

    # Sample current batch of flips
    start = batch_number * batch_size
    end = start + batch_size
    batch_flips = coin_flips[start:end]

    # Calculate number of heads and tails in the current batch
    heads = np.sum(batch_flips)
    tails = batch_size - heads
```

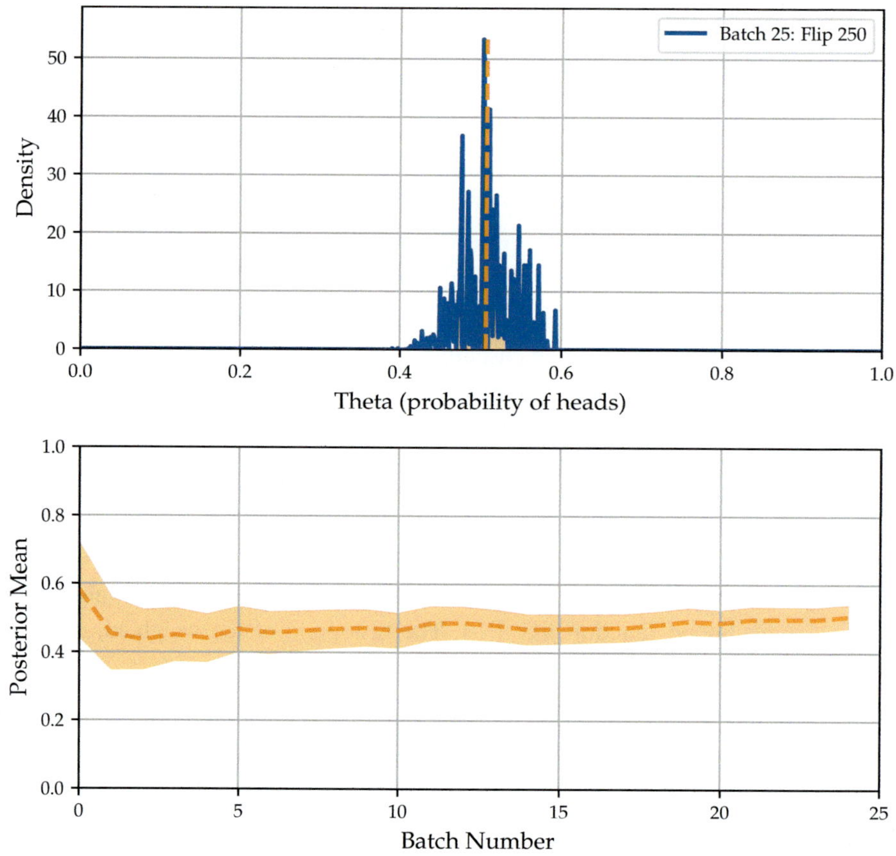

Figure 7.13 Posterior distribution (top) and posterior mean (bottom), with quantiles, for the Monte Carlo estimation. The orange shaded regions denote ±1 standard deviation.

```
# Bionomial likelihood for each prior sample given data
likelihoods = binom.pmf(heads, batch_size, prior_samples)

# Reweight prior samples based on likelihood
weights = likelihoods / np.sum(likelihoods)

# Resample weighted prior samples to form posterior
posterior_samples = np.random.choice(prior_samples, size=
    num_samples, p=weights)

# Update prior samples for next iteration (Monte Carlo)
prior_samples = posterior_samples
```

7.4 Expectation Maximization and Gaussian Mixture Models

Expectation maximization (EM) is an optimization algorithm used for parameter estimation [134, 508], for example with MLE or MAP, when there are hidden or latent variables, resulting in incomplete data. We use EM to optimize a Gaussian mixture model.

In a Gaussian mixture model, we have M data points $\{x_j\}_{j=1}^M$ that are believed to belong to one of C Gaussian clusters with unknown means $\boldsymbol{\mu}_c$ and covariance matrices $\boldsymbol{\Sigma}_c$. There is also a probability α_c associated with belonging to each of the p Gaussian clusters. The subscript c indicates an index for which *cluster* the data belongs to.

If we know which cluster each data point x_j belongs to, then it is a simple statistical estimation problem to find the means and covariance matrices, corresponding to the p independent MLE or MAP optimizations. For a single Gaussian cluster:

$$\hat{\boldsymbol{\mu}} = \frac{1}{M}\sum_{j=1}^{M} x_j \tag{7.30a}$$

$$\hat{\boldsymbol{\Sigma}} = \frac{1}{M}\sum_{j=1}^{M} \left(x_j - \hat{\boldsymbol{\mu}}\right)^{\mathrm{T}}\left(x_j - \hat{\boldsymbol{\mu}}\right). \tag{7.30b}$$

Not knowing to which cluster each data point belongs makes this a more challenging, generically nonconvex, optimization problem to learn the model parameters. For multiple Gaussian clusters, there is a hidden variable corresponding to which cluster a given data point x_j belongs to. In this case, we will use the EM algorithm below.

Formally, we consider a system with variables $\mathbf{x}$ and $\mathbf{z}$, where $\mathbf{x}$ is observed and $\mathbf{z}$ is hidden. In a GMM, $\mathbf{z}$ determines which cluster c the observation $\mathbf{x}$ belongs to, and which Gaussian parameters $\boldsymbol{\mu}_c, \boldsymbol{\Sigma}_c$ to use. The variables follow a joint probability distribution that is governed by a parameter $\boldsymbol{\theta}$ so that $(\mathbf{x}, \mathbf{z}) \sim p(\boldsymbol{\theta})$, which may be written as

$$p(\mathbf{x}, \mathbf{z} \mid \boldsymbol{\theta}). \tag{7.31}$$

The goal of a statistical estimator, such as MLE or MAP, is to find optimal parameters $\hat{\boldsymbol{\theta}}$

$$\hat{\boldsymbol{\theta}} = \underset{\boldsymbol{\theta}}{\operatorname{argmax}}\ p(\mathbf{x} \mid \boldsymbol{\theta}) \tag{7.32}$$

that maximize the marginal likelihood distribution $p(\mathbf{x} \mid \boldsymbol{\theta})$, given by

$$p(\mathbf{x} \mid \boldsymbol{\theta}) = \sum_{c=1}^{C} p(\mathbf{x} \mid \boldsymbol{\theta}, \mathbf{z}_c) p(\mathbf{z}_c). \tag{7.33}$$

The probability of being in cluster c is a constant $p(\mathbf{z}_c) = \alpha_c$, and the conditional probability $p(\mathbf{x} \mid \boldsymbol{\theta}, \mathbf{z}_c)$ is normal with parameters $\boldsymbol{\mu}_c$ and $\boldsymbol{\Sigma}_c$, so the marginal distribution is

$$p(\mathbf{x} \mid \boldsymbol{\theta}) = \sum_{c=1}^{C} \alpha_c \mathcal{N}(\boldsymbol{\mu}_c, \boldsymbol{\Sigma}_c). \tag{7.34}$$

Here $\mathcal{N}$ denotes the normal distribution. The coefficients α_c are the mixing coefficients. Thus, a GMM is a *generative* model. Given the model parameters $\{\alpha_c, \boldsymbol{\mu}_c, \boldsymbol{\Sigma}_c\}_{c=1}^C$, it is possible to generate new data that is consistent with the distribution of the training data.

We will use the EM algorithm to learn the unknown model parameters, given that there is a hidden variable corresponding to the cluster label. The EM algorithm starts with a set of clusters $\{\alpha_c, \boldsymbol{\mu}_c, \boldsymbol{\Sigma}_c\}_{c=1}^{C}$ as an initialization. There are several options for initializing a GMM, although starting with either a K-means clustering or a random initialization are common. After initializing, we iterate through the following two stages until convergence:

- Expectation (E) Step: For each data point x_j, we compute the probability r_{jc} that x_j is in cluster c:

$$r_{jc} = \frac{\alpha_c \mathcal{N}(x_j \mid \boldsymbol{\mu}_c, \boldsymbol{\Sigma}_c)}{\sum_{c=1}^{C} \alpha_c \mathcal{N}(x_j \mid \boldsymbol{\mu}_c, \boldsymbol{\Sigma}_c)}. \tag{7.35}$$

  The probability r_{jc} is called the *responsibility* value, and it is the likelihood of x_j belonging in each of the C clusters. The denominator is a normalization constant. We now fix r_{jc} and update the model parameters to maximize the likelihood.
- Maximization (M) Step: For each cluster c, we now update the parameters $\{\alpha_c, \boldsymbol{\mu}_c, \boldsymbol{\Sigma}_c\}$ to maximize the expected log-likelihood of the fit given the probabilities r_{jc} from the previous step. This amounts to recomputing each parameter using the MLE or MAP estimation, but weighted with r_{jc}. First, we compute weighting factors m_c and the probability of belonging to a cluster, α_c,

$$m_c = \sum_{j=1}^{M} r_{jc} \qquad \alpha_c = \frac{m_c}{\sum_{c=1}^{C} m_c} \tag{7.36}$$

  where the first sum is over all data points x_j. We then compute the weighted means and covariance matrices for each cluster:

$$\boldsymbol{\mu}_c = \frac{1}{m_c} \sum_{j=1}^{M} r_{jc} x_j \tag{7.37a}$$

$$\boldsymbol{\Sigma}_j = \frac{1}{m_c} \sum_{j=1}^{M} r_{jc} \left(x_j - \boldsymbol{\mu}_c\right)^{\mathrm{T}} \left(x_j - \boldsymbol{\mu}_c\right). \tag{7.37b}$$

In the E step, we compute the expectation of the log-likelihood function given the current estimates. In the M step, we update parameters to maximize this expected log-likelihood function. This is a form of bootstrapping, where we iteratively refine an initial guess. More generally, EM works well with data following any of the exponential family of distributions, discussed earlier. Each iteration increases the log-likelihood of the model and convergence to a local maximum is guaranteed. A good initialization is still required, as the algorithm will only converge to a local maximum.

The number of clusters is a hyperparameter that must be carefully chosen. As with other statistical learning and machine learning algorithms with hyperparameters, it may be possible to use information criteria, such as AIC/BIC[7], to penalize model complexity. Alternatively, it is possible to cross-validate the model with test data by choosing the number of clusters that maximize the log-likelihood of the model over test data.

[7] The Akaike information criterion [3] (AIC) prioritizes model prediction, while the Bayesian information criterion [454] (BIC) places emphasis on the model fit with a stronger penalty on complexity.

7.5 Least-Squares Regression as a Statistical Estimation Problem

We now revisit the classic linear least-squares regression problem from a statistical estimation point of view. Specifically, we reframe least-squares regression as a maximum likelihood estimation problem, which will allow us to further include prior information in the related Bayesian maximum a posteriori estimation problem.

To recast the least-squares regression problem in a statistical framework, we assume that errors to the linear regression fit follow a statistical distribution, typically a norm al distribution.

In one dimension, linear-regression finds the best-fit line

$$y = ax + b \tag{7.38}$$

that minimizes the sum-squared error of the fit. We estimate the parameters a and b. If we assume that data comes in pairs, $(x_1, y_1), (x_2, y_2), \ldots, (x_n, y_n)$, then we may formulate the data relationship as a matrix equation

$$\underbrace{\begin{bmatrix} y_1 & y_2 & \cdots & y_n \end{bmatrix}}_{\mathbf{Y}} = \underbrace{\begin{bmatrix} a & b \end{bmatrix}}_{\boldsymbol{\theta}} \underbrace{\begin{bmatrix} x_1 & x_2 & \cdots & x_n \\ 1 & 1 & \cdots & 1 \end{bmatrix}}_{\mathbf{X}}, \tag{7.39}$$

which leads to the classic least-squares regression optimization problem

$$\hat{\boldsymbol{\theta}} = \underset{\boldsymbol{\theta}}{\operatorname{argmin}} \, ||\mathbf{Y} - \boldsymbol{\theta}\mathbf{X}||_2^2. \tag{7.40}$$

Recall from Chapter 4 that this has a closed form solution in terms of the SVD.. The least-squares solution is what is known as a point estimate, since we gain no information about the distribution of y, just its expected value.

MLE Linear Regression

If the data now has measurement noise

$$y = ax + b + \eta, \tag{7.41}$$

where we have introduced the error term η, which is a normally distributed random variable: $\eta \sim \mathcal{N}(0, \sigma^2)$. This is a standard assumption that may be justified by the central limit theorem. Now we will estimate $\boldsymbol{\theta} = \begin{bmatrix} a & b \end{bmatrix}$ as the MLE based on the data and the assumption of η being a Gaussian.

Our model for y is no longer a simple point estimate, but instead follows a distribution, so it becomes a random variable Y. This is much stronger, since we can quantify the uncertainty in our estimate, for example allowing us to provide confidence intervals.

We can estimate the probability density of Y as

$$p(y \mid \mathbf{x}, \boldsymbol{\theta}) = \mathcal{N}(\boldsymbol{\theta}\mathbf{x}, \sigma^2). \tag{7.42}$$

In other words, the expected value of Y is given by the linear fit $y = \boldsymbol{\theta}\mathbf{x} = ax + b$. This estimate has a variance given by the variance of the noise, σ^2.

We now write our log-likelihood function in terms of this distribution and maximize to find the best-fit parameters $\boldsymbol{\theta}$.

The log-likelihood is

$$\ell(\boldsymbol{\theta}) = \log \prod_{j=1}^{n} p(y_j \mid x_j, \boldsymbol{\theta}) \tag{7.43a}$$

$$= \sum_{j=1}^{n} \log p(y_j \mid x_j, \boldsymbol{\theta}) \tag{7.43b}$$

$$= \sum_{j=1}^{n} \log \left[\frac{1}{\sqrt{2\pi\sigma^2}} e^{-\frac{(y_j - \boldsymbol{\theta} x_j)^2}{2\sigma^2}} \right] \tag{7.43c}$$

$$= \sum_{j=1}^{n} \Bigg[\underbrace{-\frac{(y_j - \boldsymbol{\theta} x_j)^2}{2\sigma^2}}_{\propto \sum (y_j - \boldsymbol{\theta} x_j)^2} + \underbrace{\log \left(\frac{1}{\sqrt{2\pi\sigma^2}} \right)}_{\text{constant}} \Bigg]. \tag{7.43d}$$

Thus, the likelihood function is proportional to the least-squares regression objective:

$$\ell(\boldsymbol{\theta}) \propto \|\mathbf{Y} - \boldsymbol{\theta}\mathbf{X}\|_2^2. \tag{7.44}$$

MAP Linear Regression

We may now extend the MLE linear regression above to include prior information in the Bayesian MAP regression. Recall that the MAP estimate maximizes the log-posterior, where the posterior is the Bayesian inverse, given by

$$p(\boldsymbol{\theta} \mid x, y) \propto p(y \mid x, \boldsymbol{\theta}) p(\boldsymbol{\theta}). \tag{7.45}$$

Thus the log-posterior is given by

$$\begin{aligned} \log(p(\boldsymbol{\theta} \mid x, y)) &= \log(p(y \mid x, \boldsymbol{\theta})) + \log(p(\boldsymbol{\theta})) \\ &= \underbrace{\|\mathbf{Y} - \boldsymbol{\theta}\mathbf{X}\|_2^2}_{\text{MLE}} + \underbrace{\log(p(\boldsymbol{\theta}))}_{R(\boldsymbol{\theta})}. \end{aligned} \tag{7.46}$$

The second term, $\log(p(\boldsymbol{\theta}))$ is the log-prior, which serves as a regularization term for the optimization to promote solutions that are consistent with the prior knowledge.

Gaussian Prior and Ridge Regression. As a simple example, we will assume that the prior distribution on the parameters $\boldsymbol{\theta}$ follow a Gaussian distribution, so that

$$\boldsymbol{\theta} \sim \mathcal{N}(0, \sigma^2) \quad \Longrightarrow \quad p(\boldsymbol{\theta}) = e^{-\frac{\|\boldsymbol{\theta}\|^2}{2\sigma^2}}.$$

In this case, the log-prior is given by

$$-\log p(\boldsymbol{\theta}) = \frac{\|\boldsymbol{\theta}\|^2}{2\sigma^2}$$

so the log-posteriori becomes

$$\log p(\boldsymbol{\theta} \mid x, y) = \|\mathbf{Y} - \boldsymbol{\theta}\mathbf{X}\|_2^2 + \alpha \|\boldsymbol{\theta}\|_2^2$$

Panel 7.3: Gauss and Laplace Priors

The Gaussian prior (left) and Laplace prior (right) promote different features in the parameters θ. The Laplace prior is sharply concentrated at 0, promoting sparsity, while the Gaussian prior is more distributed, promoting values that are not too large.

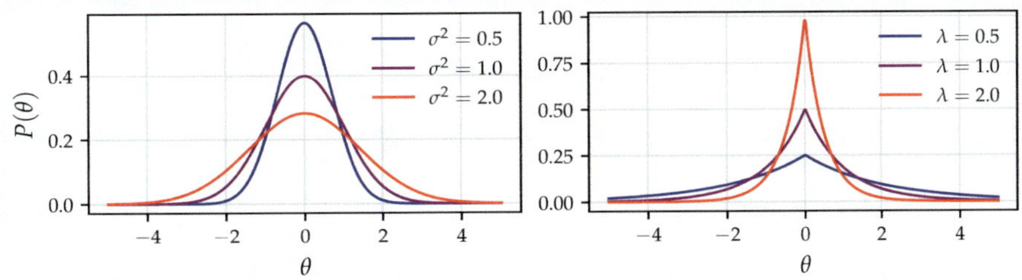

Figure 7.14: Gaussian prior (left) and Laplace prior (right).

where $\alpha \propto 1/\sigma^2$. Thus, a Gaussian prior on θ appears as a Tikhonov regularization term, which we saw earlier in Section 4.4. As in Chapter 4, there is a bias–variance tradeoff here, where the new solution is biased from the truth, but is subject to less overall variation. Note that here, σ is the standard deviation on the parameters θ rather than the data $\mathbf{X}$.

Laplace Prior and LASSO Regression. Now assume we have a Laplace prior on $\boldsymbol{\theta}$:

$$p(\boldsymbol{\theta}) = \frac{\lambda}{2} e^{-\lambda \|\boldsymbol{\theta}\|_1}.$$

In this case, the log-prior is simply $-\log p(\boldsymbol{\theta}) \propto \lambda \|\boldsymbol{\theta}\|_1$, so the log-posterior is given by

$$\log p(\boldsymbol{\theta} \mid x, y) = \|\mathbf{Y} - \boldsymbol{\theta}\mathbf{X}\|_2^2 + \lambda \|\boldsymbol{\theta}\|_1.$$

The second term is the sparsity-promoting regularization term in the LASSO regression.

Example: Linear Regression with Gaussian Prior. Figure 7.15 shows the MLE and MAP estimates for the slope, a, and intercept, b, for a one-dimensional linear regression fit. If we know that the intercept is zero, $b = 0$, then this may be incorporated into the MAP estimate through a Gaussian prior $b \sim \mathcal{N}(0, \sigma^2)$, improving the regression fit.

The following code implements MLE and MAP estimation for this example.

CodeSnip 7.f

```
from scipy.optimize import minimize

# Parameters
true_a = 2 / 3
true_b = 0
sigma_noise = 1  # Standard deviation of Gaussian noise

# Generate random x and corresponding y with Gaussian noise
n_points = 20
x = np.random.uniform(0, 5, n_points)
noise = np.random.normal(0, sigma_noise, n_points)
y = true_a * x + true_b + noise
y[19] = y[19]+3 # outlier
```

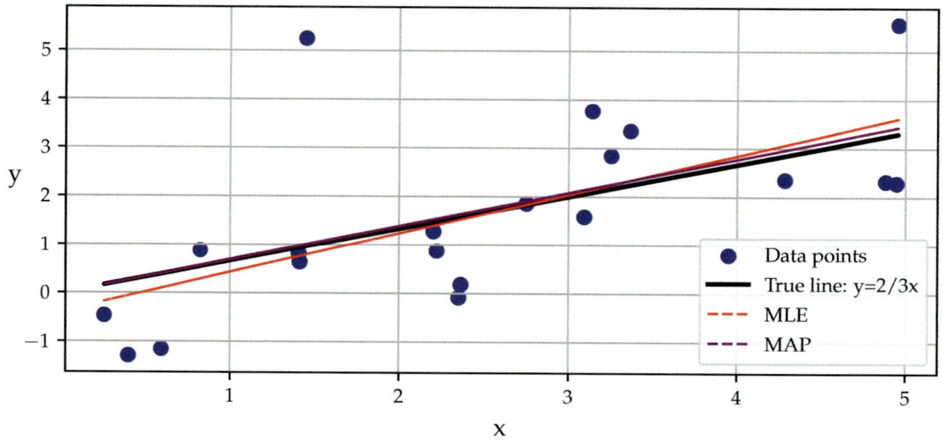

Figure 7.15 Comparison of statistical estimators in the least squares regression problem. Bayesian MAP estimation is the most accurate. The true parameters are $a = 2/3, b = 0$. The MLE is $a = 0.805, b = -0.374$, and the MAP estimate is $a = 0.694, b = -0.000$.

```
# Linear Least Squares Estimation
A = np.vstack([x, np.ones(len(x))]).T
theta_lsq, _, _, _ = np.linalg.lstsq(A, y, rcond=None)

# Maximum Likelihood Estimation (MLE)
def negative_log_likelihood(theta):
    a, b = theta
    y_pred = a * x + b
    residuals = y - y_pred
    return np.sum(residuals**2) / (2 * sigma_noise**2)

theta_mle = minimize(negative_log_likelihood, [1, 1]).x

# MAP Prior: b ~ N(0, sigma_prior^2) with sigma_prior small
sigma_prior_b = .01 # Strength of the prior on b
def negative_log_posterior(theta):
    a, b = theta
    log_prior_b = -0.5*(b**2/sigma_prior_b**2)  # Gaussian prior
        on b
    return negative_log_likelihood(theta) - log_prior_b

theta_map = minimize(negative_log_posterior, [1, 1]).x
```

Suggested Reading

Texts

(1) **Bayesian Theory**, by J. M. Bernardo, A. F. M. Smith, and M. Berliner, 1994 [47].
(2) **Bayesian Statistics: A Review**, by D. V. Lindley, 1972 [313].
(3) **Bayesian Inference in Statistical Analysis**, by G. E. P. Box and G. C. Tiao, 2011 [63].
(4) **Mathematical Statistics and Data Analysis**, by J. A. Rice, 2007 [414].
(5) **Markov Chain Monte Carlo in Practice**, by W. R. Gilks, S. Richardson, and D. Spiegelhalter, 1995 [191].

Papers and reviews

(1) **On information and sufficiency**, by S. Kullback and R. A. Leibler, *The Annals of Statistics*, 1951 [293].
(2) **A new look at the statistical model identification**, by H. Akaike, *IEEE Transactions on Automatic Control*, 1974 [4].
(3) **On the mathematical foundations of theoretical statistics**, by R. A. Fischer, *Philosophical Transactions of the Royal Society of London*, 1922 [172].
(4) **Inverse problems: A Bayesian perspective**, by A. M. Stuart, *Acta Numerica*, 2010 [476].
(5) **Bayesian inference in physics**, by U. Von Toussaint, *Reviews of Modern Physics*, 2011 [523].
(6) **Bayesian statistics and modelling**, by R. van de Schoot, S. Depaoli, R. King, B. Kramer, K. Märtens, M. G. Tadesse, M. Vannucci, A. Gelman, D. Veen, J. Willemsen, *Nature Reviews Methods Primers*, 2021 [512].
(7) **Optimal low-rank approximations of Bayesian linear inverse problems**, by A. Spantini, A. Solonen, T. Cui, J. Martin, L. Tenorio, and Y. Marzouk, *SIAM Journal on Scientific Computing*, 2015 [469].
(8) **Maximum likelihood from incomplete data via the EM algorithm**, by A. P. Dempster, N. M. Laird, and D. B. Rubin, *Journal of the Royal Statistical Society: B*, 1977 [134].

Homework

Exercise 7.1 Consider a fair coin that is flipped $N = 10$ times. What is the probability that the coin lands heads exactly 5 times? What is the probability that the coin lands heads at least 5 times? What is the probability that the coin lands tails no more than 3 times? What is the probability that the coin lands tails no more than 3 times in the first 5 coin flips? What about the last 5 coin flips?

Exercise 7.2 How many unique 5 card hands can you draw from a deck of 52 cards if the order of the cards doesn't matter? What about if the order does matter?

Exercise 7.3 How many unique p cards can you draw from a deck of n cards if the order of the cards doesn't matter? What about if the order does matter?

Exercise 7.4 You wake up in the morning in a dark room and you need to get dressed. You randomly pull two socks out of your dresser, which contains 10 black socks and 8 red socks. What is the probability that your two socks are the same color? What is the probability that both socks are red? That both socks are black?

Exercise 7.5 Write a Monte Carlo sampling code that throws darts at a square with dimensions $x, y \in [-1, 1]$. Keep track of all darts that hit inside the circle of radius 1. Compute the ratio of darts that hit inside the circle versus outside, and use this to estimate the area of the circle, and hence π. Plot your estimate of π as a function of n, the number of dart throws.

Exercise 7.6 Repeat the problem above one hundred times with different random initialization, and compute variance bars on the estimate of π.

Exercise 7.7 For how large of an n would we expect to estimate π to within 10 digits of accuracy? How did you arrive at this number?

Exercise 7.8 Now write a Monte Carlo sampling code that throws darts at a N-dimensional cube with dimensions $x_j \in [-1, 1]$. Throw $n = 100,000$ darts and compute the ratio of the volume of sphere of radius 1 with the volume of the enclosing cube. Plot this ratio for $N = 2$ through $N = 20$. Explain what this trend says about random higher-dimensional vectors $\mathbf{x}$. This is closely related to the notion of *concentration* in statistics.

Exercise 7.9 Show the second equality in the following expression

$$I(\theta) = \underbrace{\mathbb{E}\left[\frac{\partial}{\partial\theta}\log f(X\,|\,\theta)\right]^2}_{\mathbb{E}[\ell'(\theta)]^2} = \underbrace{-\mathbb{E}\left[\frac{\partial^2}{\partial\theta^2}\log f(X\,|\,\theta)\right]}_{-\mathbb{E}[\ell''(\theta)]}$$

which we introduced to explain the *normality* property of the maximum likelihood estimator.

Exercise 7.10 The kth moment of a distribution $f(x)$ is defined as

$$\mu_k = \mathbb{E}[X^k] = \int_{-\infty}^{\infty} x^k f(x)\,\mathrm{d}x.$$

Derive the moments for a normal distribution, $X \sim \mathcal{N}(\mu, \sigma^2)$. The first four moments are the mean, variance, skewness, and kurtosis.

Exercise 7.11 Derive the moments for a Poisson distribution $X \sim \text{Poisson}(\lambda)$.

Exercise 7.12 Prove that an even function will only have even moments.

Exercise 7.13 The method of moments is a statistical estimation technique, like the MLE, for fitting parameters of a distribution from data. In the method of moments we set empirical sample moments $\hat{\mu}_k = \frac{1}{n}\sum_{j=1}^{n} X_j^k$ equal to the moments of the distribution, written in the parameters of the distribution; we then invert and solve for the unknown parameters. For example, in a normal distribution, the mean and variance are related to the moments as $\mu = \mathbb{E}[X] = \mu_1$ and $\sigma^2 = \mathbb{E}[X^2] - \mathbb{E}[X]^2 = \mu_2 - \mu_1^2$. Use the method of moments to estimate λ in the Poisson example earlier in this chapter and verify it is the same as in MLE.

Exercise 7.14 The moment generating function (MGF) $M(t)$ is an important concept in probability. Much like functions can be expanded in a Taylor series, probability density functions (PDFs) can be expanded in terms of the MGF, which is defined as:

$$M(t) = \sum_x e^{tx} P(X = x)$$

for discrete X and as

$$M(t) = \int_{-\infty}^{\infty} e^{tx} f(x)\, \mathrm{d}x$$

for continuous X. The continuous version of the MGF is the the Laplace transform of the PDF. The kth moment μ_k is equal to the kth derivative of $M(t)$ evaluated at $t = 0$:

$$\mu_k = M^{(k)}(0).$$

Derive the moment generating function of the normal and Poisson distributions. Compute their moments using the MGF and verify they match the values from previous exercises.

Exercise 7.15 The central limit theorem (CLT) says that adding up independent, identically distributed (i.i.d.) random variables from the same distribution will result in a normally distributed random variable. This is a powerful result in probability and statistics.

For $X \sim \text{Uniform}[-\frac{1}{2}, \frac{1}{2}]$, draw $n \in \{5, 10, 30, 100\}$ i.i.d. samples X_j and compute the sample mean. Compute this average for 100 random realizations and plot the histogram of the sample mean. Confirm that for larger n, this distribution tends toward a normal distribution. What are the mean and standard deviation of the normal distribution for the sample mean? Repeat this for $X \sim \text{Bernoulli}(p = 0.3)$ and $X \sim \text{Exponential}(\lambda = 1)$.

Exercise 7.16 Generate $n = 200$ i.i.d. samples from an exponential distribution with $\lambda = 1$ and estimate with the MLE as $\hat{\lambda} = 1/\bar{X}$. Use the nonparametric bootstrap ($B = 1000$ resamples) to approximate the sampling distribution of $\hat{\lambda}$ and form a 95% percentile confidence interval. Compare to the confidence interval obtained via a normal approximation, $\sqrt{n}(\hat{\lambda} - \lambda) \approx \mathcal{N}(0, \lambda^2)$. Report coverage across 200 Monte Carlo trials.

Exercise 7.17 A coin is flipped 3 times and yields HHH. The MLE for p is $\hat{p} = 1$. Explain why this is problematic. Compute the MAP estimate under Beta(α, β) priors with $(\alpha, \beta) \in \{(1,1), (2,2), (20,20)\}$. Give the posterior mean and a 95% credible interval in each case. Discuss how prior strength mitigates small-sample issues.

Exercise 7.18 Continuing the previous problem, for Beta(α, β) prior and data with h heads and t tails, write the posterior $p(p \mid h, t)$ and plot it for $(h, t) = (3, 0)$ under the three priors above. Mark the posterior mean, mode (MAP), and 95% credible interval on the plot.

Exercise 7.19 To prove that the following values maximize the log-likelihood,

$$\frac{\partial \ell}{\partial \mu} = 0 \implies \hat{\mu} = \bar{X}$$

$$\frac{\partial \ell}{\partial \sigma} = 0 \implies \hat{\sigma}^2 = \frac{1}{n} \sum_{j=1}^{n} (X_j - \bar{X})^2,$$

compute their second derivatives and show that the curvature is positive.

Exercise 7.20 Suppose a coin has an unknown bias p of landing heads. We model p with a Beta prior Beta$(\alpha = 2, \beta = 2)$. After observing h heads and t tails, derive the posterior distribution for p. Plot the posterior for varying h and t and interpret.

Exercise 7.21 What is the conjugate prior to the Poisson distribution? Show that these distributions are conjugate.

Exercise 7.22 Show that a binomial distribution may be written in the exponential family of distributions.

Exercise 7.23 Show that a Poisson distribution may be written in the exponential family of distributions.

Exercise 7.24 Show that a Gaussian prior is conjugate to a Gaussian likelihood. Specifically, show that the product of these two Gaussian distributions remains a Gaussian.

Exercise 7.25 Design a Bayesian update scheme to iteratively estimate the parameters of an unknown Gaussian with new data. Test your algorithm by generating data from a known Gaussian, with specified mean μ and variance σ^2. Initialize your prior distribution with some different values of μ and σ^2, and update these parameters with new data. Show the convergence rate of μ and σ^2 with more data. Repeat the process several times, and plot the distributions of the estimated μ and σ^2 after n iterations for $n = 5, 10, 25, 50, 100$.

8 Optimization for Inverse Problems

Inverse problems involve inferring unknown parameters and/or inputs of a system from partial observations. Control theory and machine learning are both broad classes of inverse problem, and they are so important as to warrant Chapters 9 and 10 being devoted to them. There are many other examples of inverse problems in the modern world. Two of the most compelling and tangible examples are in medical imaging and computational cosmology.

Noninvasive medical imaging technologies, such as computed tomography (CT) scans and magnetic resonance imaging (MRI) make it possible to see inside the human body without damaging tissue through invasive surgery. This is fundamentally an inverse problem, and has revolutionized healthcare. These technologies image the body, typically via measuring distortion of electromagnetic radiation passing through a line, or a ray, through the body. This data is then used to reconstruct a three-dimensional model, for example via the Radon transform. Medical imaging has saved millions of lives, enabling early detection of cancer, heart disease, and brain injuries. These tools allow doctors to diagnose conditions, plan surgeries with precision, and monitor treatment progress.

Astronomy provides one of the oldest and most profound inverse problems: inferring our place in the Universe from imperfect and partial measurements, first from Earth with the naked eye, then with ground-based telescopes, and most recently with satellites and space probes. Much of our early mathematics and physics was discovered to improve the fit of models to observed data. For example, Kepler's laws, and later Newton's, were developed to fit partial data collected by Tycho Brahe. Later Einstein's theory was validated on partial measurements of the transit of Mercury. These laws now provide the foundation of our understanding of the Universe. Recently, a black hole was imaged for the first time with the Event Horizon Telescope, combining data from radio telescopes around the Earth to act as one giant telescope.

The topic of inverse problems could easily be an entire book, and so this chapter only provides a high-level overview of the field, with an emphasis on how to pose them as an optimization problem. Most inverse problems are fundamentally ill-posed, and so they involve regularization, often in the form of physics and priors. Thus, this chapter also benefits naturally from the Bayesian perspective in Chapter 7, as the two are closely related. However, inverse problems are much more general, and Bayesian inverse problems are a special class, so they are presented as separate but related topics here.

8.1 Mathematical Formulation

Inverse problems seek to infer unknown parameters or inputs that best match observed data, which is typically represented by an equation or system where the solution minimizes the discrepancy between observed and modeled data. Thus the goal is to find $\mathbf{x}$ such that

$$\mathbf{F}(\mathbf{x}) \approx \mathbf{y}_{\text{obs}}. \tag{8.1}$$

We typically reframe inversion as an optimization problem:

$$\min_{\mathbf{x}} ||\mathbf{F}(\mathbf{x}) - \mathbf{y}_{\text{obs}}||^2. \tag{8.2}$$

Here, $\mathbf{F}(\mathbf{x})$ is the forward model, and $\mathbf{x}$ are the unknown parameters. When the model $\mathbf{F}$ is linear, as in linear regression from Chapter 4 or linear optimal control in Chapter 9, then the optimization problem is convex, and we may use standard regression techniques.

Often, the observations are partial and noisy, and thus don't contain enough information to uniquely determine the unknown parameters, making the inverse problem *ill-posed* or *ill-conditioned*. Thus, we often add constraints or regularization penalties $R(\mathbf{x})$ to ensure the solution is well-posed and physical:

$$\min_{\mathbf{x}} ||\mathbf{F}(\mathbf{x}) - \mathbf{y}_{\text{obs}}||^2 + \lambda R(\mathbf{x}). \tag{8.3}$$

The regularization term $R(\mathbf{x})$ may have physical meaning, as encoded by a Bayesian prior (e.g., $||\mathbf{x}||_1$ for sparsity or $||\mathbf{x}||_2^2$ for smoothness). We have already examples of this in the case of least squares in Chapter 4, where we introduced the Tikhonov regularization to improve the conditioning of the matrix inversion. Thus, our ability to solve ill-posed and nonconvex optimization problems with regularized and relaxed formulations also improves inverse modeling.

There are a number of challenges associated with inverse problems, including ill-posedness, model uncertainty, and complexity of the forward model $\mathbf{F}$. We will try to give a flavor for these issues, with examples, and give an idea of how they are handled in practice. Importantly, there is often a mismatch between the forward model $\mathbf{F}$ and reality. For example, when optimizing the performance of an aircraft, we may rely on simulations of the aerodynamics, which only approximately match the true physics. Ill-conditioned inverse problems will be highly sensitive to these model discrepancies, necessitating careful uncertainty quantification. Aerodynamic shape optimization is also an example where the model $\mathbf{F}$ is both high-dimensional and nonlinear, which typically leads to a highly challenging, nonconvex optimization problem. Typically for these problems we use adjoint methods to approximate gradients of output quantities of interest with respect to tunable inputs, and then use constrained gradient-based optimization to find locally optimal solutions. This approach requires a good initial guess, as the problem is nonconvex.

These challenges are becoming tractable because of progress in machine learning and advanced optimization techniques. Machine learning provides improved forward models $\mathbf{F}$, that are both fast to evaluate and have computable gradients, from an increasing wealth of data. It is also possible to incorporate partial physics into these data-driven models, improving their generalization, and acting to regularize the problem for noisy and partial measurements.

8.2 Classic Inverse Problems

The line between optimization and inverse problems is especially blurred for many classic inverse problems. Here, we briefly cover several of these classical inverse problems and point out their connection to optimization. For example, Newton's method is often employed specifically to find the roots of a function, which may be viewed as an inverse problem. If Newton's method is used to find the roots of $\nabla f(\mathbf{x})$, then this is an optimization problem.

Root Finding

Finding the roots of a function $f(x) = 0$ is a deceptively challenging problem and it has a rich history in mathematics. In fact, all of complex analysis and much of modern algebra was borne out of the problem of solving for the roots of a polynomial $f(x)$.

Given a polynomial function $f(x) = x^d + c_{d-1}x^{d-1} + \cdots + c_2x^2 + c_1x + c_0$ of order d, we have well known expressions for the roots when $d \leq 4$ (e.g., the quadratic formula for $d = 2$, Cardano's formula for $d = 3$, and solutions by Ferrari and Cardano for $d = 4$) [94]. However, for $d \geq 5$ there is no general closed-form solution for the roots, making this one of the classic irreducibility results in mathematics[1].

Newton's method was originally posed to solve for the roots of a function, through numerical iteration. If instead of the roots of $f(x)$ we consider the roots of the derivative $df/dx = 0$, or more generally $\nabla f(\mathbf{x}) = \mathbf{0}$, then we have the central problem of finding local minima, maxima, or saddle points in optimization. Newton's method is typically not the best algorithm for finding roots of high-order polynomials, as it requires a good initial guess. For high-order polynomials $f(x)$, it is common to construct a companion matrix $\mathbf{C}$

$$\mathbf{C} = \begin{bmatrix} 0 & 1 & 0 & \cdots & 0 \\ 0 & 0 & 1 & \cdots & 0 \\ \vdots & \vdots & \vdots & \ddots & \vdots \\ 0 & 0 & 0 & \cdots & 1 \\ -c_0 & -c_1 & -c_2 & \cdots & -c_{d-1} \end{bmatrix}, \tag{8.4}$$

with characteristic polynomial $\det(\mathbf{C} - x\mathbf{I}) = f(x)$. We then solve for the eigenvalues of $\mathbf{C}$, which are the roots of $f(x) = 0$, using numerically stable techniques from linear algebra.

Numerical linear algebra techniques, such as QR factorization [199, 202, 500] are optimized for eigenvalue computations. They scale to high dimensions d, are numerically stable, and are included in linear algebra packages such as LAPACK [7] and NumPy [218].

Matrix Inversion

Many physical systems may be represented as a linear system of equations,

$$\mathbf{A}\mathbf{x} = \mathbf{b}, \tag{8.5}$$

[1] The irrepresentability of these roots with radicals is closely related to Galois theory, which is a cornerstone of modern abstract algebra [157, 13, 474].

where the constraint matrix $\mathbf{A}$ and vector $\mathbf{b}$ are known, and the vector $\mathbf{x}$ is unknown. Given observations $\mathbf{b}$ and a model $\mathbf{Ax} = \mathbf{b}$, the inverse problem is to solve for the $\mathbf{x}$ that is consistent with this observation. Matrix inversion is one of the most central themes in both optimization and inverse problems.

If $\mathbf{A}$ is a square, invertible matrix (i.e., $\mathbf{A}$ has non-zero determinant), then there is a unique solution $\mathbf{x}$ for every $\mathbf{b}$. However, when $\mathbf{A}$ is either singular or rectangular, there may be one, none, or infinitely many solutions, depending on the specific $\mathbf{b}$ and the column and row spaces of $\mathbf{A}$. Further, depending on the condition number of $\mathbf{A}$ the inversion may be ill-posed. We have already seen several examples of optimization solutions to matrix inversion problems in Chapter 4, including the least squares regression in Section 4.1 and the Tikhonov regularized least-squares solution in Section 4.4 for ill-conditioned problems.

Numerical Differentiation

Numerical differentiation and integration of noisy data may also be posed as optimization problems. Numerical differentiation is particularly challenging, as it typically amplifies noise. There are several optimization approaches to numerical differentiation that may be more or less suitable based on the assumptions about the noise and the underlying signal.

Given a noisy signal $\mathbf{y} = [y_1, y_2, \ldots, y_n]$, where each $y_j = y(x_j)$, we seek to approximate the derivative $\frac{dy}{dx}$ at each of these points $x_1, x_2, \ldots, x_n$. We will denote this derivative vector by $\mathbf{d}$, so that $d_j \approx \frac{dy}{dx}(x_j)$. For simplicity, we will assume that the data are evenly spaced, so that $x_{j+1} - x_j = \Delta x$ for all j.

Numerical differentiation is often accomplished using the method of finite-differences, for example approximating $d_j = \frac{1}{2\Delta x}(y_{j+1} - y_{j-1})$ using a central difference for $1 < j < n$ and $d_1 = \frac{1}{\Delta x}(y_2 - y_1)$ and $d_n = \frac{1}{\Delta x}(y_n - y_{n-1})$ using forward and backward differences at $j = 1$ and $j = n$, respectively. This may be written in terms of a matrix operator $\mathbf{D}$:

$$\mathbf{d} = \mathbf{D}\mathbf{y} = \frac{1}{2\Delta x}\begin{bmatrix} -2 & 2 & 0 & 0 & \cdots & 0 & 0 & 0 \\ -1 & 0 & 1 & 0 & \cdots & 0 & 0 & 0 \\ 0 & -1 & 0 & 1 & \cdots & 0 & 0 & 0 \\ \vdots & \vdots & \vdots & \vdots & \vdots & \vdots & \vdots & \vdots \\ 0 & 0 & 0 & 0 & \cdots & -1 & 0 & 1 \\ 0 & 0 & 0 & 0 & \cdots & 0 & -2 & 2 \end{bmatrix}\mathbf{y}.$$

Because we divide by Δx, any noise or imperfection on the data $\mathbf{y}$ is amplified.

Given the derivative $\mathbf{d}$ and y_0, we may write the integration operator in matrix form as

$$\mathbf{y} = \mathbf{H}\mathbf{d} = y_0 + \Delta x\begin{bmatrix} 0 & 0 & 0 & \cdots & 0 & 0 & 0 \\ 1 & 0 & 0 & \cdots & 0 & 0 & 0 \\ 1 & 1 & 0 & \cdots & 0 & 0 & 0 \\ \vdots & \vdots & \vdots & \vdots & \vdots & \vdots & \vdots \\ 1 & 1 & 1 & \cdots & 1 & 0 & 0 \\ 1 & 1 & 1 & \cdots & 1 & 1 & 0 \end{bmatrix}\mathbf{d}.$$

This is a matrix implementation of the simple trapezoidal integration rule.

To improve the robustness of numerical differentiation for noisy data, we pose it as an optimization problem with regularization, for example with Tikhonov regularization:

$$\mathbf{d}^* = \arg\min_{\mathbf{d}} \tfrac{1}{2}\|\mathbf{y} - \mathbf{H}\mathbf{d}\|_2^2 + \lambda\|\mathbf{d}\|_2^2. \tag{8.6}$$

Here we attempt to find the vector of derivatives $\mathbf{d}$ that best integrates to the data $\mathbf{y}$, while also having small values.

Total variation (TV) regularization [100] is another popular approach for robust numerical differentiation of noisy signals. It minimizes the trade-off between fidelity to the observed data and smoothness of the derivative, reducing the amplification of noise typically associated with differentiation. The TV-regularized derivative estimate solves:

$$\mathbf{d}^* = \arg\min_{\mathbf{d}} \tfrac{1}{2}\|\mathbf{y} - \mathbf{H}\mathbf{d}\|_2^2 + \lambda\|\mathbf{d}\|_{\mathrm{TV}}. \tag{8.7}$$

The total variation norm is given by $\|\mathbf{d}\|_{\mathrm{TV}} = \sum_{j=1}^{n-1} |\mathbf{d}_{j+1} - \mathbf{d}_j|$. The parameter $\lambda > 0$ controls the balance between the data fidelity term and the total variation regularization term. Note that the TV norm is closely related to the forward difference derivative operator $\mathbf{D}_{\mathrm{FD}}$. This method is well-suited for handling piecewise-smooth signals and mitigating noise amplification in derivative estimation.

A number of numerical differentiation schemes may be found in the open source repository of Van Breugel et al. [511].

Smoothing and Filtering Noisy Data

Numerical differentiation may be viewed as a special case of smoothing or filtering noisy data. These are closely related, as smoothing is often the first step in computing a numerical derivative. There are several approaches to smoothing noisy data, and only a few methods are discussed here.

It is common to smooth noisy data by fitting it locally to a smooth spline or other polynomial fit. Depending on the amount of data, this may be accomplished in one of several ways. For large amounts of data, it is possible to take a subset of the data for fitting, where we use a least-squares regression to fit the coefficients of a polynomial model, and the rest is used for cross-validation. This prevents overfitting, and makes it possible to determine the correct order of the fit. If the underlying model is nonlinear in the fitting coefficients, we may use nonlinear least-squares, as discussed in Chapter 4.

Savitzky–Golay smoothing [436] is a particularly useful technique, where we fit local polynomials to the data. We may then compute the analytic derivative of these polynomial fits to obtain a smoothed derivative. There are several hyperparameters, including the polynomial order, the width of the local window, and the shape of the weighting function. It is possible to formulate this in terms of a generalized linear regression, which is also amenable to regularization with closed-form solution.

If there is a model of the underlying dynamical process generating the data, such as a linear state-space ordinary differential equation, then it is possible to use a Kalman filter [266, 473], which will be discussed more in Chapter 9.

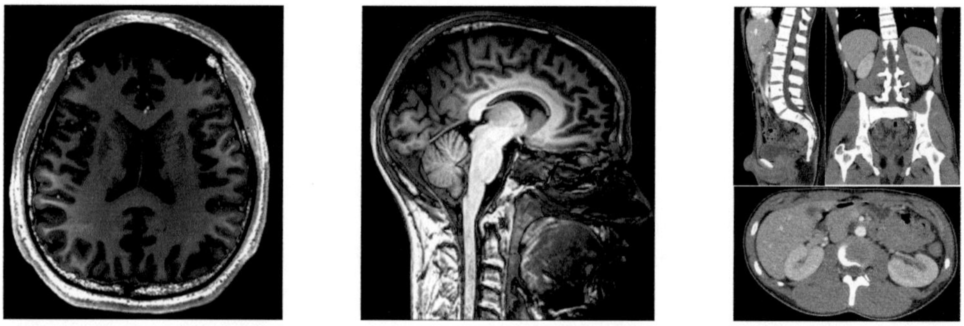

Figure 8.1 Computational imaging of the human body is an important inverse problem. Left: MRI of a human brain, image courtesy of Asnaebsa. Center: MRI of brain sagittal section, image courtesy of Chiswick Chap. Right: CT scan of normal abdomen and pelvis, image courtesy of Mikael Häggström.

8.3 Inverse Problems in Medical Imaging

One of the largest and richest classes of inverse problems involves reconstructing or improving image data from partial, noisy, or indirect measurements. The foundations of inverse problems in imaging sciences were laid well before the advent of computers and digital imaging. The field of inverse problems in imaging sciences has its roots in the 19th century, with significant developments occurring in the mid-20th century, marking the beginning of the modern era. Early contributions stem from efforts in astronomy, geophysics, and medical imaging, as researchers sought to infer hidden or indirect quantities from observable data. World War II accelerated the development of imaging technologies, such as radar and sonar, as well as related computational tools.

Two key areas that shaped the field were the mathematical work of Radon and early developments in Fourier optics. Despite their theoretical origins, both areas have become central to modern medical imaging techniques, such as X-ray computed tomography (CT) and magnetic resonance imaging (MRI); see Fig. 8.1 for examples of MRI and CT images. We will pin this discussion to problems in medical imaging, which presents some of the most impactful applications of image-based inversion. The goal is to reconstruct internal structures of the human body from indirect measurements, enabling non-invasive diagnostics. These medical applications have since driven tremendous progress in practical inversion methods.

Computed Tomography (CT) (1960s–1970s). Godfrey Hounsfield and Allan Cormack independently developed methods to reconstruct cross-sectional images from X-ray data. Their work culminated in the invention of the first CT scanner in 1972 [114, 239], which resulted in them receiving the 1979 Nobel Prize in Medicine for their contributions.

CT technology relied on the Radon transform, which was developed as a mathematical algorithm half a century earlier. At the beginning of CT, it was a major challenge to invert the Radon transform in real time on computers of the era. Techniques such as filtered backprojection (FBP) were introduced to invert the Radon transform approximately. Iterative methods, such as algebraic reconstruction techniques (ART), were also developed to handle noise and incomplete data. The Kaczmarz algorithm [260] from 1937 is an iterative method for solving linear systems of equations, and decades later it became central to tomographic reconstruction. This will be a common theme in this chapter, that mathematical progress often led technological progress by decades.

Magnetic Resonance Imaging (MRI) (1970s). MRI enables the non-invasive reconstruction of high-resolution images inside the human body by measuring radio-frequency signals emitted by hydrogen nuclei excited by a strong external magnetic field. Lauterbur and Mansfield pioneered the development of MRI, earning them the 2003 Nobel Prize in Medicine [301, 331].

MRI measurements are inherently in the frequency domain, providing Fourier coefficients. Thus, the Fourier transform, developed 150 years earlier, played a critical role in reconstructing images from frequency-domain MRI data. The main early challenges included sparse sampling in k-space and noise robustness. Recent advances have included reducing the measurement time through compressed sensing techniques.

Challenges. The birth of the field was marked by a series of profound challenges, many of which remain relevant today:

Ill-posedness: In 1902, Jacques Hadamard defined the three criteria for well-posed problems: one, that a solution exists; two, that the solution is unique; and three, that the solution depends continuously on the data. Inverse imaging problems often violate one or more of these conditions, necessitating regularization techniques.

Data Sparsity: Early imaging systems collected incomplete or noisy data, making reconstruction mathematically challenging.

Computational Complexity: The lack of computational tools in the early 20th century hindered the practical application of inverse methods. Algorithms like ART and FBP were computationally intensive before the advent of modern computers.

Noise Sensitivity: Noise in the measurement data is often amplified in the reconstructed image.

Nonlinearity: Many imaging problems, especially in MRI or ultrasound, are inherently nonlinear, resulting in expensive, nonconvex optimization problems with complex loss landscapes.

Both CT and MRI have benefited tremendously from modern advances in computational imaging and machine learning. Deep learning, in particular, is making it possible to accelerate computation by replacing iterative optimization with fast neural network inference. Further, it is possible to learn priors so that models may be trained to encode anatomical priors or expected structures, improving robustness to noise and under-sampling. These techniques allow for reduced radiation doses in CT and shorter scan times in MRI, improving patient outcomes and accessibility.

The Radon Transform and X-Ray Computed Tomography

In 1917, Johann Radon introduced the integral transform that bears his name, providing a mathematical basis for reconstructing functions from their integrals along lines or planes [401]. The name of his paper translates to "*On the determination of functions from their integral values along certain manifolds*" and it addressed a purely mathematical question. Radon was not motivated by practical imaging applications as we understand them today, but rather by theoretical problems in integral geometry and harmonic analysis. However, this work laid the foundation for CT and other imaging modalities, which were developed many decades later.

Radon's problem was to determine whether a function $f(x, y)$ could be uniquely reconstructed from its integrals along straight lines. In modern applications, the function $f(x, y)$ might indicate the density of a material, such as bone or tissue, that is being imaged with X-rays. The X-rays would pass through this material in straight lines, causing them to be attenuated by different amounts based on the integrated attenuation properties of the material along different paths. These line integrals of the attenuation field, measured via X-ray transmission, would then be measured on the other side of the object.

A specific line may be specified by s, its perpendicular distance to the reference origin $(0, 0)$, and θ, the angle of inclination of the line with respect to the x-axis. The line will then be parameterized by a path length variable z, resulting in $(x(z), y(z)) = (z\sin(\theta) + s\cos(\theta), -z\cos(\theta) + s\sin(\theta))$. These integrals form the Radon transform:

$$\mathcal{R}(f)(s,\theta) = \int_{-\infty}^{\infty} f(x(z), y(z))\, dz = \int_{-\infty}^{\infty} f(z\sin(\theta) + s\cos(\theta), -z\cos(\theta) + s\sin(\theta))\, dz.$$

Here $\mathcal{R}(f)$ is the Radon transform of f, which represents the integral of $f(x, y)$ along the line parameterized by (s, θ). This is shown in Fig. 8.2. Radon proved that a function $f(x, y)$ could be uniquely reconstructed from its line integrals under certain conditions. He also provided an explicit inversion formula, now fundamental to many imaging techniques. These transforms are closely related to the Fourier and inverse Fourier transforms.

Radon anticipated the applied challenges of ill-posedness and inversion complexity. The problem is typically ill-posed, since small errors in the measured integrals can lead to large errors in the reconstructed function. Further, Radon's explicit inversion formula, while theoretically exact, was not computationally feasible before modern computers.

Inverse Radon Transform as an Optimization Problem

There is an exact, analytic expression for the inverse Radon transform, going from $\mathcal{R}(f)(s,\theta)$ to $f(x, y)$. This is known as filtered backprojection (FBP). However, FBP is similar to numerical differentiation, making it ill-conditioned for noisy measurements or discontinuous functions $f(x, y)$. Since X-rays damage tissue, it is important to keep radiation dosages low, resulting in inherently noisy measurements. Further, interfaces between bone, tissue, and organs are all approximately discontinuous.

Thus, similar to numerical differentiation above, it is common to pose the inversion of the Radon transform as an optimization problem. Specifically, we seek the function $f(x, y)$ that minimizes the discrepancy between the observed Radon transform data $\mathcal{R}_{\text{obs}}(s, \theta)$

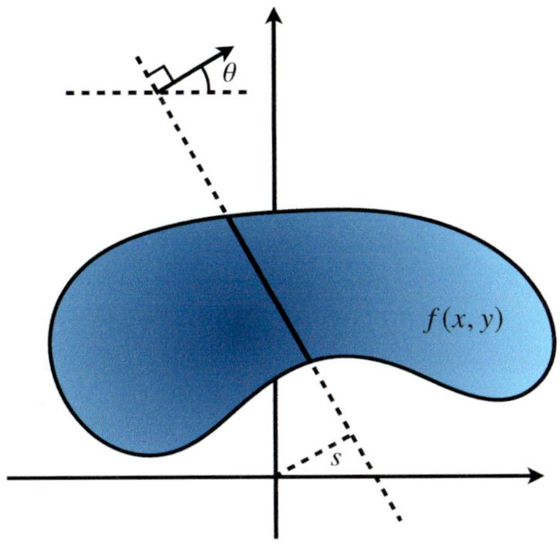

Figure 8.2 Coordinate system used to compute the Radon transform. The Radon transform is accomplished by integrating $f(x, y)$ along lines parameterized by s and θ, the shortest distance of the line to the origin, and its inclination angle from the x-axis, respectively.

and the forward Radon transform $\mathcal{R}(f)(s, \theta)$ of the reconstructed function. In addition, regularization is added to improve the inversion.

The specific algorithms are involved and are not particularly intuitive or enlightening, so we omit them here. However, the rough optimization problem is written as

$$\min_{\mathbf{x}} \|\mathcal{R}\mathbf{x} - \mathbf{y}\|_2^2 + \lambda R(\mathbf{x}). \tag{8.8}$$

Note that we have discretized the function $f(x, y)$ and the Radon transform $\mathcal{R}(f)(s, \theta)$ into the vectors $\mathbf{x}$ and $\mathbf{y}$, respectively. The operator $\mathcal{R}$ represents the discretized forward Radon transform, and $R(\mathbf{x})$ is the regularization term to address noise and ill-posedness.

Common regularization techniques include Tikhonov, total variation, and sparsity. It is also possible to impose smoothness, either in a transformed domain, such as a Fourier or wavelet basis, or by penalizing the Laplacian $|\nabla^2 f(x, y)|^2$. Finally, if prior knowledge about $f(x, y)$ is available (e.g., physical constraints, anatomical priors), regularization terms can encode this. The choice of regularization depends on the nature of the problem (e.g., noise type, expected smoothness) and the desired features of the reconstruction (e.g., sharp edges, sparsity). Further considerations include computational efficiency and robustness. The regularization strength λ is typically selected via cross-validation or other standard criteria.

Iterative methods, such as algebraic reconstruction techniques (ART), improve robustness but are computationally expensive. Deep learning approaches now incorporate priors and learned features to enhance reconstruction quality while accelerating computation.

Figure 8.3 shows the Radon transform and its inverse, applied to a simple image. The input image, referred to as a phantom, consists of a gray oval containing a circle, a square, and a triangle. Although the Radon transform yields gray-scale images corresponding to

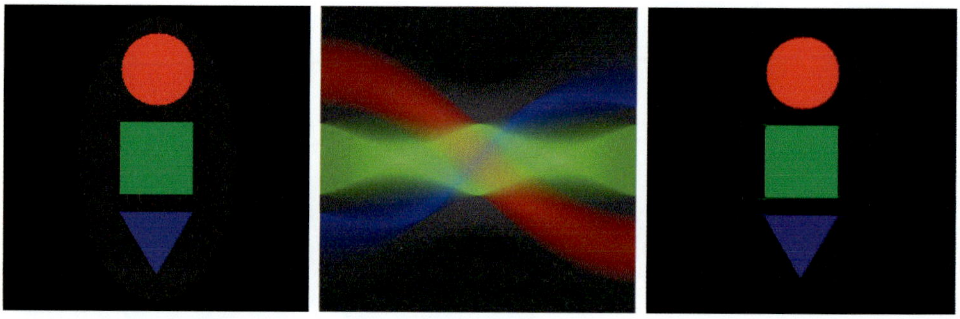

Figure 8.3 Example of Radon transform applied to simple phantom image (left). The Radon transform (middle) is inverted to achieve the reconstructed image (right). The objects are colored red, green, and blue to help illustrate which objects map to which features in the transform.

scalar intensities, here we color the circle red, the square green, and the triangle blue to make it easier to see the signature of these shapes in the transformed image.

The Fourier Transform and Magnetic Resonance Imaging

Fourier optics emerged in the 19th century as part of the broader development of wave theory, particularly the work of Joseph Fourier (1822) and Augustin-Jean Fresnel on diffraction. This field provides a framework for understanding the propagation and transformation of light as a wave, forming the basis for many imaging techniques. Inverse problems in optics are much more general than MRI, so we will briefly discuss here before focusing in on medical applications.

The central insight in Fourier optics is that a lens can act as a Fourier transform operator. When light passes through an aperture, it produces a light intensity pattern, known as a diffraction pattern, which is essentially the Fourier transform of the aperture's shape. Typically diffraction patterns are recorded after light interacts with an object or aperture. Before modern CCD cameras, light intensity patterns were captured using photographic plates or screens. The patterns recorded on a screen or photographic plate represent the Fourier transform magnitude of the object function. These patterns serve as the observable data for the inverse problem. The output is then the reconstructed object function $f(x, y)$, which could represent the structure of an aperture or an object.

The relationship between the object function f and the Fourier intensity is given by

$$I(k_x, k_y) = |\mathcal{F}[f(x, y)]|^2,$$

where $I(k_x, k_y)$ is the Fourier intensity pattern in the frequency domain, and $\mathcal{F}[f]$ is the Fourier transform of the object function $f(x, y)$.

The key challenge in Fourier optics is to reconstruct the object function $f(x, y)$ from its Fourier transform's intensity pattern. However, the intensity pattern only provides the magnitude of the Fourier transform, not the phase, making the problem nonlinear. Techniques like holography, developed in the 1940s, provided ways to recover phase information by recording interference patterns. Phase retrieval is still a challenge today, although most approaches involve computationally solving a nonlinear inverse problem . Diffraction patterns may also be noisy, requiring regularization or additional constraints for robust reconstruction. The Fourier transform has since become a foundation for many frequency-based imaging techniques, such as MRI.

Magnetic Resonance Imaging

MRI aims to reconstruct high-resolution images of internal structures by measuring radio frequency signals emitted by hydrogen nuclei in the body under a magnetic field. Unlike X-ray CT, MRI is non-ionizing, meaning it doesn't damage tissue, and it relies on frequency-domain measurements.

The raw input data in MRI are Fourier coefficients of the spatial image, collected in k-space, where k is a spatial wavenumber. The measured signals are Fourier coefficients, because the radio-frequency emissions from spinning hydrogen nuclei in a magnetic field are detected as oscillating electromagnetic waves. The frequencies of these waves correspond to spatial variations in the magnetic field, and the inverse Fourier transform naturally maps these frequency-domain measurements in k-space to spatial-domain images. This connection arises from the physics of nuclear magnetic resonance (NMR) and the spatial encoding introduced by gradient fields.

The forward model is given by the Fourier transform, which describes the relationship between the image $\mathbf{x}$ and the measured data $\mathbf{y}$

$$\mathbf{y} = \mathbf{F}\mathbf{x} + \mathbf{n}, \tag{8.9}$$

where $\mathbf{F}$ is the discrete Fourier transform operator, and $\mathbf{n}$ is noise.

The Fourier transform operator is unitary and, in general, inversion is fairly well-conditioned. However, it still helps to regularize the reconstruction problem:

$$\min_{\mathbf{x}} ||\mathbf{F}\mathbf{x} - \mathbf{y}||_2^2 + \lambda||\mathbf{F}\mathbf{x}||_1, \tag{8.10}$$

where $||\mathbf{F}\mathbf{x}||_1$ enforces sparsity in the Fourier transform of the image. Nearly all natural images are sparse in some transform basis [77], which was the principle behind compressed sensing in Section 4.3; see Panel 8.1 for details on compressed sensing in MRI. Unlike the more general problem of inversion in Fourier optics above, MRI measurements typically involve complex amplitude measurements, so phase retrieval from intensities is not a problem here.

Deep learning approaches, such as convolutional neural networks and generative adversarial networks, further enhance reconstruction speed and quality, leveraging learned priors and end-to-end optimization. Recall that the regularization terms $R(\mathbf{x})$ typically come from a prior distribution on $\mathbf{x}$, and most of the named regularization terms (e.g., Tikhonov, LASSO, etc.) come from named distributions (e.g., Gaussian, Laplace, etc.). Real-world data, like anatomical images from MRI, absolutely do follow probability distributions, but these distributions are complex and do not have names or simple parameters. Thus, learning these prior distributions from a wealth of training data, and using these priors for regularizing inverse problems, is a major advancement enabled by machine learning. These learned priors and regularizers can enable super-resolution imaging from fewer measurements.

8.4 Inverse Problems in Computational Imaging

There are several modern computational image inversion techniques that may be considered as ill-posed inverse problem, where regularization is required. These techniques have rapidly become machine learning problems, with deep learning being the method of choice. We will discuss the fundamental inverse optimization problem here and provide some general approaches; a more detailed discussion of machine learning will be provided in Chapter 10.

We will describe four basic tasks here: super-resolution, which reconstructs a high-resolution image from a low-resolution image; inpainting or occlusion removal, which fills in missing or damaged portions of an image; deblurring, which recovers a crisp image from a blurred image; and change of perspective, where we reconstruct an image from another camera angle. The first three types of images are shown schematically in Fig. 8.4.

Panel 8.1: Compressed Sensing for Faster MRI

The resolution of the image $\mathbf{x}$ is related to the resolution of k-space sampling in $\mathbf{y}$, and increasing the resolution of $\mathbf{y}$ increases the measurement time. Increased measurement time requires a patient to be very still for longer, and for infants this often involves anesthesia, which may have negative outcomes.

Therefore, modern MRI technologies seek to bypass measuring a full, high-resolution sampling of k-space, and instead measure a subset of frequencies k, given by the measurement vector $\mathbf{Cy}$, where $\mathbf{C} \in \mathbb{R}^{p\times n}$ is formed from $p \ll n$ rows of the identity matrix, indicating that only those p indices of $\mathbf{y}$ are sampled. In principle, this results in a severely underdetermined linear system to reconstruct the high-resolution image $\mathbf{x} \in \mathbb{R}^n$.

However, advances in compressed sensing [146, 92, 93] have made it possible to solve this ill-posed inverse problem with high-probability, in some instances, assuming that $\mathbf{x}$ is sufficiently sparse or compressible in an appropriate transform domain (e.g., wavelets), while the measurements are taken in Fourier space. In this case, it is possible to solve the following optimization problem

$$\min_{\mathbf{x}} \|\mathbf{CFx} - \mathbf{Cy}\|_2^2 + \lambda\|\mathbf{Fx}\|_1.$$

We may solve this compressed sensing problem using the methods discussed in Sections 5.1 and 6.5. Compressed sensing MRI [325] is now FDA-approved in several commercial scanners, and has reduced MRI scan times by factors of 2–10 in practice, enabling faster, higher-resolution, and safer imaging.

Figure 8.4 Examples of inverse problems related to image processing. From left to right: original high-resolution image; low-resolution image (goal: super resolution); occluded image (goal: inpainting and/or repair); blurred image (goal: de-blur).

Super Resolution

The goal of super resolution is to take one or more low-resolution images $\mathbf{y}$ and recover a high-resolution image $\mathbf{x}$. Applications include satellite and biomedical imaging, consumer photography, and video surveillance.

This task sounds like science fiction, as there are infinitely many high-resolution images that are consistent with a given low-resolution image. However, there is statistical structure in the space of natural images, and some high-resolution features are much more likely than others. For example, if I take off my glasses, faces appear more blurry, but I am still able to infer that eyebrows are comprised of individual hairs and that there is fine iris structure in an eye. This is because my neural network has learned these statistical features, and seamlessly fills in details. With advances in optimization and machine learning, it is becoming increasingly possible to automatically learn these statistical distributions and achieve impressively convincing super resolution.

A common single-image super-resolution model first blurs the high-resolution image $\mathbf{x}$ with an optical point-spread function $\mathbf{H}$, then down-samples by a factor s via a decimation operator $\mathbf{D}$, and finally adds sensor noise $\mathbf{n}$:

$$\mathbf{y} = \mathbf{DHx} + \mathbf{n}. \tag{8.11}$$

The super-resolution inverse optimization problem may then be stated as

$$\min_{\mathbf{x}} ||\mathbf{DHx} - \mathbf{y}||_2^2 + \lambda R(\mathbf{x}). \tag{8.12}$$

Similar to the deblurring problem, the blur filter might be assumed to be unknown, in which case it may need to be learned as well, resulting in a highly nonconvex blind super-resolution problem. There are also multi-frame versions that incorporate multiple low-resolution images to improve the super-resolution reconstruction.

Modern deep learning approaches [145, 528, 305, 559] now dominate super resolution performance, although there are also impressive classical approaches [177, 99, 544]. Today, CNNs, GANS, and transformers are widely used for super resolution. These techniques are also being used beyond image sciences, for example to provide sub-grid-scale models for turbulent fluid flows [180, 181].

Repair Damage and Fill Occlusions: Image Inpainting

Image inpainting aims to reconstruct missing or damaged regions of an image. This includes applications such as restoring damaged artwork, filling in corrupted image files, and removing objects from photographs.

Consider an observed image $\mathbf{y}$ with missing or damaged pixels specified by a binary mask $\mathbf{M}$; the entries 1 correspond to good pixels, and the entries 0 correspond to bad pixels. The goal is to reconstruct the entire image $\mathbf{x}$, filling in the values of the missing pixels.

The forward model is simply that the observed pixels match the reconstructed image:

$$\mathbf{M} \odot \mathbf{x} = \mathbf{M} \odot \mathbf{y}, \tag{8.13}$$

where the symbol $\odot$ is the Hadamard product, which is the element-wise multiplication of two matrices or vectors of the same dimensions[2].

This inversion problem may be posed as an optimization problem, where we solve for the unknown pixels in $\mathbf{x}$ by minimizing:

$$\min_{\mathbf{x}} \|\mathbf{M} \odot (\mathbf{x} - \mathbf{y})\|_2^2 + \lambda R(\mathbf{x}), \tag{8.14}$$

where $R(\mathbf{x})$ enforces priors such as smoothness (e.g., total variation), sparsity, or learned image priors from machine learning. This problem is under-determined, as the missing pixels have no direct observation. Thus it is important to encode meaningful priors via regularization to recover plausible solutions.

Total variation regularization has been widely used [98], and modern deep-learning techniques such as U-Nets [426] and GAN-based approaches achieve impressive results in complex inpainting tasks. Robust principal components (RPCA) from Section 6.5 is a useful algorithm in the case that images are cropped and aligned; i.e., the case where traditional SVD- or PCA-based image analysis is useful.

Deblurring

Another canonical imaging problem is to recover sharp images degraded by motion blur, out-of-focus optics, or atmospheric turbulence. The measurement data is a blurry image $\mathbf{y}$ and optionally a known blur kernel $\mathbf{H}$. The goal is to solve for the unknown sharp image $\mathbf{x}$, and in some cases, the blur kernel $\mathbf{H}$ if it is unknown.

The forward model convolves a sharp image with the blur kernel and adds noise $\mathbf{n}$

$$\mathbf{y} = \mathbf{H}\mathbf{x} + \mathbf{n}. \tag{8.15}$$

Deblurring may also be posed as an optimization problem. Given $\mathbf{y}$ and $\mathbf{H}$, we solve:

$$\min_{\mathbf{x}} \|\mathbf{H}\mathbf{x} - \mathbf{y}\|_2^2 + \lambda R(\mathbf{x}), \tag{8.16}$$

where $R(\mathbf{x})$ may enforce sparsity or smoothness. Again, the problem is ill-posed because there are infinitely many sharp images that give the same blurry image.

When the blur kernel $\mathbf{H}$ is unknown, the optimization becomes nonconvex

$$\min_{\mathbf{x},\mathbf{H}} \|\mathbf{H}\mathbf{x} - \mathbf{y}\|_2^2 + \lambda R(\mathbf{x}), \tag{8.17}$$

[2] If $\mathbf{A}$ and $\mathbf{B}$ are matrices of the same size, then their Hadamard product is defined as $(\mathbf{A} \odot \mathbf{B})_{ij} = \mathbf{A}_{ij} \cdot \mathbf{B}_{ij}$.

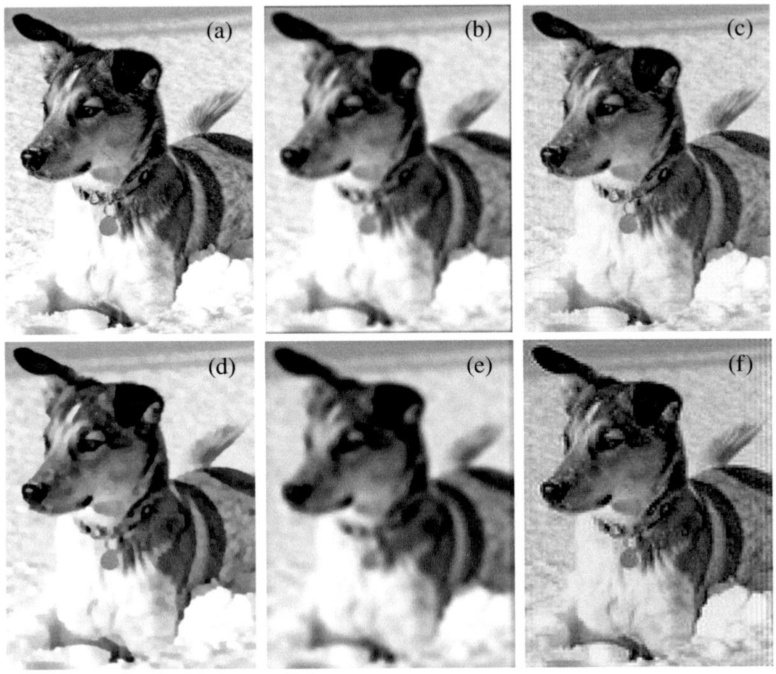

Figure 8.5 Deblurring applied to image of Mordecai the Snow Dog. (a) Original image, (b) blurred image, (c) least-squares reconstruction, (d) total variation regularized, (e) total variation regularized in Haar basis (f) FISTA regularization.

There is also the issue of computational cost, as iterative deconvolution methods can be expensive for large images. Blind deconvolution methods are widely used in photography, astronomy, and medical imaging. Here *blind* refers to the case when the the blurring kernel is unknown and must be estimated along with the sharp image.

Figure 8.5 shows the results of several deblurring algorithms on a blurred image of Mordecai the snow dog.

Change of Perspective and Inferring Three-Dimensional Shape

Recently, there has been incredible progress in our ability to infer a three-dimensional representation of a scene from multiple two-dimensional images, enabling fast and flexible change of perspective. The neural radiance field (NeRF) [349] method, based on deep learning, is particularly impressive.

Here the goal is to estimate the three-dimensional geometry of a scene from two-dimensional images captured from multiple viewpoints. The input data consists of multiple two-dimensional images $\mathbf{y}_j$. If the camera poses $\mathbf{P}_j$ are known, then reconstructing the three-dimensional representation is much simpler. However, in practice, the camera poses are also unknowns and must also be learned. A schematic of the NeRF approach is shown in Fig. 8.6.

Modern approaches leverage a feedforward neural network $\mathcal{N}(x, y, z)$ to predict the volume density and emitted radiance of (x, y, z) locations in a scene. From this, an image is

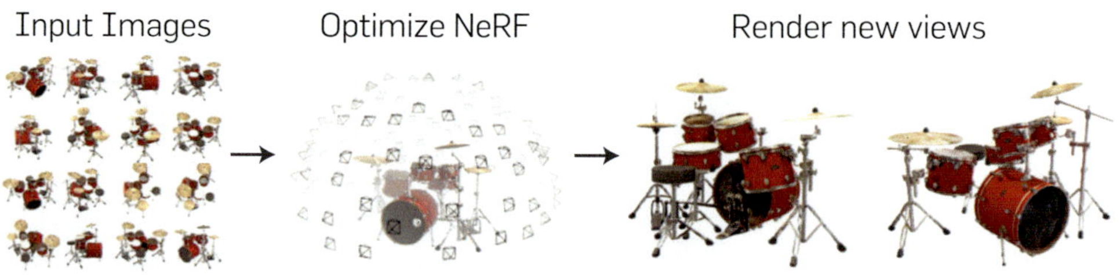

Figure 8.6 Neural Radiance Fields (NeRFs) reconstruct three-dimensional scene representations for rendering images from novel perspectives. Figure reproduced from Mildenhall et al. [349].

generated using classical volume rendering techniques, such as those used in video games. Both the neural network and the volume rendering are automatically differentiable (see Section 10.1), so it is possible to update the neural network weights to minimize the error between the predicted and actual images using gradient descent. As this is performed over multiple viewpoints, the neural network learns a coherent model of the scene.

We may formalize this as follows. Let the neural network be given by the function $\mathcal{N}_\theta(x, y, z)$ and the renderer by the function $\mathcal{R}$, so that $\mathcal{R}(\mathcal{N}_\theta, \mathbf{P}_j)$ returns the predicted two-dimensional image from perspective $\mathbf{P}_j$. The optimization problem then becomes

$$\min_{\boldsymbol{\theta}} \|\mathcal{R}(\mathcal{N}_\theta, \mathbf{P}_j) - \mathbf{y}_j\|_2^2. \tag{8.18}$$

Again, it is possible to optimize the neural network parameters $\boldsymbol{\theta}$ to minimize this loss through backpropagation. This optimization is nonlinear and high-dimensional, and solving for three-dimensional geometry from large datasets is computationally expensive. Further, a new NeRF must be trained for each new scene. In the original NeRF formulation the camera poses $\mathbf{P}_j$ are assumed known (e.g., from structure-from-motion). Recent extensions have explored jointly optimizing over $\mathbf{P}_j$ when poses are unknown [104, 275, 487].

8.5 Image-Based Inversion with Physics

Image-based inversion with physics incorporates physical models and principles into the reconstruction process, leveraging domain-specific knowledge to overcome data limitations and noise. These problems involve inferring hidden properties or structures based on observable data that interact with the underlying physics of the system, such as electromagnetic waves, acoustic waves, or X-rays. This approach is fundamental in scientific applications where the physics of the imaging system is well understood, allowing for more accurate and interpretable reconstructions. Key examples include estimating the energy yield of explosions using blast wave models, reconstructing black hole images informed by relativistic effects, and determining molecular structures from X-ray diffraction patterns. These problems are characterized by their reliance on forward models rooted in physics, computational challenges in solving associated inverse problems, and the need for regularization or physical constraints to ensure meaningful results.

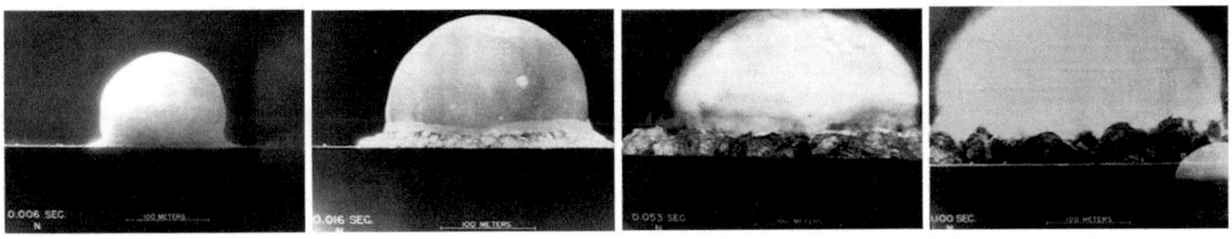

Figure 8.7 The 1945 TRINITY nuclear explosion. Rapatronic pictures taken at four times after detonation. Data of radius over time was used by G.I. Taylor to solve the inverse problem and accurately determine the bomb's yield. Trinity Site, Alamogordo Test Range, Jornada del Muerto desert. Courtesy of US Govt. Defense Threat Reduction Agency.

Yield of a Bomb from Satellite Images

The energy yield of an explosion can be estimated from satellite imagery capturing the expanding blast radius over time, a problem with significant military and scientific implications. Accurate yield estimation is critical for monitoring compliance with the comprehensive nuclear-test-ban treaty (CTBT) and understanding high-energy phenomena in controlled experiments and astrophysical contexts.

This problem dates back to the mid-20th century, when G.I. Taylor estimated the energy yield of the first nuclear tests conducted during the Manhattan Project. He was able to use a time series of images of the blast, published in a popular magazine, to back out the yield using dimensional scaling arguments. His seminal papers [490, 491] introduced the now-famous Sedov–Taylor blast wave model that connect explosion energy to the radius of the blast. See Fig. 8.7 for a sequence of images from the Trinity test.

The input data for this problem consists of time-series measurements of the blast radius, $\{R(t)\}$, derived from high-resolution satellite images taken at discrete time intervals. The unknown parameter is the explosion's energy yield, E. The forward model is based on the Sedov–Taylor blast wave theory, which relates the radius of the blast to its energy via:

$$R(t) = \alpha \left(\frac{Et^2}{\rho} \right)^{1/5}, \tag{8.19}$$

where α is a scaling constant, ρ is the ambient air density, and t is the time since detonation. For air, the constant α can be shown to be quite close to 1, and may be neglected. Using this model, the inversion problem is posed as an optimization problem:

$$\min_E \sum_k \left| R(t_k) - \alpha \left(\frac{Et_k^2}{\rho} \right)^{1/5} \right|^2. \tag{8.20}$$

This problem presents several challenges. The fifth-root dependence on E makes the optimization nonlinear and sensitive to noise in the observed data, often collected under less-than-ideal conditions. Additionally, the Sedov–Taylor model assumes isotropic explosions and requires accurate ambient density, ρ, which may not be available. Despite these challenges, this approach has been applied to estimate the yields of historical nuclear tests, such as the Trinity test, and remains relevant for modern satellite-based explosion monitoring systems. The same principles extend to astrophysical applications, including the analysis of supernova explosions; see Panel 8.2 for a discussion.

Panel 8.2: De Stella Nova: Estimating the Energy of Supernovae

The Sedov–Taylor blast wave equation (8.19) is also used to estimate the explosion energy of supernovae. After the core collapse and explosion of a particularly massive star, it may become a supernova. The expansion of the supernova through the interstellar medium (ISM) occurs in three distinct phases: 1) a free expansion phase (a few hundred years) is dominated by the mass of the supernova; 2) the Sedov–Taylor adiabatic expansion phase (up to tens of thousands of years), where the swept-up ISM dominates and energy is conserved, and 3) a radiative phase (up to tens of millions of years), where cooling becomes efficient and momentum is approximately conserved. In this second adiabatic phase, the ST equation may be used to estimate the energy of the supernova.

As the supernova remnant expands, a bright ring appears, corresponding to the dense shell of shocked ISM just behind the blast wave. During the second adiabatic expansion phase, this radius is well characterized by the ST equation for $R(t)$. We know the density ρ of the interstellar medium, which is about one atom per cubic centimeter, and the dimensionless constant $\alpha \approx 1.15$ is obtained from the self-similar solution for a strong shock in a uniform medium, depending only on the adiabatic index.

Many supernova have been observed in modern times, where we know exactly when the explosion occurred and can estimate $R(t)$ quite accurately. For example, my favorite Danish astronomer Tycho Brahe became internationally famous for discovering and documenting a supernova (*de stella nova*, known as B Cassiopeiae) in November of 1572. This supernova, SN 1572, is still in its Sedov–Taylor phase. Therefore, we can accurately estimate its explosion energy from its present radius, expansion velocity, and known age using the ST equation. The image of the supernova shows a distinct purple haze in X-ray.

The star Betelgeuse, in Orion's right shoulder, is a red supergiant nearing the end of its life. In the near astronomically speaking, future, it will go supernova, briefly shining brighter than a full moon in the night sky. I think about this every time I look up at the stars on a winter night.

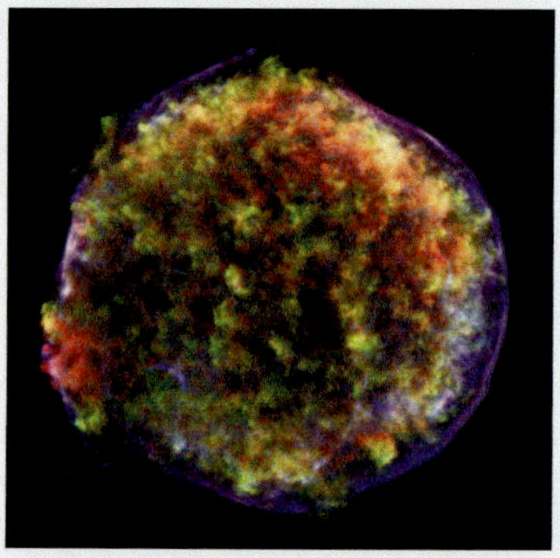

Figure 8.8: Tycho's supernova, SN 1572[a].

[a] Credit: NASA/CXC/Rutgers/J.Warren & J.Hughes et al., from `http://chandra.harvard.edu/photo/2005/tycho/`).

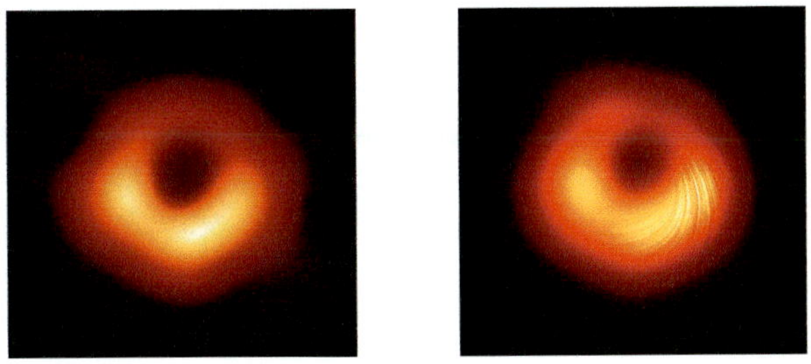

Figure 8.9 Computational images of the Messier 87 black hole, based on data collected from the Event Horizon Telescope. The image on the right is imaged using polarized light. Images courtesy of EHT Collaboration.

Mass of a Black Hole from Satellite Images

In 2019, the Event Horizon Telescope (EHT) collaboration achieved a groundbreaking milestone by producing the first image of a black hole, M87* [493]. This achievement confirmed key predictions of general relativity under extreme conditions and provided new insights into the dynamics of black holes and their accretion disks. The core problem involved reconstructing the brightness distribution surrounding a black hole and inferring its mass from sparse radio telescope measurements. Figure 8.9 shows the computationally reconstructed images of the M87 supermassive black hole.

The input data are sparse Fourier measurements of electromagnetic emissions from the black hole, obtained via very long baseline interferometry (VLBI), which combines signals from a global array of radio telescopes. The unknowns include the brightness distribution $\mathbf{I}(M, \theta)$, and the black hole's mass, M, spin, and orientation θ. General relativistic constraints relate the mass M to the brightness distribution, such as the size and shape of the black hole's shadow. It is possible to construct an image $\mathbf{x}$ from the brightness distribution $\mathbf{I}(M, \theta)$ through general relativistic ray tracing. The forward model connects the brightness distribution to the observed data via the Fourier transform $\mathbf{F}$:

$$\mathbf{y} = \mathbf{CFI}_{\mathrm{GR}}(M, \theta). \tag{8.21}$$

The matrix $\mathbf{C}$ is a sparse sampling operator, as only a few Fourier frequencies are measured. The problem is framed as a joint optimization to reconstruct estimate M and θ, from which we can derive $\mathbf{I}(M, \theta)$ and reconstruct $\mathbf{x}$:

$$\min_{M,\theta} \|\mathbf{CFI}_{\mathrm{GR}}(M, \theta) - \mathbf{y}\|_2^2. \tag{8.22}$$

This is known as the model-based fit, where we first derive the parameters of a model.

An alternative is to directly learn the image $\mathbf{x}$, which is regularized to be consistent with general relativistic considerations based on the mass M and other factors:

$$\min_{\mathbf{x}} \|\mathbf{CFx} - \mathbf{y}\|_2^2 + \lambda R(\mathbf{x}) + \mu R_{\mathrm{GR}}(\mathbf{x}, M). \tag{8.23}$$

The terms $R(\mathbf{x})$ include generic image regularizers (e.g., TV, sparsity, etc.), while $R_{\mathrm{GR}}(\mathbf{x}, M)$ include general relativistic priors.

In practice, a combination of both approaches was used to image the black hole M87* and later Sagittarius A*, the Milky Way's central black hole.

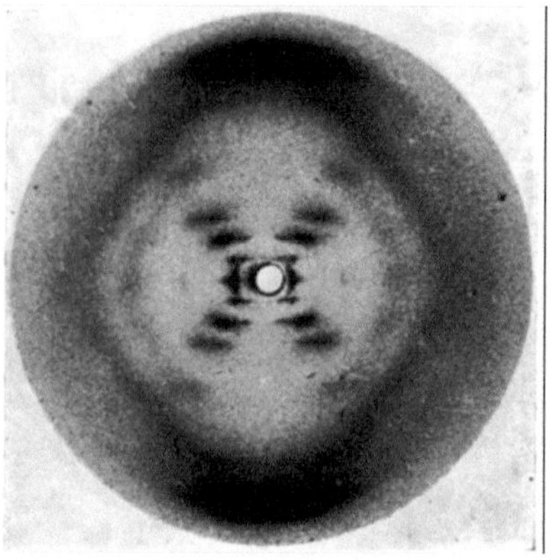

Figure 8.10 X-ray crystallography image of DNA, first taken by Franklin and Gosling [176] (reproduced with permission of Springer Nature). This image was used by Watson and Crick without Franklin's permission. They went on to win the Nobel Prize for Physiology or Medicine in 1962; Franklin died of cancer before the prize was awarded.

X-ray Crystallography

X-ray crystallography is one of the most impactful tools in structural biology and chemistry, enabling the discovery of molecular structures such as DNA, proteins, and complex drugs. This method has revolutionized our understanding of biomolecular mechanisms, from enzyme function to drug design. The origins of X-ray crystallography date back to the early 20th century, with pioneering contributions by von Laue and colleagues [178], who first demonstrated X-ray diffraction by crystals, and later by Bragg [67], who developed Bragg's law to relate diffraction patterns to crystal structures. These efforts laid the groundwork for the Nobel prize discovery of DNA's double-helix structure by Watson and Crick, based on diffraction data by Rosalind Franklin and Maurice Wilkins [532].

The input data are diffraction intensities, $I(\mathbf{k})$, corresponding to the squared magnitudes of the Fourier transform of the electron density $\rho(\mathbf{r})$:

$$I(\mathbf{k}) = |\mathcal{F}[\rho(\mathbf{r})]|^2. \tag{8.24}$$

The goal is to reconstruct $\rho(\mathbf{r})$ by solving the phase retrieval problem, which arises because only the magnitudes of the Fourier coefficients are measured, not their phases:

$$\min_{\rho} \| |\mathcal{F}[\rho]|^2 - I(\mathbf{k}) \|_2^2. \tag{8.25}$$

We denote the Fourier coefficients as $\hat{\rho}(\mathbf{k}) = \mathcal{F}[\rho(\mathbf{r})]$. This problem is challenging due to the phase ambiguity, making it nonlinear and underdetermined. Regularization techniques or prior knowledge, such as sparsity or physical constraints, are often required to make the problem well-posed. There are several ways that physics enters the optimization. First, knowing the chemical composition helps, as each atom contributes a known scattering pattern. Further, electron density is strictly non-negative, so $\rho \geq 0$. Crystal symmetries imply that many Fourier coefficients are related by group operations, so we can reduce the effective search space for phases by enforcing $\hat{\rho}(g\mathbf{k}) = \hat{\rho}(\mathbf{k})$ for $g \in G$ where G is the symmetry group of the crystal.

8.6 PDE-Constrained Inversion

Inverse problems constrained by partial differential equations (PDEs) lie at the intersection of physics, mathematics, and computational science. These problems involve inferring unknown parameters or properties of a system by fitting observations to models governed by PDEs. Applications span a wide range of scientific and engineering fields, including geophysics, fluid dynamics, and biomedical engineering. The key challenge is that solving such problems requires the repeated evaluation of computationally expensive PDE models, often leading to ill-posed and high-dimensional optimization tasks. Regularization techniques and advanced computational methods, such as adjoint-based gradient computation, play a crucial role in making these problems tractable. This section highlights two canonical examples: seismic inversion, where subsurface geophysical properties are inferred from seismic wave propagation and surface measurements, and the classic "Can you hear the shape of a drum?" problem, which explores the relationship between geometric shapes and their spectral signatures.

More generally, PDE-constrained design optimization can be seen as a related problem. For example, optimizing the shape of a wing to maximize lift and minimize drag is a complex design optimization, where we essentially have an expensive PDE-based forward model from wing shape to aerodynamic properties, but not the inverse model. The closely related field of topology optimization is useful in designing shapes for structural properties [39, 41, 40].

Seismic Inversion

Seismic inversion seeks to infer the Earth's subsurface properties, such as density and seismic velocity, from surface-based seismic wave measurements. This problem is central to geophysics, with applications in oil and gas exploration, earthquake monitoring, and carbon sequestration. By analyzing how seismic waves propagate through different subsurface materials, we can reconstruct detailed velocity models, revealing geological structures and potential resource reservoirs.

The input data consists of time-series of seismic wave amplitudes recorded at multiple receiver locations, typically from controlled sources, like explosions, or sources that may be triangulated, like earthquakes or eruptions. The unknown parameter is the spatially varying velocity model $c(\mathbf{x})$, which governs how seismic waves propagate. The forward model for this problem is based on the acoustic wave equation:

$$\frac{\partial^2 u}{\partial t^2} = c^2 \nabla^2 u, \tag{8.26}$$

where $u = u(\mathbf{x}, t)$ is the wavefield, $c = c(\mathbf{x})$ is the velocity, and the boundary conditions represent the Earth's surface and subsurface interfaces.

In seismic inversion, we aim to match observed data $\mathbf{y}$ with synthetic data generated by solving the forward model for a given $c(\mathbf{x})$. This is posed as an optimization problem:

$$\min_c \|\mathbf{F}(c) - \mathbf{y}\|_2^2, \tag{8.27}$$

where $\mathbf{F}(c)$ is a forward solver for the wave equation, typically implemented numerically using finite-difference, finite-element, or spectral methods.

A powerful tool in this setting is the adjoint state method [51, 488], which relies on a Lagrangian construction, as in Chapter 6. By introducing Lagrange multipliers (i.e., adjoint variables) to enforce the wave equation as a constraint, one can efficiently compute gradients of the objective function with respect to millions of model parameters, such as the wave speed $c(\mathbf{x})$. This comes at the cost of two PDE solves, one forward and one adjoint, per iteration. This approach underlies modern large-scale seismic inversion methods.

Seismic inversion is high-dimensional, nonconvex, and ill-posed, making it extremely challenging. The nature of the acoustic equation makes it so that small errors in the recorded data can lead to significant errors in the estimated velocity model. Regularization methods, such as Tikhonov regularization or total variation, are often employed to stabilize the inversion. Further, the velocity model $c(\mathbf{x})$ is a three-dimensional field, and higher desired resolution leads to an extremely large optimization problem with millions or billions of parameters corresponding to the estimated velocity c at each grid point. Finally, the wave equation's nonlinear relationship with c results in multiple local minima, complicating gradient-based optimization. Techniques like full waveform inversion (FWI) [519] aim to mitigate these challenges by using more accurate physics and multi-scale approaches.

Traditional seismic inversion relies heavily on solving computationally expensive PDEs, such as the wave equation, which can be prohibitively time-consuming for large-scale, high-resolution models. Machine learning mitigates these challenges by learning surrogate models, which approximate the forward solver or map seismic data directly to subsurface properties. These surrogates, trained on synthetic or real datasets, dramatically reduce computational costs while maintaining accuracy [547]. Further, automatic differentiation of these surrogates allow them to be flexibly tuned to data or be used to enable faster gradient-based optimization.

Hybrid approaches that integrate machine learning with physics-based solvers have become particularly promising. Physics-informed neural networks (PINNs)[403] incorporate physical laws, such as the wave equation, as extra terms in the objective function during training; we will discuss PINNs more in Chapter 10. This ensures that learned models respect the underlying physics, even when trained on limited data. For example, PINNs have been used to solve seismic inversion problems by embedding the wave equation into the loss function, reducing the reliance on labeled training data [555].

Another popular hybrid approach involves combining full waveform inversion (FWI) with neural networks. In these frameworks, machine learning models act as regularizers or preconditioners for traditional optimization methods, improving convergence and reducing sensitivity to local minima [556]. In addition, HPC advancements, such as GPU-accelerated solvers and parallel processing, enable large-scale simulations and optimization in realistic 3D environments, which were previously infeasible [519].

Although full wave inversion for the velocity $c(\mathbf{x})$ is already a major challenge, it is also possible to pose an optimization to solve for the velocity $c(\mathbf{x})$ as well as the density field $\rho(\mathbf{x})$. This involves using the full elastic wave equation

$$\rho(\mathbf{x})\frac{\partial^2 \mathbf{u}}{\partial t^2} = \nabla \cdot [\mathbf{C}(\mathbf{x}) : \nabla \mathbf{u}] \tag{8.28}$$

as the forward model. However, this is often more involved than it is practical. Further, the velocity field often correlates empirically with relevant lithology, porosity, and fluid saturation. Sharp velocity drops often flag gas pockets, and high velocities can indicate carbonates or cemented sands. In such cases, rock physics models can help to improve models, adding physical regularization.

Can You Hear the Shape of a Drum?

The question "Can you hear the shape of a drum?" was famously posed by Mark Kac in 1966 [259]. It refers to the mathematical problem of determining the shape of a domain from its vibrational frequencies, or eigenvalues of the Laplacian. This problem, rooted in mathematical physics, has implications for quantum mechanics, shape analysis, and modern machine learning.

The membrane of our drum is give by a bounded domain Ω, which determines the geometry of the problem. The wave equation on this domain is

$$u_{tt} = c^2 \nabla^2 u \quad u_{\partial\Omega} = 0. \tag{8.29}$$

The solution is separable in time and space, so $u(\mathbf{x}, t) = \sum_j \phi_j(\mathbf{x}) \cos(\omega_j t)$, where $f_j = \frac{\omega_j}{2\pi}$ is the audible frequency associated with the j-th spatial mode ϕ_j. The spatial mode satisfies the Helmholtz equation

$$\nabla^2 \phi_j + \frac{\omega_j^2}{c^2}\phi_j = 0 \quad \Longrightarrow \quad \nabla^2 \phi_j + \lambda_j \phi_j = 0 \tag{8.30}$$

with spatial eigenfrequencies $\lambda_j = \frac{\omega_j^2}{c^2}$ and Dirichlet boundary conditions $\phi_j(\mathbf{x})|_{\partial\Omega} = 0$. Thus, the spatial eigenfrequencies λ_j of the Laplace operator on the bounded domain Ω with Dirichlet boundary conditions are directly related to the acoustic frequencies $f_j = \frac{\omega_j}{2\pi}$.

The input data for the optimization problem are the measured eigenfrequencies λ_j, which are computed directly from the acoustic frequencies f_j. We then solve for the unknown shape of the domain, Ω. The forward model is the Helmholtz equation (8.30) from above. The inverse problem is to match the observed eigenvalues λ_j from the model eigenvalues $\hat{\lambda}_j(\Omega)$ given a particular shape Ω.

$$\min_{\Omega} \sum_j |\lambda_j - \hat{\lambda}_j(\Omega)|^2. \tag{8.31}$$

This problem is inherently non-unique. In 1992, Gordon, Webb, and Wolpert [205] showed that distinct domains can have identical eigenvalues. These isospectral domains therefore also have identical acoustic signatures. Their geometries are shown in Fig. 8.11 along with the eigenmodes of the Helmholtz equation. This introduces inherent ambiguity into the problem. However, adding physical priors or reducing the search space may help make the problem unique. It is also ill-posed, in the sense that small perturbations in the

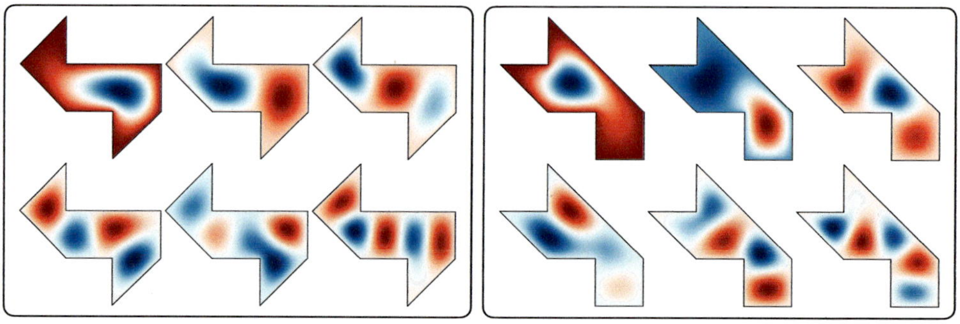

Figure 8.11 Eigenmodes of the Helmholtz equation (8.30) for two different geometries. Although the geometries are not related by simple symmetry transformations (e.g., rotations, reflections, etc.), the eigenvalue spectrum, λ_j, of the Helmholtz equation is identical for these two geometries. This means that they are *isospectral* and cannot be distinguished just by their sound. This example was discovered by Gordon, Webb, and Wolpert [205].

eigenvalues can correspond to vastly different geometries. Regularization techniques, such as penalizing high-frequency oscillations, are essential. Finally, representing and optimizing over domain shapes is computationally intensive, often requiring parameterization techniques like level sets or Fourier descriptors.

Despite these challenges, the problem of determining the shape of a domain from its spectral properties remains highly relevant, both mathematically and in practical applications. Advances in computational geometry, numerical optimization, and high-performance computing have all driven progress in this field. One notable area is spectral graph theory, where the eigenvalues of the graph Laplacian encode structural information about a graph, such as connectivity, community structure, and robustness. These spectral properties are used to solve problems in network analysis, such as detecting communities, ranking nodes, and designing efficient network flows [106, 471].

In machine learning, spectral embeddings derived from Laplacian eigenvalues are foundational for dimensionality reduction and clustering algorithms, such as spectral clustering [368]. These methods leverage the intuition that the eigenvectors corresponding to the smallest nonzero eigenvalues of the graph Laplacian capture meaningful geometric relationships in high-dimensional data. The mathematical framework provided by Kac's question has inspired new techniques for analyzing and manipulating complex shapes, including shape matching, deformation, and optimization in computer graphics [71].

The interplay between domain shape and spectral properties have far reaching implications for engineering optimization and inverse problems. For example, modern shape optimization techniques incorporate spectral properties to design objects with specific vibration characteristics, such as minimizing noise in mechanical structures or maximizing efficiency in acoustic devices.

8.7 System Identification

Another common inverse problem is to determine a dynamical system underlying an observed time series data set. These learned dynamical systems may be used for forecasting (e.g., predicting the future), estimation, and control. There is a rich history in system identification, going back over half a century, and so here we will only present a high-level sketch of the field, with a focus on recent trends. For more detail, there are several more complete references [77, 76, 444].

Although the earliest techniques in system identification were not posed explicitly as optimization problems, it was quickly realized that optimization was essential to the underlying algorithms. There are several key dichotomies in system identification that determine which optimization technique to use:

- Linear versus nonlinear dynamics: linear system identification is considerably simpler and more structured, often leading to numerically robust, convex optimization problems based on least-squares regression.
- Partial versus full measurements: system identification based on partial measurements makes the inverse problem inherently less well posed than when full-state measurements are available.
- Controlled versus autonomous systems: when a system is being actively controlled, there is ambiguity about what are internal state dynamics versus the affect of actuation, unless the actuation signal is known.
- High-dimensional versus low-dimensional systems and/or data: the dimension of the system identification problem often determines how expensive and ill-conditioned the inverse problem is.

The field of linear system identification is much more mature [253, 444, 76], with decades of progress, and so we will begin with the linear identification problem. However, nonlinear system identification is experiencing rapid progress [61, 447, 81, 116], especially because of improved optimization algorithms and function approximation techniques from machine learning. Here we will primarily discuss the simplified case of learning linear and nonlinear dynamical systems with full measurements and without control inputs, although extensions to controlled and partially measured systems will be mentioned. See Fig. 8.12 for a schematic of linear and nonlinear system identification.

Linear System Identification

A linear dynamical system, with control inputs and output measurements, is given by

$$\frac{\mathrm{d}}{\mathrm{d}t}\mathbf{x} = \mathbf{A}\mathbf{x} + \mathbf{B}\mathbf{u} \tag{8.32a}$$

$$\mathbf{y} = \mathbf{C}\mathbf{x} + \mathbf{D}\mathbf{u}, \tag{8.32b}$$

where $\mathbf{x} \in \mathbb{R}^n$ is the state of the system, $\mathbf{u} \in \mathbb{R}^q$ is the external input (e.g., a control input), and $\mathbf{y} \in \mathbb{R}^p$ is the measurement of the system. In these equations, all variables depend on

Figure 8.12 Linear system identification via DMD (left) and SINDy (right). In each case, the derivative estimates are shown in blue, the data in yellow and the unknown model parameters in red. The DMD regression is typically used for high dimensional data $\mathbf{x} \in \mathbb{R}^n$, with less snapshots $m \ll n$, so that the matrix $\mathbf{X}$ is *tall and skinny* and the matrix inversion to solve for $\mathbf{A}$ is ill-conditioned unless we add a low-rank condition. The SINDy regression is used in the opposite regime, for relatively low-dimensional $\mathbf{x} \in \mathbb{R}^n$ with more snapshots $m \gg n$. In this case, we assume a sparsity prior on $\boldsymbol{\Xi}$. Both problems essentially amount to a linear regression $\mathbf{Ax} = \mathbf{b}$ as in Chapter 4 with regularization or constraints.

time, so $\mathbf{x} = \mathbf{x}(t)$, etc., although we typically drop the explicit dependence on time, as this is clear from the fact that we are learning a differential equation.

Linear system identification is an inverse problem of identifying the system matrices $\{\mathbf{A}, \mathbf{B}, \mathbf{C}, \mathbf{D}\}$ from input–output data $\{\mathbf{u}(t), \mathbf{y}(t)\}$. More commonly, we will have data sampled discretely in time, $\{\mathbf{u}_k, \mathbf{y}_k\}$, where $\mathbf{u}_k = \mathbf{u}(k\Delta t)$ and $\mathbf{y}_k = \mathbf{y}(k\Delta t)$. We may then learn the continuous-time dynamics in Eq. (8.32), or a discrete-time dynamical system that advances the state and measurements forward by one Δt every iteration.

Typically, it is more challenging to learn the dynamical system from partial observations $\mathbf{y}$, as opposed to full-state measurements $\mathbf{x}$. However, learning linear systems from partial measurements of the inputs and outputs, $\{\mathbf{y}, \mathbf{u}\}$, has largely been a solved problem since the 1980s [231, 356, 253, 254]. The well-posedness of this inverse problem relates to whether or not the matrix pairs $(\mathbf{A}, \mathbf{B})$ and $(\mathbf{A}, \mathbf{C})$ are controllable and observable, respectively [77], and more specifically to the condition number of the controllability and observability Gramians.

If there is no control input and there are full state measurements, so that $\mathbf{y} = \mathbf{x}$, then we have a simpler inverse problem of learning the operator $\mathbf{A}$ in

$$\frac{\mathrm{d}}{\mathrm{d}t}\mathbf{x} = \mathbf{Ax}. \tag{8.33}$$

The dynamic mode decomposition (DMD) is a modern approach to full-state system identification of such systems, when the state $\mathbf{x}$ is high-dimensional [443, 428, 483, 444], as is the case for fluid flows, neuroscience, epidemiology, and many other spatio-temporal systems. DMD is deeply connected to Koopman operator theory [428, 76], where it is often possible to lift a nonlinear system into higher dimensions where the dynamics become linear.

All modern DMD algorithms are formed as optimization problems. The idea is to use data matrices and linear least-squares regression to find an approximate $\mathbf{A}$ that minimizes the error of the fit in some matrix norm.

As DMD is inherently data-driven, the first step is to collect a number of pairs of snapshots of the state of a system as it evolves in time. These snapshot pairs may be denoted by $\{(\mathbf{x}(t_k), \mathbf{x}(t'_k))\}_{k=1}^m$, where $t'_k = t_k + \Delta t$, and the time-step Δt is sufficiently small to resolve the highest frequencies in the dynamics. As before, a snapshot may be the state of a system, such as a three-dimensional fluid velocity field sampled at a number of discretized locations, which is reshaped into a high-dimensional column vector. These snapshots are then arranged into two data matrices, $\mathbf{X}$ and $\mathbf{X}'$:

$$\mathbf{X} = \begin{bmatrix} | & | & & | \\ \mathbf{x}(t_1) & \mathbf{x}(t_2) & \cdots & \mathbf{x}(t_m) \\ | & | & & | \end{bmatrix}, \quad \mathbf{X}' = \begin{bmatrix} | & | & & | \\ \mathbf{x}(t'_1) & \mathbf{x}(t'_2) & \cdots & \mathbf{x}(t'_m) \\ | & | & & | \end{bmatrix}.$$

The original formulations of Schmid [443] and Rowley et al. [428] assumed uniform sampling in time, so that $t_k = k\Delta t$ and $t'_k = t_k + \Delta t = t_{k+1}$. If we assume uniform sampling in time, we will adopt the notation $\mathbf{x}_k = \mathbf{x}(k\Delta t)$.

The DMD algorithm seeks the leading spectral decomposition (i.e., eigenvalues and eigenvectors) of the best-fit linear operator $\mathbf{A}$ that relates the snapshots in time:

$$\mathbf{X}' \approx \mathbf{A}\mathbf{X}.$$

The best-fit operator $\mathbf{A}$ then establishes a linear dynamical system that best advances snapshot measurements forward in time. If we assume uniform sampling in time, this becomes

$$\mathbf{x}_{k+1} \approx \mathbf{A}\mathbf{x}_k.$$

The best-fit operator $\mathbf{A}$ is defined as the solution to the following optimization problem:

$$\mathbf{A} = \underset{\mathbf{A}}{\operatorname{argmin}} \|\mathbf{X}' - \mathbf{A}\mathbf{X}\|_F = \mathbf{X}'\mathbf{X}^\dagger, \tag{8.34}$$

where $\|\cdot\|_F$ is the Frobenius norm and † denotes the pseudo-inverse.

There are several variants of DMD, such as DMD with control (DMDc) [397]. DMDc is closely related to linear subspace identification methods, such as the eigensystem realization algorithm (ERA) [253]. The optimized DMD algorithm generalizes exact DMD to perform a nonlinear regression to exponential-time dynamics, thus providing an improved computation of the DMD modes and their eigenvalues [14]. Physics-informed DMD (PIDMD) adds physical structure to the $\mathbf{A}$ matrix through additional terms in the objective function or via constraints. We will discuss PIDMD more in Section 10.5.

Nonlinear System Identification

Similarly, the general form of a nonlinear dynamical system is given by

$$\frac{\mathrm{d}}{\mathrm{d}t}\mathbf{x} = \mathbf{f}(\mathbf{x}, \mathbf{u}) \tag{8.35a}$$

$$\mathbf{y} = \mathbf{g}(\mathbf{x}, \mathbf{u}). \tag{8.35b}$$

In nonlinear system identification, the goal is still to learn the dynamics $\mathbf{f}$ and the output equations $\mathbf{g}$ from time series data $\{\mathbf{u}(t), \mathbf{x}(t), \mathbf{y}(t)\}$. Often, we have access to the full state $\mathbf{x}(t)$, but in the most challenging cases, we only have input–output data $\{\mathbf{u}(t), \mathbf{y}(t)\}$.

The structure of the nonlinear functions $\mathbf{f}$ and $\mathbf{g}$ is much less constrained than in linear system identification, and so typically we must first define a parametrized family of functions, $\mathbf{f}_{\boldsymbol{\xi}}(\mathbf{x}, \mathbf{u})$ and learn the parameters $\boldsymbol{\xi}$ using optimization. For simplicity we will focus on the case with full measurements $\mathbf{y} = \mathbf{x}$ and no control input $\mathbf{u}$, as above, although extensions exist to relax each of these assumptions [82].

The sparse identification of nonlinear dynamics (SINDy) algorithm [81] leverages the fact that many dynamical systems

$$\frac{\mathrm{d}}{\mathrm{d}t}\mathbf{x} = \mathbf{f}(\mathbf{x}) \tag{8.36}$$

have dynamics $\mathbf{f}$ with only a few active terms, out of all possible right-hand side functions. We then seek to approximate $\mathbf{f}$ by a generalized linear model

$$\mathbf{f}(\mathbf{x}) \approx \sum_{k=1}^{p} \theta_k(\mathbf{x})\xi_k = \boldsymbol{\Theta}(\mathbf{x})\boldsymbol{\xi}, \tag{8.37}$$

with the fewest non-zero terms in $\boldsymbol{\xi}$ as possible. It is then possible to solve for the relevant terms that are active in the dynamics using sparse regression [220, 247, 495, 564] that penalizes the number of terms in the dynamics and scales well to large problems. We have already seen examples of this in the LASSO regression from Chapter 4 and the sparse optimization algorithms in Section 5.1.

As in DMD, time-series data is collected from (8.36) and formed into a data matrix:

$$\mathbf{X} = \begin{bmatrix}\mathbf{x}(t_1) & \mathbf{x}(t_2) & \cdots & \mathbf{x}(t_m)\end{bmatrix}^{\mathrm{T}}.$$

A similar matrix of derivatives is formed:

$$\dot{\mathbf{X}} = \begin{bmatrix}\dot{\mathbf{x}}(t_1) & \dot{\mathbf{x}}(t_2) & \cdots & \dot{\mathbf{x}}(t_m)\end{bmatrix}^{\mathrm{T}}.$$

In practice, this may be computed directly from the data in $\mathbf{X}$; for noisy data, the total-variation regularized derivative tends to provide numerically robust derivatives [100]. Alternatively, it is possible to formulate the SINDy algorithm for discrete-time systems $\mathbf{x}_{k+1} = \mathbf{F}(\mathbf{x}_k)$, as in the DMD algorithm, and avoid derivatives entirely.

A library of candidate nonlinear functions $\boldsymbol{\Theta}(\mathbf{X})$ is constructed from the data $\mathbf{X}$:

$$\boldsymbol{\Theta}(\mathbf{X}) = \begin{bmatrix}\mathbf{1} & \mathbf{X} & \mathbf{X}^2 & \cdots & \mathbf{X}^d & \cdots & \sin(\mathbf{X}) & \cdots\end{bmatrix}.$$

Here, the matrix $\mathbf{X}^d$ denotes a matrix with column vectors given by all possible time series of dth-degree polynomials in the state $\mathbf{x}$. In general, this library of candidate functions is only limited by one's imagination.

The dynamical system may now be represented in terms of the data $\mathbf{X}$ and $\dot{\mathbf{X}}$ as

$$\dot{\mathbf{X}} = \boldsymbol{\Theta}(\mathbf{X})\boldsymbol{\Xi}. \tag{8.38}$$

Each column $\boldsymbol{\xi}_k$ in $\boldsymbol{\Xi}$ is a vector of coefficients determining the active terms in the k-th row of (8.36). A parsimonious model provides an accurate fit with as few terms as possible in Ξ. This is achieved by seeking solutions that minimize the following loss function

$$\min_{\Theta} \|\dot{\mathbf{X}} - \Theta(\mathbf{X})\Xi\| + \lambda\|\Xi\|_0. \tag{8.39}$$

Here λ is a sparsity-promoting hyperparameter. The term $\|\Xi\|_0$ counts the non-zero entries in Ξ, and it is neither convex nor strictly a norm. This sparse regression problem is NP-hard to find the global optimum. However, it is possible to efficiently solve Eq. (8.39) for a local optimum using the proximal gradient or SR3 method from Section 5.1. It is also possible to relax the nonconvex $\|\Xi\|_0$ to the convex $\|\Xi\|_1$ as in LASSO [495]. The approach we take is to implement sequentially thresholded least-squares (STLS) [81], which has been proven to converge [553]. The sequentially thresholded least-squares algorithm has since been demonstrated to be a formal relaxation of the ℓ_0 optimization problem, using the SR3 formulation with proximal gradient [557] discussed in Section 5.1.

There are several powerful extensions to SINDy, including extensions to include control inputs [263], to disambiguate linear and nonlinear dynamics [19], to enforce symmetry constraints [320], and to promote global stability [268]. Enforcing symmetry constraints and stability both involve modifying the optimization problem, either with hard constraints, or with additional terms, and this will be discussed more in Section 10.5. The STLS procedure is particularly appealing, because it involves a sequential series of least-squares regressions, so it is straightforward to add equality constraints at each stage [320].

Learning fully nonlinear dynamics from partial measurements is still a cutting-edge research challenge today, with advances in deep learning making it possible to learn nonlinear transformations from delay coordinates where sparse models may be learned [24]; these transformations are closely related to Takens's delay embedding theorem [485] and Koopman operator theory [75, 76]. The loss functions used to train these autoencoder neural networks are highly customized, and are essential to the learning process.

It is also possible to learn partial differential equations

$$\mathbf{u}_t = \mathcal{N}(\mathbf{u}, \mathbf{u}_x, \mathbf{u}_{xx}, \ldots), \tag{8.40}$$

from data, although this fits more into the realm of model discovery and is a much more recent field of research [430, 437, 78]. Typically, this may be viewed as an extension of the nonlinear system identification problem, where we learn $\dot{\mathbf{x}} = \mathbf{f}(\mathbf{x})$ from data.

SINDy Optimization: Sequential Thresholded Least Square. If we write the original ℓ_0-regularized problems

$$\min_{\mathbf{x}} \underbrace{\|\Theta\mathbf{x} - \dot{\mathbf{X}}\|_2^2}_{f(\mathbf{x})} + \underbrace{\lambda\|\mathbf{x}\|_0}_{g(\mathbf{x})} \tag{8.41}$$

we have several options to solve this. The sequentially thresholded least squares (STLS), advocated in the original SINDy paper, is a remarkably robust and efficient algorithm. It is generally recommended as an all-purpose SIND optimizer, and it has connections to the sparse, relaxed, regularized regression (SR3) method from Section 5.1.

However, in the broader sparse regression literature, there are several other options that are widely used. We will discuss these approaches and their relative pros and cons here.

Panel 8.3: Sequential Thresholded Least Squares, A Happy Accident

There are many algorithms for solving the SINDy regression. However, the original STLS algorithm has proven surprisingly robust and effective.

I would like to tell you that I designed this algorithm with deep optimization knowledge. I did not. It was coded in about ten lines of Matlab while I was in a hurry to go to lunch. At the time, I didn't appreciate the differences in over-determined and under-determined regression, and was trying to use compressed sensing (under-determined) on SINDy (over-determined). My initial attempts failed to handle the slightest noise. However, I noticed that even for large amounts of noise, simple linear regression revealed much of the model structure, plus some small errors:

Table 8.12:

	ξ_1	ξ_2	ξ_3
1	0.036540	-0.005219	-0.017323
x	**-10.004099**	**28.003856**	0.000611
y	**10.000864**	**-1.000628**	-0.000374
z	-0.002977	0.001258	**-2.665143**
x^2	-0.000081	0.000319	0.000036
xy	0.000058	-0.000234	**0.999894**
xz	0.000120	**-1.000051**	0.000002
y^2	-0.000021	0.000034	0.000069
yz	-0.000012	-0.000053	-0.000005
z^2	0.000059	-0.000044	-0.000032

I figured that if these small coefficients were removed, the regression onto the remaining coefficients would improve, and STLS was born. No analysis, just a hunch and some quick-and-dirty numerical experiments before lunch.

Years later, we worked with Sasha Aravkin and Peng Zheng to figure out why STLS performs so well on sparse regression tasks, even better than LASSO. It turns out we got lucky and stumbled onto an effective relaxation of the ℓ_0-regularized regression.

CodeSnip 8.a

```
%% compute Sparse regression: sequential least squares
Xi = Theta\dXdt;  % initial guess: Least-squares

% lambda is our sparsification knob.
for k=1:10
    smallinds = (abs(Xi)<lambda);    % find small
        coefficients
    Xi(smallinds)=0;                 % and threshold
    for ind = 1:n                    % n is state dimension
        biginds = ~smallinds(:,ind);
        % Regress dynamics onto remaining terms
        Xi(biginds,ind) = Theta(:,biginds)\dXdt(:,ind);
    end
end
```

The first, and most commonly advocated approach, is to relax the nonconvex ℓ_0 norm to the convex ℓ_1 norm:

$$\min_{\mathbf{x}} \|\boldsymbol{\Theta}\mathbf{x} - \dot{\mathbf{X}}\|_2^2 + \lambda\|\mathbf{x}\|_1. \tag{8.42}$$

We have been conditioned to believe that relaxing to a nearby convex problem is almost always better. However, practice has shown that theoretical convergence guarantees often fall short on real world data. For example, the LASSO regression above has well-known failure modes that occur early in the pruning optimization [477].

As an alternative, we could attempt to solve the original ℓ_0-regularized problem with proximal gradient descent. This amounts to performing a gradient descent, where we first take a gradient step on the smooth function $f(\mathbf{x})$,

$$\mathbf{z} = \mathbf{x} - \gamma\nabla f(\mathbf{x}),$$

and then perform a proximal gradient step on the nonsmooth function $g(\mathbf{x})$

$$\text{prox}_{\gamma,g}(\mathbf{z}) \equiv \underset{\mathbf{x}}{\text{argmin}} \|\mathbf{z} - \mathbf{x}\|_2^2 + g(\mathbf{x}). \tag{8.43}$$

The proximal gradient for $g(\mathbf{x}) = \lambda\|\mathbf{x}\|_0$ is the hard threshold operator, derived in Section 5.1. However, this has the downside that the gradient descent algorithm converges extremely slowly for ill-conditioned $\boldsymbol{\Theta}$ matrices, which are common in SINDy problems.

The third approach, which is very close to the STLS algorithm, is to split the objective and minimize the f and g terms with separate variables $\mathbf{x}$ and $\mathbf{z}$:

$$\min_{\mathbf{x},\mathbf{z}} f(\mathbf{x}) + g(\mathbf{z}) + \frac{\rho}{2}\|\mathbf{x} - \mathbf{z}\|_2^2. \tag{8.44}$$

These variables remain coupled via the last term $\frac{\rho}{2}\|\mathbf{x} - \mathbf{z}\|_2^2$.

It is then possible to solve this optimization problem by partially minimizing over $\mathbf{x}$, then partially minimizing over $\mathbf{z}$, and iterating until convergence. Thus, the $\mathbf{z}$ variable can be *more sparse* and the $\mathbf{x}$ variable can be a *more accurate fit* to the data, making the problem considerably less stiff. The partial minimization steps involve:

- $\min_{\mathbf{x}} \|\boldsymbol{\Theta}\mathbf{x} - \dot{\mathbf{X}}\|_2^2 + \rho\|\mathbf{x} - \mathbf{z}\|_2^2$ is solved via least squares for fixed $\mathbf{z}$.
- $\min_{\mathbf{z}} \rho\|\mathbf{x} - \mathbf{z}\|_2^2 + \lambda\|\mathbf{x}\|_0$ is the proximal gradient step, which becomes the hard threshold operator (see Section 5.1).

This essentially amounts to the sequentially thresholded least-square approach, which was formalized and generalized in the SR3 algorithm. This formulation has the benefit that we may solve the least-squares problem with efficient pre-conditioned algorithms, and solve the nonsmooth, sparse term using the proximal gradient method, instead of using gradient descent on both terms.

Technically, the original SINDy STLS algorithm does not partially minimize $\min_{\mathbf{x}} \|\boldsymbol{\Theta}\mathbf{x} - \dot{\mathbf{X}}\|_2^2 + \rho\|\mathbf{x} - \mathbf{z}\|_2^2$, but rather minimizes just the MSE error $\min_{\mathbf{x}} \|\boldsymbol{\Theta}\mathbf{x} - \dot{\mathbf{X}}\|_2^2$. Because the variables $\mathbf{x}$ and $\mathbf{z}$ are meant to be rather close to each other, the additional $\rho\|\mathbf{x} - \mathbf{z}\|_2^2$ term doesn't have much of an impact on the smooth $f(\mathbf{x})$ minimization. However, this coupling term is *essential* in formulating the $g(\mathbf{z})$ partial minimization as a proximal gradient, which

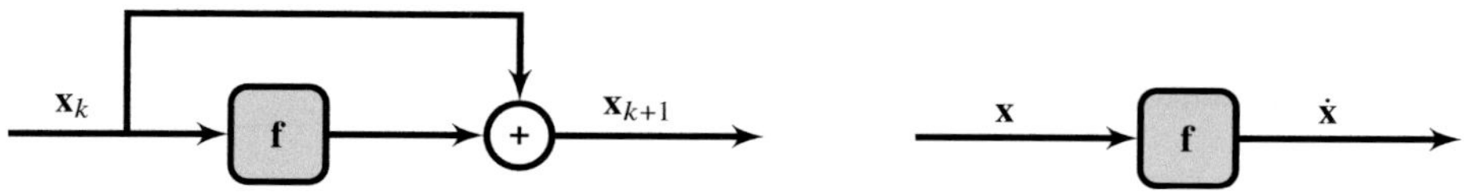

Figure 8.13 Block diagram for a residual network (left) and a neural ODE (right). The residual network mimics a forward Euler integration of the vector field **f**, while a neural ODE directly learns this function, enabling more sophisticated integration schemes.

keeps the **x** and **z** variables nearly equal. Experimentation has shown that STLS with or without including this term both perform well.

Black-Box Machine Learning Methods

More powerful, yet less interpretable methods from machine learning are increasingly being used to learn dynamical systems from data. Most of the modern techniques involve neural networks, although other techniques such as NARMAX [54] were developed for quite some time. Neural ODEs are a particularly powerful class of machine learning algorithms [101] that build on the popular residual network (ResNet) architecture [224]. Both architectures are illustrated schematically in Figure 8.13.

Residual networks were introduced in 2015 [224], primarily for image classification and super-resolution tasks, and quickly became a dominant architecture. ResNets allow for much deeper neural networks, without suffering from forgetting or vanishing/exploding gradients, by introducing jump connections that directly add a copy of the input to the output. The idea is that it may be much easier to model the difference in the input and output (e.g., in super-resolution tasks, the input should be a coarse version of the output, and so the network only needs to learn the refinement operator). However, it was recognized by Chen et al. [101] that if we think of this network as modeling a differential equation **f**, then the network is essentially mimicking a forward Euler integration scheme

$$\mathbf{x}_{k+1} = \mathbf{x}_k + \mathbf{f}(\mathbf{x}_k) \tag{8.45}$$

with time step $\Delta t = 1$. Euler schemes are known to be prone to errors and instabilities. NODEs instead learn the differential equation **f**, after which it is possible to use more sophisticated numerical integrators, such as Runge–Kutta schemes, to advance the solution.

Other prominent neural network architectures include reservoir computers [384] and long short-term memory (LSTM) networks [520]. Reservoir computers are especially simple to train, as only the last linear layer of the reservoir needs to be adjusted, which amounts to a simple least-squares regression procedure.

Open-Source Software

There are several open-source software libraries for learning dynamical systems from data:

- PyDMD [133, 242] (`https://github.com/mathLab/PyDMD`);
- PySINDy [131, 269] (`https://github.com/dynamicslab/pysindy`);
- Deeptime [233] (`https://github.com/deeptime-ml/deeptime`).

These libraries are fundamentally built on optimization. For example, in PySINDy, the optimization algorithm is explicitly defined when initializing a model:

CodeSnip 8.b

```
import pysindy as ps
import numpy as np

# Generate synthetic data: dx/dt = -x
t = np.linspace(0, 10, 100)                 # Time vector
x = np.exp(-t)                              # Data: x(t)

# Initialize and fit SINDy model using STLSQ optimizer
model = ps.SINDy(optimizer=ps.STLSQ(threshold=0.1))
model.fit(x, t=t)      # Fit the data
model.print()          # Display the discovered equation
```

It is important to note that there are several optimization techniques and regularizations that may be used for the system identification problems above, and several of them are implemented in the software packages above. While DMD uses rank to regularize the ill-conditioned inverse problem, SINDy uses sparsity. An early variant of DMD by Jovanovic et al. [251] used sparsity on the DMD mode amplitudes to downselect only the most important modes, which was useful for flows exhibiting broadband turbulence.

Suggested Reading

Texts

(1) **Applied System Identification**, by J. N. Juang, 1994 [252].

(2) **System Identification: Theory for the User**, by L. Ljung, 1999 [318].

Papers and reviews

(1) **Can one hear the shape of a drum?**, *The American Mathematical Monthly*, by M. Kac, 1966 [259].

(2) **Estimating the dimension of a model**, by G. Schwarz et al., *The Annals of Statistics*, 1978 [454].

(3) **Ensemble Kalman methods for inverse problems**, by M. A. Iglesias, K. J. H. Law, and A. M. Stuart, *Inverse Problems*, 2013 [243].

(4) **Aerodynamic shape optimization of complex aircraft configurations via an adjoint formulation**, by J. Reuther, A. Jameson, J. Farmer, L. Martinelli, and D. Saunders, *Research Institute for Advanced Computer Science, NASA Ames Research Center*, 1996 [413].

(5) **Data-driven aerospace engineering: reframing the industry with machine learning**, by S. L. Brunton, J. N. Kutz, K. Manohar, A. Y. Aravkin, K. Morgansen, J. Klemisch, N. Goebel, J. Buttrick, J. Poskin, A. W. Blom-Schieber, T. Hogan, and D. McDonald, *AIAA Journal*, 2021 [79].

(6) **Compressed sensing**, by D. L. Donoho, *IEEE Transactions on Information Theory*, 2006 [146].

(7) **Signal recovery from random measurements via orthogonal matching pursuit**, by J. A. Tropp and A. C. Gilbert, *IEEE Transactions on Information Theory*, 2007 [505].

(8) **Robust uncertainty principles: Exact signal reconstruction from highly incomplete frequency information**, by E. J. Candès, J. Romberg, and T. Tao, *IEEE Transactions on Automatic Control*, 2006 [92].

(9) **Distilling free-form natural laws from experimental data**, by M. Schmidt and H. Lipson, *Science*, 2009 [447].

(10) **Discovering governing equations from data by sparse identification of nonlinear dynamical systems**, by S. L. Brunton, J. L. Proctor, and J. N. Kutz, *Proceedings of the National Academy of Sciences*, 2016 [81].

(11) **Data-driven sparse sensor placement**, by K. Manohar, B. W. Brunton, J. N. Kutz, and S. L. Brunton, *IEEE Control Systems Magazine*, 2018 [330].

Homework

Exercise 8.1 Generate a noisy data signal $y = \sin(t) + \sigma w$ for $t = 0 : 0.001 : 10$, where w is normally distributed with mean 0 and standard deviation $\sigma = 1$.

Compare the finite difference derivative with the Tikhonov regularized derivative in Eq. (8.6). Compare both the forward difference and central difference schemes. Sweep through different values of σ and λ. Plot the results and compute the error between the estimated and true noise-free derivative.

Exercise 8.2 Repeat the problem above, now comparing the TV-regularized derivative, for various regularization strengths λ.

Exercise 8.3 Generate a piecewise smooth vector

$$y = \begin{cases} x & \text{for } x \in [0, 1] \\ 1 - x & \text{for } x \in (1, 2] \end{cases}$$

defined on $x \in [0, 2]$. Add noise to this vector, as above, and compare the various derivatives: finite difference, Tikhonov regularized, and TV-regularized. What do you notice?

Exercise 8.4 (Bomb Inversion) Estimate the yield of the Trinity nuclear bomb, shown in Fig. 8.7 in Section 8.5. In particular, use the Sedov–Taylor equation, Eq. (8.19), for the radius as a function of time, given a bomb energy E. You can extract the radius at discrete instances in time using the still frames from Fig. 8.7, just as G.I. Taylor did. How does your estimate compare to the estimate of Taylor [490, 491]?

Exercise 8.5 (Matrix inversion and SVD) Generate a random 100×100 matrix $\mathbf{A}$ (e.g., Gaussian entries). Form $\mathbf{B} = \mathbf{A}^{-1}$ (or compute the pseudoinverse if singular) and record the condition number of $\mathbf{A}$.

1. Plot the singular values of $\mathbf{A}$ and $\mathbf{B}$.
2. Discuss the effect of condition number on solving $\mathbf{Ax} = \mathbf{b}$ with noise in $\mathbf{b}$.

Exercise 8.6 (Radon Transform and Tomographic Reconstruction) Create a small 64×64 phantom image (e.g., Shepp–Logan phantom or your own shape).

1. Implement a Radon transform for a set of projection angles θ_i. Plot the resulting sinogram.
2. Add noise to the sinogram and try filtered backprojection for reconstruction. Plot the result and discuss noise artifacts.
3. Now formulate an iterative approach:

$$\min_f \|\mathcal{R}f - \hat{f}\|_2^2 + \lambda R(f),$$

where $\hat{f}$ is the noisy sinogram. Implement a simple gradient or Kaczmarz method. Compare the reconstruction with filtered backpropagation.

Exercise 8.7 (Recovering an audio signal from sparse measurements) To illustrate the use of compressed sensing to reconstruct a high-dimensional signal from a sparse set of random measurements, we consider a signal consisting of a two-tone audio signal:

$$x(t) = \cos(2\pi \times 97t) + \cos(2\pi \times 777t). \tag{8.46}$$

This signal is sparse in the frequency domain, as it is a sum of two cosine waves. The highest frequency is 777 Hz, so the Nyquist sampling rate is 1554 Hz. However, leveraging sparsity of the signal in the frequency domain, we can reconstruct the signal with random samples that are spaced at an average sampling rate of 128 Hz, which is well below the Nyquist sampling rate. In this example, the full signal is generated from $t = 0$ to $t = 1$ with a resolution of $n = 4,096$ and is then randomly sampled at $p = 128$ locations in time.

Code 8.1: Compressed sensing of two-tone cosine signal.

```
## Generate signal, DCT of signal
n = 4096 # points in high resolution signal
t = np.linspace(0,1,n)
x = np.cos(2 * 97 * np.pi * t) + np.cos(2 * 777 * np.pi * t)
xt = np.fft.fft(x) # Fourier transformed signal
PSD = xt * np.conj(xt) / n # Power spectral density

## Randomly sample signal
p = 128 # num. random samples, p = n/32
perm = np.floor(np.random.rand(p) * n).astype(int)
y = x[perm]

## Solve compressed sensing problem
Psi = dct(np.identity(n)) # Build Psi
Theta = Psi[perm,:]       # Measure rows of Psi
```

Now the objective is to solve the underdetermined system of equations

$$\mathbf{y} = \mathbf{\Theta s}$$

for the sparse vector of Fourier coefficients, **s**. Solve for **s** using several techniques:

- Directly solve the system in CVXPY with ℓ_1 regularization on **s**.
- Solve the ℓ_0 regularized system using the proximal gradient method from Section 5.1; try different initializations for **s** (e.g., random, all ones, least-squares).
- Solve the ℓ_0 regularized system using SR3.
- Solve using compressed sensing matching pursuit (CoSaMP); you may install using `pip install cosamp` and then `from cosamp_fn import cosamp`. In CoSaMP, the desired level of sparsity K must be specified.

Note that the $p = 128$ measurements are randomly chosen from the high-resolution signal. Thus, we know the precise timing of the sparse measurements at a much higher resolution than our sampling rate. If we chose $p = 128$ measurements uniformly in time, the compressed sensing algorithm fails. Try this and show the results.

Exercise 8.8 (PDE-Constrained Inverse Problem in 1D) We consider the inverse problem of estimating a forcing term $f(x)$ in the Poisson equation from sparse measurements. Let $u(x)$ solve the 1D Poisson equation

$$-u''(x) = f(x), \qquad x \in (0,1), \qquad u(0) = u(1) = 0.$$

Suppose you collect measurements $y_j \approx u(x_j)$ at $m = 10$ sensor points $\{x_j\}$ in $[0,1]$. The forcing $f(x)$ is unknown.

1. Write an optimization problem to estimate $f(x)$ from the noisy measurements. Your functional should contain a data misfit term comparing the PDE solution $u(x_j)$ with measurements y_j, and a regularization term, such as $\frac{\lambda}{2}\|f\|_{L^2}^2$ or $\frac{\lambda}{2}\|f'\|_{L^2}^2$, with $\lambda > 0$.
2. Now derive the adjoint equation by introducing a Lagrangian that enforces the PDE constraint $-u'' = f$. Derive the first-order optimality conditions by taking variations with respect to u, f, and the adjoint variable $p(x)$. Show that the adjoint PDE is

$$-p''(x) = \sum_{i=1}^{m} \left(u(x_i) - y_i\right)\delta(x - x_i), \qquad p(0) = p(1) = 0.$$

   Show that the optimality condition is $\lambda f - p = 0$ if $L = I$, or $-\lambda f'' - p = 0$ if $L = \frac{d}{dx}$.
3. Discretize $[0, 1]$ with N interior grid points. Let $\mathbf{A}$ be the tridiagonal finite-difference Laplacian with Dirichlet boundaries, so that $\mathbf{Au} = \mathbf{f}$ discretizes $-u'' = f$.

   Write the discrete inverse problem as

$$\min_{f\in\mathbb{R}^N} \quad \frac{1}{2}\|\mathbf{Mu} - \mathbf{y}\|_2^2 + \frac{\lambda}{2}\|\mathbf{L}_h\mathbf{f}\|_2^2, \quad \mathbf{Au} = \mathbf{f},$$

   where $\mathbf{M}$ selects/interpolates the m sensor locations and $\mathbf{L}_h$ is a discrete derivative or identity matrix. Eliminate $\mathbf{u}$ using $\mathbf{Au} = \mathbf{f}$ to obtain the reduced problem in $\mathbf{f}$. Derive the gradient of the reduced functional and show how it can be computed by solving one state equation $\mathbf{Au} = \mathbf{f}$ and one adjoin equation $\mathbf{Ap} = \mathbf{M}^{\mathrm{T}}(\mathbf{Mu} - \mathbf{y})$.
4. Now, implement a gradient-descent or conjugate-gradient method to update $\mathbf{f}$:

$$\mathbf{f}_{k+1} = \mathbf{f}_k - \alpha_k\left(\mathbf{p}_k + \lambda\mathbf{L}_h^{\mathrm{T}}\mathbf{L}_h\mathbf{f}_k\right),$$

   where $\mathbf{u}_k$ solves $\mathbf{Au}_k = \mathbf{f}_k$ and $\mathbf{p}_k$ solves $\mathbf{Ap}_k = \mathbf{M}^{\mathrm{T}}(\mathbf{Mu}_k - \mathbf{y})$. Explain why this only requires the sparse matrix $\mathbf{A}$ and it is never necessary to form $\mathbf{A}^{-1}$ explicitly.
5. Generate synthetic data from a known forcing $f_{\text{true}}(x) = 2 + 4\sin(2\pi x)$, solve for the corresponding $u(x)$, sample it at $m = 10$ points, and add small noise. Run your inversion with and without regularization. Plot the recovered $f(x)$ for different λ. Show that, without regularization, f exhibits spurious oscillations.
6. Why is the inverse Poisson problem ill-posed? Explain the smoothing property of the Green's function (forward map) and how it amplifies noise in inversion. Discuss the role of the adjoint equation in efficiently computing gradients, and the effect of different regularizers.

Exercise 8.9 (Can you hear the shape of a rectangular drum?) Recall that the inverse problem of reconstructing a drum's shape from its frequency spectrum (i.e., the eigenvalues of the Helmholtz/Dirichlet problem) is ill-posed. However, if we restrict the shape of the drum, the inverse problem may be solvable. Here, we restrict the drum shapes to rectangles of area 1. Let Ω be a rectangle given by:

$$0 \le x \le L, \quad 0 \le y \le W,$$

with area $LW = 1$. Thus, we can represent this rectangle using a single ratio parameter $L = r$ and $W = 1/r$, or vice versa. So the shape is determined by $r = \frac{L}{W}$. We may assume $r \ge 1$ so that $L \ge W$.

Waves on this rectangular drum are governed by the Helmholtz–Dirichlet problem:

$$\nabla^2 u + \lambda u = 0 \quad \text{in } \Omega,$$
$$u = 0 \quad \text{on all boundaries of the rectangle.}$$

This PDE models the vibrational modes, and hence acoustic frequencies of a square drum.
We will now solve the inverse problem in stages:

- Use separation of variables $u(x, y) = X(x)Y(y)$ to solve the Helmholtz–Dirichlet PDE. Specifically, plug $u(x, y) = X(x)Y(y)$ into the PDE $\nabla^2 u + \lambda u = 0$ and separate variables to yield

$$\frac{X''(x)}{X(x)} + \frac{Y''(y)}{Y(y)} = -\lambda.$$

  Use the boundary conditions to solve for the admissible frequencies λ in the x and y directions. Write down the general solution $u(x, y)$ in terms of these solutions $X(x)$, $Y(y)$ and the frequencies λ.
- Now, for $L = r$ and $W = \frac{1}{r}$, write the eigenvalues λ in terms of r.
- Discuss the identifiability of $r \geq 1$ from the spectrum λ. Is r uniquely determined? If we didn't specify that $r \geq 1$, would the shape still be uniquely identifiable?
- If only a few portions of the spectrum are available, perhaps the low wavenumbers, is is possible to solve for r?
- Bonus: Simulate the PDE for several $L = r$ and $W = \frac{1}{r}$. Compute the spectrum λ. Solve the inverse problem of estimating r from λ, with and without noise.

Exercise 8.10 Repeat the problem above, but for an elliptical drum with fixed area π. Let a be the length of the major axis and b be the length of the minor axis, so the area is $A = \pi ab$. We restrict $ab = 1$ and let $r = \frac{a}{b}$ denote the aspect ratio. The inverse problem is to solve for this aspect ratio given the measured eigenvalue spectrum λ (i.e., the acoustic signature).

Use the standard ellipse equation:

$$\frac{x^2}{a^2} + \frac{y^2}{b^2} = 1 \quad \Longrightarrow \quad \frac{x^2}{r} + ry^2 = 1.$$

Exercise 8.11 (Comparing Numerical Differentiation Techniques) Consider a noisy 1D signal $s(t) = \sin(3t) + 0.5\sin(10t)$ sampled at $t_k = k\,\Delta t$, $k = 0, \ldots, 200$, with Gaussian noise added. Compute the true derivative $\dot{s}(t)$ and implement three derivative approximations:

1. Finite-differences with a central difference scheme;
2. Tikhonov-regularized differentiation; and
3. Total variation (TV) regularization, solving

$$\min_{\dot{s}} \tfrac{1}{2}\|\mathbf{H}\dot{s} - s\|_2^2 + \lambda\|\dot{s}\|_{\text{TV}},$$

where $\mathbf{H}$ is an integration operator. Plot the estimated derivatives for each method. Which is most robust to noise? How does λ affect the TV solution?

Exercise 8.12 (SINDy on Lotka–Volterra) Generate data from a known nonlinear system, the Lotka–Volterra predator–prey model:

$$\dot{x}_1 = \alpha x_1 - \beta x_1 x_2, \quad \dot{x}_2 = \delta x_1 x_2 - \gamma x_2.$$

1. Collect trajectory data $\mathbf{x}(t_k)$ from an initial condition $[1, 1]$ for $t \in [0, 10]$, with $\Delta t = 0.01$, $\alpha = 2/3$, $\beta = 4/3$, $\delta = \gamma = 1$. Estimate $\dot{\mathbf{x}}(t_k)$ using central difference derivatives.
2. Build a library $\boldsymbol{\Theta}(\mathbf{x})$ with polynomials up to degree 2.
3. Solve $\dot{\mathbf{X}} = \boldsymbol{\Theta}(\mathbf{X})\, \boldsymbol{\Xi}$ using sequential thresholded least squares (STLS) and compare $\boldsymbol{\Xi}$ with the true $(\alpha, \beta, \delta, \gamma)$. How does the threshold λ affect your results?
4. Now solve for $\boldsymbol{\Xi}$ using LASSO. Compare the results with STLS.
5. Repeat the steps above, adding increasingly large amounts of noise to the trajectory $\mathbf{x}$, before computing the derivative. How does this change the results?
6. Repeat with a library of polynomials up to degree 3, also adding noise to $\mathbf{x}$.

Exercise 8.13 (SINDy on Lorenz) This exercise will test the data requirements for identifying an accurate SINDy model for the Lorenz system, following the work of Champion et al. [97]. The equations of motion for the Lorenz system are

$$\dot{x} = \sigma(y - x) \tag{8.47}$$

$$\dot{y} = x(\rho - z) - y \tag{8.48}$$

$$\dot{z} = xy - \beta z. \tag{8.49}$$

Use parameters $(\sigma, \rho, \beta) = (10, 28, 8/3)$ and initial condition $\mathbf{x}_0 = (x(0), y(0), z(0)) = (0, 1, 20)$. First, we will use clean data without any measurement noise.

(a) Generate a long trajectory, $t \in [0, 20]$, with a fine sampling rate, $\Delta t = 0.0001$; discard the first $t \in [0, 1]$. Generate SINDy models with increasing amounts of data T. Plot the data up to the T where SINDy learns the correct Lorenz model and summarize.

(b) Repeat this experiment for different sampling rates from $t = 0.0001$ to 0.1. How does the minimum data length T and the number of samples $T/\Delta t$ change?

(c) Repeat the experiment with small additive Gaussian noise on the trajectory. Repeat the identification for several noise realizations to obtain an average success rate. Add different noise magnitudes. Reproduce the main figure in Champion et al. [97].

Exercise 8.14 This example will use SINDy to learn the nonlinear Rossler system from data. The Rossler system is given by

$$\dot{x} = -y - z$$

$$\dot{y} = x + ay$$

$$\dot{z} = b + z(x - c)$$

with parameters $a = 0.2$, $b = 0.2$, and $c = 14$. Simulate this system with $\Delta t = 0.001$ from an initial condition $[1, 1, 0]$. Use this trajectory to identify a SINDy model. Explore different sparsifying thresholds and different trajectory lengths. Now add a small amount of noise to the trajectory and re-identify the model, again with different thresholds and lengths. When there is sufficiently long trajectory data, what happens when the threshold is too large? When it is too small?

Exercise 8.15 Now we will revisit the Lorenz system and explore the effect of different de-noising derivative algorithms on identification from noisy trajectory data.

1. First, generate a trajectory with $\Delta t = 0.01$ with the standard parameters and initial condition. Discard the first 1 time units of data, and use the next 5 time units for the trajectory (i.e., from $t = 1$ to $t = 6$).
2. Add a small amount of Gaussian noise ($\sigma = 0.001$) to the trajectory, and estimate the derivative using several techniques: forward difference, central difference, TV regularized differentiation, and polynomial smoothing.
3. Use all of these velocity estimates to identify a SINDy model. Compare the results.
4. Repeat for larger and smaller noise and report on the trends.

Exercise 8.16 Now we will compare more regularization schemes for SINDy. Using the same data from the previous exercise, compare the following SINDy models using different regularization schemes:

- L_1 regularization on Ξ with LASSO;
- L_1 regularization on Ξ with STLS;
- L_2 regularization on Ξ (i.e., ridge or Tikhonov regularization);
- L_1 and L_2 regularization on Ξ (i.e., elastic net);
- L_0 regularization on Ξ using the proximal gradient; and
- L_0 regularization on Ξ using SR3.

Plot the discovered coefficients vs. regularization weight. Which penalty yields more sparse solutions? Which is more robust to moderate noise?

Exercise 8.17 We will explore the PDE variant of SINDy, called PDE-FIND [430], to learn the viscous Burgers' equation on $x \in [0, 2\pi]$:

$$u_t + uu_x = \nu u_{xx}.$$

1. First, simulate data $u(x, t)$ from a spectral or finite-difference code.
2. Build a dictionary of candidate terms: $\{u, u^2, u_x, u_{xx}, u\, u_x, u^2 u_x, \dots\}$.
3. Perform the PDE-FIND regression:

$$\min \|\mathbf{U}_t - \Theta(\mathbf{U}, \mathbf{U}_x, \dots)\, \Xi\|_2^2 + \lambda \|\Xi\|_1.$$

4. Compare the discovered PDE with the ground truth. Plot discovered terms vs. ν.
5. Discuss how noise in $u(x, t)$ affects the results and give an idea for potentially addressing this.

Exercise 8.18 This exercise will explore identifying a PDE using PDE-FIND based on data from the Korteweg–De Vries (KdV) system

$$u_t + u_{xxx} - 6uu_x = 0.$$

This system gives rise to coherent traveling wave solutions, known as solitons, that retain their shape as they travel at a constant wave speed, despite the nonlinearity in the system. In general, a soliton solution will take the form $u(x,t) = -\frac{c}{2}\text{sech}^2\left(\frac{\sqrt{c}}{2}(x-ct)\right)$ for an arbitrary positive wave speed c. Notice that the wave speed c is linked to the amplitude and width of the wave.

(a) Generate data by initializing a single soliton solution $u(x,0) = -\frac{1}{2}\text{sech}^2\left(\frac{1}{2}(x)\right)$, with a wave speed corresponding to $c = 1$ above. Try to identify a PDE model using PDE-FIND based on this data. What is the sparsest model that supports the data?

(b) Next, use data beginning at two initial conditions. The first initial condition is the one above with $c = 1$, and the second initial condition is a soliton solution with $c = 4$. Concatenate this data and identify a PDE model using PDE-FIND. What is the sparsest model that supports the data?

(c) Discuss the two models that are identified depending on the initial data.

Exercise 8.19 Now we will explore SINDy on real-world data. Collect actual measurement data from a simple mechanical system, such as a mass-spring-damper. You may either measure the relevant position or velocity variables using an accelerometer, or from video, for example recorded on a smart phone.

First, suppose that you know the equations of motion but you do not know the damping coefficient c and spring constant k.

1. Formulate an inverse problem to estimate c and k.
2. Fit these parameters from data.
3. Discuss identifiability and how much data is required for a robust estimate.

Now, suppose that you don't know the equations of motion and must learn these as well.

1. Formulate another inverse problem to estimate the model from data.
2. Decide on an appropriate algorithm to learn the model from data.
3. Explore the data requirements to learn a good model.

Exercise 8.20 Repeat the previous example, but for a simple pendulum, for example the arm of a grandfather clock. Again, first suppose that you know the equations of motion but don't know the damping coefficient. Then assume that you don't know the equations of motion and must learn these as well.

Exercise 8.21 In this problem, we will apply dynamic mode decomposition (DMD) to analyze a simple two-dimensional fluid flow past a circular cylinder. This flow exhibits periodic vortex shedding, and it is a canonical fluid flow that is used to understand more sophisticated flows, such as over aircraft wings.

Download the cylinder flow data from `http://dmdbook.com/DATA.zip`. There are 150 vorticity field snapshots, each of dimension 199×449. Thus, reshaping each flow field into a tall skinny vector and arranging into columns of a matrix, we obtain a data matrix with dimensions 89351×150. A single flow field snapshot is shown in Fig. 8.14.

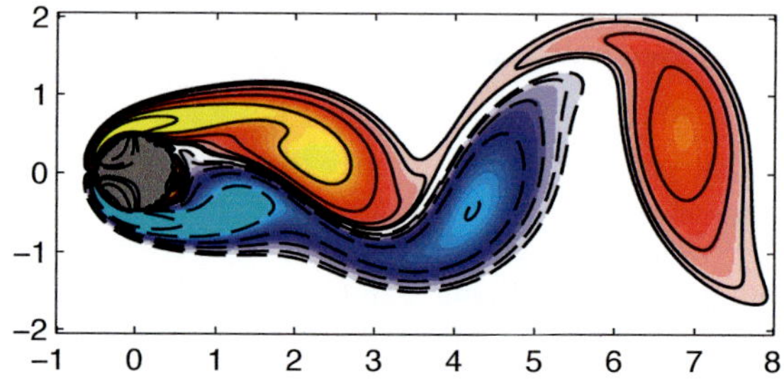

Figure 8.14 Single snapshot of the vorticity field for the fluid flow past a two-dimensional cylinder. This flow is periodic and conserves energy.

- Compute DMD on this cylinder flow data and plot the DMD eigenvalues in the complex plane. Use $r = 20$ modes by truncating and keeping only the first 20 modes in the singular value decomposition. This means you should have 20 DMD eigenvalues. Because the flow is periodic and energy conservative, the DMD eigenvalues should come in complex conjugate pairs on the unit circle in the complex plane.
- Now add a small amount of white noise to the data, and repeat the exercise above. Try 1%, 2%, 4%, etc. and plot the new DMD eigenvalues on the same plot as the DMD eigenvalues for the clean data. What trends do you notice?
- Now use the clean data, but take a subset of the flow field snapshots so that you only sample an incomplete period of vortex shedding (i.e., only sample 75% of a vortex shedding cycle). Repeat the exercise above and plot the DMD eigenvalues.

You are welcome to use `PyDMD` for any of this assignment.

In a future homework, we will use *physics -informed DMD* (PIDMD) to incorporate physics into the DMD regression to improve the results for noisy and partial data. What are some potential aspects of the physics that we might try to include?

9 Optimization for Control

Control theory is a major branch of applied optimization, where we design a control law to interact with and modify the behavior of a dynamical system. For example, a cruise controller in an automobile is a simple feedback controller that modulates the accelerator to maintain a desired reference speed, and is effective despite unmeasured factors, such as hills, wind, and cargo. Feedback control, where sensor measurements are directly used to inform a real-time control law, is an increasingly essential part of the modern engineered world. The applications are nearly endless, including autopilots, the control of construction equipment, heating ventilation and cooling systems, internet traffic routing, and many more. Indeed, progress in the use of control theory in increasingly challenging application domains has closely tracked progress in our ability to solve hard optimization problems, especially in real time.

We will define a controller in the context of optimization. An optimal controller balances the performance and robustness of the closed-loop system with the cost of control, as quantified by an objective function; in control, the objective function is often called the cost function. This results in a constrained optimization problem: the state of the system, which appears in the objective function, is constrained by the dynamics. Other constrains involve limitations on the available actuation power and desired limits on the state of the system.

Control theory is a vast topic, and a full treatment is beyond the scope of this chapter. For more details, see the *Control Bootcamp*[1]; there are also several excellent texts that provide a deeper background [156, 466, 15, 77]. This chapter presents a high-level overview of applications of optimization in control, with an emphasis on the nature of the optimization problem and solution techniques. We will almost exclusively consider closed-loop feedback control, which uses sensor measurements to inform the controller about how the system is actually responding. In contrast, open-loop control relies on a pre-programmed control sequence, perhaps informed by a model, but without sensor feedback. Closed-loop control is essential for stabilizing unstable systems and to account for an imperfect model and other forms of uncertainty.

We begin with linea optimal control, as this is the most well-developed and intuitive branch of control theory, where the optimization problem often has an analytic, closed-form solution. These solutions include the classic linear quadratic regulator (LQR) and Kalman filter, which both require solving a Riccati equation. We will derive the Riccati equation using constrained optimization and Lagrange multipliers. Linear optimal control is often

[1] `https://www.youtube.com/playlist?list=PLMrJAkhIeNNR20Mz-VpzgfQs5zrYi085m`.

also effective at regulating nonlinear systems in a neighborhood of a fixed point, where the assumption of linear dynamics is valid.

We then explore the general nonlinear optimal control problem, which involves the Hamilton–Jacobi–Bellman (HJB) equations. Here we will introduce reinforcement learning and dynamic programming, which are optimization frameworks for solving this nonlinear control problem. Reinforcement learning has been developed for a century, and it is progressing rapidly with advances in machine learning; these will be discussed in Chapter 10.

Next, we develop the model predictive control (MPC) framework as an extension to optimal control for a broader class of more complex control tasks. This framework is widely used in modern industrial control due to its flexibility, robustness, ease of implementation, and ability to respect constraints. It involves repeatedly solving an optimal control problem on a receding horizon, updating the state of the system, and reinitializing the optimization problem as new sensor information becomes available. And it is computationally intensive and has historically been used primarily for "slow" applications and for systems with simple dynamics and constraints. For example, a linear system with a quadratic objective function and convex constraints will result in a convex optimization problem that may be solved with scaleable on-line optimization methods for real-time control. Fortunately, recent advances in computational hardware and optimization algorithms have made it increasingly possible to apply MPC to larger-scale systems, with faster timescales and more complex dynamics and constraints.

Finally, we discuss linear matrix inequalities (LMIs), which allow us to reformulate many problems in control as convex optimization problems. For example, LQR and the Kalman filter can be framed in terms of LMIs and then solved efficiently using interior point methods.

Mathematical Statement of Optimal Feedback Control Problem

Figure 9.1 depicts the general feedback control framework, where sensor measurements $\mathbf{y}$ are fed into a controller, which then computes an actuation signal $\mathbf{u}$ for manipulating the system dynamics to minimize an objective function that quantifies performance, actuation cost, and robustness to model uncertainty, external disturbances, and other factors. The vector of external disturbances is typically decomposed as $\mathbf{w} = \begin{bmatrix}\mathbf{w}_d^T & \mathbf{w}_n^T & \mathbf{w}_r^T\end{bmatrix}^T$, where $\mathbf{w}_d$ are disturbances to the state of the system, $\mathbf{w}_n$ is measurement noise, and $\mathbf{w}_r$ is a reference trajectory that should be tracked by the closed-loop system.

The system dynamics are typically described as differential equations:

$$\frac{\mathrm{d}}{\mathrm{d}t}\mathbf{x} = \mathbf{f}(\mathbf{x}, \mathbf{u}, \mathbf{w}_d) \tag{9.1a}$$

$$\mathbf{y} = \mathbf{g}(\mathbf{x}, \mathbf{u}, \mathbf{w}_n), \tag{9.1b}$$

where $\mathbf{f}$ encodes the evolution of the system state $\mathbf{x}$ under actuation $\mathbf{u}$ and disturbances $\mathbf{w}_d$, and $\mathbf{y} = \mathbf{g}(\mathbf{x}, \mathbf{u}, \mathbf{w}_n)$ is a measurement equation.

The goal is to construct a control law

$$\mathbf{u} = \mathbf{k}(\mathbf{y}, \mathbf{w}_r) \tag{9.2}$$

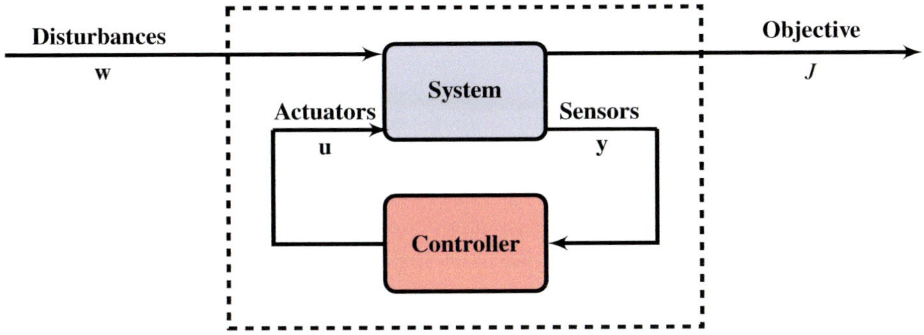

Figure 9.1 Standard framework for feedback control. Measurements $\mathbf{y}(t)$ are fed into a controller, which decides on an actuation signal $\mathbf{u}(t)$ to control the system. The control law is designed to modify the system dynamics and provide good performance, quantified by the objective J, despite external disturbances and noise in $\mathbf{w}$. The external input $\mathbf{w}$ may also include a reference trajectory $\mathbf{w}_r$ that should be tracked.

that minimizes an objective function

$$J(\mathbf{x}, \mathbf{u}, \mathbf{w}_r). \tag{9.3}$$

The sensor measurements $\mathbf{y}$ provide information about unmodeled dynamics and disturbances that would degrade the performance without closed-loop feedback control. Further, with feedback it is often possible to modify and stabilize the dynamics of the closed-loop system, something which is not possible with open-loop control. Thus, closed-loop feedback control is often able to maintain high-performance operation for systems with unstable dynamics, while being robust to model uncertainty and external disturbances.

In general, the controller in Eq. (9.2) will be a dynamical system, rather than a static function of the inputs. For example, the Kalman filter dynamically estimates the full state $\mathbf{x}$ from measurements of $\mathbf{u}$ and $\mathbf{y}$. In this case, the control law will become $\mathbf{u} = \mathbf{k}(\mathbf{y}, \hat{\mathbf{x}}, \mathbf{w}_r)$, where $\hat{\mathbf{x}}$ is an estimate of the state. Thus, optimal control typically involves optimization of a continuous function in a well-defined function space. For example, the linear quadratic regulator will involve finding the optimal control function in the space of linear functions $\mathbf{u} = \mathbf{K}\mathbf{x}$. Sometimes, the control action will be discretized in time and optimized over a finite horizon, as in MPC, resulting in a finite-dimensional optimization.

Throughout this chapter, we will consider codes that implement controller optimizations using the Python Control Systems Library (`python-control`)[2] toolbox provides extensive functionality, including Python wrappers for the SLICOT optimization libraries [44] used in Matlab's control toolboxes. It is assumed that all codes begin with the following:

CodeSnip 9.a

```
from control.matlab import *
import slycot
```

[2] Available at `https://python-control.readthedocs.io/`.

9.1 Linear Optimal Control

The most complete theory of control has been developed for linear systems [466, 156, 15]. Similar to convex optimization, linear control is essentially a solved problem, with robust and scalable numerical algorithms. In practice, linear systems almost always arise by linearizing a nonlinear system about a fixed point or periodic orbit. Even classically linear systems, such as a spring-mass-damper or a pendulum with small angle are idealizations of more complex nonlinear systems. A linear system may be written in the following form:

$$\frac{\mathrm{d}}{\mathrm{d}t}\mathbf{x} = \mathbf{A}\mathbf{x} + \mathbf{B}\mathbf{u} \tag{9.4a}$$

$$\mathbf{y} = \mathbf{C}\mathbf{x} + \mathbf{D}\mathbf{u}. \tag{9.4b}$$

Instability may quickly take a trajectory far away from the fixed point. However, an effective stabilizing controller will keep the state of the system in a small neighborhood of the fixed point where the linear approximation is valid. For example, in the case of the inverted pendulum, feedback control may stabilize the pendulum in the vertical position where the dynamics behave linearly. Note that we have neglected the disturbance and noise inputs, $\mathbf{w}_d$ and $\mathbf{w}_n$, respectively; these will be added back in the discussion on Kalman filtering.

In this section, we will formulate and solve several optimal control problems. First, we assume we have access to the full state of the system, $\mathbf{y} = \mathbf{x}$, which allows us to design the optimal full-state feedback controller, $\mathbf{u} = -\mathbf{K}\mathbf{x}$, that minimizes a quadratic objective function in terms of the state $\mathbf{x}$, and actuation input $\mathbf{u}$. This is known as the linear quadratic regulator (LQR), because the dynamics constraints are linear and the objective is quadratic. This problem is convex, with a single globally optimal solution. Next, we will relax the assumption of full-state feedback, and instead assume that we only have access to partial measurements $\mathbf{y}$, from which we must estimate the full state. This will result in the Kalman filter, which is an optimal estimator assuming Gaussian noise and disturbances; the Kalman filter may also be formulated from a Bayesian perspective, related to topics in Chapter 7. Finally, we combine these two approaches in what is known as the linear quadratic Gaussian (LQG) optimal feedback controller, given partial measurements.

Linear Quadratic Regulator and the Riccati Equation

The goal of linear, full-state feedback control is to design a control law

$$\mathbf{u} = -\mathbf{K}\mathbf{x} \tag{9.5}$$

so that the closed-loop system

$$\frac{\mathrm{d}}{\mathrm{d}t}\mathbf{x} = (\mathbf{A} - \mathbf{B}\mathbf{K})\mathbf{x} \tag{9.6}$$

is stable and has good performance, quantified by a specific objective function. Not all systems are controllable, meaning that some states of the system are not possible to influence by actuation. A system is controllable if the controllability matrix

$$\mathcal{C} = \begin{bmatrix}\mathbf{B} & \mathbf{A}\mathbf{B} & \mathbf{A}^2\mathbf{B} & \cdots & \mathbf{A}^{n-1}\mathbf{B}\end{bmatrix} \tag{9.7}$$

has n linearly independent columns. If so, then it is possible to arbitrarily manipulate the eigenvalues of the closed-loop system $(\mathbf{A} - \mathbf{BK})$ through choice of $\mathbf{K}$, and stabilize the system. The argument above assumes that full-state measurements are available (i.e., $\mathbf{C} = \mathbf{I}$ and $\mathbf{D} = \mathbf{0}$, so that $\mathbf{y} = \mathbf{x}$). Of course, full-state measurements are not always available, especially for high-dimensional systems, but we will show in the next section that if the system is observable, it is possible to build a full-state estimate from the sensor measurements via a Kalman filter.

Given a controllable system, there are many stabilizing control laws $\mathbf{u} = -\mathbf{Kx}$. A classic goal of optimal control is to choose the best controller gain matrix $\mathbf{K}$ to stabilize the system without expending too much control effort and without making the system overly sensitive to noise and disturbances. The objective function

$$J = \int_0^\infty \mathbf{x}(\tau)^\mathrm{T}\mathbf{Q}\mathbf{x}(\tau) + \mathbf{u}(\tau)^\mathrm{T}\mathbf{R}\mathbf{u}(\tau)\,\mathrm{d}\tau \tag{9.8}$$

balances the cost of regulating the state with the cost of control. The matrices $\mathbf{Q}$ and $\mathbf{R}$ weight the cost of deviations of the state from zero and the cost of actuation, respectively. The matrix $\mathbf{Q} \geq \mathbf{0}$ is positive semi-definite, and $\mathbf{R} > \mathbf{0}$ is positive definite; they are often diagonal, and the diagonal elements quantify the relative importance of the control objectives.

Adding such an objective function is essential in formulating a well-posed optimization problem. The LQR control law $\mathbf{u} = -\mathbf{K}_r\mathbf{x}$ is designed to minimize the objective function subject to the dynamics constraints:

$$\begin{aligned} \min_{\mathbf{u}=-\mathbf{K}_r\mathbf{x}} \quad & \int_0^\infty \mathbf{x}(\tau)^\mathrm{T}\mathbf{Q}\mathbf{x}(\tau) + \mathbf{u}(\tau)^\mathrm{T}\mathbf{R}\mathbf{u}(\tau)\,\mathrm{d}\tau \\ \text{s.t.} \quad & \dot{\mathbf{x}} = \mathbf{Ax} + \mathbf{Bu}. \end{aligned} \tag{9.9}$$

The LQR controller is shown schematically in Fig. 9.2. Out of all possible control laws $\mathbf{u} = \mathbf{K}(\mathbf{x})$, including nonlinear controllers, the LQR controller $\mathbf{u} = -\mathbf{K}_r\mathbf{x}$ is optimal, as we will show in the next section. The name LQR refers to the fact that we are designing a linear control law to minimize a quadratic objective function, subject to linear dynamics, that regulates the state of the system to $\lim_{t\to\infty} \mathbf{x}(t) = \mathbf{0}$.

Because the objective function in Eq. (9.8) is quadratic, there is an analytical solution for the optimal controller gain $\mathbf{K}_r$, given by

$$\mathbf{K}_r = \mathbf{R}^{-1}\mathbf{B}^\mathrm{T}\mathbf{X}, \tag{9.10}$$

where $\mathbf{X}$ is the solution to an algebraic Riccati equation, which we will derive next:

$$\mathbf{A}^\mathrm{T}\mathbf{X} + \mathbf{X}\mathbf{A} - \mathbf{X}\mathbf{B}\mathbf{R}^{-1}\mathbf{B}^\mathrm{T}\mathbf{X} + \mathbf{Q} = \mathbf{0}. \tag{9.11}$$

Solving the Riccati equation for $\mathbf{X}$, and hence for $\mathbf{K}_r$, is numerically robust and is implemented in many programming languages [300, 43]. The gain matrix $\mathbf{K}_r$ is obtained via

CodeSnip 9.b
```
>>> Kr = lqr(A,B,Q,R)[0]
```

in python-control. However, solving the Riccati equation scales as $O(n^3)$ in the state-dimension n, making it prohibitively expensive for large systems or for online computations,

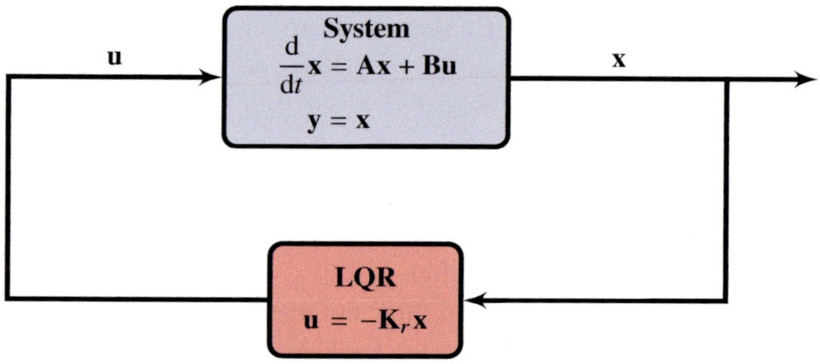

Figure 9.2 Schematic of the linear quadratic regulator (LQR) for optimal full-state feedback. The optimal controller for a linear system given measurements of the full state, $\mathbf{y} = \mathbf{x}$, is given by proportional control $\mathbf{u} = -\mathbf{K}_r\mathbf{x}$ where $\mathbf{K}_r$ is a constant gain matrix obtained by solving an algebraic Riccati equation.

for example in MPC (see Section 9.5). This motivates LMI formulations of the LQR problem (see Section 9.6), which may be solved using efficient interior point methods (see Section 6.4). Similarly, we may develop reduced-order models that capture the dominant behavior of the system with many fewer states.

Derivation of the Riccati Equation for Optimal Control

For the intrepid among us, it is worth taking a theoretical detour to derive the Riccati equation in Eq. (9.11) for the LQR optimization problem. This derivation will provide an example of how to solve convex optimization problems using the calculus of variations, and it will also provide a template for computing the optimal control solution for nonlinear systems. Because of the similarity of optimal control to the formulation of Lagrangian and Hamiltonian classical mechanics in terms of the variational principal, we adopt similar language and notation.

First, we define the LQR objective function up until a finite time t_f and add a terminal cost, introducing a factor of $1/2$ to simplify computations:

$$J = \underbrace{\int_0^{t_f} \frac{1}{2}\left(\mathbf{x}(\tau)^{\mathrm{T}}\mathbf{Q}\mathbf{x}(\tau) + \mathbf{u}(\tau)^{\mathrm{T}}\mathbf{R}\mathbf{u}(\tau)\right) \mathrm{d}\tau}_{\text{Lagrangian}\,\mathcal{L}} + \underbrace{\frac{1}{2}\mathbf{x}(t_f)^{\mathrm{T}}\mathbf{Q}_f\mathbf{x}(t_f)}_{\text{Terminal cost}}. \tag{9.12}$$

The goal is to minimize the objective J subject to the dynamics $\dot{\mathbf{x}} = \mathbf{A}\mathbf{x} + \mathbf{B}\mathbf{u}$. We solve this by introducing a Lagrangian objective function that adds an additional term to enforce the dynamics constraint s via Lagrange multipliers:

$$J_{\text{aug}} = \int_0^{t_f} \left[\frac{1}{2}\left(\mathbf{x}^{\mathrm{T}}\mathbf{Q}\mathbf{x} + \mathbf{u}^{\mathrm{T}}\mathbf{R}\mathbf{u}\right) + \boldsymbol{\lambda}^{\mathrm{T}}\left(\mathbf{A}\mathbf{x} + \mathbf{B}\mathbf{u} - \dot{\mathbf{x}}\right)\right] \mathrm{d}\tau + \frac{1}{2}\mathbf{x}(t_f)^{\mathrm{T}}\mathbf{Q}_f\mathbf{x}(t_f).$$

The Lagrange multiplier $\boldsymbol{\lambda}$ is also called the *co-state*. Taking the total variation of J_{aug} yields:

$$\delta J_{\text{aug}} = \int_0^{t_f} \left[\frac{\partial \mathcal{L}}{\partial \mathbf{x}}\delta\mathbf{x} + \frac{\partial \mathcal{L}}{\partial \mathbf{u}}\delta\mathbf{u} + \boldsymbol{\lambda}^{\mathrm{T}}\mathbf{A}\delta\mathbf{x} + \boldsymbol{\lambda}^{\mathrm{T}}\mathbf{B}\delta\mathbf{u} - \boldsymbol{\lambda}^{\mathrm{T}}\delta\dot{\mathbf{x}}\right] \mathrm{d}\tau + \mathbf{Q}_f\mathbf{x}(t_f)\delta\mathbf{x}(t_f).$$

The partial derivatives[3] of the Lagrangian are $\partial\mathcal{L}/\partial\mathbf{x} = \mathbf{x}^\mathrm{T}\mathbf{Q}$ and $\partial\mathcal{L}/\partial\mathbf{u} = \mathbf{u}^\mathrm{T}\mathbf{R}$. The last term in the integral may be modified using integration by parts:

$$-\int_0^{t_f} \boldsymbol{\lambda}^\mathrm{T}\delta\dot{\mathbf{x}}\,\mathrm{d}\tau = -\boldsymbol{\lambda}^\mathrm{T}(t_f)\delta\mathbf{x}(t_f) + \boldsymbol{\lambda}^\mathrm{T}(0)\delta\mathbf{x}(0) + \int_0^{t_f} \dot{\boldsymbol{\lambda}}^\mathrm{T}\delta\mathbf{x}\,\mathrm{d}\tau.$$

The term $\boldsymbol{\lambda}^\mathrm{T}(0)\delta\mathbf{x}(0)$ is equal to zero, since we do not let the initial condition vary with control, or else the control system would be non-causal (i.e., then future control could change the initial condition of the system).

Finally, the total variation of the augmented objective function simplifies to

$$\delta J_{\mathrm{aug}} = \int_0^{t_f} \left(\mathbf{x}^\mathrm{T}\mathbf{Q} + \boldsymbol{\lambda}^\mathrm{T}\mathbf{A} + \dot{\boldsymbol{\lambda}}^\mathrm{T}\right)\delta\mathbf{x}\,\mathrm{d}\tau + \int_0^{t_f} \left(\mathbf{u}^\mathrm{T}\mathbf{R} + \boldsymbol{\lambda}^\mathrm{T}\mathbf{B}\right)\delta\mathbf{u}\,\mathrm{d}\tau$$
$$+ \left(\mathbf{x}(t_f)^\mathrm{T}\mathbf{Q}_f - \boldsymbol{\lambda}^\mathrm{T}(t_f)\right)\delta\mathbf{x}(t_f).$$

Each variation term must equal zero for an optimal control solution that minimizes J. Thus, we may break this up into three equations:

$$\mathbf{x}^\mathrm{T}\mathbf{Q} + \boldsymbol{\lambda}^\mathrm{T}\mathbf{A} + \dot{\boldsymbol{\lambda}}^\mathrm{T} = \mathbf{0} \tag{9.13a}$$
$$\mathbf{u}^\mathrm{T}\mathbf{R} + \boldsymbol{\lambda}^\mathrm{T}\mathbf{B} = \mathbf{0} \tag{9.13b}$$
$$\mathbf{x}(t_f)^\mathrm{T}\mathbf{Q}_f - \boldsymbol{\lambda}^\mathrm{T}(t_f) = \mathbf{0}. \tag{9.13c}$$

The constraint in Eq. (9.13c) represents an initial condition for the reverse-time equation for $\boldsymbol{\lambda}$ starting at t_f. Thus, Eqs. (9.4a) and (9.13) with initial condition $\mathbf{x}(0) = \mathbf{x}_0$ and final condition $\boldsymbol{\lambda}(t_f) = \mathbf{Q}_f\mathbf{x}(t_f)$ form a two-point boundary value problem. These equations may be integrated numerically to find the optimal control solution, even for nonlinear systems.

Because the dynamics are linear, it is possible to *posit* the form $\boldsymbol{\lambda} = \mathbf{P}\mathbf{x}$, and substitute into Eq. (9.13) above. The first equation becomes:

$$\left(\dot{\mathbf{P}}\mathbf{x} + \mathbf{P}\dot{\mathbf{x}}\right)^\mathrm{T} + \mathbf{x}^\mathrm{T}\mathbf{Q} + \boldsymbol{\lambda}^\mathrm{T}\mathbf{A} = \mathbf{0}.$$

Taking the transpose, and substituting Eq. (9.4a) in for $\dot{\mathbf{x}}$, yields:

$$\dot{\mathbf{P}}\mathbf{x} + \mathbf{P}(\mathbf{A}\mathbf{x} + \mathbf{B}\mathbf{u}) + \mathbf{Q}\mathbf{x} + \mathbf{A}^\mathrm{T}\mathbf{P}\mathbf{x} = \mathbf{0}.$$

Note that $\mathbf{Q}$ is symmetric, so $\mathbf{Q}^\mathrm{T} = \mathbf{Q}$. From Eq. (9.13b), we have

$$\mathbf{u} = -\mathbf{R}^{-1}\mathbf{B}^\mathrm{T}\boldsymbol{\lambda} = -\mathbf{R}^{-1}\mathbf{B}^\mathrm{T}\mathbf{P}\mathbf{x}.$$

Again, $\mathbf{R}$ is typically symmetric and diagonal. Finally, combining yields:

$$\dot{\mathbf{P}}\mathbf{x} + \mathbf{P}\mathbf{A}\mathbf{x} + \mathbf{A}^\mathrm{T}\mathbf{P}\mathbf{x} - \mathbf{P}\mathbf{B}\mathbf{R}^{-1}\mathbf{B}^\mathrm{T}\mathbf{P}\mathbf{x} + \mathbf{Q}\mathbf{x} = \mathbf{0}. \tag{9.14}$$

This equation must be true for all $\mathbf{x}$, and so it may also be written as a matrix equation. In the infinite-horizon caes, we let time go to infinity and drop the terminal cost, in which case the $\dot{\mathbf{P}}$ term disappears and we recover the algebraic Riccati equation:

$$\mathbf{P}\mathbf{A} + \mathbf{A}^\mathrm{T}\mathbf{P} - \mathbf{P}\mathbf{B}\mathbf{R}^{-1}\mathbf{B}^\mathrm{T}\mathbf{P} + \mathbf{Q} = \mathbf{0}. \tag{9.15}$$

[3] The derivative of $\mathbf{A}\mathbf{x}$ with respect to $\mathbf{x}$ is $\mathbf{A}$, and the derivative of $\mathbf{x}^\mathrm{T}\mathbf{A}$ with respect to $\mathbf{x}$ is $\mathbf{A}^\mathrm{T}$.

Although this procedure is somewhat involved, each step is relatively straightforward. In addition, the dynamics in Eq (9.4a) may be replaced with nonlinear dynamics $\dot{\mathbf{x}} = \mathbf{f}(\mathbf{x}, \mathbf{u})$, and a similar nonlinear two-point boundary value problem may be formulated with $\partial\mathbf{f}/\partial\mathbf{x}$ replacing $\mathbf{A}$ and $\partial\mathbf{f}/\partial\mathbf{u}$ replacing $\mathbf{B}$. Therefore, this procedure is also extremely general and may be used to numerically obtain nonlinear optimal control trajectories.

Hamiltonian Formulation

It is also possible to solve the optimization problem by introducing the following dual Hamiltonian formulation:

$$\mathcal{H} = \underbrace{\frac{1}{2}\left(\mathbf{x}^{\mathrm{T}}\mathbf{Q}\mathbf{x} + \mathbf{u}^{\mathrm{T}}\mathbf{R}\mathbf{u}\right)}_{\mathcal{L}} + \boldsymbol{\lambda}^{\mathrm{T}}\left(\mathbf{A}\mathbf{x} + \mathbf{B}\mathbf{u}\right). \tag{9.16}$$

Then Hamilton's equations become:

$$\dot{\mathbf{x}} = \left(\frac{\partial \mathcal{H}}{\partial \boldsymbol{\lambda}}\right)^{\mathrm{T}} = \mathbf{A}\mathbf{x} + \mathbf{B}\mathbf{u} \qquad \mathbf{x}(0) = \mathbf{x}_0 \tag{9.17a}$$

$$-\dot{\boldsymbol{\lambda}} = \left(\frac{\partial \mathcal{H}}{\partial \mathbf{x}}\right)^{\mathrm{T}} = \mathbf{Q}\mathbf{x} + \mathbf{A}^{\mathrm{T}}\boldsymbol{\lambda} \qquad \boldsymbol{\lambda}(t_f) = \mathbf{Q}_f\mathbf{x}(t_f). \tag{9.17b}$$

Again, this is a two-point boundary value problem in $\mathbf{x}$ and $\boldsymbol{\lambda}$. Plugging in the same expression $\boldsymbol{\lambda} = \mathbf{P}\mathbf{x}$ will result in the same Riccati equation as above.

Kalman Filtering and Optimal Estimation

The optimal LQR controller relies on full-state measurements of the system, which are often unavailable. Instead of measuring the full state $\mathbf{x}$, it may be possible to estimate it from limited and noisy measurements $\mathbf{y}$. Mathematically, full-state estimation is possible as long as the pair $(\mathbf{A}, \mathbf{C})$ is observable, which is analogous to the controllability of $(\mathbf{A}^{\mathrm{T}}, \mathbf{C}^{\mathrm{T}})$. However, even if observability were technically possible, it may be a nearly ill-posed inverse problem depending on the degree of observability, as quantified by the observability Gramian [77].

The Kalman filter [266, 534, 207] is the most common full-state estimator, as it optimally balances the competing effects of measurement noise, disturbances, and model uncertainty. It is also possible to combine a Kalman filter with the LQR full-state feedback law, as explored in the next section. To derive the Kalman filter, we re-introduce disturbances $\mathbf{w}_d$ and sensor noise $\mathbf{w}_n$:

$$\frac{\mathrm{d}}{\mathrm{d}t}\mathbf{x} = \mathbf{A}\mathbf{x} + \mathbf{B}\mathbf{u} + \mathbf{w}_d \tag{9.18a}$$

$$\mathbf{y} = \mathbf{C}\mathbf{x} + \mathbf{D}\mathbf{u} + \mathbf{w}_n. \tag{9.18b}$$

The Kalman filter assumes that both the disturbance and noise are zero-mean Gaussian processes with known covariances:

$$\mathbb{E}\left(\mathbf{w}_d(t)\mathbf{w}_d(\tau)^{\mathrm{T}}\right) = \mathbf{V}_d\delta(t-\tau),$$
$$\mathbb{E}\left(\mathbf{w}_n(t)\mathbf{w}_n(\tau)^{\mathrm{T}}\right) = \mathbf{V}_n\delta(t-\tau).$$

Here $\mathbb{E}$ is the expected value and $\delta(\cdot)$ is the Dirac delta function. The matrices $\mathbf{V}_d$ and $\mathbf{V}_n$ are positive semi-definite with entries containing the covariances of the disturbance and noise terms. Extensions to the Kalman filter exist for correlated, biased, and unknown noise and disturbance terms [468, 341].

It is possible to obtain an estimate $\hat{\mathbf{x}}$ of the full-state $\mathbf{x}$ from measurements of the input $\mathbf{u}$ and output $\mathbf{y}$, via the following estimator dynamical system:

$$\frac{\mathrm{d}}{\mathrm{d}t}\hat{\mathbf{x}} = \mathbf{A}\hat{\mathbf{x}} + \mathbf{B}\mathbf{u} + \mathbf{K}_f\left(\mathbf{y} - \hat{\mathbf{y}}\right) \tag{9.19a}$$

$$\hat{\mathbf{y}} = \mathbf{C}\hat{\mathbf{x}} + \mathbf{D}\mathbf{u}. \tag{9.19b}$$

The matrices $\mathbf{A}, \mathbf{B}, \mathbf{C}$, and $\mathbf{D}$ are the same as in the systems model, and the filter gain $\mathbf{K}_f$ is determined via a similar procedure as in LQR. The Kalman filter gain $\mathbf{K}_f$ is given by

$$\mathbf{K}_f = \mathbf{Y}\mathbf{C}^{\mathrm{T}}\mathbf{V}_n^{-1}, \tag{9.20}$$

where $\mathbf{Y}$ is the solution to another algebraic Riccati equation:

$$\mathbf{Y}\mathbf{A}^{\mathrm{T}} + \mathbf{A}\mathbf{Y} - \mathbf{Y}\mathbf{C}^{\mathrm{T}}\mathbf{V}_n^{-1}\mathbf{C}\mathbf{Y} + \mathbf{V}_d = \mathbf{0}. \tag{9.21}$$

This solution is commonly referred to as the Kalman filter, and it is the optimal full-state estimator with respect to the following objective function:

$$J = \lim_{t\to\infty}\frac{1}{t}\int_0^{\mathrm{T}}\mathbb{E}\left[(\mathbf{x}(\tau) - \hat{\mathbf{x}}(\tau))^{\mathrm{T}}(\mathbf{x}(\tau) - \hat{\mathbf{x}}(\tau))\right]\mathrm{d}\tau. \tag{9.22}$$

This objective function implicitly includes the effects of disturbance and noise, which are required to determine the optimal balance between aggressive estimation and noise attenuation. The Kalman filter is referred to as linear quadratic estimation (LQE), and has a dual formulation to the LQR optimization.

The Kalman filter gain $\mathbf{K}_f$ may be determined via

CodeSnip 9.c
```
>> Kf = lqe(A,I,C,Vd,Vn); % design Kalman filter gain
```

where $\mathbf{I}$ is the $n \times n$ identity matrix. Optimal control and estimation are mathematical dual problems, so the Kalman filter may also be found using LQR:

CodeSnip 9.d
```
>> Kf = (lqr(A',C',Vd,Vn))'; % LQR and LQE are dual problems
```

The Kalman filter is shown schematically in Fig. 9.3.

Substituting the output estimate $\hat{\mathbf{y}}$ from Eq. (9.19b) into Eq. (9.19a) yields:

$$\frac{\mathrm{d}}{\mathrm{d}t}\hat{\mathbf{x}} = \left(\mathbf{A} - \mathbf{K}_f\mathbf{C}\right)\hat{\mathbf{x}} + \mathbf{K}_f\mathbf{y} + \left(\mathbf{B} - \mathbf{K}_f\mathbf{D}\right)\mathbf{u}. \tag{9.23}$$

The estimator dynamical system is expressed in terms of the estimate $\hat{\mathbf{x}}$ with inputs $\mathbf{y}$ and $\mathbf{u}$. If the system is observable, it is possible to place the eigenvalues of $\mathbf{A} - \mathbf{K}_f\mathbf{C}$ arbitrarily

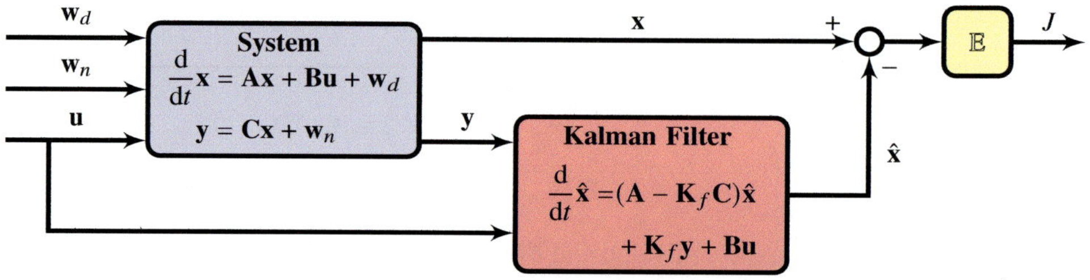

Figure 9.3 Schematic of the Kalman filter for full-state estimation from noisy measurements $\mathbf{y} = \mathbf{C}\mathbf{x} + \mathbf{w}_n$ with process noise (disturbances) $\mathbf{w}_d$. We take the expectation $\mathbb{E}$ of the output to compute the objective function J, since this system has stochastic disturbances and noise. This diagram does not have a feedthrough term $\mathbf{D}$, although it may be included.

with choice of $\mathbf{K}_f$. When the eigenvalues of the estimator are stable, then the state estimate $\hat{\mathbf{x}}$ converges to the full-state $\mathbf{x}$ asymptotically, as long as the model faithfully captures the true system dynamics. To see this convergence, consider the dynamics of the estimation error $\boldsymbol{\epsilon} = \mathbf{x} - \hat{\mathbf{x}}$:

$$\begin{aligned}\frac{\mathrm{d}}{\mathrm{d}t}\boldsymbol{\epsilon} &= \frac{\mathrm{d}}{\mathrm{d}t}\mathbf{x} - \frac{\mathrm{d}}{\mathrm{d}t}\hat{\mathbf{x}} \\ &= [\mathbf{A}\mathbf{x} + \mathbf{B}\mathbf{u} + \mathbf{w}_d] - \left[(\mathbf{A} - \mathbf{K}_f\mathbf{C})\hat{\mathbf{x}} + \mathbf{K}_f\mathbf{y} + (\mathbf{B} - \mathbf{K}_f\mathbf{D})\mathbf{u}\right] \\ &= \mathbf{A}\boldsymbol{\epsilon} + \mathbf{w}_d + \mathbf{K}_f\mathbf{C}\hat{\mathbf{x}} - \mathbf{K}_f\underbrace{[\mathbf{C}\mathbf{x} + \mathbf{D}\mathbf{u} + \mathbf{w}_n]}_{\mathbf{y}} + \mathbf{K}_f\mathbf{D}\mathbf{u} \\ &= (\mathbf{A} - \mathbf{K}_f\mathbf{C})\boldsymbol{\epsilon} + \mathbf{w}_d - \mathbf{K}_f\mathbf{w}_n.\end{aligned}$$

Therefore, the estimate $\hat{\mathbf{x}}$ will converge to the true full state when $\mathbf{A} - \mathbf{K}_f\mathbf{C}$ has stable eigenvalues. As with LQR, there is a tradeoff between over-stabilization of these eigenvalues and the amplification of sensor noise.

There are many variants of the Kalman filter for nonlinear systems [255, 256, 514], including the extended and unscented Kalman filters. The ensemble Kalman filter [8] works well for high-dimensional systems, as in geophysical data assimilation [411]. These methods still assume Gaussian noise. A particle filter is a more general, but more computationally intensive, alternative for arbitrary noise distributions [215, 418]. The unscented Kalman filter balances the efficiency of a Kalman filter and accuracy of a particle filter.

The Kalman Filter as a Bayesian Estimator. The Kalman filter provides a recursive algorithm for estimating the state of a linear dynamical system subject to Gaussian noise. Although often derived from control-theoretic or algebraic perspectives, the Kalman filter is also a special case of Bayesian estimation, which we discussed earlier in Chapter 7. At each time step, it performs an exact Bayesian update: combining a prior belief about the system state, based on a model, with a likelihood function derived from a new noisy measurement. This connection will be established rigorously in the homework.

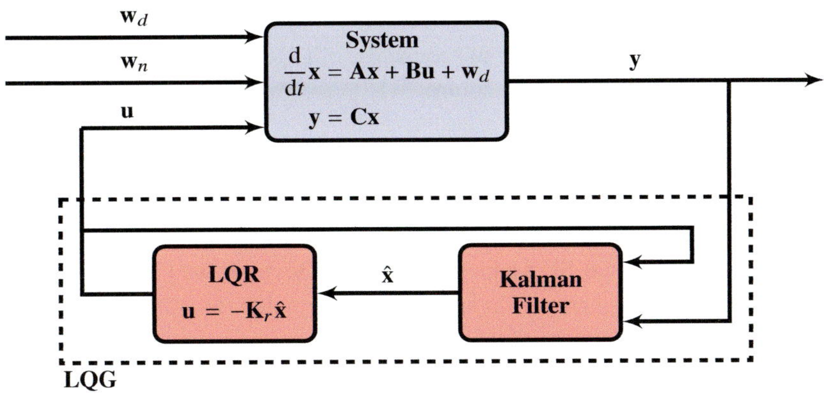

Figure 9.4 Schematic of the linear quadratic Gaussian (LQG) controller for optimal closed-loop feedback based on noisy measurements $\mathbf{y}$. The LQR and Kalman filter gain matrices $\mathbf{K}_r$ and $\mathbf{K}_f$ may be designed independently. When combined, the resulting sensor-based feedback remains optimal.

Linear Quadratic Gaussian

Combining LQR with the Kalman filter yields the linear quadratic Gaussian (LQG) controller, resulting in optimal sensor-based feedback. Remarkably, the LQR gain $\mathbf{K}_r$ and the Kalman filter gain $\mathbf{K}_f$ may be designed separately, and the resulting sensor-based feedback will remain optimal and retain the closed-loop eigenvalues when combined.

The LQG controller consists of the LQR feedback $\mathbf{u} = -\mathbf{K}_r\hat{\mathbf{x}}$ where the state $\hat{\mathbf{x}}$ is found via the estimator dynamical system

$$\frac{\mathrm{d}}{\mathrm{d}t}\hat{\mathbf{x}} = \left(\mathbf{A} - \mathbf{K}_f\mathbf{C} - \mathbf{B}\mathbf{K}_r\right)\hat{\mathbf{x}} + \mathbf{K}_f\mathbf{y}. \tag{9.24}$$

The LQG controller is optimal with respect to the following ensemble-averaged version of the objective function from Eq. (9.8):

$$J(t) = \lim_{t\to\infty}\int_0^{\mathrm{T}} \mathbb{E}\left[\mathbf{x}(\tau)^{\mathrm{T}}\mathbf{Q}\mathbf{x}(\tau) + \mathbf{u}(\tau)^{\mathrm{T}}\mathbf{R}\mathbf{u}(\tau)\right]\,\mathrm{d}\tau. \tag{9.25}$$

Applying LQR to $\hat{\mathbf{x}}$ results in the following state dynamics:

$$\begin{aligned}\frac{\mathrm{d}}{\mathrm{d}t}\mathbf{x} &= \mathbf{A}\mathbf{x} - \mathbf{B}\mathbf{K}_r\hat{\mathbf{x}} + \mathbf{w}_d\\ &= \mathbf{A}\mathbf{x} - \mathbf{B}\mathbf{K}_r\mathbf{x} + \mathbf{B}\mathbf{K}_r\left(\mathbf{x} - \hat{\mathbf{x}}\right) + \mathbf{w}_d.\end{aligned}$$

Again, if $\boldsymbol{\epsilon} = \mathbf{x} - \hat{\mathbf{x}}$, the closed-loop system may be written as

$$\frac{\mathrm{d}}{\mathrm{d}t}\begin{bmatrix}\mathbf{x}\\ \boldsymbol{\epsilon}\end{bmatrix} = \begin{bmatrix}\mathbf{A} - \mathbf{B}\mathbf{K}_r & \mathbf{B}\mathbf{K}_r\\ \mathbf{0} & \mathbf{A} - \mathbf{K}_f\mathbf{C}\end{bmatrix}\begin{bmatrix}\mathbf{x}\\ \boldsymbol{\epsilon}\end{bmatrix} + \begin{bmatrix}\mathbf{I} & \mathbf{0}\\ \mathbf{I} & -\mathbf{K}_f\end{bmatrix}\begin{bmatrix}\mathbf{w}_d\\ \mathbf{w}_n\end{bmatrix}. \tag{9.26}$$

Thus, the closed-loop eigenvalues of the LQG system are given by the eigenvalues of $\mathbf{A} - \mathbf{B}\mathbf{K}_r$ and $\mathbf{A} - \mathbf{K}_f\mathbf{C}$, which were optimally chosen by the LQR and Kalman filter gain matrices, respectively.

The LQG framework, shown in Fig. 9.4, requires an accurate model of the system and the magnitudes of the disturbances and measurement noise. In real-world systems, even small time delays and model uncertainty may destroy the robustness of LQG and result in instability [152]. The lack of robustness of LQG regulators to model uncertainty motivates the introduction of robust control [152] (see Panel 9.1).

Panel 9.1: Robust Control

In his seminal 1978 paper [152], John Doyle showed that LQG regulators can have arbitrarily small stability margins, making them *fragile* to model uncertainties, time delays, and other model imperfections.

Guaranteed Margins for LQG Regulators

JOHN C. DOYLE

***Abstract*—There are none.**

This paper motivated a new theory of *robust* control. The control formulation in Fig. 9.1 is a general optimization framework to minimize the transfer function from external inputs $\mathbf{w}$ (e.g., reference, disturbances, noise) to the cost function $\mathbf{J}$, which balances accuracy, actuation cost, performance, etc. [156, 466]. Optimal control (e.g., LQR, LQG) is optimal with respect to the $\mathcal{H}_2$-norm[a], while robust control is optimal with respect to the $\mathcal{H}_\infty$-norm. The infinity-norm is defined as

$$||\mathbf{G}||_\infty \triangleq \max_\omega \sigma_1(\mathbf{G}(i\omega)).$$

Here, σ_1 denotes the maximum singular value. Since the $||\cdot||_\infty$-norm is the maximum value of the transfer function at any frequency, it is often called a *worst-case scenario norm*; therefore, minimizing the infinity-norm provides robustness to worst-case disturbances. The robust control optimization problem is typically difficult to solve, and no general, analytic solution exists. However, there are relatively efficient iterative methods to find a controller such that $||\mathbf{G}||_\infty < \gamma$.

Selecting the objective function $\mathbf{J}$ to meet design specifications is an important part of robust control design. Considerations such as disturbance rejection, noise attenuation, controller bandwidth, and actuation cost are accounted for by a weighted sum of various transfer functions representing different sensitivities.

Robust control was especially impactful for controlling high-performance aerospace systems, such as the US Space Shuttle, where disturbances were large, physics was only partially known, and linear models were very rough. For example, when the shuttle re-entered the Earth's atmosphere to land, it passed through a wide range of extreme aerodynamic conditions. Robust control made it possible to use a relatively simple set of models and treat unmodeled dynamics as disturbances.

[a] The $\mathcal{H}_2$-norm is the bounded 2-norm on a Hardy space, consisting of stable and strictly proper transfer functions, meaning gain rolls off at high frequency. The $\mathcal{H}_\infty$ bounded infinity-norm, consisting of stable and proper transfer functions (i.e., gain does not become infinite at high frequencies).

9.2 Nonlinear Optimal Control

Now that we have explored optimal control in the context of linear dynamics and a quadratic objective function, we are ready for the more general theory of nonlinear optimal control. The goal of this section is to provide a mathematical foundation for nonlinear optimal control, which will be useful when we discuss reinforcement learning, dynamic programming, and model predictive control. Specifically, we will derive the Hamilton–Jacobi–Bellman (HJB) equations for optimal nonlinear control, which is one of the theoretical triumphs in control theory. It relies on Bellman's principle of optimality, which states that an optimal multi-step control policy must be locally optimal in every sub-sequence of steps.

Nonlinear optimal control problems do not generally have simple closed-form solutions, in contrast to linear optimal control. Instead, we typically leverage powerful optimization frameworks, such as reinforcement learning (Section 9.3) and dynamic programming (Section 9.4), to approximately solve the nonlinear, nonconvex optimization problem. This is an enormous field, and this chapter only provides a high-level overview. For more details, see the excellent text by Stengel [473].

Mathematical Statement of Nonlinear Optimal Control Problem

The goal is to optimize the control input $\mathbf{u}(t)$ to drive a nonlinear system

$$\frac{\mathrm{d}}{\mathrm{d}t}\mathbf{x} = \mathbf{f}(\mathbf{x}(t), \mathbf{u}(t), t) \tag{9.27}$$

to follow a trajectory $\mathbf{x}(t)$ that minimizes an objective function

$$J(\mathbf{x}(t), \mathbf{u}(t), t_0, t_f) = \int_{t_0}^{t_f} \mathcal{L}(\mathbf{x}(\tau), \mathbf{u}(\tau))\,\mathrm{d}\tau + Q(\mathbf{x}(t_f), t_f). \tag{9.28}$$

We have written $\mathbf{u}(t)$ as a function of time, rather than of the state $\mathbf{x}$, allowing us to consider both cases of open-loop control $\mathbf{u}(t)$ or closed-loop control $\mathbf{u}(t) = \mathbf{u}(\mathbf{x}(t))$. We have also dropped the output equation $\mathbf{y} = \mathbf{g}(\mathbf{x}, \mathbf{u}, \mathbf{w}_n)$ for simplicity, although the theory below readily generalizes to handle partial measurements.

The formulation in Eq. (9.28) generalizes the LQR objective function in Eq. (9.12), so that $\mathcal{L}(\mathbf{x}, \mathbf{u})$ and $Q(\mathbf{x}(t_f), t_f)$ may be non-quadratic. There may also be constraints on the state $\mathbf{x}$ and control $\mathbf{u}$, which determine what solutions are admissible; this is especially common in model predictive control, which we will explore in Section 9.5.

Hamilton–Jacobi–Bellman Equations and the Value Function

An optimal control $\mathbf{u}(t)$ will minimize the objective function J, starting at an initial state $\mathbf{x}_0 = \mathbf{x}(t_0)$. We reframe this optimization problem in terms of the *value* function $V(\mathbf{x}_0, t_0, t_f)$, which is the total integrated cost starting at $\mathbf{x}_0$ assuming the control law is optimal:

$$V(\mathbf{x}_0, t_0, t_f) = \min_{\mathbf{u}(t)} J(\mathbf{x}(t), \mathbf{u}(t), t_0, t_f). \tag{9.29}$$

Here $\mathbf{x}(t)$ is the solution to Eq. (9.27) for the optimal $\mathbf{u}(t)$. The value function is no longer a function of the control $\mathbf{u}(t)$, as this has been optimized over, and it is also not a function of $\mathbf{x}(t)$, but rather of an initial state $\mathbf{x}_0$, as the remainder of the trajectory is entirely specified by the dynamics and the optimal control law. The value function is often called the *cost-to-go* in control theory, as the value function evaluated at any point $\mathbf{x}(t)$ on an optimal trajectory will give the remaining cost associated with continuing to enact the optimal policy until the final time t_f. This is a statement of *Bellman's optimality principle*, that the value function V remains optimal starting with any point on an optimal trajectory.

The Hamilton–Jacobi–Bellman[4] (HJB) equation is a partial differential equation that must be satisfied by the value function V at every intermediate time $t \in [t_0, t_f]$:

$$-\frac{\partial V}{\partial t} = \min_{\mathbf{u}(t)} \left[\left(\frac{\partial V}{\partial \mathbf{x}} \right)^{\mathrm{T}} \mathbf{f}(\mathbf{x}(t), \mathbf{u}(t)) + \mathcal{L}(\mathbf{x}(t), \mathbf{u}(t)) \right]. \tag{9.30}$$

The term $\frac{\partial V}{\partial \mathbf{x}}$ is a Lagrange multiplier λ to enforce the dynamics constraint given by $\mathbf{f}$.

To derive the HJB equation, we take the total time derivative of the value function $V(\mathbf{x}(t), t, t_f)$ at some intermediate time t:

$$\begin{aligned}
\frac{\mathrm{d}}{\mathrm{d}t} V(\mathbf{x}(t), t, t_f) &= \frac{\partial V}{\partial t} + \left(\frac{\partial V}{\partial \mathbf{x}} \right)^{\mathrm{T}} \frac{\mathrm{d}\mathbf{x}}{\mathrm{d}t} \\
&= \min_{\mathbf{u}(t)} \frac{\mathrm{d}}{\mathrm{d}t} \left[\int_t^{t_f} \mathcal{L}(\mathbf{x}(\tau), \mathbf{u}(\tau))\, \mathrm{d}\tau + Q(\mathbf{x}(t_f), t_f) \right] \\
&= \min_{\mathbf{u}(t)} \underbrace{\left[\frac{\mathrm{d}}{\mathrm{d}t} \int_t^{t_f} \mathcal{L}(\mathbf{x}(\tau), \mathbf{u}(\tau))\, \mathrm{d}\tau \right]}_{-\mathcal{L}(\mathbf{x}(t), \mathbf{u}(t))} \\
\Longrightarrow \quad -\frac{\partial V}{\partial t} &= \min_{\mathbf{u}(t)} \left[\left(\frac{\partial V}{\partial \mathbf{x}} \right)^{\mathrm{T}} \mathbf{f}(\mathbf{x}(t), \mathbf{u}(t)) + \mathcal{L}(\mathbf{x}(t), \mathbf{u}(t)) \right].
\end{aligned}$$

The terminal cost does not vary with t, so it has zero time derivative. The derivative of the integral $\int_t^{t_f} \mathcal{L}(\mathbf{x}(\tau), \mathbf{u}(\tau))\, \mathrm{d}\tau$ is equal to $-\mathcal{L}(\mathbf{x}(t), \mathbf{u}(t))$ by the first fundamental theorem of calculus. Finally, the term $(\partial V / \partial \mathbf{x})^{\mathrm{T}} \mathbf{f}(\mathbf{x}, \mathbf{u})$ may be brought into the minimization with respect to $\mathbf{u}(t)$, since V is already defined as the optimal cost over $\mathbf{u}$.

The derivation above relies on Bellman's principle of optimality: at any point on an optimal trajectory $\mathbf{x}(t)$, the value function V is still optimal for the rest of the trajectory:

$$V(\mathbf{x}(t), t, t_f) = \min_{\mathbf{u}(t)} \left[\int_t^{t_f} \mathcal{L}(\mathbf{x}(\tau), \mathbf{u}(\tau))\, \mathrm{d}\tau + Q(\mathbf{x}(t_f), t_f) \right]. \tag{9.31}$$

The value function can therefore be decomposed as:

$$V(\mathbf{x}(t_0), t_0, t_f) = V(\mathbf{x}(t_0), t_0, t) + V(\mathbf{x}(t), t, t_f). \tag{9.32}$$

[4] Kalman recognized that the Bellman optimal control formulation was a generalization of the Hamilton–Jacobi equation from classical mechanics to handle stochastic input–output systems. These formulations all involve the calculus of variations, which we have seen in Section 6.6.

This makes it possible to take the total time derivative above. A more rigorous derivation is possible using the calculus of variations from Chapter 6.

The HJB equation is incredibly powerful, providing a PDE for the optimal solution of a general nonlinear control problem. Typically, the HJB equation is solved as a two-point boundary value problem using a shooting method, with boundary conditions $\mathbf{x}(0) = \mathbf{x}_0$ and $V(\mathbf{x}(t_f), t_f) = Q(\mathbf{x}(t_f), t_f)$. However, a nonlinear control problem with a three-dimensional state vector $\mathbf{x} \in \mathbb{R}^3$ will result in a three-dimensional PDE. Thus, optimal nonlinear control based on the HJB equation typically suffers from the curse of dimensionality. Phase-space clustering techniques have shown promise for systems that evolve on a low-dimensional attractor [264]. Neural ordinary differential equations are also making it possible to solve much larger HJB equations [398].

Discrete-Time HJB and the Bellman Equation

Bellman optimality is especially intuitive for discrete-time system,

$$\mathbf{x}_{k+1} = \mathbf{F}(\mathbf{x}_\mathbf{k}, \mathbf{u}_\mathbf{k}), \tag{9.33}$$

where instead of optimizing over a function, we optimize over a discrete control sequence. The objective is now given by

$$J(\mathbf{x}_0, \{\mathbf{u}_k\}_{k=0}^n, n) = \sum_{k=0}^{n} \mathcal{L}(\mathbf{x}_k, \mathbf{u}_k) + Q(\mathbf{x}_n, t_n). \tag{9.34}$$

Similarly, the value function is defined as the value of the cumulative objective function, starting at a point $\mathbf{x}_0$ assuming an optimal control policy $\mathbf{u}$:

$$V(\mathbf{x}_0, n) = \min_{\{\mathbf{u}_k\}_{k=0}^n} J(\mathbf{x}_0, \{\mathbf{u}_k\}_{k=0}^n, n). \tag{9.35}$$

Bellman's principle of optimality states that at any point along the optimal trajectory $\mathbf{x}$, the remaining control policy is optimal with respect to this new initial state

$$V(\mathbf{x}_0, n) = V(\mathbf{x}_0, k) + V(\mathbf{x}_k, n) \quad \forall k \in (0, n). \tag{9.36}$$

Thus, the value at an intermediate time step k may be written as

$$\begin{aligned} V(\mathbf{x}_k, n) &= \left[\min_{\mathbf{u}_k} \mathcal{L}(\mathbf{x}_k, \mathbf{u}_k)\right] + \underbrace{V(\mathbf{x}_{k+1}, n)}_{\text{s.t. } \mathbf{x}_{k+1} = \mathbf{F}(\mathbf{x}_k, \mathbf{u}_k)} \\ &= \min_{\mathbf{u}_k} \left[\mathcal{L}(\mathbf{x}_k, \mathbf{u}_k) + V(\mathbf{F}(\mathbf{x}_k, \mathbf{u}_k), n)\right]. \end{aligned}$$

It is also possible, given a value function $V(\mathbf{x}_k, n)$, to determine the next optimal control action $\mathbf{u}_k$ by returning the $\mathbf{u}_k$ that minimizes the above expression. This defines an optimal policy $\mathbf{u} = \pi(\mathbf{x})$. Dropping the functional dependence of V on the end time, we then have

$$V(\mathbf{x}) = \min_{\mathbf{u}} \left[\mathcal{L}(\mathbf{x}, \mathbf{u}) + V(\mathbf{F}(\mathbf{x}, \mathbf{u}))\right] \tag{9.37a}$$

$$\pi(\mathbf{x}) = \operatorname*{argmin}_{\mathbf{u}} \left[\mathcal{L}(\mathbf{x}, \mathbf{u}) + V(\mathbf{F}(\mathbf{x}, \mathbf{u}))\right]. \tag{9.37b}$$

These are the Bellman equations and they are the cornerstone of reinforcement learning.

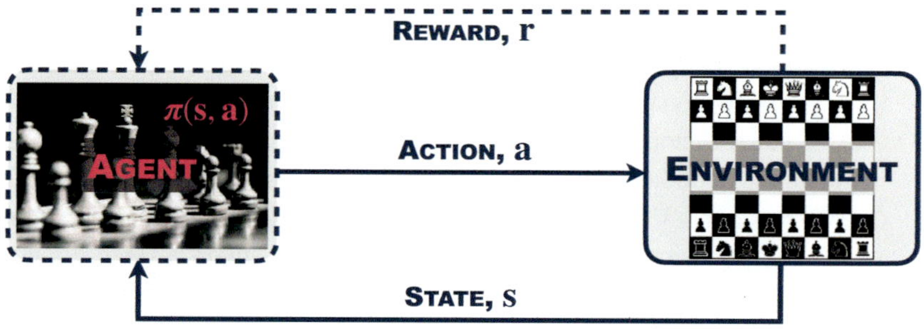

Figure 9.5 Schematic of reinforcement learning, where an agent senses its environmental state $\mathbf{s}$ and takes actions $\mathbf{a}$ according to a policy π that is optimized through learning to maximize future rewards $\mathbf{r}$.

9.3 Reinforcement Learning

Reinforcement learning (RL) is a major branch of research at the intersection of machine learning and control theory [409], and it is among the most promising fields of research towards generalized artificial intelligence and autonomy. It is concerned with how to learn control laws and policies to interact with a complex environment [480, 261]. Both machine learning and control theory fundamentally rely on optimization, and, likewise, RL involves a set of optimization techniques within a framework for learning how to interact with the environment from experience.

In reinforcement learning[5], an agent senses the state of its environment, $\mathbf{s} \in \mathcal{S}$, and learns take appropriate actions, $\mathbf{a} \in \mathcal{A}$, to optimize future rewards $\mathbf{r}$; the sets $\mathcal{S}$ and $\mathcal{A}$ denote the sets of possible states and actions, respectively. Figure 9.5 provides a schematic overview of the reinforcement learning framework.

Reinforcement learning mimics how animals learn to interact with their environment through trial and error, experience, positive and negative rewards, and feedback. Much of the history of reinforcement learning is linked to studies of animal behavior and the neurological basis of decisions, control, and learning [355, 453, 129, 128]. Pavlov's dog is an example demonstrating how animals learn environmental cues that will lead to a delayed reward [386]. The term *reinforcement* refers directly to the rewards, such as food, used to reinforce desirable behavior in humans and animals.

The Policy

The agent takes actions $\mathbf{a}$ through a control policy π that is optimized through learning to maximize future rewards $\mathbf{r}$. Rewards may be immediate or delayed, depending on the problem; in many cases, a circuitous path to the optimal solution must be learned. For example, a long term chess strategy may involve making sacrifices.

For deterministic systems, e.g. the optimal nonlinear control problem in Section 9.2, the policy may be written as $\mathbf{a} = \pi(\mathbf{s})$, directly analogous to the control law $\mathbf{u} = \mathbf{k}(\mathbf{x})$. However,

[5] Reinforcement learning uses different notation than control, although the two sets of notation can easily be mapped back and forth. The action $\mathbf{a}$ maps to the actuation $\mathbf{u}$, the state $\mathbf{s}$ maps to the full state $\mathbf{x}$ or partial measurement $\mathbf{y}$, and the policy $\pi(\mathbf{s})$ maps to the controller $\mathbf{u} = \mathbf{k}(\mathbf{x})$ or $\mathbf{u} = \mathbf{k}(\mathbf{y})$.

when the environment is stochastic, we write the policy as $\pi(\mathbf{s}, \mathbf{a})$, which is the probability of taking action $\mathbf{a}$ given state $\mathbf{s}$ to maximize future rewards. This indicates that the policy is no longer deterministic, but is rather a probability density over actions $\mathbf{a}$.

In the simplest formulation, the policy may be a look-up table that is defined on the discrete state and action spaces $\mathcal{S}$ and $\mathcal{A}$, respectively. However, for most problems, this becomes prohibitively expensive, and π must be represented as an approximate function that is parameterized by $\boldsymbol{\theta}$, for example via a deep neural network:

$$\pi(\mathbf{s}, \mathbf{a}) \approx \pi(\mathbf{s}, \mathbf{a}, \boldsymbol{\theta}). \tag{9.38}$$

Often, this parameterized function will be denoted $\pi_{\boldsymbol{\theta}}(\mathbf{s}, \mathbf{a})$. We will revisit this function approximation in the context of deep reinforcement learning in Section 10.4, where it is possible to represent these complex functions using deep neural networks.

The Environment

RL generalizes the optimal nonlinear control problem in Section 9.2. The environment may be described by the deterministic, nonlinear dynamics in Eq. (9.27), or it may represent a less structured, stochastic set of causes and effects. In a game of chess, the environment involves the rules of the game, along with the, sometimes, complex thought process of a human opponent. This is nearly impossible to write down as a tractable system of equations, and the set of states is almost innumerable. Thus, it is often the case that the model of the environment is not known. Further, the measured state of the system may be a partial measurement of a higher-dimensional state that evolves according to a stochastic, nonlinear process. Modern RL provides strategies for handling such high-dimensional, unstructured problems. We typically either use model-free RL strategies, or learn a data-driven model, for example using the techniques in Section 8.7, and proceed to model-based RL strategies. As is the case with human learning, RL often begins with an unstructured exploration, where trial-and-error are used to learn the environment and reward function, followed by exploitation, where a strategy is chosen and optimized.

For simplicity and pedagogical development of the theory, we often assume that the environment is either a nonlinear dynamical system (deterministic or stochastic), or that the state evolves according to a simple Markov decision process (MDP). Here we will develop the theory for an MDP, as this is the most common approach.

In an MDP the probability of the system occurring in the current state is determined entirely by the previous state, so that it is memory-less. An MDP consists of a set of states $\mathcal{S}$, a set of actions $\mathcal{A}$, and a set of rewards $\mathcal{R}$, along with the probability of transitioning from state $\mathbf{s}_k$ at time t_k to state $\mathbf{s}_{k+1}$ at time t_{k+1} given action $\mathbf{a}_k$,

$$P(\mathbf{s}', \mathbf{s}, \mathbf{a}) = \Pr\left(\mathbf{s}_{k+1} = \mathbf{s}' \mid \mathbf{s}_k = \mathbf{s}, \mathbf{a}_k = \mathbf{a}\right), \tag{9.39}$$

and a reward function R

$$R(\mathbf{s}', \mathbf{s}, \mathbf{a}) = \Pr\left(\mathbf{r}_{k+1} \mid \mathbf{s}_{k+1} = \mathbf{s}', \mathbf{s}_k = \mathbf{s}, \mathbf{a}_k = \mathbf{a}\right). \tag{9.40}$$

Thus, an MDP generalizes the notion of a Markov process to include actions and rewards, making it suitable for decision making and control. Sometimes[6] the transition probability $P(\mathbf{s}', \mathbf{s}, \mathbf{a})$ will be written as $P(\mathbf{s}' \mid \mathbf{s}, \mathbf{a})$.

A Markov process with a finite set of states S, may be written in terms of the vector of probabilities of being in each of the n states $\mathbf{s} \in \mathbb{R}^n$. The Markov process $P(\mathbf{s}', \mathbf{s})$ is then given by a transition matrix $\mathbf{T}$, so that $\mathbf{s}' = \mathbf{T}\mathbf{s}$. Each column of $\mathbf{T}$ must add up to 1, which is a statement that given a particular state $\mathbf{s}$, something must happen in the transition to $\mathbf{s}'$.

Similarly, for an MDP, given a policy π, the transition process may be written as

$$\mathbf{s}' = \sum_{\mathbf{a} \in \mathcal{A}} \pi(\mathbf{s}, \mathbf{a}) \mathbf{T}_{\mathbf{a}} \mathbf{s}. \tag{9.41}$$

Now for each action $\mathbf{a}$, $\mathbf{T}_{\mathbf{a}}$ is a Markov process with all columns summing to 1.

All Markov processes asymptotically approach a steady state $\boldsymbol{\mu}$, which is the eigenvector of $\mathbf{T}$ corresponding to eigenvalue 1. Given a policy π, an MDP asymptotically approaches a steady state $\boldsymbol{\mu}_\pi = \sum_{\mathbf{a}} \pi(\mathbf{s}, \mathbf{a}) \boldsymbol{\mu}_{\mathbf{a}}$.

The Value Function

We have already seen the value function in the context of the HJB equation in Section 9.2. Here we define it in the context of RL for an MDP. Given a policy π, it is possible to define a value function that quantifies the desirability of being in a given state:

$$V_\pi(\mathbf{s}) = \mathbb{E}\left(\sum_k \gamma^k \mathbf{r}_k \mid \mathbf{s}_0 = \mathbf{s} \right), \tag{9.42}$$

where $\mathbb{E}$ is the expected future total reward, subject to a discount rate γ. The optimal value function for the best policy is given by

$$V(\mathbf{s}) = \max_\pi \mathbb{E}\left(\sum_{k=0}^{\infty} \gamma^k \mathbf{r}_k \mid \mathbf{s}_0 = \mathbf{s} \right). \tag{9.43}$$

Bellman's principle of optimality states that the value may be written recursively as

$$V(\mathbf{s}) = \max_\pi \mathbb{E}\left(\mathbf{r}_0 + \sum_{k=1}^{\infty} \gamma^k \mathbf{r}_k \mid \mathbf{s}_1 = \mathbf{s}' \right), \tag{9.44}$$

which yields Bellman's equation:

$$V(\mathbf{s}) = \max_\pi \mathbb{E}\left(\mathbf{r}_0 + \gamma V(\mathbf{s}') \right), \tag{9.45}$$

where $\mathbf{s}'$ is the next state after $\mathbf{s}$. This is true for $\mathbf{s}' = \mathbf{s}_{k+1}$ and $\mathbf{s} = \mathbf{s}_k$ for all k.

Given the value function, it is possible to extract the optimal policy as

$$\pi = \underset{\pi}{\operatorname{argmax}}\, \mathbb{E}\left(\mathbf{r}_0 + \gamma V(\mathbf{s}') \right), \tag{9.46}$$

[6] We might also replace the action $\mathbf{a}$ with a policy π, which implies a sum over states:

$$P(\mathbf{s}', \mathbf{s}, \pi) = \sum_{\mathbf{a} \in \mathcal{A}} \pi(\mathbf{s}, \mathbf{a}) P(\mathbf{s}', \mathbf{s}, \mathbf{a}).$$

Optimization Techniques for Reinforcement Learning

Learning the optimal policy π, the value function V, or jointly learning both, all involve optimization. There are many approaches to learn an optimal policy π, which is the ultimate goal of RL. A major dichotomy in reinforcement learning is that of *model-based RL* versus *model-free RL*. When there is a known model for the environment, there are several strategies for learning either the optimal policy or value function through what is known as *policy* iteration or *value iteration*, which are forms of dynamic programming using the Bellman equation. We will discuss dynamic programming in the next section.

Depending on the assumed structure of π, the size and evolution dynamics of $\mathcal{S}$, and the reward landscape R, determining an optimal policy may range from a closed form optimization to a rather high-dimensional unstructured optimization. When there is no model for the environment, model-free strategies, such as Q-learning, must be employed. The reinforcement learning optimization problem may be particularly challenging for high-dimensional systems with unknown, nonlinear, stochastic dynamics and sparse and delayed rewards. All of these techniques may be combined with function approximation techniques, such as neural networks, for approximating the policy π, the value function V, or the quality function Q (discussed in Section 10.4).

One of the central challenges in reinforcement learning is that rewards are often extremely rare and may be significantly delayed from a sequence of good control actions. This leads to the so-called credit assignment problem, coined by Minsky [352] to describe the challenge of knowing what action sequence was responsible for the delayed reward. These sparse and delayed rewards have been a central challenge in RL for six decades, and they are still a focus of research today. The resulting optimization problem is computationally expensive and data intensive, requiring considerable trial and error.

Today, reinforcement learning is being used to learn sophisticated control policies for complex open-world problems in autonomy and propulsion (e.g., self-driving cars, learning to swim and fly) and as a general learning environment for rule-constrained games (e.g., checkers, backgammon, chess, Go, Atari games). Much of the history of RL may be traced through its success on increasingly challenging games, from checkers [434] to backgammon [492] and more recently to chess and Go [462]. These games serve to illustrate many of the central challenges that are still faced in RL, including the curse of dimensionality and the credit assignment problem.

The recursive nature of the value function, via Bellman's principle of optimality, is directly leveraged in reinforcement learning, where the task of learning optimal control policies is framed as estimating the value function.. While classical HJB approaches rely on solving partial differential equations, RL replaces this with iterative approximations, often through simulation or data-driven methods. This connection establishes R as a computational framework for solving Bellman-like equations in settings where models of the system dynamics may be unknown or high-dimensional. The Bellman equation reformulates the HJB's value function PDE into a recursive relationship that can be solved iteratively. This iterative structure forms the basis for RL techniques such as temporal-difference learning, which approximates the value function without explicitly solving the PDE. Modern RL techniques extend these ideas by leveraging data-driven approximations, such as neural networks.

9.4 Dynamic Programming

Dynamic programming (DP) is a recursive optimization framework introduced by Richard E. Bellman [36, 37] to solve large multi-step optimization problems, such as those found in decision making and control, by decomposing them into simpler subproblems. In the previous sections, we derived the HJB equation as a partial differential equation that must be satisfied by the value function for optimal nonlinear control problems. While powerful and surprisingly compact, the HJB equation is generally very challenging to solve analytically. Instead, we typically solve HJB approximately using dynamic programming, which provides a constructive and recursive way to compute the value function and optimal control policy.

Policy iteration and value iteration are two examples of how we use dynamic programming in optimal nonlinear control and reinforcement learning. Specifically, dynamic programming reformulates the large optimization problem as a recursive optimization in terms of smaller sub-problems, so that local improvements to the control policy may be made sequentially with new information. This approach relies on Bellman's principle of optimality, which states that a large multi-step control policy must also be locally optimal in every sub-sequence of steps. Reinforcement learning generalizes these ideas by introducing function approximation techniques to handle high-dimensional state spaces and by developing model-free approaches when the dynamics are unknown. In essence, Dynamic programming is the optimization backbone of classic reinforcement learning, while modern RL extends DP to problems with uncertainty and complex, high-dimensional dynamics.

The Bellman equation in Eq. (9.44) indicates that the large optimization problem over an entire state-action trajectory $(\mathbf{s}_k, \mathbf{a}_k)$ may be broken into a recursive optimization at each point along the trajectory. As long as the value function is known at the next point $\mathbf{s}'$, it is possible to solve the optimization at point $\mathbf{s}$ simply by optimizing the policy $\boldsymbol{\pi}(\mathbf{s}, \mathbf{a})$ at this point. Of course, this assumes that the value function is known at *all* possible next states $\mathbf{s}_{k+1}$, which is a function of the current state $\mathbf{s}_k$, the current action $\mathbf{a}_k$, and the dynamics governing the system; this becomes even more complex for non-MDP dynamics, such as the nonlinear control formulation in Section 9.2. For even moderately large problems, this suffers from the curse of dimensionality, and approximate solution methods must be employed.

When tractable, dynamic programming (i.e., the process of breaking a large problem into smaller overlapping sub-problems) provides a globally optimal solution. There are two main approaches to dynamic programming, referred to as *top down* and *bottom up*:

Top down: The top-down approach involves maintaining a table of sub-problems that are referred to when solving larger problems. For a new problem, the table is checked to see if the relevant sub-problem has been solved. If so, it is used, and if not, the sub-problem is solved. This tabular storage is called *memoization*. This approach becomes combinatorially complex for many problems.

Bottom up: The bottom-up approach involves first solving the smallest sub-problems, and then combining these to form the larger problems. This may be thought of as working backwards from every possible goal state, finding the best previous action to get there, then going back two steps, then going back three steps, etc.

Although dynamic programming still represents a brute-force search through all sub-problems, it is still more efficient than a naive brute-force search. In some cases, it reduces the computational complexity to an algorithm that scales linearly with the number of sub-problems, although this may still be combinatorially large, as in the example of the game of chess. Dynamic programming is closely related to divide-and-conquer techniques, such as quick sort, except that divide-and-conquer applies to *non-overlapping* or *non-recursive* (i.e., independent) sub-problems, while dynamic programming applies to overlapping, or recursively interdependent sub-problems.

However, the recursive strategy suggests approximate solution techniques, such as the alternating directions method, where a sub-optimal solution is initialized and the value function is iterated over. This forms the basis of policy and value iteration.

Policy Iteration

Policy iteration is a two-step optimization procedure for simultaneously finding an optimal value function V_π and the corresponding optimal policy π.

First, a candidate policy π is evaluated, resulting in the value function for this fixed policy. This typically involves a brute force calculation of the value function for this policy starting at many or all initial states. The policy may need to be run for a long time depending on the reward delay and discounting factor γ.

Next, the value function is fixed, and the policy is optimized to improve the expected rewards by taking different actions at a given state. This optimization relies on the alternative recursive formulation of the value function in Eq. (9.42) due to Bellman's Eq. (9.44):

$$
\begin{aligned}
V_\pi(\mathbf{s}) &= \mathbb{E}\left[R(\mathbf{s}', \mathbf{s}, \pi(\mathbf{s})) + \gamma V_\pi(\mathbf{s}')\right] && (9.47a)\\
&= \sum_{\mathbf{s}'} P(\mathbf{s}' \mid \mathbf{s}, \pi(\mathbf{s}))\left(R(\mathbf{s}', \mathbf{s}, \pi(\mathbf{s})) + \gamma V_\pi(\mathbf{s}')\right). && (9.47b)
\end{aligned}
$$

Note that in this expression, we have assumed a deterministic policy $\mathbf{a} = \pi(\mathbf{s})$, otherwise, Eq. (9.47a) would involve a second summation over $\mathbf{a} \in \mathcal{A}$, with the expression multiplied by $\pi(\mathbf{s}, \mathbf{a})$.

It is then possible to hold $V_\pi(\mathbf{s})$ fixed and optimize over the policy in the first term. In particular, the new deterministic optimal policy at the state $\mathbf{s}$ is given by:

$$
\pi(\mathbf{s}) = \underset{\mathbf{a} \in \mathcal{A}}{\operatorname{argmax}}\, \mathbb{E}\left[R(\mathbf{s}', \mathbf{s}, \mathbf{a}) + \gamma V_\pi(\mathbf{s}')\right]. \tag{9.48}
$$

Once the policy is updated, the process repeats, fixing this policy to update the value function, and then using this updated value function to improve the policy. The process is repeated until both the policy and the value function converge to within a specified tolerance. It is important to note that this procedure is both expensive and prone to local minima. It also resembles the alternating descent method that is widely used in optimization and machine learning.

The formulation in Eq. (9.47a) makes it clear that it may be possible to optimize backwards from a state known to give a reward with high probability. Additionally, this approach requires having a model for P and R to predict the next state $\mathbf{s}'$, making this a *model-based* approach.

Value Iteration

Value iteration is similar to policy iteration, except that at every iteration only the value function is updated, and the optimal policy is extracted from this value function at the end. First, the value function is initialized, typically either with zeros or at random. Then, for all states $\mathbf{s} \in \mathcal{S}$, the value function is updated by returning the maximum value at that state across all actions $\mathbf{a} \in \mathcal{A}$, holding the value function fixed at all other states $\mathbf{s}' \in \mathcal{S} \backslash \mathbf{s}$:

$$V(\mathbf{s}) = \max_{\mathbf{a}} \sum_{\mathbf{s}'} P(\mathbf{s}' \mid \mathbf{s}, \mathbf{a}) \left(R(\mathbf{s}', \mathbf{s}, \mathbf{a}) + \gamma V(\mathbf{s}') \right). \tag{9.49}$$

This iteration is repeated until a convergence criterion is met.

After the value function converges, it is possible to extract the optimizing policy π:

$$\pi(\mathbf{s}, \mathbf{a}) = \underset{\mathbf{a}}{\operatorname{argmax}} \sum_{\mathbf{s}'} P(\mathbf{s}' \mid \mathbf{s}, \mathbf{a}) \left(R(\mathbf{s}', \mathbf{s}, \mathbf{a}) + \gamma V(\mathbf{s}') \right). \tag{9.50}$$

Although value iteration typically requires fewer steps per iteration, policy iteration often converges in fewer iterations. This may be due to the fact that the value function is often more complex than the policy function, requiring more parameters to optimize over.

Note that the value function in RL typically refers to a discounted sum of future rewards that should be maximized, while in nonlinear control it refers to an integrated cost that should be minimized. The phrase *value function* is particularly intuitive when referring to accumulated rewards in the economic sense, as it quantifies the *value* of being in a given state. However, in the case of nonlinear control theory, the *value function* is more accurately thought of as quantifying the *numerical value* of the objective function evaluated on the optimal trajectory. This notation can be confusing, and is worth careful consideration depending on the context.

Quality Function

Polic iteration and value iteration are related by the quality function $Q(\mathbf{s}, \mathbf{a})$

$$Q(\mathbf{s}, \mathbf{a}) = \mathbb{E}\left[R(\mathbf{s}', \mathbf{s}, \mathbf{a}) + \gamma V(\mathbf{s}') \right] \tag{9.51a}$$

$$= \sum_{\mathbf{s}'} P\left(\mathbf{s}' \mid \mathbf{s}, \mathbf{a}\right) \left(R(\mathbf{s}', \mathbf{s}, \mathbf{a}) + \gamma V(\mathbf{s}') \right). \tag{9.51b}$$

In a sense, the optimal policy $\pi(\mathbf{s}, \mathbf{a})$ and the optimal value function $V(\mathbf{s}, \mathbf{a})$ contain redundant information, as one can be determined from the other via the quality function $Q(\mathbf{s}, \mathbf{a})$:

$$\pi(\mathbf{s}, \mathbf{a}) = \underset{\mathbf{a}}{\operatorname{argmax}}\ Q(\mathbf{s}, \mathbf{a}) \tag{9.52a}$$

$$V(\mathbf{s}) = \max_{\mathbf{a}} Q(\mathbf{s}, \mathbf{a}). \tag{9.52b}$$

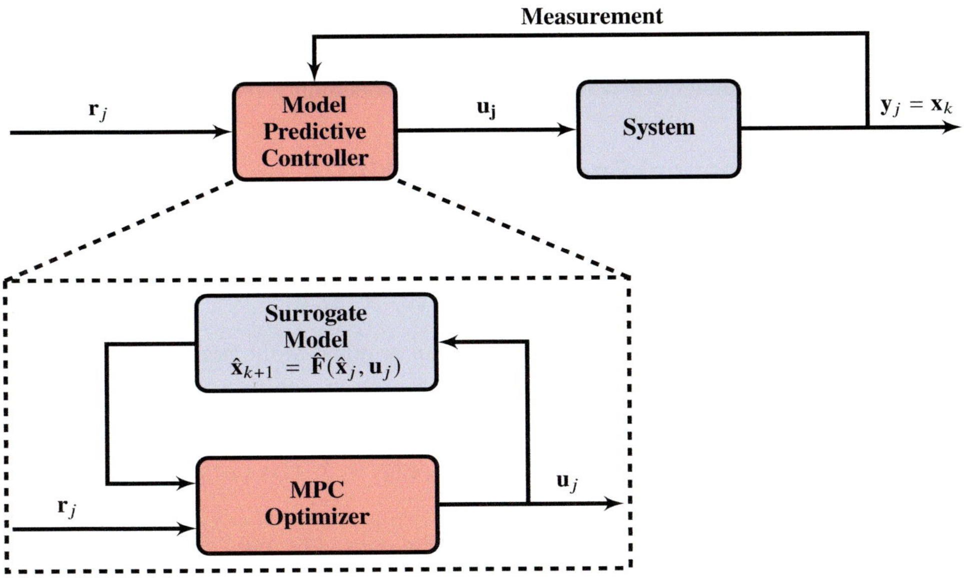

Figure 9.6 Schematic of model predictive control, where a surrogate model is used to run an optimization directly inside the control loop. This diagram assumes full state measurements $\mathbf{y} = \mathbf{x}$ for simplicity, although this is not strictly necessary for MPC.

This formulation is used for model-free Q-learning [531, 510, 164].

9.5 Model Predictive Control

Model predictive control (MPC) [89, 184] is one of the most flexible and widely used techniques in industrial control. It is also one of the most computationally intensive. It employs constrained optimization, wrapped around a predictive model, to control dynamics with constrains, time delays, non-minimum phase dynamics, and instability. This optimization loop must be real-time, which typically necessitates either expensive computational hardware or simplified models. Progress in MP has directly tracked our ability to scale the complexity of this real-time optimization. Many industrial applications of MPC have slow time scales and use linear models, because they are simple to identify and fast to optimize over. Improved computational hardware and optimization algorithms are enabling MPC for nonlinear systems with faster time scales. For example, field-programmable gate arrays (FPGAs) are being used for embedded MPC in some applications [240, 38]. Recently, deep learning and reinforcement learning have been combined effectively with MPC [306, 554].

Model predictive control is shown schematically in Fig. 9.6. MPC determines the next immediate control action by solving a constrained optimal control problem over a receding horizon. The control sequence is then implemented until the system state is updated via sensor measurements, after which the receding horizon control problem is reinitialized and solved again. This iterative procedure is repeated, balancing model prediction with sensor feedback. The predictive model is used to forecast control scenarios into the future, providing fast open-loop control, while the sensor information provides a corrective feedback signal to account for instability, uncertainty, model-mismatch, and disturbances.

Mathematical Formulation of MPC

In MPC, the open-loop control input $\mathbf{u}$ is optimized on a receding time-horizon $t_c = m_c\Delta t$ to minimize a cost J over a prediction horizon $t_p = m_p\Delta t$, based on a surrogate model

$$\hat{\mathbf{x}}_{k+1} = \hat{\mathbf{F}}(\hat{\mathbf{x}}_j, \mathbf{u}_j). \tag{9.53}$$

The control horizon is usually less than or equal to the prediction horizon, and $\mathbf{u}$ is held constant for $t \in [t_c, t_p]$. The objective function is typically similar to an LQR problem

$$J(\mathbf{x}_j) = \sum_{k=0}^{m_p-1} \|\hat{\mathbf{x}}_{j+k} - \mathbf{r}_{j+k}\|_{\mathbf{Q}}^2 + \sum_{k=1}^{m_c} \left(\|\mathbf{u}_{j+k}\|_{\mathbf{R}}^2 + \|\Delta\mathbf{u}_{j+k}\|_{\mathbf{R}_\Delta}^2 \right), \tag{9.54}$$

where $\mathbf{r}_j$ is the reference trajectory to be tracked by MPC, $\|\mathbf{x}\|_{\mathbf{Q}}^2 = \mathbf{x}^T\mathbf{Q}\mathbf{x}$, and there is an additional penalty on large changes in the control, i.e. on $\Delta\mathbf{u}_j = \mathbf{u}_j - \mathbf{u}_{j-1}$. Note that the weight matrix $\mathbf{Q}$ must be positive semidefinite and $\mathbf{R}$ and $\mathbf{R}_\Delta$ must be positive semidefinite, as in LQR. It is also possible to add a terminal cost on the final state.

The objective function is optimized over the sequence $\mathbf{U}_j = \{\mathbf{u}_{j+1}, \ldots, \mathbf{u}_{j+k}, \ldots, \mathbf{u}_{j+m_c}\}$ of control inputs, subject to the constrains of the surrogate model in Eq. (9.53), constraints on the control inputs $\mathbf{u}_j$, the rate of change of control $\Delta\mathbf{u}_j$, and the predicted state $\hat{\mathbf{x}}_j$:

$$\begin{aligned} \min_{\mathbf{U}_j} \quad & J(\mathbf{x}_j) \\ \text{s.t.} \quad & \hat{\mathbf{x}}_{k+1} = \hat{\mathbf{F}}(\hat{\mathbf{x}}_j, \mathbf{u}_j) \\ & \mathbf{u}_{\min} \leq \mathbf{u}_j \leq \mathbf{u}_{\max} \\ & \Delta\mathbf{u}_{\min} \leq \Delta\mathbf{u}_j \leq \Delta\mathbf{u}_{\max} \\ & \mathbf{x}_{\min} \leq \hat{\mathbf{x}}_j \leq \mathbf{x}_{\max}. \end{aligned} \tag{9.55}$$

This is a large-scale nonlinear program (NLP). Efficient solvers, such as IPOPT and SNOPT, can handle inequality constraints, high-dimensional state and control vectors, and complex objective function terms. Optimization details will be discussed more below.

The optimal control is then applied for one time step, or until sensor measurements are available to update the state $\mathbf{x}$, and the procedure is repeated and the receding-horizon control re-optimized at each subsequent time step. For simplicity, we will assume that sensor measurements are available at each time step. This results in the control law:

$$\mathbf{K}(\mathbf{x}_j) = \mathbf{u}_{j+1}(\mathbf{x}_j), \tag{9.56}$$

where $\mathbf{u}_{j+1}$ is the first time step of the optimized actuation starting at $\mathbf{x}_j$. This is illustrated in Fig. 9.7. For more details, see Kaiser et al. [263] and Fonzi et al. [175].

It is possible to optimize highly customized objective functions, subject to nonlinear dynamics, with constrains on the actuation and state. However, the computational requirements of re-optimizing at each time-step are considerable, putting limits on the

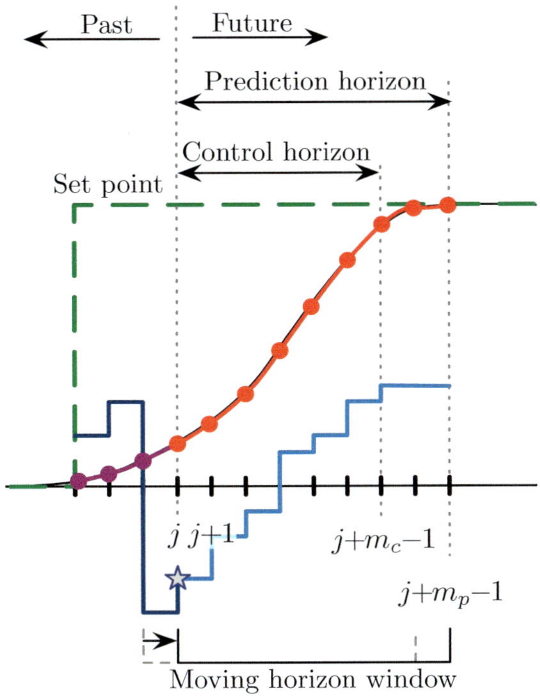

Figure 9.7 Illustration of model predictive control used to track a set point, where the actuation input **u** is iteratively optimized over a receding horizon. A receding horizon optimization is performed to compute the optimal control signal given the surrogate model, and the first step of this control sequence is enacted. After the system responds and is measured, a new optimization is initialized, allowing the controller to compensate for model errors and disturbances. The control horizon and prediction horizon may be different in MPC. *Reproduced with permission from Kaiser et al. [263].*

complexity of the model and optimization techniques. Fortunately, rapid advances in computing power and optimization are enabling MPC for real-time nonlinear control.

There have been tremendous advances in MPC, especially deep MPC, which uses deep learning for the surrogate model [306]. Deep MPC has been widely used for vision-based driving systems [153], for controlling fluid flows [358, 52], and in aeroelastic systems [175]. Deep MPC is particularly useful in data-rich applications, although deep learning requires large volumes of data and may not be readily interpretable. A complementary line of research seeks to identify models for MPC based on limited data. For example, Kaiser et al. [263] showed that it is possible to rapidly identify system models from limited data using the methods discussed in Section 8.7, and then use these models for MPC.

Tube MPC is another robust strategy that keeps the system within a certifiable tube around the reference trajectory [322, 166], which is useful for safety critical systems with model error and disturbances. More generally, developing robust and distributed MPC algorithms with guarantees [415, 416] is important in autonomy and robotics. MPC may also be used in model-based reinforcement learning [537]. Finally, recent efforts have combined differentiable programming with MPC, resulting in differentiable predictive control [6, 154].

Optimization Algorithms for MPC

Optimization is at the heart of MPC, and its feasibility in real-time applications depends on efficient solvers. Historically, the development of optimization techniques for MPC has closely followed advances in numerical methods and computing power. Mayne [343] provides a nice discussion of these trends and developments.

When the dynamics are linear and the objective function are quadratic and positive definite, and when the control, state and terminal constraints are convex, then the MPC problem is a strictly convex quadratic program (QP), which has a globally optimal solution that may be effectively computed. There are several established computational techniques that may be applied, such as interior point methods from Section 6.4. These approaches include active set methods [351, 167, 278, 88, 225], interior point methods [407, 141, 551], and gradient projection methods, for example using Nesterov's accelerated gradient [417, 283]. Early applications of MPC in the 1980s [184] primarily relied on these methods due to their simplicity and robustness. Active set methods were particularly popular for their ability to handle constrains directly and exploit the sparsity structure of the constraint and dynamics matrices, leading to faster and more memory-efficient solutions. As the scale of industrial problems grew, interior-point methods gained traction for their superior performance on large-scale QPs.

Nonlinear MPC (NMPC) requires the solution of a nonlinear programming (NLP) problem at each time step. It is common to use sequential quadratic programming (SQP), based on a local quadratic approximation to the optimal control problem at each iteration, which is then solved using the standard QP algorithms above. Each QP subproblem is constructed by approximating the Lagrangian with a second-order Taylor series, using the Hessian of the Lagrangian to form the quadratic objective of the QP subproblem. When Hessian information is unavailable, it is possible to use quasi-Newton methods, as in Section 2.6. Solvers like IPOPT [524] and SNOPT [192] have been widely adopted for NMPC due to their robustness and ability to handle large-scale problems with nonlinear constraints.

Recent developments have focused on first-order operator splitting methods, such as the alternating direction method of multipliers (ADMM) [65], which was discussed in Section 6.5. ADMM is particularly attractive for embedded MPC applications, where computational resources are limited. Its iterative nature and ability to exploit sparsity make it suitable for distributed and structured MPC, where the optimization problem is decomposed into smaller sub-problems. Robust operator-splitting solvers, such as OSQP [472], which is based on an ADMM framework tailored for QPs, has further advanced MPC. OSQP leverages the problem structure to achieve real-time performance on embedded systems, enabling MPC to be deployed in resource-constrained environments such as robotics and automotive control.

Despite these advances, challenges remain. Solvers for NMPC are computationally intensive and sensitive to the choice of initial guesses and regularization parameters.

Panel 9.2: Differentiable Control

Many control problems can be formulated as inverse problems, such as finding the precise angle of inclination for a cannon to reach a certain desired distance, despite nonlinear drag forces, variable mass, and variable wind speeds. Differentiable control is a new paradigm for such complex, nonlinear control problems, where we develop a differentiable surrogate model (e.g., a neural network) that may then be used to solve the inverse problem via gradient descent.

In this example, inspired by a trebuchet control example in FluxML [244, 245], we consider the cannon problem from Panel 1.1, except with the added complexity of drag forces that are quadratic in the velocity, a range of ambient wind speeds, and variable cannon mass. The dynamics governing this process are a nonlinear differential equation, and the final position is not easily differentiable with respect to initial cannon angle. However, in this example, we are able to train a very simple multilayer perceptron (see Chapter 10 for more on neural networks) with three inputs (angle α, cannonball mass m, and wind speed W), two hidden layers, each with 128 neurons, and a scalar output corresponding to the final position of the cannonball. The network is trained on 1000 random trajectories with the inputs randomly sampled from a uniform distribution (mass $m \in [1, 5]$, angle $\alpha \in [0, 90]$, wind speed $W \in [-10, 10]$).

Once trained, we are able to use backpropagation through the network (again, see Section 10.2) to compute the gradient of final position with respect to initial angle. In the example shown, the desired final position is 100m, and the initial guess is $\alpha_0 = 45$. After less than ten steps of gradient descent, the differentiable controller converges to the correct angle for the surrogate model to hit the target of 100m. Testing this angle by simulating the real physics, we find that our prediction is within about 1m, or 1%, of the true distance.

Working code for this example is provided in the online supplement, including the neural network used to train the surrogate model.

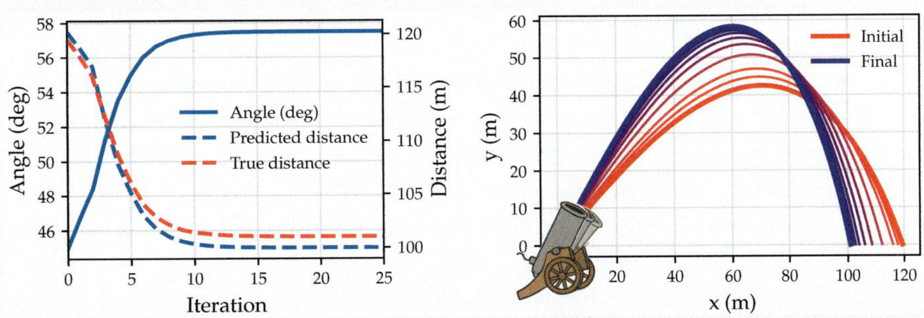

Figure 9.8: Angle and distance versus iteration (left) and trajectories (right).

Robustness to model inaccuracies and disturbances is another critical area of research, especially for safety-critical applications. Future trends include the integration of machine learning for better surrogate models and leveraging hardware accelerators, such as GPUs, to further speed up computations.

Linear Time-Varying Approximations

In practice, solving the full nonlinear MPC problem in real-time is often infeasible. Instead, linear time-varying (LTV) approximations are employed to simplify the optimization. The idea is to linearize the nonlinear dynamics $\hat{\mathbf{F}}(\hat{\mathbf{x}}, \mathbf{u})$ around the current operating point $(\hat{\mathbf{x}}_j, \mathbf{u}_j)$ at each time step:

$$\hat{\mathbf{F}}(\hat{\mathbf{x}}, \mathbf{u}) \approx \mathbf{A}\hat{\mathbf{x}} + \mathbf{B}\mathbf{u} + \mathbf{c}, \tag{9.57}$$

where $\mathbf{A} = \frac{\partial \hat{\mathbf{F}}}{\partial \hat{\mathbf{x}}}\Big|_{\hat{\mathbf{x}}_j, \mathbf{u}_j}$, $\mathbf{B} = \frac{\partial \hat{\mathbf{F}}}{\partial \mathbf{u}}\Big|_{\hat{\mathbf{x}}_j, \mathbf{u}_j}$, $\mathbf{c} = \hat{\mathbf{F}}(\hat{\mathbf{x}}_j, \mathbf{u}_j) - \mathbf{A}\hat{\mathbf{x}}_j - \mathbf{B}\mathbf{u}_j$.

The resulting optimization problem is again a QP problem:

$$\begin{aligned} \min_{\mathbf{U}_j} \quad & \sum_{k=0}^{m_p-1} \|\hat{\mathbf{x}}_{j+k} - \mathbf{r}_{j+k}\|_{\mathbf{Q}}^2 + \sum_{k=1}^{m_c} \left(\|\mathbf{u}_{j+k}\|_{\mathbf{R}}^2 + \|\Delta \mathbf{u}_{j+k}\|_{\mathbf{R}_\Delta}^2 \right) \\ \text{s.t.} \quad & \hat{\mathbf{x}}_{j+k+1} = \mathbf{A}\hat{\mathbf{x}}_{j+k} + \mathbf{B}\mathbf{u}_{j+k} + \mathbf{c}, \quad k = 0, \ldots, m_p - 1. \end{aligned} \tag{9.58}$$

By updating the linearization at each time step, LTV approximations strike a balance between computational feasibility and accuracy.

Warm-Starting and Real-Time Optimization

Warm-starting is critical to ensure that the MPC optimization can be solved in the system's sampling interval. The solution from the previous time step is used to initialize the current optimization problem, reducing the number of iterations required for convergence. In linear MPC, warm-starting involves initializing the QP solver with the previous control sequence. This approach is particularly effective when the system operates near a steady state or follows a slowly varying trajectory.

Warm-starting dates back to the early applications of MPC in the 1980s, when computational resources were limited. Active set methods naturally benefited from warm-starting because the active set of constraints at the solution tends to change gradually over time. By initializing the active set with the previous solution, the algorithm converges faster. Similarly, interior-point methods benefit from starting near the optimal trajectory, reducing the number of Newton iterations required.

Real-time optimization solvers, such as FORCES [144] and qpOASES [168], are explicitly designed to exploit warm-starting. These solvers take advantage of problem sparsity and the structure of the QP to achieve fast convergence. In embedded systems, where computational resources are constrained, warm-starting is essential for meeting the stringent timing requirements of real-time control.

One limitation of warm-starting is its reliance on the smooth evolution of the optimization problem. Abrupt changes in the system dynamics or reference trajectory can lead to slower convergence or even failure to converge. To mitigate this, robust optimization techniques and adaptive regularization strategies are often employed. Future directions include integrating warm-starting with machine learning models to better anticipate system changes and further improve solver performance.

9.6 Linear Matrix Inequalities

Linear matrix inequalities (LMIs) are a class of convex constrains that play a central role in modern control theory [64, 439]. They provide a unifying framework for solving a wide range of optimization and control problems, particularly those involving stability, robustness, and performance guarantees. LMIs are particularly appealing in control because they are convex, and efficient numerical solvers based on interior-point methods can solve them reliably. Many classical control problems, including stability, LQR, the Kalman filter, and robust control, can be reformulated as LMIs. LMI formulations are widely used in optimal control and MPC to encode constraints on states, inputs, and system dynamics. For instance, quadratic objectives can be cast into an LMI framework, as will be shown.

Motivating LMI: Lyapunov's Stability Problem

We follow Boyd et al. [64] and begin by motivating LMIs with their origin over a century ago. Lyapunov's direct method [326] states that a linear system

$$\frac{\mathrm{d}}{\mathrm{d}t}\mathbf{x} = \mathbf{A}\mathbf{x} \tag{9.59}$$

is stable if and only if there exists a positive definite matrix $\mathbf{P}$ such that

$$\mathbf{A}^{\mathrm{T}}\mathbf{P} + \mathbf{P}\mathbf{A} < \mathbf{0} \qquad \mathbf{P} > \mathbf{0}. \tag{9.60}$$

This is known as a Lyapunov inequality, and it is the original LMI, formulated in 1892.

Further, Lyapunov showed how this LMI can be solved by choosing any positive definite, self-adjoint matrix $\mathbf{Q} = \mathbf{Q}^{\mathrm{T}} > \mathbf{0}$ and solving the linear system of equations

$$\mathbf{A}^{\mathrm{T}}\mathbf{P} + \mathbf{P}\mathbf{A} = -\mathbf{Q} \tag{9.61}$$

for $\mathbf{P}$. If $\mathbf{P}$ exists, the system is stable, and $\mathbf{V}(\mathbf{x}) = \mathbf{x}^{\mathrm{T}}\mathbf{P}\mathbf{x}$ is a Lyapunov function.

LMIs were further developed in the 1960s and 1970s, resulting in many of the forms described here. Boyd et al. [64] provide a complete and interesting account of the history of LMIs in control. One notable development was by Willems in 1971 [536], who noted that LMIs can be formulated as convex optimization problems, providing computational and algorithmic benefits. The development of IPMs in the 1980s made it possible to solve LMIs efficiently at scale, and is the foundation for many modern algorithms in control.

Two Mathematical Forms of LMIs

An LMI is an inequality of the following form:

$$\mathbf{F}(\mathbf{X}) < \mathbf{Q}. \tag{9.62}$$

$\mathbf{X} \in \mathcal{X}$ is an element of a real vector space, $\mathbf{F} : \mathcal{X} \to \mathbb{H}^n$ is linear in $\mathbf{X}$, and $\mathbf{Q} \in \mathbb{H}^n$. $\mathbb{H}^n$ is the space of $n \times n$ Hermitian matrices, so $\mathbf{Q} \in \mathbb{H}^n$ implies $\mathbf{Q} = \mathbf{Q}^{\mathrm{T}}$. $\mathbf{X}$ may be a vector, $\mathcal{X} = \mathbb{R}^n$, or a symmetric $n \times n$ matrix, $\mathcal{X} = \mathbb{S}^n$. An example will help make this clear:

- $\mathbf{A}^{\mathrm{T}}\mathbf{X}\mathbf{A} - \mathbf{X} < \mathbf{Q}$, where $\mathbf{X} \in \mathbb{S}^n$ is a symmetric matrix.

In general, it is possible to write an LMI in the following form

$$\mathbf{A}_0 + \sum_{i=1}^{m} x_i \mathbf{A}_i > 0, \tag{9.63}$$

where $\mathbf{A}_0, \mathbf{A}_1, \ldots, \mathbf{A}_m$ are symmetric matrices, $x_1, x_2, \ldots, x_m$ are decision variables, and $>$ denotes positive definiteness, meaning that all eigenvalues of the matrix are nonnegative. It is also possible to have positive-semidefinite problems, which are called non-strict [64].

A simple optimization problem involving LMIs is the feasibility problem, to compute a $\mathbf{X} \in \mathcal{X}$ such that $\mathbf{F}(\mathbf{X}) < \mathbf{Q}$. The set of such $\mathbf{X}$ is convex.

Casting Quadratic Problems as LMIs

One of the strengths of the LMI framework is that it may be possible to recast seemingly nonlinear problems into an LMI. We will illustrate this with a quadratic constraint that occurs commonly in control:

$$\mathbf{X}^{\mathrm{T}}\mathbf{R}^{-1}\mathbf{X} - \mathbf{Q} < \mathbf{0}. \tag{9.64}$$

This is not an LMI, since it is quadratic in the state $\mathbf{X}$. However, using the Schur complement formula, derived below, this may be recast as the following LMI:

$$\begin{bmatrix} \mathbf{Q} & \mathbf{X}^{\mathrm{T}}\mathbf{A}^{\mathrm{T}} \\ \mathbf{A}\mathbf{X} & \mathbf{R} \end{bmatrix} > \mathbf{0}. \tag{9.65}$$

LQR as an LMI

Recall that the LQR optimal controller minimizes a quadratic cost:

$$J = \int_0^\infty \left(\mathbf{x}^{\mathrm{T}}\mathbf{Q}\mathbf{x} + \mathbf{u}^{\mathrm{T}}\mathbf{R}\mathbf{u}\right) \mathrm{d}t, \tag{9.66}$$

where $\mathbf{Q} \geq \mathbf{0}$ and $\mathbf{R} > \mathbf{0}$ are weight matrices. The optimal control input is derived from the solution of the continuous-time algebraic Riccati equation

$$\mathbf{A}^{\mathrm{T}}\mathbf{P} + \mathbf{P}\mathbf{A} - \mathbf{P}\mathbf{B}\mathbf{R}^{-1}\mathbf{B}^{\mathrm{T}}\mathbf{P} + \mathbf{Q} = \mathbf{0}. \tag{9.67}$$

This can be reformulated as an LMI using the Schur complement formula from above:

$$\begin{bmatrix} \mathbf{A}^{\mathrm{T}}\mathbf{P} + \mathbf{P}\mathbf{A} + \mathbf{Q} & \mathbf{P}\mathbf{B} \\ \mathbf{B}^{\mathrm{T}}\mathbf{P} & -\mathbf{R} \end{bmatrix} < \mathbf{0}. \tag{9.68}$$

Solving this LMI for $\mathbf{P}$ provides the optimal state feedback gain $\mathbf{K} = \mathbf{R}^{-1}\mathbf{B}^{\mathrm{T}}\mathbf{P}$.

Kalman Filteing as an LMI

The Kalman filter gain for an optimal state estimator is likewise derived from the solution of the Riccati equation:

$$\mathbf{AP} + \mathbf{PA}^{\mathrm{T}} - \mathbf{PC}^{\mathrm{T}}\mathbf{R}^{-1}\mathbf{CP} + \mathbf{Q} = \mathbf{0}. \tag{9.69}$$

This can be recast as an LMI:

$$\begin{bmatrix} \mathbf{AP} + \mathbf{PA}^{\mathrm{T}} + \mathbf{Q} & \mathbf{PC}^{\mathrm{T}} \\ \mathbf{CP} & -\mathbf{R} \end{bmatrix} < \mathbf{0}. \tag{9.70}$$

The solution $\mathbf{P}$ provides the Kalman gain $\mathbf{K} = \mathbf{PC}^{\mathrm{T}}\mathbf{R}^{-1}$.

Robust H_∞ Control

Robust control designs often require ensuring stability and performance under model uncertainties. The H_∞ control problem seeks to design a controller that minimizes the worst-case gain from disturbance to output:

$$\|\mathbf{T}_{\mathbf{w}\to\mathbf{y}}\|_\infty < \gamma, \tag{9.71}$$

Panel 9.3: The Schur Complement: It's a Schur Thing

The Schur complement formula states that the following set of matrix inequalities

$$\begin{aligned} &\mathbf{R} > \mathbf{0} \\ &\mathbf{Q} - \mathbf{XR}^{-1}\mathbf{X}^{\mathrm{T}} > \mathbf{0} \end{aligned}$$

is equivalent to

$$\begin{bmatrix} \mathbf{Q} & \mathbf{X} \\ \mathbf{X}^{\mathrm{T}} & \mathbf{R} \end{bmatrix} > \mathbf{0}.$$

To see this, we take the first set of equations, written in matrix form:

$$\begin{bmatrix} \mathbf{Q} - \mathbf{XR}^{-1}\mathbf{X}^{\mathrm{T}} & \mathbf{0} \\ \mathbf{0} & \mathbf{Q} \end{bmatrix} > \mathbf{0}$$

and apply the following change of coordinates by multiplying by

$$\begin{bmatrix} \mathbf{I} & \mathbf{XR}^{-1} \\ \mathbf{0} & \mathbf{I} \end{bmatrix}$$

on the left and by its adjoint on the right:

$$\begin{bmatrix} \mathbf{I} & \mathbf{XR}^{-1} \\ \mathbf{0} & \mathbf{I} \end{bmatrix} \begin{bmatrix} \mathbf{Q} - \mathbf{XR}^{-1}\mathbf{X}^{\mathrm{T}} & \mathbf{0} \\ \mathbf{0} & \mathbf{R} \end{bmatrix} \begin{bmatrix} \mathbf{I} & \mathbf{0} \\ \mathbf{R}^{-1}\mathbf{X}^{\mathrm{T}} & \mathbf{I} \end{bmatrix} = \begin{bmatrix} \mathbf{Q} & \mathbf{X} \\ \mathbf{X}^{\mathrm{T}} & \mathbf{R} \end{bmatrix} > \mathbf{0}.$$

This inequality is greater than zero because the change of coordinates matrix is non-singular and positive definite.

where γ is a scalar bound. This can be formulated as an LMI:

$$\begin{bmatrix} \mathbf{A}^{\mathrm{T}}\mathbf{P} + \mathbf{P}\mathbf{A} & \mathbf{P}\mathbf{B} & \mathbf{C}^{\mathrm{T}} \\ \mathbf{B}^{\mathrm{T}}\mathbf{P} & -\gamma\mathbf{I} & \mathbf{D}^{\mathrm{T}} \\ \mathbf{C} & \mathbf{D} & -\gamma\mathbf{I} \end{bmatrix} < \mathbf{0}. \tag{9.72}$$

where $\mathbf{P}$ and γ are decision variables. Solving this LMI provides the controller gains that ensure robustness, as well as the performance bound γ.

Suggested Reading

Texts

(1) **Feedback Systems: An Introduction for Scientists and Engineers**, by K. J. Aström and R. M. Murray, 2010 [15].
(2) **A Course in Robust Control Theory: A Convex Approach**, by G. E. Dullerud and F. Paganini, 200 [156].
(3) **Multivariable Feedback Control: Analysis and Design**, by S. Skogestad and I. Postlethwaite, 2005 [466].
(4) **Optimal Control and Estimation**, by R. F. Stengel, 1994 [473].
(5) **Reinforcement Learning: An Introduction**, by R. S. Sutton and A. G. Barto, 1998 [480].
(6) **Linear Matrix Inequalities in System and Control Theory**, by S. Boyd, L. El Ghaoui, E. Feron, and V. Balakrishnan, 1994 [64].

Papers and reviews

(1) **A new approach to linear filtering and prediction problems**, by R. E. Kalman, *Journal of Fluids Engineering*, 1960 [266].
(2) **Guaranteed margins for LQG regulators**, by J. C. Doyle, *IEEE Transactions on Automatic Control*, 1978 [152].
(3) **Model predictive control: theory and practice—a survey**, by C. E. Garcia, D. M. Prett, and M. Morari, *Automatica*, 1989 [184].
(4) **Model predictive control**, by E. F. Camacho and C. B. Alba, 2013 [89].
(5) **Constrained model predictive control: Stability and optimality**, by D. Q. Mayne, J. B. Rawlings, C. V. Rao, and P. O. M. Scokaert, *Automatica*, 2000 [344].
(6) **Sparse identification of nonlinear dynamics for model predictive control in the low-data limit**, by E. Kaiser, J. N. Kutz, and S. L. Brunton, *Proceedings of the Royal Society of London A*, 2018 [263].
(7) **A tour of reinforcement learning: The view from continuous control**, by B. Recht, *Annual Review of Control, Robotics, and Autonomous Systems*, 2019 [409].
(8) **Differentiable predictive control: An MPC alternative for unknown nonlinear systems using constrained deep learning**, by J. Drgona, K. Kis, A. Tuor, D. Vrabie, and M. Klauco, *arXiv*, 2020 [154].

Homework

Exercise 9.1 (LQR Design) We will explore optimal control on the inverted pendulum on a cart, shown in Fig. 9.9.

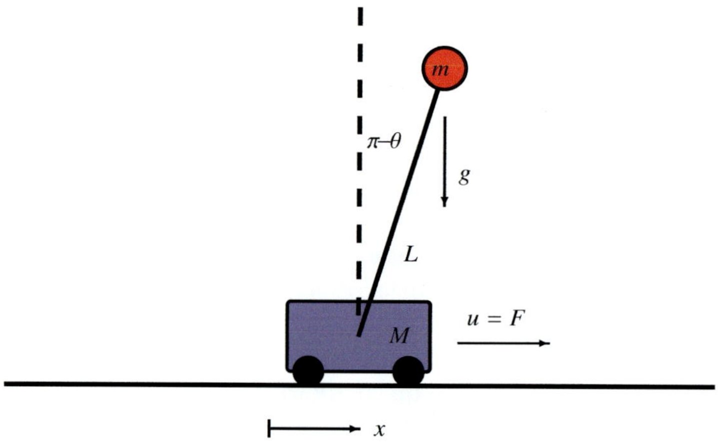

Figure 9.9 Schematic of inverted pendulum on a cart. The control forcing acts to accelerate or decelerate cart. For this example, we assume the following parameter values: pendulum mass ($m = 1$), cart mass ($M = 5$), pendulum length ($L = 2$), gravitational acceleration ($g = -10$), and cart damping ($\delta = 1$).

The full nonlinear dynamics are given by

$$\dot{x} = v \tag{9.73a}$$

$$\dot{v} = \frac{-m^2L^2g\cos(\theta)\sin(\theta) + mL^2(mL\omega^2\sin(\theta) - \delta v) + mL^2u}{mL^2(M + m(1 - \cos(\theta)^2))} \tag{9.73b}$$

$$\dot{\theta} = \omega \tag{9.73c}$$

$$\dot{\omega} = \frac{(m + M)mgL\sin(\theta) - mL\cos(\theta)(mL\omega^2\sin(\theta) - \delta v) - mL\cos(\theta)u}{mL^2(M + m(1 - \cos(\theta)^2))} \tag{9.73d}$$

where x is the cart position, v is the velocity, θ is the pendulum angle, ω is the angular velocity, m is the pendulum mass, M is the cart mass, L is the pendulum arm, g is the gravitational acceleration, δ is a friction damping on the dart, and u is a control force applied to the cart.

There are two fixed points, corresponding to either the pendulum down ($\theta = 0$) or pendulum up ($\theta = \pi$) configuration; in both cases, $v = \omega = 0$ for the fixed point, and the cart position x is a free variable, as the equations do not depend explicitly on x.

1. Linearize the equations in (9.73) about the up and the down solutions to obtain **A** and **B** matrices.
2. Confirm that the system is linearly controllable with the given **A** and **B** matrices. This may be accomplished by computing the rank of the controllability matrix using

`numpy.linalg.matrix_rank(ctrb(A,B))` in Python. The matrix should be full rank, which in this case is 4.

3. Design an LQR controller to stabilize the inverted pendulum configuration ($\theta = \pi$) assuming full-state measurements, $\mathbf{y} = \mathbf{x}$. Use gain matrices $\mathbf{Q} = \mathbf{I}$ and $R = 0.0001$. What is the LQR controller gain matrix $\mathbf{K}$, and what are the closed loop eigenvalues?
4. Plot the response of the closed loop system, starting slightly off equilibrium, at $\mathbf{x}_0 = \begin{bmatrix}-1 & 0 & \pi+.1 & 0\end{bmatrix}^T$.
5. To visually see the optimality of the LQR controller, we will now simulate the closed-loop system response obtained by non-optimal eigenvalue placement. Compute the system response and cost function for 100 randomly generated sets of stable eigenvalues, chosen in the interval $[-3.5, -.5]$; plot these in light gray. Plot the LQR solution on top of this in dark black. You can compute the $\mathbf{K}$ matrices for the random eigenvalue placement using the `place` command:

CodeSnip 9.e
```
K = place(A,B,neweigs)
```

Exercise 9.2 (Kalman Filter Design) We now revisit the pendulum on a cart, given by Eq. (9.73), but assuming that we only have partial measurements and must estimate the state with a Kalman filter. For this example, we will develop the Kalman filter for the pendulum-down condition ($\theta = 0$), since without feedback the system in the pendulum-up condition will quickly leave the fixed point where the linear model is valid. We will use the cart position as the sensor, $y = x_1$, which corresponds to an observation matrix $\mathbf{C} = \begin{bmatrix}1 & 0 & 0 & 0\end{bmatrix}$.

1. Confirm that the system is linearly observable with the given $\mathbf{A}$ and $\mathbf{C}$ matrices. This is accomplished by computing the rank of the observability matrix using `numpy.linalg.matrix_rank(obsv(A,C))` in Python. The matrix should be full rank, which in this case is 4.
2. Now design a Kalman filter with disturbance and noise covariances given by $\mathbf{V}_d = \mathbf{I}$ and $V_n = 1$. What is the Kalman filter gain matrix $\mathbf{K}_f$, and what are the eigenvalues of the estimator?
3. Simulate the augmented system, with Kalman filter, with noise disturbance. You can use code like the following:

CodeSnip 9.f
```
## Augment system with additional inputs [u I*wd 0*wn]
Baug = np.concatenate((B, np.eye(4),np.zeros_like(B)),axis=1)
Daug = np.array([0,0,0,0,0,1]) # D matrix passes noise through

sysC = ss(A,Baug,C,Daug) # Single-measurement system

# "True" system w/ full-state output, disturbance, no noise
sysTruth = ss(A,Baug,np.eye(4),np.zeros((4,Baug.shape[1])))

BKf = np.concatenate((B,np.atleast_2d(Kf).T),axis=1)
sysKF = ss(A-np.outer(Kf,C),BKf,np.eye(4),np.zeros_like(BKf))
```

4. Now modify the $\mathbf{C}$ matrix to include some imperfection in the measurement, but don't update the Kalman filter gain. How does this change the estimate?

Exercise 9.3 (LQG Design) Combine the Kalman filter and LQR controller from the exercises above, and simulate the full system response with noise and disturbances.

For this example, we use the following LQR and LQE weighting matrices: $\mathbf{Q} = \mathbf{I}_{4\times4}$, $R = 0.000001$, $\mathbf{V}_d = 0.04\,\mathbf{I}_{4\times4}$, $V_n = 0.0002$.

Start the system near the vertical equilibrium, at $\mathbf{x}_0 = \begin{bmatrix}0 & 0 & 3.14 & 0\end{bmatrix}^T$, and command a step in the cart position from $x = 0$ to $x = 1$ at $t = 10$.

Compare the system response in both the linearized system and the full nonlinear system.

Exercise 9.4 This example will design an optimal controller for a rotary inverted pendulum.

(a) Derive the equations of motion for a rotary pendulum, where the base of the pendulum is mounted to a rotating arm. The control input is a torque to the rotor arm.

(b) Identify the fixed points of the system and linearize about each point. What is the stability of each fixed point? Are they linear controllable?

(c) Design an LQR controller for the pendulum up configuration assuming full state measurements.

(d) Determine the observability of the pendulum up configuration if we cannot measure the full state, but instead measure the pendulum angle and the rotor angle. Similarly, determine the observability if we only measure the pendulum angular rate and the rotor angle. Which sensor configuration is *more* observable? Pick at least one different sensor set and assess if this configuration is observable.

(e) Assuming a measurement of the pendulum angle and rotor angle, design a Kalman filter to estimate the full state. Design for disturbance magnitude 1.e-3 and sensor noise 1.e-2. Simulate the noisy system and compare the Kalman filter estimate with the true state without added noise.

(f) Design an LQG controller for the pendulum up configuration and demonstrate this controller in simulation. How does the controller perform when you introduce a small time delay? At what point does the time delay cause the system to go unstable?

Design a full-state LQR controller to track a reference position of the cart, while minimizing the pendulum swinging. Try different LQR gain matrices, and compare the step response when setting the reference state from $x = 0$ to $x = 1$.

Exercise 9.5 Derive the equations of motion for a double pendulum on a cart.
Repeat each step above for the double pendulum.

Exercise 9.6 This exercise will design a controller to move a cart with a pendulum in the down position from one point on a track to another. The goal will be to move quickly from point $x = 0$ to point $x = 1$ while minimizing the amount the pendulum swings, which corresponds to the problem faced when designing a controller for a gantry crane. What will make this example more interesting is that it is a *non-minimum-phase* system.
Design a full-state LQR controller to track a reference position of the cart, while minimizing the pendulum swinging. Try different LQR gain matrices, and compare the step response when setting the reference state from $x = 0$ to $x = 1$.

Exercise 9.7 This example will explore how an ill-conditioned controllability Gramian can affect state-feedback control.

(a) Create a continuous-time, single-input state space system that has $n = 2$ states full-state measurements. Design the system to be technically controllable, yet with one of the directions being much more controllable than the other (i.e., 10^6 times more controllable). Compute the controllability Gramian for this system and compute its eigendecomposition. Explain what you find.

(b) Design an LQR controller for this system using weight matrices $\mathbf{Q} = \mathbf{I}$ and $R = 1$. Simulate the response of the closed loop system, and explain the results.

(c) Using this controller, initialize the system at thousands of points randomly sampled from a unit circle where $\|\mathbf{x}\| = 1$, and compute the LQR cost J for each of these trajectories. Plot the initial conditions on the sphere, color coded by the cost J. Reconcile this plot with the controllability Gramian you computed earlier.

Exercise 9.8 Consider the following 2D linear system:

$$\dot{\mathbf{x}} = \mathbf{A}\mathbf{x} + \mathbf{B}\mathbf{u}$$

with

$$\mathbf{A} = \begin{bmatrix} 0 & 1 \\ 0 & 0 \end{bmatrix}, \quad \mathbf{B} = \begin{bmatrix} 0 \\ 1 \end{bmatrix}.$$

In this problem, we will develop the LQR optimal state-feedback controller $\mathbf{u} = -\mathbf{K}\mathbf{x}$ using several different approaches. In the next exercise we will solve the same problem using LMIs. For the remainder, we will use the following LQR gain matrices:

$$\mathbf{Q} = \begin{bmatrix} 1 & 0 \\ 0 & 1 \end{bmatrix}, \quad R = 1.$$

1. First, we will explicitly solve the algebraic Riccati equation, which is necessary to derive $\mathbf{K}$. This equation is given by:

$$\mathbf{A}^T\mathbf{P} + \mathbf{P}\mathbf{A} - \mathbf{P}\mathbf{B}R^{-1}\mathbf{B}^T\mathbf{P} + \mathbf{Q} = 0.$$

   For the given $\mathbf{A}, \mathbf{B}, \mathbf{Q}$ and R, it becomes:

$$\begin{bmatrix} 0 & 0 \\ 1 & 0 \end{bmatrix}\mathbf{P} + \mathbf{P}\begin{bmatrix} 0 & 1 \\ 0 & 0 \end{bmatrix} - \mathbf{P}\begin{bmatrix} 0 \\ 1 \end{bmatrix} 1 \begin{bmatrix} 0 & 1 \end{bmatrix}\mathbf{P} + \begin{bmatrix} 1 & 0 \\ 0 & 1 \end{bmatrix} = 0.$$

   Solve for $\mathbf{P}$ explicitly by setting $\mathbf{P} = \begin{bmatrix} p_{11} & p_{12} \\ p_{12} & p_{22} \end{bmatrix}$. This yields a small system of equations in p_{11}, p_{12}, p_{22}.

2. Then compute $\mathbf{K} = R^{-1}\mathbf{B}^T\mathbf{P} = \mathbf{B}^T\mathbf{P}$ (since $R = 1$).
3. What are the closed loop eigenvalues of $\mathbf{A} - \mathbf{B}\mathbf{K}$? What is the resulting cost?
4. Now compute $\mathbf{K}$ using the `lqr` command, and compare.

Exercise 9.9 We will now solve the problem above using LMIs. We seek a feedback law $\mathbf{u} = -\mathbf{Kx}$ that minimizes the LQR integral. A standard LMI formulation is to search for $\mathbf{P} > 0$ and $\mathbf{K}$ such that the following matrix inequality holds:

$$(\mathbf{A} - \mathbf{BK})^{\mathrm{T}}\mathbf{P} + \mathbf{P}(\mathbf{A} - \mathbf{BK}) + \mathbf{Q} + \mathbf{K}^{\mathrm{T}}\mathbf{RK} < 0.$$

This condition guarantees that $V(\mathbf{x}) = \mathbf{x}^{\mathrm{T}}\mathbf{Px}$ is a valid Lyapunov function for the closed-loop system and that the derivative of V along trajectories, plus the quadratic cost, is negative definite. Additionally, we impose $\mathbf{P} > 0$ to ensure V is positive definite.

Then one can show that among all such $(\mathbf{P}, \mathbf{K})$ feasible to this inequality, the trace$(\mathbf{P})$ (or some other linear function in $\mathbf{P}, \mathbf{K}$) can serve as a surrogate for the LQR cost. Minimizing it enforces the same solution as the Riccati-based approach.

It is now possible to replicate the LQR minimization via

$$\begin{aligned} \min \quad & \mathrm{trace}(\mathbf{P}) \\ \text{s.t.} \quad & (\mathbf{A} - \mathbf{BK})^{\mathrm{T}}\mathbf{P} + \mathbf{P}(\mathbf{A} - \mathbf{BK}) + \mathbf{Q} + \mathbf{K}^{\mathrm{T}}R\mathbf{K} < 0. \end{aligned}$$

In many contexts, trace$(\mathbf{P})$ is a suitable stand-in for the LQR cost measure.

Any solution that minimizes trace$(\mathbf{P})$ under the LMI constraints above will match the $\mathbf{P}$ from standard LQR. The feedback $\mathbf{K} = (R^{-1}\mathbf{B}^{\mathrm{T}})\mathbf{P}$ is likewise identical to the LQR gain.

1. Solve this LMI by hand. Let $\mathbf{P} = \begin{bmatrix} p_{11} & p_{12} \\ p_{12} & p_{22} \end{bmatrix}$, $\mathbf{K} = (k_1, k_2)$. Write out the 2×2 LMI

$$(\mathbf{A} - \mathbf{BK})^{\mathrm{T}}\mathbf{P} + \mathbf{P}(\mathbf{A} - \mathbf{BK}) + \mathbf{Q} + \mathbf{K}^{\mathrm{T}}R\mathbf{K} < 0$$

   and solve, with the additional LMI $\mathbf{P} > 0$. This is essentially the same as rewriting the usual Riccati equation in an LMI format.
2. Now solve this using Python's `cvxpy`:

CodeSnip 9.g

```
import cvxpy as cp
import numpy as np

A = np.array([[0,1],[0,0]])
B = np.array([[0],[1]])
Q = np.eye(2)
R = 1.0

P = cp.Variable((2,2), PSD=True)    # P>0
K = cp.Variable((1,2))              # K in R^{1x2}

# Build the LMI:
M = (A - B@K).T@P + P@(A - B@K) + Q + K.T*(R)*K  # shape 2x2

constraints = [ P >> 0,
                0 > M ] # "M < 0" is '0 > M' in cvxpy
obj = cp.Minimize(cp.trace(P))

prob = cp.Problem(obj, constraints)
prob.solve(solver=cp.MOSEK) # or another solver

print("P =", P.value)
print("K =", K.value)
```

Compare the resulting **K** and **P** with the solution to the classic Riccati equation.

Exercise 9.10 We can also use LMIs to solve the simpler problem of ensuring stability (i.e., all eigenvalues of $\mathbf{A} - \mathbf{BK}$ are negative) by solving for $\mathbf{P} > 0$ such that

$$(\mathbf{A} - \mathbf{BK})^{\mathrm{T}}\mathbf{P} + \mathbf{P}(\mathbf{A} - \mathbf{BK}) < 0, \quad \mathbf{P} > 0.$$

The optimization problem is then

$$\begin{aligned} \text{find} \quad & (\mathbf{P}, \mathbf{K}) \\ \text{s.t.} \quad & \mathbf{P} > 0, \\ & (\mathbf{A} - \mathbf{BK})^{\mathrm{T}}\mathbf{P} + \mathbf{P}(\mathbf{A} - \mathbf{BK}) < 0. \end{aligned}$$

1. You may write the second inequality in LMI block form. However, in 2D you can write $\mathbf{M} = (\mathbf{A} - \mathbf{BK})^{\mathrm{T}}\mathbf{P} + \mathbf{P}(\mathbf{A} - \mathbf{BK})$ explicitly and ensure **M** is negative definite.
2. Manual approach:
   - Let $\mathbf{P} = \begin{bmatrix} p_{11} & p_{12} \\ p_{12} & p_{22} \end{bmatrix} > 0$, $\mathbf{K} = \begin{bmatrix} k_1 & k_2 \end{bmatrix}$.
   - Then $\mathbf{A} - \mathbf{BK} = \begin{bmatrix} 0 & 1 \\ 0 & 0 \end{bmatrix} - \begin{bmatrix} 0 \\ 1 \end{bmatrix} \begin{bmatrix} k_1 & k_2 \end{bmatrix} = \begin{bmatrix} 0 & 1 \\ -k_1 & -k_2 \end{bmatrix}$.
   - Evaluate $(\mathbf{A}-\mathbf{BK})^{\mathrm{T}}\mathbf{P}+\mathbf{P}(\mathbf{A}-\mathbf{BK})$ and impose negativity. Solve for unknowns p_{ij}, k_j.
3. Compare the solutions $(\mathbf{P}, \mathbf{K})$ with the ones you obtained from the Riccati approach. They should coincide, or at least be consistent.
4. Use the following Python `cvxpy` code to define the variables and constraints, initialize the optimization problem, and solve it.

CodeSnip 9.h

```
P = cp.Variable((2,2), symmetric=True)
K = cp.Variable((1,2))

constraints = [P >> 0,
               (A - B@K).T @ P + P@(A - B@K) << 0]

prob = cp.Problem(cp.Minimize(0), constraints)
prob.solve()
```

If you want only feasibility, you can use `Objective = Minimize(0)`. Or if you want to incorporate a cost on `trace(P)`, you can use `Minimize(cp.trace(P))`. You should get `P.value` and `K.value` that match, or are close to, your by-hand solution.

Exercise 9.11 Show that the LQR optimal control solution satisfies the HJB equations. First, write down the value function V for the LQR problem. Show that this function satisfies HJB. Derive an expression for the Lagrange multiplier co-state λ and explain.

Exercise 9.12 Repeat the steps in the derivation of the HJB equations more carefully using the calculus of variations.

Exercise 9.13 (Reinforcement learning vs. LQR) We will now compare RL and LQR on the linear system from Exercise 9.1 to stabilize $x = 0$.

1. Implement a basic RL algorithm, such as policy-gradient with a linear Gaussian policy ($u_t = -Kx_t + \varepsilon_t$ with $\varepsilon_t \sim \mathcal{N}(0, \sigma^2 \mathbf{I})$). Use a reward $r_t = -(x_t^{\mathrm{T}} Q x_t + u_t^{\mathrm{T}} R u_t)$. You may estimate $\nabla_K J(K)$ from rollouts using REINFORCE or actor-critic, and then perform the gradient steps on K. This is a policy search for LQR.
2. Compare the learned control law with the LQR solution.
3. Discuss sample efficiency (how many episodes or interactions are needed).
4. Now use a subset of the RL training data to learn $(\mathbf{A}, \mathbf{B})$ from data using least-squares regression. From this learned model, solve the Riccati equation and implement. How does the quality of the control vary with the amount of training data used to learn $(\mathbf{A}, \mathbf{B})$?

Exercise 9.14 (Quadratic Q-learning vs. LQR) We will now implement a reinforcement learning approach that exploits structure in the linear-quadratic regulator (LQR). It is known that the optimal Q-function for LQR has a quadratic form in $(\mathbf{x}, \mathbf{u})$:

$$Q(\mathbf{x}, \mathbf{u}) = \begin{bmatrix} \mathbf{x} \\ \mathbf{u} \end{bmatrix}^{\mathrm{T}} \mathbf{S} \begin{bmatrix} \mathbf{x} \\ \mathbf{u} \end{bmatrix},$$

for some symmetric matrix $\mathbf{S}$. We can therefore approximate Q directly using quadratic basis functions and fit using least square.

1. First, we implement a fitted Q-iteration (FQI) with quadratic features. Collect rollouts of the linear system from Exercise 9.1 using an exploratory controller (e.g., a random stabilizing gain $\mathbf{K}$ plus Gaussian noise). Build a feature vector $\phi(\mathbf{x}, \mathbf{u})$ consisting of all quadratic monomials in $(\mathbf{x}, \mathbf{u})$. Solve a least-squares regression problem to approximate the Bellman equation:

$$Q(\mathbf{x}_t, \mathbf{u}_t) \approx r_t + \gamma \min_{\mathbf{u}'} Q(\mathbf{x}_{t+1}, \mathbf{u}').$$

2. Show that the greedy policy with respect to the learned Q-function is linear state feedback $\mathbf{u} = -\mathbf{K}\mathbf{x}$. Derive $\mathbf{K}$ explicitly from the quadratic coefficients of the fitted Q-function.
3. Compare the learned feedback gain $\mathbf{K}$ with the true LQR solution from the Riccati equation. Report the closed-loop costs and trajectories.
4. Discuss sample efficiency: how many transitions are needed to learn a near-optimal controller? Compare this to the policy gradient and model-based approaches from the previous problem.

Exercise 9.15 (Model predictive control) Use a time-discretized double integrator system $x_{k+1} = x_k + \Delta t\, v_k$ and $v_{k+1} = v_k + \Delta t\, u_k$ to formulate an MPC that moves x from 0 to 10 while keeping v under some limit.

1. Write the MPC cost and constraints.
2. Implement an MPC for horizon $N = 10$. Since this double integrator is a convex quadratic program (QP), it is possible to code up the MPC optimization in CVXPY with OSQP in about 50 lines of code. Use `warm_start=True` so OSQP reuses the previous solution. Start with $q_x = 10$, $q_v = 1$, $R = 0.1$, $v_{\max} = 3$, $u_{\max} = 5$, and $\Delta t = 0.05$.
3. Compare the closed-loop trajectory for different horizons $N = 5, 10, 20$. Shorter horizons tend to be more aggressive and may need larger R or tighter $v_{\max}$.

Exercise 9.16 (Nonlinear MPC for a pendulum swing-up) Consider $\dot{\theta} = \omega$, $\dot{\omega} = -\frac{g}{L}\sin(\theta) + u$.

1. Linearize about $(\theta, \omega) = (\pi, 0)$ (upright) and see if it's controllable.
2. Set up a direct-collocation-based MPC that tries to swing the pendulum from $(0, 0)$ to $(\pi, 0)$ over finite time.
3. Compare performance with a simpler LQR that's only valid near the upright.

Hints: Implement the nonlinear MPC as a small nonlinear program in CasADi and solve it with IPOPT at each step. Discretize the pendulum with step Δt and horizon N using direct collocation (e.g., trapezoidal). Specifically, introduce decision variables $\{\theta_k, \omega_k, u_k\}_{k=0}^{N}$, and enforce the dynamics with constraints $x * k + 1 - x_k - \frac{\Delta t}{2}\left(f(x_k, u_k) + f(x_{k+1}, u_{k+1})\right) = 0$, where $x = [\theta, \omega]^{\mathrm{T}}$ and $f(x, u) = [\omega, , -(g/L)\sin\theta + u]^{\mathrm{T}}$. Minimize a quadratic stage cost toward the target $(\theta, \omega) = (\pi, 0)$ plus control effort, e.g. $\sum_{k=0}^{N-1}\left(q_\theta(\theta_k - \pi)^2 + q_\omega\omega_k^2 + ru_k^2\right)$, with optional terminal term. Add simple bounds (e.g., $|u_k| \leq u_{\max}$, wrap θ to $[-\pi, \pi]$ or use $\cos\theta/\sin\theta$ features for smoothness). In the receding-horizon loop, set the parameter x_0 to the measured state, call `opti.solver('ipopt', ...)` with warm start (i.e., reuse previous solution), apply u_0, shift, and repeat. Tune N and weights (q_θ, q_ω, r) for a reliable swing-up; start with $N \in 15, 20$, $\Delta t \approx 0.05$–0.1, and moderate $u_{\max}$.

Exercise 9.17 In this exercise, we will show that the Kalman filter is performing a Bayesian update, given prior information in the form of a system model, with a likelihood function derived from a new noisy measurement. First we consider a discrete-time linear Gaussian system model with process and measurement noise:

$$
\begin{aligned}
\mathbf{x}_{k+1} &= \mathbf{A}_k\mathbf{x}_k + \mathbf{w}_k, & \mathbf{w}_k &\sim \mathcal{N}(0, \mathbf{Q}_k),\\
\mathbf{y}_k &= \mathbf{H}_k\mathbf{x}_k + \mathbf{v}_k, & \mathbf{v}_k &\sim \mathcal{N}(0, \mathbf{R}_k).
\end{aligned}
$$

Here, $\mathbf{x}_k \in \mathbb{R}^n$ is the hidden system state, $\mathbf{y}_k \in \mathbb{R}^m$ is the observed measurement, and $\mathbf{A}_k$, $\mathbf{H}_k$ are the system and observation matrices, respectively.

The Kalman filter recursively estimates the posterior distribution $p(\mathbf{x}_k \mid \mathbf{y}_{0:k})$ as a Gaussian distribution with mean $\hat{\mathbf{x}}_{k|k}$ and covariance $\mathbf{P}_{k|k}$, using the following two steps:

Prediction step (prior):

$$
\begin{aligned}
\hat{\mathbf{x}}_{k|k-1} &= \mathbf{A}_{k-1}\hat{\mathbf{x}}_{k-1|k-1},\\
\mathbf{P}_{k|k-1} &= \mathbf{A}_{k-1}\mathbf{P}_{k-1|k-1}\mathbf{A}_{k-1}^{\mathrm{T}} + \mathbf{Q}_{k-1}.
\end{aligned}
$$

Update step (posterior):

$$
\begin{aligned}
\mathbf{K}_k &= \mathbf{P}_{k|k-1}\mathbf{H}_k^{\mathrm{T}}\left(\mathbf{H}_k\mathbf{P}_{k|k-1}\mathbf{H}_k^{\mathrm{T}} + \mathbf{R}_k\right)^{-1},\\
\hat{\mathbf{x}}_{k|k} &= \hat{\mathbf{x}}_{k|k-1} + \mathbf{K}_k\left(\mathbf{y}_k - \mathbf{H}_k\hat{\mathbf{x}}_{k|k-1}\right),\\
\mathbf{P}_{k|k} &= (\mathbf{I} - \mathbf{K}_k\mathbf{H}_k)\mathbf{P}_{k|k-1}.
\end{aligned}
$$

Show that the Kalman filter equations above arise from performing Bayesian estimation of the state $\mathbf{x}_k$ given a Gaussian prior $p(\mathbf{x}_k \mid \mathbf{y}_{0:k-1}) = \mathcal{N}(\hat{\mathbf{x}}_{k|k-1}, \mathbf{P}_{k|k-1})$ and a Gaussian likelihood $p(\mathbf{y}_k \mid \mathbf{x}_k) = \mathcal{N}(\mathbf{H}_k\mathbf{x}_k, \mathbf{R}_k)$. Specifically:

1. Derive the posterior distribution $p(\mathbf{x}_k \mid \mathbf{y}_{0:k})$ using Bayes' rule, and show that it is Gaussian.
2. Compute the mean and covariance of the posterior, and verify that they match the Kalman filter update equations for $\hat{\mathbf{x}}_{k|k}$ and $\mathbf{P}_{k|k}$.
3. Explain how this recursive update is Bayesian.

Exercise 9.18 Recall the two LMIs examples at the beginning of Section 9.6:

- $\mathbf{A}^T\mathbf{X}\mathbf{A} - \mathbf{X} < \mathbf{Q}$, where $\mathbf{X} \in \mathbb{S}^n$ is a symmetric matrix.
- $\mathbf{A}^T\mathbf{X}\mathbf{A} + \mathbf{B}\mathbf{Y} + \mathbf{Y}^T\mathbf{B}^T + \mathbf{T} < 0$, which is linear in $\mathbf{Z} = (\mathbf{X}, \mathbf{Y})$.

For each of these LMIs, show that the function $\mathbf{F}(\mathbf{X})$ is linear in $\mathbf{X}$ and maps a symmetric $\mathbf{X}$ to a self-adjoint matrix $\mathbf{F}(\mathbf{X})$.

Exercise 9.19 Show that the two forms of an LMI from Eq. (9.62) and Eq. (9.63) are equivalent. Specifically, each LMI can be transformed from one form into the other.

Exercise 9.20 (Extremum Seeking Control) An *adaptive* controller is able to modify its action based on changing dynamics, without needing a detailed system model. Extremum-seeking control (ESC) [291, 12] is a particularly attractive form of adaptive control for complex systems because it does not rely on an underlying model and it has guaranteed convergence and stability under a set of well-defined conditions. Extremum-seeking may be used to track local maxima of an objective function, despite disturbances, varying system parameters, and nonlinearities. Adaptive control may be implemented for in-time control or used for slow tuning of parameters in a working controller.

Three distinct time-scales are relevant for extremum-seeking control:

1. slow – external disturbances and parameter variation;
2. medium – perturbation frequency ω; and
3. fast – system dynamics.

In many systems, the internal system dynamics evolve on a fast time-scale. For example, turbulent fluctuations may equilibrate rapidly compared to actuation time-scales.

Extremum-seeking control may be applied to systems with nonlinear dynamics and a time varying J. The general case of extremum-seeking control of nonlinear systems is analyzed by Krstić and Wang [291], where they develop stability guarantees based on a separation of timescales and a singular perturbation analysis. The algorithm may be modified to add a phase ϕ to the sinusoidal input perturbation.

To see the ability of extremum-seeking control to handle varying system parameters, consider the time-dependent cost function given by

$$J(u) = 25 - (5 - u - \sin(t))^2. \tag{9.74}$$

The varying parameters, which oscillate at $1/2\pi$ Hz, may be considered slow compared with the perturbation frequency 10 Hz.

Design an extremum-seeking control for this slowly varying system. Plot the response and discuss how the performance varies with the various ESC parameters.

10 Optimization for Machine Learning

Machine learning is one of the most important modern applications of optimization. Every machine learning model is trained using optimization. Consider an image classification task, with training data consisting of images and their labels (e.g., dog, cat, tree). We may train a convolutional neural network architecture to extract relevant features and output the correct category when shown new images. This involves optimizing the parameters of the network using SGD and ADAM (see Section 2.3) to minimize a loss function.

More generally, developing a machine learning model involves five key stages:

The problem: Formulating what inputs and outputs to model.

The data: Collecting and curating training data for informing the model.

The model: Choosing an architecture with which to represent the model.

The loss function: Designing a loss function for assessing the performance of the model.

The optimization: Implementing an optimization algorithm for training the model.

Thus, we use optimization to tune the parameters of the model to best fit the data, by minimizing the loss function. Fundamentally, this is an inverse problem we solve using optimization. If data is the fuel for machine learning, then optimization is the engine.

The organization of steps above is only approximate, and there are overlaps and tight interconnections between each stage. For example, choosing the problem and choosing the data are two closely related decisions. Similarly, designing a custom loss function and implementing an optimization algorithm are tightly coupled.

Generally the machine learning model may be written as

$$\mathbf{y} = \mathbf{f}_\theta(\mathbf{x}), \tag{10.1}$$

where $\mathbf{x}$ are the inputs to the model, $\mathbf{y}$ are the outputs, also known as *targets*, and $\boldsymbol{\theta}$ are the parameters to be optimized over. The objective function, which we call a *loss function*, L, typically includes several terms that combine various competing objectives. As we have seen in previous chapters on regression (Chapter 4) and inverse problems (Chapter 8), the primary loss term is the mean squared error (MSE) of the model fit with the data:

$$L = \sum_{j=1}^{N} \|\mathbf{y}_j - \mathbf{f}_\theta(\mathbf{x}_j)\|^2. \tag{10.2}$$

This error term is summed over the N data points $\{\mathbf{x}_j, \mathbf{y}_j\}_{j=1}^N$. We may also add other regularization terms to the loss function, for example to promote sparsity, physics constrains, symmetries, or other prior information.

We have already seen how to solve nonlinear least-square optimization problems like this in Chapters 2 and 4, for example using stochastic gradient descent or quasi-Newton methods. However, when $\mathbf{f}_\theta(\mathbf{x})$ is a complicated, high-dimensional function, such as a neural network, it is not obvious how to obtain the gradient of this function with respect to its parameters $\boldsymbol{\theta}$. Backpropagation, powered by automatic differentiation, is the algorithm of choice for accurately and efficiently computing these gradients, providing the essential information to optimize the loss function.

In this chapter, we will explore a number of ways that optimization is used in machine learning. We will begin with background on automatic differentiation and backpropagation, as they are the core algorithmic infrastructure needed to obtain gradients for optimizing large-scale machine learning models. Automatic differentiation is a truly remarkable innovation from the 1960s, solving the issues associated with numerical differentiation or analytic differentiation of complex functions. The basic idea is to apply the chain rule to the computational graph of a function in a high-level programming language, such as Python or C++, to compute its derivative using the same order of magnitude number of calculations and the same order accuracy associated with evaluating the function itself. Automatic differentiation has been used broadly outside of neural network training, for example in inverse problems and control design.

Next, we will discuss how optimization is used for neural network training. This is such a central topic in machine learning, that it is worth taking the time to explore the various decisions one has to make associated with training and loss functions, learning rate scheduling, batch normalization, hyperparameter optimization, and vanishing and exploding gradients.

We will also revisit the topic of reinforcement learning (RL) introduced in Chapter 9, but now from the perspective of deep learning. RL agents are trained using optimization algorithms, but at the same time, a trained agent may be viewed as a way to optimize actions within an environment. In this way, we can also think of deep RL as an approach to advance the field of optimization and inverse problems.

Finally, we will explore how to incorporate and promote known physics into machine learning algorithms using various techniques from optimization. Specifically, physical knowledge typically comes in the form of known symmetries, conservation laws, and governing equations. There are two main approaches to incorporate physical knowledge into machine learning. First, it is possible to promote machine learning models respecting physics by adding custom terms to the loss function that grow when conservation laws are violated. This is the approach taken in physics-informed neural networks (PINNs), Second, it may be possible to directly enforce physical constraints, such as conservation laws and symmetries, by performing a constrained optimization when training the model. This second approach is more challenging, but it is also more effective in enforcing known constraints, such as energy conservation.

10.1 Automatic Differentiation

For high-dimensional functions, as are common in modern machine learning, symbolic differentiation is tedious, and numerical differentiation with finite differences may be slow or inaccurate. Automatic differentiation (AutoDiff, or AD) is a powerful alternative for flexibly computing derivatives and gradients of functions that are already implemented in a computer [209, 212]. It underpins the backpropagation algorithm [431], which is used to compute the gradients needed to train nearly all neural network models. AutoDiff is both more accurate than numerical differentiation and the computational cost has favorable scaling, which is related to the computational cost of the original function. In this section, we will introduce the foundations of automatic differentiation, with examples, and build towards the backpropagation algorithm in neural network training. We begin with simple examples of how to turn a function into a computational graph, and then show how this computational graph is used to compute fast and accurate derivatives via the chain rule. We will develop the two main approaches to AD: forward-mode AD and reverse-mode AD, which is most commonly used for backpropagation. At this point we will generalize from simple example functions to generic functions, and explore how these algorithms scale with the properties of the functions, such as input and output dimensions.

Finally, we will explore how automatic differentiation can be used to compute gradients in optimization problems, with examples in PyTorch and JAX. Backpropagation is a specialized, yet immensely useful instance of AutoDiff used for training neural networs, and is the subject of the next section. Automatic differentiation has a rich history [405, 210, 31], and Baydin [31] provides an exceptionally good survey and overview of this field[1].

Simple Motivating Example

The basic idea of automatic differentiation is to write a complex function $y = f(x)$ as the composition of elementary functions, such as $+, \times, \sin, \cos, \ldots$, with known derivatives, and then use the chain rule to evaluate the desired derivative $\mathrm{d}y/\mathrm{d}x$. If a function $y = f(x)$ may be written as the composition of a number of intermediate functions

$$y = f(x) = f_3(f_2(f_1(x))) := f_3 \circ f_2 \circ f_1(x) \tag{10.3}$$

[1] AD emerged in the 1950s [371, 33], and the modern forward-mode AD accumulation was introduced by Wengert in 1964 [535]; this was revived and repopularized by Griewank in 1989 [209]. Automatic differentiation was first used in control, where it is often necessary to compute adjoint information for constrained optimization [83, 84]. The earliest references to reverse-mode AD are Linnainmaa [315], although the form we know it in today was introduced by Speelpenning [470], who provided the first implementation that was actually automatic. Reverse-mode AD is closely related to backpropagation [431], which has been invented several times in the machine learning literature [381, 431, 210, 445].
Automatic differentiation enables constrained optimization without the need to explicitly compute an adjoint [6, 130, 234, 249, 154]. In fact, backpropagation appeared earlier in the context of flight path control [273]. This has the potential to replace gradient-based adjoint methods, which have been historically successful for constrained optimization problems, especially high-dimensional and nonlinear problems [248]. However, computing the adjoint is time-consuming and labor-intensive.

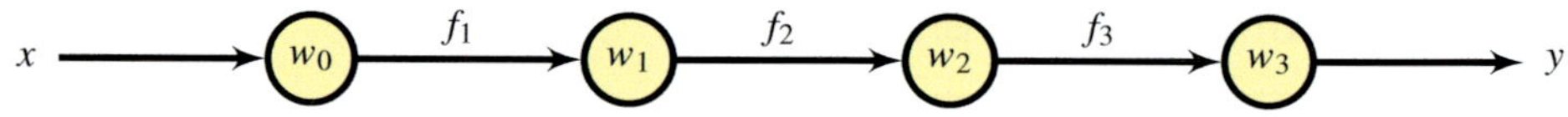

Figure 10.1 Computational graph for Eq. (10.3).

then it is possible to introduce intermediate variables

$$\begin{aligned}\text{(input)} \quad w_0 = x \longrightarrow w_1 &= f_1(w_0)\\ w_2 &= f_2(w_1)\\ w_3 &= f_3(w_2) \longrightarrow y = w_3 \quad \text{(output)}\end{aligned}$$

corresponding to the computational graph in Fig. 10.1. We may write the derivative $\mathrm{d}y/\mathrm{d}x$ using the chain rule as

$$\frac{\mathrm{d}y}{\mathrm{d}x} = \frac{\mathrm{d}f_3}{\mathrm{d}w_2} \cdot \frac{\mathrm{d}f_2}{\mathrm{d}w_1} \cdot \frac{\mathrm{d}f_1}{\mathrm{d}w_0}. \tag{10.4}$$

Each of these intermediate derivatives $\mathrm{d}f_k/\mathrm{d}w_{k-1}$ are known and are easy to compute.

But recall that the goal is not to represent this derivative analytically in a functional form[2], but rather evaluate it at a particular point x. In this case, we may further write the derivative, evaluated at $w_0 = x$ as

$$\frac{\mathrm{d}y}{\mathrm{d}x} = f_3'(w_2) \cdot f_2'(w_1) \cdot f_1'(w_0) \tag{10.5}$$

where $f_k' = \mathrm{d}f_k/\mathrm{d}w_{k-1}$. If we have computed the function $f(x)$ then we have already computed all of the intermediate values w_k in the computational graph, which are required to evaluate the terms in the chain rule. Further, these intermediate derivatives are defined analytically, and so evaluating this derivative should be approximately as accurate as evaluating the function, and with a similar order of magnitude number of operations.

Let's demonstrate this idea on a concrete example. Consider the function

$$y = \sin\left(e^{x^2}\right) = \underbrace{f_3}_{\sin} \circ \underbrace{f_2}_{\exp} \circ \underbrace{f_1}_{x^2}(x). \tag{10.6}$$

In this case the computational graph becomes

$$\begin{aligned}w_0 = x \longrightarrow w_1 &= w_0^2\\ w_2 &= e^{w_1}\\ w_3 &= \sin(w_2) \longrightarrow y = w_3\end{aligned}$$

as shown in Fig. 10.2.

[2] Why not, you might ask. If we write the derivatives $\mathrm{d}f_k/\mathrm{d}w_{k-1}$ as functions of x, so that that we may write $\mathrm{d}y/\mathrm{d}x$ in analytic form, we will experience *equation bloat*. This means that the number of terms required to write the analytic derivative grows exponentially in the complexity of the original function. However, by writing these derivatives $\mathrm{d}f_k/\mathrm{d}w_{k-1}$ as functions of w_{k-1}, we may evaluate them *numerically* by storing the intermediate values of the computational graph.

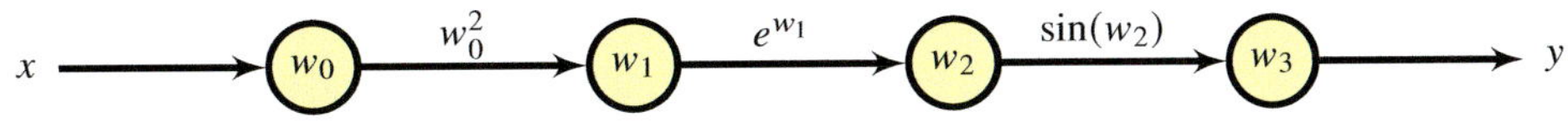

Figure 10.2 Computational graph for Eq. (10.6).

The analytic derivative is

$$\frac{\mathrm{d}y}{\mathrm{d}x} = \cos\left(e^{x^2}\right) \cdot e^{x^2} \cdot 2x. \tag{10.7}$$

Evaluating this at a specific value, such as $x = 2$, yields $\mathrm{d}y/\mathrm{d}x|_{x=2} = 4e^4 \cos(e^4) \approx -80.9$. We may also compute this using the intermediate values of the computational graph

$$\begin{aligned}
w_0 &= 2 \\
w_1 &= w_0^2 = 4 \\
w_2 &= e^{w_1} = e^4 \approx 54.6 \\
w_3 &= \sin(w_2) = \sin(54.6) \approx= -.923
\end{aligned}$$

and rewriting Eq. (10.7) as:

$$\left.\frac{\mathrm{d}y}{\mathrm{d}x}\right|_{x=2} = \cos(w_2) \cdot e^{w_1} \cdot 2w_0 = 4e^4 \cos(54.6) \approx -80.9.$$

We will now generalize this procedure for arbitrary compositions of scalar functions, introducing simple forms of the forward-mode and reverse-mode AD algorithms.

General Composition of Scalar Functions

We will now write the general expression for automatic differentiation for any composition of scalar functions (i.e., single-input, single output):

$$w_L = f(w_0) = f_L \circ f_{L-1} \circ \cdots \circ f_2 \circ f_1(w_0). \tag{10.8}$$

Here the function has L intermediate compositions, which will be helpful when we think about neural networks with L layers. Each intermediate evaluation yields the variable w_j:

$$w_j = f_j(w_{j-1}) = f_j \circ f_{j-1} \circ \cdots \circ f_1(w_0). \tag{10.9}$$

Again, we denote $f_j'(w) = \mathrm{d}f_j/\mathrm{d}w_{j-1}$ evaluated at the point w.

The derivative of the output w_L with respect to the input w_0 can now be written via the chain rule as:

$$\left.\frac{\partial w_L}{\partial w_0}\right|_{w_0} = \left.\frac{\partial f_L}{\partial w_{L-1}}\right|_{w_{L-1}} \cdot \left.\frac{\partial f_{L-1}}{\partial w_{L-2}}\right|_{w_{L-2}} \cdots \left.\frac{\partial f_2}{\partial w_1}\right|_{w_1} \cdot \left.\frac{\partial f_1}{\partial w_0}\right|_{w_0} \tag{10.10a}$$

$$= f_L'(w_{L-1}) f_{L-1}'(w_{L-2}) \cdots f_2'(w_1) f_1'(w_0). \tag{10.10b}$$

Note that we have introduced the notation of partial differentiation, since we may view each f_j as a function of w_{j-1} or any downstream w_k with $k \leq j-1$. When we differentiate with respect to the immediate argument w_{j-1} we denote this as $f_j'(w_{j-1})$ for simplicity.

We are now ready to introduce the two main modes of automatic differentiation, which traverse the chain rule in Eq. (10.10a) recursively in different directions. Forward-mode AD is the simplest conceptually, and moves from right to left in Eq. (10.10a). In the multiple-input, multiple-output case we will see that this approach is efficient when there are more outputs than inputs. Reverse-mode AD traverses Eq. (10.10a) from left to right, and is conceptually a bit more complicated. However, when there are more inputs than outputs, as is typically the case in neural network training, reverse-mode AD is more efficient. Another engineering application is the case of shape optimization of wings and other surfaces, where the shape input is high-dimensional, and the aerodynamic force outputs are low-dimensional. We will first write the recursion relationship in the single-input, single-output (SISO) case, and then will move to the more general multiple-input, multiple-output (MIMO) case.

Forward Accumulation Mode Automatic Differentiation

The basic idea here is to compute the derivative $\mathrm{d}f/\mathrm{d}w$ at some specific point w_0 by evaluating each term in Eq. (10.10a) from right to left. The major realization is that if we have computed $w_L = f(w_0)$, then we have already computed each of the intermediate nodes w_j, and so we may also evaluate each $f_j'(w_{j-1})$ as we traverse the computational graph. We don't need to store each of the intermediate w_j, as we only need them once in this forward pass for a scalar function.

In forward-mode AD, the derivative is computed recursively with the following initialization seed and recursion formula:

$$\text{seed:} \quad \dot{w}_0 := \left.\frac{\partial w_0}{\partial w_0}\right|_{w_0} = 1 \tag{10.11a}$$

$$\text{recursion:} \quad \dot{w}_k := \left.\frac{\partial w_k}{\partial w_0}\right|_{w_0} = \left.\frac{\partial w_k}{\partial w_{k-1}}\right|_{w_{k-1}} \cdot \left.\frac{\partial w_{k-1}}{\partial w_0}\right|_{w_0}. \tag{10.11b}$$

Importantly, in this simple SISO example we only need local information at each step that is already computed in the forward pass of the function evaluation, so storage is minimal. Here, we introduced $\dot{w}_k$ to denote the derivative up to the k-th function with respect to $w_0 = x$: $\dot{w}_k = \mathrm{d}w_k/\mathrm{d}w_0$, evaluated at $w_0 = x$. Since the derivative of w_0 with respect to w_0 is 1, we initialize with the seed $\dot{w}_0 = 1$.

Reverse Accumulation Mode Automatic Differentiation

Reverse-mode AD is more conceptually complicated, although in principle it isn't much more challenging to implement. However, it will be much more efficient when we have more inputs than outputs, as in backpropagation. It is easier to begin describing the method for the SISO case here, and we will generalize to MIMO functions next.

In contrast to forward-mode AD, reverse-mode AD traverses the computational graph from left to right, meaning from the later function evaluations to the earlier ones. In reverse-mode AD, the derivative is computed as:

$$\text{seed:} \quad \bar{w}_L := \left.\frac{\partial w_L}{\partial w_L}\right|_{w_L} = 1 \tag{10.12a}$$

$$\text{recursion:} \quad \bar{w}_k := \left.\frac{\partial w_L}{\partial w_k}\right|_{w_k} = \left.\frac{\partial w_L}{\partial w_{k+1}}\right|_{w_{k+1}} \cdot \left.\frac{\partial w_{k+1}}{\partial w_k}\right|_{w_k}. \tag{10.12b}$$

In reverse-mode AD, we must store all intermediate w_j, since they are used in reverse order of their computation. This can cause reverse-mode AD to have a much larger memory footprint than forward-mode AD, especially in neural network training.

Here, we introduced $\bar{w}_k$ to denote the derivative of the output $w_L = f(x)$ with respect to the k-th intermediate w_k: $\bar{w}_k = \mathrm{d}w_L/\mathrm{d}w_k$, evaluated at w_k. Since the derivative of w_L with respect to itself is 1, we initialize with the seed $\bar{w}_L = 1$. It is worth carefully going through the steps in forward-mode and reverse-mode AD and confirming that they do in fact both encode the same chain rule derivative, evaluated at a point x.

Motivating Example Revisited

Let's now reexamine the simple motivating example in Eq. (10.6). We already saw how the derivative may be evaluated at $x = 2$ using intermediate values computed in the computational graph. Now we will see how this relates to the forward-mode and reverse-mode automatic differentiation from above.

Forward-mode and reverse-mode AD may be written in tabular form, related to the computational graph of the original function $y(x)$. We begin with forward-mode AD, where we keep track of w_0, w_1, w_2, and w_3 as they are evaluated in the computational graph, along with their derivatives $\dot{w}_0, \dot{w}_1, \dot{w}_2$, and $\dot{w}_3$, defined in Eq. (10.11):

$$\begin{aligned}
w_0 &= 2 & \dot{w}_0 &= 1 \\
w_1 &= w_0^2 = 4 & \dot{w}_1 &= \frac{\partial w_1}{\partial w_0}\dot{w}_0 = 2w_0 = 4 \\
w_2 &= e^{w_1} \approx 54.6 & \dot{w}_2 &= \frac{\partial w_2}{\partial w_1}\dot{w}_1 = 4e^{w_1} \approx 218.4 \\
w_3 &= \sin(w_2) \approx -.923 & \dot{w}_3 &= \frac{\partial w_3}{\partial w_2}\dot{w}_2 = 4e^4\cos(w_2) \approx -80.9.
\end{aligned}$$

At each successive step, we use the function evaluation w_j and its derivative $\dot{w}_j$ to compute the next $\dot{w}_{j+1}$. Thus, we recover the numeric value of $\mathrm{d}y/\mathrm{d}x$ evaluated at $x = 2$.

Reverse-mode automatic differentiation proceeds very similarly, but instead of accumulating derivatives from the input w_0 to the output w_3, we *reverse* the order and accumulate derivatives from the output w_3 to the input w_0. Instead of writing our intermediate tabular values as $\dot{w}_j$, we will use the notation $\bar{w}_j$ to denote that this is an adjoint variable (i.e., the sensitivity of the output with respect to an intermediate variable). The forward function evaluation remains unchanged, and we repeat it here for clarity:

$$
\begin{aligned}
w_0 &= 2 & \bar{w}_3 &= 1 \\
w_1 &= w_0^2 = 4 & \bar{w}_2 &= \bar{w}_3 \frac{\partial w_3}{\partial w_2} = \cos(w_2) \approx -0.37 \\
w_2 &= e^{w_1} \approx 54.6 & \bar{w}_1 &= \bar{w}_2 \frac{\partial w_2}{\partial w_1} = -0.37 e^{w_1} \approx -20.24 \\
w_3 &= \sin(w_2) \approx -.923 & \bar{w}_0 &= \bar{w}_1 \frac{\partial w_1}{\partial w_0} = -20.24 \cdot 2w_0 \approx -80.9.
\end{aligned}
$$

Thus, forward-mode and reverse-mode AD arrive at the same value of the derivative dy/dx evaluated at $x = 2$. Note that we initialize $\bar{w}_3 = 1$ in the reverse-mode AD, similar to how we initialize $\dot{w}_0 = 1$ in forward-mode AD. This initialization may be thought of as pinning the subsequent sensitivities/derivatives to a unit change in the output/input.

General Scalar Functions

Although the formulation above is quite general, it only captures functions that can be written as the composition of scalar functions, each of a single input. These are called *unary* functions, and include examples like sin, exp, x^2, etc. However, we also need to consider binary functions such as + and ×: $f(x) = f_1(x) + f_2(x)$ and $f(x) = f_1(x) \times f_2(x)$. This will require us to modify and generalize the expressions in Eqs. (10.11) and (10.12).

The forward-mode and reverse-mode recursions may be modified in a relatively straightforward way to handle these more general functions. Forward-mode AD becomes

$$\text{seed:} \quad \dot{w}_0 := \left.\frac{\partial w_0}{\partial w_0}\right|_{w_0} = 1 \tag{10.13a}$$

$$\text{recursion:} \quad \dot{w}_k := \left.\frac{\partial w_k}{\partial w_0}\right|_{w_0} = \sum_{j<k} \left.\frac{\partial w_k}{\partial w_j}\right|_{w_j} \cdot \left.\frac{\partial w_j}{\partial w_0}\right|_{w_0}. \tag{10.13b}$$

Reverse-mode AD becomes

$$\text{seed:} \quad \bar{w}_L := \left.\frac{\partial w_L}{\partial w_L}\right|_{w_L} = 1 \tag{10.14a}$$

$$\text{recursion:} \quad \bar{w}_k := \left.\frac{\partial w_L}{\partial w_k}\right|_{w_k} = \sum_{j>k} \left.\frac{\partial w_L}{\partial w_j}\right|_{w_j} \cdot \left.\frac{\partial w_j}{\partial w_k}\right|_{w_k}. \tag{10.14b}$$

It is easy to verify that these reduce to Eqs. (10.11) and (10.12) in the simplified case of the composition of unary functions. We will see an example of how to compute these using tabularized forward-mode and reverse-mode AD for a simple scalar function, $f(x) = x^2 + x\sin(x)$ in the next subsection. The computational graph of this more general function is shown in Fig. 10.3, which clearly exhibits a more complex structure than the simple compositional function in Fig. 10.1.

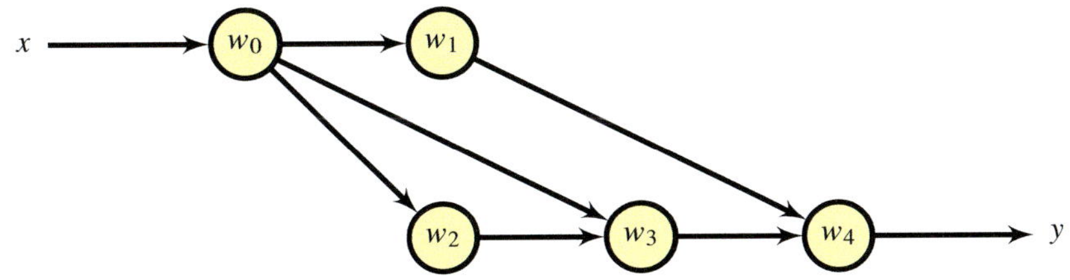

Figure 10.3 Computational graph for the SISO function in Eq. (10.15).

Generalizing Computational Graphs

We are now ready to generalize the single-input, single-output case above to handle multiple inputs and outputs. In the next subsection, we will introduce the full automatic differentiation for a general function with multiple inputs and multiple outputs. First, we explore several specific examples of how to extend the ideas above to particular functions with different inputs and outputs. Specifically, we will build concrete examples of the following types of functions: A. single-input single-output (SISO); B. multiple-input single-output (MISO); C. single-input multiple-output (SIMO); D. and multiple-input multiple-output (MIMO).

A. Single-Input, Single-Output: We've already seen how to handle simple scalar functions that are compositions of elementary unary functions. We now consider a slightly more complicated scalar function

$$y = x^2 + x\sin(x) \tag{10.15}$$

that involves the binary operations + and ×. The computational graph of this function may be derived by first introducing intermediate variables to simplify the computation:

$$\begin{aligned} w_0 = x \longrightarrow w_1 &= w_0^2 \\ w_2 &= \sin(w_0) \\ w_3 &= w_0 w_2 \\ w_4 &= w_1 + w_3 \longrightarrow y = w_4. \end{aligned}$$

This is visualized as a computational graph in Fig. 10.3.

Using the forward-mode and reverse-mode AD to compute the derivative at a specific value of x will be left as an exercise.

B. Multiple-Input, Single-Output: We now consider a function with multiple inputs and a single output. This will be especially important when considering backpropagation for neural networks, where we compute the derivatives of a scalar loss function with respect to the high-dimensional parameters of the network. A simple MISO function is

$$y = x_1 x_2 + x_1 \sin(x_1). \tag{10.16}$$

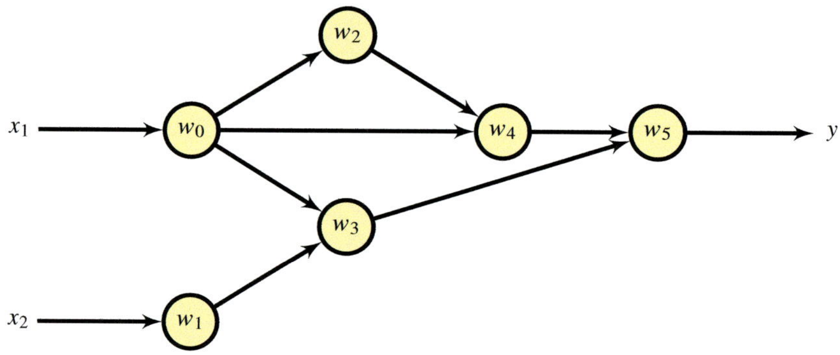

Figure 10.4 Computational graph for MISO function in Eq. (10.16).

We still introduce intermediate variables

$$\begin{aligned} w_0 = x_1 &\longrightarrow w_2 = \sin(w_0) \\ w_1 = x_2 &\longrightarrow w_3 = w_0 w_1 \\ & w_4 = w_0 w_2 \\ & w_5 = w_3 + w_4 \longrightarrow y = w_5 \end{aligned}$$

to arrive at the computational graph in Fig. 10.4.

We will demonstrate the need for multiple passes using forward-mode AD by explicitly computing $\mathrm{d}y/\mathrm{d}x_1$ and $\mathrm{d}y/\mathrm{d}x_2$ for a specific point $(x_1, x_2) = (1, 2)$. In the forward pass, we need to choose which variable we are differentiating with respect to, either $w_0 = x_1$ or $w_1 = x_2$, and initialize with either $(\dot{w}_0, \dot{w}_1) = (1, 0)$ or $(\dot{w}_0, \dot{w}_1) = (0, 1)$, respectively.

Let's begin by computing $\mathrm{d}y/\mathrm{d}x_1$. The analytic derivative is $\mathrm{d}y/\mathrm{d}x_1 = x_2 + \sin(x_1) + x_1 \cos(x_1)$, which evaluates to 3.381 at $(x_1, x_2) = (1, 2)$. We may compute this derivative using tabularized forward-mode AD by initializing $(\dot{w}_0, \dot{w}_1) = (1, 0)$:

$$\begin{aligned} w_0 &= 1 & \dot{w}_0 &= 1 \\ w_1 &= 2 & \dot{w}_1 &= 0 \\ w_2 &= \sin(w_0) = \sin(1) \approx 0.841 & \dot{w}_2 &= \frac{\partial w_2}{\partial w_0}\dot{w}_0 = \cos(w_0) = \cos(1) \approx 0.540 \\ w_3 &= w_0 w_1 = 2 & \dot{w}_3 &= \frac{\partial w_3}{\partial w_0}\dot{w}_0 + \frac{\partial w_3}{\partial w_1}\dot{w}_1 = w_1 \dot{w}_0 = 2 \\ w_4 &= w_0 w_2 \approx 0.841 & \dot{w}_4 &= \frac{\partial w_4}{\partial w_0}\dot{w}_0 + \frac{\partial w_4}{\partial w_2}\dot{w}_2 = w_2 \dot{w}_0 + w_0 \dot{w}_2 \approx 1.381 \\ w_5 &= w_3 + w_4 \approx 2.841 & \dot{w}_5 &= \frac{\partial w_5}{\partial w_3}\dot{w}_3 + \frac{\partial w_5}{\partial w_4}\dot{w}_4 = \dot{w}_3 + \dot{w}_4 \approx 3.381. \end{aligned}$$

This agrees with the analytic derivative above. However, we still need to compute $\mathrm{d}y/\mathrm{d}x_2$. The analytic derivative is $\mathrm{d}y/\mathrm{d}x_2 = x_1$, which evaluates to 1 at $(x_1, x_2) = (1, 2)$ This may also be computed using forward-mode AD, initializing $(\dot{w}_0, \dot{w}_1) = (0, 1)$ this time:

$$\begin{aligned}
w_0 &= 1 & \dot{w}_0 &= 0 \\
w_1 &= 2 & \dot{w}_1 &= 1 \\
w_2 &= \sin(w_0) = \sin(1) \approx 0.841 & \dot{w}_2 &= \frac{\partial w_2}{\partial w_0}\dot{w}_0 = 0 \\
w_3 &= w_0 w_1 = 2 & \dot{w}_3 &= \frac{\partial w_3}{\partial w_0}\dot{w}_0 + \frac{\partial w_3}{\partial w_1}\dot{w}_1 = w_0\dot{w}_1 = 1 \\
w_4 &= w_0 w_2 \approx 0.841 & \dot{w}_4 &= \frac{\partial w_4}{\partial w_0}\dot{w}_0 + \frac{\partial w_4}{\partial w_2}\dot{w}_2 = w_2\dot{w}_0 + w_0\dot{w}_2 = 0 \\
w_5 &= w_3 + w_4 \approx 2.841 & \dot{w}_5 &= \frac{\partial w_5}{\partial w_3}\dot{w}_3 + \frac{\partial w_5}{\partial w_4}\dot{w}_4 = \dot{w}_3 + \dot{w}_4 = 1.
\end{aligned}$$

Thus, the derivative $\mathrm{d}y/\mathrm{d}x_2$ at the point $(x_1, x_2) = (1, 2)$ is 1.

This is an example where reverse-mode AD is actually simpler and more computationally efficient than forward-mode AD. Computing the derivatives using forward-mode AD requires 2 passes through the computational graph, since there are 2 input variables. This is because for each input, we needed a different initialization seed for $\dot{w}_0$ and $\dot{w}_1$. Since this function has a single output, if we use reverse-mode AD, we only need one reverse pass, and we will obtain both derivatives. More generally, forward-mode AD will require m passes through the computational graph, where m is the number of input variables, and reverse-mode AD will require n passes, where n is the number of output variables.

Computing the derivatives of this function using reverse-mode AD will be left as an exercise.

C. Single-Input, Multiple-Output: We now consider the case of multiple outputs and a single input:

$$y_1 = x^2 + x\sin(x) \tag{10.17a}$$

$$y_2 = x\cos(x) - x^2. \tag{10.17b}$$

The following intermediate variables

$$\begin{aligned}
w_0 = x \longrightarrow w_1 &= w_0^2 \\
w_2 &= \sin(w_0) \\
w_3 &= \cos(w_0) \\
w_4 &= w_0 w_2 \\
w_5 &= w_0 w_3 \\
w_6 &= w_1 + w_4 \longrightarrow y_1 = w_6 \\
w_7 &= -w_1 + w_5 \longrightarrow y_2 = w_7
\end{aligned}$$

are used to arrive at the computational graph in Fig. 10.5. Again, this graph will require a single forward pass to compute both derivatives $\mathrm{d}y_1/\mathrm{d}x$ and $\mathrm{d}y_2/\mathrm{d}x$, while it will require two reverse passes. These are both left as an exercise.

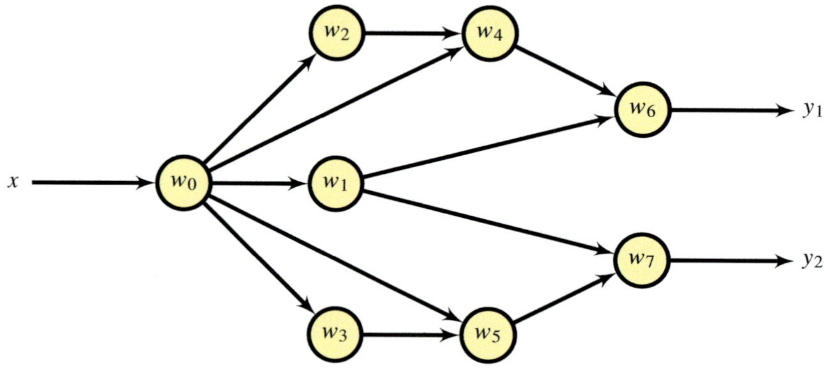

Figure 10.5 Computational graph for SIMO function in Eq. (10.17).

D. Multiple-Input, Multiple-Output: The most general case will involve functions with multiple inputs and outputs, such as the following:

$$y_1 = x_1 x_2 + x_1 \sin(x_1) \tag{10.18a}$$

$$y_2 = x_2 \cos(x_2) - x_1 x_2. \tag{10.18b}$$

It is possible to introduce the following intermediate variables

$$\begin{aligned}
w_0 = x_1 &\longrightarrow w_2 = \sin(w_0)\\
w_1 = x_2 &\longrightarrow w_3 = \cos(w_1)\\
& \quad\;\; w_4 = w_0 w_1\\
& \quad\;\; w_5 = w_0 w_2\\
& \quad\;\; w_6 = w_1 w_3\\
& \quad\;\; w_7 = w_4 + w_5 \longrightarrow y_1 = w_7\\
& \quad\;\; w_8 = -w_4 + w_6 \longrightarrow y_2 = w_8
\end{aligned}$$

to construct the computational graph in Fig. 10.6. In this case, both forward-mode and reverse-mode AD will require two passes, since there are two inputs and two outputs.

Notes on Computational Graphs. In general, computational graphs are not unique, as there may be several ways of decomposing a complex function as the composition of elementary functions.

This is closely related to the function trees in genetic programming from Chapter 5. There are many subtle details and options for constructing computational graphs, and this is a much deeper subject of computer science, which is beyond the scope of this book. However, it should be noted that constructing a computational graph is itself an optimization problem, where the goal may be to reduce redundant computations, minimize memory footprint, maintain floating point accuracy, or be amenable to downstream tasks such as automatic differentiation.

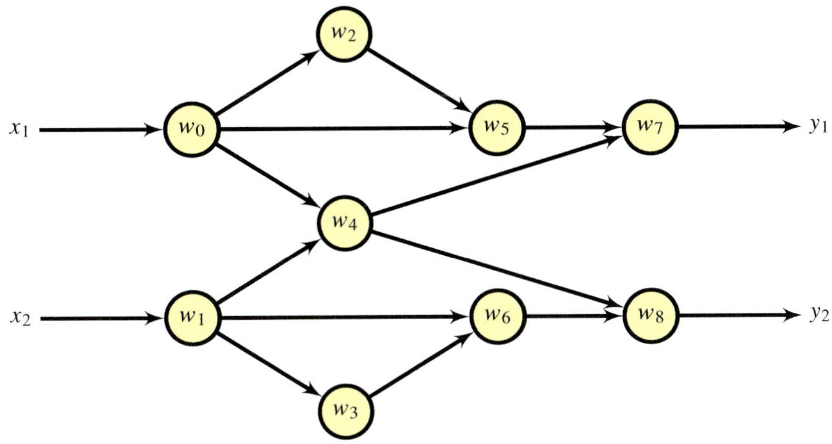

Figure 10.6 Computational graph for MIMO function in Eq. (10.18).

Multi-Input, Multi-Output Functions

We are now ready to consider the full MIMO case of automatic differentiation, which is where the main differences in forward and reverse-mode AD become apparent. Consider the function $\mathbf{w}_L = \mathbf{f}(\mathbf{w}_0)$, which may be written as the composition of several functions:

$$\mathbf{w}_L(\mathbf{w}_0) = \mathbf{f}_L \circ \mathbf{f}_{L-1} \circ \cdots \circ \mathbf{f}_2 \circ \mathbf{f}_1(\mathbf{w}_0). \tag{10.19}$$

As before, we may compute the derivative of the output with respect to the input using the chain rule and intermediate derivatives, although now we must write these in terms of Jacobian matrices of partial derivatives:

$$\frac{\mathbf{D}\mathbf{w}_L}{\mathbf{D}\mathbf{w}_0} = \left.\frac{\mathbf{D}\mathbf{f}_L}{\mathbf{D}\mathbf{w}_{L-1}}\right|_{\mathbf{w}_{L-1}} \left.\frac{\mathbf{D}\mathbf{f}_{L-1}}{\mathbf{D}\mathbf{w}_{L-2}}\right|_{\mathbf{w}_{L-2}} \cdots \left.\frac{\mathbf{D}\mathbf{f}_2}{\mathbf{D}\mathbf{w}_1}\right|_{\mathbf{w}_1} \left.\frac{\mathbf{D}\mathbf{f}_1}{\mathbf{D}\mathbf{w}_0}\right|_{\mathbf{w}_0}. \tag{10.20}$$

The Jacobian matrix of partial derivatives is defined as

$$\frac{\mathbf{D}\mathbf{f}}{\mathbf{D}\mathbf{w}} = \begin{bmatrix} \frac{\partial f_1}{\partial w_1} & \frac{\partial f_1}{\partial w_2} & \cdots & \frac{\partial f_1}{\partial w_m} \\ \frac{\partial f_2}{\partial w_1} & \frac{\partial f_2}{\partial w_2} & \cdots & \frac{\partial f_2}{\partial w_m} \\ \vdots & \vdots & \ddots & \vdots \\ \frac{\partial f_n}{\partial w_1} & \frac{\partial f_n}{\partial w_2} & \cdots & \frac{\partial f_n}{\partial w_m} \end{bmatrix}, \tag{10.21}$$

where there are m inputs and n outputs, so $\mathbf{w} \in \mathbb{R}^m$ and $\mathbf{f}(\mathbf{w}) \in \mathbb{R}^n$.

Each of these Jacobian matrices of partial derivatives must be computed, but the order in which they are multiplied might make a huge difference in the computational cost of the algorithm. We will discuss this in more detail in the next section on backpropagation.

When there are more inputs than outputs, we always want to use reverse-mode AD, since the number of forward passes through the computational graph scales with the number of inputs, while the number of reverse passes scales with the number of outputs. Consider the computational graph in Fig. 10.7 for a very structured function with a layered architecture and more inputs than outputs; this is common in many neural network architectures. The Jacobian of the output with respect to the input will involve the multiplication of the

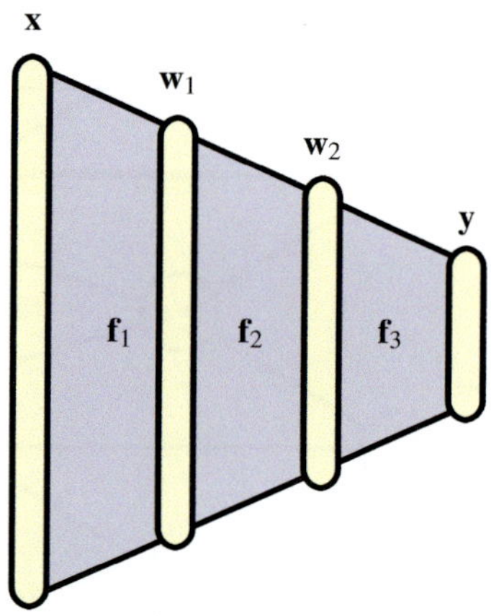

Figure 10.7 Illustration of computational graph with more inputs than outputs.

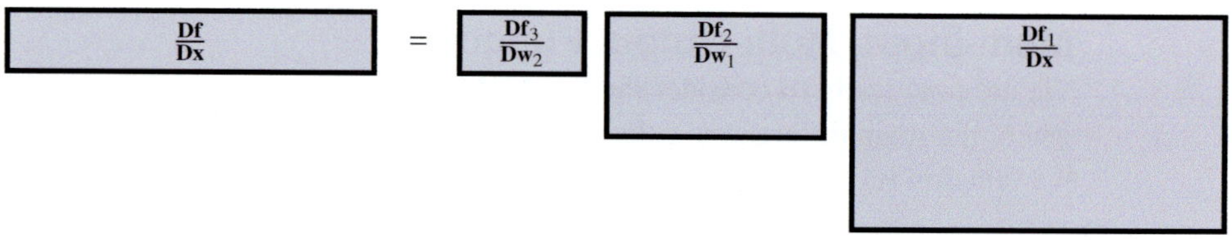

Figure 10.8 Illustration of Jacobian matrix multiplication for the example in Fig. 10.7.

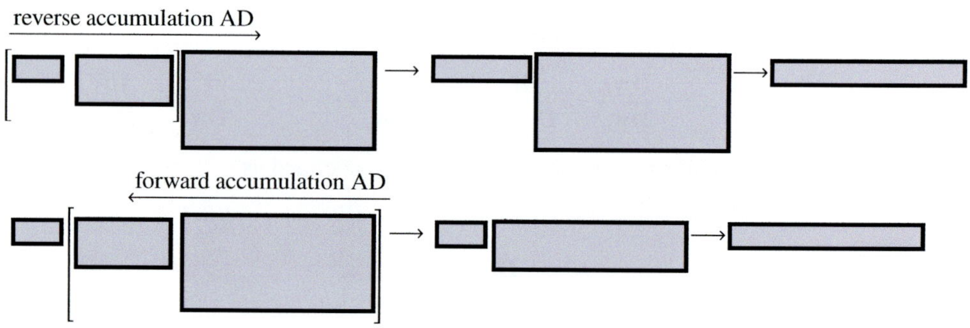

Figure 10.9 Illustration of Jacobian matrix multiplication from Fig. 10.8 and Fig. 10.7 using reverse-mode (top) and forward-mode (bottom) automatic differentiation. In reverse-mode, we traverse from **y** to **x**, so the initial matrix multiplications are small, and each subsequent matrix product involves relatively few operations. In forward mode, we begin with the largest matrix multiplications, accruing more operations in total.

Jacobians of each layer, as in Eq. (10.20) and shown schematically in Fig. 10.8. The order of operation of these matrix multiplications will dramatically change the computational complexity of the overall product, as shown in Fig. 10.9. Forward accumulation will multiply the largest matrices first, resulting in a larger number of overall computations, whereas reverse accumulation will multiply smaller matrices first, resulting in fewer computations.

Implementation in Modern ML Frameworks: PyTorch and JAX

Modern ML frameworks, such as PyTorch and JAX, computing gradients via automatic differentiation by calling functions like `.backward()` or `grad()`. Under the hood, these perform the chain rule we described earlier. Here is a demo in PyTorch to compute the derivative of $y = x^2 + x\sin(x)$ from Eq. (10.15) at $x = 2$. The analytic derivative is $\mathrm{d}y/\mathrm{d}x = 2x + \sin(x) + x\cos(x)$, which is 4.077 at $x = 2$. The PyTorch code is:

CodeSnip 10.a

```
import torch

def f(x):
    return x**2 + x*torch.sin(x)

# Evaluate at a specific point x = 2: value and derivative
x = torch.tensor(2.0, requires_grad=True)
y = f(x)              # forward pass
y.backward()          # reverse-mode AD: fills x.grad with dy/dx
print(f"x={x.item()}, f(x)={y.item()}, df/dx={x.grad.item()}")
```

It is also simple to implement automatic differentiation in JAX:

CodeSnip 10.b

```
from jax import grad
import jax.numpy as jnp

def f(x):
    return x**2 + x*jnp.sin(x)

df = grad(f) # Create derivative functions
x = jnp.array(2.0)  # Evaluate at x=2 (value and derivative)
print(f"x={float(x)}, f(x)={float(f(x))}, df/dx={float(df(x))}")
```

It is an exercise to modify this to compute the derivative of $y = \sin(\exp(x^2))$ at $x = 2$.

We now compute the derivative $\mathrm{d}f/\mathrm{d}\mathbf{w}$ of $f(\mathbf{x}, \mathbf{w}) = (\mathbf{w}^{\mathrm{T}}\mathbf{x})^2$, evaluated at $\mathbf{x} = (1, 2)$ with weights $\mathbf{w} = (0.5, -0.5)$. This is closer to backpropagation through a neural network, which we explore in the next section. It highlights the flexibility of automatic differentiation for computing derivatives of f with respect to either the input data $\mathbf{x}$ or the weights $\mathbf{w}$.

CodeSnip 10.c

```
x = torch.tensor([1.0, 2.0], requires_grad=False)
w = torch.tensor([0.5, -0.5], requires_grad=True) # grad wrt w
f = (w * x).sum()**2  # f = (w^{\rm T} x)^2

# Autograd magic: compute df/dw
f.backward()   # This populates w.grad
print("Gradient:", w.grad) # This prints df/dw evaluated at x,w
```

In JAX, the idea is similar:

CodeSnip 10.d

```
def f(w, x):
    return jnp.sum(w * x)**2

grad_f = jax.grad(f, argnums=0)    # derivative wrt w
w_init = jnp.array([0.5, -0.5])
x = jnp.array([1.0, 2.0])
print(grad_f(w_init, x))  # => [3. 6.]
```

The code constructs a computational graph and applies reverse-mode AD to find the gradient. It is an exercise to confirm that the derivative $\mathrm{d}f/\mathrm{d}\mathbf{w}$ from these codes is consistent with the analytic derivative.

Solving the Brachistochrone with Automatic Differentiation

We now revisit the Brachistochrone problem from Section 6.6. Instead of solving for the optimal cycloid solution using the Euler–Lagrange equations, we use automatic differentiation in PyTorch to compute the gradient of the travel time with respect to the curve height, $y(x)$, at various positions, x. This example was inspired by a blog post by Ben Wiener[3], who solved this problem using autograd in PyTorch. Their approach was inspired by a blog post of Declan Oller[4], who solved the same optimization problem using genetic programming and simulated annealing. Both are excellent references for further study.

First, we initialize the curve and the integration time functional:

CodeSnip 10.e

```
y0 = 1.0
n_points = 50
xpts = torch.linspace(0, 2, n_points)
ypts = torch.linspace(y0, 0, n_points, requires_grad=True)

# ----- Time Functional -----
def total_time(ypts):
    vs = torch.sqrt(torch.clamp(y0 - ypts, min=1e-6))
    dypts = ypts[1:] - ypts[:-1]
    dx = xpts[1] - xpts[0]
    lengths = torch.sqrt(dx**2 + dypts**2)
    return torch.sum((2. / (vs[:-1] + vs[1:])) * lengths)
```

Adding `requires_grad=True` to the instantiation of `ypts` informs PyTorch that gradients will be computed with respect to this variable.

Next, we iterate through gradient descent, computing the gradient of the travel time with respect to the curve shape, and then using this gradient for gradient descent:

CodeSnip 10.f

```
# ----- Gradient Descent -----
lr = 0.01
iterations = 3500
snap_interval = 100
history = []

for i in range(iterations):
    t = total_time(ypts)
    t.backward()
    if i % snap_interval == 0 or i == iterations - 1:
        history.append(ypts.detach().clone())
    with torch.no_grad():
        ypts[1:-1] -= lr * ypts.grad[1:-1]
    ypts.grad.zero_()
```

The resulting iterations are shown in Fig. 10.10.

Calling `t = total_time(ypts)` constructs the computational graph from `ypts` up to the scalar `t`. The line `t.backward()` runs backprop, initializing with $\partial t/\partial t = 1$ and then applying the chain rule backward through the computational graph, until the accumulated gradients are stored in `ypts.grad`.

[3] https://blog.benwiener.com/programming/2018/08/12/brachistochrone.html.

[4] https://declanoller.github.io/2018-12-24-solving-the-brachistochrone-and-a-cool-parallel-between-diversity-in-genetic-algorithms-and-simulated-annealing.

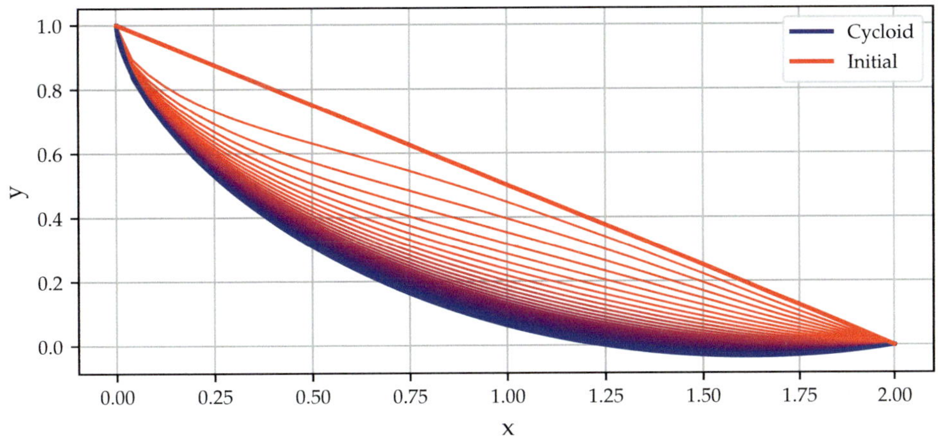

Figure 10.10 Using PyTorch to find the optimal cycloid solution to the Brachistochrone problem, starting from an initial line. The curve is plotted every 100 iterations, with color progressing from red to blue.

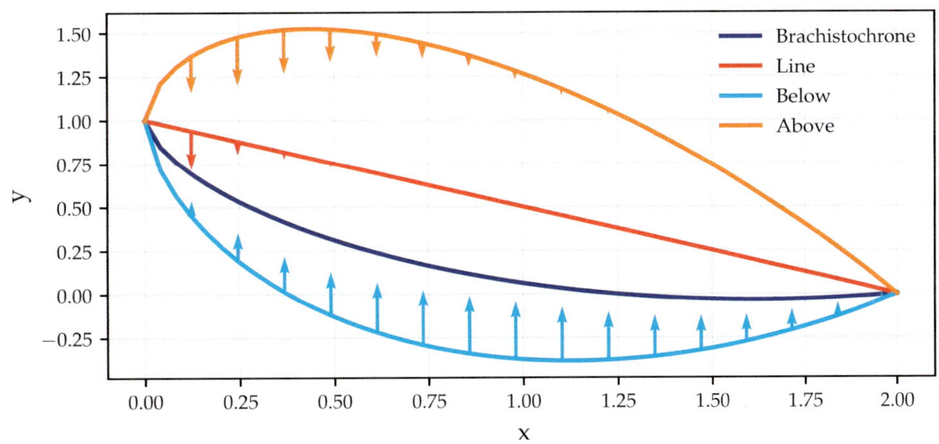

Figure 10.11 The PyTorch gradient of the travel time with respect to vertical height, computed using automated differentiation, for several different curves. All gradients push the solution towards the globally optimal Brachistochrone cycloid.

The line `with torch.no_grad(): ypts[1:-1] -= lr * ypts.grad[1:-1]` is important, since it explicitly disables autograd tracking for this this gradient descent update so it doesn't become part of the computational graph. Similarly `ypts.grad.zero()` clears the accumulated gradients at the end of the iteration, so that the next iteration starts fresh.

The gradient, computed with AD, is shown for various initial curves in Fig. 10.11.

10.2 Backpropagation

Backpropagation is the core algorithm that enables the training of neural networks with stochastic gradient descent. To implement SGD, we must be able to compute the gradient of the loss function L with respect to the neural network parameters $\boldsymbol{\theta}$. Because the loss function L is scalar, and the number of parameters in $\boldsymbol{\theta}$ is large, we have a computational graph with multiple inputs and a single output (MISO), so we use reverse-mode AD to compute this gradient. Thus, backpropagation is essentially reverse-mode AD applied to the computational graph of a neural network [431] to compute the gradient of the loss function with respect to the neural network parameters. The number of parameters $\boldsymbol{\theta}$ is typically very large, often in the millions or billions. The largest modern LLMs have hundreds of billions, if not trillions of free parameters. Table 10.1 provides a list of some of the largest neural network models with their parameter count.

Panel 10.1: A Neural Network Primer

The simplest version of a neural network is a collection of computational *neurons* that are organized in *layers*, which was originally motivated by a simplified understanding of the mammalian visual system. This type of neural network is known as a multi-layer perceptron (MLP).

In the MLP example below, information feeds forward, from left to right, through the layers. A given neuron collects all of the outputs of the previous layer in a weighted sum, weighted by $\mathbf{A}$, plus a bias term b. This weighted and biased sum is then passed through a nonlinear activation function, σ, which determines the output of the current neuron. Again, this mimics the way that biological neurons accumulate information and decide when to fire.

There are several choices for the activation function, but the two most common are the rectified linear unit (ReLU) and the sigmoidal hyperbolic tangent function. The tanh function saturates while the ReLU does not.

The network is parameterized by the weights and biases at each layer, so we may write $\boldsymbol{\theta}_k = [\mathbf{A}_k \quad \mathbf{b}_k]$ for the k-th layer. It is possible to compute gradients of outputs, such as the loss function, with respect to these parameters using backpropagation, which is a specialized form of reverse-mode AD.

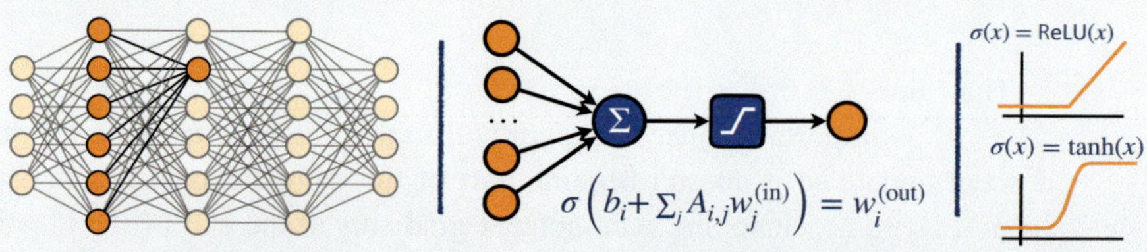

Figure 10.12: Schematic of a neural network (left), a single node and its inputs (middle), and two activation functions (right).

Table 10.1 Approximate parameter counts for representative neural network models.

Domain	Model	Parameter Count	Reference
CIFAR-10	ResNet20	~ 0.27M	[224]
CIFAR-10	ResNet56	~ 0.85M	[224]
ImageNet	AlexNet	~ 60M	[290]
ImageNet	VGG16	~ 138M	[465]
ImageNet	ResNet50	~ 25M	[224]
ImageNet	Inception v3	~ 23M	[482]
ImageNet	EfficientNet-B7	~ 66M	[486]
ImageNet	ViT-Large	~ 307M	[150]
Language Model	GPT-2	~ 1.5B	[400]
Language Model	GPT-3	~ 175B	[72]
Language Model	GPT-4	$\gg$ 175B (est.)	[375]
Language Model	LLaMA-2 (Meta)	~ 7 – 70B	[499]
Multimodal	CLIP	~ 400M	[399]
Multimodal	DALL·E 2	~ 3.5B	[406]

Throughout this section, it will help to remember that we are not taking the gradient of the neural network with respect to its inputs $\mathbf{x}$, but rather with respect to its parameters $\boldsymbol{\theta}$. Further, neural networks typically have a highly structured, layered architecture, which leads to a specialized reverse-mode AD algorithm that may be computed layer-by-layer.

The gradient of the loss function in Eq. (10.2) with respect to the neural network parameters $\boldsymbol{\theta}$ is given by

$$\nabla_{\boldsymbol{\theta}} L(\boldsymbol{\theta}) = \sum_{j=1}^{N} \nabla_{\boldsymbol{\theta}} \left\| \mathbf{y}_j - \mathbf{f}_{\boldsymbol{\theta}}(\mathbf{x}_j) \right\|^2 \tag{10.22a}$$

$$= \sum_{j=1}^{N} \nabla_{\boldsymbol{\theta}} \left(\mathbf{y}_j - \mathbf{f}_{\boldsymbol{\theta}}(\mathbf{x}_j) \right)^{\mathrm{T}} \left(\mathbf{y}_j - \mathbf{f}_{\boldsymbol{\theta}}(\mathbf{x}_j) \right) \tag{10.22b}$$

$$= -2 \sum_{j=1}^{N} \left(\mathbf{y}_j - \mathbf{f}_{\boldsymbol{\theta}}(\mathbf{x}_j) \right)^{\mathrm{T}} \nabla_{\boldsymbol{\theta}} \mathbf{f}_{\boldsymbol{\theta}}(\mathbf{x}_j). \tag{10.22c}$$

Thus, we must compute the gradient of the neural network output $\mathbf{f}_\theta$ with respect to the parameters $\boldsymbol{\theta}$ using automatic differentiation. Typically, instead of summing over all N data points, we select a smaller *mini-batch* of data to compute this gradient. This helps make the training data manageable within the memory constraints of modern GPUs. Even without memory constraints, selecting random subsets for mini-batch SGD has a regularizing effect and is akin to bootstrapping from Chapter 7.

Backpropagation and Reverse-Mode AD

In the last section, we showed that when there are more inputs than outputs, it is more efficient to use reverse-mode AD. Even neural network architectures with converging and diverging structure, such as an autoencoder network, still have a scalar loss function whose

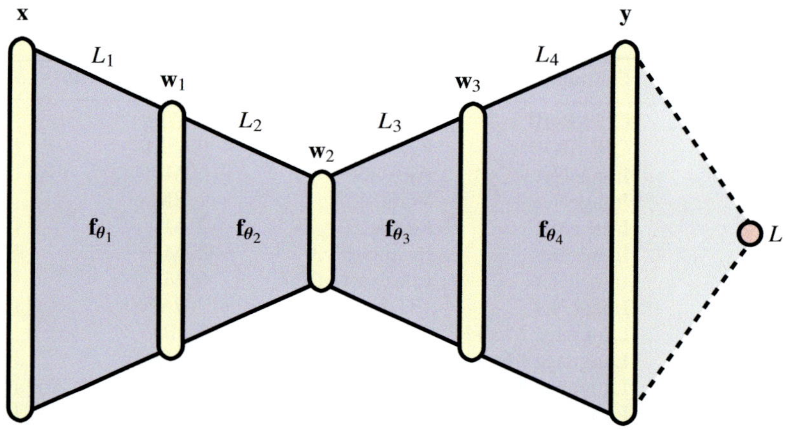

Figure 10.13 Illustration of neural network autoencoder with three hidden layers.

gradient we take. Thus, we still use reverse-mode AD since we always have a scalar output and many inputs (i.e., the parameters of the network).

Backpropagation traces backward through all layers, computing gradients via the chain rule. At each layer, we compute the gradient of the loss function with respect to the current layer and update these parameters, for example using SGD with ADAM (see Section 2.3). Because the network is compositional, to compute the gradient at a given layer, we only need the gradients of the output with respect to the previous layer. Thus, as we traverse the computational graph of the network from output to input, we only store gradient information for two layers at a time. However, we still need to store all network parameters and the values in the forward pass of the computational graph, evaluated on a mini-batch of data.

Consider a concrete example, given by the autoencoder network in Fig. 10.13. This network has three hidden layers, and we explicitly note that the true output is the scalar loss function L. We may denote the computational graph of the network as

$$\hat{\mathbf{y}} = \mathbf{f}_\theta(\mathbf{x}) = \mathbf{f}_{\theta_4} \circ \mathbf{f}_{\theta_3} \circ \mathbf{f}_{\theta_2} \circ \mathbf{f}_{\theta_1}(\mathbf{x}). \tag{10.23}$$

Recall from Eq. (10.22), to compute the gradient of L with respect to $\boldsymbol{\theta}$, we apply the chain rule and multiply $(\mathbf{y}_j - \mathbf{f}_\theta)^{\mathrm{T}}$ by $\nabla_\theta \mathbf{f}_\theta(\mathbf{x}_j)$, summing over data points in a mini-batch. Thus, we need to be able to flexibly compute the various gradients of $\mathbf{f}$ with respect to the parameters of each intermediate layer, $\boldsymbol{\theta}_j$.

To compute the gradient $\nabla_\theta \mathbf{f}_\theta(\mathbf{x})$, we leverage the highly structured, layered architecture:

$$\nabla_\theta \mathbf{f}_\theta(\mathbf{x}) = \nabla_{\theta_4} \mathbf{f}_{\theta_4}(\mathbf{w}_3) \cdot \nabla_{\theta_3} \mathbf{f}_{\theta_3}(\mathbf{w}_2) \cdot \nabla_{\theta_2} \mathbf{f}_{\theta_2}(\mathbf{w}_1) \cdot \nabla_{\theta_1} \mathbf{f}_{\theta_1}(\mathbf{x}). \tag{10.24}$$

Thus, the gradient with respect to the parameters $\boldsymbol{\theta}_j$ of any intermediate layer may be obtained by computing this product of gradients partially until the j-th gradient; this corresponds to traversing the network left, from the output $\mathbf{y}$ until the j-th layer.

These gradients of $\mathbf{f}_\theta$ may be used in Eq. (10.22) to compute the gradient of the loss function with respect to each intermediate layer. For example, the gradient of the loss function with respect to the parameters of the last computational layer $\boldsymbol{\theta}_4$ is given by

$$\nabla_{\theta_4} L(\boldsymbol{\theta}) = -2 \sum_{j=1}^{N} \left[\left(\mathbf{y}_j - \mathbf{f}_{\theta}(\mathbf{x}_j)\right)^{\mathrm{T}} \nabla_{\theta_4} \mathbf{f}_{\theta_4}(\mathbf{w}_{3,j}) \right], \tag{10.25}$$

where $\mathbf{w}_{3,j}$ denotes the computation in the $\mathbf{w}_3$ layer corresponding to the j-th data point $\mathbf{x}_j$: $\mathbf{w}_{3,j} = f_{\theta_3}(f_{\theta_2}(f_{\theta_1}(\mathbf{x}_j)))$. This is the first step in backpropagation, computing the gradient of the output with respect to the parameters $\boldsymbol{\theta}_4$ in the last computational layer.

To compute the gradient of L with respect to $\boldsymbol{\theta}_3$, we use the chain rule, multiplying the previous gradient $\nabla_{\theta_4} \mathbf{f}_{\theta_4}$ by the gradient of the third layer $\mathbf{f}_{\theta_3}$ with respect to $\boldsymbol{\theta}_3$:

$$\nabla_{\theta_3} L(\boldsymbol{\theta}) = -2 \sum_{j=1}^{N} \left[\left(\mathbf{y}_j - \mathbf{f}_{\theta}(\mathbf{x}_j)\right)^{\mathrm{T}} \nabla_{\theta_4} \mathbf{f}_{\theta_4}(\mathbf{w}_{3,j}) \cdot \nabla_{\theta_3} \mathbf{f}_{\theta_3}(\mathbf{w}_{2,j}) \right]. \tag{10.26}$$

Similarly, the gradient with respect to $\boldsymbol{\theta}_2$ and $\boldsymbol{\theta}_1$, corresponding to the second and first layers, are given by

$$\nabla_{\theta_2} L(\boldsymbol{\theta}) = -2 \sum_{j=1}^{N} \left[\left(\mathbf{y}_j - \mathbf{f}_{\theta}(\mathbf{x}_j)\right)^{\mathrm{T}} \nabla_{\theta_4} \mathbf{f}_{\theta_4}(\mathbf{w}_{3,j}) \cdot \nabla_{\theta_3} \mathbf{f}_{\theta_3}(\mathbf{w}_{2,j}) \cdot \nabla_{\theta_2} \mathbf{f}_{\theta_2}(\mathbf{w}_{1,j}) \right] \tag{10.27}$$

and

$$\nabla_{\theta_1} L(\boldsymbol{\theta}) = -2 \sum_{j=1}^{N} \left[\left(\mathbf{y}_j - \mathbf{f}_{\theta}(\mathbf{x}_j)\right)^{\mathrm{T}} \nabla_{\theta_4} \mathbf{f}_{\theta_4}(\mathbf{w}_{3,j}) \cdot \nabla_{\theta_3} \mathbf{f}_{\theta_3}(\mathbf{w}_{2,j}) \cdot \nabla_{\theta_2} \mathbf{f}_{\theta_2}(\mathbf{w}_{1,j}) \cdot \nabla_{\theta_1} \mathbf{f}_{\theta_1}(\mathbf{x}_j) \right]. \tag{10.28}$$

Differentiating Each Layer

Each $\nabla_{\theta_j} \mathbf{f}_j$ is the gradient of this layer with respect to its internal parameters. The function f_{θ_j} is usually given by an activation function σ applied to an affine transformation $\mathbf{A}_j \mathbf{w}_{j-1} + \mathbf{b}_j$ of the output of the previous layer, $\mathbf{w}_{j-1}$:

$$f_{\theta_j}(\mathbf{w}_{j-1}) = \sigma\left(\mathbf{A}_j \mathbf{w}_{j-1} + \mathbf{b}_j\right). \tag{10.29}$$

Thus, the parameters $\boldsymbol{\theta}_j$ can be written as $\boldsymbol{\theta}_j = \left[\mathbf{A}_j, \mathbf{b}_j\right]$, so the gradient of $\mathbf{f}_{\theta_j}(\mathbf{w}_{j-1})$ must be taken with respect to $\mathbf{A}_j$ and $\mathbf{b}_j$:

$$\frac{\partial \mathbf{f}_{\theta_j}}{\partial \mathbf{A}_j} = \sigma'\left(\mathbf{A}_j \mathbf{w}_{j-1} + \mathbf{b}_j\right) \cdot \mathbf{w}_{j-1}^{\mathrm{T}} \tag{10.30a}$$

$$\frac{\partial \mathbf{f}_{\theta_j}}{\partial \mathbf{b}_j} = \sigma'\left(\mathbf{A}_j \mathbf{w}_{j-1} + \mathbf{b}_j\right), \tag{10.30b}$$

where $\sigma'\left(\mathbf{A}_j \mathbf{w}_{j-1} + \mathbf{b}_j\right)$ is a column vector of elementwise derivatives.

Reverse-mode automatic differentiation can be quite memory intensive, as it stores all intermediate values of the computational graph for the reverse pass; this may be prohibitive for a large neural network. Gradient checkpointing saves memory by selectively caching intermediate states [102]. Instead of storing all activations during the forward pass, only a subset of layers are stored and the others are re-computed in the backward pass. This reduces memory usage at the cost of extra computation, since we must recompute small sections of the forwared pass, between checkpoints. However, memory is often the limiting factor, so this enables the training of larger networks.

Implementation in PyTorch

PyTorch automates backpropagation [382], making it simple to train neural networks:

CodeSnip 10.g

```
import torch
import torch.nn as nn

# Example dimensions
input_dim = 5
hidden_dim = 10
output_dim = 1

# Dummy input and target
x = torch.randn(16, input_dim)    # mini-batch of 16
y = torch.randn(16, output_dim)

# Define a simple feedforward network
net = nn.Sequential(
    nn.Linear(input_dim, hidden_dim),
    nn.ReLU(),
    nn.Linear(hidden_dim, output_dim),
    nn.Sigmoid() )

# Loss and optimizer
criterion = nn.MSELoss()
optimizer = torch.optim.SGD(net.parameters(), lr=0.01)

# Forward pass
output = net(x)
loss = criterion(output, y)

# Backward pass
optimizer.zero_grad()
loss.backward()  # << you *almost* missed the backpropagation!
optimizer.step() # grad values are used in the optimization step
```

This creates a small neural network with a single hidden layer and computes the forward pass and backward pass. The command `loss.backward()` computes the gradien using backprop. It is so easy and streamlined that you can almost miss it in the code above.

Vanishing and Exploding Gradients

One of the most significant trends in machine learning is the success of increasingly deep neural network architectures, as illustrated in Table 10.1. However, the desire to make neural networks deeper, with more layers, runs into the issue of *exploding* or *vanishing* gradients. As partial derivatives are backpropagated through the many layers of a network, large or small values may compound until they become very large or very small.

In vanishing gradients, the magnitudes of the partial derivatives shrink rapidly as they move backward through the network. This can occur when weight matrices have singular values significantly less than 1, causing repeated multiplications to push gradients toward zero. As a result, the gradient signal is very small for early layers, causing them to update slowly or not at all. The saturation of sigmoidal activation functions also causes gradiens to shrink, motivating non-saturating activations, such as rectified linear units (ReLU).

Exploding gradients occur when repeated multiplications by large singular values, or large weight components, make gradients grow through backpropagation. This typically leads to large parameter updates that can destabilize training or produce numerical overflow (NaNs). Recurrent neural networks (RNNs), especially those with long sequential inputs that are encoded in unrolled sequences, are particularly sensitive to this issue.

Proper initialization of weights can help manage exploding and vanishing gradients. Xavier initialization [194] is used for saturating activation functions,, such as sigmoid and tanh. It uses a normal distribution with standard deviation related to the number of input and output units of the layer. He initialization [223] is used for ReLU and variants. It also uses a normal distribution, with standard deviation only depending on the number of inputs.

Batch normalization of intermediate activations helps reduce extreme variance in the forward and backward passes. Gradient clipping of the norm of the gradient vector can prevent parameter updates from becoming too large in exploding-gradient scenarios, particularly RNNs. Adding residual or skip connections in the ResNet architecture can allow gradients to flow more directly through the network, reducing the tendency to vanish.

Batch Normalization

Recall that for large-scale neural network training, we typically don't compute the loss function or gradient using the entire dataset, but instead use a randomly selected mini-batch of data to compute the loss and gradient in each optimization step. Mini-batch SGD also helps improve generalization through regularization, as it introduces stochasticity in training. However, these smaller batches may lead to high variance in the output of each hidden layer of the network, causing erratic gradients and unstable training.

Batch normalization is a method for stabilizing and accelerating training in deep networks [246], by normalizing the output of each hidden layer, for each mini-batch of data. This ensures that each layer and mini-batch maintains a controlled mean and variance, reducing the sensitivity to parameter initializations and enabling higher learning rates.

Batch normalization normalizes the output of each hidden layer across the mini-batch. For a given hidden layer output $\mathbf{h} \in \mathbb{R}^{B \times d}$, where B is the size of the mini-batch and d is the dimension of the output, we compute the mean and variance:

$$\mu_j = \frac{1}{B} \sum_{i=1}^{B} h_{ij}, \quad \sigma_j^2 = \frac{1}{B} \sum_{i=1}^{B} (h_{ij} - \mu_j)^2$$

and then normalize, adding a small ϵ for numerical stability:

$$\hat{h}_{ij} = \frac{h_{ij} - \mu_j}{\sqrt{\sigma_j^2 + \epsilon}}.$$

Batch normalization reduces internal covariate shift. During training, the distribution of inputs to each layer changes as the previous layers update. This can slow training or cause instabilities, as later layers must adapt to new input distributions. Batch normalization reduces this effect by standardizing the distribution of inputs to each layer. This also improves gradient flow, preventing exploding or vanishing gradients, by keeping the scale of the inputs more consistent and reducing the saturation of nonlinear activation functions.

10.3 Neural Network Training

Training a deep neural network involves solving a high-dimensional, nonconvex optimization problem to tune the parameters of the network to minimize a loss function. This is typically an unconstrained optimization problem, and we usually rely on stochastic gradient descent with the ADAM optimizer for convergence.

Here we explore a concrete example of training a deep neural network in PyTorch, highlighting several issues that occur. We will explore the CIFAR-10 data set [289], which consists of 60,000 color images in 10 classes, with 6,000 images per class. Each image has 32×32 pixels, and example images from each class are shown in Fig. 10.14.

For this example, we will train and compare two neural network architectures: a simple multi-layer perceptron (MLP) and a convolutional neural network (CNN). The first step is to load the CIFAR-10 dataset, which is conveniently available in the `torchvision` package in PyTorch:

CodeSnip 10.h

```
import torch
import torch.nn as nn
import torch.optim as optim
from torchvision import datasets, transforms
from torch.utils.data import DataLoader, random_split

device = torch.device('cuda' if torch.cuda.is_available() else '
    cpu')

# Normalize CIFAR-10 images (using empirical mean and std of
    dataset)
transform = transforms.Compose([
    transforms.ToTensor(),  # convert to PyTorch tensor
    transforms.Normalize((0.491,0.482,0.447),(0.202,0.199,0.201)
        )
])

train_dataset = datasets.CIFAR10(root='./data', train=True,
    download=True, transform=transform)
test_dataset = datasets.CIFAR10(root='./data', train=False,
    download=True, transform=transform)

# Split train into train and validation
train_size = int(0.9 * len(train_dataset))
val_size = len(train_dataset) - train_size
train_data, val_data = random_split(train_dataset, [train_size,
    val_size])

train_loader = DataLoader(train_data, batch_size=128, shuffle=
    True)
val_loader = DataLoader(val_data, batch_size=128)
test_loader = DataLoader(test_dataset, batch_size=128)

class_names = train_dataset.classes
```

Panel 10.2: A Brief History of Neural Networks

Deep learning is the dominant paradigm in machine learning today. However, neural networks (NNs) have a rich and winding history, with deep connections to optimization and biological learning, including several false starts along the way.

Although the first artificial NN was conceived by McCulloch and Pitts in 1943 [345], the NN as we know it today was introduced in 1958 as Rosenblatt's *perceptron* [427]: the single-layer perceptron (SLP), a linear threshold unit trained by a simple gradient rule. Rosenblatt constructed perceptrons in hardware for image classification and speculated such machines may eventually translate between languages. However, interest in perceptrons declined for two decades after Minsky and Papert [353] showed that an SLP cannot represent simple nonlinear functions, such as XOR. Ironically, their criticism pointed to the solution: adding hidden layers. MLPs *can* represent XOR and, more generally, approximate arbitrary functions. The very critique that stalled NN research in the 1970s and 1980s foreshadowed the core insight of modern deep learning: depth is the source of representational power.

Despite the decline of perceptrons in the U.S., progress continued elsewhere. In Japan, Amari introduced stochastic gradient descent for MLPs in 1967 [5]. Fukushima introduced the neocognitron in 1980 [182], a precursor of today's convolutional networks, as well as the rectified linear unit (ReLU) as a biologically plausible activation function. Meanwhile, associative memory networks were being explored by Amari and later Hopfield [236], who connected neural computation with statistical mechanics. This work foreshadowed several themes of modern deep learning. Even so, neural network research remained underappreciated at the time.

Neural networks resurfaced in the 1980s with the popularization of *backpropagation* [431], which enabled efficient gradient computation in MLPs. Rumelhart, Hinton, and Williams demonstrated that MLPs could approximate nonlinear mappings, while LeCun and colleagues [304] used convolutional networks for handwritten digit recognition. Progress was still limited by computing power, small datasets, and challenging optimization. Publishing neural network research was difficult through the 1980s and 1990s, as much of the community remained skeptical.

The modern era began in the 2000s, when large datasets, GPU acceleration, and improved training methods revived interest. Hinton and collaborators introduced deep belief networks [229] and layer-wise pretraining to mitigate vanishing gradients. The turning point came in 2012, when Krizhevsky, Sutskever, and Hinton's deep convolutional network outperformed competitors on the ImageNet image recognition task [290]. This result triggered the widespread adoption of deep learning across vision and natural language. Importantly, this would not have been possible without the extensive hand-labeled ImageNet dataset [135].

At each stage, large-scale optimization has been a central challenge.

Figure 10.14 Example images from the 10 image classes in the CIFAR-10 database [289].

This code loads the training and testing data, and splits the training data into 90% for training and 10% for validation. It also converts images to `torch` tensors and normalizes by subtracting the empirical mean and dividing by the empirical standard deviation.

Next, we define the MLP architecture:

CodeSnip 10.i

```
class MLP(nn.Module):
    def __init__(self, dropout_prob=0.3):
        super().__init__()
        self.net = nn.Sequential(
            nn.Flatten(),
            nn.Linear(32*32*3, 1024),
            nn.ReLU(),
            nn.Dropout(dropout_prob),
            nn.Linear(1024, 512),
            nn.ReLU(),
            nn.Dropout(dropout_prob),
            nn.Linear(512, 10)
        )
    def forward(self, x):
        return self.net(x)
model = MLP(dropout_prob=0.3).to(device)
```

This MLP has three fully connected layers. The input layer flattens the $32 \times 32 \times 3$ image into a vector of length 3072. The first hidden layer has 1024 units and the second has 512, each followed by a ReLU activation function and dropout. The output layer has 10 units, corresponding to the 10 image classes. In the dropout layers, 30% of neurons are randomly disabled in each forward pass, regularizing the network.

Next, we define the cross entropy loss function and initialize the ADAM optimizer with a learning rate of 0.001:

CodeSnip 10.j

```
criterion = nn.CrossEntropyLoss()
optimizer = optim.Adam(model.parameters(), lr=0.001)
scheduler = optim.lr_scheduler.StepLR(optimizer, step_size=15,
    gamma=0.5)
```

We then define the `train_epoch` function, with all code to be executed in each epoch, including evaluating the model and loss function, computing the necessary gradients via backpropagation, and updating the model parameters. The `evaluate` function evaluates the model and returns necessary information for evaluation metrics and plotting:

CodeSnip 10.k

```
def train_epoch(model, loader, criterion, optimizer):
    model.train()
    total_loss, total_correct = 0, 0
    for x, y in loader:
        x, y = x.to(device), y.to(device)
        optimizer.zero_grad()
        outputs = model(x)
        loss = criterion(outputs, y)
        loss.backward()
        optimizer.step()
        total_loss += loss.item() * x.size(0)
        total_correct += (outputs.argmax(dim=1)==y).sum().item()
    return total_loss / len(loader.dataset), total_correct / len
        (loader.dataset)

def evaluate(model, loader, criterion):
    model.eval()
    total_loss, total_correct = 0, 0
    all_preds, all_labels = [], []
    with torch.no_grad():
        for x, y in loader:
            x, y = x.to(device), y.to(device)
            outputs = model(x)
            loss = criterion(outputs, y)
            total_loss += loss.item() * x.size(0)
            total_correct += (outputs.argmax(dim=1)==y).sum().
                item()
            all_preds.extend(outputs.argmax(dim=1).cpu().numpy()
                )
            all_labels.extend(y.cpu().numpy())
    return total_loss / len(loader.dataset), total_correct / len
        (loader.dataset), all_preds, all_labels
```

Finally, we run the training loop, iterating through 30 epochs:

CodeSnip 10.1

```
num_epochs = 30
train_losses, val_losses = [], []
train_accs, val_accs = [], []

for epoch in range(num_epochs):
    tr_loss, tr_acc = train_epoch(model, train_loader, criterion
        , optimizer)
    val_loss, val_acc, _, _ = evaluate(model, val_loader,
        criterion)

    train_losses.append(tr_loss)
    val_losses.append(val_loss)
    train_accs.append(tr_acc)
    val_accs.append(val_acc)

    scheduler.step()
```

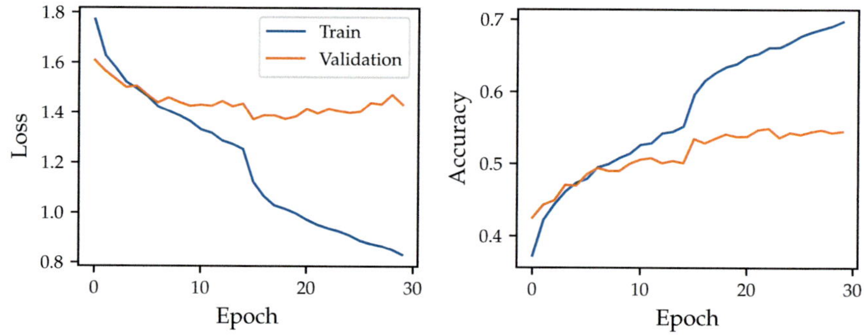

Figure 10.15 Learning curves for a multi-layer perceptron (MLP) on the CIFAR-10 database. Training and test performance are shown.

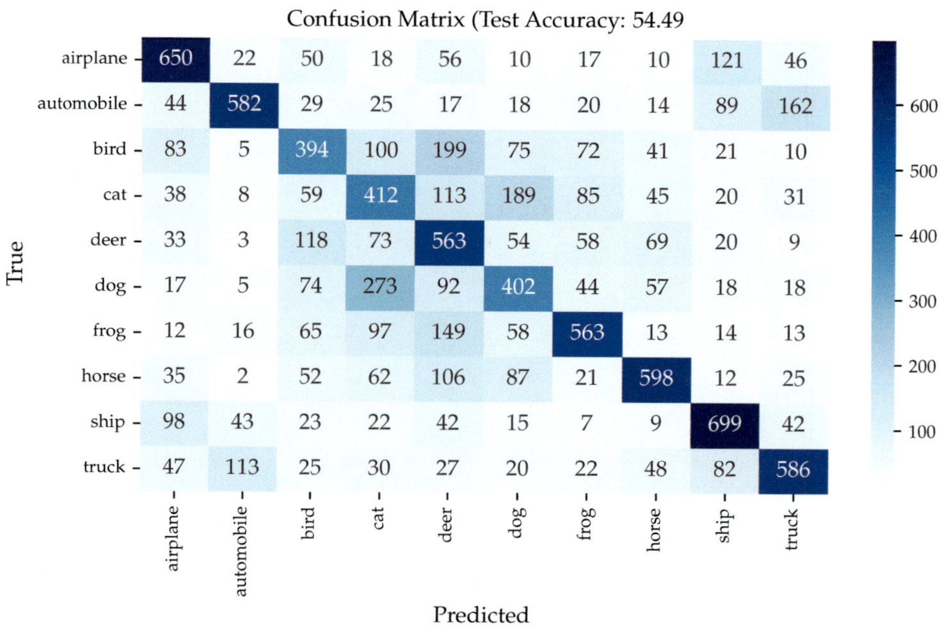

Figure 10.16 Confusion matrix for MLP classification of ten image classes in CIFAR-10.

After the model is trained, it is also customary to compute the final test error:

CodeSnip 10.m

```
test_loss, test_acc, test_preds, test_labels = evaluate(model,
    test_loader, criterion)
```

The learning curves for this MLP are shown in Fig. 10.15. Although the training accuracy continues to increase, the validation accuracy plateaus at around 50%, which is not great for a modern neural network on an image classification task. Figure 10.16 shows the rate of true classifications and misclassifications for the 10 image classes.

It is possible to dramatically improve this performance by introducing a convolutional neural network architecture. The data loading, initialization, evaluation, and epoch loop are identical, and we need only change the neural network model:

CodeSnip 10.n

```
class SimpleCNN(nn.Module):
    def __init__(self, dropout_prob=0.25):
        super().__init__()
        self.features = nn.Sequential(
            nn.Conv2d(3, 32, kernel_size=3, padding=1),  # 32
                x32x32
            nn.ReLU(),
            nn.Conv2d(32, 64, kernel_size=3, padding=1),  # 64
                x32x32
            nn.ReLU(),
            nn.MaxPool2d(2, 2),  # 64x16x16
            nn.Dropout(dropout_prob),
            nn.Conv2d(64, 128, kernel_size=3, padding=1), #128
                x16x16
            nn.ReLU(),
            nn.MaxPool2d(2, 2),  # 128x8x8
            nn.Dropout(dropout_prob)
        )
        self.classifier = nn.Sequential(
            nn.Flatten(),
            nn.Linear(128 * 8 * 8, 256),
            nn.ReLU(),
            nn.Dropout(dropout_prob),
            nn.Linear(256, 10)
        )
    def forward(self, x):
        x = self.features(x)
        return self.classifier(x)
model = SimpleCNN(dropout_prob=0.25).to(device)
```

This CNN, shown in Fig. 10.17, consists of three convolutional layers with ReLU activations, two max-pooling layers for spatial downsampling, and 25% dropout layers after pooling and between dense layers for regularization. The $32 \times 32 \times 3$ input is processed by a 3×3 conv layer with 32 filters, followed by a 3×3 conv layer with 64 filters, and finally by a 3×3 conv layer with 128 filters. Each layer is padded to preserve spatial dimensions before pooling reduces the resolution to 8×8. The flattened feature map of size $128 \times 8 \times 8$ feeds into a fully connected classifier with one hidden layer of size 256 with ReLU activation, followed by a final linear layer mapping to the 10 output classes.

This change in architecture dramatically improves performance, as shown in Fig. 10.18. The validation accuracy now plateaus closer to 80%, which is a significant improvement. The rate of misclassification is much smaller, as shown in Fig. 10.19. Interestingly, dogs and cats remain the most misclassified. The performance can be improved further by leveraging more sophisticated neural network architectures, such as the VGG-16 CNN [465] with additional convolutional layers, or the deep, recurrent residual network (ResNet) [224].

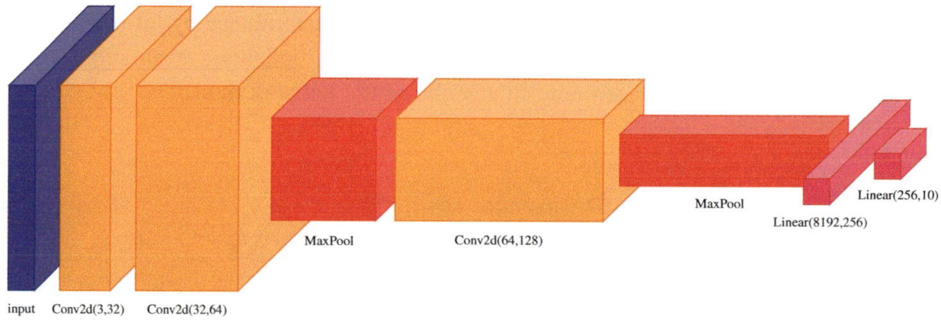

Figure 10.17 Convolutional neural network architecture used to classify CIFAR-10 dataset.

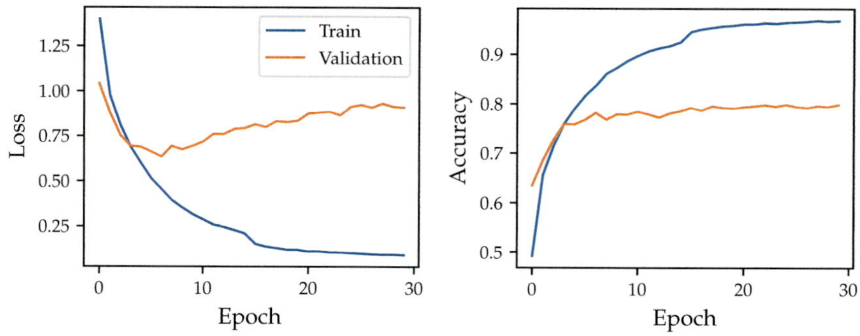

Figure 10.18 Learning curves for a convolutional neural network (CNN) on the CIFAR-10 database. Training and test performance are shown.

Hyperparameter Optimization

There are several hyperparameters involved in training a modern deep neural network, and selecting these parameters is a meta-optimization. The learning rate, the batch size, and regularization coefficients are all important. Even the network architecture, such as number and size of layers, may be considered as a hyperparameter to tune. Figure 10.20 shows the effect of batch size on the learning curves, illustrating that the training depends heavily on these parameters.

The most common approaches to optimizing hyperparameters are grid search, random search, and Bayesian optimization. A grid search is simple, as a discrete grid of possible hyperparameters are tested; however, this is computationally expensive. Random search involves sampling hyperparameters from a given distribution, as this can be more efficient in higher dimensions [45]. In both cases, it may be unclear how long to let training go before identifying "bad" or "good" parameters before zooming in and training for longer.

Bayesian optimization is a more sophisticated approach, which treats hyperparameter tuning as a black-box optimization problem, where evaluating the objective function is extremely expensive (i.e., involves training an entire model!) and should be minimized. A surrogate model (e.g., Gaussian Process) predicts performance for untested hyperparameter

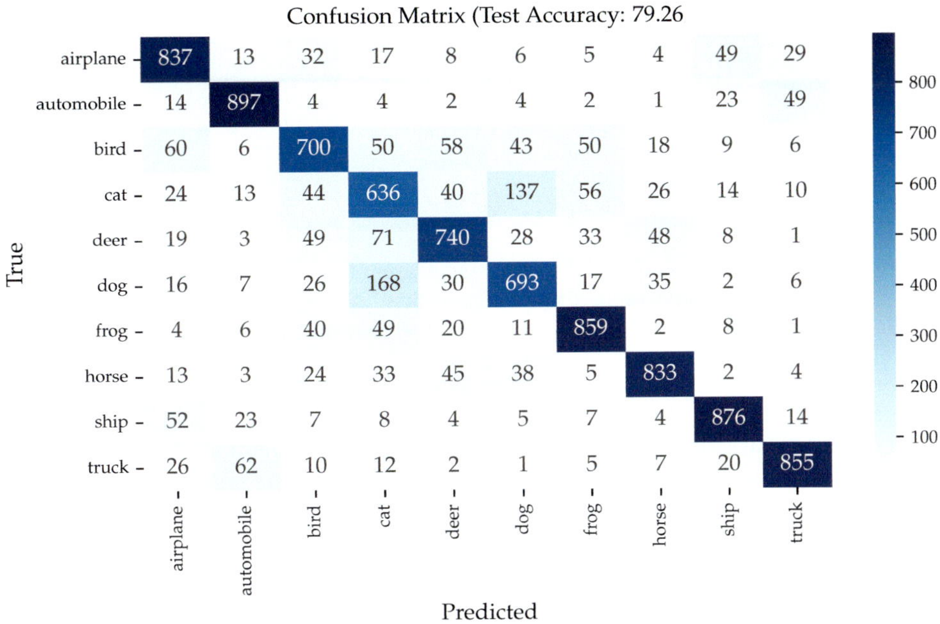

Figure 10.19 Confusion matrix for CNN classification of ten image classes in CIFAR-10.

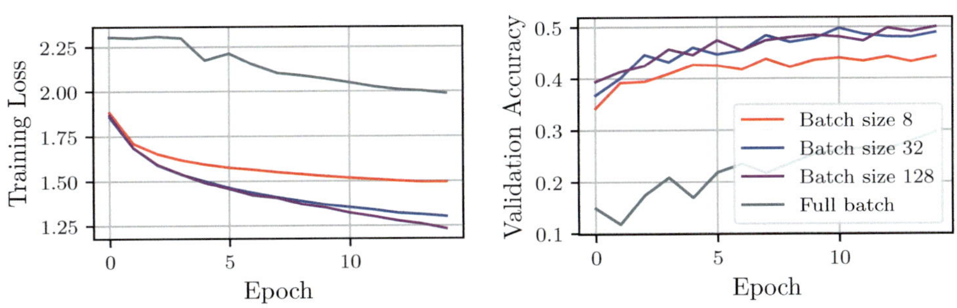

Figure 10.20 Learning curves versus batch size for the MLP on CIFAR-10. We see an improvement in performance from a small batch size of 8 to a larger size of 32 or 128, and then the performance degrades again with full batch.

points [467]. The algorithm balances exploitation (sampling regions where the model predicts good performance) and exploration (sampling uncertain regions).

Learning Rate Scheduling. The learning rate is crucial for stable and efficient convergence. Instead of using a fixed learning rate, scheduling strategies adjust the step size over time [204]. Step decay and exponential decay are two common learning rate schedules. Step decay decreases the learning rate by a factor of α every few epochs, while exponential decay continuously decrease the learning rate at an exponential rate, e.g., $\eta_t = \eta_0 \cdot e^{-kt}$.

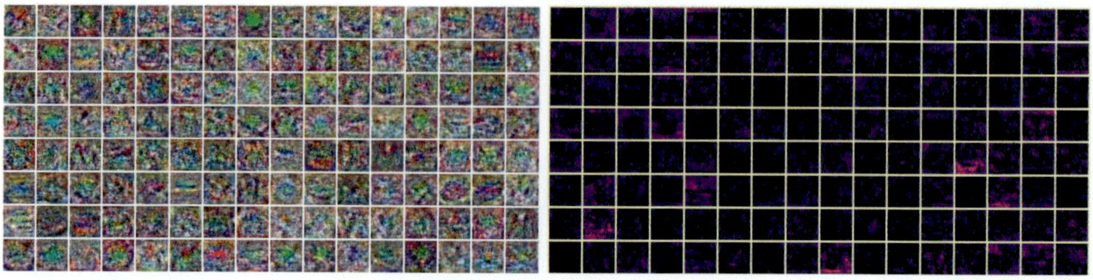

Figure 10.21 Left: Weights of the input layer of the MLP, visualized as 32×32 images. There are 1024 of these initial filters, so we plot the first 128 filters in a 8×16 grid. Right: The gradient of the loss function with respect to this input layer, computed using backprop, also visualized as a grid of images.

Cyclical learning rates allow the learning rate to oscillate between a minimum and maximum boundary, which can help the optimizer escape saddle points or poor local minima.

Visualizing the Network, Gradients, and Saliency. At first glance, neural networks may appear to be opaque and convoluted, living in a much higher dimensional space than humans can easily grasp. However, there are approach to visualize these networks based on gradients computed from backprop, providing some level of much-needed interpretability and explainability of the inner workings. Figure 10.21 shows the weights of the first input layer of the MLP, along with the gradients of the loss function with respect to this layer, computed using backpropagation. The input layer essentially consists of filters that extract features from the various images for downstream classification. The gradient of the loss function with respect to this layer provides the information needed to train and modify this layer to improve the loss function using SGD.

There is another related gradient that may be computed, which results in the *saliency map*. Specifically, we may take the gradient of the loss function with respect to the input $\mathbf{x}$ instead of the weights $\mathbf{w}$. We may evaluate this for a single image $\mathbf{x}$ or averaged over an entire category of images. Figure 10.22 (left) shows the saliency maps of the MLP architecture for 256 input images arranged in a 8×16 grid. It is possible to compute something similar for CNNs, although there is a variant, called the gradient-weighted class activation mapping (Grad-CAM), that aggregates gradients over channels to produce a spatial heatmap of regions the network relied on. Grad-CAM yields smoother, object-centric regions, compared to saliency maps, which are especially informative for CNNs. This is shown in Fig. 10.22 (right) for the CNN architecture.

10.4 Deep Reinforcement Learning

Deep reinforcement learning is one of the most exciting and rapidly developing areas of machine learning research. Some of the most impressive advances in AI have leveraged or been inspired by progress in deep RL, and it is currently one of the most promising avenues of research towards generalized artificial intelligence.

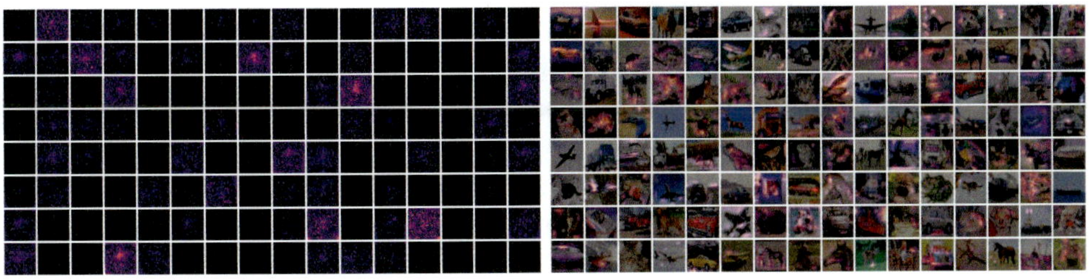

Figure 10.22 Left: Salience map of MLP input layer. Right: Grad-CAM visualization of CNN architecture, similar to saliency.

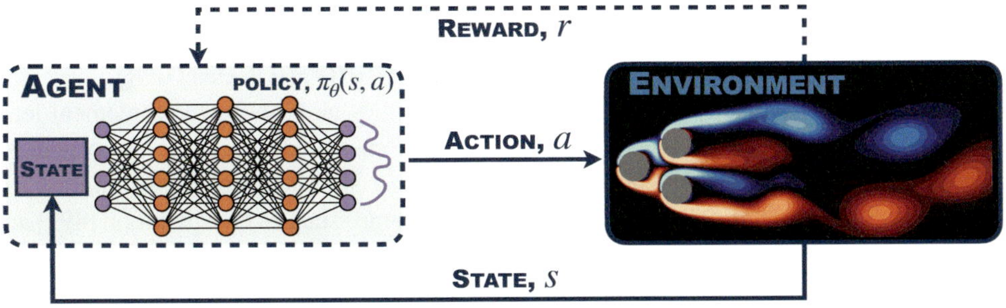

Figure 10.23 Schematic of deep reinforcement learning,, where a deep neural network is used to represent the policy π. This is known as a *deep policy* network. This figure depicts a deep RL agent controlling a fluid flow found in the HydroGym environment [295].

Classic reinforcement learning, discussed briefly in Section 9.3, is a powerful and flexible framework for learning and control. However, it suffers from a representation problem, where the policy π, the value function V, and the quality function Q, may be extremely complicated functions, defined over very high-dimensional state and action spaces. Even simple games, such as chess and Go, have exceedingly large state spaces[5]. Representing a value function or quality function, let alone sampling over these states, is infeasible with classic approaches. In the past fifteen years, deep learning has revolutionized our ability to represent such complicated functions from data, and is likewise pushing the frontiers of what is possible in RL.

It is possible to use deep learning in several different ways to approximate the various functions used in RL, or to model the environment more generally. Typically, the central challenge is in identifying and representing key features in a high-dimensional state space. For example, the policy $\pi(\mathbf{s}, \mathbf{a})$ may now be approximated by

$$\pi(\mathbf{s}, \mathbf{a}) \approx \pi(\mathbf{s}, \mathbf{a}, \boldsymbol{\theta}), \tag{10.31}$$

[5] Claude Shannon famously estimated the number of possible games of chess, known as the Shannon number, at around 10^{120} in his paper "Programming a computer for playing chess" [459]; this paper was a major inspiration for modern dynamic programming and reinforcement learning. The state space of simple video games, such as the 1972 Atari game Pong, are determined by the black-and-white screen at standard resolution 336×240, resulting in over 10^{24000} possible discrete states, making it infeasible to represent the value or quality functions exactly without approximation.

where $\boldsymbol{\theta}$ represent the weights of a neural network. This pairing of deep learning for representations with reinforcement learning for decision making and control has resulted in dramatic improvements to the capabilities of reinforcement learning.

Much of what is discussed in this section is also relevant for other function approximation techniques besides deep learning.. For example, policy gradients may be computed using other representations of the form of (10.31), and there is a long history before deep learning [265, 481]. System identification approaches in Section 8.7, such as SINDy or DMD/Koopman, have also shown promise in model-based RL [76, 429, 412, 562, 82]. That said, many of the most exciting and impressive recent demonstrations of RL leverage the full power of deep learning, and so we present these innovations in this context.

A Brief History of Deep Reinforcement Learning

Much like many kids growing up in the 1980s and 1990s, reinforcement learning got its start playing board games (e.g., checkers, chess, and Go) and video games (e.g., Atari games, StarCraft). Checkers was the earliest significant demonstration of RL by Arthur Samuel [434], followed by more complex games such as backgammon [492] and eventually chess and Go [461, 464], where RL was eventually able to reach Grandmaster status. The transition from *simple* games, like tic-tac-toe and checkers, to *hard* games, like chess and Go, was driven by deep reinforcement learning. Deep RL has similarly shown super-human performance in a range of increasingly sophisticated video games [354, 464, 462, 518]. In the case of Atari games, the pixel space is processed using a CNN architecture. A major landmark was a generalist RL agent that learned to play multiple games [462] through self-play, rather than specializing in only one task.

In recent years, RL has been increasingly used in real-world problems, for example to control the plasma dynamics in a Tokamak fusion reactor [132], to discover new pharmaceuticals [393], to improve human alignment of large language models (LLMs) [377], and to understand and predict protein folding [455], a direct precursor of the 2025 Nobel Prize work in deep protein folding [22, 257]. Deep RL has been used to improve LLMs, such as ChatGPT, using human feedback [377]. The idea is to fine tune the LLM using polic optimization to maximize the reward of aligning its output to human preferences, intentions, values, and expectations, so that it produces helpful, honest, and harmless outputs.

RL has also been used extensively in engineering applications, for example in fluid flow control [80, 186, 108, 391, 517, 53, 373, 165, 372, 21], robotics [497, 280, 550, 213], and autonomy [378, 410, 433, 457]. Physical systems require additional considerations, such as continuous state and action spaces [409], and the need for certifiable solutions, such as trust regions [451], for safety-critical applications (e.g., transportation, autonomous flight, etc.) with physical constraints [302, 489].

Progress in deep RLhas been enabled by robust benchmark environments, such as the OpenAI Gym [70] (`https://gym.openai.com`). The Gym provides a flexible simulation environment to test learning strategies on a variety of systems, including: over 100 tasks on classic Atari 2600 games; classic control benchmarks, such as balancing an inverted pendulum on a cart, swing-up of a pendulum, etc.; goal-based robotics [550], such as pushing or fetching a block to a goal position with a robot arm, with and without sliding after loss of contact, and robotic hand manipulation for reaching a pose or orienting various objects; and the MuJoCo [497] environments with multi-legged locomotion, running, hopping, swimming, etc. The HydroGym benchmark environments [295] provide a challenging set of advanced fluid flow control benchmarks.

Policy Gradient Optimization

Policy gradient optimization [265, 463, 481] is the leading technique for large-scale RL training today. Proximal policy optimization (PPO) is particularly dominant in the literature. When the policy π is parameterized by $\boldsymbol{\theta}$, it is possible to use gradient optimization on the parameters to improve the policy much faster than through traditional policy iteration.

The policy may be parameterized by a multi-layer neural network, in which case this would be a *deep policy*, although other representations and function approximations may be useful. Thus, it is possible to directly optimize the parameters $\boldsymbol{\theta}$ through gradient descent or stochastic gradient descent. The goal is still to optimize a policy to maximize future rewards, given by the following objective function $J(\boldsymbol{\theta})$:

$$J(\boldsymbol{\theta}) = \mathbb{E}\left[Q^{\pi_\theta}(\mathbf{s}, \mathbf{a})\right] = \sum_{\mathbf{s}\in\mathcal{S}} \mu_\theta(\mathbf{s}) \sum_{\mathbf{a}\in\mathcal{A}} \pi_\theta(\mathbf{s}, \mathbf{a}) Q^{\pi_\theta}(\mathbf{s}, \mathbf{a}), \tag{10.32}$$

where the expected value is taken over states and actions given a policy π_θ parameterized by $\boldsymbol{\theta}$. The states are distributed according to the asymptotic steady state distribution μ_θ; this asymptotic steady state is naturally sampled when data is collected for long times given a fixed policy. Notice that we have written the objective function J as a function of the policy parameters $\boldsymbol{\theta}$, because we are assuming that we have a fixed policy π_θ when computing the quality function.

If it is possible to compute the gradient of this objective function with respect to the parameters $\boldsymbol{\theta}$, then the polic parameters may be updated via standard gradient descent

$$\boldsymbol{\theta}^{\text{new}} = \boldsymbol{\theta}^{\text{old}} + \alpha \nabla_\theta J(\boldsymbol{\theta}), \tag{10.33}$$

with learning rate α. The weight α may be replaced with a vector of learning weights for each component of $\boldsymbol{\theta}$. Computing this gradient is the heart of gradient policy optimization, and many of the most prominent RL algorithms today.

We compute the gradient $\nabla_\theta J(\boldsymbol{\theta})$ via the chain rule as

$$\begin{aligned}
\nabla_\theta J(\boldsymbol{\theta}) &= \sum_{\mathbf{s}\in\mathcal{S}} \mu_\theta(\mathbf{s}) \sum_{\mathbf{a}\in\mathcal{A}} Q(\mathbf{s}, \mathbf{a}) \nabla_\theta \pi_\theta(\mathbf{s}, \mathbf{a}) \\
&= \sum_{\mathbf{s}\in\mathcal{S}} \mu_\theta(\mathbf{s}) \sum_{\mathbf{a}\in\mathcal{A}} \pi_\theta(\mathbf{s}, \mathbf{a}) Q(\mathbf{s}, \mathbf{a}) \frac{\nabla_\theta \pi_\theta(\mathbf{s}, \mathbf{a})}{\pi_\theta(\mathbf{s}, \mathbf{a})} \\
&= \sum_{\mathbf{s}\in\mathcal{S}} \mu_\theta(\mathbf{s}) \sum_{\mathbf{a}\in\mathcal{A}} \pi_\theta(\mathbf{s}, \mathbf{a}) Q(\mathbf{s}, \mathbf{a}) \nabla_\theta \log(\pi_\theta(\mathbf{s}, \mathbf{a})) \\
&= \mathbb{E}(Q(\mathbf{s}, \mathbf{a}) \nabla_\theta \log(\pi_\theta(\mathbf{s}, \mathbf{a}))).
\end{aligned}$$

This is known as the *policy gradient theorem* in RL. This relies on the key insight that the gradient of the policy $\nabla_\theta \pi_\theta(\mathbf{s}, \mathbf{a})$ can be rewritten in terms of the gradient of its logarithm:

$$\nabla_\theta \pi_\theta(\mathbf{s}, \mathbf{a}) = \pi_\theta(\mathbf{s}, \mathbf{a}) \nabla_\theta \log \pi_\theta(\mathbf{s}, \mathbf{a}). \tag{10.34}$$

The reformulated gradient $\nabla_\theta J(\boldsymbol{\theta})$ can now be estimated using Monte Carlo sampling. Specifically, we draw actions $\mathbf{a}_j$ from the policy $\pi_\theta(\mathbf{a}, \mathbf{s})$ and compute the empirical mean of $Q(\mathbf{s}, \mathbf{a}_j)\nabla_\theta \log \pi_\theta(\mathbf{a}_j, \mathbf{s})$ over these samples. Sampling the policy has the advantage of capturing the distribution of actions induced by π_θ, rather than sampling over all possible actions $\mathbf{a}$, which would be intractable in high-dimensional or continuous action spaces. Note that $Q^\pi(\mathbf{s}, \mathbf{a})$ is treated as a constant with respect to $\boldsymbol{\theta}$, and all gradients flow only through the policy π_θ using backpropagation.

The policy gradient method above is the foundation of the REINFORCE algorithm [538], as well as more modern proximal policy optimization (PPO) approaches [451, 452, 529, 402, 241]. Policy gradient methods also form the foundation of actor–critic algorithms, which combine value function estimation with gradient-based policy optimization. The actor learns the policy parameters, while the critic evaluates the value function to guide the updates. This tradeoff between policy and value learning helps stabilize training and accelerates convergence, especially in high-dimensional environments.

Policy gradient methods are particularly powerful in environments with continuous action spaces or high-dimensional policies, where other methods like Q-learning struggle to scale effectively. By directly optimizing the parameters of the policy, these methods avoid the need to discretize actions or maintain large tabular representations, making them well-suited for modern reinforcement learning applications.- variance gradient estimates, which can slow down convergence or lead to unstable updates. Techniques such as variance reduction, reward normalization, and entropy regularization are commonly employed to address these challenges.

Deep Q-Learning

Many of the most exciting advances in RL over the past decade have involved some variation of deep Q-learning, which uses deep neural networks to represent the quality function Q. As with the policy in (10.31), it is possible to approximate the Q function through some parameterization $\boldsymbol{\theta}$,

$$Q(\mathbf{s}, \mathbf{a}) \approx Q(\mathbf{s}, \mathbf{a}, \boldsymbol{\theta}), \tag{10.35}$$

where $\boldsymbol{\theta}$ represents the weights of a deep neural network. In this representation, the training loss function is directly related to the standard Q-learning update:

$$L = \mathbb{E}\left[(\mathbf{r}_k + \gamma \max_{\mathbf{a}} Q(\mathbf{s}_{k+1}, \mathbf{a}, \boldsymbol{\theta}) - Q(\mathbf{s}_k, \mathbf{a}_k, \boldsymbol{\theta}))^2\right]. \tag{10.36}$$

The first part of the loss function, $\mathbf{r}_k + \gamma \max_{\mathbf{a}} Q(\mathbf{s}_{k+1}, \mathbf{a}, \boldsymbol{\theta})$, is known as the temporal difference target, which is a classic RL strategy for handling delayed rewards. The second part, $Q(\mathbf{s}_k, \mathbf{a}_k, \boldsymbol{\theta})$, is the prediction.

Deep reinforcement learning based on a deep Q network (DQN) was introduced by Mnih et al. [354] for playing Atari games. Specifically, this network used a deep convolutional neural network to represent the Q function, where the inputs were pixels from the Atari screen and actions were joystick motions. In this original paper, both Q functions in (10.36) were represented by the same network weights $\boldsymbol{\theta}$. However, in a double DQN [515], different networks are used to represent the target and prediction Q functions, which reduces bias due to inaccuracies early in training. In double DQN, it may be necessary to fix the target network for multiple training iterations of the prediction network before updating to improve stability and convergence [179].

Experience replay is a critical component of training a DQN, which is possible because it is an off-policy RL algorithm. Short segments of past experiences are used in batches for SGD during training. Moreover, to place more importance on experiences with large model mismatch, it is possible to weight past experiences by the magnitude of the temporal difference (TD) error. This process is known as prioritized experience replay [438].

Dueling deep Q networks (DDQNs) [530] are another important deep Q-learning architecture that are used to improve training when actions have a marginal effect on the quality function. In particular, a DDQN splits the quality function into the sum of a value function and an *advantage* function $A(\mathbf{s}, \mathbf{a})$, which quantifies the additional benefit of a particular action over the value of being in that state:

$$Q(\mathbf{s}, \mathbf{a}, \boldsymbol{\theta}) = V(\mathbf{s}, \boldsymbol{\theta}_1) + A(\mathbf{s}, \mathbf{a}, \boldsymbol{\theta}_2). \tag{10.37}$$

Thus, separate value and advantage networks are combined to estimate the Q function.

There are a variety of other useful architectures for deep Q-learning. Deep recurrent Q networks are promising for dynamic problems [221]. Advantage actor–critic networks, discussed in the next section, combine the DDQN with deep policy networks.

Actor–Critic Networks

Actor–critic methods in reinforcement learning simultaneously learn a policy function and a value function, with the goal of taking the best of both value-based and policy-based learning. The basic idea is to have an actor, which is policy-based, and a critic, which is value-based, and to use the temporal difference signal from the critic to update the policy parameters. There are many actor–critic methods that pre-date deep learning. For example, a simple actor–critic approach would update the policy parameters $\boldsymbol{\theta}$ in (10.31) using the temporal difference error $\mathbf{r}_k + \gamma V(\mathbf{s}_{k+1}) - V(\mathbf{s}_k)$:

$$\boldsymbol{\theta}_{k+1} = \boldsymbol{\theta}_k + \alpha \nabla_{\boldsymbol{\theta}}((\log \pi(\mathbf{s}_k, \mathbf{a}_k, \boldsymbol{\theta}))(\mathbf{r}_k + \gamma V(\mathbf{s}_{k+1}) - V(\mathbf{s}_k)). \tag{10.38}$$

It is rather straightforward to incorporate deep learning into an actor–critic framework. For example, in the advantage actor–critic (A2C) network, the actor is a deep policy network, and the critic is a DDQN. In this case, the update is given by

$$\boldsymbol{\theta}_{k+1} = \boldsymbol{\theta}_k + \alpha \nabla_{\boldsymbol{\theta}}((\log \pi(\mathbf{s}_k, \mathbf{a}_k, \boldsymbol{\theta}))Q(\mathbf{s}_k, \mathbf{a}_k, \boldsymbol{\theta}_2)). \tag{10.39}$$

Challenges and Additional Techniques

Sparse and delayed rewards represent the central challenge of reinforcement learning, leading to the well-known credit assignment problem, which we have seen multiple times at this point. The following techniques, including reward shaping and hindsight experience replay, are leading techniques to overcome the credit assignment problem.

Reward Shaping. Reward shaping is a standard approach for systems with sparse rewards. This involves designing customized features that are indicative of a future reward and may be used as an intermediate reward signal. For example, in chess, the relative point count, where each piece is assigned a numeric value (e.g., a queen is worth 10 points, rooks are worth 5, knights and bishops are worth 3, and pawns are worth 1 point), is an example of a shaped reward that gives an intermediate reward signal each time a piece is taken.

Reward shaping is quite common and can be very effective. However, these rewards require expert human guidance to design, and this requires customized effort for each new task. Thus, reward shaping is not a viable strategy for a generalized artificial intelligence agent capable of learning multiple games or tasks. In addition, reward shaping generally limits the upper end of the agent's performance to that of the human expert.

Hindsight Experience Replay. In many tasks, such as robotic manipulation, the goal is to move the robot or an object from one location to another. For example, a robot arm may be required to slide an object on a table from point A to point B. Without a detailed physical model or other prior knowledge, it is unlikely that a random control policy will result in the object actually reaching the desired destination, so the rewards may be very sparse. It is possible to shape a reward based on the distance of the object to the goal state, although this is not a general strategy and suffers from the limitations discussed above.

Hindsight experience replay (HER) [9, 297] is a strategy that enriches the reward signal by taking failed trials and recasting them as successful at a different task. This approach makes the reward structure more dense, and has the benefit of enabling the simultaneous learning of a whole family of motion tasks. HER is quite intuitive in the context of human learning, for example in the case of tennis. Initially, it is difficult to aim the ball, and shots often go wild when learning. However, this provides valuable information about those muscle actions that might be useful for future tasks. After lots of practice, it then becomes possible to pick from different shots and place the ball more deliberately.

Curiosity-Driven Exploration. In large, open-world environments, an RL may get stuck in a local minimum, where it over-optimizes for a small region of state space. One approach to solve this problem is to augment the reward signal with a *novelty* reward that is large in regions of state space that are not well-modeled. This is known as curiosity-driven exploration [383], and it involves an intrinsic curiosity module (ICM), which compares a forward model of the evolution of the state, or a latent representation of the state, with the actual observed evolution. The discrepancy between the model and the actual dynamics is the novelty reward. When this difference is large, the agent becomes *curious* and explores this region more. There are similarities between this approach and TD learning. The main difference is that, in TD learning, the reward discrepancy is used as feedback to improve the value or quality function; while, in curiosity-driven exploration, the discrepancy is explicitly used as an additional reward signal. This is a clever approach to mimic the curiosity and exploration of intelligent biological learning systems.

Explainable AI. Explainable AI (XAI) is another important topic that is not only making deep RL more interpretable, but is also improving performance and generalization. For example, in fluid flow control, after the most important structures in a turbulent flow are identified [120], they may be used as the DRL reward, resulting in an improved control strategy [42] (e.g., drag reduction and lower actuation expenditure).

10.5 Physics-Constrained Machine Learning

Here we will introduce an important and growing topic in machine learning, that of enforcing and promoting known physics in the machine learning optimization. Typically this is achieved either through introducing regularizing terms in the loss function to promote physics, or directly adding physical constraints in the optimization.

We will begin this section with an overview of the five stages of machine learning, discussed briefly in the introduction to this chapter. For each stage, we will discuss how to incorporate physics into the machine learning optimization. Next, we will introduce several physics-informed or physics-constrained machine learning frameworks. We will introduce Lagrangian and Hamiltonian neural networks, which are essentially neural ODEs that introduce specific architectural choices and loss functions to preserve known physical structure. Next we will introduce physics-informed neural networks (PINNs), which include known physics, typically in the form of a partial differential equation, as regularizing terms in the loss function. Finally, we will discuss how to directly add physical constraints to machine learning optimizations to enforce known symmetries and invariances, such as energy conservation in a fluid. These ideas will be demonstrated on the sparse nonlinear system identification techniques introduced in Section 8.7.

The Five Stages of Machine Learning

Recall that applied machine learning may be broken into five steps, each of which provides an opportunity to embed prior physical knowledge:

1. Formulating the problem to model or the question to answer;
2. Collecting and curating the data used to train the model;
3. Choosing a machine learning architecture to best represent or model this data;
4. Designing loss functions to quantify performance and to guide the learning process; and
5. Implementing an optimization algorithm to train the model to minimize the loss function over the training data.

Here, we discuss these canonical stages of machine learning and investigate how to incorporate physics. This discussion is largely meant to be a high-level overview, and many more details can be found in recent papers [320, 80, 376].

The Problem. Machine learning leverages data to build models that may be deployed, ideally in real-time, to predict outputs given new data. It is critical to select the right system to model, motivated by a problem that is both important and tractable. Choosing a problem involves deciding on input data that will be readily available in the future, and output data that will represent the desired output, or prediction, of the model. The output data should be determinable from the inputs, and the functional relationship between these is precisely what the machine learning model will be trained to capture.

The nature of the problem, specifically what outputs will be modeled given what inputs, determines the large classes of machine learning algorithms: *supervised*, *unsupervised*, and *reinforcement learning*. In supervised learning, the training data will have expert labels that should be predicted or modeled with the machine learning algorithm. These output labels may be discrete, such as a categorical label of a 'dog' or a 'cat' given an input image, in which case the task is one of *classification*. If the labels are continuous, such as the average value of lift or drag given a specified airfoil geometry, then the task is one of *regression*.

In unsupervised learning, there are no expert labels, and structure must be extracted from the input data alone; thus, this is often referred to as *data mining*, and constitutes a particularly challenging field of machine learning. Again, if the structure in the data is assumed to be discrete, then the task is *clustering*. After the clusters are identified and characterized, these groupings may be used as proxy labels to then classify new data. If the structure in the data is assumed to be continuously varying, then the task is typically thought of as an *embedding* or *dimensionality reduction* task. Principal component analysis (PCA) or proper orthogonal decomposition (POD) may be thought of as unsupervised learning tasks that seek a continuous embedding of reduced dimension [77]. Reinforcement learning is a third, large branch of machine learning research, in which an *agent* learns to make control decisions to interact with an environment for some high level objective [480]. Examples include learning how to play games [354, 464], such as chess and Go.

Embedding physics. Deciding on what phenomena to model with machine learning is often inherently related to the underlying physics. Although classical machine learning has been largely applied to "static" tasks, such as image classification and the placement of advertisements, increasingly it is possible to apply these techniques to model physical systems that evolve in time according to some *rules* or *physics*. For example, we may formulate a learning problem to find and represent a conserved quantity, such as a Hamiltonian, purely from data [262]. Alternatively, the machine learning task may be to model time-series data as a differential equation, with the learning algorithm representing the dynamical system [447, 443, 81, 385, 520]. Similarly, the task may involve learning a coordinate transformation where these dynamics become simplified in some *physical* way, such as coordinate transformations to linearize or diagonalize/decouple dynamics [324, 533, 333, 484, 96].

The Data. This is the fuel that powers machine learning. Choosing data to inform a model is closely related to choosing what to model in the first place. The quality and quantity of data directly affects the resulting machine learning model. Many machine learning architectures, such as deep neural networks, are essentially sophisticated interpolation engines, and so having a diversity of training data is essential if these models are to be of use on unseen data. For example, modern deep convolutional neural networks rose to prominence with their unprecedented classification accuracy [290] on the ImageNet database [135], which contains over 14 million labeled images with over 20,000 categories, providing a sufficiently large and rich set of examples for training. This pairing of a vast labeled data set with a novel deep learning architecture is widely regarded as the beginning of the modern era of deep learning [204].

Embedding physics. The training data provides several opportunities for embedding prior physical knowledge. If a system is known to exhibit a symmetry, such translational or rotational invariance, then it is possible to augment and enrich the training data with shifted or rotated examples. More generally, it is often assumed that with an abundance of training data, these physical invariances will automatically be learned by a sufficiently expressive architecture. However, this approach tends to require considerable resources, both to collect and curate the data, as well as to train increasingly large models, making it more appropriate for industrial scale, rather than academic scale, research. In contrast, it is also possible to use physical intuition to craft new features from the training data, for example by applying a coordinate transformation that may simplify the representation or training. Physical data often comes from multiple sources with different fidelity, such as from numerical simulations, laboratory experiments, and in-flight tests. This is an important area of research for flight testing and unsteady aerodynamics [285], and recently physics-informed neural networks have been used with multifidelity data to approximate PDEs [347].

The Architecture. Once a problem has been identified, and data is collected and curated, it is necessary to choose an architecture with which to represent the machine learning model. Typically, a machine learning model is a function that maps inputs to outputs

$$\mathbf{y} = \mathbf{f}(\mathbf{x}; \boldsymbol{\theta})$$

and this function is generally represented within a specified family of functions parameterized by values in $\boldsymbol{\theta}$. For example, a linear regression model would model outputs as a linear function of the inputs, with $\boldsymbol{\theta}$ parameterizing this linear map, or matrix. Neural networks have emerged as a particularly powerful and flexible class of models to represent functional relationships between data, and they have been shown to be able to approximate arbitrarily complex functions with sufficient data and depth [238, 237]. There is a tremendous variety of potential neural network architectures [77], limited only by the imagination of the human designer.

The most common architecture is a simple feedforward network, in which data enters through an input layer and maps sequentially through a number of computational layers until an output layer. Each layer consists of nodes, where data from nodes in the previous layer are combined in a weighted sum and processed through an activation function, which is typically nonlinear. In this way, neural networks are fundamentally compositional in nature. The parameters $\boldsymbol{\theta}$ determine the network weights for how data is passed from one layer to the next, i.e. the weighted connectivity matrices for how nodes are connected in adjacent layers. The overarching network topology (i.e., how many layers, how large, what type of activation functions, etc.) is specified by the architect or determined in a meta-optimization, thus determining the family of functions that may be approximated by that class of network. Then, the network weights for the specific architecture are optimized over the data to minimize a given loss function; these stages are described next.

It is important to note that not all machine learning architectures are neural networks, although they are one of the most powerful and expressive modern architectures, powered by increasingly big data and high performance computing. Before the success of deep convolutional networks on the ImageNet dataset in 2012, neural networks were not even mentioned in the list of top ten machine learning algorithms [543]. Random forests [69] and support vector machines [448] are two other leading architectures for supervised learning. Bayesian methods are also often used, especially for dynamical systems [56]. Genetic programming has also been widely used to learn human-interpretable, yet flexible representations of data for modeling [61, 447, 119, 118] and control [560]. In addition, standard linear regression and generalized linear regression are still see widespread use for modeling time-series data, especially in fluids. The dynamic mode decomposition (DMD) [443, 483] employs linear regression with a low-rank constraint in the optimization to find dominant spatiotemporal coherent structures that evolve linearly in time. The sparse identification of nonlinear dynamics (SINDy) [81] algorithm employs generalized linea regression, with either a sparsity promoting loss function [495] or a sparse optimization algorithm [81, 557], to identify a differential equation model with as few model terms as are necessary to fit the data.

Embedding physics. Choosing a machine learning architecture with which to model the training data is one of the most intriguing opportunities for embedding physical knowledge into the learning process. Among the simplest choices are convolutional networks for translationally invariant systems, and recurrent networks, such as long-short-time memory (LSTM) networks [520] or reservoir computing [385, 384], for systems that evolve in time. LSTMs have recently been used to predict aeroelastic responses across a range of Mach numbers [308]. More generally, equivariant networks seek to encode various symmetries by construction, which should improve accuracy and reduce data requirements for physical systems [494, 350, 527, 29]. Autoencoder networks enforce the physical notion that there should be low-dimensional structure, even for high-dimensional data, by imposing an information bottleneck, given by a constriction of the number of nodes in one or more layers of the network. Such networks uncover nonlinear manifolds where the data is compactly represented, generalizing the linear dimensionality reduction obtained by PCA and POD. It is also possible to embed physics more directly into the architecture, for example by incorporating Hamiltonian [208, 171] or Lagrangian [117, 558] structure. In the field of turbulence modeling, Ling et al. [314] designed a custom neural network networks layer that enforced Galilean invariance in the Reynolds stress tensors that they were modeling. Graph neural networks have also shown the ability to learn generalizable physics in a range of challenging domains [28, 119, 435]. Deep operator networks [323] are able to learn continuous operators, such as governing partial differential equations, from relatively limited training data.

The Loss Function. The loss function is how we quantify how well the model is performing, often on a variety of tasks. For example, the ℓ_2 error between the model output and the true output, averaged over the input data, is a common term in the loss function. In addition, other terms may be added to regularize the optimization (e.g., the ℓ_1 or ℓ_2 norm of the parameters $\boldsymbol{\theta}$ to promote parsimony and prevent overfitting). Thus, the loss function typically balances multiple competing objectives, such as model performance and model complexity. The loss function may also incorporate terms used to promote a specific behavior across different sub-networks in a neural network architecture. Importantly, the loss function will provide valuable information used to approximate the gradients required to optimize the parameters.

Embedding physics. Most of the physics-informed architectures described above involve custom loss functions to promote the efficient training of accurate models. It is also possible to incorporate physical priors, such as sparsity, by adding ℓ_1 or ℓ_0 regularizing loss terms on the parameters in $\boldsymbol{\theta}$. In fact, parsimony has been a central theme in physical modeling for a century, where it is believed that balancing model complexity with descriptive capability is essential in developing models that generalize. The sparse identification of nonlinear dynamics algorithm [81] learns dynamical systems models with as few terms from a library of candidate terms as are needed to describe the training data. Sparse nonlinear modeling has been used extensively in fluid mechanics for reduced-order modeling [320, 321, 319, 136, 137] and for turbulence closure modeling [34, 35, 442]. There are several formulations involving different loss terms and optimization algorithms that promote additional physical notions, such as stability [268].

Physics-informed neural networks (PINNs) [403, 379, 545, 332, 271], are the most common approach to embedding physics via the loss function, promoting solutions that satisfy a governing physical law. For example adding the divergence of a flow field as a loss term promotes solutions that are incompressible [526]. These will be discussed in detail later in this section.

The Optimization Algorithm. If data is the fuel for machine learning, then optimization is the engine. Ultimately, machine learning models are trained using optimization algorithms to find the parameters $\boldsymbol{\theta}$ that best fit the training data. Typically, these optimization problems are both high-dimensional and nonconvex, leading to extremely challenging optimization landscapes with many local minima. While there are powerful and generic techniques for convex optimization problems [206, 66], there are few generic guarantees for convergence or global optimality in nonconvex optimization. Modern deep neural networks have particularly high-dimensional parameters $\boldsymbol{\theta}$ and require large training data sets, which necessitate stochastic gradien descent algorithms. In a sense, the optimization algorithm is the engine powering machine learning, and as such, it is often abstracted from the decision process[6]. However, developing advanced optimization algorithms is the focus of intense research efforts. It is also often necessary to explicitly consider the optimization algorithm when designing a new architecture or incorporating a novel loss term.

Embedding physics. There are several ways that the optimization algorithm may be customized or modified to incorporate prior physical knowledge. One approach is to explicitly add constrains to the optimization, for example that certain coefficients must be non-negative, or that other coefficients must satisfy a specified algebraic relationship with each other. Loiseau [320] showed that it is possible to enforce energy conservation for incompressible fluid flows directly by imposing skew-symmetry constrains on the quadratic terms of a sparse generalized linear regression (i.e. SINDy) model. These constraints manifest as equality constraints on the sparse coefficients $\boldsymbol{\theta}$ of the SINDy model. Because the standard SINDy optimization procedure is based on a sequentially thresholded least-squares procedure, it is possible to enforce these equality constraints at every stage of the regression, using the Karush–Kuhn–Tucker (KKT) conditions. The SR3 optimization package [557], discussed briefly in Section 5.1, was developed to generalize and extend these constrained optimization problems to more challenging constraints, and to more generic optimization problems. It is also possible to enforce energy stability constraints [268] in a similar way.

Another approach involves employing custom optimization algorithms required to minimize the physically motivated loss functions above, which are often nonconvex. In this way, the line between loss function and optimization algorithm are often blurred, as they are typically tightly coupled. For example, promoting sparsity with the ℓ_0 norm is nonconvex, and several relaxed optimization formulations have been developed to approximately solve this problem. The sparse relaxed regularized regression (SR3) optimization framework [557] has been developed specifically to handle challenging nonconvex loss terms that arise in physically motivated problems.

Lagrangian and Hamiltonian Neural Networks

A Hamiltonian neural network (HNN) [208] is a powerful approach in machine learning that uses structure and the loss function to promote physical solutions. An HNN is essentially a neural ODE (NODE) [101], where we learn the Hamiltonian function $\mathcal{H}$ and its flow.

Recall that the Hamiltonian is a conserved quantity, often representing the total energy of a system; we have seen the Hamiltonian in Section 6.6 as the dual of the Lagrangian. It is typically written in terms of the configuration variables $\mathbf{q}$ and their associated momenta $\mathbf{p}$. Given a Hamiltonian, the equations of motion follow the skew-symmetric dynamics given

[6] In reality, combustion engineers most certainly optimize the engine depending on the fuel source, power train, and use cases, which makes one wonder about optimization in machine learning.

by

$$\frac{\mathrm{d}\mathbf{q}}{\mathrm{d}t} = \frac{\partial \mathcal{H}}{\partial \dot{\mathbf{p}}} \tag{10.40a}$$

$$\frac{\mathrm{d}\mathbf{p}}{\mathrm{d}t} = -\frac{\partial \mathcal{H}}{\partial \dot{\mathbf{q}}}. \tag{10.40b}$$

There is a rich history of enforcing this skew-symmetric structure, for example in symplectic integrators [546], which preserve the total energy of a system. Symplectic integrators are useful for simulating chaotic systems, such as planetary dynamics in the solar system, so that the simulations are stable over long time horizons.

It is possible to promote this symplectic structure in a neural ODE model by adding the following terms to the loss function:

$$L = \left\| \frac{\mathrm{d}\mathbf{q}}{\mathrm{d}t} - \frac{\partial \mathcal{H}}{\partial \dot{\mathbf{p}}} \right\|_2^2 + \left\| \frac{\mathrm{d}\mathbf{p}}{\mathrm{d}t} + \frac{\partial \mathcal{H}}{\partial \dot{\mathbf{q}}} \right\|_2^2. \tag{10.41}$$

Note that these partial derivatives are easy to compute using autograd, similar to how the PINNs loss functions are computed.

The original HNN paper only tested the method on relatively simple example problems, such as an ideal spring-mass system and a simple pendulum. In all of these test cases, the HNN preserved total energy better than a traditional NN, similar to how a symplectic integrator improves the performance over a traditional numerical integrator.

The following code, taken from `https://github.com/greydanus/hamiltonian-nn`, develops the HNN:

CodeSnip 10.o

```
def time_derivative(self, x, t=None, separate_fields=False) :
    '''NEURAL ODE-STLE VECTOR FIELD'''
    if self. baseline:
        return self.differentiable_model (x)

    '' 'NEURAL HAMILTONIAN-STLE VECTOR FIELD'''
    F1, F2 = self. forward(x) # traditional forward pass

    conservative_field = torch.zeros_like(x) # start out with
        both components set to 0
    solenoidal_field = torch.zeros_like(x)

    if self. field_type != 'solenoidal':
        dF1 = torch-autograd grad(F1.sum(), x, create_graph=True
            ) [0] # gradients for conservative field
        conservative_field = dF1 @ torch.eye(*self.M.shape)

    if self.field_type != 'conservative':
        dF2 = torch.autograd.grad(F2.sum(), x, create_graph=True
            ) [0] # gradients for solenoidal field
        solenoidal_field = dF2 @ self.M.t()

    if separate_fields:
        return [conservative_field, solenoidal_fieldl

    return conservative_field + solenoidal_field
```

The following year, Cranmer et al. [117] introduced the Lagrangian neural network (LNN), which is a powerful extension of the HNN framework. Just as an HNN is a neural network analog of a symplectic integrator, the LNN is analogous to a variational integrator [337], which enforces the structure of the Euler–Lagrange equations to improve physical simulations. Recall that we have already seen these equations in Section 6.6 in the context of infinite-dimensional optimization. Cranmer et al. tested their method on much harder chaotic problems, such as the double pendulum. Importantly, LNNs don't need special conjugate coordinates, making them more flexible.

The idea is similar, to learn the Lagrangian function $\mathcal{L}(\mathbf{q}, \dot{\mathbf{q}})$ that satisfies the Euler–Lagrange equations:

$$\frac{\mathrm{d}}{\mathrm{d}t}\frac{\partial \mathcal{L}}{\partial \dot{\mathbf{q}}} = \frac{\partial \mathcal{L}}{\partial \mathbf{q}}. \tag{10.42}$$

This is added to the loss function via the following

$$\left\| \ddot{\mathbf{q}} - \left(\frac{\partial^2 \mathcal{L}}{\partial \dot{\mathbf{q}}^2}\right)^{-1} \left(\frac{\partial \mathcal{L}}{\partial \mathbf{q}} - \dot{\mathbf{q}}\frac{\partial^2 \mathcal{L}}{\partial \mathbf{q}\partial \dot{\mathbf{q}}}\right) \right\|_2^2. \tag{10.43}$$

Again, these terms are computed using automatic differentiation, specifically in JAX.

This loss term is derived from the Euler–Lagrange equations:

$$\frac{\mathrm{d}}{\mathrm{d}t}\frac{\partial \mathcal{L}}{\partial \dot{\mathbf{q}}} = \frac{\partial \mathcal{L}}{\partial \mathbf{q}} \quad \Longrightarrow \quad \frac{\partial^2 \mathcal{L}}{\partial \dot{\mathbf{q}}\partial \mathbf{q}}\frac{\mathrm{d}}{\mathrm{d}t}\mathbf{q} + \frac{\partial^2 \mathcal{L}}{\partial \dot{\mathbf{q}}^2}\frac{\mathrm{d}}{\mathrm{d}t}\dot{\mathbf{q}} = \frac{\partial \mathcal{L}}{\partial \mathbf{q}} \tag{10.44a}$$

$$\Longrightarrow \quad \frac{\partial^2 \mathcal{L}}{\partial \dot{\mathbf{q}}\partial \mathbf{q}}\dot{\mathbf{q}} + \frac{\partial^2 \mathcal{L}}{\partial \dot{\mathbf{q}}^2}\ddot{\mathbf{q}} = \frac{\partial \mathcal{L}}{\partial \mathbf{q}} \tag{10.44b}$$

which finally yields

$$\ddot{\mathbf{q}} = \left(\frac{\partial^2 \mathcal{L}}{\partial \dot{\mathbf{q}}^2}\right)^{-1} \left(\frac{\partial \mathcal{L}}{\partial \mathbf{q}} - \dot{\mathbf{q}}\frac{\partial^2 \mathcal{L}}{\partial \mathbf{q}\partial \dot{\mathbf{q}}}\right). \tag{10.45}$$

The matrix $\partial^2 \mathcal{L}/\partial \dot{\mathbf{q}}^2$ is given in component form as

$$\left(\frac{\partial^2 \mathcal{L}}{\partial \dot{\mathbf{q}}^2}\right)_{ij} = \frac{\partial^2 \mathcal{L}}{\partial q_j \partial \dot{q}_i}. \tag{10.46}$$

The following code, from `https://github.com/MilesCranmer/lagrangian_nns/`, builds an LNN:

CodeSnip 10.p
```
# lagrangian equation of motion
def lagrangian_eom (lagrangian, state, t=None) :
    q, q_t = jnp.split(state, 2)
    #Note: the following line assumes q is an angle. Delete it
        for problems other than double pendulum.
    q = q % (2*jnp.pi)
    q_tt = (jnp.linalg.pinv(jax.hessian(lagrangian,1)(q,q_t))
        @(jax-grad(lagrangian, 0)(q,q_t)
        - jax.jacobian(jax.jacobian(lagrangian, 1),0)(q,q_t)@q_t
          ))
    dt = 1e-1
    return dt*jnp.concatenate([q_t, q_tt])
```

```
def raw_lagrangian_eom (lagrangian, state, t=None) :
    q, q_t = jnp.split(state, 2)
    q = q % (2*jnp-pi)
    q_tt = (jnp.linalg.pinv(jax.hessian (lagrangian,1) (q,q_t))
        @ (jax.grad(lagrangian,0) (q, q_t)
        -jax.jacobian(jax.jacobian(lagrangian,1),0)(q,q_t)@q_t))
    return jp.concatenate(lq_t, q_ttl)
```

The Lagrangian is represented with a feedforward neural network as:

CodeSnip 10.q

```
from jax.experimental import stax

def mlp(args):
    return stax.serial(
        stax.Dense(args.hidden_dim),
        stax.Softplus,
        stax.Dense(args.hidden_dim),
        stax.Softplus,
        stax.Dense(args.output_dim),
    )
```

The authors use 500 hidden units, a batch size of 32, and a decaying learning rate, initialized at $\gamma = 10^{-3}$.

Physics-Informed Neural Networks

Physics-informed neural networks (PINNs) [403, 404] are one of the most straightforward ways to incorporate physics into machine learning. For systems with known governing equations, in the form of partial differential equations, it is possible to add this governing PDE as a loss function in the neural network training. Generally speaking, PDEs represent conservation laws in physics, such as conservation of mass, momentum, and energy. Thus, adding these PDE terms to the loss function penalizes models that violate these basic conservation laws. However, it is important to note that adding terms to the loss function only *promotes* the solution being physical, but typically the equations will not be satisfied exactly, as other terms in the loss function will compete with the PINNs loss terms. Alternative approaches involve constrained optimization, whereby the solution to the optimization problem is guaranteed to be on a submanifold where the physics is satisfied.

A PINN models field variables $\mathbf{u}(\mathbf{x}, t)$ that vary in space $\mathbf{x}$ and time t according to

$$\mathcal{N}(\mathbf{u}(\mathbf{x}, t)) = 0 \tag{10.47}$$

where $\mathcal{N}(\mathbf{u}) = 0$ is the partial differential equation. For example, in a three-dimensional fluid flow, the variable $\mathbf{u} = [u(x, y, z, t), v(x, y, z, t), w(x, y, z, t)]$, where (u, v, w) are the flow velocities in the x, y, and z spatial directions. For an incompressible fluid flow, the equation for mass conservation is:

$$\nabla \cdot \mathbf{u} = 0. \tag{10.48}$$

Thus, it is possible to add a loss term

$$L_{\text{div}} = \|\nabla \cdot \mathbf{u}\|_2^2 \tag{10.49}$$

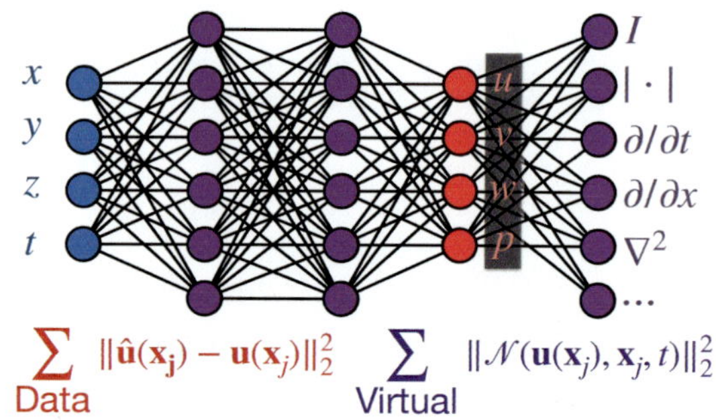

Figure 10.24 Schematic of a physics-informed neural network (PINN) [403], in which a second virtual loss function is added to the standard MSE loss function in order to enforce the satisfaction of the governing equations.

to promote solutions being incompressible [526]. Similarly, the equations for momentum conservation (i.e., Newton's second law) are:

$$\frac{\partial \mathbf{u}}{\partial t} + (\mathbf{u} \cdot \nabla)\mathbf{u} = -\nabla p + \frac{1}{\text{Re}}\nabla^2 \mathbf{u} \tag{10.50}$$

where p is the pressure and Re is the Reynolds number. Taken together, these comprise the Navier–Stokes equations. These may also be added as PINN terms in the loss function.

There are several key features of PINNs that make them convenient for scientific ML. First, the terms in the loss function involve partial derivatives of the output variables $\mathbf{u}$ with respect to the input variables $\mathbf{x}$ and t, which may be computed easily using automatic differentiation. Second, the PINNs loss terms may be evaluated on virtual, or *synthetic* test points in the space-time domain that are not in the training data. Even for spacial locations that are far away from training data, we still know that the PDE should be satisfied. Thus, it is possible to train PINNs with relatively sparse data, augmenting this *real* data with virtual data. The PINNs architecture is shown schematically in Fig. 10.24.

PINNs are particularly good at filling in partially, or sparsely measured field data. Historically this was accomplished using four-dimensional variational assimilation (4D-Var) techniques [115], which involved computationally expensive and human intensive optimization algorithms. Now, these constrained optimization problems are handled in a neural network training environment using backpropagation and automatic differentiation.

In addition to data assimilation, PINNs are also useful for accelerating computations of complex systems over complex domains, where the underlying physics is well-characterized by a partial differential equation. Modeling fluid flows in complex domains is a good example. PINNs are often powerful interpolation engines, and can handle the curse of dimensionality by exploiting low-dimensional patterns in these systems. PINNs also have fast inference time, so they may be used as a flexible surrogate model for inverse problems, design prototyping, and other applications where machine precision and exact enforcement of the physics is not necessary.

However, it is important to note that there are several well-understood drawbacks to PINNs. These algorithms often struggle with convergence, especially as the PINNs' loss may introduce a highly nonconvex loss function landscape. There are no guarantees of

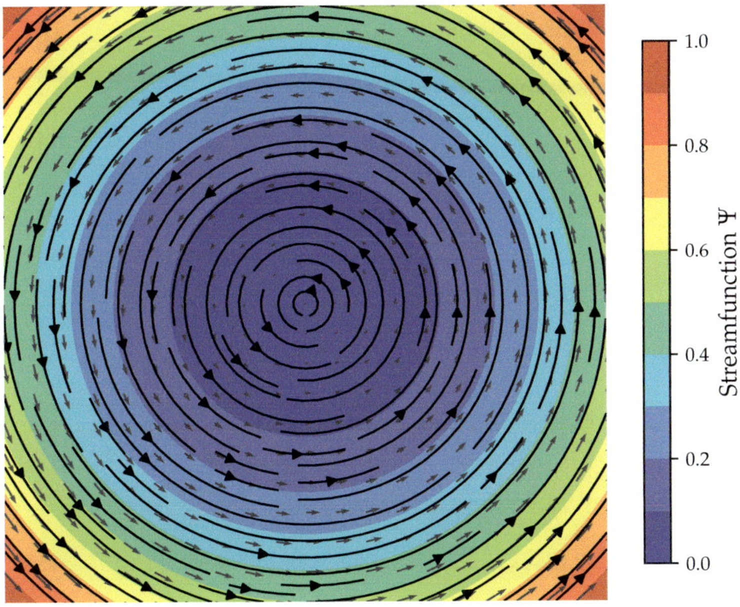

Figure 10.25 Vector field for the linear vortex in Eq. (10.51).

convergence or accuracy, unlike well-established scientific computing methods, so instead of replacing these techniques, PINNs instead complement them with another model with different fidelity and computational scaling properties. In addition, PINNs struggle with translation, advection, and discontinuities in the field variables, all of which are common in physics. Krishnapriyan et al. [288] have characterized many of the common failure modes of PINNs, with some helpful suggestions for how to fix them. Rohrhofer et al. [425] also provide an in-depth analysis on how to tune the relative loss functions in a PINN from a multi-objective optimization perspective.

Numerical Examples (PINNs)

We will explore PINNs on two example vector fields that have the incompressible property (i.e., $\nabla \cdot \mathbf{u} = 0$). We will first consider a linear flow field, given by

$$\mathbf{u}(x, y, t) = \begin{bmatrix} u(x, y, t) \\ v(x, y, t) \end{bmatrix} = \begin{bmatrix} -y \\ x \end{bmatrix}. \tag{10.51}$$

This is the vector field obtained by a simple quadratic potential $\phi(x, y) = x^2 + y^2$, where the vector field is obtained as

$$\mathbf{u}(x, y, t) = \begin{bmatrix} u(x, y, t) \\ v(x, y, t) \end{bmatrix} = \begin{bmatrix} -\partial\phi/\partial y \\ \partial\phi/\partial x \end{bmatrix}. \tag{10.52}$$

The vector field is shown in Fig. 10.25.

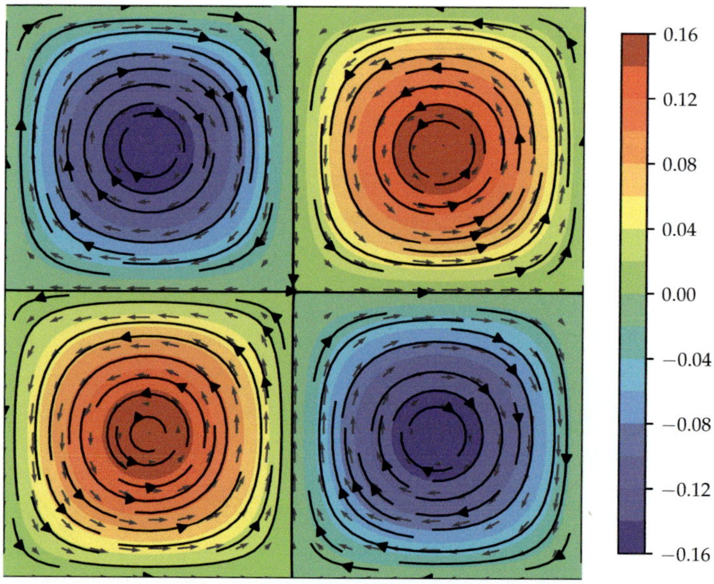

Figure 10.26 Vector field for the Taylor–Green vortex in Eq. (10.53).

The second vector field is the Taylor–Green vortex. Although this flow is often defined on a domain of $[0, 2\pi] \times [0, 2\pi]$, we will scale the x and y axes to make the domain $[0, 1] \times [0, 1]$. The vector field is given by

$$\mathbf{u}(x, y, t) = \begin{bmatrix} u(x, y, t) \\ v(x, y, t) \end{bmatrix} = \begin{bmatrix} \sin(2\pi x)\cos(2\pi y) \\ -\cos(2\pi x)\sin(2\pi y) \end{bmatrix}. \tag{10.53}$$

This is the flow field obtained from the stream function

$$\phi(x, y) = \frac{1}{2\pi}\sin(2\pi x)\sin(2\pi y). \tag{10.54}$$

The vector field is shown in Fig. 10.26.

We will design a simple neural network with 3 hidden layers and 32 units per layer. The network has 2 inputs, x and y, and 2 outputs, u and v.

CodeSnip 10.r

```
class PINNIncompressible(nn.Module):
def __init__(self, hidden_units=32, hidden_layers=3):
    super().__init__()
    layers = []
    in_dim = 2
    out_dim = 2
    # input -> hidden
    layers.append(nn.Linear(in_dim, hidden_units))
    layers.append(nn.Tanh())
    # hidden -> hidden
    for _ in range(hidden_layers-1):
        layers.append(nn.Linear(hidden_units, hidden_units))
        layers.append(nn.Tanh())
    # last hidden -> output (u, v)
```

```
        layers.append(nn.Linear(hidden_units, out_dim))
        self.net = nn.Sequential(*layers)

    def forward(self, x, y):
        # x,y in shape (N,)
        inp = torch.stack((x,y), dim=1)  # shape (N,2)
        out = self.net(inp)              # shape (N,2)
        return out[:,0], out[:,1]        # separate u,v
```

The PDE loss is the divergence term.

CodeSnip 10.s

```
# PDE loss:  div(u,v) = d/dx(u) + d/dy(v) = 0
def pde_incompressibility(model, x, y):
    # x,y:  (N,)  as torch tensors.
    # Return PDE loss: mean(  (ux + vy)^2  ).
    x.requires_grad = True
    y.requires_grad = True
    u, v = model(x,y)
    # autograd to compute partial_x u and partial_y v
    grad_u = torch.autograd.grad(u, x,
                grad_outputs=torch.ones_like(u),
                create_graph=True)[0]
    grad_v = torch.autograd.grad(v, y,
                grad_outputs=torch.ones_like(v),
                create_graph=True)[0]
    pde_val = grad_u + grad_v
    return torch.mean(pde_val**2)
```

We now test the PINN output for several different sampling strategies, comparing with a simple linea interpolation scheme as a baseline. Figure 10.27 shows the PINN performance on the linear vector field using various sampling strategies, including a coarse uniform sampling grid, random sampling, and sampling only in the left half plane. In this case, the linear interpolation is perfect, and has zero error, because the vector field is linear. For the other two sampling strategies, the limits of PINNs for extrapolation are clearly seen, as the prediction error rapidly grows away from the sampling points. This highlights the similarities between PINN and interpolation, as PINNs may be thought of as a more sophisticated neural-network interpolator.

We now explore the same comparisons, but for the nonlinear Taylor–Green vector field. Figure 10.28 shows the performance of linear interpolation versus a PINN for the same three sampling strategies. For uniform sampling, the PINN error is significantly smaller than the linear interpolation error, as linear interpolation struggles with the nonlinear vector field between sample points. However, there is still error associated with the PINN prediction in some regions of space. For the other two sampling strategies, PINNs struggle to extrapolate outside of the training data, although they are better than linear interpolation. In the homework, we will modify this example to study the double gyre vector field [456].

Physics-Constrained DMD and SINDy Models

The dynamic mode decomposition and sparse identification of nonlinear dynamics algorithms are both based on linear regression, so it is particularly straightforward to include constraints via the KKT conditions [320].

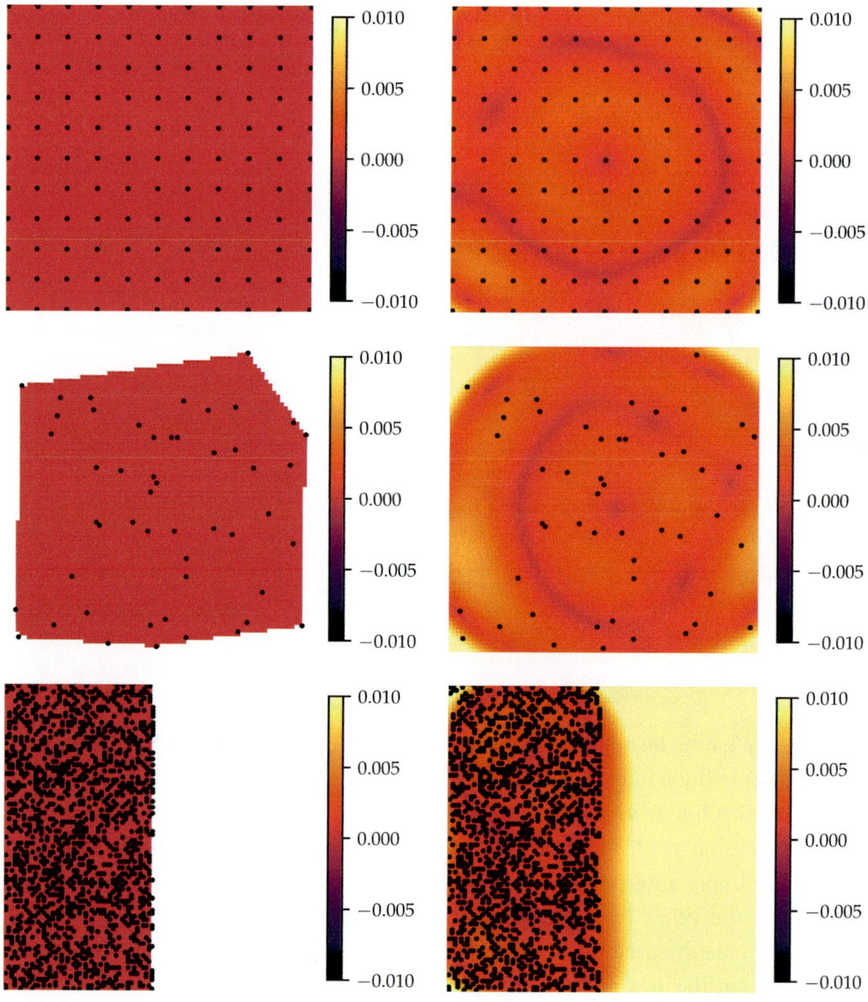

Figure 10.27 Linear interpolation (left) versus PINN prediction (right) for linear vector field with uniform sampling points (top), random sampling (middle), and sampling points only in the left half plane (bottom).

Physics-Informed DMD

The physics-informed DMD has several approaches to enforce known or partially known physics, such as symmetries, in the DMD regression [20]. The physics is either enforced as constraints in the optimization, such as manifold constraints, or promoted by adding terms to the loss function. In the case of manifold-constrained DMD, the optimization problem becomes a modified Procrustes problem:

$$\min_{\mathbf{A}} \|\mathbf{X}' - \mathbf{A}\mathbf{X}\|_F \quad \text{subject to} \quad \mathbf{C}(\mathbf{A}) = 0, \tag{10.55}$$

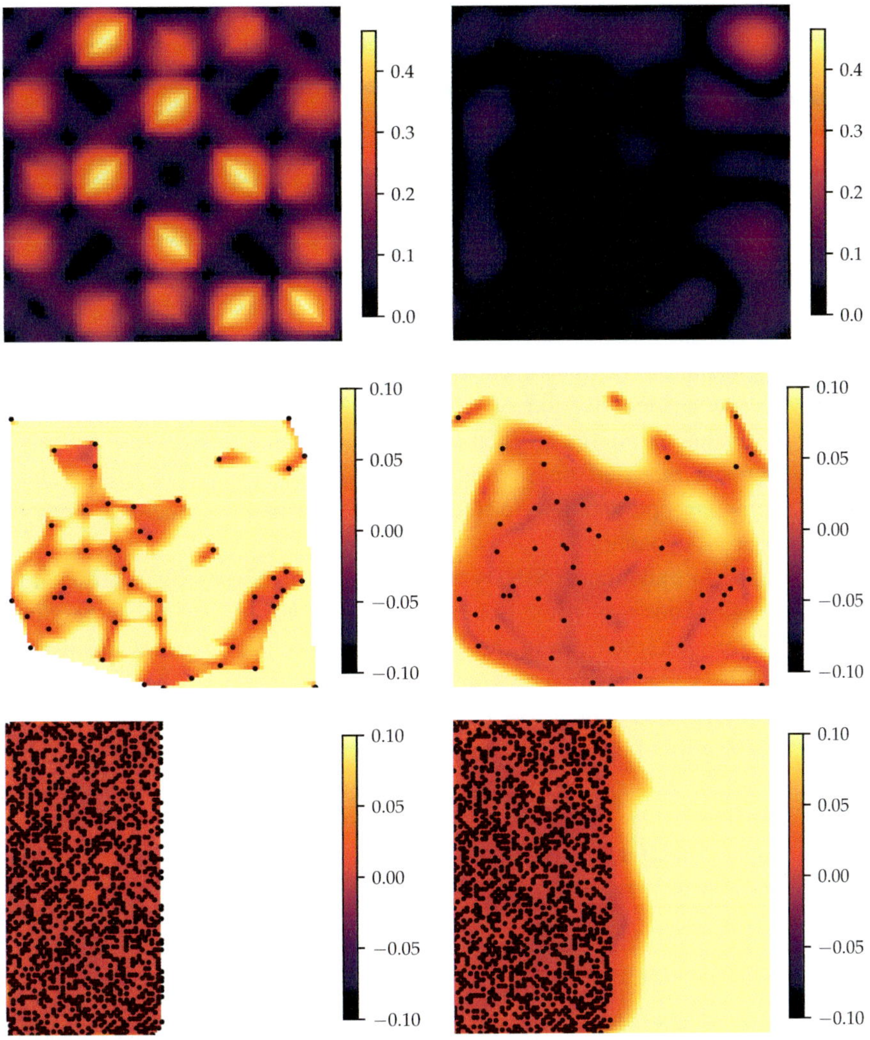

Figure 10.28 Linear interpolation (left) versus PINN prediction (right) for Taylor–Green vector field with uniform sampling points (top), random sampling (middle), and sampling points only in the left half plane (bottom).

where $\mathbf{C}(\mathbf{A})$ represents a set of constraints on $\mathbf{A}$ that encode physical properties such as conservation laws, symmetries, or invariants. Some examples of the constraints are:

Conservative: The original Procrustes problem involved searching for unitary, conservative $\mathbf{A}$ satisfying $\mathbf{A}^*\mathbf{A} = \mathbf{I}$, so $\mathbf{C}(\mathbf{A}) = \mathbf{A}^*\mathbf{A} - \mathbf{I}$.

Translationally shift-equivariant: Traveling wave physics, such as the wave equation or the Navier–Stokes equations are translationally invariant, even though the basic DMD solution is typically not. These systems require circulant or block-circulant $\mathbf{A}$ matrices.

Self-adjoint (symmetric): Solutions to the Schrödinger equation are self-adjoint, so that $\mathbf{A}^* = \mathbf{A}$, restricting our solutions to Hermitian matrices. The constraint is then $\mathbf{C}(\mathbf{A}) = \mathbf{A} - \mathbf{A}^*$.

Skew-symmetry: Similiarly, for skew-symmetric operators, the constraint $\mathbf{C}(\mathbf{A}) = \mathbf{A} + \mathbf{A}^*$ enforces this structural property.

By enforcing these constraints, piDMD ensures that the resulting operator **A** not only minimizes the data reconstruction error but also adheres to physical principles, leading to models that are more interpretable and robust. See Panel 10.3 for more details.

Constrained Sparse Nonlinear Models

In Chapter 8 we discussed system identification as an inverse problem. Specifically, the sparse identification of nonlinear dynamics (SINDy) algorithm was used to learn parsimonious nonlinear differential equation models that best fit time series data. This approach is based on solving a generalized linear regression problem with sparsity-promoting optimization, to find the fewest terms in a candidate library that describe the observed dynamics.

The formulation of SINDy as a sparse regression problem makes it possible to embed partially known physics. Loiseau [320] recognized that it was possible to constrain the SINDy regression, using the KKT constrained least-squares discussed in Chapter 4, to enforce known symmetries and relationships between the coefficients $\boldsymbol{\Xi}$. Specifically, incompressible fluid flows conserve energy through a skew-symmetry in the quadratic nonlinearity of the Navier–Stokes equations (e.g., the convective term). In a reduced-order model for a fluid flow, in terms of the amplitudes of orthogonal mode coefficients, this skew-symmetry is still present in the coefficients of the quadratic terms in the resulting ordinary differential equation.

The SINDy regression from Eq. (8.39) now becomes:

$$\begin{aligned} &\min \ \|\dot{\mathbf{X}} - \boldsymbol{\Theta}(\mathbf{X})\boldsymbol{\Xi}\| + \lambda\|\boldsymbol{\Xi}\|_0 \\ &\text{s.t.} \ \mathbf{C}\boldsymbol{\xi} = \mathbf{d} \end{aligned} \tag{10.56}$$

where $\boldsymbol{\xi}$ is the vectorized form of $\boldsymbol{\Xi}$ and **C** and **D** are matrices related to the constraint equations on $\boldsymbol{\xi}$. Because SINDy is typically solved using a sequentially thresholded least-squares algorithm, it is straightforward to incorporate these constraints directly into each of these least-squares steps using the KKT formulation discussed in Chapters 4 and 6. Specifically, each of the constrained least-squares steps may be reformulated as an unconstrained problem using Lagrange multipliers:

$$\min_{\boldsymbol{\xi},\mathbf{z}} \ \|\dot{\mathbf{X}} - \boldsymbol{\Theta}(\mathbf{X})\boldsymbol{\Xi}\| + \mathbf{z}^{\mathrm{T}}(\mathbf{C}\boldsymbol{\xi} - \mathbf{d}) \tag{10.57}$$

where **z** is the vector of Lagrange multipliers. It is then possible to solve for the optimal $\boldsymbol{\xi}$, subject to the constraints $\mathbf{C}\boldsymbol{\xi} = \mathbf{d}$, via the KKT least-square regression:

$$\begin{bmatrix} 2\hat{\boldsymbol{\Theta}}(\mathbf{X})^{\mathrm{T}}\hat{\boldsymbol{\Theta}}(\mathbf{X}) & \mathbf{C}^{\mathrm{T}} \\ \mathbf{C} & \mathbf{0} \end{bmatrix} \begin{bmatrix} \boldsymbol{\xi} \\ \mathbf{z} \end{bmatrix} = \begin{bmatrix} 2\hat{\boldsymbol{\Theta}}(\mathbf{X})^{\mathrm{T}}\dot{\mathbf{X}}(:) \\ \mathbf{d} \end{bmatrix} \tag{10.58}$$

where $\dot{\mathbf{X}}(:)$ is the vectorized form of $\dot{\mathbf{X}}$, similar to how $\boldsymbol{\xi}$ is formed by vectorizing $\boldsymbol{\Xi}$.

Panel 10.3: Physics-Informed DMD and Procrustes Optimization

The PIDMD constrained optimization may be formulated as a Procrustes problem. The Procrustes problem is to find the optimal transformation (e.g., translation, rotation, reflection, and scaling) that best aligns one set of points with another, minimizing the mismatch in the two geometries. The orthogonal Procrustes problem is the most common form, where we seek the best orthogonal transformation (e.g., rotation and reflection) to match two matrices. This involves minimizing the Frobenius norm of their difference. This is also the basis of many numerical methods, including in canonical correlation analysis (CCA). However, there are many other Procrustes problems, based on other constraints on the admissible transformations.

Another important example of Procrustes analysis is in registering point cloud data to a reference model, or CAD drawing. For example, during aircraft assembly, point cloud scans of the various parts are collected, and these then must be transformed and registered with each other and against a reference model.

This problem is named after the mythological Procrustes, a gruesome smith who was said to invite travelers to rest the night in his fortress, only to force them to sleep on an iron bed. If the travelers were too short or too tall to fit the bed – and alas, none ever did – they would be made to fit, either by stretching and breaking their limbs or by cutting them off. Procrustes was eventually killed by the hero Theseus, being forced to fit himself to his own bed.

Peter Baddoo made the connection between the constrained DMD problem and Procrustes optimization, where he outlined several major classes of physical constraints that can be included in the DMD framework using standard tools from optimization. In the examples below, the self-adjoint formulation was solved by Higham [227] and the conservative case was solved by Schönemann [449]. All other cases were new solutions by Baddoo.

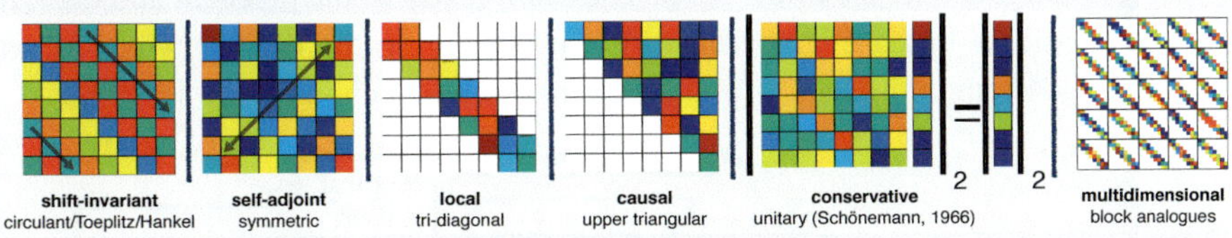

Figure 10.29: Examples of constrained DMD A matrices (Baddoo et al. [20]).

The specific constraint equations $\mathbf{C}\xi = \mathbf{d}$ may be derived from physical first principles. Broad classes of physical systems are governed by conservation laws, and these conservation laws often induce symmetries on the coefficients in ξ. When a given symmetry is known, it may be enforced on the coefficients in ξ. For incompressible fluid flows, we have a constraint on the quadratic nonlinear term

$$\mathbf{x} \cdot \mathcal{N}(\mathbf{x})\mathbf{x} = 0 \tag{10.59}$$

Panel 10.4: Digital Twins and the Future of Model-Based Engineering

Digital twins are data-enabled surrogate models for complex systems that allow for the accurate and efficient simulation of various scenarios that would be too expensive or difficult to test otherwise. For example, it may be possible to develop a digital twin for a modern commercial aircraft, making it possible to test different wing designs and engine configurations with fewer expensive wind tunnel tests. Digital twins rely on fast and flexible surrogate models that are able to incorporate physics-based models and machine learning models. These models will be synthesized from multi-fidelity data from a diversity of sources (e.g., simulations, wind tunnel testing, flight tests); thus, they also rely on a *digital thread* of the data for the entire process. Importantly, they must have an estimate of their own uncertainty, so that this can be incorporated into downstream forecasting and optimization tasks.

If the digital twin is being used effectively for design optimization, the optimization algorithm will explore new regions of parameter space that are highly performant, but poorly explored. In this case, it is necessary to update the digital twin model with new data. This involves active learning to intelligently sample the system and update the model with the right data, at the right fidelity, based on the uncertainty estimates of the model.

Consider the aerospace industry. Digital twins will enable faster and less expensive design cycles, leveraging data from previous designs more effectively. This will result in improved aircraft design, as the digital twin unifies the multi-objective design optimization landscape, which is currently too large and tightly coupled for any one group to oversee; today, individual components are optimized locally. Further, a digital twin of the factory would improve manufacturability, reliability, and reproducibility. A twin of the global supply chain would improve robustness and cost, and a twin of the operating fleet of aircraft would improve health monitoring and service. Currently these are all separate optimization problems, largely because of the scope and complexity of the problem and data. In the future, model-based engineering (MBE), leveraging the digital twin, will transform the design, optimization, and control of complex engineering systems.

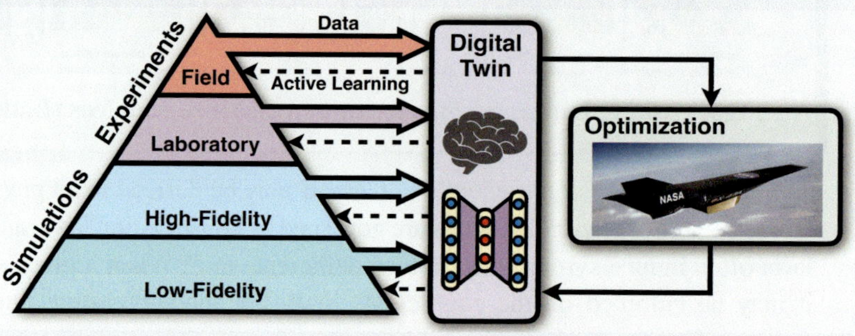

Figure 10.30: Schematic of optimization with a digital twin.

for this term to conserve energy. This may be directly translated into constraints on $\boldsymbol{\xi}$, as described in Loiseau et al. [320]. For a reduced-order state $\mathbf{x}$ with three state variables, for example the amplitudes of three orthogonal spatial modes, then this results in ten constraint equations on the coefficients in $\boldsymbol{\xi}$.

Kaptanoglu et al. [268] went on to derive more general stability conditions, based on the stability results of Schlegel and Noack [441], to promote and enforce stability in quadratic SINDy models. This idea was also used by Guan et al. [214] to derive a symmetry-preserving model for chaotic electroconvection.

More generally, Otto et al. [376] have extended these ideas to a general framework for how to promote, enforce, and discover symmetries in various machine learning algorithms, through the use of regularization and constrained optimization. Specifically, they show how a broad class of continuous Lie-group symmetries may be incorporated into machine learning algorithms, reducing data requirements and improving generalizability and robustness.

The various models described above, whether DMD models, PINNs, or SINDy models, are all approaches for developing surrogate models of real physical systems. Improved surrogate models will have a tremendous impact on improving model-based engineering via the digital twin, as discussed more in Panel 10.4.

Suggested Reading

Texts

(1) Deep Learning, by I. Goodfellow, Y. Bengio, and A. Courville, 2016 [204].

(2) Machine Learning: A Probabilistic Perspective, by K. P. Murphy, 2012 [359].

(3) Pattern Recognition and Machine Learning, by C. M. Bishop, 2006 [55].

(4) Learning with Kernels: Support Vector Machines, Regularization, Optimization, and Beyond, by B. Schölkopf and A. J. Smola, 2002 [448].

Papers and reviews

(1) Automatic differentiation in machine learning: a survey, by A. G. Baydin, B. A. Pearlmutter, A. A. Radul, and J. M. Siskind, *Journal of Machine Learning Research*, 2018 [31].

(2) Learning representations by back-propagating errors, by D. E. Rumelhart, G. E. Hinton, and R. J. Williams, *Nature*, 1986 [431].

(3) Deep learning, by Y. LeCun, Y. Bengio, and G. Hinton, *Nature*, 2015 [303].

(4) Understanding deep convolutional networks, by S. Mallat, *Philosophical Transactions of the Royal Society of London A*, 2016 [327].

(5) Constrained sparse Galerkin regression, by J. C. Loiseau and S. L. Brunton, *Journal of Fluid Mechanics*, 2018 [320].

(6) A unified framework to enforce, discover, and promote symmetry in machine learning, by S. E. Otto, N. Zolman, J. N. Kutz, and S. L. Brunton, *arXiv*, 2023 [376].

(7) Physics-informed neural networks: A deep learning framework for solving forward and inverse problems involving nonlinear partial differential equations, by M. Raissi, P. Perdikaris, and G. E. Karniadakis, *Journal of Computational Physics*, 2019 [403].

(8) Lagrangian neural networks, by M. Cranmer, S. Greydanus, S. Hoyer, P. Battaglia, D. Spergel, and S. Ho, *arXiv:2003.04630*, 2020 [117].

Homework

Exercise 10.1 Write down an example function with the computational graph given in Fig. 10.31. Use forward mode and reverse mode automatic differentiation to compute the derivative at $(x_1, x_2) = (1, 2)$.

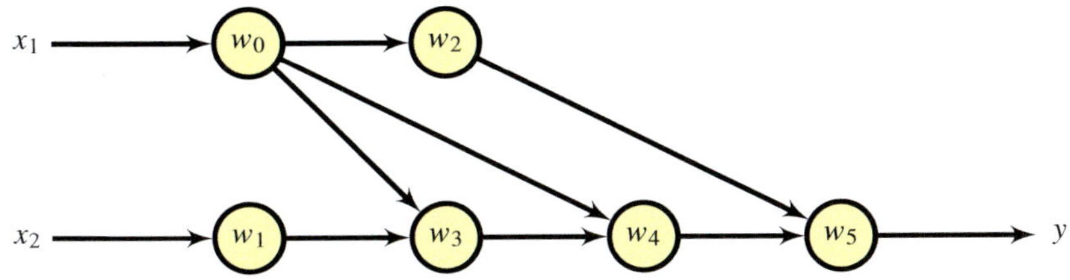

Figure 10.31 Computational graph for MISO function.

Exercise 10.2 Compute the derivative of the function in Eq. (10.6) at the point $x = 1$ using forward mode and reverse mode AD.

Exercise 10.3 Compute the derivative of the function in Eq. (10.15) at the point $x = 2$ using forward mode and reverse mode AD.

Exercise 10.4 Show that the computational graph in Fig. 10.3 can be modified to look more like the computational graph in Fig. 10.4 by making $w_2 = \sin(x)$ and $w_3 = x^2$. Draw this new computational graph.

Exercise 10.5 Compute the derivative of the function in Eq. (10.16) at the point $(x_1, x_2) = (1, 2)$ using reverse mode AD.

Exercise 10.6 Compute the derivative of the function in Eq. (10.17) at the point $x = 2$ using forward mode and reverse mode AD. Discuss which is more efficient in this case.

Exercise 10.7 Compute the derivative of the function in Eq. (10.18) at the point $(x_1, x_2) = (1, 2)$ using forward mode and reverse mode AD.

Exercise 10.8 Write down the computational graph and compute the derivative of the following functions using forward mode and reverse mode AD:

1. $y = \sin(\cos(x))$ at $x = 2$.
2. $y = \sin(x^3 + x)$ at $x = 2$.
3. $y = x^3 + x\sin(x^2)$ at $x = 1$.
4. $y = x_1^2 + x_2^2 + 2x_1x_2$ at $(x_1, x_2) = (2, 1)$.
5. $y = (x_1 - x_2)^2 + (x_1 + x_2)^2$ at $(x_1, x_2) = (1, 2)$.
6. $y = \sin(x_1^2 + x_2^2)$ at $(x_1, x_2) = (1, 1)$.
7. $y_1 = \cos(x) + \sin(x)$
   $y_2 = (x - 1)^2$ at $x = 0$.
8. $y_1 = \cos(x^2) + x^2$
   $y_2 = \sin(x^2) - x^2$ at $x = 2$.

9. $y_1 = x^2 + x^3$
   $y_2 = x - x^3$ at $x = 1$.
10. $y_1 = \exp(x_1^2 + x_2^2)$
    $y_2 = \exp(x_1^2 - x_2^2)$ at $(x_1, x_2) = (1, 0)$.
11. $y_1 = \cos(x_1 x_2) + x_1^2$
    $y_2 = \sin(x_1 x_2) - x_2^2$ at $(x_1, x_2) = (0, 1)$.
12. $y_1 = (x_2 - x_1)^2 + x_1 \cos(x_2)$
    $y_2 = x_1 x_2 - x_2^2$ at $(x_1, x_2) = (2, 0)$.

Exercise 10.9 Computational graphs are non-unique. Write down two different computational graphs for the function $y = x^3 + x \sin(x^2)$ and confirm that they yield the same forward mode and reverse mode derivatives at the point $x = 1$.

Exercise 10.10 Write down two different computational graphs for the function $y = x_1^2 + x_2^2 + 2x_1x_2$, and confirm that they yield the same forward mode and reverse mode derivatives at the point $(x_1, x_2) = (2, 1)$.

Exercise 10.11 Demonstrate that number of inputs and outputs in a function determine the number of forward and reverse mode AD passes required to compute the derivatives. Give some examples.

Exercise 10.12 Consider the computational graph in Fig. 10.7. Assume that $\mathbf{x} \in \mathbb{R}^{1024}$, $\mathbf{w}_1 \in \mathbb{R}^{512}$, $\mathbf{w}_2 \in \mathbb{R}^{256}$, and $\mathbf{y} \in \mathbb{R}^{128}$. Roughly how many multiplications are involved in forward mode AD and in reverse mode AD?

Exercise 10.13 Explain why we do not use forward mode AD to compute the gradients $\nabla_{\theta} \mathbf{f}_{\theta}$ in a neural network. Discuss this with examples, such as the CNN in Fig. 10.17 and the autoencoder in Fig. 10.13.

Exercise 10.14 Write down the formula for the six-layer autoencoder in Fig. 10.32. Write down the formula for the gradient of L with respect to the parameters of each layer, $\boldsymbol{\theta}_j$. Assume that the input and output layers have dimension 1024 and that each layer decreases or increases this size by a factor of 4. Describe the number of calculations required to traverse this computational graph in reverse-mode AD. What about forward-mode AD?

Exercise 10.15 Modify the PyTorch and JAX automatic differentiation codes to compute the derivative of $y = \sin(\exp(x^2))$ from Eq. (10.6) at $x = 2$.

Exercise 10.16 Confirm that the AD derivative $\mathrm{d}f/\mathrm{d}\mathbf{w}$ of the function $f(\mathbf{x}, \mathbf{w}) = (\mathbf{w}^{\mathrm{T}}\mathbf{x})^2$ evaluated at the point $\mathbf{x} = (1, 2)$ with weights $\mathbf{w} = (0.5, -0.5)$, matches the analytic value of the derivative evaluated at this point.

Exercise 10.17 Modify the AD code that computes $\mathrm{d}f/\mathrm{d}\mathbf{w}$ of the function $f(\mathbf{x}, \mathbf{w}) = (\mathbf{w}^{\mathrm{T}}\mathbf{x})^2$ to compute $\mathrm{d}f/\mathrm{d}\mathbf{x}$ instead. Evaluate this derivative at the point $\mathbf{x} = (1, 2)$ with weights $\mathbf{w} = (0.5, -0.5)$. Confirm that this matches the analytic value.

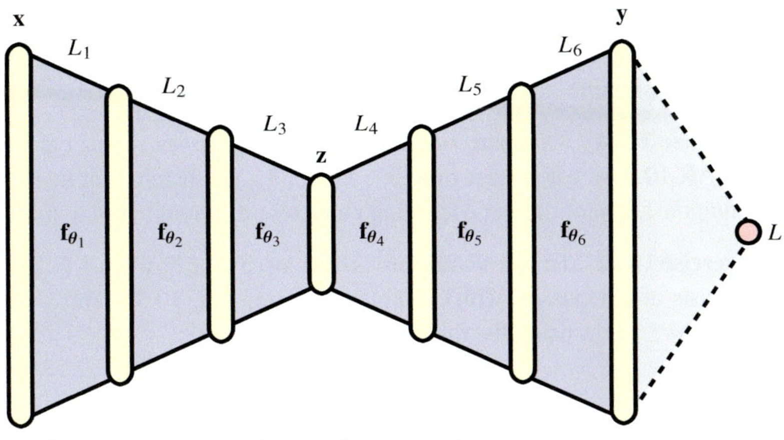

Figure 10.32 Illustration of neural network autoencoder.

Exercise 10.18 Consider the neural network in Fig. 10.33; this is a common architecture, where we expand an input before contracting to the output. For example, in a PINN, the inputs may be (x, y, z, t) and the output may be (u, v, w) evaluated at this point. We would expand the hidden layers to have more capacity to represent this input–output function.

Assuming we want to compute the gradient of $\mathbf{y}$ with respect to $\mathbf{x}$, is there any way to exploit the expanding-contracting architecture using a combination of forward mode and reverse mode AD to improve the computational complexity?

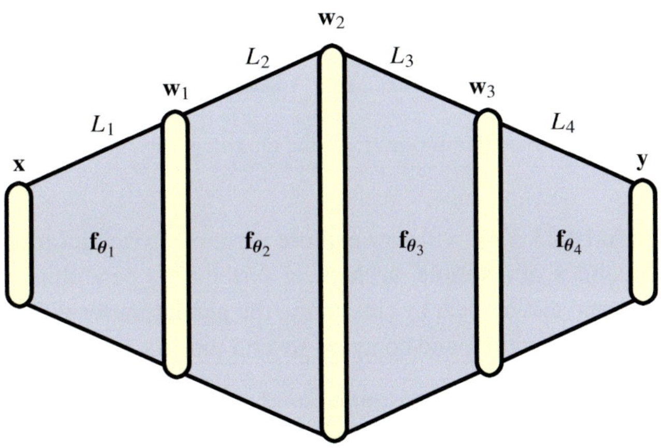

Figure 10.33 Neural network with expansion followed by contraction.

Exercise 10.19 Retrain the MLP and CNN for CIFAR-10, but without normalizing the data by the empirical mean and standard deviation. How do the results change?

Exercise 10.20 We will now examine the MLP for CIFAR-10. Instead of dropout layers, we will use ℓ_1 regularization on the weights in the loss function. Remove the dropout layers and compare the learning curves for different strengths of the ℓ_1 regularization. Examine the histogram of neural network weights for the network with dropout layers and for the various ℓ_1 regularized networks. Comment on the results.

Exercise 10.21 Compute and plot the learning curves versus batch size for the MLP on CIFAR-10. Use batch sizes of 8, 32, 128, and a full batch. Plot them together and discuss. Compute and plot the same learning curves for different learning rates. Discuss the results.

Exercise 10.22 Train a neural network to classify the Fashion-MNIST dataset. This dataset consists of ten classes, with examples shown in Fig. 10.34. This dataset can be loaded in `torchvision` using the following code

CodeSnip 10.t
```
dataset = torchvision.datasets.FashionMNIST(root='./data', train
    =True, download=True, transform=transform)
```

For this simple dataset, an MLP should reach over 90% accuracy. Plot the learning curves.

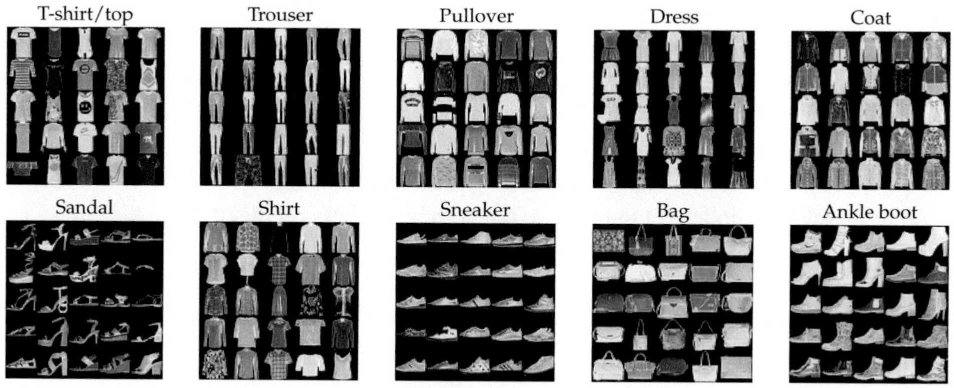

Figure 10.34 Example images from the ten categories of Fashion-MNIST.

Exercise 10.23 We will now explore various physics-informed architectures for learning the dynamics of a simple mechanical system: the pendulum. Generate a large volume of training and testing data by simulating the pendulum for different initial conditions.

You will now train and compare several models, using the following approaches:

1. Neural ODE with no constraints or physical priors;
2. Lagrangian neural network;
3. Hamiltonian neural network;
4. Symbolic regression with SINDy; and
5. Symbolic regression with PySR.

For each of these algorithms, we will compare and contrast according to several challenges:

1. Compare the data requirements and noise robustness.
2. Modify the data to have uneven time steps.

3. Try new initial conditions that were not in the training data, and report on the performance. Specifically, try initial conditions that are outside of the range of training (i.e., train for $\theta \in [-\pi/4, \pi/4]$ and test on trajectories with initial $|\theta| > \pi/4$).
4. Now extend the data set to include pendula with different arm lengths and gravity constants. Which methods can be extended. Discuss and compare.

Exercise 10.24 Repeat the above example with the chaotic double pendulum.

Exercise 10.25 Repeat the above example with rotary pendulum.

Exercise 10.26 Repeat the above example with video data from a lava lamp. Before applying these methods, first compress the data down, using an autoencoder.

Exercise 10.27 (From Loiseau et al. [321]) In this example, we will learn the Lorenz 1963 chaotic dynamical system using constrained SINDy. The governing equations are

$$\dot{x} = \sigma(y - x) \tag{10.60}$$

$$\dot{y} = x(\rho - z) - y \tag{10.61}$$

$$\dot{z} = xy - \beta z. \tag{10.62}$$

Use the parameters $(\sigma, \rho, \beta) = (10, 28, 8/3)$ and initial condition $\mathbf{x}_0 = (x(0), y(0), z(0)) = (0, 1, 20)$.

Formulate a constrained SINDy regression, following Loiseau's approach, to force the coefficients of the terms y and x in the first equation to have equal and opposite coefficients:

$$\xi_2 + \xi_3 = 0.$$

First, write this as an equality constraint $\mathbf{C}\boldsymbol{\xi} = \mathbf{d}$, and implement the KKT constrained sequentially-thresholded least-squares algorithm. Compare the learning rate versus amount of data, as well as the noise robustness, with and without this constraint.

What changes if we add the knowledge that $\xi_3 = 10$? Again, add this as a constraint equation and explore noise robustness and data requirements.

Exercise 10.28 Try to modify the PINN for the linear vortex problem to more effectively perform linear interpolation. Explore different network sizes and activation functions.

Exercise 10.29 Add symmetry conditions to the PINN for the vortex problems to try and improve the learning rate versus amount of data. See if you can improve the generalization of the PINN in the case where you only train on data in the left half of the domain.

Exercise 10.30 Modify the PINN approach for the Taylor–Green vortex problem above to the double gyre flow field. The double gyre is an unsteady potential flow, meaning that it is incompressible ($\nabla \cdot \mathbf{u} = 0$) and irrotational ($\nabla \times \mathbf{u} = 0$). The vector field is derived from the following stream function ϕ:

$$\phi(x, y, t) = A \sin(\pi f(x, t)) \sin(\pi y)$$

where the unsteadiness comes from

$$f(x, t) = a(t)x^2 + b(t)x$$

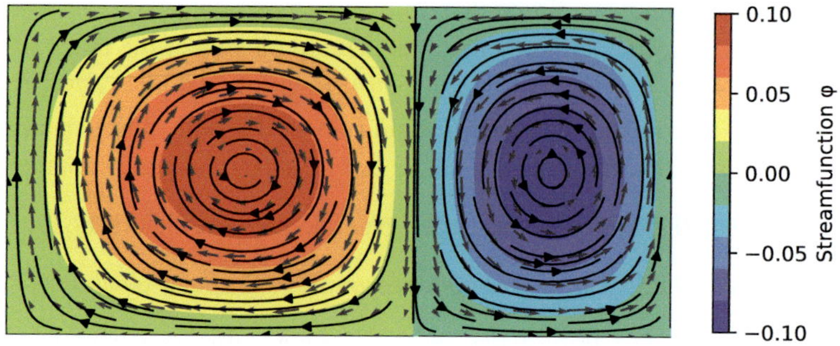

Figure 10.35 Double gyre vector field.

with time-dependent coefficients

$$a(t) = \epsilon \sin(\omega t) \quad \text{and} \quad b(t) = 1 - 2\epsilon \sin(\omega t).$$

The flow is definite on a rectangular domain of size $[0, 2] \times [0, 1]$, with the parameters $A = 0.1, \epsilon = 0.25$, and $\omega = 2\pi/10$, taken from Shadden et al. [456].

The vector field is then given by the following equations

$$\mathbf{u}(x, y, t) = \begin{bmatrix} u(x, y, t) \\ v(x, y, t) \end{bmatrix} = \begin{bmatrix} -\partial\phi/\partial y \\ \partial\phi/\partial x \end{bmatrix} = \begin{bmatrix} -\pi A \sin(\pi f(x, t)) \cos(\pi y) \\ \pi A \cos(\pi f(x, t)) \sin(\pi y) \end{bmatrix}.$$

Glossary

Active constraint – An inequality constraint $g_j(\mathbf{x}) \leq 0$ is active at an optimum solution $\mathbf{x}^*$ if $g_j(\mathbf{x}^*) = 0$. The optimal solution is at the edge of the constraint boundary.

Active Set Methods – This approach identifies a subset of the equality constraints $\mathbf{g}(\mathbf{x}) \leq 0$ to be active, and then solves the system as if they are equality constraints. Different constraints may become active or inactive at different iterations.

Adaptive momentum estimation (ADAM) – The adaptive momentum estimation method combines aspects of momentum methods and RMSProp to provide a robust, iterative algorithm for neural network training. ADAM typically uses stochastic gradient descent to compute the gradients in its updates steps.

Adjoint – For a finite-dimensional linear map (i.e., a matrix $\mathbf{A}$), the adjoint $\mathbf{A}^*$ is given by the complex conjugate transpose of the matrix. In the infinite-dimensional context, the adjoint $\mathcal{A}^*$ of a linear operator $\mathcal{A}$ is defined so that $\langle \mathcal{A} f, g \rangle = \langle f, \mathcal{A}^* g \rangle$, where $\langle \cdot, \cdot \rangle$ is an inner product.

Affine set – Affine sets are a restricted subset of convex sets, where every line passing through any two points of the set lies entirely within the set. Affine sets may be viewed as translated versions of vector subspaces to do not need to contain the origin.

Affine function – An affine function is any function $f(\mathbf{x}) = \mathbf{A}\mathbf{x} + \mathbf{b}$ that maps affine sets to affine sets. These functions include scaling, shear, rotation, translation, and all combinations thereof.

Agent – In reinforcement learning (RL), an agent senses the state $\mathbf{s}$ of its environment and learns to take appropriate actions $\mathbf{a}$ to achieve an optimal future reward $\mathbf{r}$.

Alternating direction method of multipliers – An operator-splitting optimization algorithm that solves complex constrained problems by breaking them into simpler subproblems, each of which is easier to handle. ADMM combines the decomposability of dual ascent with the robustness of the method of multipliers.

Analytic center – The minimizing solution of $-\log(-g(\mathbf{x}))$, which is used as the initialization for interior point methods.

Augmented Lagrangian – A method that modifies the classical Lagrangian by adding a quadratic penalty term to enforce constraints more robustly. This approach bridges dual methods and penalty methods for better numerical stability.

Automatic differentiation (AD) – The process of algorithmic differentiation of a function f, evaluated at a specific value, by applying the chain rule to the computational graph of f. AD is more efficient and accurate than numerical differentiation or analytic differentiation, for complex functions. The two main modes of automatic differentiation are forward-mode AD and reverse-mode AD. Forward-mode AD is used when the output dimension of f is larger than the input dimension, and reverse-mode AD is used when the output dimension is smaller than the input dimension. Reverse-mode AD underpins the backpropagation algorithm for training neural networks.

Backpropagation – Reverse-mode automatic differentiation applied to the computational graph of a neural network is referred to as backpropagation. It is used to compute the gradien of the loss function L with respect to the neural network parameters $\boldsymbol{\theta}$. This gradient is then used, for example with stochastic gradien descent, to optimize the parameters to minimize the loss function.

Barrier methods – Interior point methods that maintain iterates within the feasible region by adding a barrier term to the objective function, which becomes infinite at the constraint boundary. The influence of the barrier is gradually reduced during the optimization.

Bayes' Theorem – Bayes' theorem is a statistical identity that is used to invert conditional probabilities, and is the basis of much of statistical learning from data. Given a parameterized probability distribution $f(X \mid \theta)$ of data X given model parameters θ, it is possible to invert this to solve for the probability of the parameters given the data, $f(\theta \mid X)$, using Bayes' theorem:

$$f(\theta \mid X) = \frac{f(X \mid \theta) f(\theta)}{f(X)}. \tag{10.63}$$

Bayesian inference – The process of iteratively updating a posterior distribution given new evidence (data). The previous posterior becomes the prior for the next iteration.

Bayesian statistics – A statistical framework for balancing prior knowledge about a system with new evidence, in the form of data.

Bellman optimality – A cornerstone of dynamic programming, stating that an optimal multi-step sequence must also be locally optimal in every sub-sequence of steps.

Bias–variance tradeoff – If an ill-conditioned optimization problem or inverse problem initially has high variance, it is possible to reduce this variance with regularization, although this introduces bias. For example, Tikhonov regularization is used to reduce the variance of least-squares regression, but it introduces a bias.

Box constraints – Inequality constraints, $\mathbf{l} \leq \mathbf{x} \leq \mathbf{u}$, where each coordinate of the decision variable, x_j, must lie between a lower and upper bound, l_j and u_j, respectively.

Brachistochrone – One of the earliest variational problems, posed by Johann Bernoulli, where the goal is to find a curve connecting two points so that a particle constrained to this path will have the shortest transit time under the acceleration of gravity.

Closed-loop control – A control architecture where the actuation is informed by sensor data about the output of the system.

Compressed sensing – The process of reconstructing a high-dimensional vector signal from a random under sampling of the data using the fact that the high-dimensional signal is sparse in a known transform basis, such as the Fourier basis.

Compression – The process of reducing the size of a high-dimensional vector or array by approximating it as a sparse vector in a transformed basis. For example, MP3 and JPG compression use the Fourier basis or Wavelet basis to compress audio or image signals.

Computational graph – The representation of a function in a programming language, such as Python or C++, in terms of simple building blocks, such as +, ×, cos, sin, etc. The computational graph is used in automatic differentiation to efficiently and accurately compute the derivative of a function f using the chain rule applied to this graph.

Condition number – The condition number of a matrix $\mathbf{A}$ is the ratio of its largest singular value to its smallest singular value, and is a measure of how ill-conditioned it is to invert this matrix or solve a matrix system of equations $\mathbf{Ax} = \mathbf{b}$. Specifically, if a matrix $\mathbf{A}$ has large condition number, then the solution $\mathbf{x}$ will be highly sensitive to small changes in $\mathbf{b}$.

Conjugate functions – Also called the Fenchel dual, the conjugate $f^*(\mathbf{y})$ of a function $f(\mathbf{x})$ is given by $f^*(\mathbf{y}) = \sup_{\mathbf{x}} \left(\mathbf{y}^T\mathbf{x} - f(\mathbf{x})\right)$. Conjugate functions are used extensively in the construction of dual optimization problems.

Conjugate gradient – An iterative method for solving solve very large-scale, sparse linear systems of equations, $\mathbf{Ax} = \mathbf{b}$. The conjugate gradient method is often used to speed up Gauss–Newton methods.

Conjugate Priors – In Bayesian statistics, we obtain the posteriod distribution $P(\theta \mid X)$ by multiplying the likelihood $P(X \mid \theta)$ by the prior distribution $P(\theta)$. A conjugate prior is chosen so that when multiplied by the likelihood, it remains in the same distribution.

Constraints – Limitations on the admissible values of the decision variable $\mathbf{x}$, typically written in the form of equality constraint $\mathbf{h}(\mathbf{x}) = 0$, or inequality constraints $\mathbf{g}(\mathbf{x}) \leq 0$. Constraint equations define a feasible set of values to optimize over.

Control theory – The framework for modifying a dynamical system to conform to desired engineering specification through sensing and actuation.

Controllability – A system is controllable if it is possible to steer the system to any state with actuation. Degrees of controllability are determined by the controllability Gramian.

Convex Function – A function is convex when the line segment connecting any two points on the function never dips below the function at any intermediate points. The function is called strictly convex if the line segment is strictly greater than the function at all intermediate points.

Convex hull – The smallest convex set containing a set C is known as the convex hull, denoted $\text{hull}(C)$.

Convex Optimization – An optimization problem is convex if the set of feasible points is a convex set and the objective function is convex over the domain of feasible points. Convex optimization problems have a single global optimal point, so if a local optimum is found, then this also corresponds to the global optimum.

Convex relaxation – A technique where a non-convex optimization problem is approximated by a convex one, typically by relaxing integer or combinatorial constraints to continuous domains. Solutions of the relaxed problem can guide or bound the original non-convex problem.

Convex Set – A set is convex if the line segment connecting any two points in the set is also fully contained in the set. Intuitively, this means that the set cannot have any "fingers" or "bites" taken out of it.

Convolutional neural network (CNN) – A deep learning architecture tailored for spatial data, such as images, using convolutional layers to extract local features and reduce dimensionality with shared weights and pooling.

Cross-validation – A statistical method used to estimate the generalization performance of a model by partitioning data into training and validation sets. Common strategies include k-fold and leave-one-out cross-validation.

Dantzig's simplex method – A pivot-based algorithm for solving linear programming problems by moving along the vertices of the feasible polytope to find an optimal vertex solution.

Data matrix – A matrix $\mathbf{X} \in \mathbb{R}^{n \times m}$ representing n observations of m features, where rows correspond to samples and columns to variables.

Deep Learning – A class of machine learning methods using multi-layered neural networks to learn complex hierarchical representations from data, often trained with large datasets and backpropagation.

Density estimation – The task of estimating the probability distribution from a dataset. Techniques include parametric models, kernel density estimation, and Gaussian mixture models.

Duality – A foundational concept in optimization where every problem (the primal) has a corresponding dual problem. Duality theory provides bounds on the optimal value and informs algorithm design.

Dual space – The space of all linear functionals on a given vector space. For finite-dimensional spaces, the dual space can be identified with the space of vectors via the inner product.

Dynamical system – A mathematical model for the dynamic evolution of a system. Typically, a dynamical system is formulated in terms of ordinary differential equations on a state-space. The resulting equations may be linear or nonlinear and may also include the effect of actuation inputs and represent outputs as sensor measurements of the state.

Dynamic mode decomposition (DMD) – The leading eigendecomposition of a best-fit linear operator $\mathbf{A} = \mathbf{X}'\mathbf{X}^\dagger$ that propagates the data matrix $\mathbf{X}$ into a future data matrix $\mathbf{X}'$. The eigenvectors of $\mathbf{A}$ are DMD modes and the corresponding eigenvalues determine the time dynamics of these modes.

Dynamic programming – A powerful optimization approach used extensively for optimal nonlinear control and reinforcement learning. Dynamic programming reformulates large multi-step optimization problems into a recursive optimization of smaller sub-problems, relying on Bellman's principle of optimality.

Elastic net – A regularized regression technique that combines ℓ_1 (LASSO) and ℓ_2 (ridge) penalties to promote both sparsity and smoothness in the estimated coefficients.

Empirical priors – Priors estimated from observed data, rather than chosen subjectively or based on prior belief. They are often used to initialize Bayesian inference when domain-specific prior knowledge is limited.

Epigraph – The set of points lying on or above the graph of a function. A function is convex if and only if its epigraph is a convex set.

Equality constraints – Constraints of the form $\mathbf{h}(\mathbf{x}) = 0$, which define feasible surfaces or subspaces that the solution must be restricted to in an optimization problem.

Euler–Lagrange Equations – Necessary conditions for a function to extremize a functional in variational calculus. They arise from taking the functional derivative and setting it to zero. The EL equations are used extensively in mechanics to derive the equations of motion for mechanical systems that satisfy Hamilton's principle of least action. They are also used in optimization, for example to solve the Brachistochrone problem or problems in control theory.

Evolutionary Algorithms – Population-based optimization algorithms inspired by biological evolution. Examples include genetic algorithms, genetic programming, and particle swarm optimization. They are useful for non-convex and non-differentiable problems, although they may be quite expensive.

Expectation Maximization – An iterative method for finding maximum likelihood or MAP estimates in latent variable models. EM alternates between an expectation step (E-step) and a maximization step (M-step).

Exponential family of distributions – A large class of probability distributions that are used to construct conjugate priors for Bayesian inference. Many common distributions, such as Gaussian, Bernoulli, and Poisson distributions are in the exponential family.

Feasibility problem – An optimization problem where the objective is to find any point satisfying the constraints, rather than optimizing a cost.

Feasible set – The subset of decision space that satisfies all constraints in an optimization problem. The optimizer must lie within this region.

Feedback control – Closed-loop control where sensors measure the downstream effect of actuators, so that information is fed back to the actuators. Feedback is essential for robust control where model uncertainty and instability may be counteracted with fast sensor feedback.

Fenchel duality – A duality framework for convex functions based on their conjugates. Fenchel's inequality forms the basis for deriving strong duality results.

Fourier transform – A change of basis used to represent a function in terms of an infinite series of sines and cosines.

Fréchet Derivative – A generalization of the derivative in Banach spaces. A function f is Fréchet differentiable at $\mathbf{x}$ if there exists a bounded linear operator that best approximates f near x.

Functional derivative – The derivative of a functional with respect to a function. It determines how a functional changes when its input function is perturbed. This is central in the calculus of variations, and is used to derive the Euler–Lagrange equations.

Gauss–Newton methods – A second-order optimization technique for least-squares problems. The Hessian is approximated using the Jacobian, avoiding computation of second derivatives or storage of a large square matrix.

Gaussian mixture model (GMM) – A probabilistic model representing a distribution as a weighted sum of Gaussian components. GMMs are commonly used for clustering and density estimation.

Gâteaux derivative – A directional derivative of a functional, used when the Fréchet derivative may not exist. This derivative is defined via a limit along a direction in function space.

Genetic algorithm – An evolutionary algorithm to optimize the parameters of an expression with a pre-specified structure.

Genetic programming – An evolutionary algorithm to optimize both the structure and parameters of an expression or a function; often referred to as symbolic regression.

Global optimum – A global optimum is the best solution in the entire domain, in contrast to a local optimum.

Gradient – The vector of partial derivatives of a scalar function. The gradient points in the direction of steepest ascent and is used in gradient -based optimization methods.

Gradient descent – An iterative optimization method that updates variables in the negative gradient direction to minimize a function. The minimum of a function will have zero gradient at that point.

Gradient flow – A continuous-time dynamical system governed by the gradient of a function. Gradient flows are used to study the behavior of gradient descent in the limit of small step sizes.

Gramian – A matrix that encodes properties such as controllability or observability of a linear system. The controllability Gramian determines if a system can be steered to any state.

Hamilton–Jacobi–Bellman – A partial differential equation central to optimal control theory. The value function in dynamic programming must satisfy the HJB equations.

Hamiltonian – A function that often represents total energy in mechanics, but in optimization and control, it refers to the integrand in Pontryagin's maximum principle. Encodes both the objective and dynamics.

Hamiltonian neural network – A physics-informed neural network that preserves Hamiltonian structure to model conservative dynamical systems. HNNs are useful for learning energy-preserving systems.

Hessian – The square matrix of second derivatives of a scalar-valued function. The Hessian provides curvature information and is used in Newton's method.

Ill-conditioned – Describes problems (e.g., a linear system of equations) where small changes in the input cause large changes in the output, often due to a large condition number.

Ill-posed inverse problem – An inverse problem lacking a unique, stable solution. Such problems often require regularization to make the solution well defined.

Inequality constraints – Constraints of the form $\mathbf{g}(\mathbf{x}) \leq 0$, which define the set of feasible decision variables. Linear inequality constraints $\mathbf{Ax} - \mathbf{b} \leq 0$ define feasible half-planes and polytopes.

Integer programming – Optimization problems where some or all decision variables are constrained to be integers. These are typically combinatorially hard, although they may often be relaxed to linear programs.

Interior point methods – A class of optimization algorithms that approach the solution from the interior of the feasible region using barrier functions. Effective for large-scale linear and convex programs.

Inverse problem – A problem where one infers model parameters or inputs from observed outputs. Common inverse problems include medical imaging, seismic inversion, and system identification.

Jacobian – The matrix of all first-order partial derivatives of a vector-valued function. It describes the local linearization and is central in sensitivity analysis and gradient-based optimization.

Kalman filter – An estimator that reconstructs the full state of a dynamical system from measurements of a time-series of the sensor outputs and actuation inputs. A Kalman filter is itself a dynamical system that is constructed for observable systems to stably converge to the true state of the system. The Kalman filter is optimal for linear systems with Gaussian process and measurement noise of a known magnitude.

Kernel density estimation – A non-parametric method to estimate the probability density function of a dataset by summing localized kernels (e.g., Gaussians) at each data point.

KKT conditions – Necessary, and under certain conditions, sufficient, conditions for optimality in constrained optimization, generalizing Lagrange multipliers to inequality constraints.

Knapsack problem – A combinatorial optimization problem where the goal is to select a subset of items with given weights and values to maximize value without exceeding a weight capacity.

Lagrange multipliers – Variables introduced to convert constrained optimization problems into unconstrained problems via the Lagrangian. They quantify the sensitivity of the objective to the constraints.

Lagrangian – A scalar function combining the objective and constraints of an optimization problem using Lagrange multipliers. Stationary points of the Lagrangian satisfy optimality conditions.

Lagrangian neural network – A neural network architecture designed to learn physical systems by embedding Lagrangian mechanics into the network structure.

LASSO – A regularized least-squares regression, where we add a regularizing term, $\lambda\|\mathbf{x}\|_1$ to promote sparsity of solutions.

Least squares regression – A regression technique where a best-fit line or vector is found by minimizing the sum of squares of the error between the model and the data.

Legendre Transformation – A transformation that converts a function to its conjugate form, switching between variables and their gradients. Fundamental in convex analysis and mechanics.

Levenberg–Marquardt – A damped least-squares algorithm that interpolates between Gauss–Newton and gradient descent. Used for nonlinear least-squares problems.

Linear – A function or system that satisfies superposition: $f(ax + by) = af(x) + bf(y)$.

Linear matrix inequalities (LMIs) – Constraints of the form $\mathbf{A}(\mathbf{x}) \succeq 0$ where $\mathbf{A}(\mathbf{x})$ is an affine matrix function and $\succeq$ denotes positive semidefiniteness. LMIs are used extensively modern computational approaches to convex control problems.

Linear programming – An optimization problem where the objective function and constraint equations are linear. Because the objective function is both convex and concave, linear programs are often posed as maximization or minimization problems. As long as the feasible set is closed, the optimal solution will occur on one of the vertices of this closed, feasible polytope.

Linear quadratic regulator (LQR) – An optimal proportional feedback controller for full-state feedback, which balances the objectives of regulating the state while not expending too much control energy. The proportional gain matrix is determined by solving an algebraic Riccati equation.

Linear system – A system where superposition of any two inputs results in the superposition of the two corresponding outputs. In other words, doubling the input doubles the output. Linear time-invariant dynamical systems are characterized by linear operators, which are represented as matrices.

Line search – A technique in optimization where, at each iteration, the algorithm selects a step size that sufficiently decreases the objective along a search direction.

Low rank – A property of a matrix where the number of linearly independent rows and columns is small compared with the size of the matrix. Generally, low-rank approximations are sought for large data matrices.

Machine learning – The process of building models from data. This may involve unsupervised data mining, such as clustering, or supervised learning, such as classification.

Markov chain Monte Carlo (MCMC) – A family of sampling methods that generate samples from a target distribution by constructing a Markov chain. Includes Metropolis–Hastings and Gibbs sampling.

Maximization – The process of finding a point where a given function achieves its largest value.

Maximum a posteriori (MAP) estimation – The maximum a posteriori estimate of a parameter, obtained by maximizing the posterior distribution. The Bayesian extension of maximum likelihood estimation to include prior information.

Maximum likelihood estimation (MLE) – The maximum likelihood estimate maximizes the likelihood of observed data under a parametric model. It is asymptotically unbiased and efficient under regular conditions.

Minimization – The process of finding a point where a given function achieves its smallest value.

Minimum-norm solution – The solution with the smallest Euclidean norm among all possible solutions. Often used to select a unique solution to underdetermined systems.

Model predictive control (MPC) – A form of optimal control that optimizes a control policy over a finite-time horizon, based on a model. The models used for MPC are typically linear and may be determined empirically via system identification.

Momentum methods – A technique in optimization where updates incorporate a moving average of past gradients, accelerating convergence and damping oscillations.

Monte Carlo – A computational method relying on repeated random sampling to estimate expectations, integrals, or solutions to complex problems.

Nelder–Mead – A derivative-free optimization method that uses a simplex of points to explore the search space. Suitable for noisy or non-smooth functions.

Nesterov's accelerated gradient – A variant of gradient descent that incorporates a lookahead momentum term to accelerate convergence, particularly for convex functions.

Newton's method – A second-order optimization algorithm using the Hessian to perform quadratic local approximations of the objective. Newton's method converges quadratically near the optimum.

Non-convex – If either the feasible set or the objective function defined on this feasible set are non-convex, then the optimization problem is similarly non-convex.

Non-parametric models – Models that do not assume a fixed parametric form, such as kernel methods and Gaussian processes. Their flexibility allows them to adapt to complex, real-world data.

Norm – A function that measures the size or length of a vector **x**. Common norms include ℓ_1, ℓ_2, and ℓ_∞. Norms are used extensively in machine learning to quantify various terms in the loss function.

NP-hard – Problems for which no known polynomial-time algorithm exists. NP hard problems involve combinatorial or brute-force searches that don't scale well with Moore's law.

Objective function – The function $f(\mathbf{x})$ to be minimized or maximized in an optimization problem by choice of the decision variable $\mathbf{x}$. The objective function quantifies what is desirable or undesirable in a solution $\mathbf{x}$.

Observability – A system is observable if it is possible to estimate any system state with a time-history of the available sensors. Degrees of observability are determined by the observability Gramian.

Optimization – The process of finding the best solution (i.e., the maximum or minimum) according to a given objective and constraints.

Overdetermined system – A system $\mathbf{A}\mathbf{x} = \mathbf{b}$ where there are more equations than unknowns. Usually there is no exact solution $\mathbf{x}$ to an overdetermined system, unless the vector $\mathbf{b}$ is in the column space of $\mathbf{A}$.

Parametric models – Models that assume a fixed number of parameters, such as named distributions in statistics (e.g., the Gaussian or Poisson distributions are parametric).

Pareto front – In multi-objective optimization, the Pareto front is the set of optimal solutions where improving one objective worsens another. This represents a trade-off among competing goals. For example, in LASSO and SINDy, the competing objectives are sparsity of the solution and the model fit to data.

PDE-constrained inversion – Inverse problems where the forward model is governed by partial differential equations.

Physics-informed neural network (PINN) – These are a class of machine learning models for sequential data typically generated by a deterministic dynamical system. PINNs enforce governing equations, initial conditions, and boundary conditions in the training loss function. Thus the neural network is trained to solve a supervised learning task while respecting any given laws of physics described by general nonlinear partial differential equations.

Policy – A set of rules about what action an agent should take given the current state of the environment.

Policy iteration – A form of dynamic programming in which the policy function and value function are iteratively updated, while the other is held fixed.

Polytope – A bounded convex polyhedron defined as the intersection of a finite number of half-spaces, given by linear inequality constraints. In LP, the feasible set is a polytope.

Projected gradient – A variant of gradient descent where each step is projected onto the feasible set to enforce constraints.

Proximal gradient – A method for optimizing composite functions by combining gradient steps with proximal operators. Proximal gradients are useful for nonsmooth regularizers, such as ℓ_1.

Pseudo-inverse – A generalization of the matrix inverse, often used for solving least-squares problems. The pseudo-inverse is given by $\mathbf{A}^+ = (\mathbf{A}^T\mathbf{A})^{-1}\mathbf{A}^T$ when $\mathbf{A}$ has full column rank. For non-square or singular matrices, it is computed using the singular value decomposition.

Quadratic programming – Optimization problems with a quadratic objective function and linear constraints. QPs are efficiently solved using interior point or active set methods.

Quality function – The joint quality of being in a particular state $\mathbf{s}$ and taking a given action $\mathbf{a}$. The quality function $Q(\mathbf{s}, \mathbf{a})$ extends the value function, bypassing the need to know the next optimal state $\mathbf{s}'$ and providing the basis for Q-learning.

Quasi-Newton methods – Optimization algorithms that approximate the Hessian matrix in Newton's method, avoiding the full second derivative computation. Examples include BFGS and LBFGS.

Radon transform – An integral transform that maps a function to its integrals over hyperplanes. The Radon transform is foundational in tomography and other inverse problems.

Recurrent neural network (RNN) – A neural network architecture designed for sequential data, such as time series data, where outputs from previous time steps influence the current state.

Regression – A statistical model that represents an outcome variable in terms of indicator variables. Least-squares regression is a linear regression that finds the line of best fit to data; when generalized to higher dimensions and multi-linear regression, this generalizes to principal components regression. Nonlinear regression, dynamic regression, and functional or semantic regression are used in system identification, model reduction, and machine learning.

Regularity conditions – Conditions ensuring that the KKT conditions are sufficient for optimality. These include constrain qualifications and differentiability assumptions.

Regularization – A technique that adds a penalty to the loss function to prevent overfitting and improve generalization. Examples include LASSO, ridge regression, and elastic net.

Reinforcement learning (RL) – A major branch of machine learning that is concerned with how to learn control laws and policies to interact with a complex environment from experience.

Relaxation – The process of approximating a hard optimization problem (e.g., integer programming) with a simpler or continuous version, such as a linear program.

Riccati equation – A matrix system of equations that arises in optimal control problems, such as deriving the linear quadratic regulator (LQR) or the Kalman filter.

Ridge regression (Tikhonov) – A regularized least-squares regression for ill-conditioned problems, where we add a regularizing term, $\alpha\|\mathbf{x}\|_2^2$ to decrease the variance of solutions. This also adds a bias to the solution.

Robust control – A field of control that penalizes *worst case scenario* control outcomes, thus promoting controllers that are robust to uncertainties, disturbances, and unmodeled dynamics.

Robust statistics – Methods for producing good statistical estimates for data drawn from a wide range of probability distributions, especially for distributions that are not normal and where outliers compromise predictive capabilities.

Sequential quadratic programming (SQP) – An iterative method for nonlinear constrained optimization that solves a sequence of quadratic approximations to the problem.

Simplex tableau method – A structured tabular implementation of the simplex method for solving linear programs. The tableau method facilitates pivot operations and bookkeeping on large-scale problems.

Simulated Annealing – A gradient-free, global optimization method that explores an objective function landscape by stochastically sampling, with sample step size based on a temperature. Initially higher temperatures promote exploration, while lower temperature later in the iterations promote exploitation. Simulated annealing is inspired by the process of heating and cooling a material, such as a metal, to obtain a strong crystalline structure.

Singular value decomposition (SVD) – Given a matrix $\mathbf{X} \in \mathbb{C}^{n\times m}$, the SVD is given by $\mathbf{X} = \mathbf{U}\boldsymbol{\Sigma}\mathbf{V}^*$ where $\mathbf{U} \in \mathbb{C}^{n\times n}$, $\boldsymbol{\Sigma} \in \mathbb{C}^{n\times m}$, and $\mathbf{V} \in \mathbb{C}^{m\times m}$. The matrices $\mathbf{U}$ and $\mathbf{V}$ are unitary, so that $\mathbf{U}\mathbf{U}^* = \mathbf{U}^*\mathbf{U} = \mathbf{I}$ and $\mathbf{V}\mathbf{V}^* = \mathbf{V}^*\mathbf{V} = \mathbf{I}$. The matrix $\boldsymbol{\Sigma}$ has entries along the diagonal corresponding to the singular values that are ordered from largest to smallest. This produces a hierarchical matrix decomposition that splits a matrix into a sum of rank-1 matrices given by the outer product of a column vector (left singular vector) with a row vector (conjugate transpose of right singular vector). These rank-1 matrices are ordered by the singular value so that the first r rank-1 matrices form the *best* rank-r matrix approximation of the original matrix in a least-squares sense.

Slack variable – An auxiliary variable added to convert inequality constraints into equality constraints. The slack variable measures how far the decision variable $\mathbf{x}$ is to an inequality constraint boundary. Slack variables are used in active set methods, such as Dantzig's simplex algorithm in linear programming.

Sparse identification of nonlinear dynamics (SINDy) – A nonlinear system identification framework used to simultaneously identify the nonlinear structure and parameters of a dynamical system from data. Various sparse optimization techniques may be used to determine SINDy models.

Sparsity – A vector is *sparse* if most of its entries are zero or nearly zero. Sparsity refers to the observation that most data are sparse when represented as vectors in an appropriate transformed basis, such as Fourier or POD bases.

Splitting – A class of optimization techniques that decomposes complex problems into simpler subproblems. Splitting is common in ADMM and proximal methods.

Stationarity – A point is stationary if the gradient of the objective function or Lagrangian at this point is zero. Stationarity is necessary for optimality in unconstrained problems and for constrained problems, via the KKT conditions.

Stochastic gradient descent – A variant of gradient descent used in neural network training, where a subset of the data is used to approximate the gradient. Different batch sizes of data may be used, in the so-called mini-batch stochastic gradient descent.

Supporting hyperplanes – Hyperplanes that touch a convex set at a boundary point without cutting through the interior. These are fundamental in convex analysis and duality.

Swarm optimization – A form of distributed optimization that involves a population of agents interacting and influencing each other to find optimal solutions. This is a gradient-free, global optimization approach.

System identification – The process of learning an input–output model for a dynamical system from measurement data, possibly after perturbing the system.

Tautochrone – A classic variational problem, posed by Huygens in the 17th centry, where the goal is to solve for a curve such that every point will reach the lowest point in the same amount of time.

Total derivative – The full derivative of a multivariate function with respect to a variable, accounting for all indirect dependencies via the chain rule.

Total least squares – A least-squares regression algorithm that minimizes the error on both the inputs and the outputs. Geometrically, this corresponds to finding the line that minimizes the sum of squares of the total distance to all points, rather than the sum of squares of the vertical distance to all points.

Traveling salesman problem – Given a list of cities or nodes and the pairwise distances or costs associated with traveling between them, the traveling salesman problem seeks a route that visits each location exactly once, returning to the start, while minimizing total travel distance. This is a classic NP-hard optimization problem.

Underdetermined system – A system $\mathbf{Ax} = \mathbf{b}$ where there are fewer equations than unknowns. Generally the system has infinitely many solutions $\mathbf{x}$ unless $\mathbf{b}$ is not in the column space of $\mathbf{A}$.

Value function – A function quantifying the desirability of being in a given state $\mathbf{s}$, as calculated by the discounted sum of future rewards, given an optimal policy starting from this state. The value function is often written in a recursive form, based on Bellman's equation.

Value iteration – A form of dynamic programming that iteratively updates the value function, after which an optimal policy may be extracted.

Vanishing and exploding gradients – When training a deep neural network, partial derivatives are back-propagated through the many layers, and these partial derivatives may grow very small (i.e., vanishing) or very large (i.e., exploding), unless special care is given to manage their magnitude.

Variational Calculus – A field of analysis that involves maximizing or minimizing functionals with respect to their input functions. For example, the path of least action states that the trajectory of a physical system will minimize the Euler–Lagrange equations.

Bibliography

[1] R. Abraham, J. E. Marsden, and J. E. Marsden. *Foundations of Mechanics*, volume 36. Benjamin–Cummings Publishing Company, 1978.

[2] F. Acker and M.-A. Prestel. Convergence d'un schéma de minimisation alternée. In *Annales de la Faculté des Sciences de Toulouse: Mathématiques*, 2:1–9. Université Paul Sabatier, 1980.

[3] H. Akaike. Fitting autoregressive models for prediction. *Annals of the Institute of Statistical Mathematics*, 21(1):243–247, 1969.

[4] H. Akaike. A new look at the statistical model identification. *IEEE Transactions on Automatic Control*, 19(6):716–723, 1974.

[5] S.-I. Amari. Neural theory of association and concept-formation. *Biological Cybernetics*, 26(3):175–185, 1977.

[6] B. Amos, I. D. J. Rodriguez, J. Sacks, B. Boots, and J. Z. Kolter. Differentiable mpc for end-to-end planning and control. ArXiv:1810.13400, 2018.

[7] E. Anderson, Z. Bai, C. Bischof, S. Blackford, J. Demmel, J. Dongarra, J. Du Croz, A. Greenbaum, S. Hammerling, A. McKenney, and D. Sorensen. *LAPACK Users' Guide*, SIAM, 1999.

[8] J. L. Anderson. An ensemble adjustment Kalman filter for data assimilation. *Monthly Weather Review*, 129(12):2884–2903, 2001.

[9] M. Andrychowicz, F. Wolski, A. Ray, J. Schneider, R. Fong, P. Welinder, B. McGrew, J. Tobin, P. Abbeel, and W. Zaremba. Hindsight experience replay. ArXiv:1707.01495, 2017.

[10] A. C. Antoulas. *Approximation of Large-Scale Dynamical Systems*. SIAM, 2005.

[11] A. Y. Aravkin, J. V. Burke, D. Drusvyatskiy, M. P. Friedlander, and S. Roy. Level-set methods for convex optimization. *Mathematical Programming*, 174(1):359–390, 2019.

[12] K. B. Ariyur and M. Krstić. *Real-Time Optimization by Extremum-Seeking Control*. Wiley, 2003.

[13] E. Artin and A. N. Milgram. *Galois Rheory*. Courier Corporation, 1998.

[14] T. Askham and J. N. Kutz. Variable projection methods for an optimized dynamic mode decomposition. *SIAM Journal on Applied Dynamical Systems*, 17(1):380–416, 2018.

[15] K. J. Aström and R. M. Murray. *Feedback Systems: an Introduction for Scientists and Engineers*. Princeton University Press, 2010.

[16] H. Attouch, J. Bolte, P. Redont, and A. Soubeyran. Proximal alternating minimization and projection methods for nonconvex problems: An approach based on the Kurdyka–Lojasiewicz inequality. *Mathematics of Operations Research*, 35(2):438–457, 2010.

[17] S. D. Babacan, R. Molina, and A. K. Katsaggelos. Bayesian compressive sensing using Laplace priors. *IEEE Transactions on Image Processing*, 19(1):53–63, 2009.

[18] K. Bache and M. Lichman. UCI Machine Learning Repository, 2013.

[19] P. J. Baddoo, B. Herrmann, B. J. McKeon, and S. L. Brunton. Kernel learning for robust dynamic mode decomposition: Linear and nonlinear disambiguation optimization (LANDO).

Proceedings of the Royal Society A, 478(2260):20210830, 2022.

[20] P. J. Baddoo, B. Herrmann, B. J. McKeon, J. Nathan Kutz, and S. L. Brunton. Physics-informed dynamic mode decomposition. *Proceedings of the Royal Society A*, 479(2271):20220576, 2023.

[21] H. J. Bae and P. Koumoutsakos. Scientific multi-agent reinforcement learning for wall-models of turbulent flows. *Nature Communications*, 13(1):1443, 2022.

[22] M. Baek, F. DiMaio, I. Anishchenko, J. Dauparas, S. Ovchinnikov, G. R. Lee, J. Wang, Q. Cong, L. N. Kinch, R. D. Schaeffer, et al. Accurate prediction of protein structures and interactions using a three-track neural network. *Science*, 373(6557):871–876, 2021.

[23] R. Bai, V. Ročková, and E. I. George. Spike-and-slab meets LASSO: A review of the spike-and-slab LASSO. *Handbook of Bayesian Variable Selection*, edited by Mahlet G. Tadesse and Marina Vannucci, Chapman & Hall/CRC Press, pages 81–108, 2021.

[24] J. Bakarji, K. Champion, J. Nathan Kutz, and S. L. Brunton. Discovering governing equations from partial measurements with deep delay autoencoders. *Proceedings of the Royal Society A*, 479(2276):20230422, 2023.

[25] R. G. Baraniuk. Compressive sensing. *IEEE Sig. Proc. Mag.*, 24(4):118–120, 2007.

[26] S. Barman, X. Liu, S. C. Draper, and B. Recht. Decomposition methods for large scale LP decoding. *IEEE Transactions on Information Theory*, 59(12):7870–7886, 2013.

[27] M. S. Bartlett. An inverse matrix adjustment arising in discriminant analysis. *Annals of Mathematical Statistics*, 22(1):107–111, 1951.

[28] P. W. Battaglia, J. B. Hamrick, V. Bapst, A. Sanchez-Gonzalez, V. Zambaldi, M. Malinowski, A. Tacchetti, D. Raposo, A. Santoro, R. Faulkner, et al. Relational inductive biases, deep learning, and graph networks. ArXiv:1806.01261, 2018.

[29] S. Batzner, A. Musaelian, L. Sun, M. Geiger, J. P. Mailoa, M. Kornbluth, N. Molinari, T. E. Smidt, and B. Kozinsky. E(3)-equivariant graph neural networks for data-efficient and accurate interatomic potentials. *Nature Communications*, 13(1):2453, 2022.

[30] H. H. Bauschke, P. L. Combettes, and D. Noll. Joint minimization with alternating Bregman proximity operators. *Pacific Journal of Optimization*, 2(3):401–424, 2006.

[31] A. G. Baydin, B. A. Pearlmutter, A. A. Radul, and J. M. Siskind. Automatic differentiation in machine learning: a survey. *Journal of Machine Learning Research*, 18(153):1–43, 2018.

[32] A. Beck and M. Teboulle. A fast iterative shrinkage-thresholding algorithm for linear inverse problems. *SIAM Journal on Imaging Sciences*, 2(1):183–202, 2009.

[33] L. Beda, L. Korolev, N. Sukkikh, and T. Frolova. Programs for automatic differentiation for the machine BESM. Technical report, Institute for Precise Mechanics and Computation Techniques, 1959.

[34] S. Beetham and J. Capecelatro. Formulating turbulence closures using sparse regression with embedded form invariance. *Physical Review Fluids*, 5(8):084611, 2020.

[35] S. Beetham, R. O. Fox, and J. Capecelatro. Sparse identification of multiphase turbulence closures for coupled fluid–particle flows. *Journal of Fluid Mechanics*, 914, 2021.

[36] R. Bellman. On the theory of dynamic programming. *Proceedings of the National Academy of Sciences*, 38(8):716, 1952.

[37] R. Bellman. Dynamic programming. *Science*, 153(3731):34–37, 1966.

[38] A. Bemporad, A. Oliveri, T. Poggi, and M. Storace. Ultra-fast stabilizing model predictive control via canonical piecewise affine approximations. *IEEE Transactions on Automatic Control*, 56(12):2883–2897, 2011.

[39] M. P. Bendsøe. Optimal shape design as a material distribution problem. *Structural Optimization*, 1(4):193–202, 1989.

[40] M. P. Bendsøe. Topology optimization. In *Encyclopedia of Optimization*, edited by Christodoulos A. Floudas and Panos M. Pardalos, pages 3928–3929. Springer, 2008.

[41] M. P. Bendsøe and O. Sigmund. Material interpolation schemes in topology optimization. *Archive of Applied Mechanics*, 69(9):635–654, 1999.

[42] M. Beneitez, A. Cremades, L. Guastoni, and R. Vinuesa. Improving turbulence control through explainable deep learning. *ArXiv:2504.02354*, 2025.

[43] P. Benner, J.-R. Li, and T. Penzl. Numerical solution of large-scale Lyapunov equations, Riccati equations, and linear-quadratic optimal control problems. *Numerical Linear Algebra with Applications*, 15(9):755–777, 2008.

[44] P. Benner, V. Mehrmann, V. Sima, S. Van Huffel, and A. Varga. SLICOT – a subroutine library in systems and control theory. In *Applied and Computational Control, Signals, and Circuits*, edited by Datta, B.N. pages 499–539. Birkhäuser Boston, 1999.

[45] J. Bergstra and Y. Bengio. Random search for hyper-parameter optimization. *Journal of Machine Learning Research*, 13:281–305, 2012.

[46] J. Berkson. Examination of randomness of alpha particle emissions. In *Research Methods in Statistics*, edited by F.N. David, pages 37–54. Wiley, 1966.

[47] J. M. Bernardo, A. F. Smith, and M. Berliner. *Bayesian Theory*, Wiley, 1994.

[48] D. P. Bertsekas. *Constrained Optimization and Lagrange Multiplier Methods*. Academic Press, 2014.

[49] D. Bertsimas and J. Tsitsiklis. Simulated annealing. *Statistical Science*, 8(1):10–15, 1993.

[50] D. Bertsimas and J. N. Tsitsiklis. *Introduction to Linear Optimization*. Athena Scientific, 1997.

[51] L. T. Biegler, O. Ghattas, M. Heinkenschloss, and B. van Bloemen Waanders, editors. *Large-Scale PDE-Constrained Optimization*. Springer, 2003.

[52] K. Bieker, S. Peitz, S. L. Brunton, J. N. Kutz, and M. Dellnitz. Deep model predictive flow control with limited sensor data and online learning. *Theoretical and Computational Fluid Dynamics*, 34:577–591, 2020.

[53] L. Biferale, F. Bonaccorso, M. Buzzicotti, P. Clark Di Leoni, and K. Gustavsson. Zermelo's problem: Optimal point-to-point navigation in 2d turbulent flows using reinforcement learning. *Chaos: An Interdisciplinary Journal of Nonlinear Science*, 29(10):103138, 2019.

[54] S. A. Billings. *Nonlinear System Identification: NARMAX Methods in the Time, Frequency, and Spatio-Temporal Domains*. Wiley, 2013.

[55] C. M. Bishop et al. *Pattern Recognition and Machine Learning*. Springer, 2006.

[56] A. Blanchard and T. Sapsis. Bayesian optimization with output-weighted optimal sampling. *Journal of Computational Physics*, 425:109901, 2021.

[57] A. M. Bloch, P. Krishnaprasad, J. E. Marsden, and R. M. Murray. Nonholonomic mechanical systems with symmetry. *Archive for Rational Mechanics and Analysis*, 136(1):21–99, 1996.

[58] C. Blum and A. Roli. Metaheuristics in combinatorial optimization: Overview and conceptual comparison. *ACM Computing Surveys*, 35(3):268–308, 2003.

[59] J. Bolte, S. Sabach, and M. Teboulle. Proximal alternating linearized minimization for nonconvex and nonsmooth problems. *Mathematical Programming*, 146(1):459–494, 2014.

[60] V. Boltyanski, R. Gamkrelidze, E. Mishchenko, and L. Pontryagin. The maximum principle in the theory of optimal processes of control. *IFAC Proceedings Volumes*, 1(1):464–469, 1960.

[61] J. Bongard and H. Lipson. Automated reverse engineering of nonlinear dynamical systems. *Proceedings of the National Academy of Sciences*, 104(24):9943–9948, 2007.

[62] R. I. Bot. *Conjugate Duality in Convex Optimization*. Springer, 2009.

[63] G. E. Box and G. C. Tiao. *Bayesian inference in statistical analysis*. John Wiley & Sons, 2011.

[64] S. Boyd, L. El Ghaoui, E. Feron, and V. Balakrishnan. *Linear Matrix Inequalities in System and Control Rheory*. SIAM, 1994.

[65] S. Boyd, N. Parikh, E. Chu, B. Peleato, and J. Eckstein. Distributed optimization and statistical learning via the alternating direction method of multipliers. *Foundations and Trends® in Machine Learning*, 3(1):1–122, 2011.

[66] S. Boyd and L. Vandenberghe. *Convex Optimization*. Cambridge university press, 2009.

[67] W. H. Bragg and W. L. Bragg. The reflection of X-rays by crystals. *Proceedings of the Royal Society A*, 88:428–438, 1913.

[68] L. Breiman. Better subset regression using the nonnegative garrote. *Technometrics*, 37(4):373–384, 1995.

[69] L. Breiman. Random forests. *Machine Learning*, 45(1):5–32, 2001.

[70] G. Brockman, V. Cheung, L. Pettersson, J. Schneider, J. Schulman, J. Tang, and W. Zaremba. OpenAI Gym. ArXiv:1606.01540, 2016.

[71] M. M. Bronstein, J. Bruna, Y. LeCun, A. Szlam, and P. Vandergheynst. Geometric deep learning: Going beyond euclidean data. *IEEE Signal Processing Magazine*, 34(4):18–42, 2017.

[72] T. B. Brown et al. Language models are few-shot learners. *Advances in Neural Information Processing Systems*, 33:1877–1901, 2020.

[73] C. G. Broyden. Quasi-Newton methods and their application to function minimisation. *Mathematics of Computation*, 21(99):368–381, 1967.

[74] C. G. Broyden. The convergence of a class of double-rank minimization algorithms 1. General considerations. *IMA Journal of Applied Mathematics*, 6(1):76–90, 1970.

[75] S. L. Brunton, B. W. Brunton, J. L. Proctor, E. Kaiser, and J. N. Kutz. Chaos as an intermittently forced linear system. *Nature Communications*, 8(19):1–9, 2017.

[76] S. L. Brunton, M. Budišić, E. Kaiser, and J. N. Kutz. Modern Koopman theory for dynamical systems. *SIAM Review*, 64(2):229–340, 2022.

[77] S. L. Brunton and J. N. Kutz. *Data-Driven Science and Engineering: Machine Learning, Dynamical Systems, and Control*, 2nd edition. Cambridge University Press, 2022.

[78] S. L. Brunton and J. N. Kutz. Promising directions of machine learning for partial differential equations. *Nature Computational Science*, 4(7):483–494, 2024.

[79] S. L. Brunton, J. N. Kutz, K. Manohar, A. Y. Aravkin, K. Morgansen, J. Klemisch, N. Goebel, J. Buttrick, J. Poskin, A. W. Blom-Schieber, T. Hogan, and D. McDonald. Data-driven aerospace engineering: reframing the industry with machine learning. *AIAA Journal*, 59(8):1–26, 2021.

[80] S. L. Brunton, B. R. Noack, and P. Koumoutsakos. Machine learning for fluid mechanics. *Annual Review of Fluid Mechanics*, 52:477–508, 2020.

[81] S. L. Brunton, J. L. Proctor, and J. N. Kutz. Discovering governing equations from data by sparse identification of nonlinear dynamical systems. *Proceedings of the National Academy of Sciences*, 113(15):3932–3937, 2016.

[82] S. L. Brunton, N. Zolman, J. N. Kutz, and U. Fasel. Machine learning for sparse nonlinear modeling and control. *Annual Review of Control, Robotics, and Autonomous Systems*, 8:127–152, 2025.

[83] A. E. Bryson and W. F. Denham. A steepest-ascent method for solving optimum programming problems. *Journal of Applied Mechanics*, 29(2):247–257, 1962.

[84] A. E. Bryson and Y.-C. Ho. *Applied Optimal Control: Optimization, Estimation and Control*. Taylor & Francis, 1975.

[85] P. A. Businger and G. H. Golub. Algorithm 358: Singular value decomposition of a complex matrix [F1, 4, 5]. *Communications of the ACM*, 12(10):564–565, 1969.

[86] R. H. Byrd, S. L. Hansen, J. Nocedal, and Y. Singer. A stochastic quasi-Newton method for large-scale optimization. *SIAM Journal on Optimization*, 26(2):1008–1031, 2016.

[87] R. H. Byrd, J. Nocedal, and R. B. Schnabel. Representations of quasi-Newton matrices and their use in limited memory methods. *Mathematical Programming*, 63(1):129–156, 1994.

[88] X. Cai, M. J. Tippett, L. Xie, and J. Bao. Fast distributed MPC based on active set method. *Computers & Chemical Engineering*, 71:158–170, 2014.

[89] E. F. Camacho and C. B. Alba. *Model Predictive Control*. Springer, 2013.

[90] E. J. Candès. Compressive sensing. *Proceedings of the International Congress of Mathematicians*, Volume 3, pages 1433–1452. European Mathematical Society (EMS) Publishing House, 2006.

[91] E. J. Candès, X. Li, Y. Ma, and J. Wright. Robust principal component analysis? *Journal of the ACM*, 58(3):1–37, 2011.

[92] E. J. Candès, J. Romberg, and T. Tao. Robust uncertainty principles: exact signal reconstruction from highly incomplete frequency information. *IEEE Transactions on Information Theory*, 52(2):489–509, 2006.

[93] E. J. Candès, J. Romberg, and T. Tao. Stable signal recovery from incomplete and inaccurate measurements. *Communications in Pure and Applied Mathematics*, 59(8): 1207–1223.

[94] G. Cardano. *Ars magna or The Rules of Algebra*. Dover Publications, 1993. Originally published in 1545.

[95] C. M. Carvalho, N. G. Polson, and J. G. Scott. The horseshoe estimator for sparse signals. *Biometrika*, 97(2):465–480, 2010.

[96] K. Champion, B. Lusch, J. N. Kutz, and S. L. Brunton. Data-driven discovery of coordinates and governing equations. *Proceedings of the National Academy of Sciences*, 116(45):22445–22451, 2019.

[97] K. P. Champion, S. L. Brunton, and J. N. Kutz. Discovery of nonlinear multiscale systems: Sampling strategies and embeddings. *SIAM Journal on Applied Dynamical Systems*, 18(1):312–333, 2019.

[98] T. F. Chan and S. Esedoglu. Aspects of total variation regularized L^1 function approximation. *SIAM Journal on Applied Mathematics*, 65(5):1817–1837, 2005.

[99] H. Chang, D.-Y. Yeung, and Y. Xiong. Super-resolution through neighbor embedding. In *Proceedings of the 2004 IEEE Computer Society Conference on Computer Vision and Pattern Recognition, 2004*, pages I–I. DOI: 10.1109/CVPR.2004.1315043.

[100] R. Chartrand. Numerical differentiation of noisy, nonsmooth data. *ISRN Applied Mathematics*, doi/10.5402/2011/164564, 2011.

[101] R. T. Chen, Y. Rubanova, J. Bettencourt, and D. K. Duvenaud. Neural ordinary differential equations. In *Proceedings of the 32nd International Conference on Neural Information Processing Systems*, pages 6572–6583, 2018.

[102] T. Chen, B. Xu, C. Zhang, and C. Guestrin. Training deep nets with sublinear memory cost. In ArXiv:1604.06174, 2016.

[103] W. Cheney and A. A. Goldstein. Proximity maps for convex sets. *Proceedings of the American Mathematical Society*, 10(3):448–450, 1959.

[104] Z. Cheng, C. Esteves, V. Jampani, A. Kar, S. Maji, and A. Makadia. LU-NeRF: Scene and pose estimation by synchronizing local unposed NeRFs. In *Proceedings of the IEEE/CVF International Conference on Computer Vision*, pages 18312–18321, 2023.

[105] G. Chierchia, E. Chouzenoux, P. L. Combettes, and J.-C. Pesquet. The proximity operator repository. https://proximity-operator.net/.

[106] F. R. K. Chung. *Spectral Graph Theory*, volume 92 of *CBMS Regional Conference Series in Mathematics*. American Mathematical Society, 1997.

[107] J. F. Claerbout and F. Muir. Robust modeling with erratic data. *Geophysics*, 38(5):826–844, 1973.

[108] S. Colabrese, K. Gustavsson, A. Celani, and L. Biferale. Flow navigation by smart microswimmers via reinforcement learning. *Physical Review Letters*, 118(15):158004, 2017.

[109] P. L. Combettes and J.-C. Pesquet. Proximal splitting methods in signal processing. In *Fixed-point Algorithms for Inverse Problems in Science and Engineering*, edited by H. Bauschke, R. Burachik, P. Combettes, V. Elser, D. Luke, H. Wolkowicz, pages 185–212. Springer, 2011.

[110] P. L. Combettes and V. R. Wajs. Signal recovery by proximal forward-backward splitting. *Multiscale Modeling & Simulation*, 4(4):1168–1200, 2005.

[111] A. R. Conn. Constrained optimization using a nondifferentiable penalty function. *SIAM Journal on Numerical Analysis*, 10(4):760–784, 1973.

[112] A. R. Conn, N. I. Gould, and P. L. Toint. *Trust Region Methods*. MPS/SIAM Series on Optimization. SIAM, 2000.

[113] A. R. Conn, K. Scheinberg, and L. N. Vicente. *Introduction to Derivative-Free Optimization*. SIAM, 2009.

[114] A. Cormack. Representation of a function by its line integrals, with some radiological applications. *Journal of Applied Physics*, 34:2722–2727, 1963.

[115] P. Courtier, J.-N. Thépaut, and A. Hollingsworth. A strategy for operational implementation of 4D-Var, using an incremental approach. *Quarterly Journal of the Royal Meteorological Society*, 120(519):1367–1387, 1994.

[116] M. Cranmer. Interpretable machine learning for science with PySR and SymbolicRegression.jl. ArXiv:2305.01582, 2023.

[117] M. Cranmer, S. Greydanus, S. Hoyer, P. Battaglia, D. Spergel, and S. Ho. Lagrangian neural networks. ArXiv:2003.04630, 2020.

[118] M. Cranmer, A. Sanchez-Gonzalez, P. Battaglia, R. Xu, K. Cranmer, D. Spergel, and S. Ho. Discovering symbolic models from deep learning with inductive biases. ArXiv:2006.11287, 2020.

[119] M. D. Cranmer, R. Xu, P. Battaglia, and S. Ho. Learning symbolic physics with graph networks. ArXiv:1909.05862, 2019.

[120] A. Cremades, S. Hoyas, and R. Vinuesa. Classically studied coherent structures only paint a partial picture of wall-bounded turbulence. ArXiv:2410.23189, 2024.

[121] A. I. Dale. Bayes or Laplace? an examination of the origin and early applications of Bayes' theorem. *Archive for History of Exact Sciences*, 27(1):23–47, 1982.

[122] P. Danaher, P. Wang, and D. M. Witten. The joint graphical LASSO for inverse covariance estimation across multiple classes. *Journal of the Royal Statistical Society B*, 76(2):373–397, 2014.

[123] G. B. Dantzig. Maximization of a linear function of variables subject to linear inequalities. *Activity Analysis of Production and Allocation*, 13:339–347, 1951.

[124] G. B. Dantzig. Linear Programming and Extensions. Princeton University Press, 1959.

[125] G. Darmois. Sur les lois de probabilitéa estimation exhaustive. *CR Acad. Sci. Paris*, 260(1265):85, 1935.

[126] W. C. Davidon. Variable metric method for minimization. *SIAM Journal on Optimization*, 1(1):1–17, 1991.

[127] L. Davis et al. *Handbook of Genetic Algorithms*. Van Nostrand Reinhold, 1991.

[128] N. D. Daw, J. P. O'Doherty, P. Dayan, B. Seymour, and R. J. Dolan. Cortical substrates for exploratory decisions in humans. *Nature*, 441(7095):876–879, 2006.

[129] P. Dayan and L. F. Abbott. *Theoretical Neuroscience: Computational and Mathematical Modeling of Neural Systems*. MIT Press, 2001.

[130] F. de Avila Belbute-Peres, K. Smith, K. Allen, J. Tenenbaum, and J. Z. Kolter. End-to-end differentiable physics for learning and control. In *Proceedings of the 32nd International Conference on Neural Information Processing Systems*, pages 7178–7189, 2018.

[131] B. M. de Silva, K. Champion, M. Quade, J.-C. Loiseau, J. N. Kutz, and S. L. Brunton. PySINDy: a Python package for the sparse identification of nonlinear dynamics from data. *Journal of Open Source Software*, 5(49):2104, 2020.

[132] J. Degrave, F. Felici, J. Buchli, M. Neunert, B. Tracey, F. Carpanese, T. Ewalds, R. Hafner, A. Abdolmaleki, D. de Las Casas, et al. Magnetic control of tokamak plasmas through deep reinforcement learning. *Nature*, 602(7897):414–419, 2022.

[133] N. Demo, M. Tezzele, and G. Rozza. PyDMD: Python dynamic mode decomposition. *Journal of Open Source Software*, 3(22):530, 2018.

[134] A. P. Dempster, N. M. Laird, and D. B. Rubin. Maximum likelihood from incomplete data via the EM algorithm. *Journal of the Royal Statistical Society B*, 39(1):1–22, 1977.

[135] J. Deng, W. Dong, R. Socher, L.-J. Li, K. Li, and L. Fei-Fei. Imagenet: A large-scale hierarchical image database. In *Proceedings of the 2009 IEEE Conference on Computer Vision and Pattern Recognition*, pages 248–255. IEEE Press, 2009.

[136] N. Deng, B. R. Noack, M. Morzyński, and L. R. Pastur. Low-order model for successive bifurcations of the fluidic pinball. *Journal of Fluid Mechanics*, 884:A37. doi:10.1017/jfm.2019.959, 2020.

[137] N. Deng, B. R. Noack, M. Morzyński, and L. R. Pastur. Galerkin force model for transient and post-transient dynamics of the fluidic pinball. *Journal of Fluid Mechanics*, 918:A4. doi:10.1017/jfm.2021.299, 2021.

[138] J. E. Dennis, Jr and J. J. Moré. Quasi-Newton methods, motivation and theory. *SIAM Review*, 19(1):46–89, 1977.

[139] J. E. Dennis Jr and R. B. Schnabel. *Numerical Methods for Unconstrained Optimization and Nonlinear Equations*. SIAM, 1996.

[140] S. Diamond and S. Boyd. CVXPY: A Python-embedded modeling language for convex optimization. *Journal of Machine Learning Research*, 17(83):1–5, 2016.

[141] M. Diehl, H. J. Ferreau, and N. Haverbeke. Efficient numerical methods for nonlinear MPC and moving horizon estimation. In *Nonlinear Model Predictive Control: Towards New Challenging Applications*, edited by L. Magni, D.M. Raimondo, F. Allgöwer, pages 391–417. Springer, 2009.

[142] I. I. Dikin. Iterative solution of problems of linear and quadratic programming. In *Soviet Math. Dokl.*, 8:674–675, 1967.

[143] U. M. Diwekar. *Introduction to Applied Optimization*. Springer Nature, 2020.

[144] A. Domahidi, J. L. Jerez, N. K. Poulsen, S. Jóhannesson, A. Zanelli, and M. Diehl. FORCES: Fast optimization for real-time control on embedded systems. *IFAC Proceedings Volumes*, 45(17):145–150, 2012.

[145] C. Dong, C. C. Loy, K. He, and X. Tang. Learning a deep convolutional network for image super-resolution. In *Computer Vision – ECCV 2014*, edited by D. Fleet, T. Pajdla, B. Schiele, T. Tuytelaars, pages 184–199. Springer, 2014.

[146] D. L. Donoho. Compressed sensing. *IEEE Transactions on Information Theory*, 52(4):1289–1306, 2006.

[147] D. L. Donoho, I. M. Johnstone, J. C. Hoch, and A. S. Stern. Maximum entropy and the nearly black object. *Journal of the Royal Statistical Society B*, 54(1):41–81, 1992.

[148] D. L. Donoho and J. M. Johnstone. Ideal spatial adaptation by wavelet shrinkage. *Biometrika*, 81(3):425–455, 1994.

[149] M. Dorigo, M. Birattari, and T. Stutzle. Ant colony optimization. *IEEE Computational Intelligence Magazine*, 1(4):28–39, 2007.

[150] A. Dosovitskiy, L. Beyer, A. Kolesnikov, D. Weissenborn, X. Zhai, T. Unterthiner, M. Dehghani, M. Minderer, G. Heigold, S. Gelly, J. Uszkoreit, and N. Houlsby. An image is worth 16×16 words: Transformers for image recognition at scale. ArXiv:2010.11929, 2020.

[151] J. Douglas and H. Rachford. On the numerical solution of heat conduction problems in two and three space variables. *Transactions of the American Mathematical Society*, 82(2):421–439, 1956.

[152] J. C. Doyle. Guaranteed margins for LQG regulators. *IEEE Transactions on Automatic Control*, 23(4):756–757, 1978.

[153] P. Drews, G. Williams, B. Goldfain, E. A. Theodorou, and J. M. Rehg. Vision-based high-speed driving with a deep dynamic observer. *IEEE Robotics and Automation Letters*, 4(2):1564–1571, 2019.

[154] J. Drgoňa, K. Kiš, A. Tuor, D. Vrabie, and M. Klaučo. Differentiable predictive control: Deep learning alternative to explicit model predictive control for unknown nonlinear systems. *Journal of Process Control*, 116:80–92, 2022.

[155] J. Duchi, E. Hazan, and Y. Singer. Adaptive subgradient methods for online learning and stochastic optimization. *Journal of Machine Learning Research*, 12(7):2121–2159, 2011.

[156] G. E. Dullerud and F. Paganini. *A Course in Robust Control Theory: A Convex Approach*. Springer, 2000.

[157] D. S. Dummit and R. M. Foote. *Abstract Algebra*, volume 3. Wiley, 2004.

[158] T. Duriez, S. L. Brunton, and B. R. Noack. *Machine Learning Control: Taming Nonlinear Dynamics and Turbulence*. Springer, 2016.

[159] C. Eckart and G. Young. The approximation of one matrix by another of lower rank. *Psychometrika*, 1(3):211–218, 1936.

[160] J. Eckstein and D. P. Bertsekas. On the Douglas—Rachford splitting method and the proximal point algorithm for maximal monotone operators. *Mathematical Programming*, 55:293–318, 1992.

[161] J. Eckstein and W. Yao. Augmented Lagrangian and alternating direction methods for convex optimization: A tutorial and some illustrative computational results. *RUTCOR Research Reports*, 32(3):44, 2012.

[162] B. Efron and R. J. Tibshirani. *An Introduction to the Bootstrap*. Chapman and Hall/CRC, 1994.

[163] L. C. Evans. *Partial Differential Equations*. American Mathematical Society, revised edition, 2010.

[164] E. Even-Dar, Y. Mansour, and P. Bartlett. Learning rates for Q-learning. *Journal of Machine Learning Research*, 5(1):1–25, 2003.

[165] D. Fan, L. Yang, Z. Wang, M. S. Triantafyllou, and G. E. Karniadakis. Reinforcement learning for bluff body active flow control in experiments and simulations. *Proceedings of the National Academy of Sciences*, 117(42):26091–26098, 2020.

[166] D. D. Fan, A.-A. Agha-Mohammadi, and E. A. Theodorou. Deep learning tubes for tube MPC. ArXiv:2002.01587, 2020.

[167] H. J. Ferreau, H. G. Bock, and M. Diehl. An online active set strategy to overcome the limitations of explicit MPC. *International Journal of Robust and Nonlinear Control*, 18(8):816–830, 2008.

[168] H. J. Ferreau, C. Kirches, A. Potschka, H. G. Bock, and M. Diehl. qpOASES: A parametric active-set algorithm for quadratic programming. *Mathematical Programming Computation*, 6(4):327–363, 2014.

[169] A. V. Fiacco and G. P. McCormick. The sequential unconstrained minimization technique for nonlinear programing, a primal–dual method. *Management Science*, 10(2):360–366, 1964.

[170] A. V. Fiacco and G. P. McCormick. *Nonlinear Programming: Sequential Unconstrained Minimization Techniques*. SIAM, 1990.

[171] M. Finzi, K. A. Wang, and A. G. Wilson. Simplifying Hamiltonian and Lagrangian neural networks via explicit constraints. In *Proceedings of the 34th International Conference on Neural Information Processing Systems*, pages 13880–1388, 2020

[172] R. A. Fisher. On the mathematical foundations of theoretical statistics. *Philosophical Transactions of the Royal Society of London. Series A*, 222(594–604):309–368, 1922.

[173] R. Fletcher. A new approach to variable metric algorithms. *The Computer Journal*, 13(3):317–322, 1970.

[174] R. Fletcher. *Practical Methods of Optimization*. Wiley, 2000.

[175] N. Fonzi, S. L. Brunton, and U. Fasel. Data-driven nonlinear aeroelastic models of morphing wings for control. *Proceedings of the Royal Society A*, 476(2239):20200079, 2020.

[176] R. E. Franklin and R. G. Gosling. Molecular configuration in sodium thymonucleate. *Nature*, 171:740–741, 1953.

[177] W. T. Freeman, T. R. Jones, and E. C. Pasztor. Example-based super-resolution. *IEEE Computer Graphics and Applications*, 22(2):56–65, 2002.

[178] W. Friedrich, P. Knipping, and M. von Laue. Interferenz-erscheinungen bei Röntgenstrahlen. In *Sitzungsberichte der Kgl. Bayer. Akad. der Wiss.*, pages 303–322, 1912.

[179] S. Fujimoto, H. Hoof, and D. Meger. Addressing function approximation error in actor-critic methods. In *International Conference on Machine Learning*, pages 1587–1596. PMLR, 2018.

[180] K. Fukami, K. Fukagata, and K. Taira. Super-resolution reconstruction of turbulent flows with machine learning. *Journal of Fluid Mechanics*, 870:106–120, 2019.

[181] K. Fukami, K. Fukagata, and K. Taira. Machine-learning-based spatio-temporal super resolution reconstruction of turbulent flows. *Journal of Fluid Mechanics*, 909:A9. doi:10.1017/jfm.2020.948, 2021.

[182] K. Fukushima. Neocognitron: A self-organizing neural network model for a mechanism of pattern recognition unaffected by shift in position. *Biological Cybernetics*, 36(4):193–202, 1980.

[183] D. Gabay and B. Mercier. A dual algorithm for the solution of nonlinear variational problems via finite element approximations. *Computers and Mathematics with Applications*, 2(1):17–40, 1976.

[184] C. E. Garcia, D. M. Prett, and M. Morari. Model predictive control: theory and practice – a survey. *Automatica*, 25(3):335–348, 1989.

[185] H. P. Gavin. The Levenberg–Marquardt algorithm for nonlinear least squares curve-fitting problems. *Department of Civil and Environmental Engineering, Duke University*, `https://people.duke.edu/~hpgavin/ce281/lm.pdf`, 2019.

[186] M. Gazzola, B. Hejazialhosseini, and P. Koumoutsakos. Reinforcement learning and wavelet adapted vortex methods for simulations of self-propelled swimmers. *SIAM Journal on Scientific Computing*, 36(3):B622–B639, 2014.

[187] I. M. Gelfand and S. V. Fomin. *Calculus of Variations*. Prentice-Hall, 1963.

[188] A. C. Gilbert and P. Indyk. Sparse recovery using sparse matrices. *Proceedings of the IEEE*, 98(6):937–947, 2010.

[189] A. C. Gilbert, J. Y. Park, and M. B. Wakin. Sketched SVD: Recovering spectral features from compressive measurements. ArXiv:1211.0361, 2012.

[190] A. C. Gilbert, M. J. Strauss, and J. A. Tropp. A tutorial on fast Fourier sampling. *IEEE Signal Processing Magazine*, 25(2): 57–66, 2008.

[191] W. R. Gilks, S. Richardson, and D. Spiegelhalter. *Markov Chain Monte Carlo in Practice*. CRC Press, 1995.

[192] P. E. Gill, W. Murray, and M. A. Saunders. SNOPT: An SQP algorithm for large-scale constrained optimization. *SIAM Review*, 47(1):99–131, 2005.

[193] P. E. Gill, W. Murray, and M. H. Wright. *Practical Optimization*. SIAM, 2019.

[194] X. Glorot and Y. Bengio. Understanding the difficulty of training deep feedforward neural networks. In *Proceedings of the 13th International Conference on Artificial Intelligence and Statistics*, pages 249–256. JMLR Workshop and Conference Proceedings, 2010.

[195] R. Glowinski and A. Marrocco. Sur l'approximation, par éléments finis d'ordre un, et la résolution, par pénalisation-dualité d'une classe de problèmes de dirichlet non linéaires. *Revue Française d'Automatique, Informatique, et Recherche Opérationelle*, 9(R2):41–76, 1975.

[196] D. E. Goldberg. *Genetic Algorithms*. Pearson Education, 2006.

[197] D. Goldfarb. A family of variable-metric methods derived by variational means. *Mathematics of Computation*, 24(109):23–26, 1970.

[198] T. Goldstein, B. O'Donoghue, S. Setzer, and R. Baraniuk. Fast alternating direction optimization methods. *SIAM Journal on Imaging Sciences*, 7(3):1588–1623, 2014.

[199] G. Golub and W. Kahan. Calculating the singular values and pseudo-inverse of a matrix. *Journal of SIAM, Series B: Numerical Analysis*, 2(2):205–224, 1965.

[200] G. H. Golub, P. C. Hansen, and D. P. O'Leary. Tikhonov regularization and total least squares. *SIAM Journal on Matrix Analysis and Applications*, 21(1):185–194, 1999.

[201] G. H. Golub and C. Reinsch. Singular value decomposition and least squares solutions. *Numerical Mathematics*, 14:403–420, 1970.

[202] G. H. Golub and C. F. Van Loan. *Matrix Computations*, volume 3. JHU Press, 2012.

[203] R. E. Gomory. *Outline of an algorithm for integer solutions to linear programs and an algorithm for the mixed integer problem*. In *50 Years of Integer Programming 1958–2008*, edited by M. Jünger et al., page 77–103. Springer, 2010.

[204] I. Goodfellow, Y. Bengio, and A. Courville. *Deep Learning*. MIT Press, 2016.

[205] C. Gordon, D. Webb, and S. Wolpert. Isospectral plane domains and surfaces via Riemannian orbifolds. *Inventiones Mathematicae*, 110(1):1–22, 1992.

[206] M. Grant, S. Boyd, and Y. Ye. CVX: Matlab software for disciplined convex programming, `https://cvxr.com/cvx/`, 2008.

[207] M. S. Grewal. Kalman filtering. In *International Encyclopedia of Statistical Science*, edited by M. Lovric, M., pages 705–708. Springer, 2011.

[208] S. Greydanus, M. Dzamba, and J. Yosinski. Hamiltonian neural networks. *Advances in Neural Information Processing Systems*, 32:15379–15389, 2019.

[209] A. Griewank. On automatic differentiation. In *Mathematical Programming: Recent Developments and Applications*, edited by M. Iri and K. Tanabe, pages 83–108. Kluwer

Academic Publishers, 1989.

[210] A. Griewank. Who invented the reverse mode of differentiation. *Documenta Mathematica, Extra Volume ISMP*, 389400:26, 2012.

[211] A. Griewank and G. F. Corliss. *Automatic Differentiation of Algorithms: Theory, Implementation, and Application*. SIAM, 1991.

[212] A. Griewank and A. Walther. *Evaluating Derivatives: Principles and Techniques of Algorithmic Differentiation*. SIAM, 2008.

[213] S. Gu, E. Holly, T. Lillicrap, and S. Levine. Deep reinforcement learning for robotic manipulation with asynchronous off-policy updates. In *2017 IEEE International Conference on Robotics and Automation (ICRA)*, pages 3389–3396. IEEE, 2017.

[214] Y. Guan, S. L. Brunton, and I. Novosselov. Sparse nonlinear models of chaotic electroconvection. *Royal Society Open Science*, 8(8):202367, 2021.

[215] F. Gustafsson, F. Gunnarsson, N. Bergman, U. Forssell, J. Jansson, R. Karlsson, and P.-J. Nordlund. Particle filters for positioning, navigation, and tracking. *IEEE Transactions on Signal Processing*, 50(2):425–437, 2002.

[216] L. Guttman. Enlargement methods for computing the inverse matrix. *Annals of Mathematical Statistics*, 17(3):336–343, 1946.

[217] W. W. Hager. Updating the inverse of a matrix. *SIAM Review*, 31(2):221–239, 1989.

[218] C. R. Harris, K. J. Millman, S. J. Van Der Walt, R. Gommers, P. Virtanen, D. Cournapeau, E. Wieser, J. Taylor, S. Berg, N. J. Smith, et al. Array programming with NumPy. *Nature*, 585(7825):357–362, 2020.

[219] D. Harrison Jr and D. L. Rubinfeld. Hedonic housing prices and the demand for clean air. *Journal of Environmental Economics and Management*, 5(1):81–102, 1978.

[220] T. Hastie, R. Tibshirani, J. Friedman, T. Hastie, J. Friedman, and R. Tibshirani. *The Elements of Statistical Learning*. Springer, 2009.

[221] M. Hausknecht and P. Stone. Deep recurrent Q-learning for partially observable MDPs. In *2015 AAAI Fall Symposium Series*, 2015.

[222] B. He and X. Yuan. On the $o(1/t)$ convergence rate of the Douglas–Rachford alternating direction method. *SIAM Journal on Numerical Analysis*, 50(2):700–709, 2012.

[223] K. He, X. Zhang, S. Ren, and J. Sun. Delving deep into rectifiers: Surpassing human-level performance on imagenet classification. In *Proceedings of the IEEE International Conference on Computer Vision*, pages 1026–1034, IEEE, 2015.

[224] K. He, X. Zhang, S. Ren, and J. Sun. Deep residual learning for image recognition. In *Proceedings of the IEEE Conference on Computer Vision and Pattern Recognition*, pages 770–778, 2016.

[225] M. Herceg, C. Jones, and M. Morari. Dominant speed factors of active set methods for fast MPC. *Optimal Control Applications and Methods*, 36(5):608–627, 2015.

[226] K. K. Herrity, A. C. Gilbert, and J. A. Tropp. Sparse approximation via iterative thresholding. In *2006 IEEE International Conference on Acoustics Speech and Signal Processing Proceedings, Toulouse, France, 2006*, pages III-III, doi: 10.1109/ICASSP.2006.1660731.

[227] N. J. Higham. The symmetric Procrustes problem. *BIT Numerical Mathematics*, 28(1):133–143, 1988.

[228] F. S. Hillier and G. J. Lieberman. *Introduction to Operations Research*. McGraw-Hill, 2015.

[229] G. E. Hinton, S. Osindero, and Y.-W. Teh. A fast learning algorithm for deep belief nets. *Neural Computation*, 18(7):1527–1554, 2006.

[230] M. Hinze, R. Pinnau, M. Ulbrich, and S. Ulbrich. *Optimization with PDE Constraints*. Springer, 2008.

[231] B. L. Ho and R. E. Kalman. Effective construction of linear state-variable models from input/output data. In *Proceedings of the 3rd Annual Allerton Conference on Circuit and System Theory*, pages 449–459, 1965.

[232] A. E. Hoerl and R. W. Kennard. Ridge regression: Biased estimation for nonorthogonal problems. *Technometrics*, 12(1):55–67, 1970.

[233] M. Hoffmann, M. Scherer, T. Hempel, A. Mardt, B. de Silva, B. E. Husic, S. Klus, H. Wu, N. Kutz, S. L. Brunton, et al. Deeptime: a Python library for machine learning dynamical models from time series data. *Machine Learning: Science and Technology*, 3(1):015009, 2021.

[234] P. Holl, V. Koltun, and N. Thuerey. Learning to control PDEs with differentiable physics. ArXiv:2001.07457, 2020.

[235] J. H. Holland. *Adaptation in Natural and Artificial Systems: An Introductory Analysis with Applications to Biology, Control, and Artificial Intelligence.* University of Michigan Press, 1975.

[236] J. J. Hopfield. Neural networks and physical systems with emergent collective computational abilities. *Proceedings of the National Academy of Sciences*, 79(8):2554–2558, 1982.

[237] K. Hornik. Approximation capabilities of multilayer feedforward networks. *Neural Networks*, 4(2):251–257, 1991.

[238] K. Hornik, M. Stinchcombe, and H. White. Multilayer feedforward networks are universal approximators. *Neural Networks*, 2(5):359–366, 1989.

[239] G. Hounsfield. Computerized transverse axial scanning (tomography): Part 1. description of system. *British Journal of Radiology*, 46:1016–1022, 1973.

[240] B. Houska, H. J. Ferreau, and M. Diehl. An auto-generated real-time iteration algorithm for nonlinear MPC in the microsecond range. *Automatica*, 47(10):2279–2285, 2011.

[241] S. Huang, R. F. J. Dossa, C. Ye, J. Braga, D. Chakraborty, K. Mehta, and J. G. Araújo. CleanRL: High-quality single-file implementations of deep reinforcement learning algorithms. *Journal of Machine Learning Research*, 23(274):1–18, 2022.

[242] S. M. Ichinaga, F. Andreuzzi, N. Demo, M. Tezzele, K. Lapo, G. Rozza, S. L. Brunton, and J. N. Kutz. PyDMD: A Python package for robust dynamic mode decomposition. ArXiv:2402.07463, 2024.

[243] M. A. Iglesias, K. J. Law, and A. M. Stuart. Ensemble Kalman methods for inverse problems. *Inverse Problems*, 29(4):045001, 2013.

[244] M. Innes. Flux: Elegant machine learning with Julia. *Journal of Open Source Software*, 3(25):602, 2018.

[245] M. Innes, A. Edelman, K. Fischer, C. Rackauckas, E. Saba, V. B. Shah, and W. Tebbutt. A differentiable programming system to bridge machine learning and scientific computing. ArXiv:1907.07587, 2019.

[246] S. Ioffe and C. Szegedy. Batch normalization: Accelerating deep network training by reducing internal covariate shift. *PMLR*, 37:448-456, 2015.

[247] G. James, D. Witten, T. Hastie, and R. Tibshirani. *An Introduction to Statistical Learning*. Springer, 2013.

[248] A. Jameson, L. Martinelli, and N. Pierce. Optimum aerodynamic design using the Navier–Stokes equations. *Theoretical and Computational Fluid Dynamics*, 10(1-4):213–237, 1998.

[249] W. Jin, Z. Wang, Z. Yang, and S. Mou. Pontryagin differentiable programming: An end-to-end learning and control framework. *Advances in Neural Information Processing Systems*, 33:7979–7992, 2020.

[250] M. I. Jordan, Z. Ghahramani, T. S. Jaakkola, and L. K. Saul. An introduction to variational methods for graphical models. *Machine Learning*, 37:183–233, 1999.

[251] M. R. Jovanović, P. J. Schmid, and J. W. Nichols. Sparsity-promoting dynamic mode decomposition. *Physics of Fluids*, 26(2):024103, 2014.

[252] J. N. Juang. *Applied System Identification*. Prentice Hall, 1994.

[253] J. N. Juang and R. S. Pappa. An eigensystem realization algorithm for modal parameter identification and model reduction. *Journal of Guidance, Control, and Dynamics*, 8(5):620–627, 1985.

[254] J. N. Juang, M. Phan, L. G. Horta, and R. W. Longman. Identification of observer/Kalman filter Markov parameters: Theory and experiments. Technical Memorandum 104069, NASA, 1991.

[255] S. J. Julier and J. K. Uhlmann. New extension of the Kalman filter to nonlinear systems. In *Proc. SPIE 3068, Signal Processing, Sensor Fusion, and Target Recognition VI*, volume 3068, pages 182–193. SPIE, 1997.

[256] S. J. Julier and J. K. Uhlmann. Unscented filtering and nonlinear estimation. *Proceedings of the IEEE*, 92(3):401–422, 2004.

[257] J. Jumper, R. Evans, A. Pritzel, T. Green, M. Figurnov, O. Ronneberger, K. Tunyasuvunakool, R. Bates, A. Žídek, A. Potapenko, et al. Highly accurate protein structure prediction with AlphaFold. *Nature*, 596(7873):583–589, 2021.

[258] O. Junge, J. E. Marsden, and S. Ober-Blöbaum. Discrete mechanics and optimal control. *IFAC Proceedings Volumes*, 38(1):538–543, 2005.

[259] M. Kac. Can one hear the shape of a drum? *American Mathematical Monthly*, 73(4):1–23, 1966.

[260] S. Kaczmarz. Approximate solution of systems of linear equations. *International Journal of Control*, 57(6):1269–1271, 1993. Originally published as Angenäherte Auflösung von Systemen linearer Gleichungen in 1937.

[261] L. P. Kaelbling, M. L. Littman, and A. W. Moore. Reinforcement learning: A survey. *Journal of Artificial Intelligence Research*, 4:237–285, 1996.

[262] E. Kaiser, J. N. Kutz, and S. L. Brunton. Discovering conservation laws from data for control. In *2018 IEEE Conference on Decision and Control (CDC)*, pages 6415–6421. IEEE, 2018.

[263] E. Kaiser, J. N. Kutz, and S. L. Brunton. Sparse identification of nonlinear dynamics for model predictive control in the low-data limit. *Proceedings of the Royal Society of London A*, 474(2219), 2018.

[264] E. Kaiser, B. R. Noack, L. Cordier, A. Spohn, M. Segond, M. Abel, G. Daviller, J. Östh, S. Krajnović, and R. K. Niven. Cluster-based reduced-order modelling of a mixing layer. *Journal of Fluid Mechanics*, 754:365–414, 2014.

[265] S. M. Kakade. A natural policy gradient. In *Proceedings of the 15th International Conference on Neural Information Processing Systems*, pages 1531–1538, 2001.

[266] R. E. Kalman. A new approach to linear filtering and prediction problems. *Journal of Fluids Engineering*, 82(1):35–45, 1960.

[267] L. V. Kantorovich. *Mathematical Methods in the Organization and Planning of Production* (in Russian). Leningrad State University Publishers, 1939. Translated as translated as "Mathematical methods in the organization and planning of production", *Management Science*, 6(4):366–422, 1960.

[268] A. A. Kaptanoglu, J. L. Callaham, C. J. Hansen, A. Aravkin, and S. L. Brunton. Promoting global stability in data-driven models of quadratic nonlinear dynamics. *Physical Review Fluids*, 6(094401), 2021.

[269] A. A. Kaptanoglu, B. M. de Silva, U. Fasel, K. Kaheman, J. L. Callaham, C. B. Delahunt, K. Champion, J.-C. Loiseau, J. N. Kutz, and S. L. Brunton. PySINDy: A comprehensive

Python package for robust sparse system identification. *Journal of Open Source Software*, 7(69):3994, 2022.

[270] N. Karmarkar. A new polynomial-time algorithm for linear programming. In *Proceedings of the 16th Annual ACM Symposium on Theory of computing*, pages 302–311, 1984.

[271] G. E. Karniadakis, I. G. Kevrekidis, L. Lu, P. Perdikaris, S. Wang, and L. Yang. Physics-informed machine learning. *Nature Reviews Physics*, 3(6):422–440, 2021.

[272] W. Karush. Minima of functions of several variables with inequalities as side conditions. In *Traces and Emergence of Nonlinear Programming*, edited by G. Giorgi, T. Kjeldsen, pages 217–245. Springer, 2013.

[273] H. J. Kelley. Gradient theory of optimal flight paths. *ARS Journal*, 30(10):947–954, 1960.

[274] J. Kennedy and R. Eberhart. Particle swarm optimization. In *Proceedings of ICNN'95*, volume 4, pages 1942–1948. IEEE, 1995.

[275] I. Kim, M. Choi, and H. J. Kim. UP-NeRF: Unconstrained pose prior-free neural radiance field. *Advances in Neural Information Processing Systems*, 36:68184–68196, 2023.

[276] D. P. Kingma and J. Ba. Adam: A method for stochastic optimization. ArXiv:1412.6980, 2014.

[277] D. P. Kingma and M. Welling. Auto-encoding variational Bayes. ArXiv:1312.6114, 2013.

[278] C. Kirches, S. Sager, H. G. Bock, and J. P. Schlöder. Time-optimal control of automobile test drives with gear shifts. *Optimal Control Applications and Methods*, 31(2):137–153, 2010.

[279] S. Kirkpatrick, C. D. Gelatt Jr, and M. P. Vecchi. Optimization by simulated annealing. *Science*, 220(4598):671–680, 1983.

[280] J. Kober and J. Peters. Reinforcement learning in robotics: A survey. In *Reinforcement Learning*, edited by M. Wiering and M. Otterlo, pages 579–610. Springer, 2012.

[281] N. Komodakis and J.-C. Pesquet. Playing with duality: An overview of recent primal-dual approaches for solving large-scale optimization problems. *IEEE Signal Processing Magazine*, 32(6):31–54, 2015.

[282] B. O. Koopman. On distributions admitting a sufficient statistic. *Transactions of the American Mathematical society*, 39(3):399–409, 1936.

[283] M. Korda and C. N. Jones. Certification of fixed computation time first-order optimization-based controllers for a class of nonlinear dynamical systems. In *2014 American Control Conference*, pages 3602–3608. IEEE, 2014.

[284] B. Korte and J. Vygen. *Combinatorial Optimization: Theory and Algorithms*. Springer, 2008.

[285] J. Kou and W. Zhang. Data-driven modeling for unsteady aerodynamics and aeroelasticity. *Progress in Aerospace Sciences*, 125:100725, 2021.

[286] J. R. Koza. *Genetic Programming: On the Programming of Computers by Means of Natural Selection*. MIT Press, 1992.

[287] J. R. Koza, F. H. Bennett III, and O. Stiffelman. Genetic programming as a Darwinian invention machine. In *Genetic Programming*, pages 93–108. Springer, 1999.

[288] A. Krishnapriyan, A. Gholami, S. Zhe, R. Kirby, and M. W. Mahoney. Characterizing possible failure modes in physics-informed neural networks. *Advances in Neural Information Processing Systems*, 34:26548–26560, 2021.

[289] A. Krizhevsky, G. Hinton, et al. Learning multiple layers of features from tiny images. `https://www.cs.toronto.edu/~kriz/learning-features-2009-TR.pdf`, 2009.

[290] A. Krizhevsky, I. Sutskever, and G. E. Hinton. Imagenet classification with deep convolutional neural networks. In *Advances in Neural Information Processing Systems*, pages 1097–1105, 2012.

[291] M. Krstić and H. Wang. Stability of extremum seeking feedback for general nonlinear dynamic systems. *Automatica*, 36(4):595–601, 2000.

[292] H. W. Kuhn and A. W. Tucker. Nonlinear programming. In *Proceedings of the Second Berkeley Symposium on Mathematical Statistics and Probability*, pages 481–492, 1951.

[293] S. Kullback and R. A. Leibler. On information and sufficiency. *Annals of Mathematical Statistics*, 22(1):79–86, 1951.

[294] J. C. Lagarias, J. A. Reeds, M. H. Wright, and P. E. Wright. Convergence properties of the Nelder–Mead simplex method in low dimensions. *SIAM Journal on Optimization*, (1):112–147, 1998.

[295] C. Lagemann, L. Paehler, J. Callaham, S. Mokbel, S. Ahnert, K. Lagemann, E. Lagemann, N. Adams, and S. Brunton. HydroGym: A reinforcement learning platform for fluid dynamics. In *7th Annual Learning for Dynamics & Control Conference*, pages 497–512. PMLR, 2025.

[296] J. L. Lagrange. *Mechanique Analitique*. La Veuve Desaint, 1788.

[297] S. Lanka and T. Wu. Archer: Aggressive rewards to counter bias in hindsight experience replay. ArXiv:1809.02070, 2018.

[298] P.-S. Laplace. *Théorie Analytique des Probabilités*. Veuve Courcier, Paris, 1812.

[299] A. Larrañaga, J. Martínez, J. Míguez, and J. Porteiro. Robust optimization of heat-transfer-enhancing microtextured surfaces based on machine learning surrogate models. *International Communications in Heat and Mass Transfer*, 151:107218, 2024.

[300] A. Laub. A Schur method for solving algebraic Riccati equations. *IEEE Transactions on Automatic Control*, 24(6):913–921, 1979.

[301] P. C. Lauterbur. Image formation by induced local interactions: Examples employing nuclear magnetic resonance. *Nature*, 242:190–191, 1973.

[302] H. Le, C. Voloshin, and Y. Yue. Batch policy learning under constraints. In *International Conference on Machine Learning*, pages 3703–3712. PMLR, 2019.

[303] Y. LeCun, Y. Bengio, and G. Hinton. Deep learning. *Nature*, 521(7553):436–444, 2015.

[304] Y. LeCun, L. Bottou, Y. Bengio, and P. Haffner. Gradient-based learning applied to document recognition. *Proceedings of the IEEE*, 86(11):2278–2324, 1998.

[305] C. Ledig, L. Theis, F. Huszár, J. Caballero, A. Cunningham, A. Acosta, A. Aitken, A. Tejani, J. Totz, Z. Wang, et al. Photo-realistic single image super-resolution using a generative adversarial network. In *Proceedings of the IEEE Conference on Computer Vision and Pattern Recognition*, pages 4681–4690, 2017.

[306] I. Lenz, R. A. Knepper, and A. Saxena. DeepMPC: Learning deep latent features for model predictive control. In *Robotics: Science and Systems*, Rome, Italy. `https://www.roboticsproceedings.org/rss11/p12.pdf`, 2015.

[307] K. Levenberg. A method for the solution of certain non-linear problems in least squares. *Quarterly of Applied Mathematics*, 2(2):164–168, 1944.

[308] K. Li, J. Kou, and W. Zhang. Deep neural network for unsteady aerodynamic and aeroelastic modeling across multiple Mach numbers. *Nonlinear Dynamics*, 96(3):2157–2177, 2019.

[309] Z. Li, N. Kovachki, K. Azizzadenesheli, B. Liu, K. Bhattacharya, A. Stuart, and A. Anandkumar. Fourier neural operator for parametric partial differential equations. ArXiv:2010.08895, 2020.

[310] Z. Li, N. Kovachki, K. Azizzadenesheli, B. Liu, K. Bhattacharya, A. Stuart, and A. Anandkumar. Neural operator: Graph kernel network for partial differential equations. ArXiv:2003.03485, 2020.

[311] F. Lin, M. Fardad, and M. R. Jovanovic. Design of optimal sparse feedback gains via the alternating direction method of multipliers. *IEEE Transactions on Automatic Control*,

58(9):2426–2431, 2013.

[312] Z. Lin, M. Chen, and Y. Ma. The augmented Lagrange multiplier method for exact recovery of corrupted low-rank matrices. ArXiv:1009.5055, 2010.

[313] D. V. Lindley. *Bayesian Statistics: A Review.* SIAM, 1972.

[314] J. Ling, A. Kurzawski, and J. Templeton. Reynolds averaged turbulence modelling using deep neural networks with embedded invariance. *Journal of Fluid Mechanics*, 807:155–166, 2016.

[315] S. Linnainmaa. *The Representation of the Cumulative Rounding Error of an Algorithm as a Taylor Expansion of the Local Rounding Errors.* Master's Thesis (in Finnish), University of Helsinki, 1970.

[316] P. Lions and B. Mercier. Splitting algorithms for the sum of two nonlinear operators. *SIAM Journal on Numerical Analysis*, 16(6):964–979, 1979.

[317] D. C. Liu and J. Nocedal. On the limited memory BFGS method for large scale optimization. *Mathematical Programming*, 45(1):503–528, 1989.

[318] L. Ljung. *System Identification: Theory for the User.* Prentice-Hall, 1999.

[319] J.-C. Loiseau. Data-driven modeling of the chaotic thermal convection in an annular thermosyphon. *Theoretical and Computational Fluid Dynamics*, 34(4):339–365, 2020.

[320] J.-C. Loiseau and S. L. Brunton. Constrained sparse Galerkin regression. *Journal of Fluid Mechanics*, 838:42–67, 2018.

[321] J.-C. Loiseau, B. R. Noack, and S. L. Brunton. Sparse reduced-order modeling: sensor-based dynamics to full-state estimation. *Journal of Fluid Mechanics*, 844:459–490, 2018.

[322] B. T. Lopez, J.-J. E. Slotine, and J. P. How. Dynamic tube MPC for nonlinear systems. In *2019 American Control Conference (ACC)*, pages 1655–1662. IEEE, 2019.

[323] L. Lu, P. Jin, G. Pang, Z. Zhang, and G. E. Karniadakis. Learning nonlinear operators via DeepONet based on the universal approximation theorem of operators. *Nature Machine Intelligence*, 3(3):218–229, 2021.

[324] B. Lusch, J. N. Kutz, and S. L. Brunton. Deep learning for universal linear embeddings of nonlinear dynamics. *Nature Communications*, 9(1):4950, 2018.

[325] M. Lustig, D. L. Donoho, J. M. Santos, and J. M. Pauly. Compressed sensing MRI. *IEEE Signal Processing Magazine*, 25(2):72–82, 2008.

[326] A. M. Lyapunov. The general problem of the stability of motion. *International Journal of Control*, 55(3):531–534, 1992.

[327] S. Mallat. Understanding deep convolutional networks. *Phil. Trans. R. Soc. A*, 374(2065):20150203, 2016.

[328] R. Malouf. A comparison of algorithms for maximum entropy parameter estimation. In *COLING-02: Proceedings of the 6th Conference on Natural Language Learning*, pages 1–20, 2002.

[329] J. Mandel. Use of the singular value decomposition in regression analysis. *The American Statistician*, 36(1):15–24, 1982.

[330] K. Manohar, B. W. Brunton, J. N. Kutz, and S. L. Brunton. Data-driven sparse sensor placement: Demonstrating the benefits of exploiting known patterns. *IEEE Control Systems Magazine*, 38(3):63–86, 2018.

[331] P. Mansfield. Multi-planar image formation using NMR spin echoes. *Journal of Physics C: Solid State Physics*, 10:L55–L58, 1977.

[332] Z. Mao, A. D. Jagtap, and G. E. Karniadakis. Physics-informed neural networks for high-speed flows. *Computer Methods in Applied Mechanics and Engineering*, 360:112789, 2020.

[333] A. Mardt, L. Pasquali, H. Wu, and F. Noé. VAMPnets: Deep learning of molecular kinetics. *Nature Communications*, 9(5), 2018.

[334] D. W. Marquardt. An algorithm for least-squares estimation of nonlinear parameters. *Journal of SIAM*, 11(2):431–441, 1963.

[335] J. E. Marsden and T. S. Ratiu. *Introduction to Mechanics and Symmetry*, 2nd edition. Springer, 1999.

[336] J. E. Marsden and A. Tromba. *Vector Calculus*. Macmillan, 2003.

[337] J. E. Marsden and M. West. Discrete mechanics and variational integrators. *Acta Numerica*, 10:357–514, 2001.

[338] S. Martello and P. Toth. *Knapsack problems: algorithms and computer implementations*. Wiley, 1990.

[339] A. F. Martins, M. A. Figueiredo, P. M. Aguiar, N. A. Smith, and E. P. Xing. An augmented Lagrangian approach to constrained MAP inference. In *Proceedings of the 28th International Conference on Machine Learning (ICML)*, pages 169–176, 2011.

[340] J. R. Martins and A. Ning. *Engineering Design Optimization*. Cambridge University Press, 2021.

[341] J. L. Maryak, J. C. Spall, and B. D. Heydon. Use of the Kalman filter for inference in state-space models with unknown noise distributions. *IEEE Transactions on Automatic Control*, 49(1):87–90, 2004.

[342] H. Matthies and G. Strang. The solution of nonlinear finite element equations. *International Journal for Numerical Methods in Engineering*, 14(11):1613–1626, 1979.

[343] D. Q. Mayne. Model predictive control: Recent developments and future promise. *Automatica*, 50(12):2967–2986, 2014.

[344] D. Q. Mayne, J. B. Rawlings, C. V. Rao, and P. O. Scokaert. Constrained model predictive control: Stability and optimality. *Automatica*, 36(6):789–814, 2000.

[345] W. S. McCulloch and W. Pitts. A logical calculus of the ideas immanent in nervous activity. *The Bulletin of Mathematical Biophysics*, 5(4):115–133, 1943.

[346] K. I. M. McKinnon. Convergence of the Nelder–Mead simplex method to a nonstationary point. *SIAM Journal on Optimization*, 9(1):148–158, 1998.

[347] X. Meng and G. E. Karniadakis. A composite neural network that learns from multi-fidelity data: Application to function approximation and inverse PDE problems. *Journal of Computational Physics*, 401:109020, 2020.

[348] O. Meshi and A. Globerson. An alternating direction method for dual MAP LP relaxation. In *Joint European conference on machine learning and knowledge discovery in databases*, pages 470–483. Springer, 2011.

[349] B. Mildenhall, P. P. Srinivasan, M. Tancik, J. T. Barron, R. Ramamoorthi, and R. Ng. NeRF: Representing scenes as neural radiance fields for view synthesis. *Communications of the ACM*, 65(1):99–106, 2021.

[350] B. K. Miller, M. Geiger, T. E. Smidt, and F. Noé. Relevance of rotationally equivariant convolutions for predicting molecular properties. ArXiv:2008.08461, 2020.

[351] R. Milman and E. Davison. A fast MPC algorithm using nonfeasible active set methods. *Journal of Optimization Theory and Applications*, 139(3):591–616, 2008.

[352] M. Minsky. Steps toward artificial intelligence. *Proceedings of the IRE*, 49(1):8–30, 1961.

[353] M. Minsky and S. Papert. *Perceptrons*. MIT Press, 1969.

[354] V. Mnih, K. Kavukcuoglu, D. Silver, A. A. Rusu, J. Veness, M. G. Bellemare, A. Graves, M. Riedmiller, A. K. Fidjeland, G. Ostrovski, et al. Human-level control through deep reinforcement learning. *Nature*, 518(7540):529, 2015.

[355] P. R. Montague, P. Dayan, and T. J. Sejnowski. A framework for mesencephalic dopamine systems based on predictive hebbian learning. *Journal of Neuroscience*, 16(5):1936–1947,

1996.
[356] B. C. Moore. Principal component analysis in linear systems: Controllability, observability, and model reduction. *IEEE Transactions on Automatic Control*, AC-26(1):17–32, 1981.
[357] J. J. Moré. The Levenberg–Marquardt algorithm: Implementation and theory. In *Numerical Analysis*, edited by G.A. Watson, pages 105–116. Springer, 1978.
[358] J. Morton, F. D. Witherden, A. Jameson, and M. J. Kochenderfer. Deep dynamical modeling and control of unsteady fluid flows. *Proceedings of the 32nd Conference on Neural Information Processing Systems (NeurIPS 2018)*, pages 9278–9288, 2018.
[359] K. P. Murphy. *Machine Learning: a Probabilistic Perspective*. MIT Press, 2012.
[360] R. M. Neal. MCMC using Hamiltonian dynamics. *Handbook of Markov Chain Monte Carlo*, edited by S. Brooks, A. Gelman, G.L. Jones, X.-L. Meng, pages 113–162. Chapman and Hall/CRC, 2011.
[361] A. Nedić and A. Ozdaglar. Distributed subgradient methods for multi-agent optimization. *IEEE Transactions on Automatic Control*, 54(1):48–61, 2009.
[362] D. Needell and J. A. Tropp. CoSaMP: iterative signal recovery from incomplete and inaccurate samples. *Communications of the ACM*, 53(12):93–100, 2010.
[363] J. A. Nelder and R. Mead. A simplex method for function minimization. *The Computer Journal*, 7(4):308–313, 1965.
[364] Y. Nesterov. A method for unconstrained convex minimization problem with the rate of convergence $O(1/k^2)$. *Dokl. Akad. Nauk. SSSR*, 269(3):543, 1983.
[365] Y. Nesterov. *Introductory Lectures on Convex Optimization: A Basic Course*, Springer, 2013.
[366] Y. Nesterov. *Lectures on Convex Optimization*, 2nd edition. Springer, 2018.
[367] Y. Nesterov and A. Nemirovskii. *Interior-Point Polynomial Algorithms in Convex Programming*. SIAM, 1994.
[368] A. Y. Ng, M. I. Jordan, and Y. Weiss. On spectral clustering: Analysis and an algorithm. *Advances in Neural Information Processing Systems (NIPS)*, 14:849–856, 2002.
[369] J. Nocedal. Updating quasi-Newton matrices with limited storage. *Mathematics of Computation*, 35(151):773–782, 1980.
[370] J. Nocedal and S. J. Wright. *Numerical optimization*. Springer, 1999.
[371] J. F. Nolan. *Analytical differentiation on a digital computer*. PhD thesis, Massachusetts Institute of Technology, 1953.
[372] G. Novati, H. L. de Laroussilhe, and P. Koumoutsakos. Automating turbulence modelling by multi-agent reinforcement learning. *Nature Machine Intelligence*, 3(1):87–96, 2021.
[373] G. Novati, L. Mahadevan, and P. Koumoutsakos. Controlled gliding and perching through deep-reinforcement-learning. *Physical Review Fluids*, 4(9):093902, 2019.
[374] B. O'Donoghue, E. Chu, N. Parikh, and S. Boyd. Conic optimization via operator splitting and homogeneous self-dual embedding. *Journal of Optimization Theory and Applications*, 169(3):1042–1068, 2016.
[375] OpenAI. Gpt-4 technical report. `https://openai.com/research/gpt-4`, 2023. Accessed: 2025-07-20.
[376] S. E. Otto, N. Zolman, J. N. Kutz, and S. L. Brunton. A unified framework to enforce, discover, and promote symmetry in machine learning. ArXiv:2311.00212, 2023.
[377] L. Ouyang, J. Wu, X. Jiang, D. Almeida, C. Wainwright, P. Mishkin, C. Zhang, S. Agarwal, K. Slama, A. Ray, et al. Training language models to follow instructions with human feedback. *Advances in Neural Information Processing Systems*, 35:27730–27744, 2022.
[378] X. Pan, Y. You, Z. Wang, and C. Lu. Virtual to real reinforcement learning for autonomous driving. ArXiv:1704.03952, 2017.

[379] G. Pang, L. Lu, and G. E. Karniadakis. fPINNs: Fractional physics-informed neural networks. *SIAM Journal on Scientific Computing*, 41(4):A2603–A2626, 2019.

[380] N. Parikh, S. Boyd, et al. Proximal algorithms. *Foundations and Trends® in Optimization*, 1(3):127–239, 2014.

[381] D. Parker. Learning-logic: Casting the cortex of the human brain in silicon. Technical Report TR-47. February 1985. Center for Computational Research in Economics and Management Science, MIT. 51 pages.

[382] A. Paszke, S. Gross, S. Chintala, G. Chanan, E. Yang, Z. DeVito, Z. Lin, A. Desmaison, L. Antiga, and A. Lerer. Automatic differentiation in PyTorch. In Workshop, 31st Conference on Neural Information Processing Systems (NeurIPS), 2017.

[383] D. Pathak, P. Agrawal, A. A. Efros, and T. Darrell. Curiosity-driven exploration by self-supervised prediction. In *International Conference on Machine Learning*, pages 2778–2787. PMLR, 2017.

[384] J. Pathak, B. Hunt, M. Girvan, Z. Lu, and E. Ott. Model-free prediction of large spatiotemporally chaotic systems from data: a reservoir computing approach. *Physical Review Letters*, 120(2):024102, 2018.

[385] J. Pathak, Z. Lu, B. R. Hunt, M. Girvan, and E. Ott. Using machine learning to replicate chaotic attractors and calculate Lyapunov exponents from data. *Chaos: An Interdisciplinary Journal of Nonlinear Science*, 27(12):121102, 2017.

[386] P. I. Pavlov. *Conditioned Reflexes: An Investigation of the Physiological Activity of the Cerebral Cortex*. Oxford University Press, 1927.

[387] R. Penrose. A generalized inverse for matrices. *Mathematical Proceedings of the Cambridge Philosophical Society*, 51:406–413, 1955.

[388] R. Penrose and J. A. Todd. On best approximate solutions of linear matrix equations. *Mathematical Proceedings of the Cambridge Philosophical Society*, 52:17–19, 1956.

[389] T. Pietrzykowski. An exact potential method for constrained maxima. *SIAM Journal on Numerical Analysis*, 6(2):299–304, 1969.

[390] E. J. G. Pitman. Sufficient statistics and intrinsic accuracy. *Mathematical Proceedings of the Cambridge Philosophical Society*, 32:567–579, 1936.

[391] C. Pivot, L. Mathelin, L. Cordier, F. Guéniat, and B. R. Noack. A continuous reinforcement learning strategy for closed-loop control in fluid dynamics. In *35th AIAA Applied Aerodynamics Conference*, page 3566, 2017.

[392] B. T. Polyak. Some methods of speeding up the convergence of iteration methods. *USSR Computational Mathematics and Mathematical Physics*, 4(5):1–17, 1964.

[393] M. Popova, O. Isayev, and A. Tropsha. Deep reinforcement learning for de novo drug design. *Science Advances*, 4(7):eaap7885, 2018.

[394] F. A. Potra and S. J. Wright. Interior-point methods. *Journal of Computational and Applied Mathematics*, 124(1-2):281–302, 2000.

[395] M. Powell. On the convergence of the DFP algorithm for unconstrained optimization when there are only two variables. *Mathematical Programming*, 87:281–301, 2000.

[396] M. J. D. Powell. A hybrid method for nonlinear equations. *Numerical Methods for Nonlinear Algebraic Equations*, 7:87–114, 1970.

[397] J. L. Proctor, S. L. Brunton, and J. N. Kutz. Dynamic mode decomposition with control. *SIAM Journal on Applied Dynamical Systems*, 15(1):142–161, 2016.

[398] C. Rackauckas, Y. Ma, J. Martensen, C. Warner, K. Zubov, R. Supekar, D. Skinner, A. Ramadhan, and A. Edelman. Universal differential equations for scientific machine learning. ArXiv:2001.04385, 2020.

[399] A. Radford, J. W. Kim, C. Hallacy, A. Ramesh, G. Goh, S. Agarwal, G. Sastry, A. Askell, P. Mishkin, J. Clark, et al. Learning transferable visual models from natural language supervision. *International Conference on Machine Learning*, pages 8748–8763, 2021.

[400] A. Radford, J. Wu, R. Child, D. Luan, D. Amodei, and I. Sutskever. Language models are unsupervised multitask learners. *OpenAI Blog*, 1(8), 2019.

[401] J. Radon. Über die bestimmung von funktionen durch ihre integralwerte längs gewisser mannigfaltigkeiten. *Berichte Sächsische Akademie der Wissenschaften*, 69:262–277, 1917.

[402] A. Raffin, A. Hill, A. Gleave, A. Kanervisto, M. Ernestus, and N. Dormann. Stable-Baselines3: Reliable reinforcement learning implementations. *Journal of Machine Learning Research*, 22(268):1–8, 2021.

[403] M. Raissi, P. Perdikaris, and G. Karniadakis. Physics-informed neural networks: A deep learning framework for solving forward and inverse problems involving nonlinear partial differential equations. *Journal of Computational Physics*, 378:686–707, 2019.

[404] M. Raissi, A. Yazdani, and G. E. Karniadakis. Hidden fluid mechanics: Learning velocity and pressure fields from flow visualizations. *Science*, 367(6481):1026–1030, 2020.

[405] L. B. Rall. Perspectives on automatic differentiation: past, present, and future? In *Automatic Differentiation: Applications, Theory, and Implementations*, pages 1–14. Springer, 2006.

[406] A. Ramesh, P. Dhariwal, A. Nichol, C. Chu, and M. Chen. Hierarchical text-conditional image generation with clip latents. ArXiv:2204.06125, 2022.

[407] C. V. Rao, S. J. Wright, and J. B. Rawlings. Application of interior-point methods to model predictive control. *Journal of Optimization Theory and Applications*, 99:723–757, 1998.

[408] S. S. Rao. *Engineering Optimization: Theory and Practice*. Wiley, 2019.

[409] B. Recht. A tour of reinforcement learning: The view from continuous control. *Annual Review of Control, Robotics, and Autonomous Systems*, 2:253–279, 2019.

[410] S. Reddy, A. D. Dragan, and S. Levine. Shared autonomy via deep reinforcement learning. ArXiv:1802.01744, 2018.

[411] R. H. Reichle, D. B. McLaughlin, and D. Entekhabi. Hydrologic data assimilation with the ensemble Kalman filter. *Monthly Weather Review*, 130(1):103–114, 2002.

[412] M. Retchin, B. Amos, S. Brunton, and S. Song. Koopman constrained policy optimization: A Koopman operator theoretic method for differentiable optimal control in robotics. In *ICML 2023 Workshop on Differentiable Almost Everything: Differentiable Relaxations, Algorithms, Operators, and Simulators*, 2023.

[413] J. Reuther, A. Jameson, J. Farmer, L. Martinelli, and D. Saunders. Aerodynamic shape optimization of complex aircraft configurations via an adjoint formulation. Research Institute for Advanced Computer Science, NASA Ames Research Center, 1996.

[414] J. A. Rice. *Mathematical Statistics and Data Analysis*. Duxbury Press, 2nd edition, 1994.

[415] A. Richards and J. How. A decentralized algorithm for robust constrained model predictive control. In *Proceedings of the 2004 American Control Conference*, pages 4261–4266. IEEE, 2004.

[416] A. Richards and J. P. How. Robust distributed model predictive control. *International Journal of Control*, 80(9):1517–1531, 2007.

[417] S. Richter, C. N. Jones, and M. Morari. Computational complexity certification for real-time MPC with input constraints based on the fast gradient method. *IEEE Transactions on Automatic Control*, 57(6):1391–1403, 2011.

[418] B. Ristic, S. Arulampalam, and N. J. Gordon. *Beyond the Kalman filter: Particle filters for tracking applications*. Artech house, 2004.

[419] R. T. Rockafellar. *Convex Analysis*. Princeton University Press, 1970.

[420] R. T. Rockafellar. *Conjugate Duality and Optimization*. SIAM, 1974.

[421] R. T. Rockafellar. Lagrange multipliers and optimality. *SIAM Review*, 35(2):183–238, 1993.

[422] R. T. Rockafellar and R. J. Wets. *Variational Analysis*. Springer, 1998.

[423] V. Ročková and E. I. George. The spike-and-slab lasso. *Journal of the American Statistical Association*, 113(521):431–444, 2018.

[424] C. A. Rohde. Generalized inverses of partitioned matrices. *Journal of SIAM*, 13(4):1033–1035, 1965.

[425] F. M. Rohrhofer, S. Posch, C. Gößnitzer, and B. C. Geiger. Data vs. physics: The apparent Pareto front of physics-informed neural networks. *IEEE Access*, 11:86252–86261, 2023.

[426] O. Ronneberger, P. Fischer, and T. Brox. U-net: Convolutional networks for biomedical image segmentation. In *Medical Image Computing and Computer-Assisted Intervention–MICCAI 2015: 18th International Conference, Munich, Germany, October 5-9, 2015, Proceedings, Part III 18*, pages 234–241. Springer, 2015.

[427] F. Rosenblatt. The perceptron: a probabilistic model for information storage and organization in the brain. *Psychological Review*, 65(6):386–408, 1958.

[428] C. W. Rowley, I. Mezic, S. Bagheri, P. Schlatter, and D. Henningson. Spectral analysis of nonlinear flows. *Journal of Fluid Mechanics*, 645:115–127, 2009.

[429] P. Rozwood, E. Mehrez, L. Paehler, and S. L. Brunton. Koopman assisted reinforcement learning. In *NeurIPS Workshop on AI4Science*, 2023.

[430] S. H. Rudy, S. L. Brunton, J. L. Proctor, and J. N. Kutz. Data-driven discovery of partial differential equations. *Science Advances*, 3(e1602614), 2017.

[431] D. E. Rumelhart, G. E. Hinton, and R. J. Williams. Learning representations by back-propagating errors. *Nature*, 323(6088):533–536, 1986.

[432] Y. Saad and M. H. Schultz. GMRES: A generalized minimal residual algorithm for solving nonsymmetric linear systems. *SIAM Journal on Scientific and Statistical Computing*, 7(3):856–869, 1986.

[433] A. E. Sallab, M. Abdou, E. Perot, and S. Yogamani. Deep reinforcement learning framework for autonomous driving. ArXiv:1704.02532, 2017, 2017.

[434] A. L. Samuel. Some studies in machine learning using the game of checkers. *IBM Journal of Research and Development*, 3(3):210–229, 1959.

[435] A. Sanchez-Gonzalez, J. Godwin, T. Pfaff, R. Ying, J. Leskovec, and P. Battaglia. Learning to simulate complex physics with graph networks. In *International Conference on Machine Learning*, pages 8459–8468. PMLR, 2020.

[436] A. Savitzky and M. J. Golay. Smoothing and differentiation of data by simplified least squares procedures. *Analytical Chemistry*, 36(8):1627–1639, 1964.

[437] H. Schaeffer. Learning partial differential equations via data discovery and sparse optimization. *Proceedings of the Royal Society A*, 473(2197):20160446, 2017.

[438] T. Schaul, J. Quan, I. Antonoglou, and D. Silver. Prioritized experience replay. ArXiv:1511.05952, 2015.

[439] C. Scherer and S. Weiland. Linear matrix inequalities in control. In *The Control Systems Handbook*, 2nd edition, edited by W.S. Levine. CRC Press 2011.

[440] I. Scherl, B. Strom, J. K. Shang, O. Williams, B. L. Polagye, and S. L. Brunton. Robust principal component analysis for particle image velocimetry. *Physical Review Fluids*, 5(054401), 2020.

[441] M. Schlegel and B. R. Noack. On long-term boundedness of galerkin models. *Journal of Fluid Mechanics*, 765:325–352, 2015.

[442] M. Schmelzer, R. P. Dwight, and P. Cinnella. Discovery of algebraic reynolds-stress models using sparse symbolic regression. *Flow, Turbulence and Combustion*, 104(2):579–603, 2020.

[443] P. J. Schmid. Dynamic mode decomposition of numerical and experimental data. *Journal of Fluid Mechanics*, 656:5–28, Aug. 2010.

[444] P. J. Schmid. Dynamic mode decomposition and its variants. *Annual Review of Fluid Mechanics*, 54:225–254, 2022.

[445] J. Schmidhuber. Deep learning in neural networks: An overview. *Neural Networks*, 61:85–117, 2015.

[446] E. Schmidt. Zur theorie der linearen und nichtlinearen integralgleichungen. i teil. entwicklung willkürlichen funktionen nach system vorgeschriebener. *Math. Ann.*, 3:433–476, 1907.

[447] M. Schmidt and H. Lipson. Distilling free-form natural laws from experimental data. *Science*, 324(5923):81–85, 2009.

[448] B. Schölkopf and A. J. Smola. *Learning with Kernels: Support Vector Machines, Regularization, Optimization, and Beyond*. MIT Press, 2002.

[449] P. H. Schönemann. A generalized solution of the orthogonal Procrustes problem. *Psychometrika*, 31(1):1–10, 1966.

[450] A. Schrijver. *Theory of Linear and Integer Programming*. Wiley, 1998.

[451] J. Schulman, S. Levine, P. Abbeel, M. Jordan, and P. Moritz. Trust region policy optimization. In *International Conference on Machine Learning*, pages 1889–1897. PMLR, 2015.

[452] J. Schulman, F. Wolski, P. Dhariwal, A. Radford, and O. Klimov. Proximal policy optimization algorithms. ArXiv:1707.06347, 2017.

[453] W. Schultz, P. Dayan, and P. R. Montague. A neural substrate of prediction and reward. *Science*, 275(5306):1593–1599, 1997.

[454] G. Schwarz et al. Estimating the dimension of a model. *Annals of Statistics*, 6(2):461–464, 1978.

[455] A. W. Senior, R. Evans, J. Jumper, J. Kirkpatrick, L. Sifre, T. Green, C. Qin, A. Žídek, A. W. Nelson, A. Bridgland, et al. Improved protein structure prediction using potentials from deep learning. *Nature*, 577(7792):706–710, 2020.

[456] S. C. Shadden, F. Lekien, and J. E. Marsden. Definition and properties of Lagrangian coherent structures from finite-time Lyapunov exponents in two-dimensional aperiodic flows. *Physica D*, 212:271–304, 2005.

[457] S. Shalev-Shwartz, S. Shammah, and A. Shashua. Safe, multi-agent, reinforcement learning for autonomous driving. ArXiv:1610.03295, 2016.

[458] D. F. Shanno. Conditioning of quasi-Newton methods for function minimization. *Mathematics of Computation*, 24(111):647–656, 1970.

[459] C. E. Shannon. Xxii. programming a computer for playing chess. *The London, Edinburgh, and Dublin Philosophical Magazine and Journal of Science*, 41(314):256–275, 1950.

[460] J. Sherman and W. J. Morrison. Adjustment of an inverse matrix corresponding to a change in one element of a given matrix. *Annals of Mathematical Statistics*, 21(1):124–127, 1950.

[461] D. Silver, A. Huang, C. J. Maddison, A. Guez, L. Sifre, G. Van Den Driessche, J. Schrittwieser, I. Antonoglou, V. Panneershelvam, M. Lanctot, et al. Mastering the game of Go with deep neural networks and tree search. *Nature*, 529(7587):484–489, 2016.

[462] D. Silver, T. Hubert, J. Schrittwieser, I. Antonoglou, M. Lai, A. Guez, M. Lanctot, L. Sifre, D. Kumaran, T. Graepel, et al. A general reinforcement learning algorithm that masters chess, shogi, and Go through self-play. *Science*, 362(6419):1140–1144, 2018.

[463] D. Silver, G. Lever, N. Heess, T. Degris, D. Wierstra, and M. Riedmiller. Deterministic policy gradient algorithms. In *International Conference on Machine Learning*, pages 387–395. PMLR, 2014.

[464] D. Silver, J. Schrittwieser, K. Simonyan, I. Antonoglou, A. Huang, A. Guez, T. Hubert, L. Baker, M. Lai, A. Bolton, et al. Mastering the game of Go without human knowledge. *Nature*, 550(7676):354–359, 2017.

[465] K. Simonyan and A. Zisserman. Very deep convolutional networks for large-scale image recognition. ArXiv:1409.1556, 2014.

[466] S. Skogestad and I. Postlethwaite. *Multivariable Feedback Control: Analysis and Design*, 2nd edition. Wiley, 2005.

[467] J. Snoek, H. Larochelle, and R. P. Adams. Practical Bayesian optimization of machine learning algorithms. In *Advances in Neural Information Processing Systems*, pages 2951–2959, 2012.

[468] J. C. Spall. The Kantorovich inequality for error analysis of the Kalman filter with unknown noise distributions. *Automatica*, 31(10):1513–1517, 1995.

[469] A. Spantini, A. Solonen, T. Cui, J. Martin, L. Tenorio, and Y. Marzouk. Optimal low-rank approximations of Bayesian linear inverse problems. *SIAM Journal on Scientific Computing*, 37(6):A2451–A2487, 2015.

[470] B. Speelpenning. *Compiling Fast Partial Derivatives of Functions given by Algorithms*. Ph.D. thesis, University of Illinois at Urbana-Champaign, 1980.

[471] D. A. Spielman. Spectral graph theory and its applications. *Proceedings of the 48th Annual IEEE Symposium on Foundations of Computer Science (FOCS)*, pages 29–38, 2007.

[472] B. Stellato, G. Banjac, P. Goulart, A. Bemporad, and S. Boyd. OSQP: An operator splitting solver for quadratic programs. *Mathematical Programming Computation*, 12(4):637–672, 2020.

[473] R. F. Stengel. *Optimal Control and Estimation*. Dover Publications, 1994.

[474] I. Stewart. *Galois Theory*. Chapman and Hall/CRC, 2022.

[475] B. Strom, S. L. Brunton, and B. Polagye. Intracycle angular velocity control of cross-flow turbines. *Nature Energy*, 2(17103):1–9, 2017.

[476] A. M. Stuart. Inverse problems: a Bayesian perspective. *Acta Numerica*, 19:451–559, 2010.

[477] W. Su, M. Bogdan, E. Candes, et al. False discoveries occur early on the LASSO path. *Annals of Statistics*, 45(5):2133–2150, 2017.

[478] W. Su, S. Boyd, and E. J. Candes. A differential equation for modeling nesterov's accelerated gradient method: Theory and insights. *Journal of Machine Learning Research*, 17(153):1–43, 2016.

[479] I. Sutskever, J. Martens, G. Dahl, and G. Hinton. On the importance of initialization and momentum in deep learning. In *Proceedings of the 30th International Conference on Machine Learning (ICML)*, pages 1139–1147, 2013.

[480] R. S. Sutton and A. G. Barto. *Reinforcement Learning: An Introduction*. MIT Press, 1998.

[481] R. S. Sutton, D. A. McAllester, S. P. Singh, and Y. Mansour. Policy gradient methods for reinforcement learning with function approximation. In *Proceedings of the 13th International Conference on Neural Information Processing Systems*, pages 1057–1063, 1999.

[482] C. Szegedy, V. Vanhoucke, S. Ioffe, J. Shlens, and Z. Wojna. Rethinking the inception architecture for computer vision. In *Proceedings of the IEEE Conference on Computer Vision and Pattern Recognition*, pages 2818–2826. IEEE, 2016.

[483] K. Taira, S. L. Brunton, S. Dawson, C. W. Rowley, T. Colonius, B. J. McKeon, O. T. Schmidt, S. Gordeyev, V. Theofilis, and L. S. Ukeiley. Modal analysis of fluid flows: An overview. *AIAA Journal*, 55(12):4013–4041, 2017.

[484] N. Takeishi, Y. Kawahara, and T. Yairi. Learning Koopman invariant subspaces for dynamic mode decomposition. In *Advances in Neural Information Processing Systems*, pages 1130–1140, 2017.

[485] F. Takens. Detecting strange attractors in turbulence. *Lecture Notes in Mathematics*, 898, pages 366–381. Springer, 1981.

[486] M. Tan and Q. V. Le. EfficientNet: Rethinking model scaling for convolutional neural networks. In *International Conference on Machine Learning*, pages 6105–6114. PMLR, 2019.

[487] Z. Tan, Z. Zhou, Y. Ge, Z. Wang, X. Chen, and D. Hu. TD-NeRF: Novel truncated depth prior for joint camera pose and neural radiance field optimization. In *2024 IEEE/RSJ International Conference on Intelligent Robots and Systems (IROS)*, pages 372–379. IEEE, 2024.

[488] A. Tarantola. *Inverse Problem Theory and Methods for Model Parameter Estimation*. SIAM, 2005.

[489] A. Taylor, A. Singletary, Y. Yue, and A. Ames. Learning for safety-critical control with control barrier functions. In *Learning for Dynamics and Control*, pages 708–717. PMLR, 2020.

[490] G. I. Taylor. The formation of a blast wave by a very intense explosion. *Proceedings of the Royal Society A*, 186(1006):273–292, 1941.

[491] G. I. Taylor. The formation of a blast wave by a very intense explosion. II. The atomic explosion of 1945. *Proceedings of the Royal Society A*, 201(1065):175–186, 1950.

[492] G. Tesauro et al. Temporal difference learning and td-gammon. *Communications of the ACM*, 38(3):58–68, 1995.

[493] The Event Horizon Telescope Collaboration. First M87 event horizon telescope results. I. the shadow of the supermassive black hole. *Astrophysical Journal Letters*, 875:L1, 2019.

[494] N. Thomas, T. Smidt, S. Kearnes, L. Yang, L. Li, K. Kohlhoff, and P. Riley. Tensor field networks: Rotation-and translation-equivariant neural networks for 3D point clouds. ArXiv:1802.08219, 2018.

[495] R. Tibshirani. Regression shrinkage and selection via the Lasso. *Journal of the Royal Statistical Society B*, pages 267–288, 1996.

[496] A. N. Tikhonov et al. On the stability of inverse problems. *Dokl. Akad. Nauk SSSR*, 39(5):195–198, 1943.

[497] E. Todorov, T. Erez, and Y. Tassa. MuJoCo: A physics engine for model-based control. In *2012 IEEE/RSJ International Conference on Intelligent Robots and Systems*, pages 5026–5033. IEEE, 2012.

[498] J. F. Toland. Duality in nonconvex optimization. *Journal of Mathematical Analysis and Applications*, 66(2):399–415, 1978.

[499] H. Touvron and et al. Llama 2: Open foundation and fine-tuned chat models. ArXiv:2307.09288, 2023.

[500] L. N. Trefethen and D. Bau. *Numerical Linear Algebra*. SIAM, 2022.

[501] J. A. Tropp. Greed is good: Algorithmic results for sparse approximation. *IEEE Transactions on Information Theory*, 50(10):2231–2242, 2004.

[502] J. A. Tropp. Recovery of short, complex linear combinations via l_1 minimization. *IEEE Transactions on Information Theory*, 51(4):1568–1570, 2005.

[503] J. A. Tropp. Algorithms for simultaneous sparse approximation. Part II: Convex relaxation. *Signal Processing*, 86(3):589–602, 2006.

[504] J. A. Tropp. Just relax: Convex programming methods for identifying sparse signals in noise. *IEEE Transactions on Information Theory*, 52(3):1030–1051, 2006.

[505] J. A. Tropp and A. C. Gilbert. Signal recovery from random measurements via orthogonal matching pursuit. *IEEE Transactions on Information Theory*, 53(12):4655–4666, 2007.

[506] J. A. Tropp, A. C. Gilbert, and M. J. Strauss. Algorithms for simultaneous sparse approximation. Part I: Greedy pursuit. *Signal Processing*, 86(3):572–588, 2006.

[507] J. A. Tropp, J. N. Laska, M. F. Duarte, J. K. Romberg, and R. G. Baraniuk. Beyond Nyquist: Efficient sampling of sparse bandlimited signals. *IEEE Transactions on Information Theory*, 56(1):520–544, 2010.

[508] P. Tseng. An analysis of the EM algorithm and entropy-like proximal point methods. *Mathematics of Operations Research*, 29(1):27–44, 2004.

[509] J. N. Tsitsiklis. *Problems in Decentralized Decision Making and Computation*. PhD thesis, Massachusetts Institute of Technology, 1984.

[510] J. N. Tsitsiklis. Asynchronous stochastic approximation and Q-learning. *Machine Learning*, 16(3):185–202, 1994.

[511] F. Van Breugel, J. N. Kutz, and B. W. Brunton. Numerical differentiation of noisy data: A unifying multi-objective optimization framework. *IEEE Access*, 8:196865–196877, 2020.

[512] R. van de Schoot, S. Depaoli, R. King, B. Kramer, K. Märtens, M. G. Tadesse, M. Vannucci, A. Gelman, D. Veen, J. Willemsen, et al. Bayesian statistics and modelling. *Nature Reviews Methods Primers*, 1(1):1, 2021.

[513] E. Van Den Berg and M. P. Friedlander. Probing the Pareto frontier for basis pursuit solutions. *SIAM Journal on Scientific Computing*, 31(2):890–912, 2009.

[514] R. Van Der Merwe. *Sigma-point Kalman filters for probabilistic inference in dynamic state-space models*. Ph.D. thesis, Oregon Health & Science University, 2004.

[515] H. Van Hasselt, A. Guez, and D. Silver. Deep reinforcement learning with double Q-learning. In *Proceedings of the 30th AAAI Conference on Artificial Intelligence*, pages 2094–2100, 2016.

[516] P. J. Van Laarhoven and E. H. Aarts. *Simulated Annealing*. Springer, 1987.

[517] S. Verma, G. Novati, and P. Koumoutsakos. Efficient collective swimming by harnessing vortices through deep reinforcement learning. *Proceedings of the National Academy of Sciences*, 115(23):5849–5854, 2018.

[518] O. Vinyals, I. Babuschkin, W. M. Czarnecki, M. Mathieu, A. Dudzik, J. Chung, D. H. Choi, R. Powell, T. Ewalds, P. Georgiev, et al. Grandmaster level in Starcraft II using multi-agent reinforcement learning. *Nature*, 575(7782):350–354, 2019.

[519] J. Virieux and S. Operto. An overview of full-waveform inversion in exploration geophysics. *Geophysics*, 74(6):WCC1–WCC26, 2009.

[520] P. R. Vlachas, W. Byeon, Z. Y. Wan, T. P. Sapsis, and P. Koumoutsakos. Data-driven forecasting of high-dimensional chaotic systems with long short-term memory networks. *Proceedings of the Royal Society, A*, 474(2213):20170844, 2018.

[521] B. Völker, V. Maier-Kiener, K. Werbach, T. Müller, S. Pilz, M. Calin, J. Eckert, and A. Hohenwarter. Influence of annealing on microstructure and mechanical properties of ultrafine-grained Ti45Nb. *Materials & Design*, 179:107864, 2019.

[522] J. Von Neumann. *Functional Operators, Volume 2: The Geometry of Orthogonal Spaces*. Princeton University Press, 1950.

[523] U. Von Toussaint. Bayesian inference in physics. *Reviews of Modern Physics*, 83(3):943–999, 2011.

[524] A. Wächter and L. T. Biegler. On the implementation of an interior-point filter line-search algorithm for large-scale nonlinear programming. *Mathematical Programming*, 106(1):25–57, 2006.

[525] N. J. Walkington. Nesterov's method for convex optimization. *SIAM Review*, 65(2):539–562, 2023.

[526] R. Wang, K. Kashinath, M. Mustafa, A. Albert, and R. Yu. Towards physics-informed deep learning for turbulent flow prediction. In *Proceedings of the 26th ACM SIGKDD International Conference on Knowledge Discovery & Data Mining*, pages 1457–1466, 2020.

[527] R. Wang, R. Walters, and R. Yu. Incorporating symmetry into deep dynamics models for improved generalization. ArXiv:2002.03061, 2020.

[528] X. Wang, K. Yu, S. Wu, J. Gu, Y. Liu, C. Dong, Y. Qiao, and C. Change Loy. ESRGAN: Enhanced super-resolution generative adversarial networks. In *Proceedings of the European Conference on Computer Vision (ECCV) workshops*, 2018.

[529] Y. Wang, H. He, X. Tan, and Y. Gan. Trust region-guided proximal policy optimization. In *Proceedings of the 33rd International Conference on Neural Information Processing Systems*, pages 626–636, 2019.

[530] Z. Wang, T. Schaul, M. Hessel, H. Hasselt, M. Lanctot, and N. Freitas. Dueling network architectures for deep reinforcement learning. In *International Conference on Machine Learning*, pages 1995–2003. PMLR, 2016.

[531] C. J. Watkins and P. Dayan. Q-learning. *Machine Learning*, 8(3-4):279–292, 1992.

[532] J. D. Watson and F. H. C. Crick. Molecular structure of nucleic acids: A structure for deoxyribose nucleic acid. *Nature*, 171:737–738, 1953.

[533] C. Wehmeyer and F. Noé. Time-lagged autoencoders: Deep learning of slow collective variables for molecular kinetics. *Journal of Chemical Physics*, 148(241703):1–9, 2018.

[534] G. Welch and G. Bishop. An introduction to the Kalman filter. Technical Report, University Of North Carolina, Chapel Hill, 1995.

[535] R. E. Wengert. A simple automatic derivative evaluation program. *Communications of the ACM*, 7(8): 463–464, 1964.

[536] J. Willems. Least squares stationary optimal control and the algebraic Riccati equation. *IEEE Trans. Automatic Control*, 16(6):621–634, 1971.

[537] G. Williams, N. Wagener, B. Goldfain, P. Drews, J. M. Rehg, B. Boots, and E. A. Theodorou. Information theoretic MPC for model-based reinforcement learning. In *2017 IEEE International Conference on Robotics and Automation (ICRA)*, pages 1714–1721. IEEE, 2017.

[538] R. J. Williams. Simple statistical gradient-following algorithms for connectionist reinforcement learning. *Machine Learning*, 8(3):229–256, 1992.

[539] D. M. Witten and R. Tibshirani. Penalized classification using fisher's linear discriminant. *Journal of the Royal Statistical Society B*, 73(5):753–772, 2011.

[540] L. A. Wolsey and G. L. Nemhauser. *Integer and Combinatorial Optimization*. Wiley, 1999.

[541] M. A. Woodbury. The stability of out-input matrices. Technical Report, Chicago, 1949.

[542] S. J. Wright. *Primal-Dual Interior-Point Methods*. SIAM, 1997.

[543] X. Wu, V. Kumar, J. R. Quinlan, J. Ghosh, Q. Yang, H. Motoda, G. J. McLachlan, A. Ng, B. Liu, S. Y. Philip, et al. Top 10 algorithms in data mining. *Knowledge and Information Systems*, 14(1):1–37, 2008.

[544] J. Yang, J. Wright, T. S. Huang, and Y. Ma. Image super-resolution via sparse representation. *IEEE Transactions on Image Processing*, 19(11):2861–2873, 2010.

[545] L. Yang, D. Zhang, and G. E. Karniadakis. Physics-informed generative adversarial networks for stochastic differential equations. *SIAM Journal on Scientific Computing*, 42(1):A292–A317, 2020.

[546] H. Yoshida. Construction of higher order symplectic integrators. *Physics letters A*, 150(5-7):262–268, 1990.

[547] S. Yu and J. Ma. Deep learning for geophysics: Current and future trends. *Reviews of Geophysics*, 59(3):e2021RG000742, 2021.

[548] S. Yu, S. M. Won, H. W. Baac, D. Son, and C. Shin. Quantitative evaluation of line-edge roughness in various FinFET structures: Bayesian neural network with automatic model

selection. *IEEE Access*, 10:26340–26346, 2022.

[549] X. Yuan and J. Yang. Sparse and low-rank matrix decomposition via alternating direction methods. Pacific Journal of Optimization, 9(1):167–180, 2009.

[550] I. Zamora, N. G. Lopez, V. M. Vilches, and A. H. Cordero. Extending the OpenAI Gym for robotics: a toolkit for reinforcement learning using ROS and Gazebo. ArXiv:1608.05742, 2016.

[551] M. N. Zeilinger, D. M. Raimondo, A. Domahidi, M. Morari, and C. N. Jones. On real-time robust model predictive control. *Automatica*, 50(3):683–694, 2014.

[552] C. Zhang, J. Bütepage, H. Kjellström, and S. Mandt. Advances in variational inference. *IEEE Transactions on Pattern Analysis and Machine Intelligence*, 41(8):2008–2026, 2018.

[553] L. Zhang and H. Schaeffer. On the convergence of the SINDy algorithm. *Multiscale Modeling & Simulation*, 17(3), 2019.

[554] T. Zhang, G. Kahn, S. Levine, and P. Abbeel. Learning deep control policies for autonomous aerial vehicles with MPC-guided policy search. In *IEEE Robotics and Automation (ICRA)*, pages 528–535, 2016.

[555] Y. Zhang, X. Zhu, and J. Gao. Seismic inversion based on acoustic wave equations using physics-informed neural network. *IEEE Trans. Geosci. Remote Sens.*, 61:1–11, 2023.

[556] Z. Zhang and T. Alkhalifah. Regularized elastic full-waveform inversion using deep learning. In *Advances in Subsurface Data Analytics*, pages 219–250. Elsevier, 2022.

[557] P. Zheng, T. Askham, S. L. Brunton, J. N. Kutz, and A. Y. Aravkin. Sparse relaxed regularized regression: SR3. *IEEE Access*, 7(1):1404–1423, 2019.

[558] Y. D. Zhong and N. Leonard. Unsupervised learning of Lagrangian dynamics from images for prediction and control. *Advances in Neural Information Processing Systems*, 33, 2020.

[559] S. Zhou, K. Chan, C. Li, and C. C. Loy. Towards robust blind face restoration with codebook lookup transformer. *Advances in Neural Information Processing Systems*, 35:30599–30611, 2022.

[560] Y. Zhou, D. Fan, B. Zhang, R. Li, and B. R. Noack. Artificial intelligence control of a turbulent jet. *Journal of Fluid Mechanics*, 897:A27. doi:10.1017/jfm.2020.392, 2020.

[561] S. Zlobec. An explicit form of the moore-penrose inverse of an arbitrary complex matrix. *SIAM Review*, 12(1):132–134, 1970.

[562] N. Zolman, U. Fasel, J. N. Kutz, and S. L. Brunton. SINDy-RL: Interpretable and efficient model-based reinforcement learning. ArXiv:2403.09110, 2024.

[563] F. Zou, L. Shen, Z. Jie, W. Zhang, and W. Liu. A sufficient condition for convergences of Adam and RMSProp. In *Proceedings of the IEEE/CVF Conference on Computer Vision and Pattern Recognition*, pages 11127–11135, 2019.

[564] H. Zou and T. Hastie. Regularization and variable selection via the elastic net. *Journal of the Royal Statistical Society B*, 67(2):301–320, 2005.

Index

Printed in the United Kingdom by TJ Clays Ltd.